Immune Aspects of Biopharmaceuticals and Nanomedicines

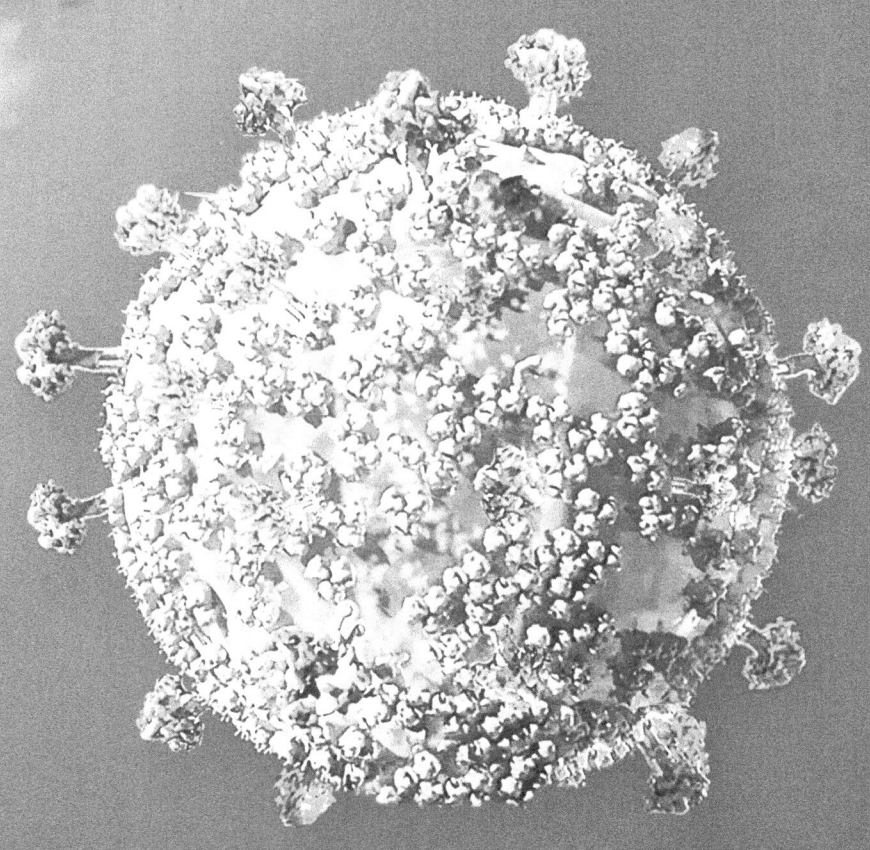

Endorsed by the American Society for Nanomedicine

Pan Stanford Series on Nanomedicine

Series Editor
Raj Bawa

Titles in the Series

Published

Vol. 1
Handbook of Clinical Nanomedicine: Nanoparticles, Imaging, Therapy, and Clinical Applications
Raj Bawa, Gerald F. Audette, and Israel Rubinstein, eds.
2016
978-981-4669-20-7 (Hardcover)
978-981-4669-21-4 (eBook)

Vol. 2
Handbook of Clinical Nanomedicine: Law, Business, Regulation, Safety, and Risk
Raj Bawa, ed., Gerald F. Audette and Brian E. Reese, asst. eds.
2016
978-981-4669-22-1 (Hardcover)
978-981-4669-23-8 (eBook)

Vol. 3
Immune Aspects of Biopharmaceuticals and Nanomedicines
Raj Bawa, János Szebeni, Thomas J. Webster, and Gerald F. Audette, eds.
2018
978-981-4774-52-9 (Hardcover)
978-0-203-73153-6 (eBook)

Forthcoming

Vol. 4
The Road from Nanomedicine to Precision Medicine
Shaker A. Mousa and Raj Bawa, eds., Gerald F. Audette, Thomas J. Webster, and Gert Storm, asst. eds.
2019

Pan Stanford Series on Nanomedicine Vol. 3

Immune Aspects of Biopharmaceuticals and Nanomedicines

edited by

Raj Bawa, MS, PhD
Patent Agent, Bawa Biotech LLC, Ashburn, Virginia, USA
Adjunct Professor, Rensselaer Polytechnic Institute, Troy, New York, USA
Scientific Advisor, Teva Pharmaceutical Industries, Ltd., Israel

János Szebeni, MD, PhD, DSc
Director, Nanomedicine Research and Education Center
Department of Pathophysiology, Semmelweis University, Budapest, Hungary
Professor of Biology and Immunology, Miskolc University, Miskolc, Hungary
President and CEO, SeroScience Ltd., Budapest, Hungary

Thomas J. Webster, MS, PhD
The Art Zafiropoulo Professor and Department Chair
Department of Chemical Engineering
Northeastern University, Boston, Massachusetts, USA

Gerald F. Audette, PhD
Associate Professor, Department of Chemistry
York University, Toronto, Canada

Published by

Pan Stanford Publishing Pte. Ltd.
Penthouse Level, Suntec Tower 3
8 Temasek Boulevard
Singapore 038988
Email: editorial@panstanford.com
Web: www.panstanford.com

Note from the Series Editor and Publisher

It is important to note that knowledge and best practices in the various fields represented in this book (medicine, immunology, biologics, nanomedicine, protein science, patent law, FDA law, regulatory science, toxicology, and pharmaceutical sciences) are constantly evolving. As new research and experience broaden our knowledge base, changes in research methods, legal and business practices, diagnostics, assays, tools and techniques, and/or medical treatments may become necessary. Therefore, the reader is advised to consult the most current information regarding: (i) various products, therapies, assays, diagnostics, technologies, and companies featured; (ii) FDA regulatory guidance documents provided; and (iii) legal, business, or commercial information provided. It is imperative that the reader does not solely rely on the information presented herein and always carefully reviews product labels, warnings, and directions before using or consuming a product. For additional information about a product, please contact the manufacturer, FDA, physician, pharmacist, or other licensed health-care professional, as appropriate. Similarly, careful evaluation of any information, procedures, formulations, legal ideas, medical protocols, regulatory guidances, or assays described herein is warranted. To the fullest extent of the law, neither the publisher nor the editors or the authors make any representations or warranties, express or implied, with respect to the information presented in this book. In this regard, they assume no liability for any injury and/or damage to persons or property as a matter of product liability, negligence or otherwise, or from any use or operation of any medical, therapeutic, immunological, or legal opinions, business methods, medical procedures, therapeutic products, assays, techniques, instructions, or ideas presented or highlighted in this book.

A catalogue record for this book is available from the Library of Congress and the British Library.

Immune Aspects of Biopharmaceuticals and Nanomedicines

Copyright © 2018 Pan Stanford Publishing Pte. Ltd. All rights reserved. This book, or parts thereof, may not be reproduced in any form or by any means, electronic or mechanical, including photocopying, recording or by any information storage and retrieval system now known or to be invented, without written permission from the publisher. For photocopying of material in this volume, please pay a copying fee through the Copyright Clearance Center, Inc., 222 Rosewood Drive, Danvers, MA 01923, USA. In this case, permission to photocopy is not required from the publisher.

ISBN 978-981-4774-52-9 (Hardcover)
ISBN 978-0-203-73153-6 (eBook)

Printed in the USA

This book is dedicated to patients of autoimmune diseases, who endure pain with grace and await new treatments with patience.

—Raj Bawa

I dedicate this book to my sons, hoping that one day they will enjoy science and then understand what kept their dad living enchanted, working at night.

—János Szebeni

I dedicate this book to our next generation (particularly, my daughters, Mia, Zoe, and Ava), who will push the boundaries of science, engineering, and all fields—let this book inspire you to a life with no boundaries.

—Thomas Webster

I dedicate this book to my family, for their unflagging support, and to my students, who continually inspire me with their curiosity.

—Gerald Audette

Dr. Bawa's (above, second from left) personal inspiration for this volume resulted from medical events in the lives of his loved ones. His wife (above left) suffers from rheumatoid arthritis (RA) and Sjögren's syndrome (SjS), both chronic autoimmune diseases. His mother (above, second from right) recently had the "red man syndrome" (a hypersensitivity reaction) while in a hospital emergency room following intravenous administration of vancomycin to treat aspiration pneumonia. She also suffers from Type 1 diabetes (T1D), now considered an autoimmune disease that results from the destruction of the insulin-producing beta cells of the pancreas. His father (above right), a former professor and dean, apart from reviewing numerous chapters of this book, recommended that a volume on the immune aspects of biotherapeutics and nanomedicines was critically needed for regulators, clinicians, and pharma. Note that RA, SjS, and T1D are not caused by immune reactions of biotherapeutics or nanodrugs—autoantibodies are to blame for all. Vancomycin, a glycopeptide antibiotic, is not classified as a biotherapeutic or nanomedicine.

Also Read
Handbook of Clinical Nanomedicine
(Two-Volume Set)

Vol. 1: 9789814669207 (Hardcover), 9789814669214 (eBook),
1662 pages
Vol. 2: 9789814669221 (Hardcover), 9789814669238 (eBook),
1448 pages
Set: 9789814316170 (Hardcover), 9789814411660 (eBook),
3,110 pages

Handbook of Clinical Nanomedicine. Vol. 1. Nanoparticles, Imaging, Therapy, and Clinical Applications, Raj Bawa, PhD, Gerald F. Audette, PhD, and Israel Rubinstein, MD (Editors)

This handbook (55 chapters) provides a comprehensive roadmap of basic research in nanomedicine as well as clinical applications. However, unlike other texts in nanomedicine, it not only highlights current advances in diagnostics and therapeutics but also explores related issues like nomenclature, historical developments, regulatory aspects, nanosimilars and 3D nanofabrication. While bridging the gap between basic biomedical research, engineering, medicine and law, the handbook provides a thorough understanding of nano's potential to address (i) medical problems from both the patient and health provider's perspective, and (ii) current applications and their potential in a healthcare setting.

"*Dr. Bawa and his team have meticulously gathered the distilled experience of world-class researchers, clinicians and business leaders addressing the most salient issues confronted in product concept development and translation.*"

Gregory Lanza, MD, PhD
Professor of Medicine and Oliver M. Langenberg Distinguished Professor
Washington University Medical School, USA

"*This is an outstanding, comprehensive volume that crosscuts disciplines and topics fitting individuals from a variety of fields looking to become knowledgeable in medical nanotech research and its translation from the bench to the bedside.*"

Shaker A. Mousa, PhD, MBA
Vice Provost and Professor of Pharmacology
Albany College of Pharmacy and Health Sciences, USA

"*Masterful! This handbook will have a welcome place in the hands of students, educators, clinicians and experienced scientists alike. In a rapidly evolving arena, the authors have harnessed the field and its future by highlighting both current and future needs in diagnosis and therapies. Bravo!*"

Howard E. Gendelman, MD
Margaret R. Larson Professor and Chair
University of Nebraska Medical Center, USA

"*It is refreshing to see a handbook that does not merely focus on preclinical aspects or exaggerated projections of nanomedicine. Unlike other books, this handbook not only highlights current advances in diagnostics and therapies but also addresses critical issues like terminology, regulatory aspects and personalized medicine.*"

Gert Storm, PhD
Professor of Pharmaceutics
Utrecht University, The Netherlands

Handbook of Clinical Nanomedicine. Vol. 2. *Law, Business, Regulation, Safety, and Risk,* Raj Bawa, PhD (Editor), Gerald F. Audette, PhD, and Brian E. Reese, PhD, MBA, JD (Assistant Editors)

This unique handbook (60 chapters) examines the entire "product life cycle," from the creation of nanomedical products to their final market introduction. While focusing on critical issues relevant to nanoproduct development and translational activities, it tackles topics such as regulatory science, patent law, FDA law, ethics, personalized medicine, risk analysis, toxicology, nano-characterization and commercialization activities. A separate section provides fascinating perspectives and editorials from leading experts in this complex interdisciplinary field.

"The distinguished editors have secured contributions from the leading experts in nanomedicine law, business, regulation and policy. This handbook represents possibly the most comprehensive and advanced collections of materials on these critical topics. An invaluable standard resource."

Gregory N. Mandel, JD
Peter J. Liacouras Professor of Law and Associate Dean
Temple University Beasley School of Law, USA

"This is an outstanding volume for those looking to become familiar with nanotechnology research and its translation from the bench to market. Way ahead of the competition, a standard reference on any shelf."

Shaker A. Mousa, PhD, MBA
Vice Provost and Professor of Pharmacology
Albany College of Pharmacy, USA

"The editors have gathered the distilled experience of leaders addressing the most salient issues confronted in R&D and translation. Knowledge is power, particularly in nanotechnology translation, and this handbook is an essential guide that illustrates and clarifies our way to commercial success."

Gregory Lanza, MD, PhD
Professor of Medicine and Oliver M. Langenberg Distinguished Professor
Washington University Medical School, USA

"The title of the handbook reflects its broad-ranging contents. The intellectual property chapters alone are worthy of their own handbook. Dr. Bawa and his coeditors should be congratulated for gathering the important writings on nanotech law, business and commercialization."

Richard J. Apley, JD
Chief Patent Officer
Litman Law Offices/Becker & Poliakoff, USA

"It is clear that this handbook will serve the interdisciplinary community involved in nanomedicine, pharma and biotech in a highly comprehensive way. It not only covers basic and clinical aspects but the often missing, yet critically important, topics of safety, risk, regulation, IP and licensing. The section titled 'Perspectives and Editorials' is superb."

Yechezkel (Chezy) Barenholz, PhD
Professor Emeritus of Biochemistry and Daniel Miller Professor of Cancer Research
Hebrew University-Hadassah Medical School, Israel

About the Editors

Raj Bawa, MS, PhD, is president of Bawa Biotech LLC, a biotech/pharma consultancy and patent law firm based in Ashburn, Virginia, that he founded in 2002. He is an inventor, entrepreneur, professor and registered patent agent licensed to practice before the U.S. Patent & Trademark Office. Trained as a biochemist and microbiologist, he has extensive expertise in pharmaceutical sciences, biotechnology, nanomedicine, drug delivery, microbial biodefense, FDA regulatory issues, and patent law. Since 1999, he has held various positions at Rensselaer Polytechnic Institute in Troy, NY, where he is an adjunct professor and where he received his doctoral degree in three years (biophysics/biochemistry). Currently, he serves as a scientific advisor to Teva Pharma, Israel, is a visiting research scholar at the Pharmaceutical Research Institute of Albany College of Pharmacy in Albany, NY, and is vice president of Guanine, Inc., based in Rensselaer, NY. He has served as a principal investigator of SBIRs and reviewer for both the NIH and NSF. Currently, he is principal investigator of a CDC SBIR Phase 1 grant to develop an assay for carbapenemase-resistant bacteria. In the 1990s, Dr. Bawa held various positions at the US Patent & Trademark Office, including primary examiner from 1996–2002. He is a life member of Sigma Xi, co-chair of the nanotech and personalized medicine committees of the American Bar Association and founding director of the American Society for Nanomedicine. He has authored over 100 publications, co-edited four texts and serves on the editorial boards of 14 peer-reviewed journals, including serving as an associate editor of *Nanomedicine* (Elsevier). Some of Dr. Bawa's awards include the Innovations Prize from the Institution of Mechanical Engineers, London, UK (2008); Appreciation Award from the Undersecretary of Commerce, Washington, DC (2001); the

Key Award from Rensselaer's Office of Alumni Relations (2005); and Lifetime Achievement Award from the American Society for Nanomedicine (2014).

Janos Szebeni, MD, PhD, DSc, is director of the Nanomedicine Research and Education Center at Semmelweis University in Budapest, Hungary. He is also the founder and CEO of a contract research SME, SeroScience, Ltd., and a full professor of immunology and biology at the University of Miskolc in Hungary. He has held various academic positions in Hungary and abroad, including at the National Institute of Hematology and Blood Transfusion and at the First Clinics of Internal Medicine, Semmelweis University in Budapest, the University of Arizona in Tucson, AZ, the National Institute of Health in Bethesda, Maryland, Massachusetts General Hospital, Harvard University in Boston, MA, the Walter Reed Army Institute of Research in Silver Spring, Maryland. After residing in the United States for over two decades, Dr. Szebeni returned to Hungary in 2006. His research on various themes in hematology, membrane biology, nanotechnology, and immunology has produced 160+ peer-reviewed papers, book chapters, patents, etc. (citations: ≈6000, H index: 39), and a book titled *The Complement System: Novel Roles in Health and Disease* (Kluwer, 2004). He has made significant contributions to three fields: artificial blood, liposomes, and the complement system. His original works led to the "CARPA" concept, i.e., that complement activation underlies numerous drug-induced (pseudo) allergic (infusion) reactions.

Thomas J. Webster, MS, PhD (H index: 77), has degrees in chemical engineering from the University of Pittsburgh (BS, 1995) and in biomedical engineering from Rensselaer Polytechnic Institute (MS, 1997; PhD, 2000). He was appointed department chair of Chemical Engineering at Northeastern University, Boston, in 2012. He is the current director of the Nanomedicine Laboratories (currently at 25 members) and has completed extensive studies on the use of nanophase materials in medicine. In his 17+ years in academics, Dr. Webster has graduated or supervised over 109 visiting faculty, clinical fellows, post-doctoral students, and thesis completing BS, MS, and PhD students. To date, his lab group has generated over 9 textbooks, 48 book chapters, 306 invited presentations, at least 403 peer-reviewed literature articles (222) and/or conference proceedings (181), at least 567 conference presentations, and 32 provisional or full patents. He is the founding editor-in-chief of the *International Journal of Nanomedicine* (the first international journal in nanomedicine that has a 5-year impact factor of 5.034). Dr. Webster has received numerous honors including: 2012, Fellow, American Institute for Medical and Biological Engineering (AIMBE, representing the top 2% of all medical and biological engineers); 2013, Fellow, Biomedical Engineering Society; 2015, Wenzhou 580 Award; 2015, Zheijang 1000 Talent Program; 2016, International Materials Research Chinese Academy of Science Lee-Hsun Lecture Award; and 2016, International College of Fellows, Biomaterials Science and Engineering. He was recently elected president of the U.S. Society For Biomaterials. He has appeared on BBC, NBC, ABC, Fox News, National Geographic, the Weather Channel, and many other news outlets talking about science and medicine.

Gerald F. Audette, PhD, has been a faculty member at York University in Toronto, Canada, in the Department of Chemistry since 2006. Currently he is associate professor in the department and a member of the Centre for Research on Biomolecular Interactions at York University. He received his doctorate in 2002 from the Department of Biochemistry at the University of Saskatchewan, Canada. Working with Drs. Louis T. J. Delbaere and J. Wilson Quail (1995–2001), Dr. Audette's research focused on the elucidation of the protein-carbohydrate interactions that occur during blood group recognition, in particular, during the recognition of the O-blood type, using high-resolution X-ray crystallography. Dr. Audette conducted his postdoctoral research at the University of Alberta, Canada (2001–2006). Working with Drs. Bart Hazes and Laura Frost, his research again utilized high-resolution protein crystallography to examine the correlation between protein structure and biological activity of type IV pilins from *P. aeruginosa* and the type 4 secretion system from the conjugative F-plasmid of *Escherichia coli*. His current research interests include structure/function studies of proteins involved in bacterial conjugation systems, as well as the type IV pilins and assembly systems from several bacterial pathogens, and exploring the adaptation of these protein systems for applications in bionanotechnology and nanomedicine. Dr. Audette is the co-editor of volumes 1–4 of the *Pan Stanford Series on Nanomedicine* and is a subject editor of structural chemistry and crystallography for the journal *FACETS*.

Contents

List of Corresponding Authors	xxvii
Foreword	xxix
Howard E. Gendelman, MD	
My Life with Biologicals and Nanodrugs: A Twenty-Year Affair	xxxvii
Raj Bawa, MS, PhD	

1. Current Immune Aspects of Biologics and Nanodrugs: An Overview — 1

Raj Bawa, MS, PhD

1.1	Introduction	3
1.2	Biologics versus Small-Molecule Drugs	21
1.3	What Are Nanodrugs?	24
1.4	Are Biologics and Nanodrugs Adversely Immunogenic?	32
1.5	Immunogenicity Assessment of Biologics and Nanodrugs	51
1.6	Entering the Era of Biosimilars	56
1.7	Immune Aspects of Biosimilars and Nanosimilars: The Copaxone® Example	62
1.8	Concluding Remarks and Future Directions	67

2. Immunological Issues with Medicines of Nano Size: The Price of Dimension Paradox — 83

János Szebeni, MD, PhD, DSc, and Raj Bawa, MS, PhD

2.1	Adverse Immune Effects of Nanodrugs	84
2.2	Issues of Terminology	94
2.3	Adverse Immune Effects of Nanodrugs: The Dimension Paradox	96
2.4	Vicious Cycle between Specific and Nonspecific Immune Responses to Nanodrugs	111

2.5	CARPA as Blood Stress	112
2.6	CARPA Testing	114

3. Immunotherapy and Vaccines — 123
Johanna Poecheim, PhD, and Gerrit Borchard, PhD

3.1	Introduction	123
3.2	The Immune System	124
3.3	Nanotechnology in Vaccines	128
3.4	Conclusions	145

4. Site-Specific Antibody Conjugation for ADC and Beyond — 155
Qun Zhou, PhD

4.1	Introduction	155
4.2	Site-Specific ADC through Specific Amino Acids	159
4.3	Site-Specific ADC through Unnatural Amino Acids	161
4.4	Site-Specific ADC through Glycans	164
4.5	Site-Specific ADC through Short Peptide Tags	165
4.6	Site-Specific Antibody Conjugation for Diagnosis	166
4.7	Site-Specific Antibody Conjugation for Other Therapeutic Applications	169
4.8	Conclusions	173

5. Current Understanding of Interactions between Nanoparticles and the Immune System — 183
Marina A. Dobrovolskaia, PhD, Michael Shurin, MD, PhD, and Anna A. Shvedova, PhD

5.1	Introduction	184
5.2	Achievements	188
5.3	Disappointments	199
5.4	Lessons Learned	204
5.5	Conclusions	208

6. Auto-antibodies as Biomarkers for Disease Diagnosis — 229
Angelika Lueking, Heike Göhler, and Peter Schulz-Knappe

6.1	Introduction	229
6.2	Auto-antibodies as Biomarkers	230

	6.3	Auto-antibodies for Companion Diagnostics Enabling Personalized Medicine	233
	6.4	Biomarker Discovery Strategies	235
	6.5	Antigen/Auto-antibody Interactions as Biomarker Candidates	239
	6.6	Diagnostic Assays Based on Antigen/Auto-antibody Interactions	242
	6.7	Conclusion	243

7. The Acceleated Blood Clearance Phenomenon of PEGylated Nanocarriers — **249**

Amr S. Abu Lila, PhD, and Tatsuhiro Ishida, PhD

	7.1	Introduction	249
	7.2	Mechanism of ABC Phenomenon	251
	7.3	Correlation Between Complement Activation and ABC Phenomenon	254
	7.4	Factors That Affect the Magnitude of the ABC Phenomenon	256
	7.5	Strategies to Abrogate/Attenuate Induction of the ABC Phenomenon	270
	7.6	Clinical Implications of ABC Phenomenon	275
	7.7	Conclusion	276

8. Anti-PEG Immunity Against PEGylated Therapeutics — **289**

Amr S. Abu Lila, PhD, and Tatsuhiro Ishida, PhD

	8.1	Introduction	289
	8.2	PEG Immunogenicity in Animal Models	290
	8.3	PEG Immunogenicity in Humans	293
	8.4	Properties of Anti-PEG Antibody Epitope	298
	8.5	Strategies to Avert Anti-PEG Antibody Responses	299
	8.6	Conclusion	301

9. Complement Activation: Challenges to Nanomedicine Development — **311**

Dennis E. Hourcade, PhD, Christine T. N. Pham, MD, and Gregory M. Lanza, MD, PhD

	9.1	Introduction	311

9.2	C Activation Pathways and Downstream Effectors	312
9.3	Role of C in Human Health and Disease	314
9.4	Complement Activation by Biomaterials	316
9.5	Nanomedicine-Mediated C Activation	317
9.6	Current Methods to Measure Nanomedicine-Mediated C Activation	319
9.7	Conclusions	328

10. Intravenous Immunoglobulin at the Borderline of Nanomedicines and Biologicals: Antithrombogenic Effect via Complement Attenuation — 341

Milan Basta, MD, PhD

10.1	Introduction	341
10.2	Atherosclerosis	343
10.3	Antiphospholipid Syndrome	347
10.4	Sickle Cell Disease	351
10.5	Mechanism of IVIG Modulation of Vaso-Occlusive Disorders	353
10.6	Summary and Outlook	355

11. Lessons Learned from the Porcine CARPA Model: Constant and Variable Responses to Different Nanomedicines and Administration Protocols — 361

Rudolf Urbanics, MD, PhD, Péter Bedőcs, MD, PhD, and János Szebeni, MD, PhD

11.1	Introduction: CARPA as an Immune-Mediated Stress Reaction in Blood Triggered by Nanomedicines	361
11.2	*In vitro* Tests for CARPA	363
11.3	Animal Models of Immune Toxicity: Which Is Good for CARPA Evaluation?	363
11.4	Non-Standard Immunotoxicity Tests of CARPA in Different Animals	364
11.5	The Use of Pigs as Disease Models	366
11.6	Technical Details of the Porcine CARPA Model	366
11.7	The Symptom Tetrad	367

11.8	Uniqueness of the Porcine CARPA Model	369
11.9	Invariable Parameters of Porcine CARPA	369
11.10	Variable Parameters of Porcine CARPA	371
11.11	Summary and Future Directions	376

12. Blood Cell Changes in Complement Activation-Related Pseudoallergy: Intertwining of Cellular and Humoral Interactions — 389

Zsófia Patkó, MD, PhD, and János Szebeni, MD, PhD, DSc

12.1	Introduction	389
12.2	Blood Cell Changes in CARPA: Human and Animal Data	390
12.3	Platelets and Their Role in CARPA	393
12.4	Zooming on Interactions, Receptors and Mediators in CARPA-Associated Blood Cell Changes	399
12.5	Conclusions	407

13. Rodent Models of Complement Activation-Related Pseudoallergy: Inducers, Symptoms, Inhibitors and Reaction Mechanisms — 417

László Dézsi, PhD, László Rosivall, MD, PhD, Péter Hamar, MD, PhD, János Szebeni, MD, PhD, and Gábor Szénási, PhD

13.1	Introduction	417
13.2	The CARPA genic Effects of CVF, Zymosan and LPS in Rodents and Their Modulation with Complement Antagonists	418
13.3	CARPA in Pregnancy	420
13.4	Characteristics of Liposome-Induced CARPA in Rats	420
13.5	Effects of Complement Components C3a and C5a in the Guinea Pig	426
13.6	Effects of Complement Components C3a and C5a in the Rat	428
13.7	Effects of Complement Components C3a and C5a in the Rabbit, Hamster and Mouse	429
13.8	Conclusions	431

14. Immune Reactions in the Delivery of RNA Interference-Based Therapeutics: Mechanisms and Opportunities   441

Kaushik Thanki, PhD, Emily Falkenberg, Monique Gangloff, PhD, and Camilla Foged, PhD

14.1	Background	441
14.2	Overview of Approaches for Efficient Intracellular Delivery of siRNA	445
14.3	Immune Reactions Elicited during RNAi Therapy	452
14.4	Recent Advancements in Predicting Immunological Complications	456
14.5	Conclusions and Future Perspectives	462

15. Lipid Nanoparticle Induced Immunomodulatory Effects of siRNA   473

Ranjita Shegokar, PhD, and Prabhat Mishra, PhD

15.1	Introduction	473
15.2	Discovery of siRNA	474
15.3	Strategies for Delivering siRNA	477
15.4	Immune Response to siRNA Payload and Its Modulation	479
15.5	Clinical Status	491
15.6	Conclusions	494

16. Nanovaccines against Intracellular Pathogens Using *Coxiella burnetii* as a Model Organism   507

Erin J. van Schaik, PhD, Anthony E. Gregory, PhD, Gerald F. Audette, PhD, and James E. Samuel, PhD

16.1	Introduction	507
16.2	Using Nanomedicine to Tackle Intracellular Pathogens	508
16.3	*C. burnetii* Bacteriology	509
16.4	Epidemiology and Clinical Manifestations	509
16.5	Virulence	510
16.6	Immune Evasion Strategies	512
16.7	Qvax® and Correlates of Protective Immunity	514

16.8 Nanovaccines	516
16.9 Why Are Nanoparticle Vaccines a Good Strategy for *C. burnetii*?	522

17. Immunogenicity Assessment for Therapeutic Protein Products — 537

17.1 Introduction	538
17.2 Background	539
17.3 Clinical Consequences	540
17.4 Recommendations for Immunogenicity Risk Mitigation in the Clinical Phase of Development of Therapeutic Protein Products	545
17.5 Patient- and Product-Specific Factors That Affect Immunogenicity	549
17.6 Conclusion	567

18. Assay Development and Validation for Immunogenicity Testing of Therapeutic Protein Products — 585

18.1 Introduction	586
18.2 Background	587
18.3 General Principles	588
18.4 Assay Design Elements	591
18.5 Assay Development	608
18.6 Assay Validation	615
18.7 Implementation of Assay Testing	621
18.8 Documentation	623

19. The "Sentinel": A Conceptual Nanomedical Strategy for the Enhancement of the Human Immune System — 627

Frank J. Boehm and Angelika Domschke, PhD

19.1 Introduction	627
19.2 Brief Survey of Current Nanomedical Research: Toward Immune System Augmentation	628
19.3 Conceptual "Sentinel" Nanomedical Platform for the Significant Enhancement of the Human Immune System	631
19.4 Conclusion	638

20. Immunotherapy for Gliomas and Other Intracranial Malignancies   643

Mario Ganau, MD, PhD, Gianfranco K. I. Ligarotti, MD, Salvatore Chibbaro, MD, PhD, and Andrea Soddu, PhD

    20.1 Regional Immunotherapy: A Rising Trend in Nanomedicine   643
    20.2 Primary and Secondary Brain Tumors   645
    20.3 Current Approaches to Immunotherapy for Brain Tumors   646
    20.4 Review of Ongoing Clinical Trials   648
    20.5 Neuro-Oncology and Immunotherapy: An Outlook for the Next 10 Years   649

21. Engineering Nanoparticles to Overcome Barriers to Immunotherapy   657

Randall Toy, PhD, and Krishnendu Roy, PhD

    21.1 Introduction   657
    21.2 Engineering Nanoparticles to Manipulate Transport and the Immune Response   664
    21.3 Improving Nanoparticle Design to Enhance Immunotherapy Efficacy   670
    21.4 Conclusions   685

22. Metal-Based Nanoparticles and the Immune System: Activation, Inflammation, and Potential Applications   699

Yueh-Hsia Luo, PhD, Louis W. Chang, PhD, and Pinpin Lin, PhD

    22.1 Introduction   699
    22.2 Nanoparticles and Immune System   707
    22.3 Conclusion and Future Perspectives   719

23. Silica Nanoparticles Effects on Hemostasis   731

Volodymyr Gryshchuk, PhD, Volodymyr Chernyshenko, PhD, Tamara Chernyshenko, Olha Hornytska, PhD, Natalya Galagan, PhD, and Tetyana Platonova, DrSc

    23.1 Introduction   731
    23.2 Materials and Methods   732

23.3	Results	735
23.4	Discussion	745
23.5	Conclusions	746

24. Valproate-Induced Rodent Model of Autism Spectrum Disorder: Immunogenic Effects and Role of Microglia — 753

Prabha S. Awale, PhD, James C. K. Lai, PhD, Srinath Pashikanthi, PhD, and Alok Bhushan, PhD

24.1	Introduction	753
24.2	Autism Spectrum Disorders: Etiology and Pathogenesis	754
24.3	Valproate-Induced Rodent Model of Autism Spectrum Disorders	755
24.4	Mechanism of Action of VPA	763
24.5	Conclusions and Future Prospects	764

25. Accelerated Blood Clearance Phenomenon and Complement Activation-Related Pseudoallergy: Two Sides of the Same Coin — 771

Amr S. Abu Lila, Janos Szebeni, and Tatsuhiro Ishida

25.1	Introduction	771
25.2	Immunogenicity of Liposomal Drug Delivery Systems	772
25.3	Features Distinguishing CARPA from Classical IgE-Mediated Immunity	777
25.4	Mechanism of CARPA	777
25.5	Factors Affecting Complement Activation by Liposomes	781
25.6	Predictive Tests for CARPA	784
25.7	Strategies to Attenuate/Abrogate CARPA	787
25.8	Conclusions	789

26. Current and Rising Concepts in Immunotherapy: Biopharmaceuticals versus Nanomedicines — 801

Matthias Bartneck, PhD

| 26.1 | Immunity in Inflammatory Disease and Cancer | 801 |
| 26.2 | Nanomedicine | 804 |

26.3 Therapeutic Modulation of Immunity 808
26.4 Conclusions 824

27. Characterization of the Interaction between Nanomedicines and Biological Components: *In vitro* Evaluation 835

Cristina Fornaguera, MSc, PhD

27.1 Introduction 835
27.2 Experimental Techniques for the Analysis of Nanoparticle Interaction with Biological Components 842
27.3 Conclusions and Future Prospects 860

28. Unwanted Immunogenicity: From Risk Assessment to Risk Management 867

Cheryl Scott

28.1 Introduction 867
28.2 Cause and Effect 870
28.3 Evaluating Immunogenicity 872
28.4 Predict and Prevent 878
28.5 From Start to Finish—and Beyond 886

29. Emerging Therapeutic Potential of Nanoparticles in Pancreatic Cancer: A Systematic Review of Clinical Trials 893

Minnie Au, MBBS, Theophilus I. Emeto, PhD, Jacinta Power, MBBS, Venkat N. Vangaveti, PhD, and Hock C. Lai, MBBS

29.1 Introduction 894
29.2 Methods Section 896
29.3 Results 898
29.4 Synthesis of Study Results 907
29.5 Discussion 913
29.6 Conclusions 919

30. SGT-53: A Novel Nanomedicine Capable of Augmenting Cancer Immunotherapy 929

Joe B. Harford, PhD, Sang-Soo Kim, PhD, Kathleen F. Pirollo, PhD, Antonina Rait, PhD, and Esther H. Chang, PhD

30.1 Introduction 929

30.2	The Role of p53 in Cancer	932
30.3	Cancer Therapeutics Based on p53	937
30.4	The Role of p53 as Guardian of Immune Integrity	938
30.5	SGT-53, A Novel Nanomedicine for *TP53* Gene Therapy	939
30.6	SGT-53 Augments Cancer Immunotherapy Based on an Anti-PD1 Monoclonal Antibody	948
30.7	Summary and Perspectives	954

Index   971

List of Corresponding Authors

Matthias Bartneck Department of Medicine III, Medical Faculty, RWTH Aachen, Pauwelsstr. 30, 52074 Aachen, Germany, Email: mbartneck@ukaachen.de

Milan Basta Biovisions, Inc., 9012 Wandering Trail Dr, Potomac, MD 20854, USA, Email: Basta.Milan@gmail.com

Raj Bawa Bawa Biotech LLC, 21005 Starflower Way, Ashburn, Virginia, USA, Email: bawa@bawabiotech.com

Alok Bhushan Department of Pharmaceutical Sciences, Jefferson College of Pharmacy, Thomas Jefferson University, 901 Walnut St., Suite 915, Philadelphia, PA 19107, USA, Email: alok.bhushan@jefferson.edu

Frank Boehm 1987 W14th Avenue, Vancouver, BC, Canada V6J 2K1, Email: frankboehm@nanoappsmedical.com

Gerrit Borchard School of Pharmaceutical Sciences, University of Geneva, University of Lausanne, 30 quai Ernest-Ansermet, 1211 Geneva, Switzerland, Email: Gerrit.Borchard@unige.ch

László Dézsi Nanomedicine Research and Education Center, Semmelweis University, 1089 Budapest, Nagyvárad tér 4, Hungary, Email: dezsi.laszlo@med.semmelweis-univ.hu

Marina A. Dobrovolskaia Nanotechnology Characterization Laboratory, Cancer Research Technology Program, Leidos Biomedical Research Inc., Frederick National Laboratory for Cancer Research, NCI at Frederick, 1050 Boyles Street, Frederick MD 21702, USA, E-mail: marina@mail.nih.gov

Theophilus I. Emeto Public Health and Tropical Medicine, College of Public Health, Medical and Veterinary Sciences, James Cook University, James Cook Drive, Douglas, Townsville QLD 4811, Australia, Email: Theophilus.emeto@jcu.edu.au

Camilla Foged Department of Pharmacy, University of Copenhagen, Building: 13-4-415b, Universitetsparken 2, DK-2100 Copenhagen, Denmark, Email: camilla.foged@sund.ku.dk

Cristina Fornaguera Group of Materials Engineering (GEMAT), Bioengineering Department, IQS School of Engineering, Ramon Llull University, Via Augusta 390, 08017 Barcelona, Spain, Email: cristina.fornaguera@iqs.url.edu

Mario Ganau Suite 2204—70 Temperance St, MH5 0B1 Toronto, Canada, Email: mario.ganau@alumni.harvard.edu

Volodymyr Gryshchuk ESC "Institute of Biology", Kyiv National Taras Shevchenko University, 64/13 Volodymyrska Street, Kyiv 01601, Ukraine, Email: gryshchukv@gmail.com

Joe B. Harford SynerGene Therapeutics, Inc., 9812 Falls Rd., Suite 114, Potomac, MD, USA

Dennis E. Hourcade Division of Rheumatology, Washington University School of Medicine, 660 South Euclid Avenue, Campus Box 8045St. Louis, MO 63110, USA, Email: dhourcade@wustl.edu

Tatsuhiro Ishida Department of Pharmacokinetics and Biopharmaceutics, Subdivision of Biopharmaceutical Sciences, Institute of Biomedical Sciences, Tokushima University, 1-78-1, Sho-machiTokushima 770-8505, Japan, Email: ishida@tokushima-u.ac.jp

Pinpin Lin National Environmental Health Research Center, National Health Research Institutes, 35 Keyan Road, Zhunan 35053, Miaoli County, Taiwan, Email: pplin@nhri.org.tw

Angelika Lueking Protagen AG, Otto-Hahn-Str. 15, 44227, Dortmund, Germany, Email: angelika.lueking@protagen.com

Zsófia Patkó Department of Nuclear Diagnostics and Therapy, Borsod-Abaúj-Zemplén County Hospital, 3526 Miskolc, Szentpeteri Kapu 72-76, Hungary, Email: patko.zsofia@gmail.com

Krishnendu Roy The Wallace H. Coulter Department of Biomedical Engineering, at Georgia Tech and Emory, The Parker H. Petit Institute for Bioengineering and Biosciences, Georgia Institute of Technology, EBB 3018, 950 Atlantic Dr. NW, Atlanta, GA 30318, USA, Email: krish.roy@gatech.edu

James E. Samuel Department of Microbial Pathogenesis and Immunology, College of Medicine Texas A&M University, 3107 Medical Research and Education Building, 8447 State Hwy 47, Bryan, TX 77807, USA, Email: jsamuel@medicine.tamhsc.edu

Cheryl Scott 340 South 46th Street, Springfield, OR 97478, USA, Email: cscott@knect365.com

Ranjita Shegokar Free University of Berlin, Department of Pharmaceutics, Biopharmaceutics and NutriCosmetics, 12169 Berlin, Germany, Email: ranjita@arcslive.com

János Szebeni Nanomedicine Research and Education Center Semmelweis University, 1089 Nagyvárad tér 4, Budapest, Hungary, Email: jszebeni2@gmail.com

Qun Zhou Protein Engineering, Biologics Research, Sanofi, 49 New York Avenue, Framingham, MA 01701, USA, Email: qun.zhou@sanofi.com

Foreword

It is a pleasure to write the foreword for the third volume of the Series on Nanomedicine, titled *Immune Aspects of Biopharmaceuticals and Nanomedicines*. I have had the pleasure of knowing Dr. Raj Bawa, as a friend and colleague, for more than a decade and have seen firsthand his leadership in the legal, business, translational, and regulatory matters of nanomedicine. These are at the forefront of a global effort to see nanoformulations deployed in the diagnosis and therapy of human disease for the betterment of humankind and the focus of the current volume.

The chapters herein and my own research are congruent as both focus on how the immune system may be harnessed to influence nanomedicine treatment outcomes. My journey in the field centers on harnessing immunity to improve the delivery of biopharmaceuticals. While innate "nonspecific" immunity provides a first line of defense against foreign microbes, cancer, and a spectrum of particulates, nanoparticles potentiate such activities by inciting inflammatory responses. These affect the function and control of neutrophils and macrophages in their fight against a spectrum of diseases. However, it is the size, shape, and charge of the nanoparticles that affect their positive control over immune responses. Thus, as is highlighted in this book, preclinical safety tests for therapeutic nanoparticles need to include the assessment of immune control. A principal part of this control revolves around complement activation, which affects dynamic interactions between the nanoparticle surface composition and the clearance and safety of the formulations themselves. Nanoparticles stimulate immunity and can mimic what is seen by the bacterial lipopolysaccharide. These all can affect the delivery and depot formations of the nanoparticle containing drug, protein, or nucleic acid as well as alter the adverse outcomes of immune-based toxicological assessments. For example, nanoparticles themselves, in severe cases, can induce adverse cardiopulmonary distress. In this instance, the complement system and reactive macrophages become the common effectors. Rapid macrophage clearance of nanoparticles is the known mediator of these adverse cardiopulmonary reactions and as such delays particle macrophage recognition and attenuates adverse reactions. Indeed, the immune system facilitates the controlled release of drugs to the site of disease or injury and is helped by specific nanocarriers that enable drug efficiency in disease or injury.

Interestingly, it is the macrophage that acts as the conductor of immune repair. It has been my lifelong quest to understand and apply the role of the immune system, specifically the macrophage, in combating disease. Many believe that the macrophage leads the orchestra that makes up various components of innate and adaptive immunity. The macrophage is a versatile cell. This versatility is realized by its abilities to sense its environment, engulf toxic tissue products and microbes, and rid the body of all of them. The elimination of harmful factors occurs in tandem with a cell's abilities to maintain the tissue microenvironment. This is made possible through the secretion of bioactive products enabling the continuance of general homeostatic functions. Macrophages perform specific adaptive activities and serve also as the major body armor protecting it against infectious, cancerous, and chemical insults. The past decade has seen a realization that macrophage function could be harnessed for biopharmaceuticals and nanomedicines to promote drug delivery and to sustain drug depots. Moreover, the cell's mobility function enables it to carry payloads of active biopharmaceuticals to action sites to curb microbial growth or sustain tissue and neighboring cell function. Receptor-based targeting facilitates drug-carried nanoparticles resulting in improved outcome measures. All are made possible through high loading capacity, reduced toxicity, and nanoparticle drug stability when comparisons are made against native medicines. Macrophages serve as therapeutic carriers, facilitate tissue repair, and rid the body of cancer, toxic cells, debris, and proteins. All are anchored to nanotherapeutics and create a footprint for pharmaceutical developments. Indeed, state-of-the-art "instruments," including functionalizing ligands and targeting modalities, serve to facilitate cell-based delivery. Nanotechnology promotes cell-to-cell interactions improving a diseased microenvironment, an especially important outcome for cancer treatment.

This volume, forged by experts in the field, demonstrates that whether it is cell-based delivery or direct targeting of cells by biomaterials, harnessing immune control for nanomedicines is promising for diagnostic and therapeutic gain. This, at the end of the day, is a singular goal of all nanoscientists engaged in health care. However, the immune system may prove to be *friend or foe* in such an effort. As a friend, it serves to facilitate the actual formulations and their development, is used as a drug carrier, enables probes for imaging, improves targets for chemotherapy, serves as a means to modulate immune responses after vaccines to prevent infections, or repairs disease-associated tissue injuries. Nanocarriers serve as the enablers of improved drug distribution that occurs by slow effective release of bioactive agents at specific disease sites either by altering the hydrolysis of drug or by serving as particle-drug depots. Drugs or bioagents most amenable for drug action are by definition poorly soluble and evolve through improved bioavailability and lipophilicity. The cells' membranes are penetrated by nanoparticles and as such can overcome a lack of drug specificity. Drug specificity can be improved by nanoparticle decoration, by extending drug circulation time, by altering dissolution rates, or by encasing prodrugs into particles and affecting hydrolysis. All this can result in improved outcomes for the immune clearance of infectious agents and tumor cells. The foe, of course, are the harmful effects that occur as a result of immune engagement of the particles themselves. Particles will nearly always engage the innate immune system and stimulate this function resulting in tissue injuries. This can also lead to the blockade of functionalized particles to specific parts of the immune system that would otherwise metabolize and destroy them. Allergic reactions may also occur as a result of the particles. Thus, to act as immune stimulators or to facilitate adaptive immune responses, a balance is struck between the clearance of infectious and tumor agents and the generalized immune dysfunction. This could preclude therapeutic outcomes if the balance between help and harm is not achieved.

The process of simulating an infectious agent to aid in the clearance of infectious agents and for cancer and other diseases typifies the positive aspects of nanotechnology and medicine. Without question, nanoparticles can provide successful vaccination outcomes with virus-like particles mimicking natural infection and are able to elicit effective immune responses. The events linked to immune control not only improve drug biodistribution and drug action but can affect their rapid polymer degradation.

Such polymers can also enable such active proteins to assemble at appropriate times and under specified biological conditions by using the nanocarrier as a delivery vehicle and shield of the protein of interest. Proteins may serve as adjuvants for vaccines, maintain tissue homeostasis, and stimulate the conversion of somatic cells to inducible pluripotential stem cells (iPSCs). The iPSC technology has been aided greatly by nanomedicines and currently is undergoing a renaissance strongly dependent on the use of recombinant proteins for cell growth, differentiation, and replacement. The enablers are proteins that provide the required growth factors, and as such nanomaterials hold numerous prospective applications and therapeutic opportunities, including a plethora of anti-infective and anti-cancerous medicines.

On balance, vaccine approaches have had considerably less success for latent microbial infections. Organisms that are latent or that can change their molecular coat are singled out as they are much less effectively eliminated by the immune system. These include, but are not limited to, the human immunodeficiency virus, protozoal infections, and tuberculosis. Underlying the lack of pathogen clearance is the capability of the innate and adaptive immune systems. For example, macrophages often become reservoirs of such infections, and replication can continue unabated or the organisms can be stored with subcellular organelles. Alternatively, B and T cells, while producing large amounts of antibodies and cytotoxic T cells, become ineffective in their responses to antigens. Neutralizing antibodies and cytotoxic effector cells mounted against the invading pathogens fall short of eliminating them. While neutralizing antibodies elicit protective immunity, later microbial exposures fail to curtail disease symptoms. Moreover, during these types of chronic infections and notably in cancer, the host fails to be protected, as the spectrum of antigens changes and antibody affinity is not adequate to clear the organism or the cancerous cells. There is simply no protective immunity against individual pathogen strains, and any re-exposure leads to disease and, in the more severe cases, death. There are numerous explanations for the lack of protective immunity, and many of them could be overcome through the use of novel nanoparticle strategies for antigen delivery. Apart from the traditional and conjugate vaccines, recombinants were developed where genetically engineered nucleic acid encodes a specific antigen. This antigen is used to elicit specific immunological responses. DNA-induced immunity results from dendritic cell induction of cytotoxic lymphocytes through MHC and enhances specific adaptive immunity. Nanoparticles have proved to be helpful in vaccine development based on improved delivery and in the generation of potent immune responses. Increased antigen stability and immunogenicity with improved delivery to dendritic cells can be accomplished through nanoformulations. Once the nanocarrier is delivered in polymeric, liposomal, virus-like nanoparticles it can deliver the antigen in conjunction with the MHC II complex and facilitates effector T cell activation. This leads to humoral and cell-mediated immunity responses. Therapeutic nanoparticle vaccines are also being developed to target tumor cells as well as to suppress those elements of the immune system (innate or adaptive) that affect potent immune activities against cancer cells.

Micro- and nanoparticles have also been developed to improve diagnostic endpoints. For example, paramagnetic microparticles, when coated with specific antigens, can enhance magnetic resonance imaging and can be used for precise cancerous or microbial diagnoses. Rapid detection methods are being developed with antibody-immobilized fluorescent nanoparticles for point-of-care diagnostics. Positron emission tomography and single-photon emission computed tomography imaging have led to the development of new nanoparticle drug delivery systems, and at the same time, afford new diagnostic and therapeutic radiopharmaceuticals. These particles offer diagnostic and therapeutic (theranostic pharmaceuticals) approaches for delivering drugs, ferrite, and radionuclides and can be completed using the identical biological and pharmacological mechanisms. The discovery and development of innovative nanomedicines will certainly improve the delivery of therapeutic and diagnostic agents. The "next generation" therapies will deliver drugs, therapeutic proteins, and recombinant DNA to focal areas of disease or to tumors to maximize clinical benefits while limiting untoward side effects. The use of nanoscale technologies to design novel drug delivery systems and devices in biomedical research promises breakthrough advances in immunology and cell biology. All are facilitated by the engagement of the innate immune system.

In summary, the scholarly chapters presented in this book represent a rich undertaking of the roles by which nanoparticles can engage the immune system to improve health as well as cautionary notes for notable untoward reactions. The work is both comprehensive and well written and surely will occupy a place for experts and students alike seeking to better understand the consequences of the immune control of nanotechnologies. Hearty congratulations to the editors and the contributors for a job well done.

Howard E. Gendelman, MD
Margaret R. Larson Professor of Infectious Diseases and Internal Medicine
Professor and Chair
Department of Pharmacology and Experimental Neuroscience
Editor-in-Chief, the *Journal of Neuroimmune Pharmacology*
University of Nebraska Medical Center, Omaha, Nebraska, USA

Howard E. Gendelman is Margaret R. Larson Professor of Internal Medicine and Infectious Diseases, and Chairman of the Department of Pharmacology and Experimental Neuroscience at the University of Nebraska Medical Center. He is credited for unraveling how functional alterations in brain immunity induce metabolic changes and ultimately lead to neural cell damage for a broad range of infectious, metabolic, and neurodegenerative disorders. These discoveries have had broad implications in developmental therapeutics aimed at preventing, slowing, or reversing neural maladies. He is also the first to demonstrate that AIDS dementia is a reversible metabolic encephalopathy. His work has led to novel immunotherapy and nanomedicine strategies for Parkinson's and viral diseases recently tested in early clinical trials. Dr. Gendelman obtained his bachelor's degree in natural sciences and Russian studies with honors from Muhlenberg College and his MD from the Pennsylvania State University-Hershey Medical Center, where he was the 1999 Distinguished Alumnus. He completed a residency in internal medicine at Montefiore Hospital, Albert Einstein College of Medicine, and was a clinical and research fellow in Neurology and Infectious Diseases at the Johns Hopkins University Medical Center. He occupied senior faculty and research positions at the Johns Hopkins Medical Institutions, the National Institute of Allergy and Infectious Diseases, the Uniformed Services University of the Health Sciences Center, the Walter Reed Army Institute of Research, and the Henry Jackson Foundation for the Advancement in Military Medicine before joining the University of Nebraska Medical Center faculty in March 1993. He retired from the US Army as a lieutenant colonel. Dr. Gendelman has authored over 450 peer-reviewed publications, edited 11 books and monographs, holds 12 patents, is the founding editor-in-chief of the *Journal of Neuroimmune Pharmacology*, and serves on many editorial boards international scientific review and federal and state committees. He has been an invited lecturer to more than 250 scientific seminars and symposia and is a recipient of numerous honors, including the Henry L. Moses Award in Basic Science, the Carter-Wallace Fellow for Distinction in AIDS Research, the David T. Purtilo Distinguished Chair of Pathology and Microbiology, the UNMC Scientist Laureate, NU Outstanding Research and Creativity, the 2013 UNMC Innovator of the Year, the 2014 Outstanding Faculty Mentor of Graduate Students, the 2016 Pioneer in Neurovirology, the 2017 Humanitarian of the Year from the Jewish Federation of Omaha, and the Joseph Wybran Distinguished Scientist. Dr. Gendelman was named a J. William Fulbright Research Scholar at the Weizmann Institute of Science in Israel. In 2001, he received the prestigious Jacob Javits Neuroscience Research Award from the National Institute of Neurological Disorders and Stroke and the Career Research Award in Medicine from the Department of Internal Medicine, UNMC. He is included among a selective scientific group listed on http://hcr.stateofinnovation.com/ as one of the top cited scientists in his field. He has trained more than 50 scientists and his leadership is credited with the growth of the department to be among the top-like ranked and federally funded departments (top seven) nationwide—a particularly noted feat as its position was 89 when he assumed its leadership.

My Life with Biologicals and Nanodrugs: A Twenty-Year Affair

Twenty years now
Where'd they go?
Twenty years
I don't know
Sit and I wonder sometimes
Where they've gone[1]

Twenty years ago, as Primary Examiner at the US Patent Office, I reviewed and granted many US patents on biotechnology-based drug products and nanoparticulate drug formulations, a small fraction of which were approved by drug regulatory agencies[2] and eventually commercialized. Most of these first-generation, early drug products are still on the market. Since then, I have been involved in all aspects of biotherapeutics and nanomedicines—research, patent practitioner, professor, journal editor, FDA regulatory filings, conference organizer, keynote speaker, and advisor to the drug industry. I have seen the evolution of anything and everything "biotherapeutic" and "nanomedicine." I have marveled at the cutting-edge discoveries and inventions in these emerging fields. I have also stood up to criticize inept governmental regulatory policies, spotty patent examination at patent offices, hyped-up press releases from eminent university professors taunting translation potential of their basic research and development (R&D), inadequate safety policies, and inaccurate

---

Copyright © 2018 Raj Bawa. All rights reserved. The copyright holder permits unrestricted use, distribution, online posting, and reproduction of this article or excerpts therefrom in any medium, provided the author and source are clearly identified and properly credited.

[1]This is an excerpt from the classic song titled *Like A Rock* by music legend Bob Seger, in which the aging songwriter laments the loss of his youth once filled with vim and vigor and wonders where time went.

[2]The primary drug agencies are the European Medicines Agency (EMA), the US Food and Drug Administration (FDA), Health Canada (HC), and the Japanese Pharmaceuticals and Medical Devices Agency (PMDA).

depiction of these drug products by scientists, media, government agencies, and politicians.

Twenty years later, there is a wave of "newer" therapeutics sweeping the world of medicines. Specifically, there is a rapid introduction of two somewhat distinct but overlapping categories of drugs into the pharmaceutical landscape: (1) biotherapeutics ("biologics," "biologicals," "biological products," "biopharmaceuticals," "biomolecular drugs," or "protein products")[3] and (2) nanomedicines ("nanodrugs," "nanoparticulate drug formulations," or "nanopharmaceuticals").[4] For example, biotherapeutics alone have grown from 11% of the total global drug market in 2002 to around 20% in 2017.[5] I estimate that there are over 225 approved biotherapeutics and around 75 approved nanomedicines for various clinical applications. Similarly, by my estimate, hundreds of companies globally are engaged in nanomedicine R&D; the majority of these have continued to be startups or small- to medium-sized enterprises rather than big pharma.

---

[3]Biologicals, including those made by biotechnology, are a special category of "drugs" or medicines. They differ from conventional small-molecule drugs derived by chemical means in that they are derived biologically from microorganisms (generally engineered) or cells (often mammalian, including human cells). In other words, these are human health products generated or produced by modern molecular biological methods, and differ from traditional biological products that are directly extracted from natural biological sources such as proteins obtained from plasma or plants. Most biologicals are large, complex molecules as compared to small-molecule pharmaceuticals. Slight variations between manufactured lots of the same biological product are normal and expected within the manufacturing process. As part of its review, the FDA assesses this and the manufacturer's strategy to control within-product variations. See: Walsh, G. (2002). Biopharmaceuticals and biotechnology medicines: an issue of nomenclature. *Eur. J. Pharm. Sci.*, **15**, 135–138: *"A biopharmaceutical is a protein or nucleic acid-based pharmaceutical substance used for therapeutic or in vivo diagnostic purposes, which is produced by means other than direct extraction from a native (non-engineered) biological source."*

[4]There is no formal or internationally accepted definition for a nanodrug. The following is my definition (see: Bawa, R. (2016). What's in a name? Defining "nano" in the context of drug delivery. In: Bawa, R., Audette, G., Rubinstein, I., eds. *Handbook of Clinical Nanomedicine: Nanoparticles, Imaging, Therapy, and Clinical Applications*, Pan Stanford Publishing, Singapore, chapter 6, pp. 127–169): *"A nanomedicine is (1) a formulation, often colloidal, containing therapeutic particles (nanoparticles) ranging in size from 1–1,000 nm; and (2) either (a) the carrier(s) is/are the therapeutic (i.e., a conventional therapeutic agent is absent), or (b) the therapeutic is directly coupled (functionalized, solubilized, entrapped, coated, etc.) to a carrier."*

[5]Data from the IMS Institute for Healthcare Informatics.

This book will focus on those biologics, biotechnology products, nanomedicines, nanodrug products, and nanomaterials that are employed for medicinal purposes for humans. Many terms used in this book are definitions that come from specific regulations or compendia, but others are being defined as they are used here. The terms "product," "drug formulation," "therapeutic product," or "medicinal product" will be used in the manner the FDA defines a "drug," encompassing both small-molecule pharmaceutical drugs, biologicals, and nanomedicines in the context of describing the final "drug product."[6] Some of the terms will be used synonymously. For example, biotherapeutics, biologicals, biological products, and biologics are equivalent terms. Similarly, nanomedicines, nanodrugs, nanopharmaceuticals, nanoparticulate drug formulations, and nanotherapeutics are the same.

Although there are major benefits touted for these "newer" therapeutics, including a reduction in unwanted side effects, their use does not guarantee the absence of side effects. For example, studies have shown that these therapeutic agents can interact with various components of the immune system to various immunological endpoints, interactions that are fast, complex, and poorly understood. These interactions with the immune system play a leading role in the intensity and extent of side effects occurring simultaneously with their therapeutic efficacy. In fact, when compared to conventional small-molecule pharmaceutical drugs, both biologics and nanomedicines have biological and synthetic entities of a size, shape, reactivity, and structure that are *often* recognized by the human immune system, *sometimes* in an adverse manner. This obviously can negatively affect their effectiveness and safety, and thereby limit their therapeutic application.[7] Some of the undesired immune responses

---

[6]Branded drugs are referred to as "pioneer," "branded" or "reference" drugs. Small molecule drugs approved by the FDA are known as New Chemical Entities (NCEs) while approved biologics are referred to as New Biological Entities (NBEs). As a result, a new drug application for an NCE is known as a New Drug Application (NDA), whereas a new drug application for an NBE is called a Biologic License Application (BLA). Note that prior to the 1980s there were very few marketed biologics, so the very term "pharmaceutical" or "drug" implied a small molecule drug.

[7]10–20% of the medicinal products removed from clinical practice between 1969 and 2005 were withdrawn due to immunotoxic effects. See: Wysowski, D. K., Swartz, L. (2005). Adverse drug event surveillance and drug withdrawals in the United States, 1969–2002: The importance of reporting suspected reactions. *Archiv. Intern. Med.*, **165**(12), 1363–1369.

include complement activation, tissue inflammation, leucocyte hypersensitivity and formation of antibodies associated with clinical conditions. This has highlighted the critical need to evaluate, assay, and devise strategies to overcome adverse immunogenicity of both biotherapeutics[8] and nanotherapeutics[9].

---

[8]Early developers of biologics assumed that as many of these drugs were based on human genes and proteins, the human immune system would not treat them as foreign and not produce antidrug antibodies (ADAs). However, this optimistic view has turned into alarm as some biologics elicit a vigorous immune response that *may sometimes* neutralize, block, or destroy them. Also, most biotherapeutics are engineered to enable dual or multiple binding sites (e.g., conjugated proteins, functionalized antibodies)—all of which could lead to them being recognized as foreign and therefore immunogenic. Specifically, ADAs may (i) neutralize the activity of the biotherapeutic, (ii) reduce half-life by enhancing clearance, (iii) result in allergic reactions, and/or (iv) cross-react with endogenous counterparts to result in "autoimmune-like" reactions. Such effects are rarely observed with conventional small-molecule drug products. For example, some studies have shown that AbbVie's HUMIRA® (adalimumab) does not work in ~20% of patients. Similarly, in 2016, Pfizer had to withdraw a promising anticholesterol biologic (bococizumab) after testing it in more than 25,000 persons. In 2016, the Netherlands Cancer Institute reported that >50% of the anticancer biologics in 81 clinical trials worldwide were generating ADAs, although they could not confirm that this always negatively affected the drug candidate being tested.

Another issue with some biologics is that they show a concentration-dependent propensity for self-association. This can induce adverse immune responses in patients that may affect drug safety and efficacy. See: Ratanji, K. D., Derrick, J. P., Dearman, R. J., Kimber, I. (2014). Immunogenicity of therapeutic proteins: influence of aggregation. *J. Immunotoxicol.*, **11**(2), 99–109.

[9]Clinical application of nanomedicines and nanocarriers is also dogged by safety and nanotoxicity concerns (undesirable adverse effects), especially about their long-term use. In the case of nanomedicines, therapeutic particles are engineered to break tissue physiological barriers for entry and to escape immune surveillance, thereby persisting in body fluids and delivering their active pharmaceutical ingredients (APIs) to the right tissue site. However, this persistence in the body may trigger immune responses. Novel "immune-toxicity" from nanomedicines may result from the unique combinations of shape, size, surface charge, porosity, reactivity, and chemical composition—all aspects to which the immune system may not have adapted to. Often, intravenously administered nanomedicines prime the immune system, leading to adverse reactions and/or loss of efficacy of the drug product. For example, it is now well established that intravenous administration of nanomedicines and nanocarriers *may* provoke "hypersensitivity reactions" (HSR) or "anaphylactoid reactions" that are referred to as complement (C) activation-related pseudoallergy (CARPA). See: Szebeni, J. (2005). Complement activation-related pseudoallergy: A new class of drug-induced acute immune toxicity. *Toxicology*, **216**, 106–121 and Szebeni, J. (2018). Mechanism of nanoparticle-induced hypersensitivity in pigs: complement or not complement? *Drug Discov.*

Not all biotherapeutics, nanoformulations, and nanomaterials are created equal. Given this scientific fact, the risks for immunogenicity should be assessed on a case-by-case basis. In fact, while some biologics, particularly glycoproteins, cause the body to produce antidrug antibodies (ADAs),[8] few elicit immunogenicity in a manner that induces any clinically relevant reaction. Similarly, the diversity of nanomedicines makes it impossible to extrapolate or generalize the immunologic findings from one class of nanomedicines (e.g., nanoliposomes, solid nanoparticles, carbon nanotubes) to another. Nevertheless, the degree of risk for eliciting immune responses from biotherapeutics, nanoformulations, or nanomaterials is considered a major issue during drug R&D and administration to patients. It is now well established that any biotherapeutic, nanoformulation, or nanomaterial can *potentially* exert an immunogenic effect ("immunogenicity risks") depending on a patient's immunologic status, prior history, route/dose/frequency of delivery and unique characteristics of the administered therapeutic product. Therefore, regulatory agencies, particularly the FDA and the EMA, recommend that drug developers employ a risk-based approach to evaluate and reduce adverse immune events related to the administration of these therapeutics that could affect safety and efficacy. These must be carefully evaluated at the earliest stages of drug formulation/development as well as throughout the product lifecycle, including during phase IV. Biotherapeutic drug products containing a non-biologic nanomaterial component are on the rise and may have different immunogenic properties compared with those that contain the biologic alone. Consequently, it is also important that immunogenicity aspects and risks of biotherapeutic drug products

Today, **23**(3), 487–492. These hypersensitivity reactions typically occur directly at first exposure to the nanocarriers without prior sensitization, and the symptoms usually lessen and/or disappear on later treatment. That is why these reactions are labelled as "pseudoallergic" or "nonspecific hypersensitivity." Nanomedicines causing CARPA include radio-contrast media, liposomal drugs (Doxil®, Ambisome® and DaunoXome®, Abelcet®, Visudyne®), micellar solvents (e.g., Cremophor EL, the vehicle of Taxol®), PEGylated proteins and monoclonal antibodies. Drug products other than biologics and nanomedicines such as nonsteroidal anti-inflammatory medicinal products, analgesics and morphine can also trigger CARPA. Also, see: Szebeni, J, Bawa, R. (2018). Immunological issues with medicines of nano size: The price of dimension paradox. In: Bawa, R., et al., eds. *Immune Aspects of Biopharmaceuticals and Nanomedicines*, Pan Stanford Publishing, Singapore, chapter 2, pp. 83–122.

containing non-biologic nanomaterial components be assessed with a focus on whether the nanomaterial components possess adjuvant properties. Similarly, carriers may exhibit inherent immunologic activity that is not related to the loaded active pharmaceutical ingredient (API); this could also affect the safety and effectiveness of the drug product. Another important issue involves the approval of follow-on versions of both biologics and nanomedicines.[10] I wonder how often cost considerations drive the approval process. I suspect that there are enormous pressures on drug regulatory agencies (e.g., the Trump administration's FDA) to grant these drug products. It is no secret that in certain countries these follow-on versions are the preferred drug products and driven by government-controlled healthcare programs. However, it is critical that immune aspects of these so-called "biosimilars" and "nanosimilars" be transparently evaluated and reported during the drug approval process: *Lower drug prices should not supplant patient safety and efficacy.* The recent FDA approval of follow-on versions of Copaxone® is an example that highlights this troubling trend. I believe that accelerating the approval of follow-on versions of biologics and nanomedicines should be science-based and undertaken on a case-by-case basis.[11]

---

[10]Since the replication of biologics is complex and less precise as compared to small molecule drug products, the term generic has been deemed inappropriate.

[11]See: Conner, J. B., Bawa, R., Nicholas, J. M., Weinstein, V. (2016). Copaxone® in the era of biosimilars and nanosimilars. In: Bawa, R., Audette, G., Rubinstein, I., eds. *Handbook of Clinical Nanomedicine: Nanoparticles, Imaging, Therapy, and Clinical Applications,* Pan Stanford Publishing, Singapore, chapter 28, pp. 783–826; Bawa, R. (2018). Immunogenicity of biologics and nanodrugs: An overview. In: Bawa, R., Szebeni, J., Webster, T. J. and Audette, G. F. eds. *Immune Aspects of Biopharmaceuticals and Nanomedicines,* Pan Stanford Publishing, Singapore, chapter 1, pp. 1–82.

Copaxone® is a non-biologic (synthetic) complex drug ("NBCD") and can be considered a first-generation nanomedicine. It is composed of an uncharacterized mixture of immunogenic polypeptides in a colloidal solution. The complexity of glatiramer acetate is amplified by several aspects: (1) the active moieties in glatiramer acetate are unknown; (2) the mechanisms of action are not completely elucidated; (3) pharmacokinetic testing is not indicative of glatiramer acetate bioavailability; (4) pharmacodynamic testing is not indicative of therapeutic activity and there are no biomarkers available as surrogate measures of efficacy; and (5) small changes in the glatiramer acetate mixture can change its immunogenicity profile. There is one aspect of Copaxone® that raises special safety and effectiveness

In our rapidly changing yet interconnected and globalized world, biologics and nanomedicines will continue to surprise and expand. There are numerous second- and third-generation biologics and nanomedicines at the basic research stage. Hopefully, despite enormous bottlenecks, we will find a greater number of these translated into practical patient applications. In the meantime, we need to temper our expectations yet continue to hope for paradigm-shifting advances in the bio-nano world.

Against this backdrop, the editors felt that enormous advances in the past 20 years in immunology of biologics and nanomedicines warranted an authoritative and comprehensive reference resource that can be relied upon by immunologists, biomedical researchers, clinicians, pharmaceutical companies, formulation scientists, regulatory agencies, technology transfer officers, venture capitalists, and policy makers alike. Hence, this volume aims to provide a broad survey of theoretical and experimental knowledge currently available and presents a framework that is readily applicable to develop strategies for clinical applications. Each chapter contains key words, tables and figures in color, future predictions, and an extensive list of references. The focus is on the current, most relevant information, all accomplished in a user-friendly format.

Assorted topics pertain to the immune effects of biologics and nanomedicines, both beneficial and adverse. A thorough understanding of immunology, therapeutic potential, clinical applications, adverse reactions and approaches to overcoming

---

concerns that merit heightened vigilance with respect to the approval of any potentially interchangeable follow-on glatiramer acetate product: Glatiramer acetate is an immunomodulator. In other words, Copaxone® is intended to achieve its therapeutic effects by interacting with and modulating a patient's immune system over an extended period. For this reason, Copaxone®'s package insert warns that chronic use has the potential to alter healthy immune function as well as induce pathogenic immune mechanisms, although no such effects have been observed with Copaxone®. Due to the complexity and inexorable link between the manufacturing process and quality, any follow-on product almost certainly will differ from Copaxone®'s structure and composition of active ingredients because it will be made using a different manufacturing process than that developed by the branded product developer (Teva). Although it is not possible to fully characterize and compare these complex mixtures, differences are revealed via sophisticated analytical techniques. Despite these immunological concerns, the FDA in 2017 approved so-called follow-on versions of Copaxone®.

immunotoxicity of biologics and nanomedicines is presented. For instance, chapters are devoted to immune stimulatory and suppressive effects of antibodies, peptides and other biologics, as well as various nanomedicines. The state of the art in therapeutic and preventive vaccines along with their potential molecular mechanisms underlying immunogenicity is also highlighted. Adverse immune effect of certain biologics and nanomedicines, namely, complement (C) activation-related pseudoallergy (CARPA), is discussed in unprecedented detail in terms of occurrence, prediction, prevention, and mechanism. Furthermore, critical, yet often overlooked topics such as immune aspects of nano-bio interactions, current FDA regulatory guidance, immunogenicity testing of therapeutic protein products, and engineering bio/nanotherapeutics to overcome barriers to immunotherapy are also covered.

I express my sincere gratitude to the authors, coeditors, and reviewers for their excellent effort in undertaking this project with great enthusiasm. I thank my father, Dr. S. R. Bawa, for meticulously reviewing various chapters of this book. Finally, I also thank Mr. Stanford Chong and Ms. Jenny Rompas of Pan Stanford Publishing for commissioning me to edit this volume. Mr. Arvind Kanswal of Pan Stanford Publishing and my staff at Bawa Biotech LLC are acknowledged for their valuable assistance with publication coordination.

Raj Bawa, MS, PhD
Series Editor
Ashburn, Virginia, USA
June 7, 2018*

---

*The day my beloved *Washington Capitals* ice hockey team won the Stanley Cup for the first time in franchise history!

Chapter 1

Current Immune Aspects of Biologics and Nanodrugs: An Overview

Raj Bawa, MS, PhD

Patent Law Department, Bawa Biotech LLC, Ashburn, Virginia, USA
The Pharmaceutical Research Institute,
Albany College of Pharmacy and Health Sciences, Albany, New York, USA
Department of Biological Sciences,
Rensselaer Polytechnic Institute, Troy, New York, USA

> Copyright 2018 Raj Bawa. All rights reserved. As a service to authors and researchers, the copyright holder permits unrestricted use, distribution, online posting and reproduction of this article or unaltered excerpts therefrom, in any medium, provided the author and original source are clearly identified and properly credited. The figures in this chapter that are copyrighted to the author may similarly be used, distributed, or reproduced in any medium, provided the author and the original source are clearly identified and properly credited. A copy of the publication or posting must be provided via email to the copyright holder for archival.

Immune Aspects of Biopharmaceuticals and Nanomedicines
Edited by Raj Bawa, János Szebeni, Thomas J. Webster, and Gerald F. Audette
Copyright © 2018 Raj Bawa
ISBN 978-981-4774-52-9 (Hardcover), 978-0-203-73153-6 (eBook)
www.panstanford.com

Keywords: biotherapeutics, biologics, biologicals, biological products, biopharmaceuticals, biomolecular drugs, protein products, nanomedicine, nanodrugs, nanoparticulate drug formulations, nanopharmaceuticals, nanotechnology, nanomaterial, nanoscale, patents, commercialization, research and development (R&D), US Food and Drug Administration (FDA), European Medicines Agency (EMA), drug delivery systems (DDS), site-specific delivery, nanoparticles (NPs), protein aggregation, small-molecule drug, New Chemical Entities (NCEs), New Biological Entities (NBEs), New Drug Application (NDA), Biologic License Application (BLA), Bayh–Dole Act, Hatch–Waxman Act, Biologics Price Competition and Innovation Act (BPCI Act), immunotoxic effects, complement activation, immunogenicity, antidrug antibodies (ADAs), antibody–drug conjugates (ADCs), adverse drug reaction (ADR), conjugated proteins, functionalized antibodies, Federal Food, Drug, and Cosmetic Act (FD&C Act), target mediated drug disposition (TMDD), pharmacodynamic (PD), Public Health Service (PHS) Act, pharmacokinetics (PK), Humira®, protein aggregates, active pharmaceutical ingredient (API), hypersensitivity reactions (HSR), anaphylactoid reactions, complement activation-related pseudoallergy (CARPA), Doxil®, Ambisome®, DaunoXome®, Abelcet®, Visudyne®, Cremophor EL, PEGylated proteins, monoclonal antibodies (mABs), Humulin®, PEGylated liposomes, accelerated blood clearance (ABC), reticuloendothelial system (RES), immune complexes (ICs), biosimilar, generic drugs, bioequivalent, interchangeable product, nanosimilars, nonbiologic complex drug (NBCD), NBCD similar, glatiramer acetate, Copaxone®, immunomodulator, clinical trials, immunopharmacology, immunomodulatory effects, iTope™, TCED™, Epibase®, EpiMatrix™, EpiScreen™, immunogenic epitopes, artificial intelligence (AI), single-cell genomics, user fees, druggable genome, cryo-electron microscopy (cryo-EM), epitope mapping analysis, bench-to-bedside, translation, drug-like molecule, CRISPR-Cas9

1.1 Introduction

A wave of "newer" therapeutics is sweeping the drug world. Specifically, there is a rapid introduction of two somewhat distinct yet overlapping classes of drugs into the pharmaceutical landscape: (1) biologics[1] and (2) nanodrugs.[2] Biologics have already entered an era of rapid growth due to their wider applications, and in the near future they will replace many existing organic based small-molecule drugs. According to one drug analysis firm, biologics have grown from 11% of the total global drug market in 2002 to around 20% in 2017.[3] On the other hand, nanodrugs have sputtered along a somewhat different trajectory with greater challenges to their translation [1–3]. I estimate that since the approval of the first recombinant biologic (recombinant human insulin, in 1982), there are over 225+ marketed biologics and at least 75 nanodrugs for various clinical applications approved by various regulatory agencies.[4] According to the Pharmaceutical Research and Manufacturers of America (PhRMA) website, as of 2013, there are over 900 biologic medicines and vaccines in development. I estimate that hundreds of companies globally are engaged in nanomedicine research and development (R&D), the clear majority of these have continued to be startups or small- to medium-sized enterprises rather than big pharma. Despite immature regulatory mechanisms, follow-on versions of these two drug classes, namely biosimilars and nanosimilars, respectively, have also started to trickle into the marketplace.

According to the US Food and Drug Administration (FDA), the products it regulates represent around 20% of all products sold in the United States, representing more than $2.4 trillion. The FDA regulates products according to specific categories: food, dietary supplements, cosmetics, drugs, biologics, medical devices, veterinary products, and tobacco. The Center for Biologics

---

[1]Analogous terms include biotherapeutics, biologicals, biological products, biopharmaceuticals, biomolecular drugs, therapeutic protein product (TPP), and protein products.
[2]Analogous terms include nanomedicines, nanoparticulate drug formulations, and nanopharmaceuticals.
[3]Data from the IMS Institute for Healthcare Informatics.
[4]My estimate for nanodrugs is based on my broader definition of a nanodrug that appears in Section 1.3.

Evaluation and Research (CBER) regulates what are often referred to as traditional biologics, such as vaccines, blood and blood products, allergenic extracts, and certain devices and test kits. CBER also regulates gene therapy products, cellular therapy products, human tissue used in transplantation, and the tissue used in xenotransplantation—the transplantation of nonhuman cells, tissues, or organs into a human. On the other hand, the Center for Drug Evaluation and Research (CDER) regulates branded and generic drugs, over-the-counter (OTC) drugs, and most therapeutic biologics (Fig. 1.1a). Food, dietary supplements, and cosmetics fall under the jurisdiction of the Center for Food Safety and Nutrition (CFSAN). Since dietary supplements are intended to supplement the diet, they are classified under the "umbrella" of foods and do not require premarket authorization from the FDA. Cosmetics containing sunscreen components are regulated as drugs. In these cases, the products must be labeled as OTC drugs and meet OTC drug requirements. Tobacco products are subject to a unique regulatory framework as they only pose risks without providing any health benefits. They are regulated by the Center for Tobacco Products (CTP). Medical devices are regulated by the Center for Devices and Radiological Health (CDRH), and veterinary products by the Center for Veterinary Medicine (CVM). Drugs that have high potential for abuse with no accepted medical use are illegal and cannot be imported, manufactured, distributed, possessed, or used. The Drug Enforcement Administration (DEA) is the US agency tasked with overseeing these dangerous products and enforcing the controlled substances laws. The Office of Combination Products (OCP) has authority over the regulatory life cycle of combination products. Combination products are therapeutic and diagnostic products that combine drugs, devices, and/or biological products. As technological advances continue to merge product types and blur the historical lines of separation between various FDA centers, I expect that more products in the near future will fall into the category of combination products. Naturally, this will present unique regulatory, policy, and review management challenges.

    The main law that governs various products in the United States is the Federal Food, Drug, and Cosmetic Act (FD&C Act). It was established in 1938 and has been amended numerous times since. The laws are passed as Acts of Congress and organized/codified into United States Code (USC). Of the 53 titles in the USC, title 21 corresponds to the FD&C Act. To operationalize the law for enforcement, federal agencies, including the FDA, are authorized to create regulations. The Code of Federal Regulations (CFR) details how the law will be enforced. The CFR is divided into 50 titles according to subject matter. Therefore, there are three types of references for regulatory compliance: FD&C Act, 21USC, and 21CFR. The FD&C Act provides definitions for the different product categories along with allowable claims. For example, drugs, biologics, and medical devices can make therapeutic claims like "treatment of a particular disease" or "reduction of symptoms associated with a particular disease." Therapeutic claims also include implied statements like "relieves nausea" or "relieves congestion." It is illegal for nonmedical products like pharma-cosmetics, dietary supplements, and cosmetics to make therapeutic claims. Even if a product lacks any therapeutic ingredient, its intended use may cause it to be categorized as a drug.

This chapter focuses on those biologics, biotechnology products, nanomedicines, nanodrug products, and nanomaterials that are used for medicinal purposes in humans. Many biologics (e.g., monoclonal antibodies or drug–protein conjugates) are of nanoscale and hence can also be considered to be nanodrugs. Conversely, many nanodrugs are biologics according to standard definitions (Sections 1.2 and 1.3). For example, Copaxone® (Section 1.7) is a biologic (Section 1.2) but also falls within the definition of a nanodrug (Section 1.3). Many terms used here are definitions that come from specific regulations or compendia. The terms "product," "drug formulation," "therapeutic product," or "medicinal product" will be used in the manner the FDA defines a "drug," encompassing pharmaceutical drugs, biologics, and nanomedicines in the context of describing the final "drug product." Some of the terms will be used synonymously. For example, biotherapeutics, protein drugs, biologicals, biological products, and biologics are equivalent terms.[1] Similarly, nanomedicines, nanodrugs, nanopharmaceuticals, nanoparticulate drug formulations, and nanotherapeutics are the same.[2] Branded drugs are referred to as "pioneer," "originator," "branded," or "reference" drugs. Small-molecule drugs approved by the FDA are known as New Chemical Entities (NCEs) while approved biologics are referred to as New Biological Entities (NBEs) (Fig. 1.1a, Table 1.1, and Box 1.1). As a result, a new drug application for an

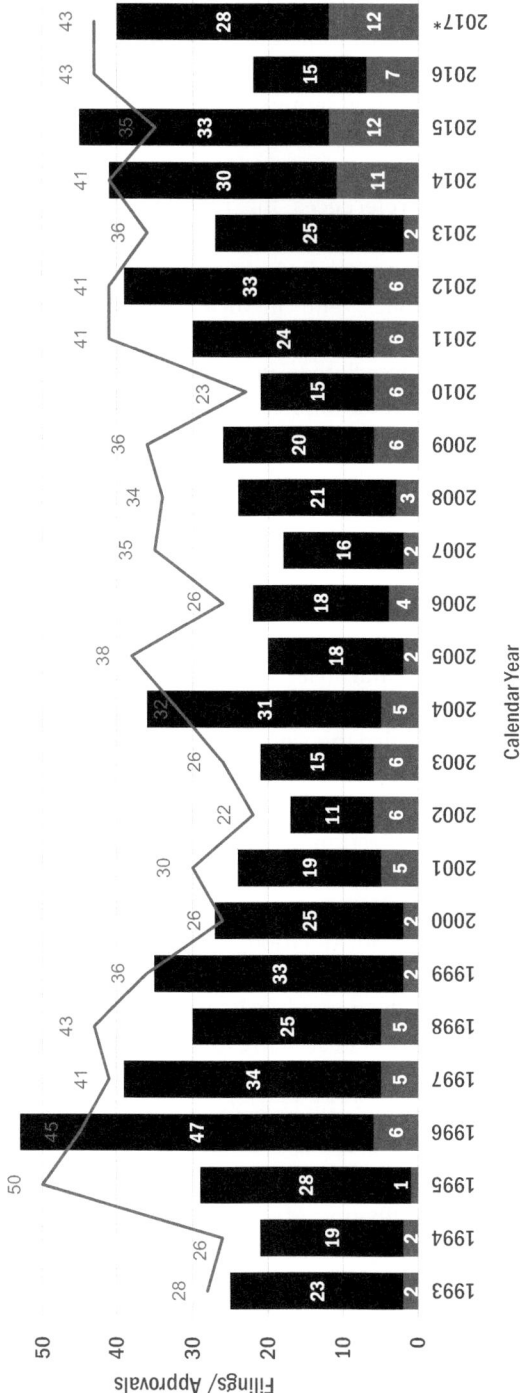

*The data in this figure for 2017 reflect 43 drug filings and are accurate as of November 30, 2017. However, the final count for 2017 (as of December 31, 2017) is 46.

Figure 1.1a Drug filings and approvals by the FDA's Center for Drug Evaluation and Research (CDER) since 1993. Overall, new drugs approved by the FDA have risen in recent years. Strong numbers were posted in 2017, rebounding from a steep drop in 2016. In fact, the 2017 bounty of 46 new drugs was only second to the peak of 59 drugs approved in 1996. Emphasizing its focus on orphan drugs, the FDA approved 18 drugs with orphan designation, 7 potential blockbusters candidates, 12 cancer drugs, and 17 products with breakthrough designation.

CDER approves novel drugs, either as New Molecular Entities (NMEs) under New Drug Applications (NDAs), or as New Biological Entities (NBEs) under Biologics License Applications (BLAs). In recent years, many products were known more for their breathtaking price tags and rapidity of approval rather than innovation. The FDA defines "novel drugs" as innovative products that serve previously unmet medical needs or otherwise significantly help to advance patient treatments and the active ingredient or ingredients in a novel drug have never before been approved in the United States.

In rare instances, it may be necessary for the FDA to change a drug's NME designation or the status of its application as a new BLA. This applies to all references to NMEs/BLAs in this figure. For instance, new information may become available which could lead to a reconsideration of the original designation or status. Note that approvals by the Center for Biologics Evaluation and Research (CBER) are excluded from this figure. Since applications are received and filed throughout a calendar year, the filed applications in a given calendar year do not necessarily correspond to approvals in the same calendar year. Certain applications are within their 60-day filing review period and may not be filed upon completion of the review. Original BLAs that do not contain a new active ingredient are excluded.

Data courtesy of the FDA and various drug companies. For more information about the approved drugs in 2017 and for their complete risk information, refer to the drug approval letters and FDA-approved labeling at Drugs@FDA. Also see CDER's Novel Drug Approvals for 2017 on the FDA website for the approval dates, nonproprietary names and what each drug is used for: https://www.fda.gov/Drugs/DevelopmentApprovalProcess/DrugInnovation/ucm537040.htm.

Figure is courtesy of the FDA while the legend is by the author.

Table 1.1 Properties of biologics versus small-molecule drugs

Property	Biologics	Small-Molecule Drugs
Size and MW	generally large and high MW; MW >700 Da; complex structure	generally small and low; MW <700 Da; simple and defined structure
Manufacturing	numerous critical process steps; highly susceptible to slight alterations in production process; lengthy and complex purification; great possibility of contamination and detection/removal often impossible	fewer critical process steps; not affected by slight alterations in production process; easy to purify; contamination can generally be avoided and detection/removal easy
Composition	protein-based; amino acids; heterogenous mixture that may include variants; may involve post-translational modifications	chemical-based; synthetic organic compound(s); homogenous drug substance (single entity)
Origin	isolated from living cells or recombinantly produced	chemical synthesis
Toxicity	more consistent with exaggerated pharmacology than off-target toxicity; much greater contact surface area for binding allows access to a much wider range of protein targets as well as a more specific binding interaction, decreasing the potential for off-target effects	drug product or metabolites that are generated can be toxic; target binding results in the small-molecule drug being nearly completely buried within a hydrophobic pocket of the protein target to maximize hydrophobic contact plus create a more stable complex, thereby effectively limiting targets to those that possess solvent accessible pockets

Property	Biologics	Small-Molecule Drugs
Dosing Frequency	increased blood circulation time can allow far less frequent dosing	greater dosing frequency
Half-Life	variable; longer half-life (hours, days, weeks)	variable; mostly shorter half-life (hours to days)
Clearance	slow	rapid
Pharmacokinetic (PK) and Distribution	target can affect PK behavior (TMDD); larger molecule(s) and hence reach blood via lymphatics; subject to proteolysis during interstitial and lymphatic transit; distribution generally limited to plasma and/or extracellular fluid	mostly linear PK; nonlinearity from saturation of metabolic pathways; rapid entry into systemic circulation via capillaries; distributed to any combination of organ/tissue
Cost	high, often extremely high	generally low
Drug–Drug Interaction (DDI)	rare or few examples, mostly pharmacodynamic (PD)-related	possible and many examples; metabolic and/or PD related
Off-target Action	rare; mostly "on-target" effects	often "off-target" effects
Mode of Action	regulatory or enzyme activity to replace/augment cell action; may target cell surface to induce action; binding to cell-surface receptors and other markers specifically associated with or overexpressed; limited to extracellular and cell surface interactions	antagonistic/agonistic activity on intracellular and extracellular targets
Storage and Handling Risk	variable; sensitive to environmental conditions (heat and shear)	relatively stable

(Continued)

Table 1.1 (Continued)

Property	Biologics	Small-Molecule Drugs
Contamination Risk	high	low
Structure	may or may not be precisely elucidated or known; inherent variability due to complex manufacturing	precisely defined structure (or structures, e.g, racemic mixtures)
Delivery	generally parenteral (e.g., IV and SC)	various routes; generally oral
Dispensed By	physicians (often specialists) or hospitals	general practitioner or retail pharmacies
Duration of Action	long; days to weeks	short; hours
Characterization	less easily characterized; cannot always be fully characterized	can be fully characterized
Immunogenicity	low to high; usually antigenic and hence potential exists	often non-antigenic and hence low to none
Toxicity	receptor-mediated toxicity	specific toxicity
FDA Approval	licensed under the provisions of both the FD&C Act and the PHS Act (for exceptions see Box 1.1); biologics approved by the FDA are referred to as New Biological Entities (NBEs); a new drug application for an NBE is called a Biologic License Application (BLA) (see Fig. 1.1a)	licensed under the FD&C Act; small-molecule drugs approved by the FDA are known as New Molecular Entities (NMEs); a new drug application for an NCE is known as a New Drug Application (NDA) (see Fig. 1.1a)

Property	Biologics	Small-Molecule Drugs
Compilation	*Purple Book* published by the FDA lists biologics, their biosimilars and interchangeable generic equivalents	*Orange Book* published by the FDA lists drugs and their generic equivalents
Follow-on Versions	biosimilars (see Section 1.6); high barriers to entry; follow-ons will not be identical to the reference innovator product; preclinical and clinical (i.e., safety/efficacy) studies are needed to demonstrate comparability	generics (see Section 1.6); preclinical analytical methods can be used to validate and demonstrate comparability; full clinical studies not needed; follow-ons have identical API(s), strength, dosage form, route, and purity
Patent Issues	patent prosecution and litigation are often more complex; patents and legal exclusivities may delay the FDA approval of applications for biosimilars	patent prosecution and litigation generally less complex; patents and legal exclusivities may delay the FDA approval of applications for generics
Selectivity	high species selectivity (affinity/potency)	generally low species selectivity
Targets	multiple target binding	mostly a single or few targets

Abbreviations: BLA, Biologic License Application; Da, Daltons; DDI, drug–drug interaction; FD&C Act, Federal Food, Drug, and Cosmetic Act; IV, intravenous; MW, molecular weight; NBE, New Biological Entity; NME, New Molecular Entity; NDA, New Drug Application; TMDD, target mediated drug disposition; PD, pharmacodynamic; PHS Act, Public Health Service Act; PK, pharmacokinetic; SC, subcutaneous; API, active pharmaceutical ingredient. Copyright 2018 Raj Bawa. All rights reserved.

NCE is known as a New Drug Application (NDA) while a new drug application for an NBE is known as a Biologic License Application (BLA). Note that prior to the 1980s, there were very few marketed biologics, so the very term "pharmaceutical" or "drug" implied a small-molecule drug. Although biologics are subject to federal regulation under the Public Health Service (PHS) Act, they also meet the definition of "drugs" and are considered a subset of drugs. Hence, biologics are regulated under the provisions of both PHS Act and FD&C Act. Table 1.2 shows the different regulatory routes for therapeutic products.

Table 1.2 FDA regulatory routes for therapeutic products

	Medical Devices	Drugs	Biologics
FDA Center Jurisdiction	CDRH	CDER	CBER/CDER
Regulatory Route(s)	510(k) waived 510(k) notification PMA	OTC ANDA NDA	BLA
Clinical Trial Initiation	IDE	IND	IND

Abbreviations: CBER, Center for Biologics Evaluation and Research; CDER, Center for Drug Evaluation and Research; CDRH, Center for Devices and Radiological Health; NDA, New Drug Application; BLA, Biologic License Application; OTC, over-the-counter; ANDA, Abbreviated New Drug Application; PMA, Premarket Approval Application; IND, Investigational New Drug; IDE, Investigational Device Exemption. Copyright 2018 Raj Bawa. All rights reserved.

As the boundaries between big pharma and biotech companies have further blurred, big pharma has adapted its operational strategy, employing outside collaborations with respect to research, technology, workforce, and marketing. Obviously, big pharma's evolving role has resulted partly from the "biotech boom" and the "genomics boom," where enormous advances resulted from molecular biology and DNA technology, but also from advances in information and computer technology. In addition, two important pieces of legislation in the 1980s have had a major impact on the drug industry in the United States. The first was the Bayh–Dole (or Patent and Trademark Law Amendments) Act of 1980, which allowed universities, hospitals, nonprofit organizations and small businesses to patent and retain ownership arising from federally funded research [4]. The second was the Hatch–Waxman (or Drug Price Competition and Patent Term Restoration) Act of 1984, which established abbreviated pathways for the approval of small-molecule drug products [5]. It set up the modern system of generic drug regulations in the United States by amending the FD&C Act. Section 505(j) of the Hatch–Waxman Act, codified as 21 USC § 355(j), outlines the process for pharmaceutical manufacturers to file an Abbreviated New Drug Application (ANDA) for approval of a generic drug by the FDA.

In addition to the Bayh–Dole Act and Hatch–Waxman Act, the more recent Biologics Price Competition and Innovation Act of 2009 (BPCI Act), which is included in the Patient Protection and Affordable Care Act signed into law by President Obama in 2010, pertains specifically to biologics. This Act created an abbreviated approval pathway for biologics proven to be "highly similar" (biosimilar) to or "interchangeable" with an FDA-licensed reference biologic product [6]. In concept, the goal of the BPCI Act is similar to the Hatch–Waxman Act.

The prohibitive costs of most biologics and some small-molecule drugs has led to increased scrutiny in understanding the US government's role in the development of costly novel drug products. For example, for almost all of the biosimilars approved by the FDA so far, the associated brand-name drug (among the top-selling drugs in the world) was originally formulated by scientists at public-sector research institutions. Hence, like most US tax payers, I question the logic behind allowing sky-rocketing drug prices, especially for branded biologics. Should there be more robust governmental controls on this front? Should the US taxpayer have significant leverage to affect the process? Based on two recent US Court of Appeals for the Federal Circuit (CAFC) decisions and imperfections in the BPCI Act itself, some argue that the law impairs the potential for a flourishing generic market for biologics [7a]. Moreover, since around 90% of the global biosimilar sales come from the European Union (EU), compared to just 2% from the United States, some have questioned whether the US biosimilar industry is falling behind [7b]. The global biosimilars market in 2017 was $4.49 billion and is expected to grow with a compound annual growth rate (CAGR) of 31.7% to $23.63 billion by 2023.[5] Biosimilars are discussed in Section 1.6. Figures 1.1b and 1.1c represent the FDA drug approval process.

---

[5]Data from MarketsandMarkets.com

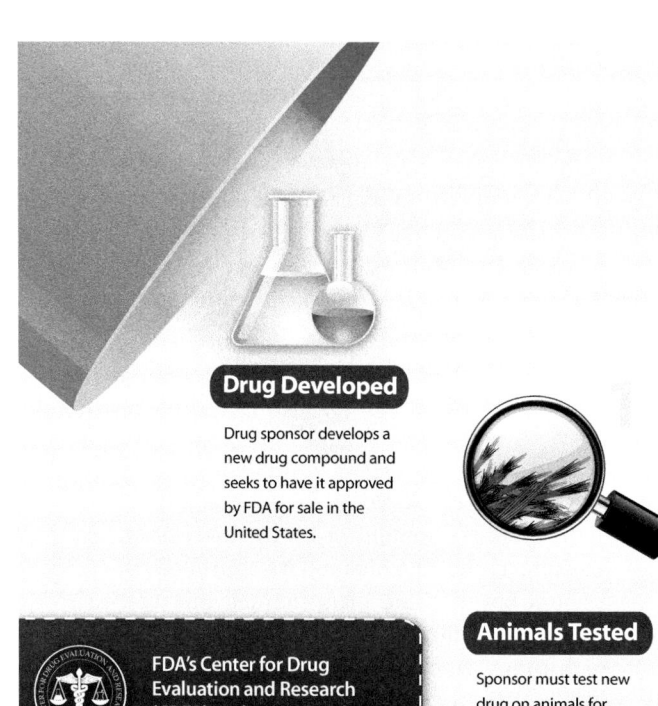

Figure 1.1b Drug sponsor's discovery and screening phase (preclinical).

Figure 1.1c Drug sponsor's clinical studies/trials.

> **Box 1.1 The FDA's view of biological products**
> (courtesy of the FDA, with modifications by the author)
>
> **1. What is a biological product?**
>
> Biological products, like other drugs, are used for the treatment, prevention, or cure of disease in humans. In contrast to chemically synthesized small-molecular-weight drugs, which have a well-defined structure and can be thoroughly characterized, biological products are generally derived from living material—human, animal, or microorganism—are complex in structure, and thus are usually not fully characterized. Section 351 of the *Public Health Service (PHS) Act* defines a biological product as a "virus, therapeutic serum, toxin, antitoxin, vaccine, blood, blood component or derivative, allergenic product, or analogous product, ... applicable to the prevention, treatment, or cure of a disease or condition of human beings." FDA regulations and policies have established that biological products include blood-derived products, vaccines, *in vivo* diagnostic allergenic products, immunoglobulin products, products containing cells or microorganisms, and most protein products. Biological products subject to the *PHS Act* also meet the definition of *drugs* under the *Federal Food, Drug and Cosmetic Act (FDC Act)*. Note that hormones such as insulin, glucagon, and human growth hormone are regulated as drugs under the *FDC Act*, not biological products under the *PHS Act*.
>
> **2. What Center has the regulatory responsibility for therapeutic biological products?**
>
> Both the FDA's Center for Drug Evaluation and Research (CDER) and Center for Biologics Evaluation and Research (CBER) have regulatory responsibility for therapeutic biological products, including premarket review and oversight. The categories of therapeutic biological products regulated by CDER (under the *FDC Act* and/or the *PHS Act*, as appropriate) are the following:
>
> • Monoclonal antibodies for *in vivo* use.

- Most proteins intended for therapeutic use, including cytokines (e.g., interferons), enzymes (e.g., thrombolytics), and other novel proteins, except for those that are specifically assigned to the Center for Biologics Evaluation and Research (CBER) (e.g., vaccines and blood products). This category includes therapeutic proteins derived from plants, animals, humans, or microorganisms, and recombinant versions of these products. Exceptions to this rule are coagulation factors (both recombinant and human plasma derived).
- Immunomodulators (non-vaccine and non-allergenic products intended to treat disease by inhibiting or down-regulating a pre-existing, pathological immune response).
- Growth factors, cytokines, and monoclonal antibodies intended to mobilize, stimulate, decrease, or otherwise alter the production of hematopoietic cells *in vivo*.

3. **Are the biologic development requirements different from the requirements for a new drug product?**

Biological products are a subset of drugs; therefore, both are regulated under provisions of the *FDC Act*. However, only biological products are licensed under section 351 of the *PHS Act*. (As previously noted, some therapeutic protein products are approved under section 505 of the *FDC Act*, not under the *PHS Act*.) Following initial laboratory and animal testing that shows that investigational use in humans is reasonably safe, biological products (like other drugs) can be studied in clinical trials in humans under an investigational new drug application (IND) in accordance with the regulations at *21 CFR 312*. If the data generated by the studies demonstrate that the product is safe and effective for its intended use, the data are submitted as part of a marketing application. Whereas a new drug application (NDA) is used for drugs subject to the drug approval provisions of the *FDC Act*, a biologics license application (BLA) is required for biological products subject to licensure under the *PHS Act*. FDA form 356h is used for both NDA and BLA submissions.

FDA approval to market a biologic is granted by issuance of a biologics license (see Fig. 1.1a).

4. What are the requirements for licensing a biologic?

Issuance of a biologics license is a determination that the product, the manufacturing process, and the manufacturing facilities meet applicable requirements to ensure the continued safety, purity, and potency of the product. Among other things, safety and purity assessments must consider the storage and testing of cell substrates that are often used to manufacture biologics. A potency assay is required due to the complexity and heterogeneity of biologics. The regulations regarding BLAs for therapeutic biological products include *21 CFR parts 600, 601, and 610*.

5. What does safety mean?

The word safety means the relative freedom from harmful effects, direct or indirect, when a product is prudently administered, taking into consideration the character of the product in relation to the condition of the recipient at the time.

6. What is purity?

Purity means relative freedom from extraneous matter in the finished product, whether or not harmful to the recipient or deleterious to the product. Purity includes but is not limited to relative freedom from residual moisture or other volatile substances and pyrogenic substances.

7. What is potency?

The word potency is interpreted to mean the specific ability or capacity of the product, as indicated by appropriate laboratory tests, to yield a given result.

8. Does FDA issue license certificates upon approval of a BLA?

Approval to market a biologic is granted by issuance of a biologics license (including US license number) as part of the approval letter. The FDA does not issue a license certificate. The US License number must appear on the product labeling.

9. Why are biologics regulated under the *PHS Act*?

As mentioned above, biologics are subject to provisions of both the *FD&C Act* and the *PHS Act*. Because of the complexity of manufacturing and characterizing a biologic, the *PHS Act* emphasizes the importance of appropriate manufacturing control for products. The *PHS Act* provides for a system of controls over all aspects of the manufacturing process. In some cases, manufacturing changes could result in changes to the biological molecule that might not be detected by standard chemical and molecular biology characterization techniques yet could profoundly alter the safety or efficacy profile. Therefore, changes in the manufacturing process, equipment, or facilities may require additional clinical studies to demonstrate the product's continued safety, identity, purity, and potency. The *PHS Act* also provides authority to immediately suspend licenses in situations where there exists a danger to public health.

10. How is the manufacturing process for a biological product usually different from the process for drugs?

Because, in many cases, there is limited ability to identify the identity of the clinically active component(s) of a complex biological product, such products are often defined by their manufacturing processes. Changes in the manufacturing process, equipment, or facilities could result in changes in the biological product itself and sometimes require additional clinical studies to demonstrate the product's safety, identity, purity, and potency. Traditional drug products usually consist of pure chemical substances that are easily analyzed after manufacture. Since there is a significant difference in how biological products are made, the production is monitored by the agency from the initial stages to make sure the final product turns out as expected.

11. What is comparability testing of biologics?

A sponsor may be able to demonstrate product comparability between a biological product made after a manufacturing change and a product made before implementation of the change through different types of analytical and functional testing without additional clinical studies. The agency may determine that the two products are comparable if the results of the comparability testing demonstrate that the manufacturing change does not affect safety, identity, purity, or potency. For more information, see Chapter 17, titled "Immunogenicity Assessment for Therapeutic Protein Products" (FDA), and Chapter 18, titled "Assay Development and Validation for Immunogenicity Testing of Therapeutic Protein Products" (FDA).

12. Where can I find additional information about therapeutic biologics?

There are several guidances that may be helpful:

- "Changes to an Approved Application for Specified Biotechnology and Specified Synthetic Biological Products" (PDF-33 KB) (http://www.fda.gov/downloads/Drugs/GuidanceComplianceRegulatoryInformation/Guidances/UCM124805.pdf)

- "Content and Format of INDs for Phase I Studies of Drugs Including Well Characterized, Therapeutic, Biotechnology-Derived Products" (PDF-42 KB) (http://www.fda.gov/downloads/Drugs/GuidanceComplianceRegulatoryInformation/Guidances/UCM071597.pdf)

- "Providing Clinical Effectiveness of Human Drugs and Biological Products" (http://www.fda.gov/BiologicsBloodVaccines/GuidanceComplianceRegulatoryInformation/Guidances/default.htm)

- "Points to Consider in the Manufacture and Testing of Monoclonal Antibody Products for Human Use" (PDF-140 KB) (http://www.fda.gov/downloads/BiologicsBloodVaccines/GuidanceComplianceRegulatoryInformation/OtherRecommendationsforManufacturers/UCM153182.pdf)

Pharmaceutical versus Biotechnology Companies

The demarcations between pharmaceutical and biotechnology (and between branded and generic) companies are no longer that clear. For example, Genentech (owned by Roche) and Medimmune (owned by AstraZeneca), although operate independently, are technically part of big pharma. Many biotechs are developing therapeutics that are traditional small-molecule drugs rather than biotech products. Conversely, big pharma is developing biotech products along with traditional small molecules. Often, branded companies are developing generics and *vice versa*. Currently, there is a symbiotic relationship between all these diverse players. For example, big pharma (which is well versed in clinical trials and commercialization) often turns to biotech companies (that are generally low on funds, lack a robust sales force or lack regulatory expertise) to license compounds or to develop platform technologies with the promise to yield multiple molecules.

1.2 Biologics versus Small-Molecule Drugs

Biologics are a distinct regulatory category of drugs that differ from conventional small-molecule drugs by their manufacturing processes (i.e., biological sources vs. chemical/synthetic manufacturing). They are biologically derived from microorganisms (generally engineered) or cells (often mammalian, including human). In other words, biologics are drugs produced via modern molecular biological methods, and they are distinguished from traditional biological products that are directly extracted from natural biological sources (such as proteins derived from plasma or plants). Biologics include a diverse range of therapeutics, including blockbuster monoclonal antibodies (mAbs) (e.g., Avastin® (bevacizumab) and Humira® (adalimumab)), Fc fusion proteins, anticoagulants, blood factors, hormones, cytokines, growth factors, engineered protein scaffolds, and cell-based gene therapies (e.g., chimeric antigen receptor T-cell therapy (CAR-T)) to treat various diseases—cancers, rheumatoid arthritis (RA), multiple sclerosis (MS), inflammatory bowel disease (IBD), hemophilia, anemia, etc. Most biologics are large, complex molecules as compared to small-molecule drugs (Fig. 1.2) and are often more difficult to characterize than small-molecule drugs (Table 1.1).

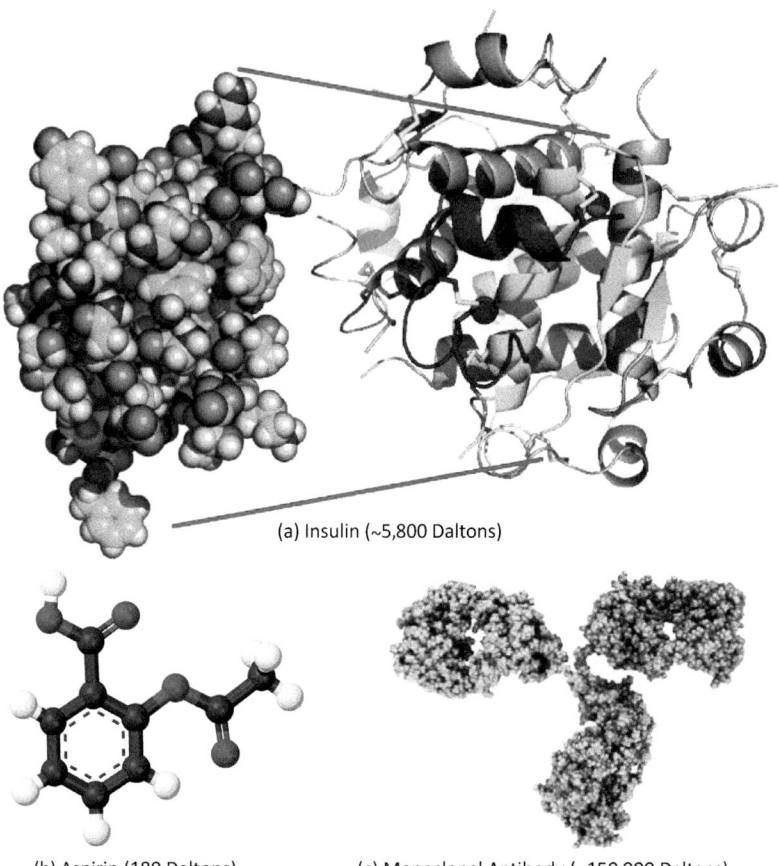

(a) Insulin (~5,800 Daltons)

(b) Aspirin (180 Daltons)   (c) Monoclonal Antibody (~150,000 Daltons)

Figure 1.2 Comparing biologics to small-molecule drugs. The molecular model of two biologics (insulin and monoclonal antibody) and the molecular structure of a small-molecule drug (acetylsalicylic acid or aspirin) are shown to demonstrate the differences in size and molecular complexity associated with these two overlapping drug classes. The molecular weight (MW) of insulin is ~5,800 Daltons and that of a monoclonal antibody is ~150,000 Daltons. The MW of aspirin is 180 Daltons. Structures shown are not to scale. (a) The left side is a space-filling model of the insulin monomer, believed to be biologically active. Carbon atoms are shown in green, hydrogen in gray, oxygen in red, and nitrogen in blue. On the right side is a ribbon diagram of the insulin hexamer, believed to be the stored form. A monomer unit is highlighted with the A chain in blue and the B chain in cyan. Yellow denotes disulfide bonds, and magenta spheres are zinc ions (courtesy of Wikipedia). (b) Ball-and-stick model of the aspirin molecule. (c) X-ray crystallographic structure of a monoclonal antibody shown as a space-filling model (courtesy of the FDA).

The FDA's statutory definition of a "biological product" is listed in Box 1.1.[6] This definition has important regulatory and commercial ramifications as it determines which regulatory pathway governs the approval/licensure of an innovator product and any subsequent follow-on competitor products (i.e., biosimilars, see Section 1.6) that seek to rely on that product's approval. Note that some protein drug products (hormones such as insulin and human growth hormone) are regulated by the FDA as drugs under the FD&C Act, not biological products under the PHS Act. In fact, when human insulin (Humulin®) was approved as the world's first recombinant protein therapeutic in 1982, it was approved under the FD&C Act. This bizarre dichotomy continues today, with some proteins licensed under the PHS Act and some approved under the FD&C Act. Thankfully, this mess is set to clear up in March 2020, when an approved application for a biological product under section 505 of the FD&C Act "shall be deemed to be a license for the biological product under section 351 of the PHS Act."[7]

Growth of Biologics: Technological Drivers

Advances built on two seminal technologies (recombinant DNA technology and hybridoma technology) have been the driving forces behind the expansion of biologics. Specifically, the development of recombinant DNA technology in the 1970s revolutionized the production of biologics. In 1982, human insulin (brand name Humulin® and manufactured by Genentech in partnership with Eli Lilly) was the first recombinant protein therapeutic approved by the FDA. Since Humulin® was fully human and produced via genetically engineered *Escherichia coli*, issues with immunogenicity were minimized. In the 1980s, modified biologics joined recombinant versions of natural proteins as a major new class of biologics. In 1975, Köhler and Milstein's hybridoma technology established a continuous immortal culture of cells secreting an antibody of predefined specificity (monoclonal antibody (mAb)) by fusing antibody-producing B cells with myeloma cells.

---

[6] In December 2017, the FDA formally announces rulemaking to amend the definition of a biologic to conform to the statutory definition (21 U.S.C. 262) adopted in the Biologics Price Competition and Innovation Act of 2009. See: Definition of the Term "Biological Product." Available at: https://www.reginfo.gov/public/do/eAgendaViewRule?pubId=201710&RIN=0910-AH57 (accessed on May 1, 2018).

[7] Federal Register (2016). Notices, vol. 81, no. 49, p. 13373, Docket No. FDA–2015–D–4750.

Below appears a well-accepted definition of a biologic [8]:

"A biopharmaceutical is a protein or nucleic acid-based pharmaceutical substance used for therapeutic or in vivo diagnostic purposes, which is produced by means other than direct extraction from a native (non-engineered) biological source."

Since most biologics are very complex molecules and cannot be fully characterized by existing scientific technologies, they are often characterized via their manufacturing processes. However, due to their structural complexity, the manufacturing processes are also often complex, very sensitive, and proprietary. In fact, minor variations in temperature or other production factors can profoundly change the final biologic drug product. Naturally, this can affect product performance and patient safety. Hence, even minor alterations in the manufacturing process or facility may require clinical studies to demonstrate safety (including immune-related), purity, and potency of the synthesized biologic. According to the FDA [9], "[t]he nature of biological products, including the inherent variations that can result from the manufacturing process, can present challenges in characterizing and manufacturing these products that often do not exist in the development of small-molecule drugs. Slight differences between manufactured lots of the same biological product (i.e., acceptable within-product variations) are normal and expected within the manufacturing process."

1.3 What Are Nanodrugs?

Optimists tout nanotechnology as an enabling technology, a sort of next industrial revolution that could enhance the wealth and health of nations. They promise that many areas within nanomedicine (nanoscale drug delivery systems, theranostics, imaging, etc.) will soon be a healthcare game-changer by offering patients access to personalized or precision medicine. Pessimists, on the other hand, take a cautionary position, preaching instead a go-slow approach and pointing to lack of sufficient scientific data on health risks, general failure on the part of regulatory agencies to provide clearer guidelines and issuance of patents of dubious scope by patent offices. As usual, the reality is somewhere between such extremes. Whatever your stance, nanomedicine has already permeated virtually every sector of the global economy. It continues to evolve and play a pivotal role in various industry segments, spurring new directions in research, product development, and translational efforts [1–3].

Nano Frontiers: Dreams, Hype and Reality

The rush to celebrate "eureka" moments is overshadowing the research enterprise. Some blame the current pervasive culture of science that focuses on rewarding eye-catching and positive findings. Others point to an increased emphasis on making provocative statements rather than presenting technical details or reporting basic elements of experimental design. "Fantastical claiming" is nothing new to academia and start-ups where exaggerated basic research developments are often touted as revolutionary and translatable advances. Claims of early-stage discoveries are highlighted as confirmation of downstream novel products and applications to come. Even distinguished professors at reputable universities are guilty of such hype. In this context, nano's potential benefits are also often overstated or inferred to be very close to application when clear bottlenecks to commercial translation persist.

In the nanoworld, many have desperately and without scientific basis thrown around the "nano" prefix to suit their selfish purpose, whether it is to obtain research funds, gain patent approval, raise venture capital, run for public office, or seek publication of a manuscript. Sadly, many fall prey to such outrageous hype and are even willing to provide venture funds. An extreme example of this is the recent Theranos case where the blood-testing company concocted fantastical claims of doing hundreds of tests from a single drop of human blood and raised billions in the process (market valuation of $9 billion). See: Carreyrou, J. (2018). *Bad Blood: Secrets and Lies in a Silicon Valley Startup*, Alfred A. Knopf, New York. There are also a few cautionary tales from the world of nanomedicine. Consider, for example, the recent demise and bankruptcy of BIND Therapeutics Inc. See: WTF happened to BIND Therapeutics? Available at: https://www.nanalyze.com/2017/08/wtf-happened-bind-therapeutics/ (accessed on August 5, 2018).

Obviously, the Holy Grail of any drug delivery system, whether it is nanoscale or not, is to deliver to a patient the correct dose of an active agent to a specific disease or tissue site while simultaneously minimizing toxic side effects and optimizing therapeutic benefit. This is mostly unachievable via conventional small-molecule formulations and drug delivery systems. However, the potential to do so may be greater now via nanodrugs. The prototype of targeted drug delivery can be traced back to the concept of a "magic bullet" that was postulated by Nobel laureate Paul Ehrlich in 1908 (*magische Kugel*, his term for an ideal therapeutic agent) wherein a drug could selectively target a pathogenic organism or diseased tissue while leaving healthy cells unharmed [10]. Half a century later, this concept of the magic bullet was realized by the development of antibody–drug conjugates (ADCs) when in 1958 methotrexate was linked to an antibody targeting leukemia cells wherein the antibody component provided specificity for a target antigen and the active agent portion conferred cytotoxicity. (Technically, ADCs are nanodrugs.) Half a century since ADCs, various classes of nanoscale drug delivery systems are in early development though first-generation nanodrugs have been commercialized (Fig. 1.3). However, the arrival of revolutionary nanodrugs are just promises at this stage. There are many second- and third-generation nanodrugs at various stages of R&D (Fig. 1.3). Obviously, advanced nanodrugs will be (i) those that can specifically deliver active agents to target tissue, specific cells or even organelles (site-specific drug delivery); or (ii) offer simultaneous controlled delivery of active agents with concurrent real-time imaging (theranostic drug delivery).

Data obtained from industry and the FDA show that most of the approved or pending nanodrugs are oncology-related and based on protein–polymer conjugates or liposomes. The first FDA-approved nanodrug was Doxil® (doxorubicin hydrochloride liposome injection) in 1995 while AmBisome® (amphotericin B liposome injection) was the first one approved by the EMA in 1997. The first protein-based nanodrug to receive regulatory approval was albumin-bound paclitaxel (Abraxane®), approved by the FDA in 2005. However, note that a nanoparticulate iron oxide intravenous solution that was marketed in the 1960s and certain nanoliposomal products that were approved in the 1950s should, in fact, be considered true first-generation nanodrugs. Polymer–drug conjugates (with a short peptide spacer between the two that prolonged release) were also prepared back in the 1950s, when a polyvinylpyrrolidone–mescaline conjugate was produced.

Nanodrugs: Relabeling of Earlier Terms?

"The new concept of nanomedicine arose from merging nanoscience and nanotechnology with medicine. Pharmaceutical scientists quickly adopted nanoscience terminology, thus "creating" "nanopharmaceuticals." Moreover, just using the term "nano" intuitively implied state-of-the-art research and became very fashionable within the pharmaceutical science community. Colloidal systems reemerged as nanosystems. Colloidal gold, a traditional alchemical preparation, was turned into a suspension of gold nanoparticles, and colloidal drug-delivery systems became nanodrug delivery systems. The exploration of colloidal systems, i.e., systems containing nanometer sized components, for biomedical research was, however, launched already more than 50 years ago and efforts to explore colloidal (nano) particles for drug delivery date back about 40 years. For example, efforts to reduce the cardiotoxicity of anthracyclines via encapsulation into nanosized phospholipid vesicles (liposomes) began at the end of the 1970s. During the 1980s, three liposome-dedicated US start-up companies (Vestar in Pasadena, CA, USA, The Liposome Company in Princeton, NJ, USA, and Liposome Technology Inc., in Menlo Park, CA, USA) were competing with each other in developing three different liposomal anthracycline formulations. Liposome technology research culminated in 1995 in the US Food and Drug Administration (FDA) approval of Doxil®, "the first FDA-approved nanodrug". Notwithstanding, it should be noted that in the liposome literature the term "nano" was essentially absent until the year 2000."

Source: Weissig, V., Pettinger, T. K., Murdock, N. (2014). Nanopharmaceuticals (part 1): Products on the market. *Int. J. Nanomed.*, **9**, 4357–4373.

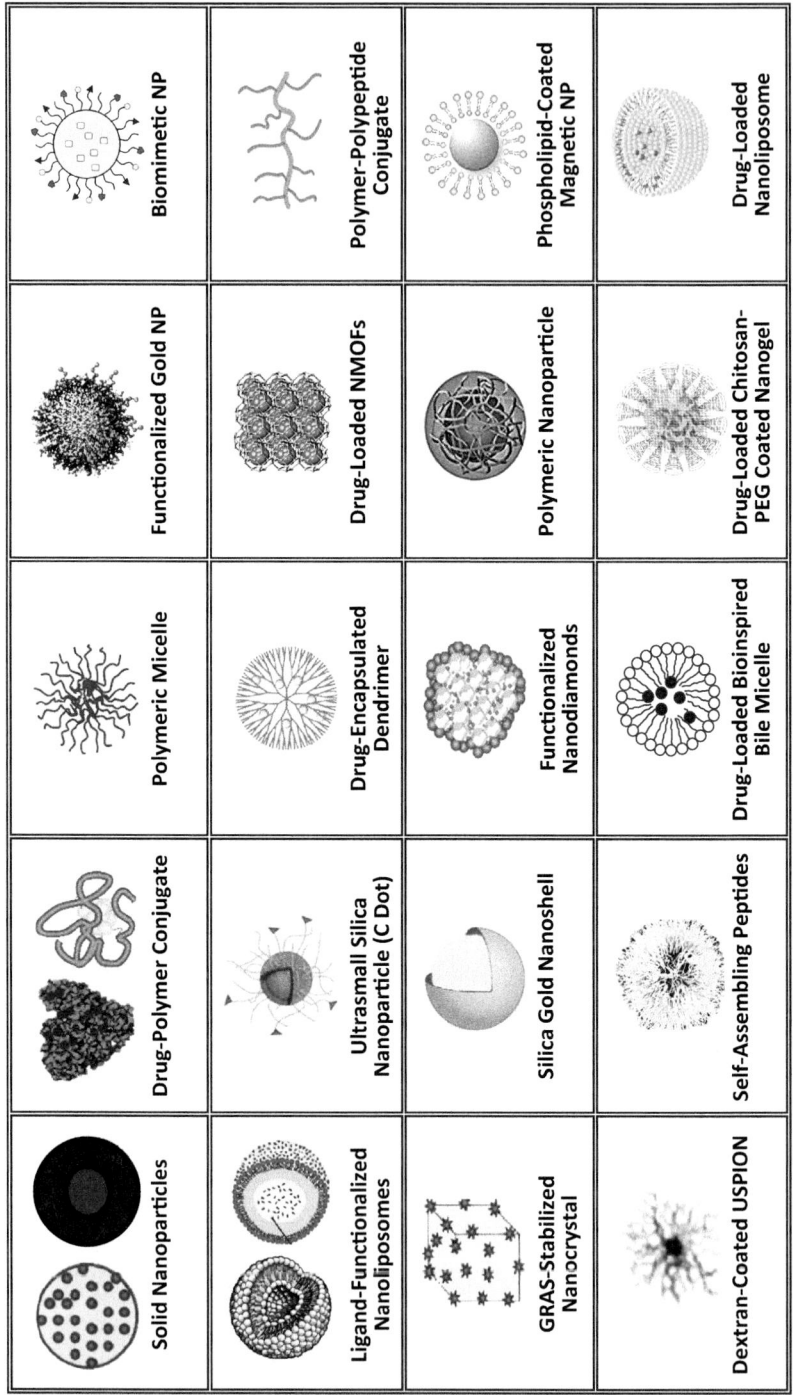

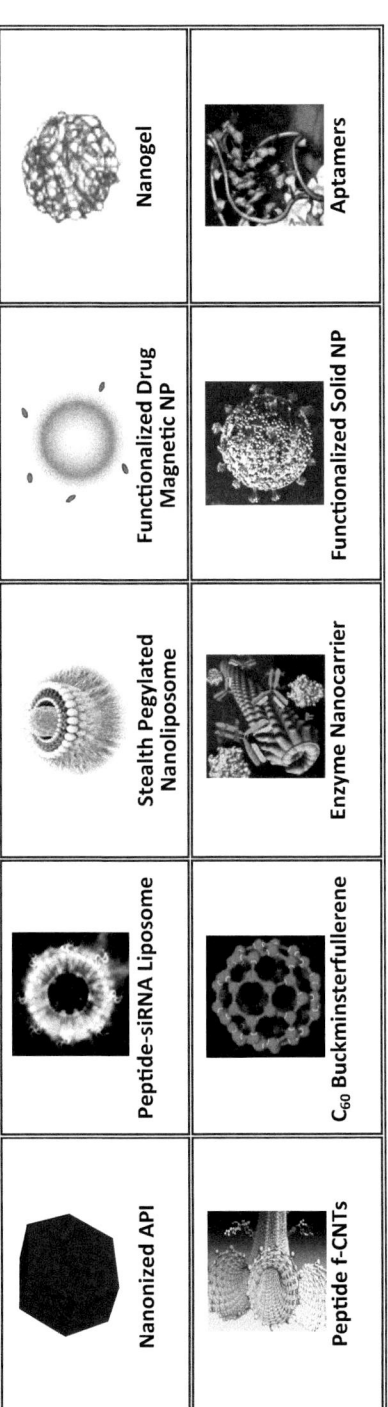

Figure 1.3 Nanoscale Drug delivery system platforms (nanodrug products). Schematic representation of selected engineered nanoparticles (NPs) used in drug delivery that are either approved by regulatory bodies, are in preclinical development or are in clinical trials. In most cases shown above, they are considered as first or second generation multifunctional engineered NPs, ranging in average diameter from one nanometer (1 nm) to a micron (1000 nm). Active bio-targeting of the NP is often achieved via conjugating ligands or functional groups (antibodies, peptides, aptamers, folate, hyaluronic acid). These molecules are tagged to the NP surface with or without spacers/linkers such as PEG. Many nanodrugs depicted above (e.g., metal-based NPs, f-CNTs, NMOFs, etc.), although extensively advertised for drug delivery, will pose enormous drug approval and commercialization challenges and will not appear in the clinic this century. Non-engineered antibodies, biological motors (e.g., sperms), engineered nanomotors, and naturally occurring NPs (natural protein nanotubes) are specifically excluded here. Antibody–drug conjugates (ADCs) are encompassed by the cartoons labeled "Polymer-Polypeptide Conjugate" or "Drug-Polymer Conjugate." Therapeutic monoclonal antibodies (TMAbs), polymer-polypeptide conjugates, and aptamers shown above are classic biologics but they are also nanodrugs as they fall within the widely accepted standard definition of nanodrugs. The list of NPs depicted here is not meant to be exhaustive, the illustrations do not reflect precise three-dimensional shape or configuration, and the NPs are not drawn to scale. *Abbreviations*: NPs, nanoparticles; PEG, polyethylene glycol; GRAS, Generally Recognized As Safe; C dot, Cornell dot; API, active pharmaceutical ingredient; ADCs, antibody–drug conjugates; NMOFs, nanoscale metal organic frameworks; f-CNTs, functionalized carbon nanotubes; siRNA, small interfering ribonucleic acid; USPION, ultrasmall superparamagnetic iron oxide nanoparticle. Copyright 2018 Raj Bawa. All rights reserved.

In 2011, drug shortages were such a pressing issue in the United States that an executive order from the President was issued directing the FDA to streamline the approval process for new therapeutics that could fill the voids. One of the major drugs whose supply was deficient in the United States was Doxil®, and to curb this shortage, the FDA on February 21, 2012, authorized the temporary importation of Lipodox® (doxorubicin hydrochloride liposome injection, Sun Pharmaceutical Industries Ltd., India), a generic version of Doxil®. Following this, the FDA evaluated and approved Lipodox® within a year on February 4, 2013, in roughly one-third of the time it takes for an average generic to receive premarket regulatory approval. Hence, Lipodox® became the first generic nanodrug (i.e., nanosimilar) approved in the United States. Obviously, this helped alleviate the Doxil® shortage and reduced the cost of care (Fig. 1.4). However, a recent study [11] concluded that "*the data available from this study and in the peer-reviewed literature are compelling suggesting that Lipodox for treatment of recurrent ovarian cancer does not appear to have equal efficacy compared to Doxil. It raises many concerns how to balance the challenges of drug shortages with maintaining the standards for drug approval. A prospective clinical study to compare the two products is warranted before Lipodox can be deemed equivalent substitution for Doxil.*"

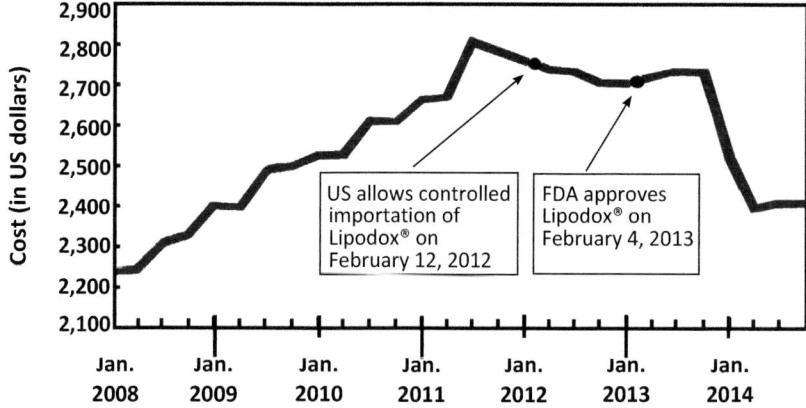

Figure 1.4 Cost for treatment of AIDS-related Kaposi sarcoma (KS) from January 2008 to September 2014.

What Is Nanotechnology?

My definition of nanotechnology omits any strict size limitation. See: Bawa, R. (2007). Patents and nanomedicine. *Nanomedicine (London)*, **2**(3), 351–374: *"The design, characterization, production, and application of structures, devices, and systems by controlled manipulation of size and shape at the nanometer scale (atomic, molecular, and macromolecular scale) that produces structures, devices, and systems with at least one novel/superior characteristic or property."*

This flexible definition has four key features: (i) It recognizes that the properties and performance of the synthetic, engineered *"structures, devices, and systems"* are inherently rooted in their nanoscale dimensions. The definition focuses on the unique physiological behavior of these *"structures, devices, and systems"* that is occurring at the nanoscale; it does not focus on any shape, aspect ratio, specific size or dimension; (ii) it focuses on "technology" that has commercial potential, not "nanoscience" or "basic R&D" conducted in a lab setting; (iii) the *"structures, devices, and systems"* that result from or incorporate nano must be *"novel/superior"* compared to their bulk/conventional counterparts; and (iv) the concept of *"controlled manipulation"* (as compared to "self-assembly") is critical.

The prefix "nano" in the SI measurement system denotes 10^{-9} or one-billionth. There is no firm consensus over whether the prefix "nano" is Greek or Latin. While the term "nano" is often linked to the Greek word for "dwarf," the ancient Greek word for "dwarf" is spelled "nanno" (with a double "n") while the Latin word for dwarf is "nanus" (with a single "n"). While a nanometer refers to one-billionth of a meter in size (10^{-9} m = 1 nm), a nanosecond refers to one billionth of a second (10^{-9} s = 1 ns), a nanoliter refers to one billionth of a liter (10^{-9} l = 1 nl) and a nanogram refers to one billionth of a gram (10^{-9} g = 1 ng). The diameter of an atom ranges from about 0.1–0.5 nm, a DNA molecule about 2–3 nm, and a gold atom about 1/3rd of a nm.

Given that this specific example of generic approval is a problem, I believe that while the development of generics is important to facilitate patient access to vital medications at a reasonable price, generic approvals should be science-based, data-driven, and reported transparently. Another example of the issuance of generics is discussed in detail in Section 1.7.

There is no formal or internationally accepted definition for anything "nano." In this regard, a harmonized definition and nomenclature is urgently needed. For example, there is no standard definition for a nanodrug. The following is my definition for a nanodrug [12]:

> "A nanodrug is: (1) a formulation, often colloidal, containing therapeutic particles (nanoparticles) ranging in size from 1–1,000 nm; and (2) either (a) the carrier(s) is/are the therapeutic (i.e., a conventional therapeutic agent is absent), or (b) the therapeutic is directly coupled (functionalized, solubilized, entrapped, coated, etc.) to a carrier."

Nanodrugs cannot merely be defined by their size. Nanodrugs may have unique properties (nanocharacter) that can be beneficial for various clinical applications but there is no specific size range or dimensional limit to which superior properties are confined to. In fact, size limitation below 100 nm, frequently recited in journals and talks as well as touted as the definition of anything "nano" by US federal agencies like the National Nanotechnology Initiative (NNI), cannot serve as an arbitrary basis of novel properties of nanodrugs [12]. For instance, such a bizarre definition falls on its face in all these scenarios: Larger materials may contain nanostructures with size-specific properties; nanomaterials may be employed during a manufacturing step but not found in the final finished product; or nanoparticles may aggregate/dissociate in a dynamic equilibrium state and therefore the formulation may contain a mixed population of size ranges. *In conclusion, viable sui generis definition of nanodrugs having a bright-line size limit below 100 nm has no scientific or legal basis.*

1.4 Are Biologics and Nanodrugs Adversely Immunogenic?

Almost all small-molecule drug-induced allergic reactions may be easily classified into one of four classic Gell and Coombs hypersensitivity categories.[8] However, many others with an immunologic component, including biologics and nanodrugs, are difficult to classify in such a manner because of a lack of mechanistic information [13]. Adverse clinical events (sometimes referred to as Adverse Drug Reactions or ADRs) can not only occur due to primary factors such as off-target toxicity or exaggerated pharmacologic effects, but also due to secondary drug effects such as immune reactions to the drug product. While approximately 80% of human adverse drug reactions are directly related to an effect of the drug or a metabolite, around 6–10% are immune-mediated and unpredictable [13]. One study showed that 10–20% of the medicinal products removed from clinical practice between 1969 and 2002 were withdrawn due to immunotoxic effects [14]. Some claim that the actual number of serious adverse events like hospitalizations and death from FDA-approved drugs, vaccines, and medical devices is grossly underreported by the FDA [15].[9] Suspected ADRs can be reported to the FDA at 1-800-FDA-1088 or www.fda.gov/medwatch.

Immune-mediated side effects of small molecules are unpredictable. Most small molecules that have a MW <1 KDa do not elicit an immune response in their native state, becoming immunogenic only when they act as a hapten, bind covalently to high-molecular-weight proteins, and undergo antigen processing and presentation. On the other hand, "newer" larger molecule drugs can be inherently immunogenic. For example, protein-based biologics and nanodrugs can be digested and processed for presentation by antigen-presenting cells (APCs); this can sometimes cause ADRs. The very untested nature of these therapeutics that make them so revolutionary in some respects also makes them problematic and potentially dangerous. For example, major benefits touted for nanodrugs—a reduction in unwanted side effects, increased specificity, fewer off-target effects, generation of fewer harmful metabolites, slower clearance from the body, longer

---

[8]Gell and Coombs developed their widely accepted classification of hypersensitivity reactions in the context of deleterious connotation. See: Gell, P. G. H., Coombs, R. R. A. (1963). The classification of allergic reactions underlying disease. In: Coombs, R. R. A., Gell, P. G. H., eds., *Clinical Aspects of Immunology*, Blackwell Science.

[9]Also, see the FDA Death Meter: http://www.anh-usa.org/microsite-subpage/fda-death-meter/.

What Is Immunogenicity?

Immunogenicity is the ability of an antigen or epitope to provoke an immune response, i.e., to induce a humoral and/or cell-mediated immune response. Put differently, it is the propensity of a therapeutic (e.g., biologic or nanodrug) to generate immune responses to itself and to related products. These responses can either (i) induce immunologically related nonclinical effect(s); (ii) provide beneficial or protective effect(s); or (iii) result in adverse clinical events (Adverse Drug Reactions or ADRs). Immunogenicity can be one of two types: (a) wanted or (b) unwanted. Wanted immunogenicity is typically related to vaccines where the injection of the vaccine (the antigen) stimulates an immune response against a pathogen. On the other hand, unwanted immune responses are adverse events (i.e., ADRs). The meaning of immunogenicity in this chapter is the latter, namely, an adverse immune response to the therapeutic. The detection of antidrug antibodies (ADAs) (Section 1.4.1(a)) has generally been equated as a measure of immunogenicity. ADAs may neutralize a therapeutic and inhibit its efficacy or cross-react to endogenous counterparts, leading to loss of physiological function. An example of unwanted immunogenicity is the generation of neutralizing antibodies against recombinant erythropoietin (EPO) in patients receiving EPO for chronic kidney disease (CKD) and resulting pure red cell aplasia (PRCA)–related anemia due to the neutralization of the endogenous EPO. Immunogenicity associated with protein drugs was first observed more than a century ago in 30% of diphtheria patients treated with antitoxin administered in whole horse serum. See: Weaver, G. H. (1900). Serum disease. *Arch. Intern. Med. (Chic.)*, **5**, 485–513.

duration of effect, reduced intrinsic toxicity, etc.—do not guarantee the absence of adverse immune side effects: *An inherent risk of introducing these drug products into the human body is the potential to provoke an unwanted immune response.* Thus, managing their immunogenicity profile is critical during drug R&D and later. Studies have shown that these drug products can interact with various components of the immune system to various immunological endpoints, interactions that are fast, complex, and poorly understood. These interactions with the immune system play a leading role in the intensity and extent of side effects occurring simultaneously with their therapeutic efficacy. In fact, when compared to conventional small-molecule drugs, both biologics and nanodrugs have biological and synthetic entities of a size, shape, reactivity, and structure that are *often* recognized by the human immune system, *sometimes* in an adverse manner. This can obviously negatively affect their effectiveness and safety, and thereby, limit their therapeutic application. This also poses challenges for regulatory agencies and patent offices, all serving as bottlenecks to effective translation of these therapeutics.

Multiple risk factors influencing the immunogenicity of biologics and nanodrugs include patient-, clinical use-, manufacturing-, and product-related factors (Fig. 1.5). Some of the ADRs include complement activation, tissue inflammation, leucocyte hypersensitivity, and formation of antibodies associated with clinical conditions. However, detailed mechanisms and causal linkages between various risk factors and immunogenicity induction onset as shown in Fig. 1.5 have yet to be fully elucidated. This is primarily due to the limited amount of data from mechanistic studies, a lack of multi-factorial analysis and a lack of standard immunogenicity assessment methods.

1.4.1   Immune Aspects of Biologics

(a) Antidrug Antibodies (ADAs) and Immune Complexes (ICs)

Early developers of biologics assumed that as many of these drugs were based on human genes, the human immune system would not treat them as foreign and not produce antibodies. However, this optimistic view has turned into alarm as some biologics can elicit a vigorous immune response that *may* sometimes neutralize, block or destroy the administered biologic. Also, most biologics are engineered to enable dual or multiple binding sites (via conjugated proteins, functionalized antibodies, etc.) — all of which could result in them being recognized as foreign and therefore immunogenic. In most cases, immunogenicity manifests itself

DRUG PRODUCT
origin
formulation, handling
aggregation/degradation of excipient(s) and/or active(s)
drug conjugates
mode of action/nature[1]
molecular structural differences from native active
proportion of "non-self" protein sequences/epitopes[2]
presence of foreign proteins
misfolding related to oxidation/deamidation
glycosylation patterns in proteins, protein mutations
nanoscale dimensions/nanoparticle size[3]
surface functionality, surface charge
protein size[4]
topology, shape, geometry, protein conformation

MANUFACTURING
production protocol variations
denaturation and/or alteration of structure
chemical modifications[5]
post translational modifications of proteins
impurities, contaminants, degradants, fragments[6]
aggregates, agglomerates[7]
leachables from containers[8]

CLINICAL USE
dose level
mechanism of action
dosing regimen (procedure, concentration)
delivery route[9]
frequency of administration[10]
duration of treatment[11]
use of DEHP or other plasticizers in plastic components[12]

PATIENT
patient genetics, predisposition, genetic deficiency[13]
age[14]
immunocompetency[15]
preexisting antibodies and CD4+T cells reactive to drug[16]
extended drug residence time[17]
presence of chronic conditions
disease state being treated, concurrent illness
prior exposure to related or cross-reacting drug products
in vivo modifications of endogenous proteins
interruptions in therapy
concomitant therapies[18]
binding to specific cell surface versus soluble targets and/or determinants
"superagonist" formation by cross-linking with ADAs

[1] immunomodulatory versus immunosuppressive, or agonist versus antagonist
[2] proportion of endogenous versus non-endogenous protein sequences; monoclonal antibody-based therapeutics have low immunogenicity
[3] a high surface area to volume ratio when compared to their corresponding bulk counterpart
[4] immunogenicity increases with size
[5] oxidation, deamidation, isomerization has varying effects
[6] host cell proteins, DNA and excipients from formulations are highly immunogenic
[7] unique conformational epitopes may be present
[8] introduction or exposure of new epitopes
[9] immunogenicity order: inhalation > subcutaneous > intraperitoneal > intramuscular > intravenous
[10] repeat administration increases immunogenicity
[11] prolonged exposure increases immunogenicity
[12] di(2-ethylhexyl) phthalate (DEHP) is a manufactured chemical that is commonly added to plastics to make them flexible
[13] certain MHC alleles, polymorphisms in cytokine genes, autoimmune or proinflammatory predisposition has a higher immunogenicity risk
[14] pediatric versus adult immune system
[15] if the patient is immunosuppressed, then may be more immunotolerant
[16] examples include cross-reacting auto-antibodies, preexisting anti-PEG antibodies
[17] at a specific site of action, within specific targeted tissue or in systemic circulation
[18] co-medicated immunosuppressive drugs (e.g., methotrexate or steroids) reduce immunogenicity

Figure 1.5 Key risk factors contributing to adverse immunogenicity of biologics and nanodrugs. *Abbreviations*: DEHP, di-(2-ethylhexyl) phthalate; ADAs, anti-drug antibodies; CD4$^+$T cell, cluster of differentiation 4 T cell; MHC, major histocompatibility complex. Copyright 2018 Raj Bawa. All rights reserved.

as the generation of polyclonal neutralizing and non-neutralizing antidrug antibodies (ADAs) directed against the biologic, rendering it less effective. The detailed mechanisms leading to ADA formation are still not fully understood and characterized. Examples of ADA formation against biologics observed in clinical practice include the treatment of Crohn's disease and rheumatoid arthritis (RA) with anti-TNF adalimumab (Humira®), hemophilia A treatment with recombinant Factor VIII and multiple sclerosis (MS) patients receiving interferon-beta therapy, although the incidence rate of ADA varies among studies, even while using the same drug. Some studies have shown that Humira® does not work in ~20% of patients (the extensive warning list for Humira® includes "immune reactions, including a lupus-like syndrome"). Similarly, in 2016, Pfizer had to pull a promising anticholesterol biologic (bococizumab) after testing it in more than 25,000 persons. In six trials, ~50% of those who received the formulation developed ADAs, spelling doom for the drug candidate. According to Pfizer, this potential biologic was "not likely to provide value to patients, physicians or shareholders" [16]. In 2016, the Netherlands Cancer Institute reported that >50% of the anticancer biologics in 81 clinical trials worldwide were generating ADAs, although they could not confirm that this always negatively affected the drug candidate being tested.

Specifically, ADAs *may* (i) neutralize the activity of the biologic drug product, (ii) reduce half-life by enhancing clearance, (iii) result in allergic reactions, (iv) alter the drug's pharmacokinetic profile, (v) abrogate the pharmacological activity of the drug, and/or (vi) cross-react with endogenous counterparts to result in "autoimmune-like" reactions. Furthermore, antibody responses can potentially affect the interpretation of toxicology studies. As indicated earlier, such effects are much less frequently observed with conventional small-molecule drug products (Table 1.1).

Most biologics are administered to patients as repeated doses. This can elicit ADAs that can form antidrug immune complexes (ICs) with the biologic, which in turn can drive more ADA formation. In general, formation of ICs is a normal immunologic process. For example, binding of antibodies to their respective antigens forms ICs. Most formed ICs, even those that develop due to ADAs, are small and cleared from circulation. It is only when systems to clear or degrade ICs are impaired, clinical or immune consequences may be observed. The role of ADAs is relevant to a discussion of ICs [17]: *"ADAs can be elicited in vivo to a therapeutic and their detection has generally been equated as a measure of immunogenicity. The detection, reporting, and characterization of the ADA are done in a tiered manner after careful consideration of immunogenic risk factors. Most adverse effects consequential to ADA formation, such as pharmacological abrogation, impact on therapeutic exposure, or hypersensitivity reactions, are a consequence of formation of immune complexes (ICs) between the ADA and therapeutic protein. Their levels, kinetics of interaction, size, polyclonal diversity, distribution, and Fc-mediated physiological effects can be potentially translated to clinically observable adverse effects. This leads to the paradigm of immunogenicity where therapeutic exposure leads to ADA generation that in turn forms ICs that mediate adverse effects related to immunogenicity. While the detection of such therapeutic specific IC from in vivo samples has remained analytically challenging, there are other biomarkers that mediate the interplay of the innate and adaptive immune responses and are potentially amenable to analysis. Such markers can reflect either the formation or the downstream effects of ICs... Clinical consequences of ADA make a compelling case for early IC formation that is an important consideration whether or not a long-lasting, pharmacologically meaningful ADA response will form. With the advent of personalized treatment, there will be a greater need to monitor underlying differences between individuals who are reactive to a therapeutic and how they impact either their response to treatment or their manifestation of any immunological adverse effects. Clinical decisions in routine practice rarely make use of information on the patient's immune response to a therapeutic as a basis to understand poor therapeutic response or an unexpected adverse effect; to some extent, this has been due to limitations to identify the right dose of the drug required to neutralize the target in the presence of ADA, challenges in ascertaining total amount of ADA, and a general lack of immunogenicity assessments in patients to investigate failure of response after a drug has been approved for market."*

Immunotoxicology of Biological Response Modifiers

"[F]or human biopharmaceuticals, the immune system is often the intended target of the therapy and the immunotoxicity observed may be exaggerated pharmacology. The intended effects of biotherapeutics on the immune system can be classified as immunopharmacology or as immunomodulatory effects. Adverse events can result from the intended immunomodulatory mechanism of action. For example, excessive downregulation of the immune system can result in recrudescence of a previously inactive virus. Immunotoxicity, on the other hand, refers to adverse immune effects that occur with products that are not targeting the immune system or with unintended effects on the immune system. These effects include inflammatory reaction at the injection site and autoimmunity due to altered expression of surface antigens. Although immunogenicity is an immune response of the animal to a foreign protein, it is not viewed as immunotoxicity *per se*."

Source: Bussiere, J. L. (2016). Immunotoxicology of biological response modifiers. *Reference Module in Biomedical Sciences.* Elsevier.

The immunogenicity risk profile of a biologic is characterized by measurement of ADA levels in patients and correlation with therapeutic outcomes. An immune response to a biologic can occur in animal species, in clinical trial subjects or in patients. This is well recognized by regulatory agencies and hence it is mandatory to test immunogenicity of biologicals in clinical trials as well as to monitor patients after drug approval. This minimizes an unnecessary safety risk for the patient while saving time, resources and effort. It is imperative that drug and biotechnology companies develop both novel tools as well as improve upon existing ADA-testing technologies to look for ADAs before and during clinical trials of biologics. In fact, multiple assay formats, technology platforms and sample preparation protocols are available to measure ADA responses including the enzyme-linked immunosorbent assay (ELISA), pH-shift anti-idiotype antigen-binding test (PIA), surface plasmon resonance (SPR), radioimmunoassay (RIA), electrochemiluminescence assay (ECLA), and homogenous mobility shift assay (HMSA). Obviously, the incidence of such reactions and their action on drug efficacy and patient safety must be transparently and promptly reported.

(b) Species Origin of Biologics

The species origin of biologics has been identified as a significant factor in determining immunogenicity. For example, nonhuman proteins tend to elicit a prolonged and more pronounced immune responses than biologics developed from human or humanized molecules (Fig. 1.5). This may be because of amino acid sequence and glycosylation differences in the proteins such that the immune system sees them as self versus non-self [18]. Glycosylated proteins are generally less immunogenic than nonglycosylated proteins, possibly due to fewer exposed antigenic sites on the protein's tertiary structure. The greater the structural and amino acid sequence homology of the biologic with native human protein(s), the lesser the immunogenic potential. However, induction of antibody responses has been observed with biological products that are identical or nearly identical to native human proteins. This shows that other factors (Fig. 1.5) may be involved.

(c) Aggregation of Biologics

Another issue with some biologics is that they show a concentration-dependent propensity for self-association, which often leads to the formation of aggregates that range in size from nanometers (oligomers) to microns (subvisible and visible particles). Aggregation[10] can occur throughout the life cycle of a biologic product: during upstream and downstream processing, during shipping, shelf-storage, and during handling in the clinic. The presence of aggregates in biologic drug products can induce adverse immune responses in patients that may affect drug safety and efficacy, cause infusion reactions, cytokine release syndrome, anaphylaxis, or even death [19, 20]. Hence, just like ADAs and ICs discussed above, aggregates are of concern to manufacturers, clinicians, patients, and regulatory agencies. Aggregation of biologics is a challenging phenomenon to mitigate due to knowledge gaps of the molecular mechanisms underlying aggregation as well as a lack of standard and reliable aggregation prediction tools.

---

[10]Many diseases are characterized by protein aggregation *in vivo*, including Alzheimer's disease, prion disorders, amyotrophic lateral sclerosis (ALS), Huntington's disease and Parkinson's disease.

However, in recent years, regulators and drug industry experts have spearheaded development of novel techniques to detect and characterize aggregates, increase research into the role of protein aggregates of all sizes in immunogenicity, aid in revising pharmacopoeia monographs to improve subvisible particle testing, and clarify terminology like "practically" or "essentially free of particles."

1.4.2 Immune Aspects of Nanodrugs

The clinical application of nanodrugs and nanocarriers is dogged by safety and toxicity concerns, especially about their long-term use. As discussed earlier (Fig. 1.5), immunogenicity of nanodrugs may result from a unique combination of physicochemical properties, such as shape, size, surface charge, porosity, reactivity, and composition. Many nanodrugs are engineered to break tissue physiological barriers for entry and to escape immune surveillance, thereby persisting in body fluids and delivering their active pharmaceutical ingredients (APIs). However, this persistence in the body may trigger immune responses.

(a) A well-studied but poorly understood immune issue with nanodrugs is the formation of the so-called "protein corona" (Fig. 1.6) at the interface between nanodrugs and blood (bio-nano interface). Protein corona refers to the adsorption of proteins onto the nanodrug surface, thereby reducing their stability and facilitating their rapid *in vivo* clearance. Obviously, this phenomenon has important implications on immune safety, biocompatibility, and the use of nanodrugs in medicine [21, 22]. This formation of protein corona may be one factor that has contributed to the inefficient accumulation of nanodrugs (<10% accumulation [23]) in diseased tissues despite the oft-highlighted advantages of "targeted" nanodrug delivery. However, this area of research has suffered from a mechanistic understanding of the bio-nano interface.

(b) Since the surface area-to-volume ratio is very high at the nanoscale [12], the surface properties of nanodrugs dictate their interactions with the bioenvironment. This enormous surface area of nanoparticles can in turn cause increased biological activity, including immunogenicity. Adverse effects can lead to either suppressed or stimulated immune functions and they can involve various blood and immune cells (Fig. 1.7). Evaluation of the interaction of nanodrugs with blood components (Fig. 1.8) is, therefore, critical as most administered nanodrugs will end up being distributed by the bloodstream [24]. Hence, experimental techniques for the analysis of nanodrug interaction with biological components are critical [24].

Figure 1.6 Protein Corona. A nanoparticle gains a new biological identity upon its dynamic interactions with biological fluids, giving rise to the protein corona (shown as adsorbed green, blue, and cyan globules), which consequently influences drug delivery and targeting of the functionalized nanoparticle (illustrated as aqua blue fibrils). Reproduced with permission from [22]. Copyright 2017 American Chemical Society.

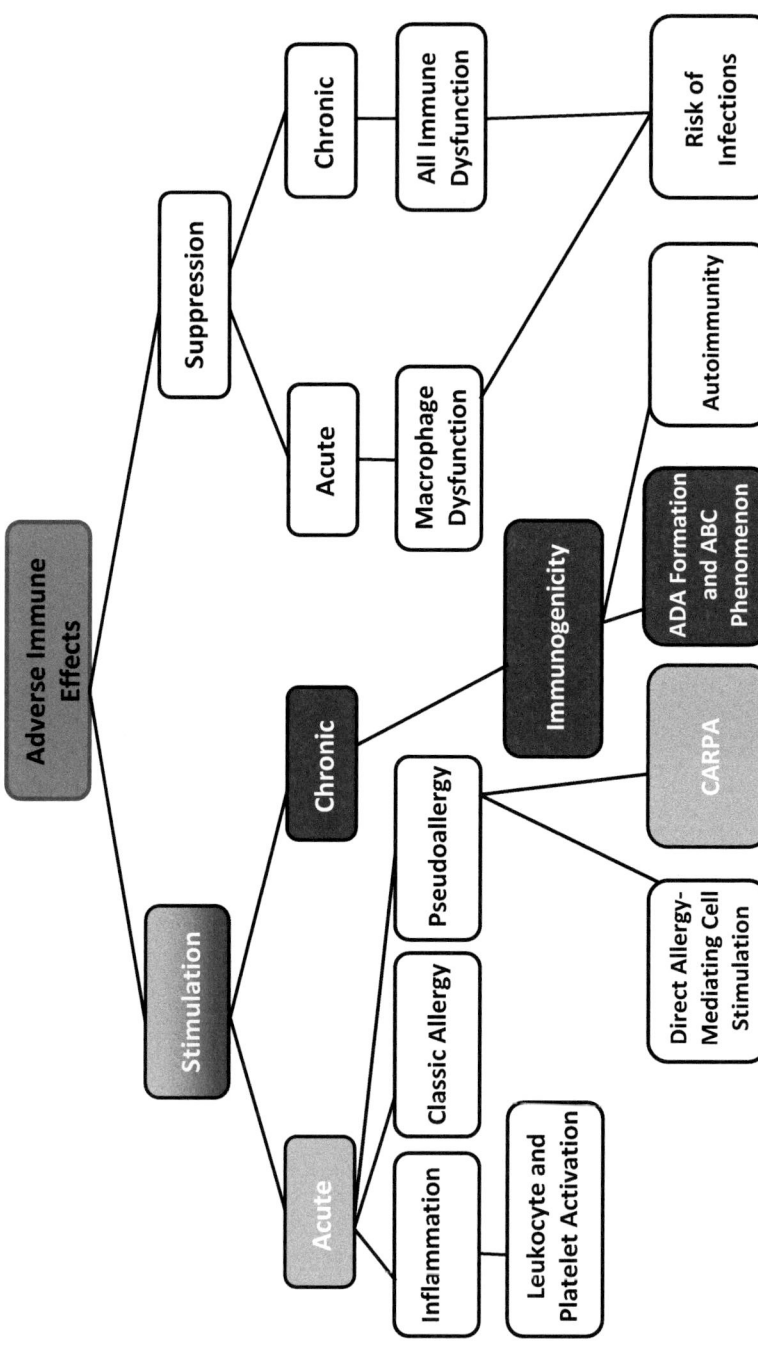

Figure 1.7 Adverse immune effects of nanodrugs, classified according to their impact and time course. *Abbreviations:* ABC, accelerated blood clearance; ADA, antidrug antibody; CARPA, complement activation-related pseudoallergy.

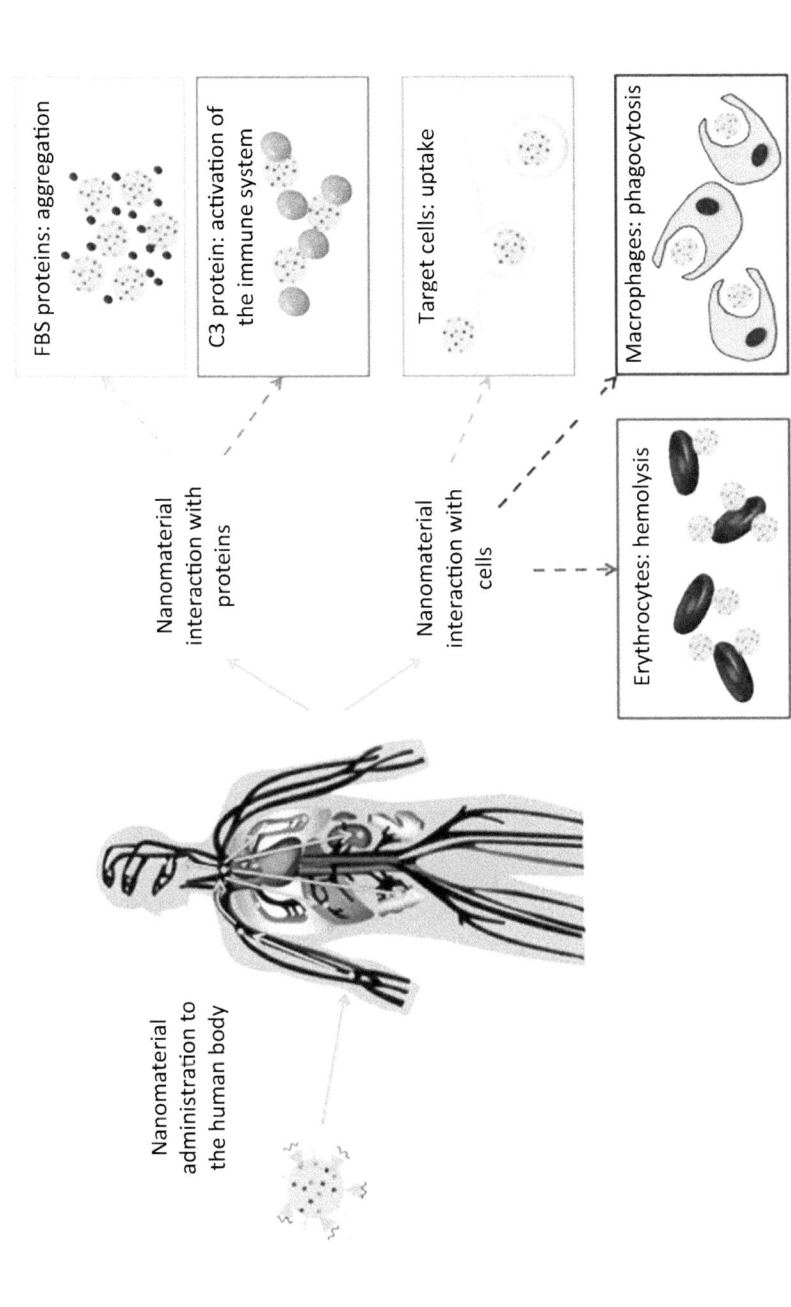

Figure 1.8 Schematic representation of possible interactions of some nanosystems with biological components, namely cells and proteins. Courtesy of Dr. Cristina Fornaguera, Sagetis-Biotech, Barcelona, Spain.

(c) Often, intravenously administered therapeutics (certain nanodrugs, biologics, NBCDs, etc.) prime the immune system, leading to adverse reactions and/or the loss of efficacy of the drug product. It is now well established that these therapeutics may provoke "hypersensitivity reactions" (HSRs), also known as "infusion" or "anaphylactoid" reactions. Due to the association of complement activation with many of these adverse reactions, the term "complement activation-related pseudoallergy" (CARPA) was coined in the late 1990s [25–27] (Tables 1.3 and 1.4). CARPA was based on pig studies involving intravenously administered liposomes; the model is now known as the "porcine CARPA model" (Fig. 1.9).

These hypersensitivity reactions typically occur directly at first exposure to the drug without prior sensitization, and the symptoms usually lessen and/or disappear upon later treatment. The rapidly arising symptoms, namely, shortness of breath, facial redness and swelling, chest pain, back pain, flashing, rash, chills, panic, and fever are also typical of acute or Type 1 hypersensitivity. However, a role of IgE has not been implicated in most of these reactions. Therefore, these HSRs are labeled as "pseudoallergic" or "nonspecific hypersensitivity." Nanodrugs causing CARPA (Table 1.3) include radio-contrast media, liposomal drugs (Doxil®, Ambisome®, DaunoXome®, Abelcet®, Visudyne®), nanoparticulate iron, micellar solvents (Cremophor EL, the vehicle of Taxol®), PEGylated proteins, and monoclonal antibodies (mABs).[11] Drug products other than nanodrugs such as nonsteroidal anti-inflammatory drugs (NSAIDs), analgesics, and morphine can also trigger CARPA (Table 1.3). Now, CARPA is a well-established cornerstone of pharmacotherapy and liposomal chemotherapy. The clinical relevance of CARPA is highlighted by notes in industry guidances issued by the FDA and the EMA. The FDA recommends detection of complement activation by-products in animals showing signs of anaphylaxis [28a], while the EMA refers to CARPA tests as

---

[11]Monoclonal antibodies (mABs) (Fig. 1.2) are the largest group of biologics. They include in their names the type of target (immune system, renal system, cancer, cardiovascular system, bone) and their origin (chimeric, humanized, human). Classification for different biologics includes the prefix of the name (generally provided by the pharmaceutical company), and the suffix defines the type of biologic, namely, a monoclonal antibody (mab), a soluble receptor (cept), or a kinase inhibitor (inib).

a potentially useful preclinical safety test for liposomal drug R&D [28b]. The World Health Organization (WHO) also emphasizes evaluating complement binding and activation for biologics [29]: *"Unless otherwise justified, the ability for complement-binding and activation, and/or other effector functions, should be evaluated even if the intended biological activity does not require such functions."* The FDA has approved a few drugs for inhibiting various complement proteins, while many others are in preclinical and clinical stages of drug development (Table 1.4).

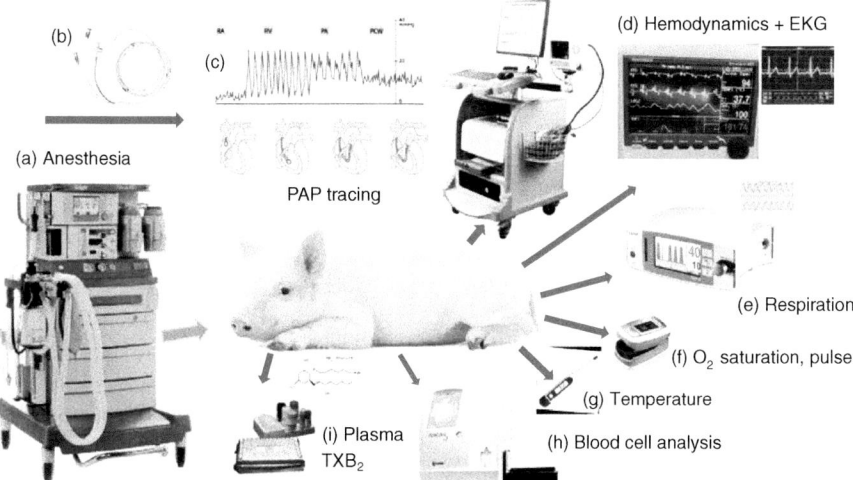

Figure 1.9 Instruments and parameters measured in the porcine CARPA model. (a) anesthesia machine; (b) Swan–Ganz catheter; (c) blood pressure wave forms directing the passage of the tip of the Swan–Ganz catheter via the right atrium (RA), right ventricle (RV) and pulmonary artery (PA) until being wedged into the pulmonary capillary bed; (d) computerized multiple parameter hemodynamic monitoring system tracing the systemic and pulmonary pressures, heart rate, and the EKG; (e) capnograph connected to the tracheal tube to measure respiratory rate (RR), etCO$_2$ and inCO$_2$; (f) pulse oximeter (fixed on the tail) measures O$_2$ saturation in blood and pulse rate; (g) temperature is measured with a thermometer placed in the rectum; (h) veterinary hematology analyzer measuring all blood cell counts and WBC differential; (i) enzyme-linked immunosorbent assay (ELISA) for measuring biomarkers of allergic/inflammatory reactions, e.g., TXB2, histamine, leukotrienes, adenosine, tryptase, PAF and C3 levels, etc. Courtesy of Dr. János Szebeni, Semmelweis University School of Medicine, Hungary.

Table 1.3 Drugs linked to CARPA.

Liposomal Drugs	Micelle-Solubilized Drugs	Antibodies	Pegylated Proteins	Contrast Media	Enzymes/Proteins/ Peptides	Miscellaneous
Abelcet	Cyclosporine	Avastin	Adagen	Diatrizoate	Abbokinase	ACE inhibitors
AmBisome	Elitec	Campath	Neulasta	Iodipamide	ACH	AR blockers
Amphotec/ Amphocyl	Etoposide	Erbitux	Oncaspar, Pegaspargas	Iodixanol	Actimmune	Aspirin
DaunoXome	Fasturec	Herceptin		Iohexol	Activase	Cancidas
Doxil, Caelyx	Taxol	Infliximab		Iopamidol	Aldurazyme	Copaxone
Myocet	Taxotere	Muronomab		Iopromide	Avonex	Corticosteroids
Visudyne	Vumon	Mylotarg		Iothalamate	Fasturtec	Cyclofloxacin
		Remicade		Ioversol	Neulasta	Eloxatin
		Rituxan		Ioxaglate	Neupogen	Intralipid
		Vectibix		Ioxilan	Plenaxis	Opiates
		Xolair		Magnevist	Protamine	Orencia
				Metrizamide	Urokinase	Salicylates
				SonoVue	Zevalin	Vancomycin

Table 1.4 FDA-approved complement inhibitors and new drugs under development

Company/Reference	Drug/Class	Target	Structure/Derivation	Route	Biochemical Data	Stage
Alexion	Eculizumab	C5	Humanized monoclonal antibody	Intravenous	K_d = 120 pM	FDA approved for PNH and aHUS
ViroPharma, CSC Behring	C1 esterase inhibitor: plasma derived	C1 esterase	Human protein	Intravenous		FDA approved for hereditary angioedema
Pharming-Salix	C1 esterase inhibitor: recombinant	C1 esterase	Protein analogue, produced in rabbits	Intravenous		FDA approved for hereditary angioedema
Alexion (Taligen)	TT30	C3 convertase	Factor H-CR2 fusion	Intravenous/Subcutaneous	IC_{50} = 0.5 μm	Human studies
Norvartis	LFG 316	C5	mAb	Intravenous/Intravitreal		Human studies
Amyndas, Apellis, Potentia	Compstatin analogues	C3/C3b	Cyclic peptide/Phage display	Intravitreal, Subcutaneous, Inhaled	IC_{50} = 62 nM, K_d for C3b = 2.3 nM	Human studies

(Continued)

Table 1.4 (Continued)

Company/Reference	Drug/Class	Target	Structure/Derivation	Route	Biochemical Data	Stage
Volution Akari	Coversin	C5	Peptide/tick saliva	Subcutaneous	Maximal inhibition at 10 μg/mL	Human studies
Achillion	Small molecule	Complement factor D	X-ray crystallography	Oral	K_d < 1 nm, IC_{50} = 17 nm (protease inhibition)	Preclinical studies
Amyncas	Mini factor H	C3 convertase	Derived from factor H		IC_{50} = 0.22 μM (for C3 deposition)	Preclinical studies
Alnylam	ALN-CC5	C5 RNA	RNAi conjugate	Subcutaneous		Preclinical studies
Lindofer et al.	3E7/H17	C3b	mAb		100% blockage of lysis at 1 μM	Preclinical studies
Ra Pharmaceuticals	Several	C5	Cyclic peptide	Subcutaneous	K_d = 2.6 nM, IC_{50} = 8.1 nm (% RBC lysis)	Preclinical studies

Source: Reprinted with permission from [20]. Copyright (2018) Elsevier.

Abbreviations: aHUS, atypical hemolytic uremic syndrome; IC_{50}, half maximal inhibitory concentration; K_d, dissociation constant; PNH, paroxysmal nocturnal hemoglobinuria; RBC, red blood cell.

(d) Another immunologic issue specific to PEGylated liposomes is referred to as the accelerated blood clearance (ABC) phenomenon [30–31] (Fig. 1.10). Liposomes are the most widely used nanodrugs and PEGylation is a common strategy involved in designing stealth liposomes to shield them from reticuloendothelial system (RES) uptake. However, a repeated-dose injection of PEGylated liposomes affects their clearance rate and bioavailability. The delivery of the first dose of PEGylated liposomes ("priming the system") accelerates subsequent dose elimination as compared to the initial dose, mainly mediated through specific anti-PEG IgM. This finding is clinically significant as well as concerning if PEGylated liposome therapy is involved because it decreases the therapeutic efficacy upon repeated administration. Therefore, repeated-dosage PK studies are critical to prevent immunogenicity of PEGylated liposome drug products without hampering their efficacy or safety. Table 1.5 lists some PEGylated nanodrugs that have adverse immune effects.

1.5 Immunogenicity Assessment of Biologics and Nanodrugs

There is a crucial need to evaluate, assay, and devise strategies to overcome adverse immunogenicity aspects of both biologics and nanodrugs. Not all biologics and nanodrugs are created equal. Given this scientific fact, the risks for immunogenicity should be assessed on a case-by-case basis. Contrary to widely held belief, few ADAs elicit any clinically relevant issues. In fact, while some biologics, particularly glycoproteins, cause the body to produce ADAs, the safety and efficacy of most is unaffected during clinical use. Similarly, the diversity of nanodrugs makes it impossible to extrapolate or generalize the immunologic findings from one class of nanodrugs to another. Nevertheless, the degree of risk for eliciting immune responses from biologics and nanodrugs is considered a major issue during drug R&D and use in patients. Any biologic or nanodrug can *potentially* exert an immunogenic effect depending on a patient's immunologic status, prior history, route/dose/frequency of delivery and unique characteristics of the administered therapeutic product (Fig. 1.5).

Therefore, regulatory agencies, particularly the FDA and the EMA, recommend that drug developers employ a risk-based approach for immunogenicity evaluation and reduction of adverse immune events related to the administration of these therapeutics.

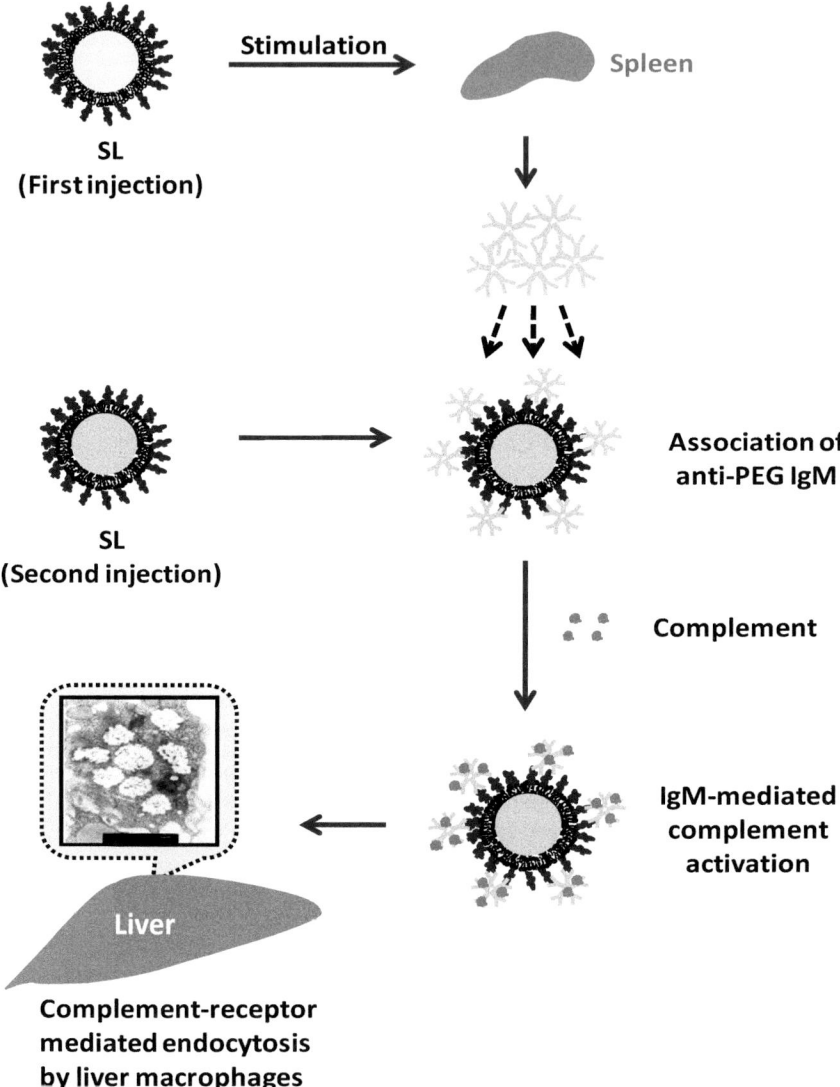

Figure 1.10 Mechanism of the ABC phenomenon. Courtesy of Dr. Tatsuhiro Ishida, Tokushima University, Japan.

Table 1.5 PEGylated nanodrugs with documented adverse immune effects

Generic Name	Trade Name	Api or Vehicle	Company
PEGylated liposomal doxorubicin	Doxil®/Caelyx®	Liposome	ALZA/Janssen
Pegaspargase	Oncaspar®	Enzyme: asparaginase	Enzon
Pegfilgrastim	Neulasta®	Protein (GCSF)	Amgen
Pegaptanib	Macugen®	Aptamer (anti-VEGF)	Eye Tech/Pfizer
Mono-mPEG-epoetin-β	Mircera®	Protein (EPO)	Hoffmann-LaRoche
Certolizumab pegol	Cimzia®	Fab of anti-TNF mAb	UCB, Inc, Smyrna
Withdrawn from the market			
Pegvisomant	Somavert®	Peptid (somatotropin antagonist)	Pfizer
Pegloticase	Krystexxa®	Enzyme: Urate oxidase	Horizon Pharma
Peginesatide	Omontys®/Hematide®	Peptide (EPO-mimetic)	Affymax/Takeda
Pegnivacogin+Anivamersen	Revolixys kit	F-IXa blocker RNA aptamer + reverse agent	Regado/Tobira

These must be carefully evaluated at the earliest stages of drug formulation/development as well as throughout the product lifecycle, including phase IV. Biologic drug products containing a nonbiologic component or nanomaterial component are on the rise and may have different immunogenic properties compared with those that contain the biologic alone. Consequently, it is also important that immunogenicity aspects and risks of these biologics be assessed with a focus on whether the nonbiologic component or nanomaterial component possesses adjuvant properties. Also, immunogenic potential of drug carriers and other adjuvants cannot be overlooked either as these drug components may exhibit inherent immunologic activity unrelated to the loaded API.

Immunogenicity could be measured by experimental approaches or predicted via mathematical models and *in vitro/in vivo/in silico* assays. Therefore, few tools have been developed to access potential immunogenicity of biologics and nanodrugs (Table 1.6). The key methods for preclinical measurement of immunogenicity use *in silico*, *in vitro*, and *in vivo* models to predict CD4+ T cell responses as well as conventional mouse models, immune-tolerant transgenic mice, HLA-immune-tolerant transgenic mice, and nonhuman primate models (Table 1.6).

The immunogenicity for biologics has been primarily assessed by monitoring the presence and amount (titer) of ADA responses and *in vitro* neutralizing ability of ADA following biologic administration. Such assessment strategies are often driven by indication-specific, product-specific or risk assessment/performance-based goals.

On the other hand, the immunogenicity assessment of nanodrugs is less well developed. There are very few detailed regulatory guidance documents specifically dedicated to evaluating immunogenicity. In fact, immunogenicity or immunotoxicity assessment of nanodrugs is often performed based on existing guidelines for conventional therapeutic drug products. However, due to various unique properties of nanodrugs as compared to conventional therapeutic drug products, the currently prescribed set of tests or assays may provide insufficient information for an adequate evaluation of potential immunogenicity or immunotoxicity of nanodrugs.

Table 1.6 Standard industry immunogenicity prediction tools and models

In silico	In vitro	In vivo
iTope™ TCED™ Epibase® EpiMatrix™	EpiScreen™—*Ex vivo* assessment of immunogenicity ➢ EpiScreen™ time course T cell assay ➢ EpiScreen™ DC:T cell assay ➢ EpiScreen™ T Cell Epitope Mapping ➢ EpiScreen™ MAPPS—MHC Class II—Associated Peptide Proteomics Epibase® REVEAL®	conventional mouse models immune-tolerant transgenic mice HLA-immune-tolerant transgenic mice nonhuman primate models

Abbreviations: DCs, dendritic cells; MHC, Major Histocompatibility Complex; MAPPS, MHC Class II Associated Peptide Proteomics; TCED™, T Cell Epitope Database; HLA, human leukocyte antigen.

Note: Although these tests are widely used for biologic immunogenicity prediction, they could pertain to both biologics and nanodrugs because of considerable overlap in their definitions (Sections 1.2 and 1.3). Copyright 2018 Raj Bawa. All rights reserved.

Although the complex field of immunogenicity assessment is still evolving, numerous hurdles persist. One major issue is the so-called "immunogenicity testing dilemma" for biologics and nanodrugs due to the recognized fact that the phylogenetic distance between laboratory animals and humans limits the predictive value for testing. For example, immune responses to biologics in conventional animal models has been rarely predictive of the response in humans. This fact is critical when evaluating human immunogenicity due to pronounced species-specific differences in antigen recognition, in immune reactivity of nonlymphoid/lymphoid cells, and in the systemic immunity at the organ level. Efforts to overcome this immunogenicity testing dilemma have not been particularly successful. For example, employing a broad spectrum of 2D *in vitro* assays in conventional culture plates based on suspension or matrix-assisted human immune cell cultures for evaluation of immunogenicity prior to human testing is fraught with problems and still not an industry standard. Similarly, overprediction of immunogenicity risk via *in silico* methods may occur as these models depend heavily on how well computational algorithms have been created in the first place. However, some tests are slowly gaining ground and may become standard in due course. For instance, the fact that CARPA (Section 1.4.2(c)) is a major immunologic issue with intravenous nanodrug formulations, has recently prompted the FDA to list testing for complement activation *in vitro* and/or *in vivo* as one of the immunotoxicology tests [28a].

1.6 Entering the Era of Biosimilars

1.6.1 What Are Biosimilars?[12]

A biosimilar (Fig. 1.11) is a biological product that is "highly similar" to and has no clinically meaningful differences from an existing FDA-approved reference product. A "reference product" is the single biological product, already approved by the FDA, against which a proposed biosimilar product is compared (Fig. 1.12). A reference product is approved based on, among other things, a full complement of safety and effectiveness data. A proposed biosimilar product is compared to and evaluated against a reference product to ensure that the product is highly similar and has no clinically meaningful differences.

Biosimilars and generic drugs are versions of brand name drugs and may offer more affordable treatment options to patients. Biosimilars and generics are each approved through different abbreviated pathways that avoid duplicating costly clinical trials. But biosimilars are not generics, and there are crucial differences between biosimilars and generic drugs. For example, the active ingredients of generic drugs are the same as those of brand name drugs. In addition, the manufacturer of a generic drug must show that the generic is *bioequivalent* to the brand name drug. By contrast, biosimilar manufacturers must demonstrate that the biosimilar is *highly similar* to the reference product (Fig. 1.12), except for minor differences in clinically inactive components.

---

[12]This US perspective on biosimilars was kindly provided by the FDA. The figures in this section have been modified by the author.

European Medicines Agency - A biosimilar is a biological medicine that is developed to be similar to an existing biological medicine (the 'reference medicine'). When approved, a biosimilar's variability and any differences between it and its reference medicine will have been shown not to affect safety or effectiveness.

United States Food and Drug Administration - A biosimilar is a biological product that is highly similar to a US licensed reference biological product notwithstanding minor differences in clinically inactive components, and for which there are no clinically meaningful differences between the biological product and the reference product in terms of safety, purity and potency of the product.

World Health Organization - A biosimilar is a biotherapeutic product which is similar in terms of quality, safety and efficacy to an already licensed reference biotherapeutic product.

Figure 1.11 Official global definitions for biosimilars.

Reference Product
A reference product is the single biological product, already approved by FDA, against which a proposed biosimilar product is compared.

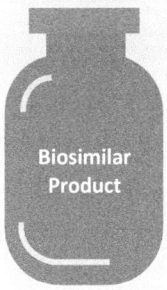
Biosimilar Product
A biosimilar is a biological product that is highly similar or has no clinically meaningful differences from an existing FDA-approved reference product.

Interchangeable Product
An interchangeable product is a biosimilar product that meets additional requirements.

Figure 1.12 FDA Terminology regarding biosimilars.

Biosimilar manufacturers must also prove that there are no clinically meaningful differences between the biosimilar and the reference product in terms of safety and effectiveness.

A manufacturer developing a proposed biosimilar demonstrates that its product is highly similar to the reference product by extensively analyzing (i.e., characterizing) the structure and function of both the reference product and the proposed biosimilar. State-of-the-art technology is used to compare characteristics of the products, such as purity, chemical identity, and bioactivity. The manufacturer uses results from these comparative tests, along with other information, to demonstrate that the biosimilar is highly similar to the reference product.

Minor differences between the reference product and the proposed biosimilar product in clinically inactive components are acceptable. For example, these could include minor differences in the stabilizer or buffer compared to what is used in the reference product. As mentioned above, slight differences (i.e., acceptable within-product variations) are expected during the manufacturing process for biological products, regardless of whether the product is a biosimilar or a reference product. For both reference products and biosimilars, lot-to-lot differences (i.e., acceptable within-product differences) are carefully controlled and monitored. A manufacturer must also demonstrate that its proposed biosimilar product has no clinically meaningful differences from the reference product in terms of safety, purity, and potency (safety and effectiveness). This is generally demonstrated through human pharmacokinetic (exposure) and pharmacodynamic (response) studies, an assessment of clinical immunogenicity, and, if needed, additional clinical studies (Fig. 1.13).

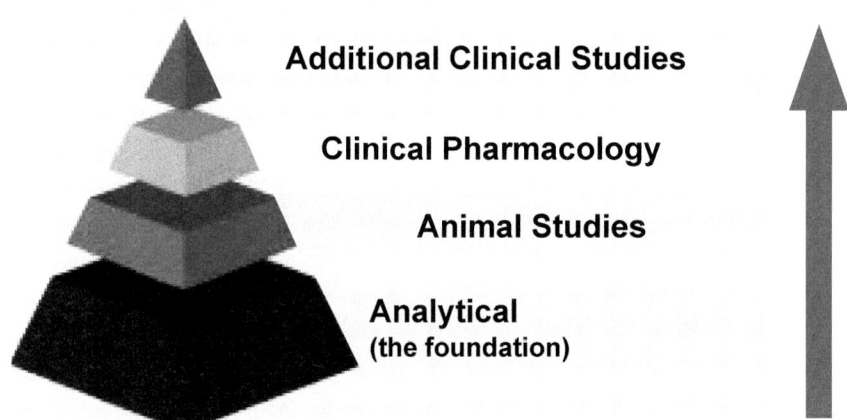

Figure 1.13 The FDA's review for licensure of a biosimilar product.

When considering licensure of a biosimilar product, the FDA reviews the totality of the data and information, including the foundation of detailed analytical (structural and functional) characterization, animal studies if necessary, then moving on to clinical pharmacology studies and, as needed, other comparative clinical studies (Fig. 1.13).

An "interchangeable product" (Fig. 1.12) is a biosimilar product that meets additional requirements outlined by the BPCI Act (Section 1.1). As part of fulfilling these additional requirements, information is needed to show that an interchangeable product is expected to produce the same clinical result as the reference product in a patient. A manufacturer of a proposed interchangeable product will need to provide additional information to show that an interchangeable product is expected to produce the same clinical result as the reference product in any given patient. Also, for a product that is administered to a patient more than once, a manufacturer will need to provide data and information to evaluate the risk, in terms of safety (including immunogenicity) and decreased efficacy, of alternating or switching between the products. As a result, a product approved as an interchangeable product means that the FDA has concluded it may be substituted for the reference product without consulting the prescriber. For example, say a patient self-administers a biological product by injection to treat their RA. To receive the biosimilar instead of the reference product, the patient may need a prescription from a health care prescriber written specifically for that biosimilar. However, once the FDA approves a product as interchangeable, that patient may be able to take a prescription for the reference product to the pharmacy and, depending on the state, the pharmacist could substitute the interchangeable product for the reference product without consulting the prescriber. Note that pharmacy laws and practices vary from state to state.

1.6.2 FDA Challenges Regarding Biosimilar Approval

According to a 2018 speech[13] by the FDA commissioner, about a third of new drugs approved by the FDA are now biologics while they account for about 40% of all US drug spending, and 70% of spending growth from 2010–2015. Developing a generic

---

[13]Gottlieb, S. (2018). Capturing the benefits of competition for patients. Available at: https://www.fda.gov/NewsEvents/Speeches/ucm599833.htm (accessed on July 14, 2018).

version of a small-molecule drug can cost ~$10 million. Due to the complexity of manufacturing and testing biosimilars, more significant outlays by sponsors are required: typically, $100–$250 million per program.

Since 2007, 31 biosimilar products have been approved by the EMA while 5 have been refused or withdrawn. On the other hand, the FDA has struggled with biosimilar approval. Since the passage of the BPCI Act, as of May 2018, the FDA has licensed only nine biosimilar products. The FDA has been justifiably criticized for the slow entrance of biosimilars into the US market. It is obvious to me that the steep cost (~$150+ million) and lengthy development (~7–9 years) of biosimilars are untenable and need urgent addressing, possibly via appropriate regulatory adjustments. Table 1.7 lists suggested modifications to the FDA's current biosimilar guidelines.

Table 1.7 Recommendations to the FDA for faster development and licensing of biosimilar products[14]

- The FDA should remove the current default requirements of conducting bridging studies between a US-licensed product and a non-US approved comparator to establish biosimilarity.

- The FDA should present clear and open scientific views to the public, more particularly, to the prescribers that a biosimilar product has "no clinically meaningful difference" from the originator product and thus suitable for naïve patients.

- The FDA should encourage the development of *in vitro* immunogenicity testing methods to reduce exposure of test subjects on ethical grounds.

- The FDA should revise some of the specific statistical testing methodologies in establishing analytical similarity to remove certain contradictions in the guidance.

- The FDA should take a fresh look at the clinical relevance of the protocols and statistical methods used to establish PK/PD similarity, and to make these studies more clinically relevant while reducing their cost.

---

[14]Based on the Citizen Petition (CP) of Dr. S. K. Niazi of the University of Illinois College of Pharmacy to the FDA (dated May 11, 2018; docket number FDA-2018-P-1876) that focuses on reducing human testing to establish bioequivalence. It was accepted by the FDA and as of June 2018 was under the comment period. In the past, I have filed CPs on behalf of Teva pertaining to Copaxone®.

1.7 Immune Aspects of Biosimilars and Nanosimilars: The Copaxone® Example

Many veteran drug industry experts, including this author, believe that there are enormous pressures on drug regulatory agencies to approve follow-on versions (i.e., generic equivalents) of both biologics and nanodrugs. Frankly, judging from the rapid pace of biosimilars that were approved in the past year, the Trump administration seems to be pushing for an increase in biosimilar approvals at the FDA. Concurrently, the increase in the number of drug companies targeting generic opportunities and seeking US market exclusivity for generic versions of major branded products is on the rise. There are many factors for this, including governmental drug policy, price pressures, and statutes. However, it is critical that immune aspects of these follow-on versions of branded products be transparently evaluated in a science-based context and reported during all phases of drug R&D (from preclinical to post-marketing): *Lower drug prices, a priority for the Trump Administration,*[15] *should not supplant patient safety and drug efficacy.*

The following discussion regarding biosimilar therapeutic monoclonal antibodies (TMAbs) highlights the fact that such follow-on biologic approval by a regulatory agency must be carefully evaluated on a case-by-case basis for clinical data based on the "totality-of-evidence" [32]:

"By contrast with generic small-molecule drugs, clinical performance of a biologic pharmaceutical is a function of its structural complexity and higher-order structure (HOS). Biomanufacturing controls of such complex products cannot fully ensure chemical similarity between an innovator product and putative biosimilar because minor differences in chemical modifications and HOS can significantly alter a product's safety and efficacy. Therefore, to substantiate claims of clinical functionality, a demonstration of bioequivalence is inadequate for biosimilar pharmaceuticals. This is different from regulatory approval for generic drugs, in which bioequivalence demonstration is adequate. The overall challenge in approving biosimilar pharmaceuticals is to enable scientific inference of similarity

---
[15]Sterling, J. (2018). President reveals plan to cut drug prices. *Genet. Eng. Biotechnol. News*, **38**(12), 1, 30.

in safety and efficacy for a new biologically derived product compared with an innovator without repeating burdensome clinical studies…. So although they are helpful, biological and/ or functional assays may not fill a gap in analytical assay sensitivity to detect minor conformational differences between biosimilar TMAbs and innovator products. It is important to note that no analytical test or combination for HOS has yet been sufficiently validated for analytical testing as a substitute for clinical studies in the development of a biosimilar TMAbs drug substance."

In this context, the recent FDA approval of multiple generic versions of Copaxone® is an example that merits discussion as it highlights this problematic issue [33]. Copaxone® is a nonbiologic complex drug (NBCD) [34] but can also be considered a nanodrug (Section 1.3). However, it also shares features with biologics and given the loose definition of biologics (Section 1.2), it can be classified as a biologic as well. In this chapter, it will be considered a NBCD, a nanodrug, and a biologic. Owing to the complexity of NBCDs and nanodrugs, showing equivalence is more challenging for their follow-on versions. Therefore, the interchangeability or substitutability of nanosimilars and their listed reference product(s) cannot be taken for granted. In the past, nanosimilars have been approved via generic pathways but differences in clinical efficacy and safety have been reported in the scientific literature following approval [35].

What Is a Nonbiologic Complex Drug (NBCD)?

"A medicinal product, not being a biological medicine, where the active substance is not a homomolecular structure, but consists of different (closely) related and often nanoparticulate structures that cannot be isolated and fully quantitated, characterized, and/or described by physicochemical analytical means. It is also unknown which structural elements might affect the therapeutic performance. The composition, quality, and *in vivo* performance of NBCDs are highly dependent on the manufacturing processes of both the active ingredient and the formulation. Examples of NBCDs include liposomes, iron-carbohydrate (iron-sugar) drugs, and glatiramoids."

Source: [35]

Copaxone® is composed of an uncharacterized mixture of immunogenic polypeptides in a colloidal solution. The active ingredient in Copaxone®—glatiramer acetate—is a heterogeneous synthetic mixture of polypeptides comprising four amino acids found in myelin basic protein (L-glutamic acid, L-alanine, L-lysine, and L-tyrosine) in a defined molar ratio. Glatiramer acetate has immunomodulatory effects on innate and acquired immunity and is indicated for the treatment of patients with relapsing forms of multiple sclerosis (MS). Copaxone® is not a single molecular entity but a heterogenous mixture of potentially millions of distinct, synthetic polypeptides of varying lengths, some containing up to 200 amino acids with structural complexity comparable to that of proteins, or even more complex than proteins. It is presently impossible to isolate and identify its pure components even via the most technologically sophisticated multidimensional separation techniques. The complexity of glatiramer acetate is amplified by several aspects: (1) The active moieties in glatiramer acetate are unknown; (2) the mechanisms of action are not completely elucidated; (3) pharmacokinetic testing is not indicative of glatiramer acetate bioavailability; (4) pharmacodynamic testing is not indicative of therapeutic activity and there are no biomarkers available as surrogate measures of efficacy; and (5) small changes in the glatiramer acetate mixture can change its immunogenicity profile. There is one aspect of Copaxone® that raises special safety and effectiveness concerns that merit heightened vigilance with respect to the approval of any potentially interchangeable follow-on glatiramer acetate product: *Glatiramer acetate is an immunomodulator* [33]. In other words, Copaxone® is intended to achieve its therapeutic effects by interacting with and modulating a patient's immune system over an extended period. For this reason, Copaxone®'s package insert warns that chronic use has the potential to alter healthy immune function as well as induce pathogenic immune mechanisms, although no such effects have been observed with Copaxone®.

For small-molecule drugs, regulatory approval of generic versions is based on factors like molecular identity of the active ingredient, identity in strength, purity and quality, and bioequivalence. In other words, a demonstration of bioequivalence can result in regulatory approval of a small-molecule generic without having to conduct the full set of clinical trials that prove clinical safety and efficacy. However, this strategy cannot be followed for biologics like Copaxone®. Even if a biosimilar were to have the exact same primary amino acid sequence as the innovator, the innovator's manufacturing process is usually proprietary and not in the public domain. *Hence, biosimilars are, by definition, manufactured using different processes than the innovator (Section 1.6). Obviously, these differences in manufacturing process, no matter how subtle, can generate unique heterogeneities within a potential biosimilar product as compared to the branded product. This can have different pharmacologic effects or adverse immune effects on the patient. Therefore, biosimilars necessitate careful consideration for safety and efficacy.* With this backdrop, it is clear that due to the complexity and inexorable link between the manufacturing process and quality, any Copaxone® biosimilar almost certainly will differ from Copaxone®'s structure and composition of active ingredients because it will be made using a different manufacturing process than that developed by the branded product developer (Teva Pharmaceutical Industries Ltd., Israel) [33]. Although it is not possible to fully characterize and compare these complex mixtures, differences are revealed via sophisticated analytical techniques. In the past few years, purported generic glatiramer acetate follow-on versions have been approved in India, Argentina, and Mexico. More recently, the FDA has also approved substitutable generic glatiramer acetate formulations.[16] A variety of physicochemical tests performed by Teva have been done on these generic products and they have been proven to be similar to Copaxone® in some basic features [33]. However, they are different in the bulk composition of constituents when analyzed via methods for analysis of complex closely related molecules [33]. In this regard, a widely used analytical tool for characterization of complex mixtures of biologics in the context of biosimilars is ion mobility mass spectrometry (IMMS). The ion mobility method applies multidimensional separation techniques based on size, shape, charge, and mass of the molecules in the sample mixture and can

---

[16] Teva Pharmaceuticals USA, Inc. v. Sandoz, Inc., 574 U.S.___(2015), is a landmark Supreme Court patent law case pertaining to Teva's Copaxone® patent. Available at: https://www.supremecourt.gov/opinions/14pdf/13-854_o7jp.pdf (accessed on June 21, 2018).

separate isomeric peptides that chromatographic techniques cannot. The analysis produces a three-dimensional "heat map" to highlight intensity differences of peptides at various mass/charge ratios and drift times. The difference between the intensities of heat maps for the generics tested by Teva as compared to Copaxone® (result of subtraction of generic heat map from that of Copaxone®) show highlighted areas indicating different polypeptide populations compared to those of Copaxone® lots tested. Clearly, these results indicate a profound difference in size, shape, and charge of the constituent polypeptides in Copaxone® as compared to the purported generic products tested by Teva [33].

What does this mean in the context of immune aspects of Copaxone®? Because Copaxone® is an immunomodulator, a follow-on product characterized by different constituent population could have significant and unpredictable differences from Copaxone® in its immunological mechanisms, raising major safety and efficacy concerns. The potential risks associated with such follow-on products include increased immunogenicity, immunotoxicity, induction of additional autoimmune disorders, lack of efficacy, and exacerbation of the MS disease processes. Moreover, because of the nature of both RRMS and Copaxone®, these risks may not develop for months or years and, once apparent, may be irreversible. Since the active amino acid sequences in the glatiramer acetate mixture responsible for its efficacy are unknown, it is impossible to predict whether already-approved and future follow-on products will have the same efficacy as Copaxone®. They could have a weaker anti-inflammatory effect and/or enhance a pro-inflammatory environment, further exacerbating MS pathogenic processes. A reduced anti-inflammatory effect may provide less effective control of MS relapses, which would be difficult to detect in the post-marketing environment because MS relapses and progression of disability are not completely abolished by any MS therapy. On the other hand, creation or amplification of a pro-inflammatory environment would likely increase relapse rate and progression of disability or worse (e.g., have a profound encephalitogenic effect).

Finally, the potential for the development of cross-reactive neutralizing antibodies must be assessed before any regulatory authority approves any follow-on glatiramer acetate product intended to be used interchangeably with Copaxone®. Switching between two complex polypeptide products with subtle differences in structure and/or composition may increase the chance of cross-reactivity, a phenomenon that has been observed with interferon beta products. Upon switching from Copaxone® to a follow-on product or using them interchangeably, antibodies formed against Copaxone® may neutralize the activity of the proposed generic product and *vice versa*. If this were the case, patients would be left without any effective treatment. Again, there is no evidence that progression of neurologic disability associated with untreated MS can ever be reversed.

It is thus critical to ensure that any proposed follow-on product has a long-term immunogenicity profile that is comparable to Copaxone®'s before approval. This can only be done based upon data from appropriate clinical testing.[17] Surprisingly, despite these immunological concerns, the FDA recently approved so-called generic versions of Copaxone®.

1.8 Concluding Remarks and Future Directions

Immune regulation is mediated by a highly complex network of cells and signaling pathways, massive and dynamically interacting gene networks, host–pathogen interactions, and nutrition–microbiota–host interplay. Therefore, dysregulation of immune pathways (i.e., when immunoregulatory mechanisms deviate or fail) is central to many diseases. In fact, immune-mediated diseases are often multifactorial, exhibit enormous patient-to-patient variability, and are often hard to treat via traditional therapies. There continues to be a lack of understanding of the physicochemical determinants underlying immune mechanisms as they relate to biologics and nanodrugs. Despite enormous advances in medicine in the last hundred years, there exist major

---

[17]Tyler, R. S. (2013). The goals of FDA regulation and the challenges of meeting them. *Health Matrix*, **22**(2), 423–431: "[W]ith respect to drugs, there is no substitute for a well-controlled clinical trial to establish a drug's safety and effectiveness and conducting such a trial is beyond the competence of individual consumers. Consumers, unprotected by regulations requiring such trials, are unable to judge the safety and effectiveness of a drug…Nevertheless, the regulatory framework is unsettled and there are now, as there have been in the past, demands in Congress and elsewhere to change the laws under which FDA operates."

gaps in our current understanding of immunological responses and immune mechanisms. We are on a steep learning curve with respect to fully comprehending the extremely complex mechanisms, side reactions, and interactions of various immune cells.

Unwanted immunogenicity of biologics and nanodrugs is a major safety and efficacy concern during drug development and clinical use. Hence, assessment of immunogenicity remains a key element during drug R&D. Unfortunately, extensive testing during drug development does not guarantee that the approved product will be free of immune issues, including immunogenicity, that could adversely affect drug effectiveness and patient safety. There are basic underlying reasons responsible for this unpredictability. For example, the medical and/or scientific concepts related to immunogenicity are incompletely understood. The pressure to develop effective and safe drugs for disease states with unmet medical needs adds another wrinkle to the mix.

Pharmacovigilance is, therefore, critical for biologics and nanodrugs. Due to unpredictability of their immunogenicity profile, managing it is essential not only during the drug R&D phases but all the way to postmarketing surveillance (PMS). In this context, a multidisciplinary approach is called for to better understand and minimize immune issues associated with biologics and nanodrugs. The assessment of unwanted immunogenicity can be improved by using immuno-prediction tools, optimizing immunoassays, and monitoring patients receiving these drug products. In fact, routine immunogenicity and drug level assessment in patients receiving biologics and nanodrugs should become a healthcare standard to better understand their underlying immune mechanisms. Basically, we need to identify various modulating factors that could reduce drug immunogenicity below clinically significant levels. An early indicator of a potentially highly immunogenic drug, before it enters clinical phase testing, will avoid an unnecessary safety risk to patients and save time and resources. Although the etiology of immunogenicity is still not fully understood for these drug products, advances in approaches to mitigate immunogenicity that are currently underway involve rigorous immunogenicity characterization, advances in animal models, and *in silico, in vitro*, and *in vivo* prediction tools. Future biomedical research must expand and standardize analytical methods.

Another broader problem impacting immunogenicity assessment is that there are many defects in the current drug research environment. The "evidence" from clinical studies of drug effects, including immune potential, and why such evidence might fail in the prediction of the clinical utility of drugs is an issue of much concern to me. Although the standards used by the regulatory agencies have evolved and expanded over the past two decades, serious concerns persist with the current approach [36]:

"Problems in clinical studies are an indication of missed opportunities to successfully define the real-world effectiveness and safety of drugs. Driven largely by commercial interests, many clinical studies generate more noise than meaningful evidence to guide clinical decision making. Greater involvement of nonconflicted bodies is needed in the design and conduct of clinical studies, along with more head-to-head comparisons, representative patient populations, hard clinical outcomes, and appropriate analytical approaches. Documenting, registering, and publishing study protocols at the outset and sharing participant-level data at study completion would help ensure transparency and enhance public trust in the clinical research enterprise. Such an approach is needed to generate evidence that is better suited to the tasks of predicting the clinical utility of drugs and providing the information needed by patients and clinicians. Future efforts should focus on engaging the industry, researchers, regulators, clinicians, patients, and other decision makers in discussions to develop transformative ideas with the aim of tackling the numerous defects in the current research environment. Emerging ideas should be piloted and subjected to scientific scrutiny before they are widely implemented and touted as solutions."

Many concerned experts highlight another key issue that affects the entire pharma enterprise. It is referred to as the "institutional corruption of pharmaceuticals" and is due to an interplay of key players with often-serious conflicts of interest: physicians, Congress, and the drug industry. Naturally, this jeopardizes the safety and effectiveness of all drug products, not only biologics and nanodrugs. One apparent consequence for patients are serious ADRs [37]:

"Institutional corruption is a normative concept of growing importance that embodies the systemic dependencies and informal practices that distort an institution's societal mission. An extensive range of studies and lawsuits already documents strategies by which pharmaceutical companies hide, ignore, or misrepresent evidence about new drugs; distort the medical literature; and misrepresent products to prescribing physicians... First, through large-scale lobbying and political contributions, the pharmaceutical industry has influenced Congress to pass legislation that has compromised the mission of the Food and Drug Administration (FDA). Second, largely as a result of industry pressure, Congress has underfunded FDA enforcement capacities since 1906, and turning to industry-paid "user fees" since 1992 has biased funding to limit the FDA's ability to protect the public from serious adverse reactions to drugs that have few offsetting advantages. Finally, industry has commercialized the role of physicians and undermined their position as independent, trusted advisers to patients."

Advances in immune aspects of biologics and nanodrugs over the past decade have created tremendous opportunity to accelerate the discovery and development of these novel therapeutic agents to treat devastating human diseases. However, despite enormous advances, wide gaps persist. *So, what to expect in the next decade in this vast field regarding efforts to blunt adverse immune reactions and design safer biologics and nanodrugs? What tools, techniques, and analytical methods will be leveraged? Will these advances to come leave us poised on a threshold of innovation?*

I expect that in the next decade there will be an intense competition for targets, introduction of second- and third-generation biologics and nanodrugs, more follow-on versions on validated targets, expiration of blockbuster patents, spotty patent examination at patent offices, nomenclature confusion, poor regulatory guidelines from regulatory agencies, third-party payor pressures, sky-rocketing prices of biologics, and governmental pricing pressures—all impacting and reshaping the drug industry landscape. I also expect that due to limited current experience with the evaluation of biologics and first-generation nanodrugs, manufacturers, regulatory agencies, clinicians, patients, and patent offices will face challenges not only regarding second- and third-generations of these two drug classes but also on the biosimilars, nanosimilars, and NBCD similars front.

Immunogenic effects are likely to be especially challenging to evaluate for highly complex biologics, combination products such as theranostics, and later generation nanoformulations. So, for the time being (this decade), immune reactions to biologics and nanodrugs will be common and regulatory agencies will continue to approve drugs based on an analysis of the risk–benefit ratio that changes significantly depending on the treatment modality. However, as more drug products are developed, information will accumulate on the structure and function of biologics and nanodrugs. As a result, the description and understanding of these drug products and their functionality will be revised, as applicable, and supported with characterization data. Moreover, as the intricacies of the human immune system are further elucidated, we will learn more about the interactions of these therapeutics with immune cells. In the meantime, all medicinal products, including biologics and nanodrugs, will continue to be evaluated by regulatory agencies on a case-by-case basis.

Academic immunology research is generally lagging industry and other medical research fields in incorporating modeling approaches. Due to the high failure rates, long time line (10–17 years), high attrition rate, and enormous R&D costs (estimated at $2.6 billion total)[18] involved in the approval of a new drug, pharma has increasingly turned to computational and mathematical modeling at all levels—modeling drug–receptor interactions, PK and pharmacodynamic (PD) modeling, *in silico* clinical trials. Given this trend, I predict that we will glean greater information regarding the immune aspects of biologics and nanodrugs as we expand our arsenal of both *in vitro, in silico,* and *in vivo* analytical methods as well as instrumentation to evaluate potential immunotoxic effects. Computer-driven computational methods followed by *in vitro* and/or *in vivo*

---

[18]DiMasi, J. A., Grabowski, H. G., Hansen, R. W. (2016). Innovation in the pharmaceutical industry: New estimates of R&D costs. *J. Health Econ.*, **47**, 20–33.

testing of any potentially immunogenic epitopes will help in minimizing immune responses. In future, due to the great cost and time needed for comprehensive animal studies, researchers will increasingly develop various *ex vivo* mimics of *in vivo* biological environments to study the interactions of these drug products with the immune system. Artificial intelligence (AI) is expected to change the drug discovery process as machine learning and other technologies are likely to make the hunt for new drugs quicker, cheaper and more effective [38]. Specifically, AI will be employed in this arena to analyze large data sets from clinical trials, health records, gene profiles, and preclinical studies. Technically, a sufficiently large medicinal chemistry database of transformations could provide novel approaches to improving drug discovery [39].

Drug Discovery Technologies: Current and Future Trends

"[D]rug discovery remains perhaps the most challenging applied science largely due to the complexity of human biology, the vastness of chemical space, the discontinuous impact of functional group changes on molecular properties, and the inability to optimize a single variable (potency, selectivity, permeability, metabolic stability, solubility) without having simultaneous and sometimes detrimental effects on other critical parameters. For these reasons, a successful drug discovery campaign often emerges after investigating dozens of pharmacological targets, with each one requiring thousands of chemical hits to be triaged and hundreds of close-in synthetic analogues to be evaluated. A recent 2016 publication based on 10^6 new drugs from 10 pharmaceuticals firms estimated that the overall investment in discovery and clinical development approaches $2.6 billion for each successful launch...Technologies that enable more effective selection of productive biomolecular targets provide novel ways to engage targets, or appropriately guide design to the most effective regions of chemical space will lead to transformative improvements in drug discovery efficiency."

Source: Noe, M. C., Peakman, M-C. (2017). Drug discovery technologies: Current and future trends. In: Chackalamannil, S, Rotella, D, Ward, S, eds. *Comprehensive Medicinal Chemistry III*, vol. 2, pp. 1–32, Elsevier.

Single-cell genomics involving cell capture and accurate analysis of DNA, RNA, and protein of single cells will certainly transform our understanding of the immune system. Single-cell genomic analysis of blood samples or biopsies will be routine in the next decade, and the entire immune composition of patients will be analyzed and compared with all known healthy and diseased states [40].

I am not a fan of the various accelerated approaches currently underway and on the rise at global regulatory agencies, primarily at the FDA, EMA, and PMDA. For serious or life-threatening disease, the FDA can approve drugs through its accelerated approval review track based on surrogate end-points (rather than hard clinical end-points) that are "reasonably likely to predict clinical benefit." This pathway was designed in the early 1990s to speed drug development. Various accelerated approaches include breakthrough therapy designation, accelerated approval, and conditional marketing authorization—collectively referred to as "facilitated regulated pathways" (FRPs). A greater uncertainty is introduced into the regulatory approval process via FRPs. This could translate into unwanted immunogenicity.

A comprehensive map of molecular drug targets is currently lacking. Gaps and opportunities need to be identified to shed light on the so called "druggable genome"—the subset of genes (~3,000 of the ~20,000 total protein-coding genes in the human genome) encoding proteins that could bind drug-like molecules.[19] In fact, out of ~3,000 of these druggable genes, less than 700 are currently targeted by FDA-approved drugs [41]. This is because big pharma focuses on relatively well-characterized proteins as targets for drug development to mitigate risk. However,

---

[19]The phrase "drug-like molecule" implies that certain properties of a chemical compound (drug candidate) confer on it a greater propensity to become a successful drug product. The standard method to evaluate "druglikeness" or to determine if a chemical compound with a certain pharmacological or biological activity has chemical properties and physical properties that would make it a likely orally active drug in humans is to check compliance of "Lipinski's Rule of Five" that covers certain features or properties of the compound: the numbers of hydrophilic groups, molecular weight, and hydrophobicity. See: Lipinski, C. A., Lombardo, F., Dominy, B. W., Feeney, P. J. (1997). Experimental and computational approaches to estimate solubility and permeability in drug discovery and development settings. *Adv. Drug Deliv. Rev.*, **23**(1–3), 3–25.

it is hard to fault it for following this path: There is a lack of consolidated information on the druggable genome and also a scarcity of high-quality technologies to characterize the function of protein family members. Hence, there is a critical unmet need to expand our basic knowledge of the druggable genome and to increase our catché of potential drug targets by studying druggable gene families [42]. This will aid in determining the relevance of drug targets to human health and disease as well as in the identification of off-target effects of existing drugs and drug candidates. An important long-term outcome of this would be the development of new drugs for immune targets. Also, rapid and precise gene editing technologies, including CRISPR-Cas9,[20] can be applied to build systems of greater physiological relevance and disease significance.

The impact of noncrystalline single-particle cryo-electron microscopy (cryo-EM) on structural biology cannot be understated in the context of immunogenicity. Although, cryo-EM has been used to determine the structure of biological macromolecules and assemblies, its potential for application in drug discovery has been limited by two issues: the minimum size of the structures that can be used to study and the resolution of the images [43]. However, recent technological advances, including the development of direct electron detectors and improved computational image analysis techniques, are leading to high-resolution structures of large macromolecular assemblies [43]. These improvements should further enable structural determination for "intractable" targets that are still not accessible to X-ray crystallographic analysis. Therefore, negative staining techniques and cryo-EM, which have both been employed previously for both linear and conformational epitope mapping analysis, should further enable epitope mapping for designing novel biologics and nanodrugs as well as for determining epitopes at the amino acid level that are critical to immune aspects of these therapeutics. This could also aid in anti-ADA vaccine design in future.

---

[20]What are genome editing and CRISPR-Cas9? U.S. National Library of Medicine. Available at: https://ghr.nlm.nih.gov/primer/genomicresearch/genomeediting (accessed on July 7, 2018).

In the coming years, the study of immune complex (IC) biology specific to biologics will shed more light on the role and relationship of ADA to clinical outcome measures. The formation and contribution of ICs (Section 1.4.1(a)) is central to most of the downstream sequelae that are seen following development of ADA [17]. IC formation and corresponding risks could persist if the same treatment continues unabated even with symptomatic remediation of adverse effects. One central question for now is: *Why do some individuals develop clinically significant ADA titers while others do not?* Attention is also warranted to address the discrepancies currently seen when measuring ADAs with different assays. This can lead to biased clinical interpretation and treatment modalities. Hence, accurate immunogenicity measurement, as reflected by the presence and magnitude (titre) of ADAs, is essential towards assessing, predicting, and mitigating unwanted immunogenicity in a clinical setting. Ultimately, this can lead to safer and more effective drug products.

Compared to conventional small-molecule drugs, further understanding will be essential about the interactions of biologics, nanodrugs and their carriers with biological tissues. Even if these drug products are declared nontoxic according to standard regulatory assays, more robust testing of their interaction with the immune system needs to be performed. Specifically, the impact of intrinsic (e.g., disease, age, sex) and extrinsic factors (e.g., co-administered drugs, presence of impurities, dosing frequency, disease state of the patient) exposure and response, the role of enzymes and transporters in their disposition and their immunogenic potential will be essential to advancing the safe use of these drug products (Fig. 1.5). *In future, drug companies will need to increasingly prove to regulators that neither their manufacturing processes nor later use of the final drug product generates CARPA, immunogenicity, ADAs, or ICs in a manner that causes adverse reactions impacting safety or efficacy.* Regulatory agencies must hold biologics and nanodrugs to strict safety and efficacy standards now so that corresponding follow-on versions later (biosimilars, nanosimilars, NBCD similars [34, 35, 44–46]) are also safe and efficacious. The FDA and the EMA, in particular, should formulate regulatory pathways that are science-based and follow the "totality-of-the-evidence approach" for highly complex drugs like biologics and nanodrugs.

The ever-expanding landscape of innovative technology, techniques, and assays makes it critical for immunologists, protein chemists, drug formulators, nanotechnologists, medicinal chemists, analytical chemists, structural biologists, screening biologists, and computational scientists to expand and integrate their efforts into cross-disciplinary collaborations and to become more familiar with a multitude of areas outside their expertise. Only then can we provoke transformative change in this complex field and address issues regarding immune aspects of biologics and nanodrugs. Ultimately, developing biologics and nanodrugs that have minimal, or no adverse immune aspects, will only be addressed in a comprehensive manner with firm commitment and cooperation between all stakeholders—the public, researchers, pharmaceutical and biotechnology companies, government policymakers, patients, and regulatory agencies. After all, our common mission of building a bridge from "bench-to-bedside" is quite simple: *enhancing translation of biologics, nanodrugs as well as their follow-on versions.*

Disclosures and Conflict of Interest

The author declares that he has no conflict of interest. No writing assistance was used in the production of this chapter and the author has received no payment for its preparation. The author is a scientific advisor to Teva Pharmaceutical Industries Ltd. (Israel). The author is not aware of any affiliations, memberships or funding that might be perceived as affecting the objectivity of this chapter.

Corresponding Author

Dr. Raj Bawa
Bawa Biotech LLC
21005 Starflower Way
Ashburn, Virginia, USA
bawa@bawabiotech.com

About the Author

Raj Bawa, MS, PhD, is president of Bawa Biotech LLC, a biotech/pharma consultancy and patent law firm based in Ashburn, Virginia, that he founded in 2002. He is an inventor, entrepreneur, professor, and registered patent agent licensed to practice before the U.S. Patent & Trademark Office. Trained as a biochemist and microbiologist, he has extensive expertise in pharmaceutical sciences, biotechnology, nanomedicine, drug delivery, microbial biodefense, FDA regulatory issues, and patent law. Since 1999, he has held various positions at Rensselaer Polytechnic Institute in Troy, NY, where he is an adjunct professor and where he received his doctoral degree in three years (biophysics/biochemistry). Currently, he serves as a scientific advisor to Teva Pharma, Israel, is a visiting research scholar at the Pharmaceutical Research Institute of Albany College of Pharmacy in Albany, NY, and is vice president of Guanine, Inc., based in Rensselaer, NY. He has served as a principal investigator of SBIRs and reviewer for both the NIH and NSF. Currently, he is principal investigator of a CDC SBIR Phase 1 grant to develop an assay for carbapenemase-resistant bacteria. In the 1990s, Dr. Bawa held various positions at the US Patent & Trademark Office, including primary examiner from 1996–2002. He is a life member of Sigma Xi, co-chair of the nanotech and personalized medicine committees of the American Bar Association, and founding director of the American Society for Nanomedicine. He has authored over 100 publications, co-edited four texts, and serves on the editorial boards of 14 peer-reviewed journals, including serving as an associate editor of *Nanomedicine* (Elsevier). Some of Dr. Bawa's awards include the Innovations Prize from the Institution of Mechanical Engineers, London, UK (2008); Appreciation Award from the Undersecretary of Commerce, Washington, DC (2001); the Key Award from Rensselaer's Office of Alumni Relations (2005); and Lifetime Achievement Award from the American Society for Nanomedicine (2014).

References

1. Bawa, R. (2017). A practical guide to translating nanomedical products. In: Cornier, J., et al., eds. *Pharmaceutical Nanotechnology: Innovation and Production,* 1st ed., Wiley-VCH Verlag, chapter 28, pp. 663–695.

2. Bawa, R., Barenholz, Y., Owen, A. (2016). The challenge of regulating nanomedicine: Key issues. In: Braddock, M., ed. *Nanomedicines: Design, Delivery and Detection,* Royal Society of Chemistry, UK, RSC Drug Discovery Series No. 51, chapter 12, pp. 290–314.

3. Bawa, R., Bawa, S. R., Mehra, R. (2016). The translational challenge in medicine at the nanoscale. In: Bawa, R., ed.; Audette, G. F., Reese, B. E., asst. eds. *Handbook of Clinical Nanomedicine: Law, Business, Regulation, Safety and Risk,* Pan Stanford Publishing, Singapore, chapter 58, pp. 1291–1346.

4. Markel, H. (2013). Patents, profits, and the American people: The Bayh–Dole Act of 1980. *N. Engl. J. Med.,* **369**, 794–796.

5. Kesselbaum, A. S. (2011). An empirical view of major legislation affecting drug development: Past experiences, effects, and unintended consequences. *Milbank Q.,* **89**, 450–502.

6. Johnson, J. A. (2017). Biologics and biosimilars: Background and key issues. *Congressional Research Service Report* R44620.

7a. Koballa, K. E. (2018). The Biologics Price Competition and Innovation Act: Is a generic market for biologics attainable? *Wm. & Mary Bus. L. Rev.,* **9**, 479–520.

7b. ISR (2018). Is the U.S. biosimilar industry falling behind. *Life Sci. Lead.,* **10**(6), 17.

8. Walsh, G. (2002). Biopharmaceuticals and biotechnology medicines: An issue of nomenclature. *Eur. J. Pharm. Sci.,* **15**, 135–138.

9. FDA. Biological Product Definitions. Available at: https://www.fda.gov/downloads/Drugs/DevelopmentApprovalProcess/HowDrugsareDevelopedandApproved/ApprovalApplications/TherapeuticBiologicApplications/Biosimilars/UCM581282.pdf (accessed on April 30, 2018).

10. Ehrlich, P. (1913). Address in pathology. On chemiotherapy. Delivered before the 17th International Congress of Medicine. *Br. Med. J.,* **16**, 353–359.

11. Smith, J. A., Costales, A. B., Jaffari, M., Urbauer, D. L., Frumovitz, M., Kutac, C. K., Tran, H., Coleman, R. L. (2016). Is it equivalent? Evaluation

of the clinical activity of single agent Lipodox® compared to single agent Doxil® in ovarian cancer treatment. *J. Oncol. Pharm. Practice*, **22**(4), 599–604.

12. Bawa, R. (2016). What's in a name? Defining "nano" in the context of drug delivery. In: Bawa, R., Audette, G., Rubinstein, I., eds. *Handbook of Clinical Nanomedicine: Nanoparticles, Imaging, Therapy, and Clinical Applications,* Pan Stanford Publishing, Singapore, chapter 6, pp. 127–169.

13. Gruchalla, R. S. (2003). Drug allergy. *J. Allergy Clin. Immunol.*, **111**(2), suppl. 2, S548–S559.

14. Wysowski, D. K., Swartz, L. (2005). Adverse drug event surveillance and drug withdrawals in the United States, 1969–2002: The importance of reporting suspected reactions. *Archiv. Intern. Med.*, **165**(12), 1363–1369.

15. Bouley, J. (2016). Exposing that murderous ole FDA. *Drug Deliv. News*, **12**(7), 10.

16. Pfizer scraps cholesterol fighter, trims profit forecast. Reuters. November 1, 2016. Available at: https://www.reuters.com/article/us-pfizer-results/pfizer-cholesterol-drug-fizzles-hitting-shares-idUSKBN12W3S8 (accessed on February 18, 2018).

17. Krishna, M., Nadler, S. G. (2016). Immunogenicity to biotherapeutics: The role of anti-drug immune complexes. *Front. Immunol.*, **7**, 21. Available at: https://www.ncbi.nlm.nih.gov/pubmed/26870037 (accessed on April 29, 2018).

18. Hermeling, S., Crommelin, D. J. A., Schellekens, H., Jiskoot, W. (2004). Structure–immunogenicity relationships of therapeutic proteins. *Pharm. Res.* **21**(6), 897–903.

19. Rosenberg, A. S. (2006). Effects of protein aggregates: An immunologic perspective. *AAPS J.*, **8**(3), E501–E507.

20. Ratanji, K. D., Derrick, J. P., Dearman, R. J., Kimber, I. (2014). Immunogenicity of therapeutic proteins: Influence of aggregation. *J. Immunotoxicol.*, **11**(2), 99–109.

21. Barberoa, F., Russoa, L., Vitalic, M., Piellaa, J., Salvoc, I., Borrajoc, M. L., Busquets-Fité, M., Grandori, R., Bastús, N. G., Casals, E., Puntes, V. (2017). Formation of the protein corona: The interface between nanoparticles and the immune system. *Semin. Immunol.*, **34**, 52–60.

22. Ke, P. C., Lin, S., Parak, W. J., Davis, T. P., Caruso, F. (2017). A decade of the protein corona. *ACS Nano*, **11**(12), 11773–11776.

23. Wilhelm, S., Tavares, A. J., Dai, Q., Ohta, S., Audet, J., Dvorak, H. F., Chan, W. C. W. (2016). Analysis of nanoparticle delivery to tumours. *Nat. Rev. Mater.*, **1**(5), 16014.

24. Fornaguera, C. (2018). Characterization of the interaction between nanomedicines and biological components: *In vitro* Evaluation. In: Bawa, R., Szebeni, J., Webster, T. J., Audette, G. F., eds. *Immune Aspects of Biopharmaceuticals and Nanomedicines*, Pan Stanford Publishing, Singapore, chapter 27, pp. 833–862.

25. Szebeni, J., Fontana, J., Wassef, N., Mongan, P., Morse, D., Stahl, G., et al. (1998). Liposome-induced and complement-mediated cardiopulmonary distress in pigs as a model of pseudo-allergic reactions to liposomal drugs. *Mol. Immunol.*, **35**(6), 401 (abstract).

26. Szebeni, J., Fontana, J. L., Wassef, N. M., Mongan, P. D., Morse, D. S., Dobbins, D. E., et al. (1999). Hemodynamic changes induced by liposomes and liposome-encapsulated hemoglobin in pigs: A model for pseudoallergic cardiopulmonary reactions to liposomes. Role of complement and inhibition by soluble CR1 and anti-C5a antibody. *Circulation*, **99**(17), 2302–2309.

27. Szebeni, J., Bawa, R.. (2018). Immunological issues with medicines of nano size: The price of dimension paradox. In: Bawa, R., Szebeni, J., Webster, T. J., Audette, G. F., eds. *Immune Aspects of Biopharmaceuticals and Nanomedicines*, Pan Stanford Publishing, Singapore, chapter 2, pp. 83–122.

28a. Food and Drug Administration (2002). Immunotoxicology evaluation of investigational new drugs, page 14. Available at: https://www.fda.gov/downloads/drugs/guidancecomplianceregulatoryinformation/guidances/ucm079239.pdf (accessed on July 9, 2018).

28b. European Medicines Agency (2013). Reflection paper on the data requirements for intravenous liposomal products developed with reference to an innovator liposomal product. Available at: http://www.ema.europa.eu/docs/en_GB/document_library/Scientific_guideline/2013/03/WC500140351.pdf (accessed on April 25, 2018).

29. World Health Organization (2013). Guidelines on the quality, safety, and efficacy of biotherapeutic protein products prepared by recombinant DNA technology. Available at: http://www.who.int/biologicals/biotherapeutics/rDNA_DB_final_19_Nov_2013.pdf (accessed on April 27, 2018).

30. Dams, E. T., Laverman, P., Oyen, W. J., et al. (2000). Accelerated blood clearance and altered biodistribution of repeated injections of sterically stabilized liposomes. *J. Pharmacol. Exp. Ther.*, **292**(3), 1071–1079.

31. Abu Lila, A. S., Ishida, T. (2018). The accelerated blood clearance (ABC) phenomenon of PEGylated nanocarriers. In: Bawa, R., Szebeni, J., Webster, T. J., Audette, G. F., eds. *Immune Aspects of Biopharmaceuticals and Nanomedicines*, Pan Stanford Publishing, Singapore, chapter 8, pp. 289–310.
32. Kaur, S. J., Sampey, D., Schultheis, L. W., Freedman, L. P., Bentley, W. E. (2016). Biosimilar therapeutic monoclonal antibodies. *BioProcess Int.* **14**(9), 12–21.
33. Conner, J. B., Bawa, R., Nicholas, J. M., Weinstein, V. (2016). Copaxone® in the era of biosimilars and nanosimilars. In: Bawa, R., Audette, G., Rubinstein, I., eds. *Handbook of Clinical Nanomedicine: Nanoparticles, Imaging, Therapy, and Clinical Applications*, Pan Stanford Publishing, Singapore, chapter 28, pp. 783–826.
34. Crommelin, D. J. A., de Vlieger, J. S. B. (2015). *Non-Biologic Complex Drugs: The Science and the Regulatory Landscape*, Springer, Switzerland.
35. Astier, A., Pai, A. B., Bissig, M., Crommelin, D. J. A., Flühmann, B., Hecq, J.-D., Knoeff, J., Lipp, H.-P., Morell-Baladrón, A., Mühlebach, S. (2017). How to select a nanosimilar. *Ann. N. Y. Acad. Sci.*, **1407**(1), 50–62.
36. Naci, H., Ioannidis, J. P. A. (2015). How good is "evidence" from clinical studies of drug effects and why might such evidence fail in the prediction of the clinical utility of drugs? *Annu. Rev. Pharmacol. Toxicol.*, **55**, 169–189.
37. Light, D. W., Lexchin, J., Darrow, J. J. (2013). Institutional corruption of pharmaceuticals and the myth of safe and effective drugs. *J. Law Med. Ethics*, **14**(3), 590–610.
38. Fleming, N. (2018). Computer-calculated compounds. *Nature*, **557**, 555–557.
39. Griffen, E. J., Dossetter, A. G., Leach, A. G., Montague, S. (2018). Can we accelerate medicinal chemistry by augmenting the chemist with Big Data and artificial intelligence? *Drug Discov. Today* **23**(7), 1373–1384.
40. Giladi, A., Amit, I. (2018). Single-cell genomics: A stepping stone for future immunology discoveries. *Cell*, **172**(1–2), 14–21.
41. Santos, R., Ursu, O., Gaulton, A., Bento, A. P., Donadi, R. S., Bologa, C. G., Karlsson, A., Al-Lazikani, B., Hersey, A., Oprea, T. I., Overington, J. P. (2017). A comprehensive map of molecular drug targets. *Nat. Rev. Drug Discov.* **16**, 19–34.
42. Rodgers, G., Austin, C., Anderson, F., Pawlyk, A, Colvis, C., Margolis, R., Baker, J. (2018). Glimmers in illuminating the druggable genome. *Nat. Rev. Drug Discov.*, **17**, 301–302.

43. Renaud, J.-P., Chari, A., Ciferri, C., Liu, W., Rémigy, H.-W., Stark, H., Wiesmann, C. (2018). Cryo-EM in drug discovery: Achievements, limitations and prospects. *Nat. Rev. Drug Discov.*, **17**, 471–492.
44. Crommelin, D. J., de Vlieger, J. S., Weinstein, V., Mühlebach, S., Shah, V. P., Schellekens, H. (2014). Different pharmaceutical products need similar terminology. *AAPS J.*, **16**, 11–14.
45. Schellekens, H., Stegemann, S., Weinstein, V., de Vlieger, J. S., Flühmann, B., Mühlebach, S., Gaspar, R., Shah, V. P., Crommelin, D. J. (2014). How to regulate nonbiological complex drugs (NBCD) and their follow-on versions: Points to consider. *AAPS J.*, **16**(1), 15–21.
46. Mühlebach, S., Borchard, G., Yildiz, S. (2015). Regulatory challenges and approaches to characterize nanomedicines and their follow-on similars. *Nanomedicine (Lond.)*, **10**(4), 659–674.

Chapter 2

Immunological Issues with Medicines of Nano Size: The Price of Dimension Paradox[1]

János Szebeni, MD, PhD, DSc,[a,b,c] and Raj Bawa, MS, PhD[d,e,f]

[a]*Nanomedicine Research and Education Center, Department of Pathophysiology, Semmelweis University School of Medicine, Budapest, Hungary*
[b]*Department of Nanobiotechnology and Regenerative Medicine, Faculty of Health, Miskolc University, Miskolc, Hungary*
[c]*SeroScience Ltd., Budapest, Hungary*
[d]*Patent Law Department, Bawa Biotech LLC, Ashburn, Virginia, USA*
[e]*Department of Biological Sciences, Rensselaer Polytechnic Institute, Troy, New York, USA*
[f]*The Pharmaceutical Research Institute, Albany College of Pharmacy and Health Sciences, Albany, New York, USA*

Keywords: nanomedicine, nanopharmaceutical, nanodrug, nanoparticles (NPs), biologics, small-molecule drug (SMD), immunogenicity, hypersensitivity reaction (HSR), complement (C) activation-related pseudoallergy (CARPA), adverse drug reactions (ADRs), antidrug antibodies (ADAs), immune complexes (ICs), protein corona, accelerated blood clearance (ABC) phenomenon, aggregation of biologic nanodrugs, non-biologic complex drug (NBCD), Copaxone®, polyethylene glycol (PEG), anti-PEG antibodies, liposomes, micelles, anaphylaxis, anaphylactoid reactions,

---

[1]Copyright 2018 János Szebeni and Raj Bawa. All rights reserved. The copyright holders permit unrestricted use, distribution, online posting and reproduction of this article, figures, or unaltered excerpts therefrom, in any medium, provided the authors and original source are clearly identified and properly credited. A copy must be provided to the copyright holders for archival at bawa@bawabiotech.com.

Immune Aspects of Biopharmaceuticals and Nanomedicines
Edited by Raj Bawa, János Szebeni, Thomas J. Webster, and Gerald F. Audette
Copyright © 2018 János Szebeni and Raj Bawa
ISBN 978-981-4774-52-9 (Hardcover), 978-0-203-73153-6 (eBook)
www.panstanford.com

opsonization, anaphylactic shock, thromboxane, macrophages, pulmonary intravascular macrophages (PIM) cells, histamine, opsonins, C3a, C3b, C5a, cardiac anaphylaxis, polystyrene nanoparticles

2.1 Adverse Immune Effects of Nanodrugs

While there are thousands of nanotechnology-related products in the marketplace, nanodrugs ("nanomedicines," "nanoparticulate drug formulations," or "nanopharmaceuticals") are gradually entering the drug landscape (Fig. 2.1). In fact, they have sputtered along a slower trajectory with greater challenges to their translation and commercialization [1–2]. This is partly due to scarcity of venture funds, toxicity concerns, gaps in regulatory guidance, disagreements over nomenclature (Box 2.1), and issuance of patents of dubious scope [1–8]. The comparison of nanodrugs is often made to biologics, a distinct category of therapeutics. However, in comparison to nanodrugs, biologics are entering an era of rapid growth and are expected to overtake small-molecule drugs in global sales in the near future. Despite bottlenecks and challenges, follow-on versions of both biologics and nanodrugs, namely biosimilars and nanosimilars, respectively, are also trickling into the marketplace.

Most small-molecule drug-induced allergic reactions may be easily classified into one of four classic Gell and Coombs hypersensitivity categories.[2] However, many others with an immunologic component, including biologics and nanodrugs, cannot be classified in such a manner because of a lack of mechanistic information [9]. Adverse clinical events (sometimes referred to as adverse drug reactions, or ADRs) can occur not only due to primary factors such as off-target toxicity or exaggerated pharmacologic effects, but also due to secondary drug effects, such as immune reactions to the drug product [10, 11]. While approximately 80% of human ADRs are directly related to an effect of the drug or a metabolite, around 6–10% are immune-mediated and unpredictable [9]. One study showed that 10–20% of the medicinal products removed from clinical practice between 1969 and 2005 were withdrawn due to immunotoxic effects [12].

---

[2]Gell and Coombs developed their widely accepted classification of hypersensitivity reactions in the context of deleterious connotation. See: Gell, P. G. H., Coombs, R. R. A. (1963) The classification of allergic reactions underlying disease. In: Coombs, R. R. A., Gell, P. G. H., eds., *Clinical Aspects of Immunology*, Blackwell Science.

Some claim that the actual number of serious adverse events such as hospitalizations and death from US Food and Drug Administration (FDA)-approved drugs, vaccines, and medical devices is grossly under-reported by the FDA [13].[3]

Studies have shown that nanodrugs can interact with various components of the immune system, interactions that are fast, complex, and poorly understood. These interactions with the immune system play a leading role in the intensity and extent of side effects occurring simultaneously with their therapeutic efficacy. Compared with conventional small-molecule drug products, nanodrugs have biological and synthetic entities of a shape, size, surface charge, porosity, reactivity, chemical composition, and topology that are often recognized by the human immune system, sometimes in an adverse manner. This can obviously negatively affect their effectiveness and safety, thereby limiting their therapeutic application.

Most small molecules of molecular weight (MW) <1 KDa usually are not able to elicit an immune response in their native state, only becoming immunogenic when they act as a hapten, bind covalently to high-MW proteins, or undergo antigen processing and presentation. On the other hand, "newer" large molecule drugs like biologics, nanodrugs, non-biologic complex drugs (NBCDs), biosimilars, and nanosimilars can be inherently immunogenic. For example, protein-based biologics and nanodrugs can be digested and processed for presentation by antigen-presenting cells (APCs); this can sometimes cause ADRs.

The very untested nature of nanodrug products that make them so revolutionary in some respects also makes them problematic and potentially dangerous. The major benefits touted for nanodrugs—a reduction in unwanted side effects, increased specificity, fewer off-target effects, generation of fewer harmful metabolites, slower clearance from the body, longer duration of effect, reduced intrinsic toxicity, etc.—does not guarantee the absence of adverse immune effects. In fact, some of these perceived advantages may contribute to immune reactivity and/or immunogenicity (defined in Section 2.1.1). For example, the fact that nanodrugs are engineered to break down physiological barriers to entry and escape immune surveillance for longer persistence may also trigger adverse immune responses.

---

[3]Also, see the FDA Death Meter: http://www.anh-usa.org/microsite-subpage/fda-death-meter/.

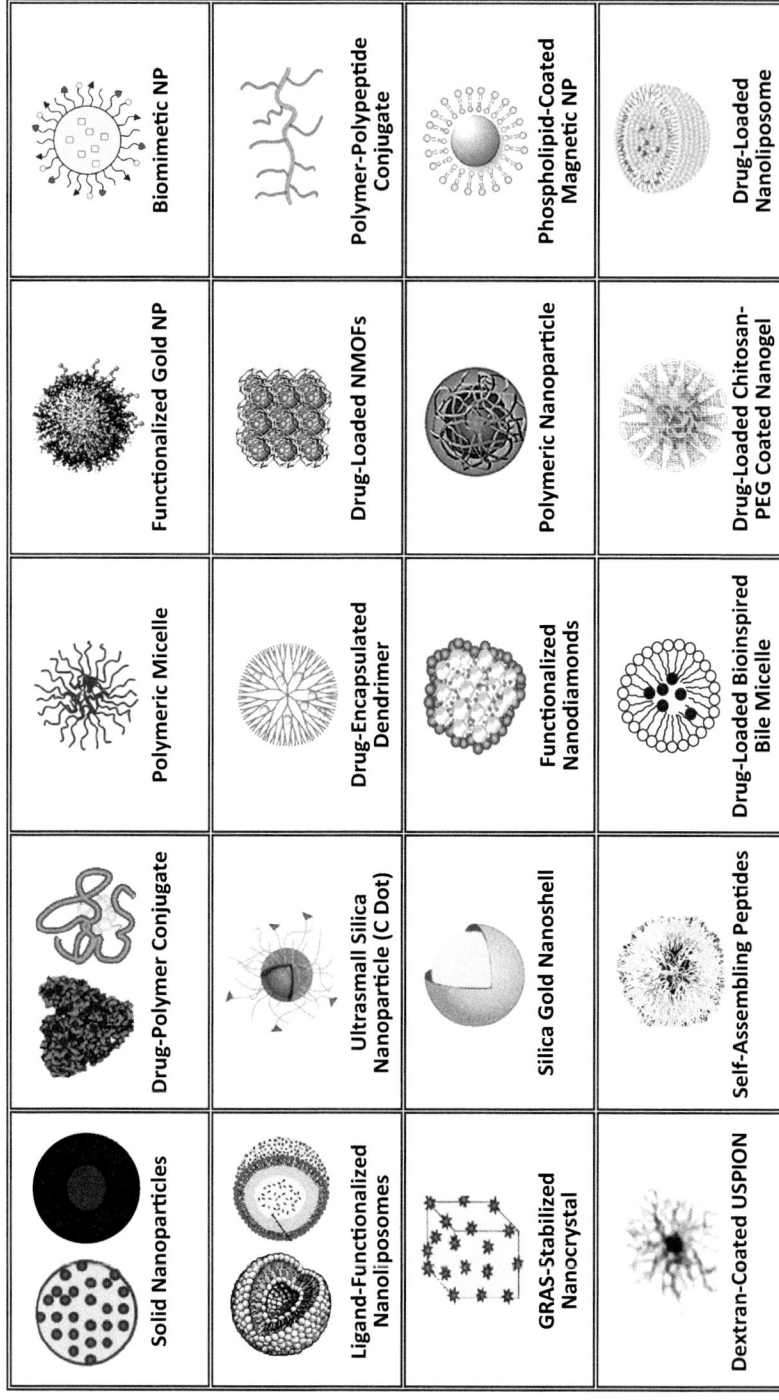

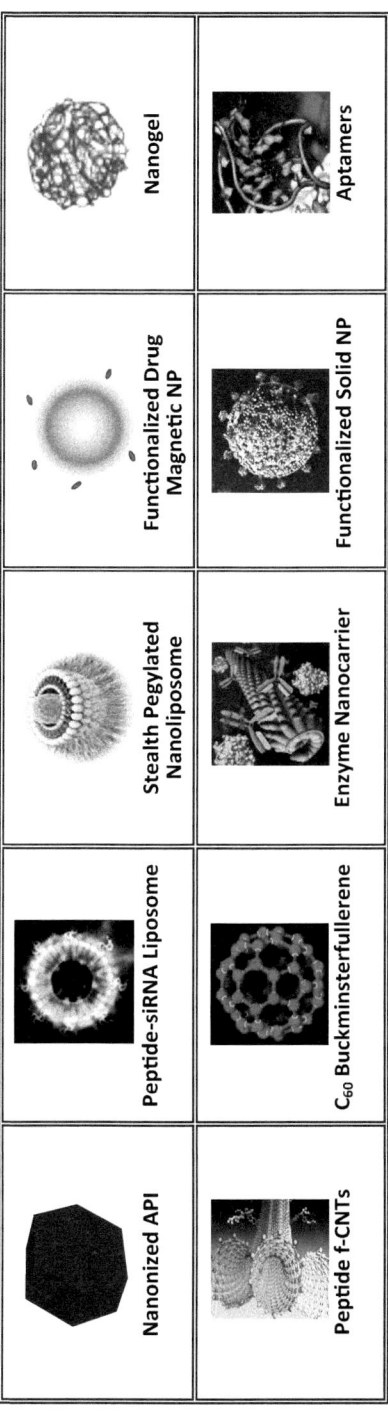

Figure 2.1 Nanoscale Drug delivery system platforms (nanodrug products). Schematic representation of selected engineered nanoparticles (NPs) used in drug delivery that are either approved by regulatory bodies, are in preclinical development or are in clinical trials. In most cases shown above, they are considered as first or second generation multifunctional engineered NPs, ranging in average diameter from one nanometer (1 nm) to a micron (1000 nm). Active bio-targeting of the NP is often achieved via conjugating ligands or functional groups (antibodies, peptides, aptamers, folate, hyaluronic acid). These molecules are tagged to the NP surface with or without spacers/linkers such as PEG. Many nanodrugs depicted above (e.g., metal-based NPs, f-CNTs, NMOFs, etc.), although extensively advertised for drug delivery, will pose enormous drug approval and commercialization challenges and will not appear in the clinic this century. Non-engineered antibodies, biological motors (e.g., sperms), engineered nanomotors, and naturally occurring NPs (natural protein nanotubes) are specifically excluded here. Antibody–drug conjugates (ADCs) are encompassed by the cartoons labeled "Polymer-Polypeptide Conjugate" or "Drug-Polymer Conjugate." Therapeutic monoclonal antibodies (TMAbs), polymer–polypeptide conjugates, and aptamers shown above are classic biologics but they are also nanodrugs as they fall within the widely accepted standard definition of nanodrugs. The list of NPs depicted here is not meant to be exhaustive, the illustrations do not reflect precise three-dimensional shape or configuration, and the NPs are not drawn to scale. *Abbreviations*: NPs, nanoparticles; PEG, polyethylene glycol; GRAS, Generally Recognized As Safe; C dot, Cornell dot; API, active pharmaceutical ingredient; ADCs, antibody–drug conjugates; NMOFs, nanoscale metal organic frameworks; f-CNTs, functionalized carbon nanotubes; siRNA, small interfering ribonucleic acid; USPION, ultrasmall superparamagnetic iron oxide nanoparticle. Copyright 2018 Raj Bawa. All rights reserved.

2.1.1 What Is Immunogenicity?

We define immunogenicity as the ability of an antigen or epitope to provoke an immune response, i.e., to induce a humoral and/or cell-mediated immune response. Put differently, it is the propensity of a therapeutic to generate immune responses to itself or to related drug products. These responses can either (i) induce immunologically related nonclinical effect(s) or (ii) result in ADRs. Immunogenicity can be one of two types: (a) wanted or (b) unwanted. Wanted immunogenicity is typically related to vaccines where the injection of a vaccine (the antigen) stimulates an immune response against a pathogen. On the other hand, unwanted immune responses are adverse events (i.e., ADRs).[4]

At the molecular level, immunogenicity manifests itself as the generation of antidrug antibodies (ADAs) directed against the drug product, though the detailed mechanisms leading to ADA formation are poorly understood and characterized. ADAs may (i) neutralize the activity of the therapeutic, (ii) reduce half-life by enhancing clearance, (iii) result in allergic reactions, (iv) alter the therapeutics's pharmacokinetic profile, (v) abrogate its pharmacological effect, and/or (vi) cross-react with endogenous cell components to result in "autoimmune-like" reactions. Furthermore, ADAs can potentially affect the interpretation of toxicology studies. Factors such as contamination with endotoxin or with synthetic chemicals, incompletely characterized nanomaterials, nanoparticle aging effects, assay interference by nanoparticles, or batch-to-batch variability between nominally identical particle preparations may all contribute to immunogenicity. Immunogenicity is less often observed with conventional small-molecule drugs but more frequently with biologics and sometimes nanodrugs, particularly those of biologic origin. Immunogenicity associated with protein drugs was first seen more than a century ago in diphtheria patients treated with antitoxin horse serum.

---

[4]An example of unwanted immunogenicity is the generation of neutralizing antibodies against recombinant erythropoietin (EPO) in patients receiving EPO for chronic kidney disease (CKD) and resulting pure red cell aplasia (PRCA)-related anemia due to neutralization of endogenous EPO.

2.1.2 Formation of Protein Corona

A well-studied but poorly understood immune issue is the formation of the so-called "protein corona" at the interface between nanodrugs and blood. Protein corona refers to the adsorption of proteins onto nanodrugs, thereby reducing their stability and facilitating their fast *in vivo* clearance. Obviously, this phenomenon has important implications on immune safety, biocompatibility, and the use of nanodrugs in medicine [14–16]. This may be one factor that has contributed to the inefficient accumulation of nanodrugs (<10% accumulation [17]) in diseased tissues despite the oft-highlighted advantages of "targeted" nanodrug delivery. However, this area of research has suffered from a lack of mechanistic understanding of the bio–nano interface.

2.1.3 The Accelerated Blood Clearance (ABC) Phenomenon

Another immunologic phenomenon observed with PEGylated nanoliposomes is referred to as the "accelerated blood clearance (ABC) phenomenon" [18, 19] (Fig. 2.2). Liposomes are the most widely used nanodrugs and PEGylation is a common strategy involved in designing stealth liposomes to shield them from reticuloendothelial system (RES) uptake. However, repeated-dose injection of PEGylated liposomes affects their clearance rate and bioavailability: The delivery of the first dose of PEGylated liposomes ("priming the system") accelerates later dose elimination as compared to the initial dose, mainly mediated through specific anti-PEG IgM. This finding is clinically significant if PEGylated nanoliposome therapy is involved because it decreases the therapeutic efficacy upon repeated administration. Therefore, repeated-dosage pharmacokinetic (PK) studies are critical to prevent immunogenicity of these drug products without hampering their efficacy or safety. Table 2.1 lists some PEGylated nanodrugs that have adverse immune effects.

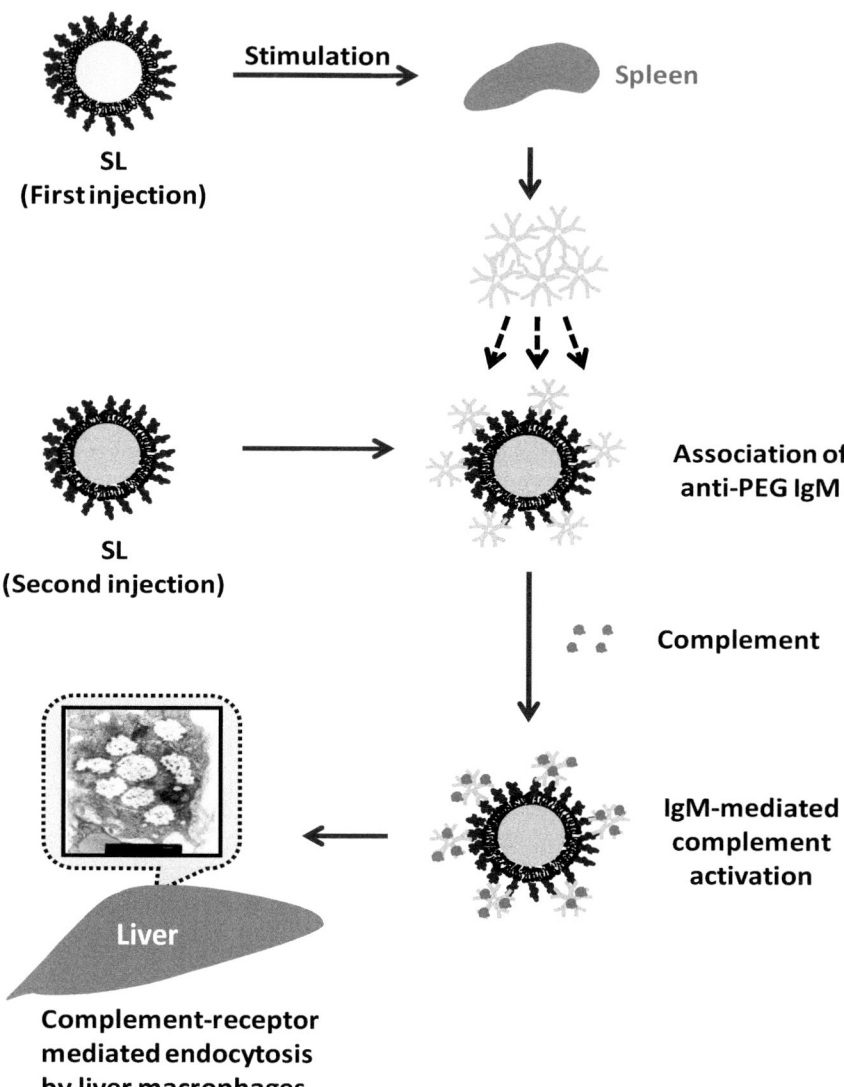

Figure 2.2 Mechanism of the ABC phenomenon. Courtesy of Dr. Tatsuhiro Ishida (Tokushima University, Japan).

Table 2.1 PEGylated nanodrugs with documented adverse immune effects

Generic Name	Trade name	API or vehicle	Company
PEGylated liposomal doxorubicin	Doxil®/Caelyx®	Liposome	ALZA/Janssen
Pegaspargase	Oncaspar®	Enzyme: asparaginase	Enzon
Pegfilgrastim	Neulasta®	Protein (GCSF)	Amgen
Pegaptanib	Macugen®	Aptamer (anti-VEGF)	Eye Tech/Pfizer
Mono-mPEG-epoetin-β	Mircera®	Protein (EPO)	Hoffmann-LaRoche
Certolizumab pegol	Cimzia®	Fab of anti-TNF mAb	UCB, Inc., Smyrna
Withdrawn from the market			
Pegvisomant	Somavert®	Peptid (somatotropin antagonist)	Pfizer
Pegloticase	Krystexxa®	Enzyme: Urate oxidase	Horizon Pharma
Peginesatide	Omontys®/Hematide®	Peptide (EPO-mimetic)	Affymax/Takeda
Pegnivacogin+Anivamersen	Revolixys kit	F-IXa blocker RNA aptamer + reverse agent	Regado/Tobira

2.1.4 Aggregation of Biologic Nanodrugs

An issue seen with some biologic nanodrugs is that they show a concentration-dependent propensity for self-association, which often leads to the formation of aggregates ranging in size from a few nanometers (oligomers) to microns (subvisible and visible particles). Aggregation can occur throughout the life cycle of a nanodrug product: during manufacturing, shipping, storage, reconstitution, and delivery in the clinic. Potential immunogenicity of such nanoaggregates has been implicated in triggering adverse immune responses in patients that may affect drug safety and efficacy and cause infusion reactions, cytokine release syndrome, anaphylaxis, or even death. Hence, just like ADAs and anti-drug immune complexes (ICs), nanodrug aggregates are of concern to manufacturers, clinicians, patients, and regulators. Yet, aggregation of nanodrugs remains a challenging phenomenon to mitigate due to knowledge gaps in molecular mechanisms underlying aggregation, and a lack of standard and reliable aggregation prediction tools. However, in recent years, regulators and drug industry experts have spearheaded development of novel techniques and assays to detect and characterize aggregates, increase research into the role of aggregates in immunogenicity, aid in revising pharmacopoeia monographs to improve subvisible-particle testing and clarify terminology like "practically" or "essentially free of particles."

2.1.5 Complement-Mediated Hypersensitivity

The complement (C) system is an ancient and conserved network of ~40 blood and cell surface proteins [20, 21]. Its limited and highly coordinated cascadic proteolysis plays a critical role in host defense even in invertebrates that are incapable of adaptive immune response [22]. In the 1880s, factors involved in the host-defense mechanism to pathogens were placed into two categories based on sensitivity to heat: (a) a heat-stable component (i.e., antibody) that was "specific" for the invading pathogens and also arose following immunization; and (b) a heat-labile (>56°C) fraction ("the complement system") that acted in a "nonspecific" manner to "complement" the antibody-mediated pathogen lysis.

In addition to its lytic role, the complement system serves several other functions: (i) clearance of targeted pathogens via opsonization[5]; and (ii) inflammatory processes due to the release of small peptide fragments from complement proteins. These peptides cause mast cell degranulation, smooth muscle contraction, and chemotaxis of motile cells to inflammatory sites. Eleven complement proteins (C1q, r, s, and C2 to C9) circulate in blood in an inactive form, and their engagement in a proteolytic chain, known as complement activation, was thought until recently to proceed via three pathways: classical, alternative, and lectin. However, now a fourth pathway is recognized and called "extrinsic" pathway (Fig. 2.3) [20]. Although all four activation pathways depend upon different molecules for initiation, they converge to generate identical effector molecules, most importantly C3a and C5a anaphylatoxins and C5b-9, the membrane attack complex (MAC) [23a].

Often, intravenously administered nanodrugs (as well as many other biologics and NBCDs) prime the immune system, leading to adverse reactions and/or the loss of efficacy of the drug product. It is now well proven that these agents may provoke "hypersensitivity reactions" (HSRs), also known as "infusion" or "anaphylactoid" reactions [23b]. Based on the association of complement activation with many of these adverse reactions, the term "complement activation-related pseudoallergy" (CARPA) was coined. CARPA is discussed in detail in Section 2.3. This chapter will relate solely to nanodrugs and not biologics or small-molecule drugs *per se*.

> Various immune issues discussed in Section 2.1 pose serious challenges with respect to developing and delivering nanodrug products. In effect, they serve as bottlenecks to effective translation of nanodrugs [1–8]. The bottom line is this: *An inherent risk of introducing nanodrugs into the human body is the potential to provoke an unwanted adverse immune response.*

---

[5]Opsonization (from the Greek opsōneîn, to prepare for eating) is the process by which a pathogen is coated (marked for destruction) that enhances phagocytosis via tissue macrophages and activated follicular dendritic cells (FDCs), as well as binding by receptors on peripheral blood cells.

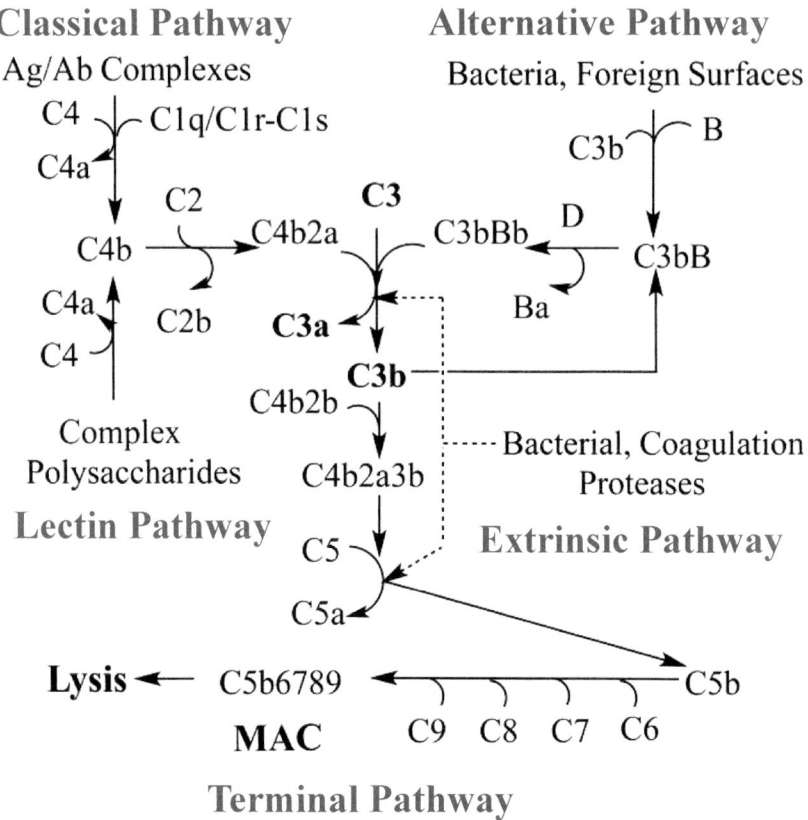

Figure 2.3 The four pathways of C activation. Note that the distinction of the fourth pathway in addition to the three "canonical" pathways is relatively recent. It involves direct cleavage of C3 and C5 by proteases in the coagulation cascade, such as thrombin, factor IXa, factor Xa, factor XIa, and plasmin. FXa can also promote membrane attack complex (MAC) formation and cell lysis. Reprinted with permission from [20]. Copyright 2018 American Chemical Society.

2.2 Issues of Terminology

Many biologics are on the nanoscale and hence can also be considered nanodrugs. For example, monoclonal antibodies or drug protein conjugates are biologics, but they also fall within the definition of nanodrugs (Fig. 2.1). Conversely, many nanodrugs are biologics according to standard definitions (Box 2.1).

Box 2.1 Standard Nomenclature

"A nanodrug is: (1) a formulation, often colloidal, containing therapeutic particles (nanoparticles) ranging in size from 1–1,000 nm; and (2) either (a) the carrier(s) is/are the therapeutic (i.e., a conventional therapeutic agent is absent), or (b) the therapeutic is directly coupled (functionalized, solubilized, entrapped, coated, etc.) to a carrier."

Source: Bawa, R. (2016). What's in a name? Defining "nano" in the context of drug delivery. In: Bawa, R., Audette, G., Rubinstein, I., eds. *Handbook of Clinical Nanomedicine: Nanoparticles, Imaging, Therapy, and Clinical Applications*, Pan Stanford Publishing, Singapore, chapter 6, pp. 127–169.

"A biopharmaceutical is a protein or nucleic acid-based pharmaceutical substance used for therapeutic or in vivo diagnostic purposes, which is produced by means other than direct extraction from a native (non-engineered) biological source."

Source: Walsh, G. (2002). Biopharmaceuticals and biotechnology medicines: An issue of nomenclature. *Eur. J. Pharm. Sci.* **15**, 135–138.

A small-molecule drug (SMD) is a chemically synthesized pharmaceutical compound of precise structure and low molecular weight (<700 Daltons) used for therapy or in vivo diagnosis, that lacks immunogenicity in the patient but may produce off-target effects.

Source: Raj Bawa, unpublished Work, 2018.

A non-biologic complex drug (NBCD) is "[a] medicinal product, not being a biological medicine, where the active substance is not a homomolecular structure, but consists of different (closely) related and often nanoparticulate structures that cannot be isolated and fully quantitated, characterized, and/or described by physicochemical analytical means. It is also unknown which structural elements might affect the therapeutic performance. The composition, quality, and in vivo performance of NBCDs are highly dependent on the manufacturing processes of both the active ingredient and the formulation. Examples of NBCDs include liposomes, iron–carbohydrate (iron–sugar) drugs, and glatiramoids."

Source: Astier, A., Pai, A. B., Bissig, M., Crommelin, D. J. A., Flühmann, B., Hecq, J.-D., Knoeff, J., Lipp, H.-P., Morell-Baladrón, A., Mühlebach, S. (2017). How to select a nanosimilar. *Ann. N.Y. Acad. Sci.*, **1407**(1), 50–62.

Nanotechnology is "[t]he design, characterization, production, and application of structures, devices, and systems by controlled manipulation of size and shape at the nanometer scale (atomic, molecular, and macromolecular scale) that produces structures, devices, and systems with at least one novel/superior characteristic or property."

Source: Bawa, R. (2007). Patents and nanomedicine. *Nanomedicine* (London), **2**(3), 351–374.

Another drug class has gained prominence in the literature in recent years: NBCDs, which have properties of biologics but are also considered to be nanodrugs (Box 2.1). An example of an NBCD is the blockbuster drug, Copaxone® [24]. Furthermore, just as the demarcations between pharmaceutical and biotechnology (and between branded and generic) companies are becoming increasingly blurred, it is hard to always place a specific drug product into a regulatory-based therapeutic category (i.e., small-molecule drug versus biologic). Both the FDA and the European Medicines Agency (EMA) review nanodrugs on a case-by-case basis on preexisting drug laws and not as a separate drug category with any specific nano-characteristic.

2.3 Adverse Immune Effects of Nanodrugs: The Dimension Paradox

As discussed in considerable detail in Section 2.1, nanodrugs, like all drugs in human therapy, can have adverse effects on the immune system (Fig. 2.4). However, these adverse effects are caused not because nanodrugs are very small (which explains their unique physicochemical characteristics), but because they are too large compared to the traditional small-molecule drugs. This represents a "dimension paradox" in the sense that the uniqueness of nanodrugs lies in their smallness, yet their adverse effects are due to their relatively large size. In fact, most nanodrugs happen to be in the size range or dimension of pathogenic viruses (i.e., 40–300 nm) (Fig. 2.5), against which the immune system has effective clearing mechanisms developed over millions of years of biological evolution.

2.3.1 Complement Activation-Related Pseudoallergy (CARPA)

Focusing on acute immune stimulation, intravenously (I.V.) administered nanodrugs (as well as many other biologics and non-biologic complex drugs, NBCDs) can "prime" the innate immune system, leading to adverse reactions and/or the loss of efficacy of the drug product. The adverse events, also known as "hypersensitivity" or "infusion" reactions (HSRs, IRs) are often

Adverse Immune Effects of Nanodrugs | 97

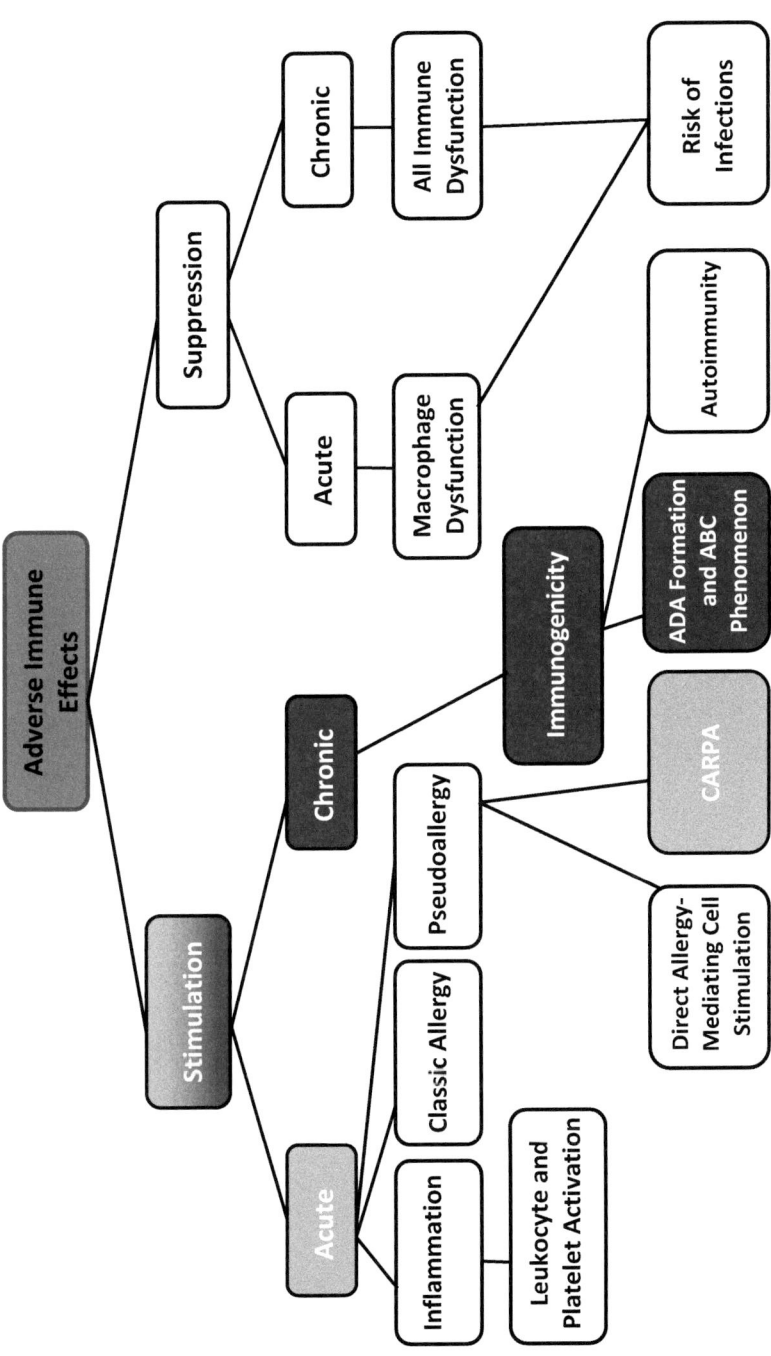

Figure 2.4 Adverse immune effects of nanodrugs, classified according to their impact and time course. *Abbreviations*: ABC, accelerated blood clearance; ADA, antidrug antibodies; CARPA, complement activation-related pseudoallergy.

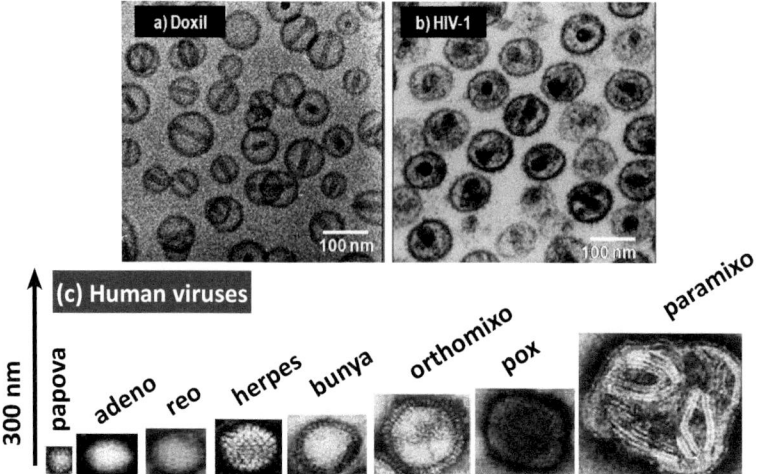

Figure 2.5 Size range of pathogenic virus strains in the 40–300 nm range.

non-IgE-mediated, and are therefore called "pseudoallergies." Pseudoallergic HSRs are often associated with complement activation, rationalizing the term "complement activation-related pseudoallergy" (CARPA). The following section focuses only on the latter phenomenon, i.e., abnormal induction of nonspecific immune response via complement activation, resulting in specific physiological changes known as CARPA syndrome. Table 2.2 lists various drug categories and drugs within them that are known to cause CARPA.

The term CARPA was introduced in the late 1990s [25, 26] in the context of pig experiments pursued to understand the adverse physiological effects of I.V. liposomes. A niche in immune toxicology model systems, the phenomenon in pigs, and to lesser extent in other species, represents a unique over-stimulation of the humoral and cellular arms of innate immunity by certain nanoparticles directly exposed to blood. It involves an unprecedented cross talk between the immune and other major organ systems, most importantly the cardiovascular system. Strong reactions along the latter, immune-cardiovascular axis can lead to lethal anaphylaxis, making CARPA a safety issue, an increasingly recognized cornerstone of nanodrug pharmacotherapy. To illustrate these unique features of CARPA, Fig. 2.6 puts the phenomenon on a map of clinically relevant nanotoxicities categorized according to particle source, route of exposure, organ involvement, and immune abnormality [27].

Table 2.2 Drugs linked to CARPA

Liposomal drugs	Micelle-solubilized drugs	Antibodies	PEGylated proteins	Contrast media	Enzymes/proteins/ peptides	Miscellaneous
Abelcet	Cyclosporine	Avastin	Adagen	Diatrizoate	Abbokinase	ACE inhibitors
AmBisome	Elitec	Campath	Neulasta	Iodipamide	ACH	AR blockers
Amphotec/ Amphocyl	Etoposide	Erbitux	Oncaspar, Pegaspargas	Iodixanol	Actimmune	Aspirin
DaunoXome	Fasturec	Herceptin		Iohexol	Activase	Cancidas
Doxil, Caelyx	Taxol	Infliximab		Iopamidol	Aldurazyme	Copaxone
Myocet	Taxotere	Muronomab		Iopromide	Avonex	Corticosteroids
Visudyne	Vumon	Mylotarg		Iothalamate	Fasturtec	Cyclofloxacin
		Remicade		Ioversol	Neulasta	Eloxatin
		Rituxan		Ioxaglate	Neupogen	Intralipid
		Vectibix		Ioxilan	Plenaxis	Opiates
		Xolair		Magnevist	protamine	Orencia
				Metrizamide	Urokinase	Salicylates
				SonoVue	Zevalin	Vancomycin

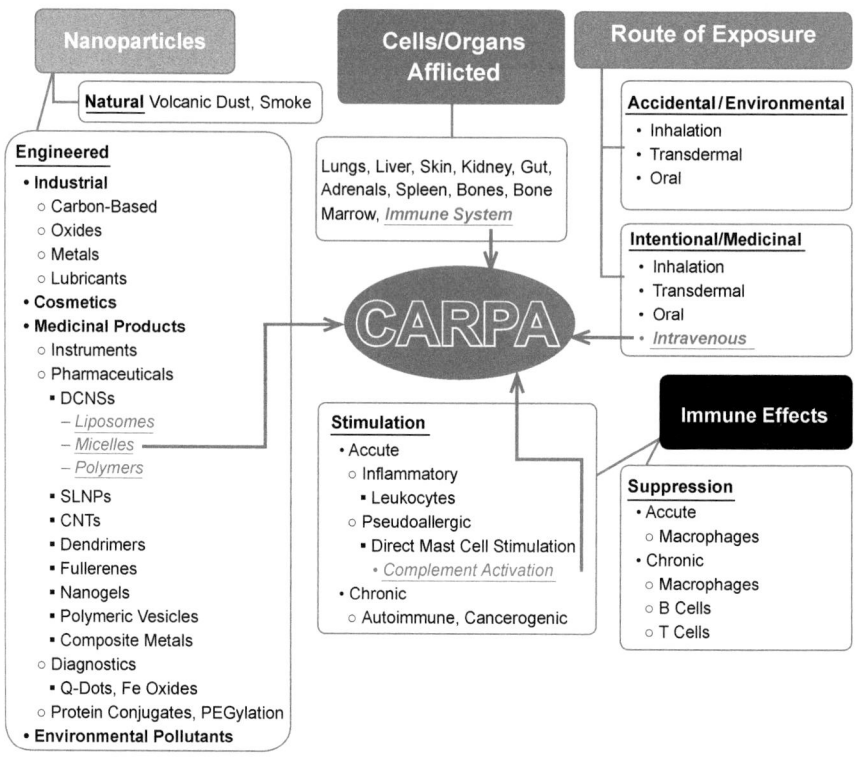

Figure 2.6 Niche position of CARPA on a map of nanotoxicity determinants. The scheme shows different variables that determine the toxic effects of nanoparticles. Variables in the same categories are placed in boxes of different colors, with the best recognized CARPAgenic entries shown in red. *Abbreviations:* DCNSs, drug carrier nanosystems; SLNPs, solid lipid nanoparticles; CNTs, carbon nanotubes; Q-dots, quantum dots. Modified from the author's original in [27].

The clinical relevance of CARPA is highlighted by notes in industry guidances issued by the FDA and the EMA. The FDA recommends detection of complement activation by-products in animals showing signs of anaphylaxis [28], while EMA refers to CARPA tests as a potentially useful preclinical safety tests for liposomal drug R&D [29]. The World Health Organization (WHO) also emphasizes evaluating complement binding and activation for biologics [30]: *"Unless otherwise justified, the ability for complement-binding and activation, and/or other effector functions,*

Table 2.3 FDA-approved complement inhibitors and new drugs under development

Company/reference	Drug/class	Target	Structure/derivation	Route	Biochemical data	Stage
Alexion	Eculizumab	C5	Humanized monoclonal antibody	Intravenous	K_d = 120 pM	FDA approved for PNH and aHUS
ViroPharma, CSC Behring	C1 esterase inhibitor: plasma derived	C1 esterase	Human protein	Intravenous		FDA approved for hereditary angioedema
Pharming-Salix	C1 esterase inhibitor: recombinant	C1 esterase	Protein analogue, produced in rabbits	Intravenous		FDA approved for hereditary angioedema
Alexion (Taligen)	TT30	C3 convertase	Factor H-CR2 fusion	Intravenous/Subcutaneous	IC_{50} = 0.5 μm	Human studies
Norvartis	LFG 316	C5	mAb	Intravitreal/Intravitreal		Human studies
Amyndas, Apellis, Potentia	Compstatin analogues	C3/C3b	Cyclic peptide/Phage display	Intravitreal, Subcutaneous, Inhaled	IC_{50} = 62 nM, K_d for C3b = 2.3 nM	Human studies

(Continued)

Table 2.3 (Continued)

Company/reference	Drug/class	Target	Structure/derivation	Route	Biochemical data	Stage
Volution Akari	Coversin	C5	Peptide/tick saliva	Subcutaneous	Maximal inhibition at 10 µg/mL	Human studies
Achillion	Small molecule	Complement factor D	X-ray crystallography	Oral	$K_d < 1$ nm, $IC_{50} = 17$ nm (protease inhibition)	Preclinical studies
Amyndas	Mini factor H	C3 convertase	Derived from factor H		$IC_{50} = 0.22$ µM (for C3 deposition)	Preclinical studies
Alnylam	ALN-CC5	C5 RNA	RNAi conjugate	Subcutaneous		Preclinical studies
Lindofer et al.	3E7/H17	C3b	mAb		100% blockage of lysis at 1 µM	Preclinical studies
Ra Pharmaceuticals	Several	C5	Cyclic peptide	Subcutaneous	$K_d = 2.6$ nM, $IC_{50} = 8.1$ nm (% RBC lysis)	Preclinical studies

Source: Reprinted with permission from [20]. Copyright (2018) Elsevier.

Abbreviations: aHUS, atypical hemolytic uremic syndrome; IC_{50}, half maximal inhibitory concentration; K_d, dissociation constant; PNH, paroxysmal nocturnal hemoglobinuria; RBC, red blood cell.

should be evaluated even if the intended biological activity does not require such functions." CARPA caused by nanodrugs represents a double harm in that it hurts not only the patient in whom it develops but also the drug, in the sense that the hyper-reactive nanoparticles undergo rapid clearance from blood and therefore lose their therapeutic benefit. The FDA has approved a few drugs for inhibiting various complement proteins, while many others are in preclinical and clinical stages of drug development (Table 2.3).

2.3.2 The Porcine CARPA Model: Features and Human Relevance

The first detailed [26] as well as later studies addressing various aspects of CARPA describe pigs as being extremely sensitive to adverse hemodynamic and other toxic effects of liposomes and various I.V.-administered nanodrugs (Fig. 2.7) [31–38]. The symptoms in pigs resemble the clinical picture of drug-induced HSRs in patients; they also proved to be consistent and quantitatively reproducible. These facts led to the proposal that pigs may be used as a model of nanodrug-induced HSR in humans, enabling preclinical assessment of safety [38].

The latter claim, i.e., that normal healthy pigs provide a disease model, namely, allergy to nanodrugs, is important to recognize, as critics of the model [39, 40] point to its overt sensitivity leading to false overestimation of HSR risk in humans. In fact, while reactogenic liposomal and other drugs cause HSR in essentially all pigs, the frequency of such reactions in humans is in the 2–10% range. While recognizing the danger of overkill, the discrepancy in reaction frequency can also be considered as a uniquely fortunate condition, since a few animals can provide important information about a rare, hardly reproducible acute illness. There is no need to use a large number of pigs to obtain statistically analyzable experimental data. Note that the conclusions obtained in pigs have to be applied only to hypersensitive patients and not to normal individuals [38].

When comparing the HSRs in humans and pigs, the symptoms in both species arise within minutes after intravenous infusion of reactogenic nanoparticles (i.e., reactive NPs) and the reactions subside within 15–60 min, depending on severity. The reactions typically arise at first exposure to the nanodrug without prior exposure, and the symptoms usually lessen and/or disappear on later treatment [26, 31–38]. Table 2.4 shows the symptoms of CARPA in humans involving different organ systems. Figure 2.8 illustrates the typical symptoms of severe HSRs in pigs involving major hemodynamic (panel A), blood cell (B), plasma thromboxane A2 (C), and skin changes (D) [36].

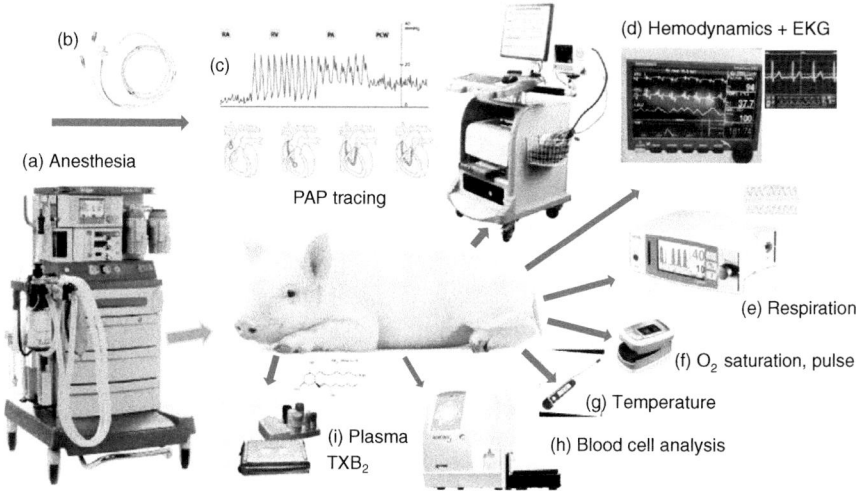

Figure 2.7 Parameters measured and equipment used in the porcine CARPA model: (a) Anesthesia machine; (b) Swan–Ganz balloon catheter, used for the measurement of pulmonary arterial pressure; (c) blood pressure wave forms during passage of the tip of the Swan–Ganz catheter via the right atrium, right ventricle, and pulmonary artery until being wedged into the pulmonary capillary bed; (d) computerized hemodynamic monitoring system tracing the systemic and pulmonary pressures, heart rate, and the EKG; (e) capnograph measuring the respiratory rate (RR) and end-tidal carbon dioxide ($EtCO_2$); (f) pulse oximeter measuring oxygen saturation and pulse rate; (g) rectal temperature probe; (h) blood cell analyzer; and (i) enzyme linked immunosorbent assay for measuring plasma mediators, such as TxB2. Reprinted with permission from [36].

Table 2.4 Symptoms of pseudoallergy

Cardiovascular	Broncho-pulmonary	Hematological	Mucocutaneous	Gastro-intestinal	Neuro-psycho-somatic	Systemic
Angioedema	Apnea	Granulopenia	Cyanosis	Bloating	Back pain	Chills
Arrhythmia	Bronchospasm	Leukopenia	Erythema	Cramping	Chest pain	Diaphoresis
Cardiogenic shock	Coughing	Lymphopenia	Flushing	Diarrhea	Chest tightness	Feeling of warmth
Edema	Dyspnea	Rebound leukocytosis	Nasal congestion	Metallic taste	Confusion	Fever
Hypertension	Hoarseness	Rebound granulocytosis	Rash	Nausea	Dizziness	Loss of consciousness
Hypotension	Hyperventilation	Thrombocytopenia	Rhinitis	Vomiting	Feeling of imminent death	Rigors
Hypoxia	Laryngospasm		Swelling		Fright	Sweating
Myocardial infarction	Respiratory distress		Tearing		Headache	Wheezing
Tachycardia	Shortness of breath		Urticaria		Panic	
Ventricular fibrillation	Sneezing					
Syncope	Stridor					

Note: The most dangerous symptoms are shown in bold.

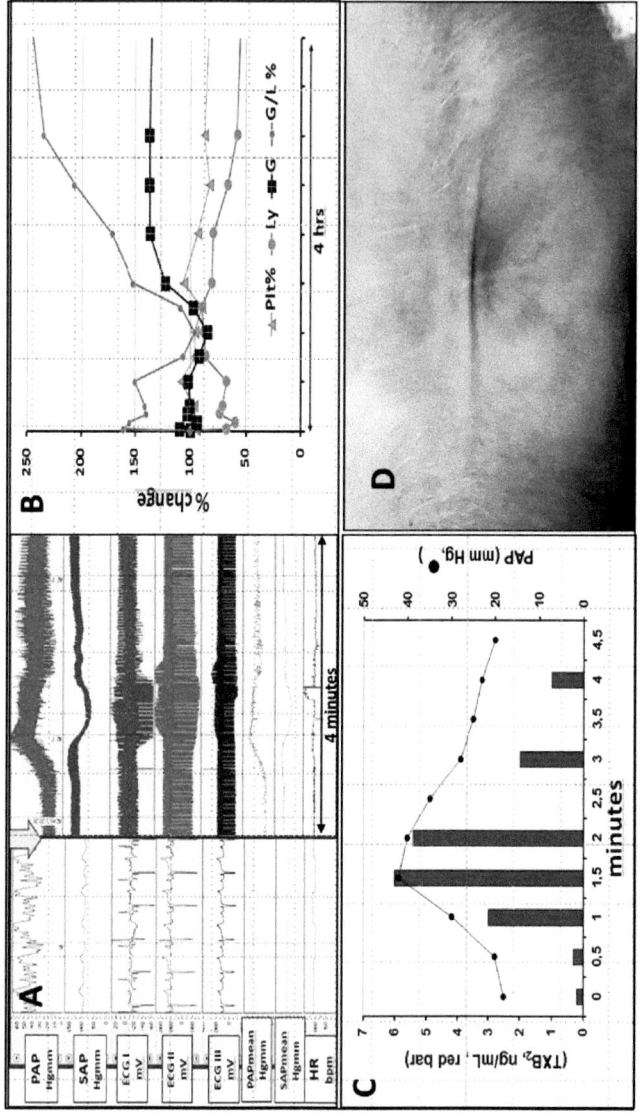

Figure 2.8 The main symptoms of CARPA in pigs: the "porcine CARPA tetrad." (A) Hemodynamic changes, involving rise of PAP (red), drop of SAP (blue), ECG changes (green, purple, and dark blue). Real-time readings. PAP and SAP mean (bottom second and third curves) show averaged PAP and SAP readings, while HR (bottom) is the heart rate, beats/min). Typical readings during an AmBisome® reaction. (B) Typical hematological changes during infusion of PEgylated liposomes. (C) Rise of TXB2 during a liposome reaction, showing remarkable correlation with the rise of PAP. (D) A skin reaction, observed 2 min after bolus injection of liposomes. Reproduced with permission from [36].

In addition to reproducing the severe, life-threatening cardiovascular symptoms and skin alterations, what makes the pig model relevant to human HSRs is that the dose eliciting the reaction is identical or very similar to the doses that trigger life-threatening reactions in humans. This statement is based on calculations that the bolus dose of PEGylated liposomal doxorubicin (Doxil®) triggering pulmonary hypertension in pigs is identical to the amount of Doxil® reaching the circulation of reactive patients within the first 10–20 seconds of infusion, the earliest time when symptoms can start [41]. Although similar calculations have not been made for other reactive nanoparticles (NPs), the reactogenic dose of phospholipid-containing reactive NPs is in the 0.01–0.1 mg/kg concentration range on a phospholipid weight basis, a value that may guide similar calculations for other liposomal drugs. In rats and mice hemodynamic changes are triggered only by doses that are orders of magnitude larger [42].

Another unique property of the pig model is tachyphylaxis, i.e., self-induced tolerance arising to certain reactogenic NPs after repetitive administration. Typically, the reactions caused by PEGylated liposomes are tachyphylactic [35], while those caused by multilamellar dimyristoyl phosphatidylglycerol (DMPG)-containing liposomes [26] or AmBisome® [34] are not. This implies that when testing tachyphylactic drugs, only the first administration will reflect the response of a hypersensitive individual, and the rest of the injections will underestimate the drug's reactivity. In contrast, if the agent is non-tachyphylactic, the model enables quantitation of multiple injections over hours, enabling dose–effect relationship and inhibition studies in individual animals.

2.3.3 Hemodynamic Alterations in CARPA

Focusing on the initial hemodynamic changes and their likely mechanism, the transient but significant rise of PAP is the most reproducible measure of adverse immuno-circulatory response to reactogenic NPs, which we quantify as the primary endpoint of an HSR. Interestingly, while the PAP almost always rises, the extent and direction of changes of SAP are highly variable. The HR usually increases, or it does not change, while the most intense reactions may entail paradoxical bradycardia [32].

Pulmonary hypertension is most likely due to the release of TxA2, a known pulmonary vasoconstrictor eicosanoid. This assumption is supported by the remarkable parallelism between the rises of PAP and TxB2, the stable metabolite of TxA2 measured by enzyme linked immunosorbent assay (ELISA), and the observation that indomethacin, a cyclooxygenase blocker of TxA2 release, inhibits both processes [26].

As for the source of TxA2, the primary suspects are pulmonary intravascular macrophages (PIM cells), such as resident macrophages adhered to the endothelium of pulmonary capillaries. The abundant presence of PIM is observed only in a few species, such as sheep, cattle, horse, and cat [43]. Their function is to screen the blood from particulate pathogens [43]. PIM cells express anaphylatoxin C5a receptors (ATR, C5aR) on their surface as well as Fc, toll-like and C receptors (CR1, CR3, and CR4) and can secrete vasoactive mediators, including TxA2, histamine, leukotrienes, PAF, interleukin-6, IL-8, and IL-1β. The combination of different vasoactivity of all secretion products explains the versatility of systemic blood pressure changes in CARPA). However, TxA2 and many of these mediators may also be released by other ATR+ cells, including mast cells, leukocytes, and activated platelets [44]. The relative contribution of these cells to TxA2 production in pig CARPA has not yet been clarified.

The key role of PIM cells in the cardiopulmonary distress of pigs following reactogenic NP administration was supported by a recent study showing close parallelism of the time courses of pulmonary hypertension caused by spherical polystyrene NPs (PS-NPs), their clearance from blood in mice, and their uptake by cultured macrophages [37]. Although rapid phagocytosis of PS-NPs by PIM cells was considered to be the main mechanism of the pigs' pulmonary response independent of C activation [39] a follow-up review [45] and a recent study [46] argued against the premature exclusion of the role of the complement. It was pointed out that the *in vitro* ELISA results conducted in whole blood [37] could not provide adequate evidence for the absence of C activation *in vivo*, and the question needs to be further studied [45]. In doing so, we found that other methodical approaches, namely FACS analysis of NP-bound C fragments

(C5b-9 and iC3b) and Western blot detection of C3 degradation to C3dg, did indicate C activation in pig serum by PS-NPs *in vitro* [46]. Since the detected iC3b and C3dg are potent opsonins, and opsonization is a well-known trigger for enhanced phagocytic uptake, it is very likely that C activation played a role in the rapid clearance and "robust" uptake of PS-NPs by macrophages via its opsonic ability. Whether or not the increased opsonization of NPs by C3b byproducts is "complemented" by concurrent C3a/C5a production and stimulation of cell via anaphylatoxin receptors is not yet clear, but in any case, these data emphasize the complexity of CARPA, and the existence of two or more activation mechanisms ("double hit") [35, 38, 45–47] rather than rapid phagocytosis representing an alternative mechanism of HSRs, competing with CARPA [39]. On top of advancing this academic debate, the latter study [46] highlighted that 500 nm PS-NPs are the most potent inducers of HSR in pigs studied to date, possibly due to their high negative surface charge and hydrophobicity. Also, despite the difficulties in projecting *in vitro* C assay data to *in vivo* physiological changes, the study presented significant correlation between C activation by different sized PS-NPs in human serum and pulmonary hypertensive effect of these NPs *in vivo* [46], providing strong support for the CARPA background of PS-NP reactions in pigs.

2.3.4 The Complex Mechanism of CARPA

CARPA is not limited to the immune system but involves many different cells from different organ systems in a highly coordinated manner. The multilevel signaling among these organ systems from the molecular recognition of nanomedicines to clinical symptoms is illustrated in Fig. 2.9.

Regarding the role of allergy mediator secretory cells, the double-hit theory mentioned above proposes that NPs can induce HSRs by simultaneous stimulation of anaphylatoxin receptors (mainly C5aR, CD88) and other surface receptors on PIM cells, which can be linked to release reactions directly or indirectly. Potential receptors include pathogen recognition receptors, also known as pathogen-associated molecular pattern receptors, mannose-binding lectin receptor, C receptors (CR2 and CR3 and Fc receptors) and many others typically present on the surface of mast cells [38].

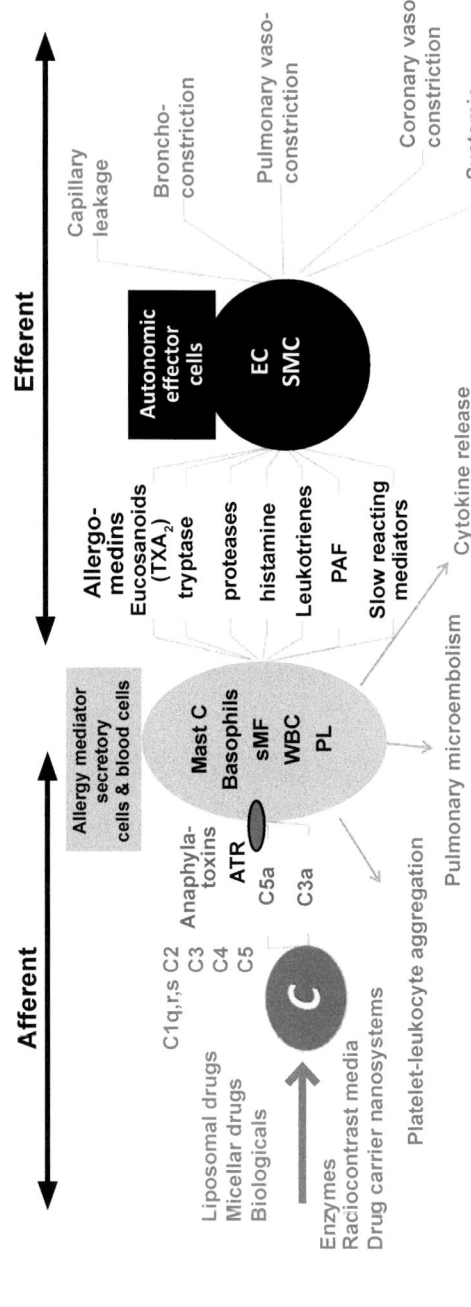

Figure 2.9 Multilevel signaling among cells of different organ systems in CARPA: Double-hit theory. The hypothetical scheme illustrates the steps and interactions among cells and mediators involved in CARPA. *Abbreviations*: AR, anaphylatoxin receptors; Mast C, mast cells; sMF, secretory macrophages; WBC, white blood cells; PL, platelets; CR2 and CR3, C receptors; FcR, Fc receptor; PRR, pathogen recognition receptors; EC, endothelial cells; SMC, smooth muscle cells. The different types of systems, cells, mediators, and effects are color coded. Modified from the author's original figure in [47].

These different activation pathways are likely to represent different degrees of stimulation, and exhaustion of one or more pathways upon repetitive exposures might explain tachyphylaxis. It also follows from the above "multiple hit" concept that if one or the other activation pathway is sufficiently strong, or dominates, they alone might trigger the release reaction. For example, in case of very strong C activation, the C5aR (or C3aR)-mediated anaphylatoxin "pathway" might dominate, while direct stimulants of mast cells, such as opioids, neuromuscular blocking agents, quinolones, compound 48/80, or physical stimuli of cold and trauma might induce pseudoallergy directly, without C activation. Such reactions can be referred to as C-independent pseudoallergies [38, 45].

2.4 Vicious Cycle between Specific and Nonspecific Immune Responses to Nanodrugs

It should be pointed out that the specific and non-specific immune responses to nanoparticles can act in a vicious cycle (Fig. 2.10),

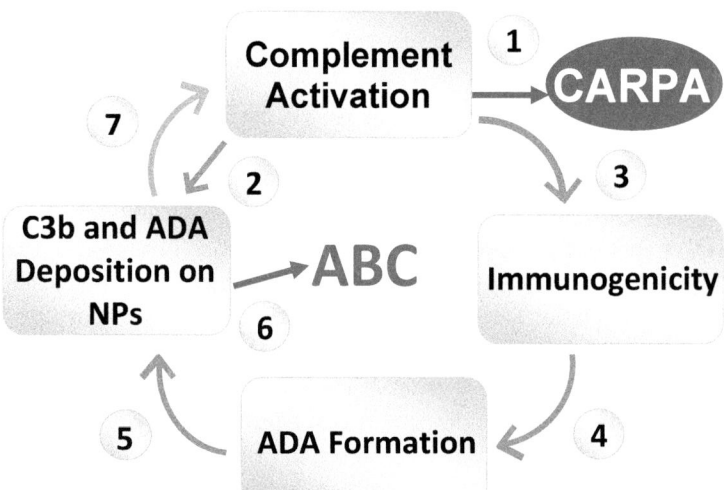

Figure 2.10 Vicious cycle between specific and nonspecific immune responses to nanodrugs. The numbers indicate different steps in the cycle. *Abbreviations:* ABC, accelerated blood clearance; CARPA, complement activation-related pseudoallergy; ADA, antidrug antibody.

inasmuch as C activation enhances immunogenicity at the level of antigen presentation and B cell maturation, while antibody formation and binding to nanoparticles is one of the most efficient mechanisms of their C activation [27].

2.5 CARPA as Blood Stress

An unconventional perception of CARPA is that it is a homeostatic defense mechanism fighting "blood stress" [47]. The proposal is based on the fact that the types of symptoms (cardiovascular/pulmonary, skin, hematological, and blood chemistry) are essentially the same in all animals, with, of course, species and drug-dependent qualitative and quantitative differences. The Doxil® and AmBisome® reactions in pigs are very similar to the reactions obtained after I.V. bolus injections of a large variety of nanoparticulate drugs or agents, including other types of liposomes, micelles, polymers, dendrimers, solid lipid and metal nanoparticles, carbon nanotubes, microbubbles, etc. [36]. Although the composition, inner and outer constituents, the presence of surface ligands or conjugates, and the method of administration have a major influence on the reaction, the basic phenomenon (transient physiological derangement) starting within minutes remains the same in most cases. This inter- and intra-species uniformity of alarm reaction to particulate agents exposed to blood is remarkably similar to the classic stress reaction to physical and psychologic harms described in detail in [48]. The theory was based on the finding that injection of a variety of organ extracts in mice produced the same symptoms (swelling of the adrenal cortex, atrophy of the thymus, gastric and duodenal ulcers), which János Selye called the "general adaptation syndrome," the common response to "noxious agents" or stressors [48].

Table 2.5 presents a comprehensive comparison of CARPA with classical stress, leading to the suggestion that CARPA may be considered a "blood stress," and that the stress theory may be expanded with another axis of uniform self-defense. An important difference is that while classic stress proceeds on the hypothalamic-pituitary-adrenal axis (HPA axis), CARPA proceeds on the C-blood and immune secretory cell-bronchopulmonary/

cardiovascular autonomic regulatory cell axis. In other words, it represents a crosstalk between the immune and cardiovascular systems, also affecting many other organs and vegetative functions. CARPA may be perceived as a misguided immune defense with the difficult-to-understand biological sense that acute hemodynamic changes ("cardiovascular distress") and other allergic changes

Table 2.5 Comparison of CARPA with classic stress

	Similarities	Differences
Stress	• Homeostatic function	• Function: fighting physical and psychical harm
	• Rapid response	• Noxious agents: physical harm, emotional stress
	• Nonspecific response to variable stimuli	• Time course: from hyperacute to chronic
	• Leads to adaptation	• Mediators: neurotransmitters, hormones (ACTH, cortisol)
	• Mediators are secretory products	• Adverse impact: adrenal swelling, thymus atrophy, ulcers
		• Organs involved: neuroendocrine, cardiovascular, muscle
CARPA	• Can lead to death	• Axis: hypothalamo-pituitary-adrenal (HPA axis)
	• Function is to fight harmful impacts	• Function: fighting infections, contributes to blood clearance
	• Error-prone, causing adverse effects	• Noxious agents: viruses, intravenous drugs and agents
	• Stages: Alarm → resistance → exhaustion	• Time course: from hyperacute to subacute
		• Mediators: anaphylatoxins, allergomedins, cytokines
		• Adverse impact: HSRs, anaphylaxis, CARPA tetrad
		• Organs involved: cardiovascular system, skin, blood
		• Axis: C system-mast cell/cardiovascular-macrophage system (CMC axis)

would help in antimicrobial defense. Thus, CARPA may represent an evolutionarily ancient, basic physiological reaction whose biological sense needs to be understood. The pig CARPA model will hopefully give a clue in this regard.

2.6 CARPA Testing

Considering the increased attention to CARPA by drug regulatory authorities discussed earlier, the development of standard, validated *in vitro* and *in vivo* tests for the quantitation and prediction of CARPA is an important goal in experimental toxicology. The C assays mandated by regulatory agencies for the approval of medical devices (e.g., endovascular grafts, shunts, rings, patches, heart valves, balloon pumps, stents, pacemakers, hemopheresis filters) are not applicable for CARPA without adaptation for (nano)particle dispersions. In an attempt to fill this gap, a decision tree was suggested (Fig. 2.11) [47] that may guide the development of CARPA free drug candidates. According to this scheme, the test agent (drug candidate) is first incubated with a few NHS to explore possible major C activation. If the result is positive, the agent is likely to carry an elevated risk for CARPA *in vivo*. As for the threshold for considering C activation as "major," an activation factor (for example a rise of sC5b-9 above baseline over 20–30 min incubation at 37°C) of 5–10-fold may be a realistic predictor of a clinical reaction, as such increases (of sC5b-9) were shown to correlate with clinical symptoms of patients treated with Doxil® [49]. However, the correlation between C activation by a drug *in vitro* and the clinical symptoms it causes *in vivo* remains to be established in the future with higher precision. If the *in vitro* C assay in NHS is not showing C activation, based on the substantial individual variation of C response, testing in a much larger number of NHS (in the range of 10–100) can be recommended.

The reactogenicity of test agents in the porcine CARPA model can be quantitated, among others, by using the cardiac abnormality score (CAS) [32]. This score combines all symptoms of CARPA into one parameter, i.e., CAS, whereupon the scores represent categories of associated symptoms of increasing severity on a 0–5 scale [32]. In case of low reactogenicity (CAS score 1–2), the test agent may carry a small but not negligible risk for CARPA in a small percentage of hypersensitive individuals. In case of strong reactivity (CAS = 3–5), the risk of CARPA is great(er). If none of the highly sensitive animal models indicate reactivity to any therapeutically relevant bolus doses, or the test drug does not cause C activation in many NHS, it may be considered as possessing low or minimal risk for CARPA, although obviously the experimental conditions need to be relevant and the tests technically valid. However, even if bolus administration leads to HSR in the animal model, or C is activated in NHS, desensitization, premedications, and inhibition of C activation can be considered and used to decrease the risk of clinical reactions [21]. Alternatively, slow administration protocols can be developed that secure safe human use of (mild C activator) drug candidates. The ultimate decision needs the consideration of other contraindications of the therapeutic application of CARPAgenic drugs, as certain conditions (e.g., atopic constitution, heart disease, autoimmune processes) represent risk factors for aggravated reactions.

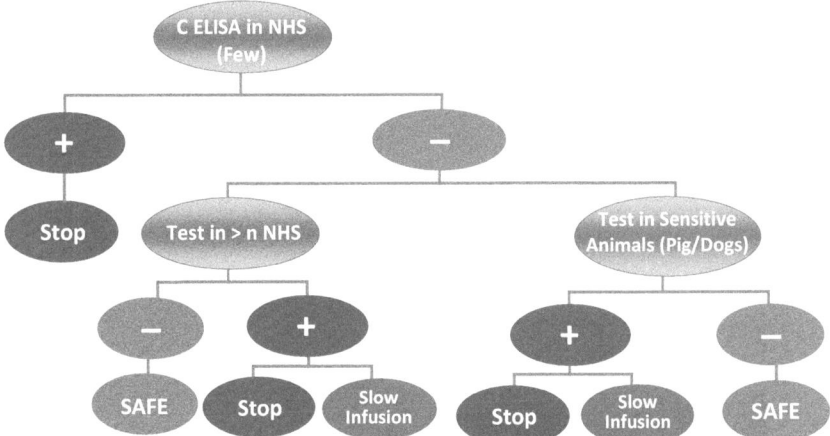

Figure 2.11 Decision tree for CARPA testing. NHS, normal human sera; C ELISA, ELISA of C activation byproducts (C3a, C5a, sC5b-9, Bb, C4d). The blue entries are tests. The + and – symbols mean major reaction and no reaction, respectively, where major is defined in the text; >n, large number of human NHS. SAFE means that the tested drug candidate is unlikely to cause CARPA, while "slow infusion" means the possibility to develop a safe administration protocol by slow infusion. STOP means high risk for CARPA in sensitive large animals (pig and/or dogs) with bolus administration.

Disclosures and Conflict of Interest

The authors declare that they have no conflict of interest. No writing assistance was used in the production of this chapter and the authors have received no payment for its preparation. Dr. Bawa is a scientific advisor to Teva Pharmaceutical Industries Ltd. (Israel). Dr. Szebeni holds shares in SeroScience Ltd. (Budapest, Hungary) and acknowledges the financial support of EU projects FP7-NMP-2012-LARGE-6-309820 (NanoAthero); FP-7-HEALTH-2013-Innovation-1 (2013) (602923-2) (TheraGlio) and the Applied Materials and Nanotechnology Center of Excellence at Miskolc University, Hungary.

Corresponding Authors

Dr. János Szebeni
Nanomedicine Research and Education Center
Dept of Pathophysiology, Semmelweis University
1089 Nagyvárad tér 4
Budapest, Hungary
Email: jszebeni2@gmail.com

Dr. Raj Bawa
Bawa Biotech LLC
21005 Starflower Way
Ashburn, Virginia, USA
Email: bawa@bawabiotech.com

About the Authors

Janos Szebeni, MD, PhD, DSc, is director of the Nanomedicine Research and Education Center at Semmelweis University in Budapest, Hungary. He is also the founder and CEO of a contract research SME, SeroScience, Ltd., and a full professor of immunology and biology at the University of Miskolc in Hungary. He has held various academic positions in Hungary and abroad, including at the National Institute of Hematology and Blood Transfusion and at the First Clinics of Internal Medicine, Semmelweis University in Budapest, the University of Arizona in Tucson, AZ, the National Institute of

Health in Bethesda, Maryland, Massachusetts General Hospital, Harvard University in Boston, MA, the Walter Reed Army Institute of Research in Silver Spring, Maryland. After residing in the United States for over two decades, Dr. Szebeni returned to Hungary in 2006. His research on various themes in hematology, membrane biology, nanotechnology, and immunology has produced 160+ peer-reviewed papers, book chapters, patents, etc. (citations: ≈6000, H index: 39), and a book titled *The Complement System: Novel Roles in Health and Disease* (Kluwer, 2004). He has made significant contributions to three fields: artificial blood, liposomes, and the complement system. His original works led to the "CARPA" concept, i.e., that complement activation underlies numerous drug-induced (pseudo) allergic (infusion) reactions.

Raj Bawa, MS, PhD, is president of Bawa Biotech LLC, a biotech/pharma consultancy and patent law firm based in Ashburn, Virginia, that he founded in 2002. He is an inventor, entrepreneur, professor, and registered patent agent licensed to practice before the U.S. Patent & Trademark Office. Trained as a biochemist and microbiologist, he has extensive expertise in pharmaceutical sciences, biotechnology, nanomedicine, drug delivery, microbial biodefense, FDA regulatory issues, and patent law. Since 1999, he has held various positions at Rensselaer Polytechnic Institute in Troy, NY, where he is an adjunct professor and where he received his doctoral degree in three years (biophysics/biochemistry). Currently, he serves as a scientific advisor to Teva Pharma, Israel, is a visiting research scholar at the Pharmaceutical Research Institute of Albany College of Pharmacy in Albany, NY, and is vice president of Guanine, Inc., based in Rensselaer, NY. He has served as a principal investigator of SBIRs and reviewer for both the NIH and NSF. Currently, he is principal investigator of a CDC SBIR Phase 1 grant to develop an assay for carbapenemase-resistant bacteria. In the 1990s, Dr. Bawa held various positions at the US Patent & Trademark Office, including primary examiner from 1996–2002. He is a life member of Sigma Xi, co-chair of the nanotech and personalized medicine committees of the American Bar Association, and founding director of the American Society for Nanomedicine. He has authored over 100 publications, co-edited four texts, and serves on the editorial boards of 14 peer-reviewed journals, including serving as an associate editor of *Nanomedicine* (Elsevier). Some of Dr. Bawa's awards include the Innovations Prize from the Institution of Mechanical Engineers, London, UK (2008); Appreciation Award from the Undersecretary of Commerce, Washington, DC (2001); the Key Award from Rensselaer's Office of Alumni Relations (2005); and Lifetime Achievement Award from the American Society for Nanomedicine (2014).

References

1. Bawa, R., Bawa, S. R., Mehra, R. (2016). The translational challenge in medicine at the nanoscale. In: Bawa, R., ed.; Audette, G. F., Reese, B. E., asst. eds. *Handbook of Clinical Nanomedicine: Law, Business, Regulation, Safety, and Risk*, Pan Stanford Publishing, Singapore, chapter 58, pp. 1291–1346.
2. Bawa, R. (2017). A practical guide to translating nanomedical products. In: Cornier, J., et al., eds. *Pharmaceutical Nanotechnology: Innovation and Production*, 1st ed., Wiley-VCH Verlag, chapter 28, pp. 663–695.
3. Bawa, R. (2016). Racing ahead: Nanomedicine under the microscope. In: Bawa, R., ed.; Audette, G. F., Reese, B. E., asst. eds. *Handbook of Clinical Nanomedicine: Law, Business, Regulation, Safety, and Risk*, Pan Stanford Publishing, Singapore, Preface, pp. xxxvii–li.
4. Bawa, R. (2016). What's in a name? Defining "nano" in the context of drug delivery. In: Bawa, R., Audette, G., Rubinstein, I., eds. *Handbook of Clinical Nanomedicine: Nanoparticles, Imaging, Therapy, and Clinical Applications*, Pan Stanford Publishing, Singapore, chapter 6, pp. 127–169.
5. Bawa, R. (2016). FDA and nano: Baby steps, regulatory uncertainty and the bumpy road ahead. In: Bawa, R., ed.; Audette, G. F., Reese, B. E., assst. eds. *Handbook of Clinical Nanomedicine: Law, Business, Regulation, Safety, and Risk*, Pan Stanford Publishing, Singapore, chapter 58, pp. 339–383.
6. Bawa, R., Barenholz, Y., Owen, A. (2016). The challenge of regulating nanomedicine: Key issues. In: Braddock, M., ed. *Nanomedicines: Design, Delivery and Detection*, Royal Society of Chemistry, UK, RSC Drug Discovery Series No. 51, chapter 12, pp. 290–314.
7. Tinkle, S., McNeil, S. E., Mühlebach, S., Bawa, R., Borchard, G., Barenholz, Y., Tamarkin, L., Desai, N. (2014). Nanomedicines: Addressing the scientific and regulatory gap. *Ann. N. Y. Acad. Sci.*, **1313**, 35–56.
8. Bawa, R. (2007). Nanotechnology patent proliferation and the crisis at the US Patent Office. *Albany Law J. Sci. Technol.* **17**(3), 699–735.

9. Gruchalla, R. S. (2003). Drug allergy. *J. Allergy Clin. Immunol.*, **111**(2), suppl. 2, S548–S559.
10. Bawa, R. (2018). My life with biologicals and nanodrugs: A twenty-year affair. In: Bawa, R., Szebeni, J., Webster, T. J., Audette, G. F., eds. *Immune Aspects of Biopharmaceuticals and Nanomedicines,* Pan Stanford Publishing, Singapore, Preface, pp. xxix–xxx.
11. Bawa, R. (2018). Current immune aspects of biologics and nanodrugs: An overview. In: Bawa, R., Szebeni, J., Webster, T. J., Audette, G. F., eds. *Immune Aspects of Biopharmaceuticals and Nanomedicines,* Pan Stanford Publishing, Singapore, chapter 1, pp. 1–82.
12. Wysowski, D. K., Swartz, L. (2005). Adverse drug event surveillance and drug withdrawals in the United States, 1969–2002: The importance of reporting suspected reactions. *Archiv. Intern. Med.*, **165**(12), 1363–1369.
13. Bouley, J. (2016). Exposing that murderous ole FDA. *Drug Deliv. News*, **12**(7), 10.
14. Ke, P. C., Lin, S., Parak, W. J., Davis, T. P., Caruso, F. (2017). A decade of the protein corona. *ACS Nano*, **11**(12), 11773–11776.
15. Barbero, F., Russo, L., Vitali, M., Piella, J., Salvo, I., Borrajo, M. L., Busquets-Fité, M., Grandori, R., Bastús, N. G., Casals, E., Puntes, V. (2017). Formation of the protein corona: The interface between nanoparticles and the immune system. *Semin. Immunol.*, **34**, 52–60.
16. Cai, K., Wang, A. Z., Yin, L., Cheng, J. (2017). Bio-nano interface: The impact of biological environment on nanomaterials and their delivery properties. *J. Control. Release*, **263**, 211–222.
17. Wilhelm, S., Tavares, A. J., Dai, Q., Ohta, S., Audet, J., Dvorak, H. F., Chan, W. C. W. (2016). Analysis of nanoparticle delivery to tumours. *Nat. Rev. Mater.*, **1**, 16014.
18. Dams, E. T., Laverman, P., Oyen, W. J., Storm, G., Scherphof G. L., van Der Meer, J. W., Corstens, F. H., Boerman, O. C. (2000). Accelerated blood clearance and altered biodistribution of repeated injections of sterically stabilized liposomes. *J. Pharmacol. Exp. Ther.*, **292**(3), 1071–1079.
19. Abu Lila, A. S., Ishida, T. (2018). The accelerated blood clearance (ABC) phenomenon of PEGylated nanocarriers. In: Bawa, R., Szebeni, J., Webster, T. J., Audette, G. F., eds. *Immune Aspects of Biopharmaceuticals and Nanomedicines,* Pan Stanford Publishing, Singapore, chapter 8, pp. 289–310.
20. Iyer, A., Xu, W., Reid, R. C., Fairlie, D. P. (2018). Chemical approaches to modulating complement-mediated diseases. *J. Med. Chem.*, **61**, 3253–3276.

21. Ricklin, D., Barratt-Due, A., Mollnes, T. E. (2017). Complement in clinical medicine: Clinical trials, case reports and therapy monitoring. *Mol. Immunol.*, **89**, 10–21.
22. López-Lera, A., Corvillo, F., Nozal, P., Regueiro, J. R., Sánchez-Corral, P., López-Trascasa, M. (2018). Complement as a diagnostic tool in immunopathology. *Semin. Cell Dev. Biol.*, pii: S1084-9521(17)30146-5.
23a. Hourcade, D. E., Pham, C. T. N., Lanza, G. M. (2018). Complement activation: Challenges to nanomedicine development. In: Bawa, R., Szebeni, J., Webster, T. J., Audette, G. F., eds. *Immune Aspects of Biopharmaceuticals and Nanomedicines*, Pan Stanford Publishing Pte. Ltd., Singapore, chapter 9, pp. 311–340.
23b. Szebeni, J., Dmitri Simberg, D., González-Fernández, A., Yechezkel Barenholz, Y., Dobrovolskaia, M. (2018). Roadmap and strategy for overcoming infusion reactions to nanomedicines. *Nat. Nanotech.*, October 22.
24. Conner, J. B., Bawa, R., Nicholas, J. M., Weinstein, V. (2016). Copaxone® in the era of biosimilars and nanosimilars. In: Bawa, R., Audette, G., Rubinstein, I., eds. *Handbook of Clinical Nanomedicine: Nanoparticles, Imaging, Therapy, and Clinical Applications*, Pan Stanford Publishing, Singapore, chapter 28, pp. 783–826.
25. Szebeni, J., Fontana, J., Wassef, N., Mongan, P., Morse, D., Stahl, G., et al. (1998). Liposome-induced and complement-mediated cardiopulmonary distress in pigs as a model of pseudo-allergic reactions to liposomal drugs. *Mol. Immunol.*, **35**(6), 401 (abstract).
26. Szebeni, J., Fontana, J. L., Wassef, N. M., Mongan, P. D., Morse, D. S., Dobbins, D. E., Stahl, G. L., Bünger, R., Alving, C. R. (1999). Hemodynamic changes induced by liposomes and liposome-encapsulated hemoglobin in pigs: A model for pseudoallergic cardiopulmonary reactions to liposomes. Role of complement and inhibition by soluble CR1 and anti-C5a antibody. *Circulation*, **99**(17), 2302–2309.
27. Szebeni, J. (2017), Challenges and risks of nanotechnology in medicine: An immunologist's point of view. In: Bert Mueller, B., Van de Voorde, M., eds. *Nanotechnology for Human Health*, 1st ed., Wiley-VCH Verlag GmbH & Co. KGaA., pp. 97–123.
28. Food and Drug Administration (2002). Immunotoxicology evaluation of investigational new drugs, page 14. Available at: https://www.fda.gov/downloads/drugs/guidancecomplianceregulatoryinformation/guidances/ucm079239.pdf (accessed on July 9, 2018).
29. European Medicines Agency (2013). Reflection paper on the data requirements for intravenous liposomal products developed with reference to an innovator liposomal product. Available at: http://

www.ema.europa.eu/docs/en_GB/document_library/Scientific_ guideline/2013/03/WC500140351.pdf (accessed on April 25, 2018).

30. World Health Organization (2013). Guidelines on the quality, safety, and efficacy of biotherapeutic protein products prepared by recombinant DNA technology. Available at: http://www.who.int/ biologicals/biotherapeutics/rDNA_DB_final_19_Nov_2013.pdf (accessed on April 27, 2018).

31. Szebeni, J., Baranyi, B., Savay, S., Bodo, M., Morse, D. S., Basta, M., Stahl, G. L., Bunger, R., Alving, C. R. (2000). Liposome-induced pulmonary hypertension: Properties and mechanism of a complement-mediated pseudoallergic reaction. *Am. J. Physiol.*, **279**, H1319–H1328.

32. Szebeni, J., Baranyi, L., Savay, S., Bodo, M., Milosevits, J., Alving, C. R., Bunger, R. (2006). Complement activation-related cardiac anaphylaxis in pigs: Role of C5a anaphylatoxin and adenosine in liposome-induced abnormalities in ECG and heart function. *Am. J. Physiol. Heart Circ. Physiol.*, **290**, H1050–H1058.

33. Szebeni, J., Bedocs, P., Csukas, D., Rosivall, L., Bunger, R., Urbanics, R. (2012). A porcine model of complement-mediated infusion reactions to drug carrier nanosystems and other medicines. *Adv. Drug Deliv. Rev.*, **64**, 1706–1716.

34. Szebeni, J., Bedocs, P., Rozsnyay, Z., Weiszhar, Z., Urbanics, R., Rosivall, L., Cohen, R., Garbuzenko, O., Bathori, G., Toth, M., Bunger, R., Barenholz, Y. (2012). Liposome-induced complement activation and related cardiopulmonary distress in pigs: Factors promoting reactogenicity of Doxil and AmBisome. *Nanomed. Nanotechnol. Biol. Med.*, **8**, 176–184.

35. Szebeni, J., Bedocs, P., Urbanics, R., Bunger, R., Rosivall, L., Toth, M., Barenholz, Y. (2012). Prevention of infusion reactions to PEGylated liposomal doxorubicin via tachyphylaxis induction by placebo vesicles: A porcine model. *J. Control. Release*, **160**, 382–387.

36. Urbanics, R., Bedocs, P., Szebeni, J. (2015). Lessons learned from the porcine CARPA model: Constant and variable responses to different nanomedicines and administration protocols. *Eur. J. Nanomed.*, **7**, 219–231.

37. Wibroe, P. P., Anselmo, A. C., Nilsson, P. H., Sarode, A., Gupta, V., Urbanics, R., Szebeni, J., Hunter, A. C., Mitragotri, S., Mollnes, T. E., Moghimi, S. M. (2017). Bypassing adverse injection reactions to nanoparticles through shape modification and attachment to erythrocytes. *Nat. Nanotechnol.*, **12**, 589–594.

38. Szebeni, J., Bedocs, P., Dézsi, L., Urbanics, R. (2018). A porcine model of complement activation-related pseudoallergy to nano-

biopharmaceuticals: Pros and cons of translation to a preclinical safety test. *J. Precision Nanomed.*, **1**(1), 63–75.

39. Moghimi, S. M. (2018). Nanomedicine safety in preclinical and clinical development: Focus on idiosyncratic injection/infusion reactions. *Drug Discov. Today*, **23**, 1034–1042.

40. Skotland, T. (2017). Injection of nanoparticles into cloven-hoof animals: Asking for trouble. *Theranostics*, **7**, 4877–4878.

41. Szebeni, J., Baranyi, B., Savay, S., Lutz, L. U., Jelezarova, E., Bunger, R., Alving, C. R. (2000). The role of complement activation in hypersensitivity to pegylated liposomal doxorubicin (Doxil®). *J. Liposome Res.*, **10**, 347–361.

42. Dézsi, L., Fülöp, T., Mészáros, T., Szénási, G., Urbanics, R., Vázsonyi, C., Őrfi, E., Rosivall, L., Nemes, R., Jan Kok, R., Metselaare, J. M., Storm, G., Szebeni, J. (2014). Features of complement activation-related pseudoallergy to liposomes with different surface charge and PEGylation: Comparison of the porcine and rat responses. *J. Control. Release*, S0168–3659.

43. Csukas, D., Urbanics, R., Weber, G., Rosivall, L., Szebeni, J. (2015). Pulmonary intravascular macrophages: Prime suspects as cellular mediators of porcine CARPA. *Eur. J. Nanomed.*, **7**, 27–36.

44. Patkó, Z., Szebeni, J. (2015). Blood cell changes in complement activation-related pseudoallergy. *Eur. J. Nanomed.*, **7**, 233–244.

45. Szebeni, J. (2018). Mechanism of nanoparticle-induced hypersensitivity in pigs: Complement or not complement? *Drug Discov. Today*, **23**, 487–492.

46. Mészáros, T., Kozma, G. T., Shimizu, T., Miyahara, K., Turjeman, K., Ishida, T., Barenholz, Y., Urbanics, R., Szebeni, J. (2018). Involvement of complement activation in the pulmonary vasoactivity of polystyrene nanoparticles in pigs: Unique surface properties underlying alternative pathway activation and instant opsonization. *Int. J. Nanomed.*, **13**, 6345–6357.

47. Szebeni, J. (2014). Complement activation-related pseudoallergy: A stress reaction in blood triggered by nanomedicines and biologicals. *Mol. Immunol.*, **61**, 163–173.

48. Selye, H. (1955). Stress and disease. *Science*, **122**, 625–631.

49. Chanan-Khan, A., Szebeni, J., Savay, S., Liebes, L., Rafique, N. M., Alving, C. R., Muggia, F. M. (2003). Complement activation following first exposure to pegylated liposomal doxorubicin (Doxil®): Possible role in hypersensitivity reactions. *Ann. Oncol.*, **14**, 1430–1437.

Chapter 3

Immunotherapy and Vaccines

Johanna Poecheim, PhD, and Gerrit Borchard, PhD

School of Pharmaceutical Sciences, University of Geneva, University of Lausanne, Geneva, Switzerland

Keywords: adaptive immune response, adjuvant, antigen, antigen-presenting cells (APCs), antibody, CD4+ cells, CD8+ cells, cytotoxic T lymphocyte (CTL), dendritic cell (DC), formulation, humoral immunity, immune-stimulating complexes, immunogenicity, innate immune system, infection, infectious disease, inflammatory response, lymph node (LN), macrophages, MHC class I, MHC class II, nanoemulsion, nanoparticles, natural killer (NK) cells, neutrophils, opsonin, opsonization, phagocyte, pathogen-associated molecular patterns (PAMPs), pattern recognition receptors (PRRs), particle-mediated epidermal delivery (PMED), polyethylene glycol (PEG), reactive oxygen species (ROS), shelf-life stability, superparamagnetic iron oxide nanoparticles (SPIONs), surface charge, T helper (Th) cell, vaccine, virus-like nanoparticles

3.1 Introduction

The human immune system has evolved to recognize antigenic properties of pathogens and initiate an immune response resulting in the restriction of the pathogen from entering the body or in the clearance of the pathogen after infection. Among the signals leading to an activation of the immune system are the pathogen's surface properties, its size, and its shape. It is therefore not

---

Immune Aspects of Biopharmaceuticals and Nanomedicines
Edited by Raj Bawa, János Szebeni, Thomas J. Webster, and Gerald F. Audette
Copyright © 2018 Pan Stanford Publishing Pte. Ltd.
ISBN 978-981-4774-52-9 (Hardcover), 978-0-203-73153-6 (eBook)
www.panstanford.com

surprising that the general paradigm of vaccine development demands the preparation of vaccines combining both antigen and adjuvant in the same—particulate—formulation. Vaccine development today, though, is fraught with the failure to achieve the required sufficient level of protection, possibly due to the lack of appropriate adjuvants—especially for mucosal vaccines—and delivery systems (Doroud and Rafati, 2011). However, and unfortunately, there is a bias between current knowledge on specific immune activation and its translation into real products. Whereas groundbreaking discoveries in the field of pathogenic pattern recognition by dedicated receptor families have been made over the last decades, only few ligands specifically interacting with these receptors have been introduced into clinical studies. One reason for this pause in vaccine development may have been the reluctance of manufacturers to introduce these novel adjuvants into the pipeline; another lies in the fact that they need to be presented to the cells of the immune system in an appropriate way to exploit their full potential. It may even be necessary to include combinations of novel adjuvants in one vaccine formulation to achieve synergistic effects.

Nanotechnology may offer the possibility to design novel, more effective vaccines. The nanoparticulate carrier systems should be of appropriate size, have favorable surface parameters for immune recognition, be targeted to antigen-presenting cells (APCs), and be suitable for the inclusion of the respective antigen. In addition, such novel vaccines may prove to have enhanced shelf-life stability, rendering refrigeration unnecessary, and to be applicable by mucosal pathways, allowing for the avoidance of needles for injection and the health risks related.

This chapter discusses the current influence of nanotechnology on the development of vaccines.

3.2   The Immune System

Vaccines are meant to elicit an immune response and to create long-lasting immune memory against a pathogen-specific antigen by staging an "artificial" infection. To rationally design novel vaccines it is essential to understand the function of the immune system itself. The vertebrate immune system consists of the innate and adaptive branches that cooperate to protect against infection and subsistence of pathogenic agents in the host. How both arms of the immune response are orchestrated is described in the following sections and illustrated in Fig. 3.1.

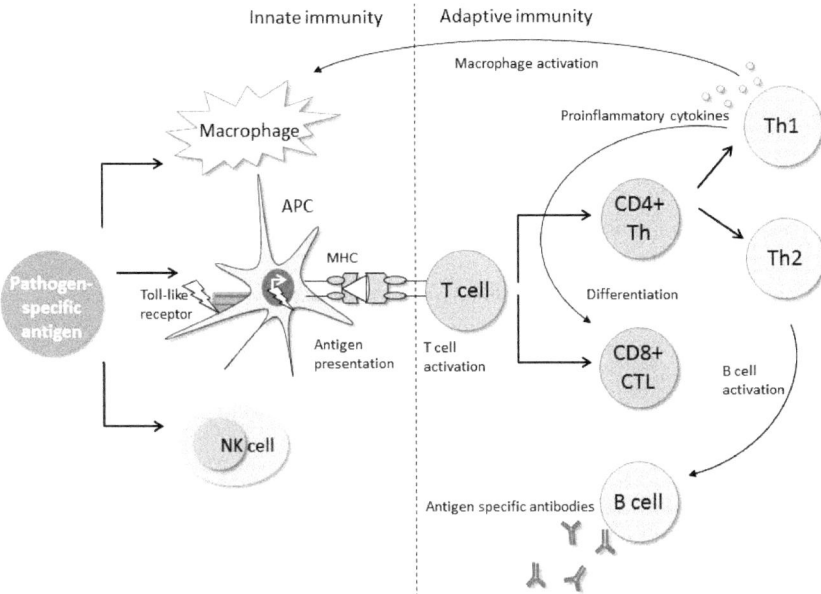

Figure 3.1 The role of the innate and adaptive immune responses following vaccination with a pathogen-specific antigen. Antigen-presenting cells (APCs), which activate naive T cells through antigen presentation, activate the adaptive immune system and induce T cell differentiation. Activated CD4+ T helper cells of type 1 activate macrophages and cytotoxic T lymphocytes (CTLs) through proinflammatory cytokine release. Type 2 helper T cells activate B cells to release specific antibodies against the antigen. Additionally to those antigen-specific B and T cells spread as effector cells, long-lived clones of memory cells are produced that form the immunological memory.

3.2.1   The Innate Immune System

When pathogens overcome the barriers formed by the skin and the mucosae and gain access to the body's soft tissues, the innate immune system detects the invading agent and is activated as a fast and early-stage immune response. Innate immunity provides immune surveillance and immediate defenses that are always available and do not improve with repeated exposure to the same antigen. The immune system has to identify danger signals related to the pathogen by detecting antigenic properties or patterns of a pathogen in the absence of prior immune recognition (first infection). Such danger signals are, for example, the recognition of pathogen-associated molecular patterns (PAMPs) by pattern recognition receptors (PRRs), missing components, or "missing self" signals distinguishing between bacterial cells and healthy human cells, tissue damage caused by pathogens, and alarm signals such as the release of certain cytokines. Upon recognition, complement activation is triggered, consisting in a system of plasma proteins and cell surface molecules that mark pathogens for destruction. These interactions assist resident macrophages to phagocytose the microbiological invaders, as well as to induce macrophages to secrete inflammatory cytokines that in turn attract neutrophils and natural killer (NK) cells to the site of infection (Parham, 2000) (Fig. 3.1).

3.2.2 The Adaptive Immune System

Under normal conditions, the adaptive immune system is silent and is only called into action directly when pathogens evade or overcome the innate immune defense. Thus, the adaptive immune response is initiated only if the innate immune system is activated and signals the presence of a pathogen. It "adapts" to the presence of a pathogen by activating, proliferating, and creating potent mechanisms to neutralize and/or eliminate pathogens.

Two types of adaptive immune responses are to be distinguished. Humoral immunity is mediated by antibodies, also known as immunoglobulins (Igs), which are produced by B lymphocytes. The most abundant isotypes are IgG, found in blood serum and lymph, and IgA, as the main effector of the mucosal immune system. Ratios of IgG2/IgG1 subclasses >1 are associated with type 1 helper (Th1) responses, whereas any ratio <1 corresponds to type 2 helper (Th2)-biased immune responses. Simultaneously cell-mediated immunity is induced, mediated by T lymphocytes. The main difference between these lymphocyte subtypes is that Ig receptors of B cells bind whole molecules and intact pathogens, whereas T cells recognize only short peptide antigens (Kumar et al., 2012b), the so-called T cell epitopes. At the site of infection APCs internalize, process, and present T cell epitopes by the major histocompatibility complex (MHC or CD1) molecules expressed at the cell surface. These cells subsequently migrate to the draining lymph nodes (LNs), a collection point where APCs interact with naive T cells. The T cell receptor (TCR) recognizes peptide antigens presented at the surface of these immune cells and, thus, activates the T cell (Von Andrian and Mackay, 2000). Antigen structures from intracellular infections are presented by MHC class I molecules and antigens from extracellular pathogens by MHC class II molecules. Depending on these two classes, T cells differentiate into one of two types of effector T cells: T helper cells (CD4+ cells) to fight off extracellular pathogens and cytotoxic T cells (CD8+ cells) to eliminate infected cells. The CD4+ T helper (Th) lymphocytes are commonly divided into Th1 and Th2 subtypes. Th1 cells release pro-inflammatory cytokines to activate macrophages and cytotoxic T lymphocytes (CTLs) that are a part of the cellular immune system. Th2 cells help B cells to mature to antibody-producing plasma cells, thereby supporting humoral immunity (Fig. 3.1).

3.2.3 Immunological Memory

In the course of the adaptive immune response to infection, clones of pathogen-specific B and T cells spread as effector cells. Additionally, long-lived clones of memory T cells that form the immunological memory are produced. Subsequent immune responses to the same pathogen will be faster and stronger, since memory cells are more quickly activated than naive cells. Unlike naive T cells, memory T cells can patrol nonlymphoid tissues, such as mucosae, and detect infection at an earlier stage. The greater power of a secondary immune response supports the generation of vaccine-mediated protection. By applying antigenic structures of the pathogen to the body without inducing the disease, immunological memory is elicited. Additionally, Th1- and Th2-polarizing adjuvants may be introduced to direct the desired immune response (Parham, 2000; Plotkin et al., 2013).

3.3 Nanotechnology in Vaccines

3.3.1 Particle Characteristics Interacting with the Immune System

In the following section some of the key parameters of nanoparticles suitable for targeting to specific cells of the immune system will be discussed. Nanoparticles are of great scientific interest as they are effectively a bridge between bulk materials and atomic or molecular structures (Pignataro, 2010). To design an optimal vaccine carrier, physicochemical properties of nanoparticles such as size, surface charge, functional material(s) and their composition, hydrophilicity/hydrophobicity, and biodegradability have to be taken into consideration. Nanoparticles activate APCs, which phagocytose them and travel from the application site, that is, lung, skin, or nose, to the LNs where immune reactions are initiated.

3.3.1.1 Particle size

3.3.1.1.1 Size-dependent uptake

Particle size is a critical parameter that influences uptake by phagocytic immune cells via either direct penetration through the cell membrane (i.e., energy-independent mechanism) or by endocytosis (i.e., energy-dependent mechanism). Various types of endocytosis have been identified: virus-sized particles (20–200 nm) are taken up by clathrin-dependent endocytosis (particles < 150 nm), through caveolae-mediated endocytosis (particles within 50–80 nm), or via clathrin- and caveolae-independent pathways. Endocytosis of bacteria and larger-sized particles (>0.5 μm) occurs mainly via macropinocytosis and phagocytosis, effected only by macrophages and immature dendritic cells (DCs) (Xiang et al., 2006). It was verified in vitro that APCs are able to ingest particles up to 5 μm (Tabata and Ikada, 1988). Although nanoparticles' size seems to deeply influence the uptake mechanisms inside the cells, many reported data are still controversial. Some studies have shown that the most efficient uptake is achieved for particles in the nanosize range (Joshi et al., 2012; Yue et al., 2010), other studies have demonstrated a similar or even preferred internalization of microparticles of a size of up to 5 µm (Chua et al., 2011; Kobiasi et al., 2012).

It has been suggested that the immune system has evolved to react to particles on the scale of viruses (<0.5 µm) and bacteria (>0.5 µm) (Fifis et al., 2004). Uptake and immunostimulating mechanisms of nanoparticles and microparticles may be related to their similarity in size to these pathogens, and the immune system reacts to particles that fall within this size range (Chua et al., 2011). In vivo, either the particles are phagocytosed by macrophages at the site of application and transported to the LNs or they diffuse to the LNs by interstitial flow where they become phagocytosed by LN-resident APCs. Studies have shown that only nanoparticles below 100 nm were taken up into the lymphatic system directly and activated LN-resident APCs more efficiently than nanoparticles of larger sizes (Reddy et al., 2006, 2007; Xie et al., 2009). This size dependency of absorption into the LNs is likely to be related to the process of particle transport through the interstitium. Large particles (>100 nm) will penetrate the interstitium less easily and will remain at the site of injection, targeting peripheral rather than LN-resident APCs (Oussoren and Storm, 2001). Manolova et al. reported on the impact of particle size on the specific targeting of distinct DC populations. Whereas small nanoparticles (20 nm) were found in LN-resident APCs, larger particles (500–2000 nm) were mainly associated with DC at the injection site. This size-dependent manner of transport to the LNs implicates a delayed appearance of nanoparticles > 500 nm in the LNs after subcutaneous injection and is consistent with the active transport of these particles by skin-resident DCs. Small particles of 20 nm were detected in the LNs within two hours after injection and taken up by LN-resident APCs, suggesting free drainage (Manolova et al., 2008). This explains the findings in other studies, where a more rapid appearance of chitosan nanoparticles in the LNs compared to chitosan microparticles after subcutaneous injection was observed (Chua et al., 2011, Kobiasi et al., 2012). From these findings it can be concluded that both nano- and microparticles are internalized by cells at comparable efficiency, but nanoparticles are transported to LNs much faster.

These results, however, do not necessarily implicate an increased immunogenicity for nanoparticles. Besides investigation of particle uptake into the LNs, their ability to elicit immune responses was examined. Both nano- and microparticles showed the ability to induce antigen-specific antibody responses. Polystyrene particles of 40 nm were found in more LN cells and induced higher levels of interferon gamma (IFN-γ) and antibody titers in mice than other smaller or larger-sized beads (Fifis et al., 2004). Particles of 1 µm elicited higher serum IgG levels than smaller particles (Gutierro et al., 2002). Some studies revealed no significant differences in IgG production between different-sized particles after parenteral administration (Chua et al., 2011; Gutierro et al., 2002; Nagamoto et al., 2004). However, it has to be mentioned that nanoparticles up to 1 µm applied intranasally elicited a significantly higher production of mucosal IgA when compared to microparticles (Nagamoto et al., 2004). Intranasal immunization with 200 nm nanoparticles enhanced CD4+ T cell responses in the lungs compared to 30 nm nanoparticles (Stano et al., 2012). Regarding the route of administration, following intranasal application, higher serum IgG2a/IgG1 ratios were found for 500 nm poly(lactic-co-glycolic) acid (PLGA) nanoparticles than for 200 nm particles, indicating a Th1 polarized response (Gutierro et al., 2002).

In general, these findings suggest that local immunization may be preferably induced by nanoparticles at a size range between 200 and 1000 nm. However, there is not always a clear size dependency with regard to particle uptake and immunogenicity. Extent and type of immune responses are presumably also associated with other physicochemical properties of the particles, the various materials used, the antigen type transferred, and the route of administration, as well as the vaccination regimen (Chua et al., 2011; Kobiasi et al., 2012).

3.3.1.1.2 *Nanoparticle size and Th response*

The concept of immune activation can be further extended by investigating the type of immune response induced. Particle size may influence Th cell differentiation, whereas Th1 cells are mainly developed following infections by intracellular bacteria and some viruses. Th1 cells produce cytokines that activate macrophages and are responsible for cell-mediated immunity and phagocyte-dependent protective responses. Th2 cells are predominant in responding to large extracellular pathogens and are responsible for antibody production and activation of eosinophils (Romagnani, 1999).

Seen the conflicting data available in the literature, it is difficult to accurately predict particle size ranges that will induce a Th1 or a mixed Th1/Th2 immune response outcome (Oyewumi et al., 2010). Some data are suggesting that nanoparticles promote cellular immune responses. For instance, codelivery of the hepatitis B viral protein HBcAg and monophosphoryl lipid A (MPLA) in the copolymer PLGA nanoparticles of around 300 nm promoted antigen-specific Th1 immune responses, including IFN-γ production (Chong et al., 2005). In a different study, mice vaccinated with 300 nm sized PLGA particles loaded with a model antigen (ovalbumin, OVA) generated the highest fraction of OVA-specific CTLs. They also induced more than a 50-fold increase in the IgG2a/IgG1 ratio compared to microparticles, suggesting polarization toward a Th1-type immune response (Joshi et al., 2012).

Unlike these observations, Henriksen-Lacey et al. described how DDA:TDB liposomes, regardless of their size, stimulated a characteristic Th1 immune response with production of IgG2 antibodies and IFN-γ (Henriksen-Lacey et al., 2011). Other reports showed that 200–600 nm particles favored induction of Th1 responses associated with higher levels of IFN-γ production, whereas microparticles of 2–8 μm in size promoted interleukin-4 (IL-4, a cytokine that induces differentiation of naive helper T cells (Th0 cells) to Th2 cells) secretion inducing humoral immunity (Kanchan and Panda, 2007).

It has also been reported that as the particle size increases from the nanometer to the micrometer range, antibody titers increase, as well. Gutierro et al. showed that 1 μm particles elicited higher total serum IgG levels than nanoparticles. IgG2a/IgG1 ratios typical for a Th1 response, on the other hand, were similar for all particle sizes (Gutierro et al., 2002). In addition, according to Katare et al., microparticles in the size range of 2–8 μm elicited the highest antibody titers, whereas decreasing the particles size resulted in lower-peak antibody titers (Katare et al., 2005).

Table 3.1 List of representative studies demonstrating the effects of particle size on immune responses

Materials	Particle size (nm)	Route	Parameters measured	Comments
PS (Manolova et al., 2008), Chitosan (Chua et al., 2011; Kobiasi et al., 2012)	20/500/1000 150/1300, 163/2100	s.c.	Particle trafficking	Faster appearance of NPs in LNs than MPs.
PS (Fifis et al., 2004)	20/40/100/500/ 1000/2000	i.d.	IFN-γ, IgG	40 nm beads induced the highest levels of IFN-γ and IgG.
PLGA (Gutierro et al., 2002)	200/500/1000	s.c., p.o., i.n.	IgG	1 µm particles elicited higher total serum IgG levels than smaller particles for all tested routes.
Chitosan (Chua et al., 2011; Nagamoto et al., 2004)	163/2100 700/1300/3000	s.c, ip., i.n.	IgG, IgA	Similar ability in inducing IgG after parenteral, IgA levels higher for smaller particles after i.n. administration.
PPS sulfide (Stano et al., 2012), PLGA (Gutierro et al., 2002)	30/200, 200/500/1000	i.n.	IFN-γ, TNF-α, IL-2, IgG1, IgG2a	Higher Th1 responses were obtained with NP of 200–500 nm after i.n. administration.
PLGA (Chong et al., 2005)	633	s.c.	IFN-γ, IL-4, IgG	Only NPs induced Th1 responses with IFN-γ production.

Materials	Particle size (nm)	Route	Parameters measured	Comments
Liposomes (Henriksen-Lacey et al., 2011)	<200/700/1500/ 2000–3000	i.m.	IL-1β, IL-2, IL-6, IFN-γ, IgG1, IgG2a, IL-10, IL-5	High levels of IFN-γ and IgG2a and low levels of IL-5 and IL-10 were noted regardless of the size. Particles of ~700–1000 nm produced the highest levels of IFN-γ and IL-1β.
PLA (Kanchan and Panda, 2007)	200–600/2000–8000	i.p.	IFN-γ, IL-4	NPs favored IFN-γ secretion, MPs IL-4 secretion and higher IgG1/IgG2a ratios.
PLA (Katare et al., 2005)	<2000/2–8 μm/>8	i.m.	IgG	MPs of 2–8 μm elicited the highest antibody titers.
DEP, CBP, SIP (van Zijverden and Granum, 2000)	<500/1000–5000	s.c.	IgE, IgG1, IgG2a	NPs induced a strong Th2 response, MPs induced Th1.
PS (Samuelsen et al., 2009)	64/202/1053/4646	i.t.	TNF-α, IL-1β	MPs induced higher cellular responses than NPs.
Bilosomes (Mann et al., 2009)	250/980	p.o.	IFN-γ, IgG1, IgG2a	Large vesicles are better inducers of Th1 responses, assessed by IFN-γ and IgG2a production.

Abbreviations: CBP: carbon black particle; DEP: diesel exhaust particle; i.d.: intradermal; i.m.: intramuscular; i.n.: intranasal; i.p.: intraperitoneal; i.t.: intratracheal; p.o.: per os; s.c.: subcutaneous; MP: microparticles; NP: nanoparticle; PLGA: poly(lactic-co-glycolic) acid; PPS: polypropylene sulfide; PS: polystyrene; SIP: silica particle.

By contrast, subcutaneous application of several particle types present in the environment led to different results. Diesel exhaust particles below a size of 500 nm increased the production of IgE and IgG1 antibodies as indicators for Th2-like responses, while IgG2a titers remained low. Carbon black particles, on the other hand, induced a mixed Th1/Th2 response, and the larger silica particles (1–5 µm) induced a Th1 bias (van Zijverden and Granum, 2000). In addition, data from Samuelsen et al. applying polystyrene particles of different diameters intratracheally, as well as from Mann et al. using orally applied bile salt stabilized vesicles (named bilosomes), showed that larger particles (≥1 µm) generated stronger Th1 responses than nanoparticles (Mann et al., 2009; Samuelsen et al., 2009).

Not only nanoparticle size as a parameter has been determined to initiate immunostimulatory reactions through mediating release of pro-inflammatory cytokines, but also the surface charge and the material used for nanoparticles production itself may modulate immune responses.

The studies described here are listed in Table 3.1 summarizing the effects of particle size on the resultant immune responses.

3.3.1.2 Charge

Charged nanoparticles have been shown to be more likely taken up by phagocytes than neutral particles of the same size. Polyethylene glycol (PEG)-modified neutral nanoparticles had the lowest percentage of uptake when compared to particles with cationic or anionic surface charges (Dobrovolskaia et al., 2008; Zahr et al., 2006). Nanoparticles rendered negatively charged, for example, by adsorption of mucus components after mucosal administration, again may interact to a much lesser extent with the negatively charged cell membrane surface. Consequently, it is more likely that positively charged particles are internalized by immune cells and induce mucosal immune responses, which was confirmed in a study showing a positive correlation between particle uptake and increasing zeta potential (Kwon et al., 2005). Usually particles displaying a cationic surface charge are thought to be more toxic, since they can penetrate into cells more easily.

However, some studies revealed a higher uptake of negatively charged particles compared to cationic ones, which was explained by nonspecific binding and clustering of the particles at cationic cell membrane domains and subsequent endocytosis (Verma and Stellacci, 2010). Another possibility in this case would be that these positively charged particles become neutralized by adsorption of negatively charged components of the mucus to their surface and thus reduced in their targeting ability (Rajapaksa and Lo, 2010).

Not only surface charge, but also surface charge density is crucial for interactions between nanoparticle surface and proteins, that is, opsonins. Nanoparticles have been shown to be covered with serum proteins in the bloodstream shortly after injection. This process is called opsonization and enables macrophages to recognize and subsequently phagocytose foreign particles. How different coating patterns of particles may change the pathway of opsonization (classic, alternative, lectin mediated) has recently been described by Moghimi et al. (2011, 2012). Particles of neutral surface potential may have slower opsonization kinetics, a much lower opsonization rate, and very possibly a different opsonization pattern than charged particles (Nie, 2010). Such particles avoiding opsonization are used when a "stealth effect" for prolonged circulation in the bloodstream and reduced immunogenicity is desired. Serum proteins bound to charged particles, on the other hand, have been shown to activate immune cell receptors upon unfolding of the protein (Deng et al.).

Last but not least, surface charge may also have an effect on the intracellular transport and processing following phagocytosis. As shown, for instance, by Harush-Frenkel et al. (2008), anionic nanoparticles are targeted to the degradative lysosomal route, whereas cationic particles, bound to the transcytotic pathway, are not found in lysosomes suggesting a longer intracellular residence time.

3.3.1.3 Immunogenicity of nanoparticle materials

Nanoparticles are classified as transport vehicles and their adjuvant activity is mainly related to enhanced delivery of antigens to APCs or to the LNs. However, mere transport function and interaction with target cells determining overall effect cannot be separated for these complex systems. Different types of bulk materials used for nanoparticle preparation may induce cytokine release, as has been shown in several studies (Plotkin et al., 2013). Materials used for nanoparticle preparation include (bio)polymers (e.g., PLGA and chitosan), lipids, and metals, of which certain substances have been shown to be immunogenic and/or have adjuvant properties themselves.

3.3.1.3.1 Polymers

Response to polymeric particles is characterized by mild inflammation, differentiation of immune cells, and degradation of the particles. (Chakravarthi et al., 2007). Biomaterials used in nanoparticulate vaccines have the potential to control the host response in order to reduce or enhance immune responses. Babensee et al. investigated the effect of biomaterial-induced DC phenotype changes and observed differential levels of DC maturation, depending on the biomaterial used. PLGA or chitosan films induced maturation of DCs, as seen in augmentation of pro-inflammatory cytokine production and expression of costimulatory molecules, such as CD86, providing increased interaction with T cells. By contrast, this effect was not observed when DCs were treated with alginate or agarose films, and was even inhibited by hyaluronic acid (Babensee, 2008; Rogers and Babensee, 2011; Yoshida et al., 2007).

In addition, a different effect on TNF-α secretion by the shape of biomaterials was observed, showing stronger effects for particles than for films of the same material (Yoshida et al., 2007). PLGA is considered to be safe and biocompatible at low immunogenicity, and is therefore approved by the US Food and Drug Administration (FDA). However, several studies exhibited the induction of weak inflammatory responses to PLGA, which in vaccine carriers could lead to an adjuvant effect enhancing the overall immunogenicity of the vaccine. After tracheal application of PLGA, an acute inflammatory response was demonstrated by the increase of neutrophils when compared with animals treated with saline solution. By contrast, the inflammatory response seen in animals instilled with saline and polystyrene was similar. This suggests that polystyrene under these conditions behaves as an inert material and does not induce a sustained inflammatory response in the airways (Avital et al., 2002; Springer et al., 2005).

Fiore et al. suggested that the degradation of PLGA into acidic by-products could be responsible for its potential to induce inflammatory responses. Having assessed macrophage and lymphocyte numbers, PLGA-treated lungs of mice displayed a slight increase in inflammation. By contrast, upon intratracheal administration of polyketal particles, which degrade into neutral compounds, no significant inflammatory reactions were detected (Fiore et al., 2010).

The adjuvant properties of chitosan, a biodegradable biopolymer, have also been subjected to investigations. Chitosan potently activated the NLRP3 (NOD-like receptor family, pyrin domain containing 3) inflammasome (a multiprotein oligomer that is a component of the innate immune system) in a phagocytosis-dependent manner. Chitosan is the deacetylated derivative of chitin, which was shown to be relatively inert, suggesting the influence of charge on the immunostimulatory properties of chitosan because of the presence of a secondary amino group (Bueter et al., 2011). Also Tokura et al., who studied the immunological aspects of chitin and chitin derivatives administered to animals, pointed out that chitosan itself, as an adjuvant, can induce polarized Th2 responses (Borchard et al., 2012; Tokura et al., 1999). Whether this activation is based on the interaction of chitosan with PRRs in a lectin-like fashion is still under discussion.

3.3.1.3.2 *Lipids*

Lipid-based delivery systems have been subjected to investigation as vaccine adjuvants, as well. It is generally assumed that hydrophobic surfaces of particles enhance their phagocytosis (Thiele et al., 2003). Among lipid-based vaccine deliver systems, such as solid lipid nanoparticles (SLNs), liposomes, polymerized and nonpolymerized liposomes, and new classes of lipid drug carriers such as immune-stimulating complexes (ISCOMs) and multilamellar vesicles (Alonso-Romanowski et al., 2003; Cai et al., 2011; Moon et al., 2011; Wilson et al., 2012), oil-in-water emulsions are the most extensively studied particulate vaccine carriers. In Europe, nanoscale emulsion-based adjuvants are licensed for influenza vaccines, containing MF59 (Novartis; 1997) or AS03 (GlaxoSmithKline; 2009). AF03 (Sanofi Pasteur) is another emulsion containing squalene, a naturally occurring intermediate metabolite of cholesterol. AF03 has been tested in humans, but is not yet licensed for use in humans (Montomoli et al., 2011; Salvador et al., 2011). The potential to improve vaccine performance has been shown, yet the mechanism of adjuvanticity of these oil-in-water emulsions remains unknown or undisclosed. However, as described above, several studies suggest that these nanoemulsions do not act only as antigen delivery systems but also as immune modulators.

In a study on the influence of squalene-based nanoemulsions on the activation of the innate immune system, it was hypothesized that changes in lipid metabolism and innate immunity are closely linked. Uptake of squalene-based nanoemulsions induced accumulation of neutral lipids that have been shown to have pro-inflammatory properties (Kalvodova, 2010; Lorentzen, 1999; Prieur et al., 2010). Other findings show that the nonionic surfactant polysorbate 80, used in the above-mentioned emulsions, may have an immunomodulatory effect, inducing Th2 responses (Kozutsumi et al., 2006). The systemic expression of IL-5 is in agreement with the Th2 immune response elicited by the polysorbate 80–containing nanoemulsion MF59 (Mosca et al., 2008). The contribution of vitamin E to the overall adjuvanticity in AS03 was investigated by Morel et al. (2011), where higher antibody responses with the inclusion of α-tocopherol than with the plain squalene-in-water emulsion were observed.

3.3.1.3.3 *Inorganic materials*

As an alternative to organic compounds, inorganic materials such as metals and metal oxides are also known to form nanoparticles. Due to their rigid nondeformable shape and much smaller size, they exhibit many distinct biological properties compared to polymeric or lipid nanoparticles (Huang et al., 2010). Being nonbiodegradable, these nanoparticles remain at the site of injection for extended periods of time, resulting in long-lasting presentation and thus enhanced immunogenicity (Maquieira et al., 2012). Among the most widely used metal nanoparticles, gold, silver, or copper and as magnetic metals iron or cobalt is taken into consideration for drug delivery; however, few attempts have been made to develop solid inorganic nanoparticles as vaccine platforms (Bhattacharyya et al., 2011; Pusic et al., 2012).

The main advantage of this type of nanoparticles is that they can be manufactured in fine-tunable sizes between 5 and 100 nm. In addition, they show a high affinity for sulfhydryl groups, which facilitates the binding of various biomolecules through chemisorption of thiol residues onto the metal nanoparticles

(Algar et al., 2011; Cruz et al., 2012). Gold, silver and copper are known for being antimicrobial materials, probably due to their ability to react with the –SH groups of enzymes and hence inactivating these proteins (Bhattacharya and Mukherjee, 2008; Yoon et al., 2007). As an example, inhibitory effects on HIV-1 infection in CD4+MT-2 cells (human T cell leukemia [HTLV-1] virus carrier cell line) and cMAGI HIV-1 reporter cells (HeLa cell clone expressing human CD4 and HIV-long terminal repeat-coupled genes) due to the interaction of silver nanoparticles with the virus was reported by Elechiguerra et al. (2005).

Furthermore, the inflammatory potential of metals could be of interest in vaccine development. Goebel et al. (1995) showed that the T cell response in mice treated with gold(I) compounds for 12 weeks is directed against the oxidized gold(III), which is a metabolite generated in vivo by macrophages. In another study the inflammatory potential of colloidal silver on macrophages was evaluated through assessing the release of IL-8, cell viability, and induction of oxidative stress. While reactive oxygen species (ROS) are generated as part of the normal oxidative metabolism, their overproduction can lead to cell death. Early-stage oxidative stress, on the contrary, can stimulate inflammatory responses, and the generated ROS were found to trigger the release of IL-8 in macrophages (Park et al., 2011). While different types of metallic nanoparticles without any specific antigen loading reportedly exert antimicrobial activity, other studies revealed a compromise of innate immune response and clearance of bacterial pathogens after exposure. Shevedova et al. suggested that elimination of microbes depends to a larger extent on nitric oxide production than ROS from the oxidative burst (Nathan and Hibbs Jr., 1991; Shvedova et al., 2008). This could explain the controversial findings of decreased pulmonary bacterial clearance in mice after exposure to copper nanoparticles, despite induction of robust inflammatory responses (Kim et al., 2011).

Also metal oxide nanoparticles and their soluble ions were investigated with regard to their immunostimulatory/inflammatory effects. In vitro, nanoparticles made of nickel, copper, and zinc oxide all showed toxic effects, but only zinc and copper oxide nanoparticles induced IL-8 production and activation of transcription factors. Aqueous extracts of Zn and Cu oxide nanoparticles showed similar profiles of cytotoxicity and inflammatory effects, whereas nickel ions did not induce any cytotoxicity or immune responses. The inflammatory response assessed in rats given by nanoparticles was also different to the aqueous extracts, with zinc and copper oxide nanoparticles leading to the recruitment of eosinophils but not their soluble ions (Cho et al., 2012b).

3.3.2 Immune Receptor Targeting and Antigen Delivery

Currently, intense research efforts are aimed at the development of novel formulations and delivery systems for vaccines. By the use of nanoparticles, antigens are expected to be adequately presented to the immune system to elicit immune responses appropriate for sufficient protection against infectious diseases. Ideally, the nanoparticulate vaccine carrier system should have sufficient antigen- and adjuvant-loading capacity, through either incorporation or surface adsorption.

Formulation strategies can facilitate the capture by and the entry of the antigen into APCs. T cell antigens, in the form of peptides, proteins, plasmid DNA, or RNA formulated into nanoparticles appear to increase CTL responses in vivo. Such particulate formulations significantly increase efficiency as compared to the soluble antigen alone, as they are in general readily recognized as antigenic and ingested by immune cells, which subsequently leads to antigen processing and presentation to other cells of the immune system (Moingeon et al., 2002). Ligands of APC-specific PRRs, such as Toll-like receptors (TLRs), can be grafted onto the surface of particle-based vaccines to increase specific delivery to immune cells (Heuking and Borchard, 2012).

Application of vaccines in particulate form not only allows moving the antigen within the tissue to individual cells, it also facilitates its penetration through the cell membrane. This transfection process is a major requirement for making use of intracellular mechanisms, such as signaling pathways and antigen processing, to activate the immune system. Many biomolecules, used as antigens for vaccines, are not able to diffuse within tissues and through the plasma membrane, due to their size and other physicochemical properties. Particle-based strategies to overcome this problem have been developed for genetic vaccination relying on chemical materials used as carriers and physical treatment. In the "gene gun technique" DNA is coated on gold particles, energy is transferred to the DNA carriers to accelerate and propel the particles into the cells (Villemejane and Mir, 2009). This method is limited to low-depth penetration of particles in tissues but is sufficient for skin DNA vaccination (Luz Garcia-Hernandez et al., 2008). Magnetic nanoparticles have attracted attention because of their potential use for direct targeting to diseased tissues and organs by applying an external magnetic field. "Magnetofection" describes a technique of gene transfection involving naked plasmid DNA or DNA vectors associated to the surface of magnetic nanoparticles. Superparamagnetic iron oxide nanoparticles (SPIONs) are applied for magnetofection by delivering genes to the target cells, hence increasing their local concentration (Al-Deen et al., 2011; Vasir and Labhasetwar, 2007). Application of an external alternating magnetic field leads to the production of energy in the form of heat. Hyperthermia has been demonstrated to stimulate the innate immune response through the release of heat-shock proteins that activate neighboring immune cells (Baronzio et al., 2006; Colombo et al., 2012). As such, SPIONs may have adjuvant properties activating the immune system in a nonspecific manner. These two methods are still under investigation. Although a few clinical trials have already been performed or are in progress, long-term safety and potential risks have not been sufficiently evaluated (Villemejane and Mir, 2009). Another approach is impalefection by functionalizing nanoscale carbon fibers with DNA to impale cells. It is a useful tool for exploring gene delivery in vitro but has yet to be used in the therapeutic field (Pearce et al., 2013).

The types of effective immune responses against infectious diseases depend on the localization of the pathogens after infection. Generally, extracellular pathogens are combated by antibodies mediated by humoral immune responses, whereas protection against intracellular pathogens depends on cell-mediated immunity via CD4+ Th1 cells. In addition, CD8+ CTLs play an important role in the protection against these pathogens by directly killing infected cells (Nagata and Koide, 2010). Antigen recognition by T lymphocytes depends on the ability of APCs to process protein antigen into peptides and to present them via MHC molecules at the cell surface. The MHC I presentation pathway allows CD8+ T cells to identify and eliminate cells infected by intracellular pathogens. By contrast, MHC II and CD1 molecules present peptides derived from captured exogenous antigens to CD4+ T cells and NK cells (Paul, 2008) (Fig. 3.1).

3.3.3 Current Nanoparticle Vaccines on the Market and in Clinical Studies

Nanotechnology is enabling novel vaccines to enter clinical trials that often show superiority of the particulate systems over traditional vaccine types. We will discuss here several examples of nanoparticle-based vaccines.

On the basis of antigen presentation by DCs and innate immune system stimulation via TLRs, a vaccine against melanoma was designed. Melanoma-specific peptide Melan-A/Mart-1 was linked to virus-like nanoparticles (VLPs) loaded with CpG oligonucleotides as activators of TLR 9 located in the endosomal compartment of target cells (Speiser et al., 2005). Already vaccination of HLA-A2 transgenic mice in preclinical studies showed strong Melan-A-specific CD8 T cell responses. A phase I/II study in stage II–IV melanoma patients confirmed good tolerability of the vaccine and ex vivo T cell responses. Increased activated Melan-A-specific CD8 T cell populations were detected directly ex vivo as well as cytokine production of INF-γ, TNF-α, and IL-2, and LAMP-1 expression by vaccine-induced T cells. Furthermore, a central memory phenotype of those specific T cells was frequently observed and an enhancement of T cell responses was achieved by subsequent vaccination with the peptide emulsified in incomplete Freund's adjuvant (Speiser et al., 2010). Other vaccines using nanotechnology based on VLPs have already been licensed for the prevention of cervical and anogenital infections, that is, by GlaxoSmithKline (Engerix® and Cervarix®) and Merck and Co. (Recombivax HB and Gardasil®) (Buonaguro and Buonaguro, 2013). These products are noninfectious viral subunit vaccines with proteins of the respective pathogens assembled in nanosized VLPs, structures resembling the virus itself and thus recognized as antigens but lacking the genetic repertoire necessary for replication.

A novel liposomal influenza subunit vaccine is currently under clinical investigation. It consists of liposomes containing the viral surface proteins hemagglutinin (HA) and neuraminidase

(NA) derived from various influenza strains (Babai et al., 2001). An IL-2-supplemented liposomal formulation injected intramuscularly has proven to be both safe and effective in inducing strong anti-HA and anti-NA antibody responses in mice and humans (Ben-Yehuda et al., 2003).

DNA vaccines with one specific or even several pathogenic antigenic epitopes encoded by the plasmid are known to drive cellular and humoral immune responses at least in the animal model. To increase transfection efficiency, and to enhance and direct immune responses to DNA vaccines, particle-based delivery systems have been explored (Bivas-Benita et al., 2004, 2009).

FDA-controlled phase I studies to assess safety and immunogenicity of particle-mediated DNA vaccines against influenza and hepatitis B viruses (HBVs) have been completed only for microparticles. These studies were a part of the clinical development of a particle-mediated epidermal delivery (PMED) DNA vaccine or "gene gun" based on a dry powder formulation with DNA plasmid adsorbed onto microscopic gold particles (PowderMed, 2007, 2008). Particle-mediated DNA administration permits the use of small quantities of DNA and application via the skin as an immunological active tissue increases the efficacy of the vaccine. HBV-specific protective antibody responses of at least 10 mIU/mL and antigen-specific CD8+ cells were detected. Hepatitis B surface antigen–specific IFN-γ-secreting Th cells were measured in the majority of individuals treated, indicating Th1-biased responses. Current data demonstrate that particle-mediated DNA administration with this needle-free method of administration was safe and well tolerated. Since Th1 and CD8+ T cell responses are associated with resolution of HBV and other chronic viral infections, the results reported suggest a potential for particle-mediated DNA vaccine delivery (Roy et al., 2000).

Another topical vaccine, DermaVir Patch for skin delivery of the DNA vaccine, has already entered clinical trials for the treatment of HIV infections (Immunity, 2013). Treatment of HIV-1/AIDS requires a vaccine to induce robust and long-lasting HIV-1-specific T cells to control viral replication. A plasmid DNA encoding the majority of HIV-1 genes has been chemically formulated into polyethylenimine mannose (PEIm), a cationic polymer that forms nanoparticles with the negatively charged plasmid through condensation. The mannose moiety targets the vaccine to receptors on the surface of APCs (Lisziewicz et al., 2007). On the basis of the potent induction of Gag, Tat, and Rev antigen-specific memory T cells, it is assumed that DermaVir boosts T cell responses specific to all the 15 HIV antigenic epitopes expressed by a single plasmid applied. Furthermore, a dose-dependent expansion of HIV-specific memory T cells with high proliferation capacity was detected (Lisziewicz et al., 2012).

Just recently a phase II clinical trial evaluated the safety and immunogenicity of a respiratory syncytial virus fusion (RSVF) protein vaccine, formulated as nanoparticles. The protein was extracted and purified from insect cell membranes and assembled into 40 nm nanoparticles composed of multiple RSVF oligomers arranged in the form of rosettes (Smith et al., 2012). The primary immune response measured confirmed the vaccine to be a potent antigen eliciting immune responses at levels that would be predicted to protect infants through maternal immunization (AZoNano, Novavax, 2013). A study using a polypeptide vaccine against malaria that self-assembles into spherical nanoparticles, displaying repetitive epitopes of the Plasmodium falciparum circumsporozoite protein, showed promising results in preclinical testing. The outcome was high-titer, long-lasting protective antibody and long-lived central memory CD8+ T cells and could provide sterile protection against a lethal challenge of the transgenic parasites in mice (Kaba et al., 2012).

From a soluble type of chitosan, Viscosan, a hydrogel was formed, which was further mechanically processed into gel particles. This so-called ViscoGel was mixed with a commercial vaccine against Haemophilus influenza type b (Act-HIB). ViscoGel-Act-HIB with a tenth of the vaccine dose is as efficient in eliciting a humoral response in mice as the high-dose vaccine alone (Neimert-Andersson et al., 2011).

Mixing saponin, cholesterol, and phospholipid under controlled conditions, 40 nm spherical negatively charged ISCOMs of the ISCOMATRIX® adjuvant are formed. A range of ISCOMATRIX® vaccines have been tested in clinical trials, generating both antibody and T cell responses. The adjuvant has been combined with influenza viral proteins or recombinant antigens for two cancer/neoplasia vaccines and one hepatitis C virus (HCV) vaccine to be tested in humans. Higher and faster antibody responses compared to conventional influenza vaccines and virus-specific CTL responses were observed. Furthermore, strong virus-specific humoral and cellular immune responses were detected in vaccinated subjects of the two types of human papilloma virus (HPV) and of the HCV core ISCOMATRIX® vaccines (Pearse and Drane, 2005).

3.4 Conclusions

Prevention is better than treatment. Preventing a pathogen from gaining access to the body is better than to fight it off after infection. Antibiotics are failing, occurrence of multidrug resistance is on the rise, and new pathogens are emerging: we are in dire need of novel vaccines that elicit long-lasting and protective immune responses. By our growing understanding of how adjuvants may work, the "dirty little secrets" of vaccinologists may be no more, and through nanotechnology we may have the opportunity to design vaccine delivery systems as nature intended them to be.

In practical terms, which antigens would be good candidates to be used in a novel (nanoparticulate) formulation, possibly using novel adjuvants? Using known antigens, for which successful vaccines already exist, might not be a good choice: the danger of nonacceptance by regulatory authorities and/or the public may be too high. Therefore, one would choose novel antigens that are being identified through sequencing of pathogen genomes, or a combination thereof. The concept of "new formulations for new vaccines" might rightfully apply, as tolerance toward undesired side effects is certainly higher for first-in-man vaccines than for new formulations of vaccines that have been tried and tested. The use of adjuvants, on the other hand, is fraught with their sometimes difficult handling; the skepticism of regulatory authorities toward new, not well-defined excipients/active principles; and the ongoing discussion in a public that remains often closed to scientific argumentation.

The way out of this dilemma would, in our opinion, be to enhance the immunogenicity of protein and peptide vaccines themselves, to use particulate systems to better target and present the antigen to the immune system, and to finally use well-characterized small-molecular-weight entities of a defined mechanism of action and known toxicity. These avenues are followed in modern vaccine development and will result in highly effective vaccines against old and new threats to our health. Nanotechnology is playing a prominent part in this endeavor.

Disclosures and Conflict of Interest

The authors declare that they have no conflict of interest and have no affiliations or financial involvement with any organization or entity discussed in this chapter. This includes employment, consultancies, honoraria, grants, stock ownership or options, expert testimony, patents (received or pending) or royalties. No writing assistance was utilized in the production of this manuscript and the authors have received no payment for the preparation of this chapter. This chapter is a revised and updated version of the chapter that originally appeared in *Controlled Release Systems: Advances in Nanobottles and Active Nanoparticles*, 2016, Alexander van Herk et al. (Editors), Pan Stanford Publishing, Singapore, and appears here with kind permission of the copyright holder.

Corresponding Author

Dr. Gerrit Borchard
School of Pharmaceutical Sciences
University of Geneva, University of Lausanne
30 quai Ernest-Ansermet, 1211 Geneva, Switzerland
Email: Gerrit.Borchard@unige.ch

About the Authors

Johanna Poecheim graduated in pharmaceutical sciences at the University of Innsbruck, Austria, where she worked on the improvement of storage stability of drugs in aqueous parenteral formulations for her master thesis. She obtained her PhD from the University of Geneva, Switzerland. Within her doctoral thesis, performed in the Biopharmacy group of Prof. Gerrit Borchard, she focused on nanoparticle-based formulation development, including novel immunological adjuvants, to improve immunogenicity of a tuberculosis DNA vaccine. The project included formulation development, physicochemical characterization, and in vitro testing. In vivo studies with mice were performed in collaboration with Dr. Nicolas Collin of the WHO Vaccine Formulation Laboratory at the University of Lausanne. She then started a postdoc within the Early-Stage Pharmaceutical Development division of F. Hoffmann-La Roche in Basel, Switzerland. Dr. Poecheim's current research interest there focuses on the characterization of nucleic acid molecules and stabilizing effects of various excipients and buffer systems with the objective to develop a platform for oligonucleotide formulations.

Gerrit Borchard is a licensed pharmacist and obtained his PhD in pharmaceutical technology from the University of Frankfurt, Germany, for his thesis on the interaction of colloidal drug carrier systems with the immune system. After holding several academic posts, including a lecturer position at Saarland University, Germany, and assistant and associate professorships at Leiden University, the Netherlands, he joined Enzon Pharmaceuticals, Inc., USA, as vice president research. In 2005, he was appointed full professor of Biopharmaceutics at the University of Geneva, Switzerland, and scientific director of the Centre Pharmapeptides in Archamps, France, an international center for biopharmaceutical research and training. Prof. Borchard is author or coauthor of over 130 scientific publications and book chapters and coeditor of 1 book and has been named as an inventor on 7 patents. From 2008 to 2013, he served as vice president of the School of Pharmaceutical Sciences Geneva-Lausanne and from 2013 to 2014 as acting president. In 2012 Prof. Borchard joined the Non-Biological Complex Drugs working group hosted at Lygature (former Top Institute Pharma, Leiden, the Netherlands) and became a member of the steering committee in 2015. He was nominated chair of the NBC working party at the European Directorate for the Quality of Medicines & Health Care by Swissmedic in 2013. Prof. Borchard was nominated fellow of the Swiss Society of Pharmaceutical Sciences in 2010, and has been president of the

Swiss Academy of Pharmaceutical Sciences since 2014. He also served as vice president of the European Federation of Pharmaceutical Sciences from 2013 to 2015. Owing to his working in both academia and industry, and living in four countries, Prof. Borchard has acquired extensive experience in diverse working and cultural environments and speaks fluent Dutch, English, French, and German. Time allowing, he enjoys roaming the trails and by-roads of the Jura Mountains on foot and bike.

References

Al-Deen, F. N., Ho, J., Selomulya, C., Ma, C., Coppel, R. (2011). *Langmuir*, **27**, 3703–3712.

Algar, W. R., Prasuhn, D. E., Stewart, M. H., Jennings, T. L., Blanco-Canosa, J. B., Dawson, P. E., Medintz, I. L. (2011). *Bioconjug. Chem.*, **22**, 825–858.

Alonso-Romanowski, S., Chiaramoni, N. S., Lioy, V. S., Gargini, R. A., Viera, L. I., Taira, M. C. (2003). *Chem. Phys. Lipids*, **122**, 191–203.

Avital, A., Shapiro, E., Doviner, V., Sherman, Y., Margel, S., Tsuberi, M., Springer, C. (2002). *Am. J. Respir. Cell Mol. Biol.*, **27**, 511–514.

AzoNano (2013). *Positive Data from Novavax Phase II RSV F Protein Nanoparticle Vaccine Clinical Trial*, May 8, 2013, Available from: http://www.azonano.com/news.aspx?newsID=27000 (accessed on March 6, 2017).

Babai, I., Barenholz, Y., Zakay-Rones, Z., Greenbaum, E., Samira, S., Hayon, I., Rochman, M., Kedar, E. (2001). *Vaccine*, **20**, 505–515.

Babensee, J. E. (2008). *Semin. Immunol.*, **20**, 101–108.

Baronzio, G., Gramaglia, A., Fiorentini, G. (2006). *In Vivo*, **20**, 689–695.

Ben-Yehuda, A., Joseph, A., Zeira, E., Even-Chen, S., Louria-Hayon, I., Babai, I., Zakay-Rones, Z., Greenbaum, E., Barenholz, Y., Kedar, E. (2003). *J. Med. Virol.*, **69**, 560–567.

Bhattacharya, R., Mukherjee, P. (2008). *Adv. Drug Deliv. Rev.*, **60**, 1289–1306.

Bhattacharyya, S., Kudgus, R., Bhattacharya, R., Mukherjee, P. (2011). *Pharm. Res.*, **28**, 237–259.

Bivas-Benita, M., Lin, M. Y., Bal, S. M., van Meijgaarden, K. E., Franken, K. L. M. C., Friggen, A. H., Junginger, H. E., Borchard, G., Klein, M. R., Ottenhoff, T. H. M. (2009). *Vaccine*, **27**, 4010–4017.

Bivas-Benita, M., van Meijgaarden, K. E., Franken, K. L. M. C., Junginger, H. E., Borchard, G., Ottenhoff, T. H. M., Geluk, A. (2004). *Vaccine*, **22**, 1609–1615.

Borchard, G., Esmaeili, F., Heuking, S. (2012). Chitosan-based delivery systems for mucosal vaccination. In: Sarmento, B., das Neves, J., eds. *Chitosan-Based Systems for Biopharmaceuticals*, Wiley, UK, pp. 211–224.

Bueter, C. L., Lee, C. K., Rathinam, V. A. K., Healy, G. J., Taron, C. H., Specht, C. A., Levitz, S. M. (2011). *J. Biol. Chem.*, **286**, 35447–35455.

Buonaguro, F., Buonaguro, L. (2013). *Expert Rev. Vaccines*, **12**, 99–99.

Cai, S., Yang, Q., Bagby, T. R., Forrest, M. L. (2011). *Adv. Drug Deliv. Rev.*, **63**, 901–908.

Chakravarthi, S. S., Robinson, D. H., De, S. (2007). Nanoparticles prepared using natural and synthetic polymers. In: Thassu, D., Deleers, M., Pathak, Y., eds. *Nanoparticulate Drug Delivery Systems*, CRC Press, USA, Vol. 166, pp. 51–60.

Cho, W.-S., Duffin, R., Poland, C. A., Duschl, A., Oostingh, G. J., MacNee, W., Bradley, M., Megson, I. L., Donaldson, K. (2012b). *Nanotoxicology*, **6**, 22–35.

Chong, C. S. W., Cao, M., Wong, W. W., Fischer, K. P., Addison, W. R., Kwon, G. S., Tyrrell, D. L., Samuel, J. (2005). *J. Control. Release*, **102**, 85–99.

Chua, B. Y., Al Kobaisi, M., Zeng, W., Mainwaring, D., Jackson, D. C. (2011). *Mol. Pharm.*, **9**, 81–90.

Colombo, M., Carregal-Romero, S., Casula, M. F., Gutierrez, L., Morales, M. P., Bohm, I. B., Heverhagen, J. T., Prosperi, D., Parak, W. J. (2012). *Chem. Soc. Rev.*, **41**, 4306–4334.

Cruz, L. J., Tacken, P. J., Rueda, F., Domingo, J. C., Albericio, F., Figdor, C. G. (2012). *Methods Enzymol.*, **509**, 143–163.

Deng, Z. J., Liang, M., Toth, I., Monteiro, M., Minchin, R. F. (2012). *Nanotoxicology*, **7**(3), 314–322.

Dobrovolskaia, M. A., Aggarwal, P., Hall, J. B., McNeil, S. E. (2008). *Mol. Pharm.*, **5**, 487–495.

Doroud, D., Rafati, S. (2011). *Expert Rev. Vaccines*, **11**, 69–86.

Elechiguerra, J., Burt, J., Morones, J., Camacho-Bragado, A., Gao, X., Lara, H., Yacaman, M. (2005). *J. Nanobiotechnol.*, **3**, 6.

Fifis, T., Gamvrellis, A., Crimeen-Irwin, B., Pietersz, G. A., Li, J., Mottram, P. L., McKenzie, I. F. C., Plebanski, M. (2004). *J. Immunol.*, **173**, 3148–3154.

Fiore, V. F., Lofton, M. C., Roser-Page, S., Yang, S. C., Roman, J., Murthy, N., Barker, T. H. (2010). *Biomaterials*, **31**, 810–817.

Goebel, C., Kubickamuranyi, M., Tonn, T., Gonzalez, J., Gleichmann, E. (1995). *Arch. Toxicol.*, **69**, 450–459.

Gutierro, I., Hernández, R. M., Igartua, M., Gascón, A. R., Pedraz, J. L. (2002). *Vaccine*, **21**, 67–77.

Harush-Frenkel, O., Rozentur, E., Benita, S., Altschuler, Y. (2008). *Biomacromolecules*, **9**, 435–443.

Henriksen-Lacey, M., Devitt, A., Perrie, Y. (2011). *J. Control. Release*, **154**, 131–137.

Heuking, S., Borchard, G. (2012). *J. Pharm. Sci.*, **101**, 1166–1177.

Huang, Y., Yu, F., Park, Y.-S., Wang, J., Shin, M.-C., Chung, H. S., Yang, V. C. (2010). *Biomaterials*, **31**, 9086–9091.

Immunity, G. (2013). *Single DermaVir Immunization in HIV-1 Infected Patients on HAART (GIHU004)*, NCT00712530, May 6, 2013, Available from: https://clinicaltrials.gov/ct2/show/NCT00712530 (accessed on March 6, 2017).

Joshi, V., Geary, S., Salem, A. (2012). *AAPS J.*, **15**, 85–94.

Kaba, S. A., McCoy, M. E., Doll, T. A. P. F., Brando, C., Guo, Q., Dasgupta, D., Yang, Y., Mittelholzer, C., Spaccapelo, R., Crisanti, A., Burkhard, P., Lanar, D. E. (2012). *PLoS One*, **7**, 1–11.

Kalvodova, L. (2010). *Biochem. Biophys. Res. Commun.*, **393**, 350–355.

Kanchan, V., Panda, A. K. (2007). *Biomaterials*, **28**, 5344–5357.

Katare, Y. K., Muthukumaran, T., Panda, A. K. (2005). *Int. J. Pharm.*, **301**, 149–160.

Kim, J. S., Adamcak, A. K. (2005). *Int. J. Pharm.*, **301**, 149–160.

Kim, J. S., Adamcakova-Dodd, A., O'Shaughnessy, P., Grassian, V., Thorne, P. (2011). *Part. Fibre Toxicol.*, **8**, 29.

Kobiasi, M. A., Chua, B. Y., Tonkin, D., Jackson, D. C., Mainwaring, D. E. (2012). *J. Biomed. Mater. Res. Part A*, **100A**, 1859–1867.

Kozutsumi, D., Tsunematsu, M., Yamaji, T., Murakami, R., Yokoyama, M., Kino, K. (2006). *Immunology*, **118**, 392–401.

Kwon, Y. J., Standley, S. M., Goh, S. L., Fréchet, J. M. J. (2005). *J. Control. Release*, **105**, 199–212.

Lisziewicz, J., Bakare, N., Calarota, S. A., Bánhegyi, D., Szlávik, J., Újhelyi, E., Tőke, E. R., Molnár, L., Lisziewicz, Z., Autran, B., Lori, F. (2012). *PLoS One*, **7**, e35416.

Lisziewicz, J., Calarota, S. A., Lori, F. (2007). *Expert Opin. Biol. Ther.*, **7**, 1563–1574.

Lorentzen, J. C. (1999). *Scand. J. Immunol.*, **49**, 45–50.

Luz Garcia-Hernandez, M., Gray, A., Hubby, B., Klinger, O. J., Kast, W. M. (2008). *Cancer Res.*, **68**, 861–869.

Mann, J. F. S., Shakir, E., Carter, K. C., Mullen, A. B., Alexander, J., Ferro, V. A. (2009). *Vaccine*, **27**, 3643–3649.

Manolova, V., Flace, A., Bauer, M., Schwarz, K., Saudan, P., Bachmann, M. F. (2008). *Eur. J. Immunol.*, **38**, 1404–1413.

Maquieira, Á., Brun, E. M., Garcés-García, M., Puchades, R. (2012). *Anal. Chem.*, **84**, 9340–9348.

Moghimi, S. M., Andersen, A. J., Ahmadvand, D., Wibroe, P. P., Andresen, T. L., Hunter, A. C. (2011). *Adv. Drug Deliv. Rev.*, **63**, 1000–1007.

Moghimi, S. M., Hunter, A. C., Andresen, T. L. (2012). *Annu. Rev. Pharmacol. Toxicol.*, **52**, 481–503.

Moingeon, P., de Taisne, C., Almond, J. (2002). *Br. Med. Bull.*, **62**, 29–44.

Montomoli, E., Piccirella, S., Khadang, B., Mennitto, E., Camerini, R., De Rosa, A. (2011). *Expert Rev. Vaccines*, **10**, 1053–1061.

Moon, J. J., Suh, H., Bershteyn, A., Stephan, M. T., Liu, H., Huang, B., Sohail, M., Luo, S., Ho Um, S., Khant, H., Goodwin, J. T., Ramos, J., Chiu, W., Irvine, D. J. (2011). *Nat. Mater.*, **10**, 243–251.

Morral-Ruiz, G., Melgar-Lesmes, P., Solans, C., Garcia-Celma, M. J. (2013). *J. Control. Release*, **171**, 163–171.

Mosca, F., Tritto, E., Muzzi, A., Monaci, E., Bagnoli, F., Iavarone, C., O'Hagan, D., Rappuoli, R., De Gregorio, E. (2008). *Proc. Natl. Acad. Sci. U. S. A.*, **105**, 10501–10506.

Nagamoto, T., Hattori, Y., Takayama, K., Maitani, Y. (2004). *Pharm. Res.*, **21**, 671–674.

Nagata, T., Koide, Y. (2010). *J. Biomed. Biotechnol.*, **2010**, 1–11.

Nathan, C. F., Hibbs Jr, J. B. (1991). *Curr. Opin. Immunol.*, **3**, 65–70.

Neimert-Andersson, T., Hällgren, A.-C., Andersson, M., Langebäck, J., Zettergren, L., Nilsen-Nygaard, J., Draget, K. I., van Hage, M., Lindberg, A., Gafvelin, G., Grönlund, H. (2011). *Vaccine*, **29**, 8965–8973.

Nie, S. (2010). *Nanomed.*, **5**, 523–528.

Novavax (2013). *A Study to Evaluate the Immune Response and Safety of a Seasonal Virus-Like Particle Influenza Vaccine in Healthy Young Adults*, NCT01561768, May 8, 2013, Available from: https://clinicaltrials.gov/

ct2/show/NCT01561768?term=NCT01561768&rank=1 (accessed on March 6, 2017).

Oussoren, C., Storm, G. (2001). *Adv. Drug Deliv. Rev.*, **50**, 143–156.

Oyewumi, M. O., Kumar, A., Cui, Z. (2010). *Expert Rev. Vaccines*, **9**, 1095–1107.

Parham, P. (2009). *The Immune System*, 3rd ed., Garland Science, Taylor & Francis Group, LLC, USA.

Park, J., Lim, D.-H., Lim, H.-J., Kwon, T., Choi, J.-S., Jeong, S., Choi, I.-H., Cheon, J. (2011). *Chem. Commun.*, **47**, 4382–4384.

Paul, W. E. (2008). *Fundamental Immunology*, 6th ed., Lippincott Williams and Wilkins, USA.

Pearce, R. C., Railsback, J. G., Anderson, B. D., Sarac, M. F., McKnight, T. E., Tracy, J. B., Melechko, A. V. (2013). *ACS Appl. Mater. Interfaces*, **5**, 878–882.

Pearse, M. J., Drane, D. (2005). *Adv. Drug Deliv. Rev.*, **57**, 465–474.

Pignataro, B. (2010). *Ideas in Chemistry and Molecular Sciences: Advances in Nanotechnology, Materials and Devices*, 1st ed., Wiley, Weinheim.

Plotkin, S. A., Orenstein, W. A., Offit, P. A. (2013). *Vaccines*, 6th ed., Elsevier Saunders, Edinburgh.

PowderMed (2007). *A Safety and Immunology Study of a DNA Trivalent Influenza Vaccine*, NCT00375206, May 6, 2013, Available from: https://clinicaltrials.gov/ct2/show/NCT00375206?term=NCT00375 206&rank=1 (accessed on March 7, 2017).

PowderMed (2008). *Safety Study of HBV DNA Vaccine to Treat Patients With Chronic Hepatitis B Infection*, NCT00277576, May 6, 2013, Available from: https://clinicaltrials.gov/ct2/show/NCT00277576?term=NCT 00277576&rank=1 (accessed on March 7, 2017).

Prieur, X., Rőszer, T., Ricote, M. (2010). *Biochim. Biophys. Acta (BBA): Mol. Cell Biol. Lipids*, **1801**, 327–337.

Pusic, K., Aguilar, Z., McLoughlin, J., Kobuch, S., Xu, H., Tsang, M., Wang, A., Hui, G. (2012). *FASEB J.*, **27**, 1153–1166.

Rajapaksa, T. E., Lo, D. D. (2010). *Curr. Immunol. Rev.*, **6**, 29–37.

Reddy, S. T., Rehor, A., Schmoekel, H. G., Hubbell, J. A., Swartz, M. A. (2006). *J. Control. Release*, **112**, 26–34.

Reddy, S. T., van der Vlies, A. J., Simeoni, E., Angeli, V., Randolph, G. J., O'Neil, C. P., Lee, L. K., Swartz, M. A., Hubbell, J. A. (2007). *Nat. Biotechnol.*, **25**, 1159–1164.

Rogers, T. H., Babensee, J. E. (2011). *Biomaterials*, **32**, 1270–1279.

Romagnani, S. (1999). *Inflamm. Bowel Dis.*, **5**, 285–94.

Roy, M. J., Wu, M. S., Barr, L. J., Fuller, J. T., Tussey, L. G., Speller, S., Culp, J., Burkholder, J. K., Swain, W. F., Dixon, R. M., Widera, G., Vessey, R., King, A., Ogg, G., Gallimore, A., Haynes, J. R., Heydenburg Fuller, D. (2000). *Vaccine*, **19**, 764–778.

Salvador, A., Igartua, M., Hernández, R. M., Pedraz, J. L. (2011). *J. Drug Deliv.*, **2011**, 1–23.

Samuelsen, M., Nygaard, U. C., Løvik, M. (2009). *Scand. J. Immunol.*, **69**, 421–428.

Shvedova, A. A., Fabisiak, J. P., Kisin, E. R., Murray, A. R., Roberts, J. R., Tyurina, Y. Y., Antonini, J. M., Feng, W. H., Kommineni, C., Reynolds, J., Barchowsky, A., Castranova, V., Kagan, V. E. (2008). *Am. J. Respir. Cell Mol. Biol.*, **38**, 579–590.

Smith, G., Raghunandan, R., Wu, Y., Liu, Y., Massare, M., Nathan, M., Zhou, B., Lu, H., Boddapati, S., Li, J., Flyer, D., Glenn, G. (2012). *PLoS One*, **7**, e50852.

Speiser, D. E., Liénard, D., Rufer, N., Rubio-Godoy, V., Rimoldi, D., Lejeune, F., Krieg, A. M., Cerottini, J.-C., Romero, P. (2005). *J. Clin. Invest.*, **115**, 739–746.

Speiser, D. E., Schwarz, K., Baumgaertner, P., Manolova, V., Devevre, E., Sterry, W., Walden, P., Zippelius, A., Conzett, K. B., Senti, G. (2010). *J. Immunother.*, **33**, 848–858.

Springer, C., Benita, S., Sherman, Y., Gursoy, N., Gilhar, D., Avital, A. (2005). *Pediatr. Res.*, **58**, 537–541.

Stano, A., Nembrini, C., Swartz, M. A., Hubbell, J. A., Simeoni, E. (2012). *Vaccine*, **30**, 7541–7546.

Tabata, Y., Ikada, Y. (1988). *Biomaterials*, **9**, 356–362.

Thiele, L., Diederichs, J. E., Reszka, R., Merkle, H. P., Walter, E. (2003). *Biomaterials*, **24**, 1409–1418.

Tokura, S., Hiroshi, T., Ichiro, A. (1999). Immunological aspects of chitin and chitin derivatives administered to animals. In: Jollès, P., ed. *Chitin and Chitinases*, Birkhäuser, Basel, Vol. **87**, pp. 279–292.

van Zijverden, M., Granum, B. (2000). *Toxicology*, **152**, 69–77.

Vasir, J. K., Labhasetwar, V. (2007). Nanoparticles for gene delivery: formulation characteristics. In: *Nanoparticulate Drug Delivery Systems*, pp. 291–304.

Verma, A., Stellacci, F. (2010). *Small*, **6**, 12–21.

Villemejane, J., Mir, L. M. (2009). *Br. J. Pharmacol.*, **157**, 207–219.

Wilson, N. S., Yang, B., Morelli, A. B., Koernig, S., Yang, A., Loeser, S., Airey, D., Provan, L., Hass, P., Braley, H., Couto, S., Drane, D., Boyle, J., Belz, G. T., Ashkenazi, A., Maraskovsky, E. (2012). *Immunol. Cell Biol.*, **90**, 540–552.

Xiang, S. D., Scholzen, A., Minigo, G., David, C., Apostolopoulos, V., Mottram, P. L., Plebanski, M. (2006). *Methods*, **40**, 1–9.

Xie, Y., Bagby, T. R., Cohen, M., Forrest, M. L. (2009). *Expert Opin. Drug Deliv.*, **6**, 785–792.

Yoon, K.-Y., Hoon Byeon, J., Park, J.-H., Hwang, J. (2007). *Sci. Total Environ.*, **373**, 572–575.

Yoshida, M., Mata, J., Babensee, J. E. (2007). *J. Biomed. Mater. Res. Part A*, **80A**, 7–12.

Yue, H., Wei, W., Yue, Z., Lv, P., Wang, L., Ma, G., Su, Z. (2010). *Eur. J. Pharm. Sci.*, **41**, 650–657.

Zahr, A. S., Davis, C. A., Pishko, M. V. (2006). *Langmuir*, **22**, 8178–8185.

Chapter 4

Site-Specific Antibody Conjugation for ADC and Beyond

Qun Zhou, PhD

Protein Engineering, Biologics Research, Sanofi, Framingham, Massachusetts, USA

Keywords: site-specific conjugation, specific amino acids, unnatural amino acids, glycans, short peptide tags, ADC, antibody–drug conjugates, acute myelogenous leukemia, acute lymphoblastic leukemia, monomethyl auristatin E, pyrrolobenzodiazepine, antibody-antibiotic conjugate, microbial transglutaminase, strain-promoted azide-alkyne cycloaddition, aminoacyl-tRNA synthetase, p-acetylphenylalanine, pyrrolysyl-tRNA synthetase, p-azidomethyl-L-phenylalanine, formylglycine-generating enzyme, optical imaging, near-infrared fluorescent, peripheral blood mononuclear cells, antibody–polymer conjugation, human epidermal growth factor receptor 2, THIOMAB, Sec incorporation sequence, positron emission tomography, prostate-specific membrane antigen, bispecific antibody, C-type lectin-like molecule-1

4.1 Introduction

Therapeutic antibodies have become a major class of biologics in treating many challenging diseases, including cancer. Due to their high specificity and affinity, monoclonal antibodies have

---

Immune Aspects of Biopharmaceuticals and Nanomedicines
Edited by Raj Bawa, János Szebeni, Thomas J. Webster, and Gerald F. Audette
Copyright © 2018 Pan Stanford Publishing Pte. Ltd.
ISBN 978-981-4774-52-9 (Hardcover), 978-0-203-73153-6 (eBook)
www.panstanford.com

less off-target side effects compared to small-molecular-weight compounds. Currently, many monoclonal antibodies are under accelerated clinical developments [1]. It was predicted that by 2020 about 70 of them will be approved by regulatory agencies and used in clinics in treating patients [2]. Besides the traditional immunoglobulin G (IgG), many different antibody formats are being developed to enhance therapeutic efficacy, such as antibody-drug conjugates (ADCs) and bispecific antibodies [3–11]. ADCs contain both antibody and coupled cytotoxins or chemotherapeutic drugs. They combine the advantages of high affinity and specificity related to antibody with those of small molecular drugs, such as the high penetration toward intracellular targets for inducing apoptosis.

As one of the targeted therapies or the "magic bullet" proposed by Paul Ehrlich more than a century ago [12], the ADC is the targeted delivery of a highly cytotoxic drug for selective (as opposed to systemic) chemotherapy, resulting in an improved therapeutic index and enhanced efficacy relative to traditional chemotherapies or naked monoclonal antibodies. Thus, this unique format would allow the use of certain chemotherapeutic drugs that are too potent or toxic to be applied systemically. Since these cytotoxins need to be delivered into the tumor cells targeting intracellular microtubules or DNA, the chemotherapeutic drugs carrying antibodies should be efficiently internalized into lysosomes once binding to the antigen on the tumor surface (Fig. 4.1). Within the lysosome, cytotoxins would be efficiently released from antibodies through proteolytic cleavage or disulfide reduction before they diffuse into the cytoplasm for cytotoxicity. Although the ADC idea has been around for decades, the medicine used in clinics was not available until 2000 with the regulatory approval of the first ADC, gemtuzumab ozogamicin (anti-CD33 ADC). Currently, four ADCs have been approved by regulatory agencies for cancer treatment: gemtuzumab ozogamicin (anti-CD33 ADC) for acute myelogenous leukemia (AML), brentuximab vedotin (anti-CD30 ADC) for treating anaplastic large cell lymphoma/Hodgkin's lymphoma, trastuzumab emtansine (Anti-HER2 ADC) for advanced HER2 (human epidermal growth factor receptor 2)-positive breast cancer, and inotuzumab ozogamicin (anti-CD22 ADC) for the treatment of relapsed or refractory acute lymphoblastic leukemia (ALL) [13–22].

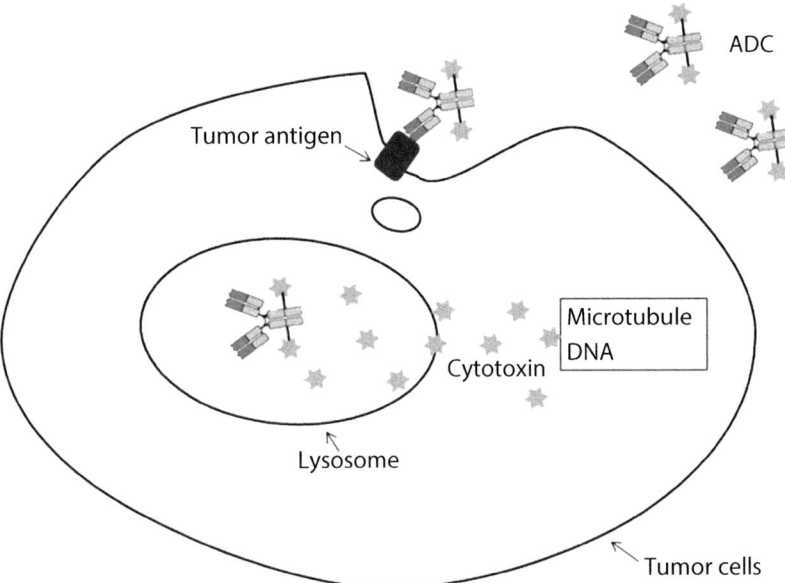

Figure 4.1 Binding and internalization of an antibody–drug conjugate (ADC) followed by the release of its cytotoxins inside the cell.

All these ADCs are prepared using conventional conjugation methods that couple antibodies through either surface-exposed lysines (~70 to 90) or cysteines from interchain disulfides (8 in IgG1) after partial reduction. These lysine and cysteine methods generate heterogeneous products with varied numbers of drugs coupled across several possible sites although the average drug-to-antibody ratio (DAR) appears around 4 to 6, creating significant challenges for process consistency and product characterization [23–25]. There is also limited understanding of the relationships between the site/extent of drug loading and ADC attributes such as efficacy, safety, pharmacokinetics, and immunogenicity. The non-specific conjugation would also result in more off-target side effects, leading to relatively low maximum tolerated dose. To improve the technology aiming for homogeneous molecules with higher therapeutic index, several site-specific ADC technologies have been developed as next-generation methods [26].

In contrast to the conventional conjugations through lysines or cysteines, which are abundant in an antibody, the site-specific conjugations couple the antibodies with cytotoxins through the unique and defined sites based on antibody engineering. There are four major categories of methods based on the conjugation sites in the antibody molecules: specific amino acids, unnatural amino acids, short peptide tags, or glycans (Fig. 4.2 and Table 4.1).

Table 4.1 The four categories of the site-specific antibody–drug conjugation (ADC)

Technologies	Specific amino acids	Unnatural amino acids	Glycans	Short peptide tags
Conjugation sites	C, Q	pAcF, pAMF, Sec, etc.	SA, Gal, Fuc, etc.	LLQG, LCxPxR, etc.
Cell line engineering	–	+	–	±
Metabolic labeling	–	+	±	–
In vivo protein engineering	+	+	–	+
In vitro enzymatic modification	±	±	+	±
Chemical modification	+	–	±	–
Selected references	[27, 28, 29, 30, 31, 32, 33, 34, 35]	[36, 37, 38, 39, 40, 41]	[42, 43, 44, 45, 46, 47, 48, 49]	[50, 51, 52, 53]

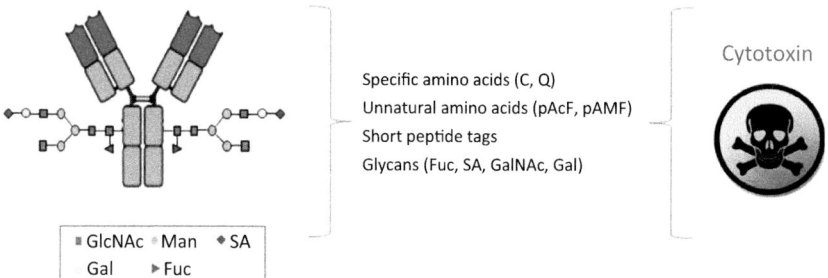

Specific amino acids (C, Q)
Unnatural amino acids (pAcF, pAMF)
Short peptide tags
Glycans (Fuc, SA, GalNAc, Gal)

Cytotoxin

GlcNAc Man SA
Gal Fuc

A monoclonal antibody IgG

Figure 4.2 The categories of the site-specific ADC with cytotoxin coupled at unique and defined sites in an antibody molecule.

4.2 Site-Specific ADC through Specific Amino Acids

Several native or engineered amino acids, including cysteines and glutamines, are selected as the sites for conjugation.

The unpaired cysteine-mediated conjugation was the first described site-specific ADC as THIOMAB or TDC [27]. A cysteine residue was engineered into different positions of the antibody heavy chain (HC) or light chain (LC) for coupling. Since engineered cysteines are always capped with glutathione or something else during expression, the antibodies need to be partially reduced to remove the cap. The uncapped cysteines were then coupled with thiol-reactive linkers containing cytotoxins, such as monomethyl auristatin E (MMAE), using thiol-maleimide chemistry. The ADCs generated through drug-linker conjugation at the cysteine residue (HC-A114C) appeared to be nearly homogeneous conjugates with an improved therapeutic index. The anti-MUC16 TDC displayed a twofold improved *in vivo* efficacy in a mouse xenograft model of ovarian cancer over the same MMAE containing ADC prepared using the conventional cysteine method at an equivalent drug (or cytotoxin) dose, while both TDC (DAR at 1.6) and conventional ADC (DAR at 3.1) had similar efficacy at an equivalent antibody dose. The TDC was tolerated at higher doses in rats and cynomolgus monkeys than the conventional ADC. Similar results were obtained with anti-HER2 TDC prepared with another tubulin inhibitor containing the drug-linker, mertansine (DM1), compared to the same DM1 containing ADC prepared using the conventional lysine method [28]. In a safety study with cynomolgus monkey, the TDC was tolerated at a higher antibody dose than the conventional ADC (48 mg/kg vs. 30 mg/kg). The THIOMAB approach was also applied to other antibodies, such as anti-CD70, with a cytotoxic DNA cross-linking pyrrolobenzodiazepine (PBD) linker coupled at an engineered cysteine (S239C) in HC [29]. The TDC showed low aggregation, high stability in plasma, and strong *in vivo* and *in vitro* antitumor activities.

Moreover, cysteine insertion was also described in the antibody before and after the selected sites in either IgG HC or LC, including LC-V205, HC-A114, and HC-S239 [30]. There was no difference in conjugation efficiency between cysteine-inserted and cysteine-substituted antibodies in coupling to a PBD drug-linker. The ADCs prepared with a drug-linker coupled through the cysteine-insertion after site HC-S239 of anti-5T4 (trophoblast glycoprotein, an oncofetal antigen on a breast tumor) demonstrated potent dose-dependent antitumor activity in a mouse xenograft model [30].

In addition to conjugation through unpaired cysteine, thiol bridge methods were developed to conjugate a bifunctional drug-linker to both cysteines from each of the interchain disulfide, instead of conjugation of drug-linker to only one cysteine using the conventional method, after partial reduction and re-oxidation [31, 32]. Thus, four drugs were coupled to eight cysteines from four interchain disulfides in IgG1 with low heterogeneity. In a method developed, a reduction bis-alkylation approach was applied to rebridge the reduced interchain disulfide bonds in the antibody [31]. Anti-HER2 was conjugated with bisAlk-vc-MMAE with DAR at 4 as a major product. There was improved stability of the ADC prepared using this approach compared to the ADC prepared using the conventional cysteine method. The ADC also showed higher *in vivo* antitumor efficacy than trastuzumab-DM1 prepared using the conventional lysine approach. Behrens et al. reported the conjugation of the antibody with a bifunctional dibromomaleimide (DBM) linker instead of a conventional maleimide linker [32]. They found ~70% of the DBM-MMAF derivative cross-linked both cysteines from interchain disulfides, while 30% of the drug-linker was half of the antibody conjugate due to intrachain cross-linking of cysteines originally from interchain disulfides. The ADCs demonstrated improved pharmacokinetics and reduced toxicity *in vivo* compared to analogous conventional cysteine ADCs. Anti-HER2 ADC prepared using this method showed better *in vivo* efficacy than the ADC generated using conventional cysteine conjugation. It significantly delayed the tumor growth, while treatment with the conventional ADC did not result in any significant inhibition at a similar dose in a mouse xenograft model of ovarian cancer.

The site-specific conjugation through glutamine was also reported [33, 34]. Instead of using reducing and oxidizing reagents, the method was designed by using microbial transglutaminase (mTG) to transfer an amine containing drug-linker or a reactive spacer into HC-Q295 in a deglycosylated antibody. The conjugation was optimized using a two-step chemoenzymatic approach whereby a reactive spacer containing a bioorthogonal azido or thiol functional linker was attached to the antibody by mTG and subsequently reacted with either dibenzocyclooctynes (DBCO) or maleimide containing MMAE. By using strain-promoted azide-alkyne cycloaddition (SPAAC) or thiol-maleimide chemistry, homogeneous ADCs were generated with DAR at ~2. The Anti-HER2-MMAE showed strong *in vitro* potency against tumor cells.

The site-specific conjugation through unpaired cysteine is relatively simple and scalable. The drug coupling is done without the need of special reagents. The method has been applied to prepare multiple site-specific ADCs for preclinical or clinical developments. As described above, the ADCs prepared through site-specific cysteines showed twofold stronger *in vivo* antitumor activities and were better tolerated than the conventional conjugates. However, the stability of the conjugate generated through unpaired cysteine varied depending on where the cysteine was introduced in an antibody molecule [35]. It appears that the TDC prepared with conjugation at highly solvent-accessible sites in a relatively neutral environment (HC-S396C) rapidly lost the drug in plasma due to maleimide exchange with other reactive thiols, such as albumin, glutathione, or cysteine. The TDC prepared by conjugating drug-linker to a partially accessible site with a positively charged environment (LC-V205C) showed the least thiol maleimide exchange, while the conjugate at the site with partial solvent-accessibility and neutral charge (HC-A114C) displayed the intermediate drug stability in plasma.

4.3 Site-Specific ADC through Unnatural Amino Acids

The coupling of drug-linker to unnatural amino acid residues in the antibody is another approach for site-specific conjugation [36]. An orthogonal amber suppressor tRNA/aminoacyl-tRNA synthetase (aaRS) pair from *Methanococcus jannaschii* was expressed in the cells to site-specifically incorporate *p*-acetylphenylalanine (pAcF) in response to an amber nonsense codon engineered in the antibodies. The keto group of pAcF was then selectively coupled to a drug-linker with aminooxy functionality. The pAcF-containing anti-HER2 was produced from Chinese hamster ovary (CHO) cells co-expressing antibody with an amber codon at the heavy chain (HC-A121) and an orthogonal amber suppressor tRNA/aaRS pair. The pAcF residue was coupled with a non-cleavable auristatin linker, which is a potent tubulin inhibitor, through oxime chemistry with DAR at ~2. The anti-HER2 ADC showed strong *in vitro* and *in vivo* antitumor activities. The method was further optimized, and a stable CHO cell line was generated to express an orthogonal amber suppressor tRNA/aaRS pair [37]. Multiple pAcF containing antibodies, including anti-5T4, anti-EGFR (epidermal growth factor receptor), anti-HER2, and anti-PSMA (prostate-specific membrane antigen), were expressed in the cell line with titers over 1 g/L in fed-batch processes. The unnatural amino acid was introduced into different sites in these antibodies, such as HC-S115 for anti-5T4 or HC-A114 for anti-HER2. It was then conjugated with the drug-linker, aminooxy containing monomethyl auristatin D (MMAD), with DAR at ~2. Interestingly, the ADC prepared with this site-specific technology showed superior *in vivo* antitumor efficacy compared to the ADC prepared using either conventional conjugation through hinge disulfides or site-specific conjugation through unpaired cysteine as described above. Although all three ADCs caused complete regression of the tumor at 3 mg/kg dose, only anti-HER2 HC-A114pAcF ADC showed complete tumor regression at 1 mg/kg in a mouse tumor xenograft established from breast cancer cell lines [37, 38].

CHO cells expressing the pyrrolysyl-tRNA synthetase (pylRS) and its cognate tRNA (tRNA pyl) were also generated to genetically encode an unnatural amino acid containing an azido moiety in response to an amber stop codon [39]. Anti-HER2 antibodies were expressed from the engineered cells to contain N6-((2-azidoethoxy)carbonyl)-L-lysine at four different positions in either the heavy or light chain for DAR at 2 and a combination of two sites in both the heavy and light chains for DAR at 4. The azido group introduced at position HC-H274 of the antibody enabled click cycloaddition chemistry that generated a stable heterocyclic triazole linkage of the toxin auristatin F or PBD with over 95% conjugation efficiency. The ADCs were potent and specific in *in vitro* cytotoxicity assays. They demonstrated stability *in vivo* and a PBD containing ADC with DAR of 1.8 showed similar efficacy in the sustained regression of tumor growth in a mouse tumor xenograft model compared to the ADC with DAR of 3 prepared using conventional cysteine conjugation.

In addition, a cell-free expression system was established to produce ADCs through site-specific incorporation of the optimized unnatural amino acid, *p*-azidomethyl-L-phenylalanine (pAMF) [40]. A novel variant of the *Methanococcus jannaschii* tyrosyl tRNA synthetase (TyrRS) was discovered through library screening with a high activity and specificity toward pAMF. The site-specific incorporation of pAMF at HC-S136 of anti-HER2 facilitated near complete conjugation of a DBCO-PEG-monomethyl auristatin F (DBCO-PEG-MMAF) through SPAAC using copper-free click chemistry. The resultant ADCs showed *in vitro* antitumor potency.

The ADC was also site-specifically generated using selenocysteine (Sec) residues engineered at the C-terminus of the antibody with iodoacetamide containing monomethyl auristatin F (MMAF) [41]. In eukaryotes, Sec is encoded by the stop codon UGA, and its translational incorporation requires the presence of a Sec incorporation sequence (SECIS) in the UTR of the mRNA. Since the selenol group of Sec is more nucleophilic than the thiol group of cysteine, the antibody was conjugated under mildly acidic and reducing conditions without the antibody re-oxidation required for THIOMAB conjugation. The ADC showed strong antitumor activities *in vitro* and *in vivo*. Significant tumor growth inhibition and regression were observed for the anti-HER2 scFv-Fc-Sec conjugate. Four of the five mice treated with the ADC at a high dose were tumor free at six weeks after the last treatment.

The ADCs prepared through conjugation of drug-linkers to the unnatural amino acids were more efficacious *in vivo* than the conjugates generated using conventional or THIOMAB methods. The unnatural amino acid–containing antibodies were expressed in a bioreactor at the gram scale, and the conjugates were stable. However, cell line engineering is required for optimal expression of an orthogonal amber suppressor tRNA/aaRS pair in addition to antibody engineering. The potential immunogenicity of these unnatural amino acids containing the antibody in a human is currently unknown.

4.4 Site-Specific ADC through Glycans

The glycan-mediated conjugation provides a unique site-specific method by conjugating the drug-linker to N297 glycans located on the CH2 domain instead of coupling the relatively hydrophobic cytotoxins directly into amino acid residues. Since there are several different monosaccharides present at the non-reducing terminus of the glycans, various approaches were developed to conjugate the drug-linkers to these sugars, including fucose, galactose, N-acetylgalactosamine (GalNAc), N-acetylglucosamine (GlcNAc), and sialic acid (SA). It was reported by Okeley et al. that 6-thiofucose, a fucose analogue, could be metabolically incorporated into anti-CD30 or anti-CD70 antibodies. The thiofucose in the antibodies was then conjugated with a maleimide containing MMAE drug-linker with DAR at ~1.3 [42]. The ADCs generated through the thiofucose maintained good plasma stability and showed strong *in vitro* antitumor activity.

The galactose or GalNAc analogues were also introduced *in vitro* by a mutant galactosyltransferase, GalT (Y289L) which was discovered by Qasba et al. [43, 44], for preparing site-specific ADCs with a DAR close to 2. The ADC, which was generated using C-2 keto galactose labeling and conjugation with aminooxy containing auristatin F, showed strong *in vitro* antitumor potency [45]. The same enzyme, GalT (Y289L), was also used to introduce azido-GalNAc to core GlcNAc exposed after pre-treatment of antibodies with endoglycosidases, such as endo F3, endo S, or endo S2, which release most of N-297 glycans except for the innermost GlcNAc [46]. The drug-linkers, such as bicyclononyne (BCN) containing MMAE, MMAF, maytansine, or doxorubicin, were efficiently conjugated to the introduced azido-GalNAc through SPAAC using copper-free click chemistry. The process was scaled up for preparing 5 g ADCs with excellent homogeneity. The anti-HER2 ADCs prepared using this method with BCN containing maytansine or MMAF showed better *in vivo* antitumor efficacy than the anti-HER2 coupled with DM1 using a conventional lysine conjugation approach. In a breast cancer xenograft mouse model, anti-HER2 glycoconjugate containing a cleavable or noncleavable maytansine linker demonstrated complete tumor regression, while both the MMAE containing glycoconjugate and conventional lysine conjugate showed partial tumor suppression.

Moreover, sialic acid (SA) was also used as a site for conjugation. SA was first transferred to the antibody before being oxidized with periodate for conjugation to aminooxy-containing drug-linkers [47]. Several different antibody ADCs were prepared with two different drug-linkers. Anti-HER2 ADC prepared using this approach showed strong *in vitro* and *in vivo* antitumor activities. A similar method was developed by transferring C9-azido-modified SA into an antibody for conjugation with DBCO containing doxorubicin using copper-free click chemistry [48]. Finally, a report showed the use of a mutant endoglycosidase in preparing site-specific ADC containing homogeneous glycans [49].

The glycoengineering approach is unique in that the drug-linkers are coupled to glycans without a need to engineer the amino acid sequence and they are coupled far away from the amino acid residues. It was demonstrated in a study as described above that the ADC made through glycoconjugation was more efficacious *in vivo* than conventional lysine conjugate. However, the method needs special reagents and enzymes for glycoengineering.

4.5 Site-Specific ADC through Short Peptide Tags

There are several site-specific conjugation methods being developed through coupling of cytotoxins to specific short peptide tags that contain four to six amino acid residues.

Strop et al. engineered a glutamine tag (LLQG) into an antibody molecule, so the glutamine in the tag can be recognized by mTG for transferring an amine containing drug [50]. Twelve out of 90 sites in an anti-EGFR antibody were found to be efficiently conjugated with mTG. They demonstrated that the drug-linker, MMAD, was efficiently transferred by the enzyme to the glutamine tags, including LLQGA in the C-terminal heavy chain or GGLLQGA in the C-terminal light chain. The ADCs generated were homogeneous with DAR at ~2 and they showed strong *in vitro* and *in vivo* antitumor activities. Interestingly, the ADC with the drug conjugated at the antibody light chain had better pharmacokinetics and serum stability than those at the heavy chain in a species-dependent manner probably due to a different mechanism other than the chemical instability associated with cysteine conjugates.

Another conjugation platform was also reported by using sortase A-mediated transpeptidation reaction, generating ADCs with cytotoxins coupled to pre-defined sites [51]. This method includes C-terminal modification of antibody heavy and light chains with sortase A recognition motif, LPETG, which make them suitable substrates for sortase A-mediated transpeptidation of a pentaglycine peptide containing either MMAE or maytansine. Anti-CD30 ADC containing MMAE and anti-HER2 ADC containing maytansine were generated with sortase A from *S. aureus*. The ADCs displayed strong *in vitro* and *in vivo* antitumor activities similar to the ADCs generated using the conventional approaches.

Wu et al. described a site-specific antibody conjugation using a genetically encoded aldehyde tag [52, 53]. A short peptide tag, LCxPxR, which was recognized by formylglycine-generating enzyme (FGE), was inserted into the N or C terminal region of the antibodies. After co-expression of the antibody with FGE, the cysteine in the short peptide tag was oxidized by the enzyme inside the cells to aldehyde-bearing formylglycine, which can be coupled with aminooxy-functionalized reagents. Although the efficiency of conversion of cysteine to formylglycine varied among different locations of the inserted short peptide tag (44% to 91%), the site-specific conjugation was found to be efficient when the tag was introduced in either the N or C terminus of the antibodies.

The methods in this category rely on introducing unique short peptide tags into antibodies for enzyme modification either *in vivo* or *in vitro*. They allow specific amino acids in the peptide tags to be functionalized and coupled to the drug-linkers. Although the approaches are straightforward, the potential immunogenicity of these short peptide tags located at different regions of antibodies is currently unknown, nor the scalability.

4.6 Site-Specific Antibody Conjugation for Diagnosis

The recent technical advancements in positron emission tomography (PET) and optical imaging (OI) resulted in great interest in the development of multimodal PET/OI probes that can be employed during the diagnosis, staging, and surgical treatment of cancer [54]. The combination of PET/OI agents with antibodies using site-specific conjugations would enhance both the sensitivity and selectivity (Fig. 4.3). A chemoenzymatic strategy for the construction of multimodal PET/OI and radiolabeled immunoconjugates was developed by the site-specific labeling of antibody through N297 glycans [55–57]. The method includes the removal of terminal galactose, followed by enzymatic incorporation of azido-GalNAc using GalT (Y289L). Instead of being coupled to cytotoxins as described above, the azido sugar in the antibodies was conjugated with chelator- or chelator plus fluorophore–modified DBCO through SPAAC before the antibodies were radiolabeled. In one study, an anti-PSMA antibody was site-specifically conjugated with the chelator desferrioxamine-modified DBCO through antibody glycans using copper-free click chemistry [55]. The chelator-modified antibody was then radiolabeled with the position-emitting radiometal ^{89}Zr. The radiolabeled antibody displayed high selective tumor uptake and tumor-to-background contrast in mice bearing PSMA expressing tumor. In another study, a colorectal cancer–targeting antibody was GalT (Y289L) modified with azido-GalNAc, which was then conjugated with two reporters [56]. These reporters include DBCO-containing near-infrared fluorescent dye Alexa Fluor 680 and DBCO containing desferrioxamine, which subsequently reacted with the positron-emitting radiometal ^{89}Zr. In *in vivo* PET and fluorescence imaging experiments, a hybrid ^{89}Zr-and Alexa Fluor 680-labeled antibody conjugate displayed high levels of specific uptake in the tumor in mouse xenograft models. The data suggest that the site-specific conjugation strategy is robust and reproducible, producing well-defined immunoconjugates.

An antibody was also site-specifically conjugated with PET, near-infrared fluorescent (NIRF), and dual-modal (PET/NIRF) imaging agents [57]. The N297 glycans of anti-CA19.9 (a tumor associated carbohydrate antigen) was remodeled and incorporated with azido-GalNAc using GalT (Y289L). The azido sugar was then coupled using copper-free click chemistry with DBCO-containing desferrioxamine for ^{89}Zr radiolabeling for PET imaging used in noninvasive whole-body imaging and/or a NIRF dye for guided delineation of surgical margins. The antibodies conjugated with single or dual imaging agents showed specific uptake and contrast in antigen-positive tumors with negligible nonspecific uptake in antigen-negative tumor in a mouse xenograft model using human pancreatic cancer cell lines.

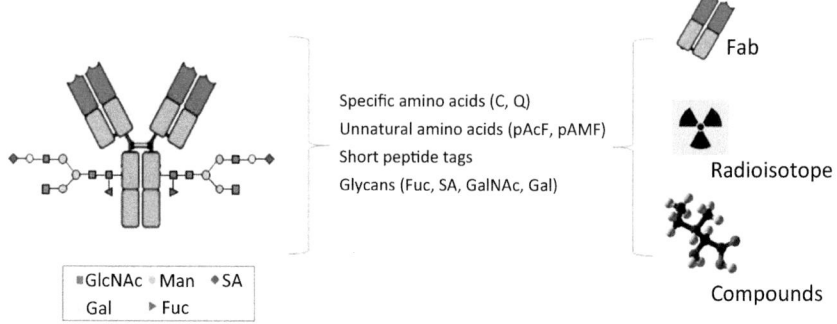

Figure 4.3 The application of the site-specific ADCs in coupling an antibody with a small protein, radioisotope, and non-cytotoxic compounds.

In another study, a colorectal cancer-targeting antibody was conjugated with trans-cyclooctene through N297 glycans [58]. The immunoconjugate was injected into human colorectal carcinoma xenografts after the administration of a pretargeted PET imaging agent, ^{64}Cu-labeled tetrazine radioligand. The antibody and radioligand reacted *in vivo* in mice via strain-promoted azide-alkyne click chemistry. PET imaging and biodistribution studies revealed that this strategy clearly delineated tumor tissue, producing images with excellent contrast and high tumor-to-background ratio.

Kazane et al. developed a site-specific DNA-antibody conjugation method for specific and sensitive immuno-polymerase chain reaction (PCR) used as diagnosis and imaging [59]. An unnatural amino acid, pAcF, was introduced into the anti-HER2 antibody Fab at LC-K169 or LC-S202 and it was coupled with an aminooxy-modified ssDNA primer (32 nt) to produce the oligobody. After antigen binding of the Fab, the oligonucleotide was amplified, ligated, and hybridized with complementary fluorophore, which was detected with either a fluorescence or confocal microscope. The immunoconjugates were tested in immuno-PCR assays to detect HER2 expressing tumor cells. They showed greater sensitivity and specificity as well as a lower background signal than nonspecifically coupled fragments and could detect extremely rare tumor cells in a complex cellular environment. Thus, the site-specific antibody-oligonucleotide conjugates should provide sensitive and specific reagents for diagnostics.

The use of site-specific conjugation for radiolabeling or coupling of an antibody with fluorescence/oligonucleotides could potentially improve sensitivity and specificity. It could reduce false positive outcomes in cancer diagnosis. However, there is no *in vivo* data yet that demonstrates its superior selectivities compared to the immunoconjugates prepared using conventional methods.

4.7 Site-Specific Antibody Conjugation for Other Therapeutic Applications

The site-specific conjugation approaches have also been applied to other small-molecular-weight proteins or compounds in addition to cytotoxins or radioisotopes (Fig. 4.3) [60].

Antibody fragments have been coupled using a site-specific method as bispecific antibody Fab conjugates. In one study, a bispecific antibody Fab conjugate was generated using genetically encoded unnatural amino acids with orthogonal chemical reactivity as described above for cytotoxin containing ADC [61]. A tRNA/aaRS pair derived from *Methanococcus jannaschii* was co-expressed to site-specifically incorporate pAcF at defined sites in each of two Fab fragments in response to an amber nonsense codon. The unnatural amino acids were incorporated into LC-S202 of the Fab fragment of anti-HER2 and subsequently conjugated with a PEG linker containing azide, while pAcF was incorporated in the HC-K138 of a Fab fragment of anti-CD3 before being conjugated with a PEG linker containing BCN. Both Fab fragments were subsequently coupled using copper-free click chemistry. In an *in vitro* cytotoxicity assay, the bispecific antibody Fab conjugate efficiently recruited T cells from human peripheral blood mononuclear cells (PBMCs) to kill tumor cells at picomolar concentrations. Moreover, the same site-specific antibody conjugation approach was applied to prepare bispecific antibody Fab conjugates against both CD3 and C-type lectin-like molecule-1 (CLL1) as well as both CD3 and CD33 [62]. CLL1 and CD33 are cell surface antigens overexpressed in acute myeloid leukemia, and the antibody Fab fragments were coupled to either azido-PEG3-aminooxy or BCN-PEG3-aminooxy linkers, respectively. The bispecific antibody Fab conjugate against CD3 and CLL1 displayed strong *in vitro* and *in vivo* antitumor activity compared to the Fab conjugate against CD3 and CD33. In addition to use of a PEG linker for coupling, either oligonucleotides or peptide nucleic acids of defined sequences were site-specifically conjugated to unnatural amino acids introduced in an antibody for the preparation of bispecific antibody Fab conjugates or multimeric antibody Fab conjugates [63]. As described above, pAcF was incorporated into HC-K138 of anti-CD3 Fab, while pAcF was introduced into LC-S202 of anti-HER2. Complementary peptide nucleic acid strands were then coupled to both Fab fragments, respectively. The bispecific Fab conjugates were self-assembled based on Watson-Crick base pairing properties of oligonucleotides. They were shown to recruit cytotoxic T cells to kill cancer cells *in vitro*. Tetrameric Fab conjugates were also generated using a similar approach.

Besides the bispecific antibody Fab conjugation, chemically programmed bispecific antibody conjugation is another strategy for site-specifically preparing a bispecific antibody. Kim et al. reported the generation of a bispecific small molecule-antibody conjugate for targeting prostate cancer [64]. They incorporated an unnatural amino acid, pAcF, into different locations of the anti-CD3 Fab that are distal to the antigen-binding site based on the crystal structure. The unnatural amino acid in the Fab was conjugated to a synthetic small molecule ligand, 2-[3-dicarboxy propyl]-ureido] pentanedioic acid (DUPA), that selectively binds PSMA with high affinity. A bivalent Fab was also prepared by introducing pAcF in two different positions (LC-S202 and HC-K138) and subsequently coupled with two DUPA ligands to a single Fab. The anti-CD3 DUPA conjugate showed potent *in vitro* cytotoxicity against prostate cancer cells and strong *in vivo* efficacy in mouse xenograft models. Cui et al. also reported another approach in preparing chemically programmed bispecific antibodies [65]. A C-terminal selenocysteine (Sec) was cotranslationally introduced into the anti-CD3 antibody Fab. The Sec containing Fab was conjugated with maleimide containing LLP2A, a high affinity ligand for integrin α4β1 overexpressed in malignant B cells, or maleimide-containing folate, a high affinity ligand for folate receptor α overexpressed in many cancer cells. The bispecific small molecule-antibody conjugates showed potent and specific *in vitro* and *ex vivo* cytotoxicity against tumor cell lines and primary tumor cells in the presence of T cells. In another study, a diabody containing both anti-hapten and anti-CD3 Fv (disulfide linked polypeptides containing either variable heavy or variable light chains) was coupled with hapten-derivatized small molecule folate through a reactive lysine introduced in one of the polypeptides of anti-hapten [66]. The chemically programmed diabody demonstrated high selectivity and potency against folate receptor α-expressing ovarian cancer cells both *in vitro* and *in vivo*.

Lehar et al. reported the generation of a novel antibody-antibiotic conjugate (AAC) using the THIOMAB approach [67]. They found a virulent subset of bacteria *Staphylococcus aureus* that can establish infection even in the presence of antibiotics such as vancomycin. They prepared an AAC in which an anti-S. aureus antibody was conjugated to highly efficacious antibiotics such as rifalogue, which is activated only after being released in the lysosome. A cysteine residue was engineered at the V205 position of the anti-bacteria antibody light chain and the thiol containing antibody was conjugated to MC-vc-PAB-rifalogue. The AAC was superior to vancomycin for the treatment of bacteremia *in vivo*.

A site-specific antibody–polymer conjugation (APC) approach was also reported [68]. An unnatural amino acid, pAcF, was introduced into anti-HER2 Fab (LC-S202) or IgG (HC-Q389) using an evolved orthogonal tRNA/aaRS pair. The engineered ketone-containing unnatural amino acid was conjugated with aminooxy-derivatized cationic block copolymer. The cationic polymer on the antibody specifically delivered siRNAs to HER2-positive tumor cells and mediated potent gene silencing at both the mRNA and protein levels *in vitro*.

The site-specific antibody conjugation has been applied for preparing immunoconjugates against other diseases such as autoimmune diseases and atherosclerosis. A highly potent phosphodiesterase 4 (PDE4) inhibitor, GSK256066, was site-specifically coupled to a human anti-CD11a through an unnatural amino acid, pAcF, introduced at HC-A122 with DAR at ~2

[69]. PDE4 is a cAMP phosphodiesterase widely expressed in a variety of cells and some small molecule PDE4 inhibitors showed wide-ranging preclinical efficacy in autoimmune diseases with a few being approved by regulatory agencies for the treatment of some moderate to severe inflammatory conditions. However, dose-limiting side effects have impeded their broader therapeutic application. The site-specific conjugation of pan-immune cell targeting human anti-CD11a with GSK256066 resulted in an immunoconjugate that was rapidly internalized into immune cells and suppressed lipopolysaccharide (LPS)-induced TNFα secretion in primary human monocytes. In another study, a liver X receptor (LXR) agonist was site-specifically conjugated to pAcF at HC-A122 of anti-CD11a [70]. Liver X receptor agonists have been explored as potential treatments for atherosclerosis and other diseases based on their ability to induce reverse cholesterol transport and suppress inflammation. However, this therapeutic potential has been limited by on-target adverse effects in the liver mediated by excessive lipogenesis after the interaction of the ligand with LXR-α. To prevent the adverse effect, the aminoooxy-modified LXR agonist was coupled to pAcF in anti-CD11a for selective delivery of the agonist to monocytes/macrophages while sparing hepatocytes. The anti-CD11a IgG-LXR agonist immunoconjugate induced LXR activation specifically in human THP-1 monocyte/macrophage cells *in vitro* with EC50 at nM range, but had no significant effect in hepatocytes, indicating that the payload delivery was CD11a-mediated. This approach represents a fundamentally different strategy that uses tissue targeting to overcome the limitations of LXR agonists for potential use in treating atherosclerosis.

As next-generation technologies, the site-specific antibody conjugations are likely to be applied to different therapeutic areas for preparing homogeneous immunoconjugates. The combination of therapeutic antibodies with a wide variety of small-molecular-weight proteins or drugs could potentially expand the current treatment options for many challenging diseases. Although it is unknown which methods are superior to others, each category of these methods may have their unique advantages related to the *in vivo* and *in vitro* properties of the conjugates.

4.8 Conclusions

Site-specific antibody drug conjugation technologies have been developed by coupling the cytotoxins to engineered specific amino acids, unnatural amino acids, short peptide tags, or N297 glycans. As next-generation methods, these approaches generated homogeneous ADCs with a high therapeutic index compared to the conventional conjugations. Some of the site-specific ADCs even showed better antitumor efficacy *in vivo* than the ADCs prepared using conventional methods. Moreover, there are trends in applying these site-specific antibody conjugations to other therapeutic areas or diagnosis. All those studies have provided promising results that suggest the usefulness of the next-generation methods in the coupling of small proteins, small-molecular-weight compounds, DNAs, and RNAs. It is not surprising that these new technologies will lead to important therapeutic platforms for many unmet medical needs.

Abbreviations

ADC	Antibody–drug conjugates
AML	Acute myelogenous leukemia
ALL	Acute lymphoblastic leukemia
MMAE	Monomethyl auristatin E
PBD	Pyrrolobenzodiazepine
LC	Light chain
mTG	Microbial transglutaminase
SPAAC	Strain-promoted azide-alkyne cycloaddition
aaRS	Aminoacyl-tRNA synthetase
EGFR	Epidermal growth factor receptor
pylRS	Pyrrolysyl-tRNA synthetase
pAMF	*p*-Azidomethyl-L-phenylalanine
DBCO-PEG-MMAF	DBCO-PEG-monomethyl auristatin F
Sec	Selenocysteine
GlaNAc	*N*-acetylgalactosamine

BCN	Bicyclononyne
FGE	Formylglycine-generating enzyme
OI	Optical imaging
NIRF	Near-infrared fluorescent
PBMCs	Peripheral blood mononuclear cells
DUPA	2-[3-Dicarboxy propyl]-ureido] pentanedioic acid
APC	Antibody–polymer conjugation
LPS	Lipopolysaccharide
IgG	Immunoglobulin G
HER2	Human epidermal growth factor receptor 2
DAR	Drug-to-antibody ratio
DM1	Mertansine
HC	Heavy chain
DBM	Dibromomaleimide
DBCO	Dibenzocyclooctynes
pAcF	*p*-Acetylphenylalanine
CHO	Chinese hamster ovary
MMAD	Monomethyl auristatin D
tRNA pyl	Pyrrolysyl-tRNA
TyrRS	Tyrosyl tRNA synthetase
MMAF	Monomethyl auristatin F
SECIS	Sec incorporation sequence
GlcNAc	*N*-acetylglucosamine
SA	Sialic acid
PET	Positron emission tomography
PSMA	Prostate-specific membrane antigen
PCR	Polymerase chain reaction
CLL1	C-type lectin-like molecule-1
AAC	Antibody-antibiotic conjugate
PDE4	phosphodiesterase 4
LXR	Liver X receptor

Disclosures and Conflict of Interest

This chapter was originally published as an open-access article distributed under the terms and conditions of the Creative Commons Attribution (CC BY) license (http://creativecommons.org/licenses/by/4.0/) and originally appeared in *Biomedicines* 2017, 5, 64; doi:10.3390/biomedicines5040064. The permission of the author and publisher has been obtained. The author has not received any payment for the preparation of this chapter and declares no conflict of interest.

Corresponding Author

Dr. Qun Zhou
Protein Engineering, Biologics Research, Sanofi
49 New York Avenue, Framingham, MA 01701, USA
Email: qun.zhou@sanofi.com

References

1. Reichert, J. M. (2017). Antibodies to watch in 2017. *Money Adv. Budg. Serv.*, **9**, 167–181.
2. Ecker, D. M., Jones, S. D., Levine, H. L. (2015). The therapeutic monoclonal antibody market. *Money Adv. Budg. Serv.*, **7**, 9–14.
3. Alley, S. C., Okeley, N. M., Senter, P. D. (2010). Antibody-drug conjugates: Targeted drug delivery for cancer. *Curr. Opin. Chem. Biol.*, **14**, 529–537.
4. Beck, A. (2014). Review of antibody-drug conjugates, methods in molecular biology series: A book edited by Laurent Ducry. *Money Adv. Budg. Serv.*, **6**, 30–33.
5. Carter, P. J., Senter, P. D. (2008). Antibody-drug conjugates for cancer therapy. *Cancer J.*, **14**, 154–169.
6. Teicher, B. A. (2009). Antibody-drug conjugate targets. *Curr. Cancer Drug Targets*, **9**, 982–1004.
7. Okeley, N. M., Alley, S. C., Senter, P. D. (2014). Advancing antibody drug conjugation: From the laboratory to a clinically approved anticancer drug. *Hematol. Oncol. Clin. N. Am.*, **28**, 13–25.
8. Lambert, J. M. (2012). Drug-conjugated antibodies for the treatment of cancer. *Br. J. Clin. Pharmacol.*, **76**, 248–262.

9. Senter, P. D., Sievers, E. L. (2012). The discovery and development of brentuximab vedotin for use in relapsed Hodgkin lymphoma and systemic anaplastic large cell lymphoma. *Nat. Biotechnol.*, **30**, 631–637.
10. Sievers, E. L., Senter, P. D. (2013). Antibody-drug conjugates in cancer therapy. *Annu. Rev. Med.*, **64**, 15–29.
11. Brinkmann, U., Kontermann, R. E. (2017). The making of bispecific antibodies. *Money Adv. Budg. Serv.*, **9**, 182–212.
12. Strebhardt, K., Ullrich, A. (2008). Paul Ehrlich's magic bullet concept: 100 years of progress. *Nat. Rev. Cancer*, **8**, 473–480.
13. Godwin, C. D., Gale, R. P., Walter, R. B. (2017). Gemtuzumab ozogamicin in acute myeloid leukemia. *Leukemia*, **31**, 1855–1868.
14. Dhillon, S. (2014). Trastuzumab emtansine: A review of its use in patients with HER2-positive advanced breast cancer previously treated with trastuzumab-based therapy. *Drugs*, **74**, 675–686.
15. Doronina, S. O., Toki, B. E., Torgov, M. Y., Mendelsohn, B. A., Cerveny, C. G., Chace, D. F., DeBlanc, R. L., Gearing, R. P., Bovee, T. D., Siegall, C. B., et al. (2003). Development of potent monoclonal antibody auristatin conjugates for cancer therapy. *Nat. Biotechnol.*, **21**, 778–784.
16. Gualberto, A. (2012). Brentuximab Vedotin (SGN-35), an antibody-drug conjugate for the treatment of CD30-positive malignancies. *Expert Opin. Investig. Drugs*, **21**, 205–216.
17. Kantarjian, H. M., DeAngelo, D. J., Stelljes, M., Martinelli, G., Liedtke, M., Stock, W., Gökbuget, N., O'Brien, S., Wang, K., Wang, T., et al. (2016). Inotuzumab ozogamicin versus standard therapy for acute lymphoblastic leukemia. *N. Engl. J. Med.*, **375**, 740–753.
18. Lambert, J. M., Chari, R. V. (2014). Ado-trastuzumab emtansine (T-DM1): An antibody-drug conjugate (ADC) for HER2-positive breast cancer. *J. Med. Chem.*, **57**, 6949–6964.
19. Haddley, K. (2012). Brentuximab vedotin: Its role in the treatment of anaplastic large cell and Hodgkin's lymphoma. *Drugs Today (Barc.)*, **48**, 259–270.
20. Horwitz, S. M., Advani, R. H., Bartlett, N. L., Jacobsen, E. D., Sharman, J. P., O'Connor, O. A., Siddiqi, T., Kennedy, D. A., Oki, Y. (2014). Objective responses in relapsed T-cell lymphomas with single agent brentuximab vedotin. *Blood*, **123**, 3095–3100.
21. Krop, I., Winer, E. P. (2014). Trastuzumab emtansine: A novel antibody-drug conjugate for HER2-positive breast cancer. *Clin. Cancer Res.*, **20**, 15–20.

22. Krop, I. E., Kim, S. B., Gonzalez-Martin, A., Lorusso, P. M., Ferrero, J. M., Smitt, M., Badovinac-Crnjevic, T., Hoersch, S., Wildiers, H. (2014). Trastuzumab emtansine versus treatment of physician's choice for pretreated HER2-positive advanced breast cancer (TH3RESA): A randomised, open-label, phase 3 trial. *Lancet Oncol.*, **15**, 689–699.

23. Stephan, J. P., Kozak, K. R., Wong, W. L. (2011). Challenges in developing bioanalytical assays for characterization of antibody-drug conjugates. *Bioanalysis*, **3**, 677–700.

24. Wakankar, A. A., Feeney, M. B., Rivera, J., Chen, Y., Kim, M., Sharma, V. K., Wang, Y. J. (2010). Physicochemical stability of the antibody-drug conjugate Trastuzumab-DM1: Changes due to modification and conjugation processes. *Bioconjug. Chem.*, **21**, 1588–1595.

25. Wang, L., Amphlett, G., Blattler, W. A., Lambert, J. M., Zhang, W. (2005). Structural characterization of the maytansinoid-monoclonal antibody immunoconjugate, huN901-DM1, by mass spectrometry. *Protein Sci.*, **14**, 2436–2446.

26. Panowski, S., Bhakta, S., Raab, H., Polakis, P., Junutula, J. R. (2014). Site-specific antibody drug conjugates for cancer therapy. *Money Adv. Budg. Serv.*, **6**, 34–45.

27. Junutula, J. R., Raab, H., Clark, S., Bhakta, S., Leipold, D. D., Weir, S., Chen, Y., Simpson, M., Tsai, S. P., Dennis, M. S., et al. (2008). Site-specific conjugation of a cytotoxic drug to an antibody improves the therapeutic index. *Nat. Biotechnol.*, **26**, 925–932.

28. Junutula, J. R., Flagella, K. M., Graham, R. A., Parsons, K. L., Ha, E., Raab, H., Bhakta, S., Nguyen, T., Dugger, D. L., Li, G., et al. (2010). Engineered thio-trastuzumab-DM1 conjugate with an improved therapeutic index to target human epidermal growth factor receptor 2-positive breast cancer. *Clin. Cancer Res.*, **16**, 4769–4778.

29. Jeffrey, S. C., Burke, P. J., Lyon, R. P., Meyer, D. W., Sussman, D., Anderson, M., Hunter, J. H., Leiske, C. I., Miyamoto, J. B., Nicholas, N. D., et al. (2013). A potent anti-CD70 antibody-drug conjugate combining a dimeric pyrrolobenzodiazepine drug with site-specific conjugation technology. *Bioconjug. Chem.*, **24**, 1256–1263.

30. Dimasi, N., Fleming, R., Zhong, H., Bezabeh, B., Kinneer, K., Christie, R. J., Fazenbaker, C., Wu, H., Gao, C. (2017). Efficient preparation of site-specific antibody-drug conjugates using cysteine insertion. *Mol. Pharm.*, **14**, 1501–1516.

31. Bryant, P., Pabst, M., Badescu, G., Bird, M., McDowell, W., Jamieson, E., Swierkosz, J., Jurlewicz, K., Tommasi, R., Henseleit, K., et al. (2015). In vitro and in vivo evaluation of cysteine rebridged trastuzumab-

MMAE antibody drug conjugates with defined drug-to-antibody ratios. *Mol. Pharm.*, **12**, 1872–1879.

32. Behrens, C. R., Ha, E. H., Chinn, L. L., Bowers, S., Probst, G., Fitch-Bruhns, M., Monteon, J., Valdiosera, A., Bermudez, A., Liao-Chan, S., et al. (2015). Antibody-drug conjugates (ADCs) derived from interchain cysteine cross-linking demonstrate improved homogeneity and other pharmacological properties over conventional heterogeneous ADCs. *Mol. Pharm.*, **12**, 3986–3998.

33. Dennler, P., Schibli, R., Fischer, E. (2013). Enzymatic antibody modification by bacterial transglutaminase. *Methods Mol. Biol.*, **1045**, 205–215.

34. Dennler, P., Chiotellis, A., Fischer, E., Bregeon, D., Belmant, C., Gauthier, L., Lhospice, F., Romagne, F., Schibli, R. (2014). Transglutaminase-based chemo-enzymatic conjugation approach yields homogeneous antibody-drug conjugates. *Bioconjug. Chem.*, **25**, 569–578.

35. Shen, B. Q., Xu, K., Liu, L., Raab, H., Bhakta, S., Kenrick, M., Parsons-Reponte, K. L., Tien, J., Yu, S. F., Mai, E., et al. (2012). Conjugation site modulates the in vivo stability and therapeutic activity of antibody-drug conjugates. *Nat. Biotechnol.*, **30**, 184–189.

36. Axup, J. Y., Bajjuri, K. M., Ritland, M., Hutchins, B. M., Kim, C. H., Kazane, S. A., Halder, R., Forsyth, J. S., Santidrian, A. F., Stafin, K., et al. (2012). Synthesis of site-specific antibody-drug conjugates using unnatural amino acids. *Proc. Natl. Acad. Sci. U.S.A.*, **109**, 16101–16106.

37. Tian, F., Lu, Y., Manibusan, A., Sellers, A., Tran, H., Sun, Y., Phuong, T., Barnett, T., Hehli, B., Song, F., et al. (2014). A general approach to site-specific antibody drug conjugates. *Proc. Natl. Acad. Sci. U.S.A.*, **111**, 1766–1771.

38. Jackson, D., Atkinson, J., Guevara, C. I., Zhang, C., Kery, V., Moon, S. J., Virata, C., Yang, P., Lowe, C., Pinkstaff, J., et al. (2014). In vitro and in vivo evaluation of cysteine and site specific conjugated herceptin antibody-drug conjugates. *PLoS One*, **9**, e83865.

39. VanBrunt, M. P., Shanebeck, K., Caldwell, Z., Johnson, J., Thompson, P., Martin, T., Dong, H., Li, G., Xu, H., D'Hooge, F., et al. (2015). Genetically encoded azide containing amino acid in mammalian cells enables site-specific antibody-drug conjugates using click cycloaddition chemistry. *Bioconjug. Chem.*, **26**, 2249–2260.

40. Zimmerman, E. S., Heibeck, T. H., Gill, A., Li, X., Murray, C. J., Madlansacay, M. R., Tran, C., Uter, N. T., Yin, G., Rivers, P. J., et al. (2014). Production of site-specific antibody-drug conjugates using optimized non-

natural amino acids in a cell-free expression system. *Bioconjug. Chem.*, **25**, 351–361.

41. Li, X., Nelson, C. G., Nair, R. R., Hazlehurst, L., Moroni, T., Martinez-Acedo, P., Nanna, A. R., Hymel, D., Burke, T. R., Jr., Rader, C. (2017). Stable and potent selenomab-drug conjugates. *Cell Chem. Biol.*, **24**, 433–442.

42. Okeley, N. M., Toki, B. E., Zhang, X., Jeffrey, S. C., Burke, P. J., Alley, S. C., Senter, P. D. (2013). Metabolic engineering of monoclonal antibody carbohydrates for antibody-drug conjugation. *Bioconjug. Chem.*, **24**, 1650–1655.

43. Ramakrishnan, B., Qasba, P. K. (2002). Structure-based design of beta 1,4-galactosyltransferase I (beta 4Gal-T1) with equally efficient *N*-acetylgalactosaminyltransferase activity: Point mutation broadens beta 4Gal-T1 donor specificity. *J. Biol. Chem.*, **277**, 20833–20839.

44. Ramakrishnan, B., Boeggeman, E., Pasek, M., Qasba, P. K. (2011). Bioconjugation using mutant glycosyltransferases for the site-specific labeling of biomolecules with sugars carrying chemical handles. *Methods Mol. Biol.*, **751**, 281–296.

45. Zhu, Z., Ramakrishnan, B., Li, J., Wang, Y., Feng, Y., Prabakaran, P., Colantonio, S., Dyba, M. A., Qasba, P. K., Dimitrov, D. S. (2014). Site-specific antibody-drug conjugation through an engineered glycotransferase and a chemically reactive sugar. *Money Adv. Budg. Serv.*, **6**, 1190–1200.

46. Van Geel, R., Wijdeven, M. A., Heesbeen, R., Verkade, J. M., Wasiel, A. A., van Berkel, S. S., van Delft, F. L. (2015). Chemoenzymatic conjugation of toxic payloads to the globally conserved N-glycan of native mAbs provides homogeneous and highly efficacious antibody-drug conjugates. *Bioconjug. Chem.*, **26**, 2233–2242.

47. Zhou, Q., Stefano, J. E., Manning, C., Kyazike, J., Chen, B., Gianolio, D. A., Park, A., Busch, M., Bird, J., Zheng, X., et al. (2014). Site-specific antibody-drug conjugation through glycoengineering. *Bioconjug. Chem.*, **25**, 510–520.

48. Li, X., Fang, T., Boons, G. J. (2014). Preparation of well-defined antibody-drug conjugates through glycan remodeling and strain-promoted azide-alkyne cycloadditions. *Angew. Chem. Int. Ed. Engl.*, **53**, 7179–7182.

49. Tang, F., Wang, L. X., Huang, W. (2017). Chemoenzymatic synthesis of glycoengineered IgG antibodies and glycosite-specific antibody-drug conjugates. *Nat. Protoc.*, **12**, 1702–1721.

50. Strop, P., Liu, S. H., Dorywalska, M., Delaria, K., Dushin, R. G., Tran, T. T., Ho, W. H., Farias, S., Casas, M. G., Abdiche, Y., et al. (2013).

Location matters: Site of conjugation modulates stability and pharmacokinetics of antibody drug conjugates. *Chem. Biol.*, **20**, 161–167.

51. Beerli, R. R., Hell, T., Merkel, A. S., Grawunder, U. (2015). Sortase enzyme-mediated generation of site-specifically conjugated antibody drug conjugates with high in vitro and in vivo potency. *PLoS One*, **10**, e0131177.

52. Wu, P., Shui, W., Carlson, B. L., Hu, N., Rabuka, D., Lee, J., Bertozzi, C. R. (2009). Site-specific chemical modification of recombinant proteins produced in mammalian cells by using the genetically encoded aldehyde tag. *Proc. Natl. Acad. Sci. U.S.A.*, **106**, 3000–3005.

53. Rabuka, D., Rush, J. S., deHart, G. W., Wu, P., Bertozzi, C. R. (2012). Site-specific chemical protein conjugation using genetically encoded aldehyde tags. *Nat. Protoc.*, **7**, 1052–1067.

54. Moek, K. L., Giesen, D., Kok, I. C., de Groot, D. J. A., Jalving, M., Fehrmann, R. S. N., Lub-de Hooge, M. N., Brouwers, A. H., de Vries, E. G. E. (2017). Theranostics using antibodies and antibody-related therapeutics. *J. Nucl. Med. Off. Publ. Soc. Nucl. Med.*, **58**, 83s–90s.

55. Zeglis, B. M., Davis, C. B., Aggeler, R., Kang, H. C., Chen, A., Agnew, B. J., Lewis, J. S. (2013). Enzyme-mediated methodology for the site-specific radiolabeling of antibodies based on catalyst-free click chemistry. *Bioconjug. Chem.*, **24**, 1057–1067.

56. Zeglis, B. M., Davis, C. B., Abdel-Atti, D., Carlin, S. D., Chen, A., Aggeler, R., Agnew, B. J., Lewis, J. S. (2014). Chemoenzymatic strategy for the synthesis of site-specifically labeled immunoconjugates for multimodal PET and optical imaging. *Bioconjug. Chem.*, **25**, 2123–2128.

57. Houghton, J. L., Zeglis, B. M., Abdel-Atti, D., Aggeler, R., Sawada, R., Agnew, B. J., Scholz, W. W., Lewis, J. S. (2015). Site-specifically labeled CA19.9-targeted immunoconjugates for the PET, NIRF, and multimodal PET/NIRF imaging of pancreatic cancer. *Proc. Natl. Acad. Sci. U.S.A.*, **112**, 15850–15855.

58. Cook, B. E., Adumeau, P., Membreno, R., Carnazza, K. E., Brand, C., Reiner, T., Agnew, B. J., Lewis, J. S., Zeglis, B. S. (2016). Pretargeted PET imaging using a site-specifically labeled immunoconjugate. *Bioconjug. Chem.*, **27**, 1789–1795.

59. Kazane, S. A., Sok, D., Cho, E. H., Uson, M. L., Kuhn, P., Schultz, P. G., Smider, V. V. (2012). Site-specific DNA-antibody conjugates for

specific and sensitive immuno-PCR. *Proc. Natl. Acad. Sci. U.S.A.*, **109**, 3731–3736.

60. Rader, C. (2014). Chemically programmed antibodies. *Trends Biotechnol.*, **32**, 186–197.
61. Kim, C. H., Axup, J. Y., Dubrovska, A., Kazane, S. A., Hutchins, B. A., Wold, E. D., Smider, V. V., Schultz, P. G. (2012). Synthesis of bispecific antibodies using genetically encoded unnatural amino acids. *J. Am. Chem. Soc.*, **134**, 9918–9921.
62. Lu, H., Zhou, Q., Deshmukh, V., Phull, H., Ma, J., Tardif, V., Naik, R. R., Bouvard, C., Zhang, Y., Choi, S., et al. (2014). Targeting human C-type lectin-like molecule-1 (CLL1) with a bispecific antibody for immunotherapy of acute myeloid leukemia. *Angew. Chem. Int. Ed. Engl.*, **53**, 9841–9845.
63. Kazane, S. A., Axup, J. Y., Kim, C. H., Ciobanu, M., Wold, E. D., Barluenga, S., Hutchins, B. A., Schultz, P. G., Winssinger, N., Smider, V. V. (2013). Self-assembled antibody multimers through peptide nucleic acid conjugation. *J. Am. Chem. Soc.*, **135**, 340–346.
64. Kim, C. H., Axup, J. Y., Lawson, B. R., Yun, H., Tardif, V., Choi, S. H., Zhou, Q., Dubrovska, A., Biroc, S. L., Marsden, R., et al. (2013). Bispecific small molecule-antibody conjugate targeting prostate cancer. *Proc. Natl. Acad. Sci. U.S.A.*, **110**, 17796–17801.
65. Cui, H., Thomas, J. D., Burke, T. R., Jr., Rader, C. (2012). Chemically programmed bispecific antibodies that recruit and activate T cells. *J. Biol. Chem.*, **287**, 28206–28214.
66. Walseng, E., Nelson, C. G., Qi, J., Nanna, A. R., Roush, W. R., Goswami, R. K., Sinha, S. C., Burke, T. R., Jr., Rader, C. (2016). Chemically programmed bispecific antibodies in diabody format. *J. Biol. Chem.*, **291**, 19661–19673.
67. Lehar, S. M., Pillow, T., Xu, M., Staben, L., Kajihara, K. K., Vandlen, R., DePalatis, L., Raab, H., Hazenbos, W. L., Morisaki, J. H., et al. (2015). Novel antibody-antibiotic conjugate eliminates intracellular *S. aureus*. *Nature*, **527**, 323–328.
68. Lu, H., Wang, D., Kazane, S., Javahishvili, T., Tian, F., Song, F., Sellers, A., Barnett, B., Schultz, P. G. (2013). Site-specific antibody-polymer conjugates for siRNA delivery. *J. Am. Chem. Soc.*, **135**, 13885–13891.
69. Yu, S., Pearson, A. D., Lim, R. K., Rodgers, D. T., Li, S., Parker, H. B., Weglarz, M., Hampton, E. N., Bollong, M. J., Shen, J., et al. (2016). Targeted delivery of an anti-inflammatory PDE4 inhibitor to immune

cells via an antibody-drug conjugate. *Mol. Ther. J. Am. Soc. Gene Ther.*, **24**, 2078–2089.

70. Lim, R. K., Yu, S., Cheng, B., Li, S., Kim, N. J., Cao, Y., Chi, V., Kim, J. Y., Chatterjee, A. K., Schultz, P. G., et al. (2015). Targeted delivery of LXR agonist using a site-specific antibody-drug conjugate. *Bioconjug. Chem.*, **26**, 2216–2222.

Chapter 5

Current Understanding of Interactions between Nanoparticles and the Immune System

Marina A. Dobrovolskaia, PhD,[a] Michael Shurin, MD, PhD,[b,c] and Anna A. Shvedova, PhD[d,e]

[a]*Nanotechnology Characterization Laboratory, Cancer Research Technology Program, Leidos Biomedical Research Inc., Frederick National Laboratory for Cancer Research, NCI at Frederick, Frederick, Maryland, USA*
[b]*Department of Pathology, University of Pittsburgh Medical Center, Pittsburgh, Pennsylvania, USA*
[c]*Department of Immunology, University of Pittsburgh Medical Center, Pittsburgh, Pennsylvania, USA*
[d]*Health Effects Laboratory Division, National Institute of Occupational Safety and Health, Centers for Disease Control and Prevention, Morgantown, West Virginia, USA*
[e]*Department of Physiology and Pharmacology, West Virginia University, Morgantown, West Virginia, USA*

Keywords: anti-inflammatory, antiretroviral, biodegradation, cancer, CARPA, CD4+ T-cell, CD8+ T-cell, drug delivery, immunology, immunosuppressive, immunotherapy, immunotoxicity, mononuclear phagocytic system (MPS), nanoformulations, nanoparticles, preclinical, structure-activity relationship, total protein binding, tumor necrosis factor α (TNF-α), vaccine, endotoxin, inflammation, immune cells, protein corona, cytokines

---

Immune Aspects of Biopharmaceuticals and Nanomedicines
Edited by Raj Bawa, János Szebeni, Thomas J. Webster, and Gerald F. Audette
Copyright © 2018 Pan Stanford Publishing Pte. Ltd.
ISBN 978-981-4774-52-9 (Hardcover), 978-0-203-73153-6 (eBook)
www.panstanford.com

5.1 Introduction

The immune system's function in the maintenance of tissue homeostasis is to protect the host from environmental agents such as microbes or chemicals, and thereby preserve the integrity of the body. This is done through effective surveillance and elimination of foreign and abnormal self cells and structures from the body. It is well known that certain environmental contaminants and xenobiotics, as well as other drugs, may alter the immune system's normal function. Therefore, screening for immunotoxicity is a generally accepted step in toxicological research related to both environmental factors and pharmaceutical products (Luebke, 2012).

The interactions between nanoparticles and various components of the immune system have become an active area of research in bio- and nanotechnology because the benefits of using nanotechnology in industry and medicine are often questioned over concerns regarding the safety of these novel materials. The past decade of research has shown that, while nanoparticles can be toxic, nanotechnology engineering can modify these materials to either avoid or specifically target the immune system. Avoiding interaction with the immune system is desirable when the nanoparticles are being used for medical applications not intended to stimulate or inhibit the immune system, as well as when they are used for industrial and environmental applications. Specific targeting of the immune system, on the other hand, provides an attractive option for vaccine delivery, as well as for improving the quality of anti-inflammatory, anticancer, and antiviral therapies (Mallipeddi and Rohan, 2010; Gonzalez-Aramundiz et al., 2012; Zaman et al., 2013; Tran and Amiji, 2015). Moreover, nanotechnology-based carriers can be used to reduce the immunotoxicity of traditional drugs (Libutti et al., 2010).

Some nanomaterials, metal colloids and liposomes, for example, were in use more than a decade ago (Gregoriadis et al., 1974). However most active research in this field began in early 2000, fueled by the attention paid by regulatory agencies, such as the United States Environmental Protection Agency (EPA) and the U.S. Food and Drug Administration (FDA), to the rapidly growing number of applications containing various types of engineered nanomaterials. The increase in submissions was expected since innovative research in this area had been progressing for years, culminated by the establishment of several breakthrough technologies that led to the discovery of fullerenes (Benning et al., 1992), carbon nanotubes (Ramirez et al., 1994), dendritic polymers (Tomalia, 1991; Newkome et al., 2002), and quantum dots (Takagahara, 1987). In 2005–2006, many worldwide initiatives were launched to improve the understanding of nanoparticle safety and included, among others, the establishment of the Nanotechnology Task Force by the FDA (http://www.fda.gov/ScienceResearch/SpecialTopics/Nanotechnology/ucm2006658.htm), several nanotechnology research programs by the EPA (http://www2.epa.gov/chemical-research/research-evaluating-nanomaterials-chemical-safety), the E56 committee by the American Society for Testing and Materials (ASTM) International (http://www.astm.org/COMMITTEE/E56.htm), and the TC229 Nanotechnologies Technical Committee by the International Organization for Standardization (ISO) (http://www.iso.org/iso/iso_technical_committee?commid=381983). In addition to these efforts, the U.S. National Cancer Institute established the Nanotechnology Characterization Laboratory (NCL) to accelerate the translation of nanotechnology-based concepts intended for medical applications in the area of cancer diagnosis and therapy from bench to bedside (http://ncl.cancer.gov/). One of the initial goals of the NCL was to support the nanotechnology community by developing a so-called assay cascade that would include, among other tests, a battery of immunological assays. This assay cascade contributed to the initial understanding of the interactions between nanoparticles and the immune system and created a framework for stimulating discussions in the area of nano-immunotoxicology (Dobrovolskaia and McNeil, 2007; Marx, 2008; Dobrovolskaia et al., 2009a; Pantic, 2011; Smith et al., 2013). Recently, the European Commission has established the European Nanomedicine Characterization Laboratory (EU-NCL), which shares several objectives with those of the NCL (https://ec.europa.eu/jrc/en/news/eu-ncl-launched).

The rapid growth of this field becomes obvious when one compares the number of publications searchable in PubMed using the key words "nanoparticles" and "immune system" between years 2000 and 2015 (Fig. 5.1). Reviewing these data reveals many advances, as well as disappointments. Moreover, delving into the mechanisms of nanoparticle immunotoxicity uncovered many challenges in material characterization. Due to the wide variety of nanomaterials available, the characterization of their physicochemical properties is directed toward addressing parameters specific to certain type of particles (e.g., porosity is applicable to silicon nanoparticles, but is not informative for liposomes and dendrimers). The grand challenge in the particle characterization that precedes immunotoxicity studies relates to the estimation of immunoreactive contaminants, such as synthesis byproducts (e.g., iron catalysts in carbon nanotubes, cetyltrimethylammonium bromide [CTAB] in gold nanorods), and bacterial endotoxins, as well as excipients (e.g., Cremophor EL, polysorbate 80), and linkers (e.g., certain linkers used to attach poly(ethylene glycol) [PEG] to the nanoparticle surface) (Crist et al., 2013).

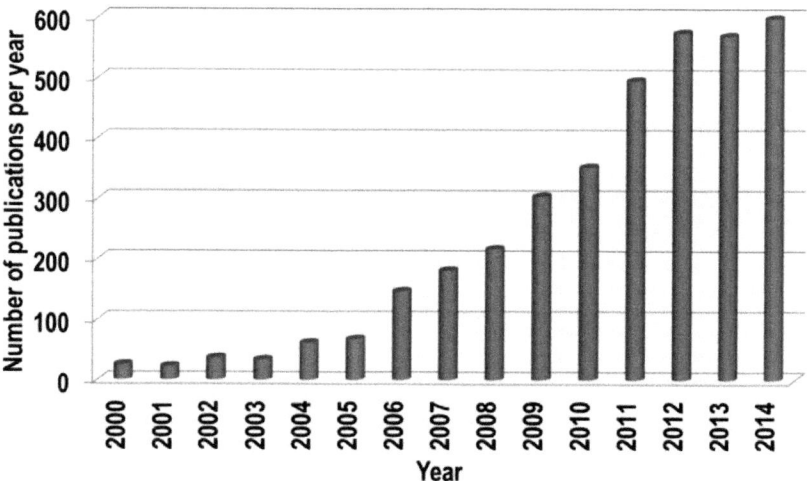

Figure 5.1 Publications statistics. The PubMed data base was searched using the keywords "nanoparticles" and "immune system" for the years 2000–2015. The data for 2015 were excluded from the analysis because the publication year was incomplete at the time of the search. Each bar shows the total publication number per year.

The challenges related to the physicochemical characterization (Clogston and Patri, 2013) and estimation of endotoxin contamination have been recently reviewed elsewhere (Crist et al., 2013; Dobrovolskaia, 2015).

The immunotoxicity of environmental materials has also been reviewed elsewhere (Kagan et al., 2010b).

Introduction | 187

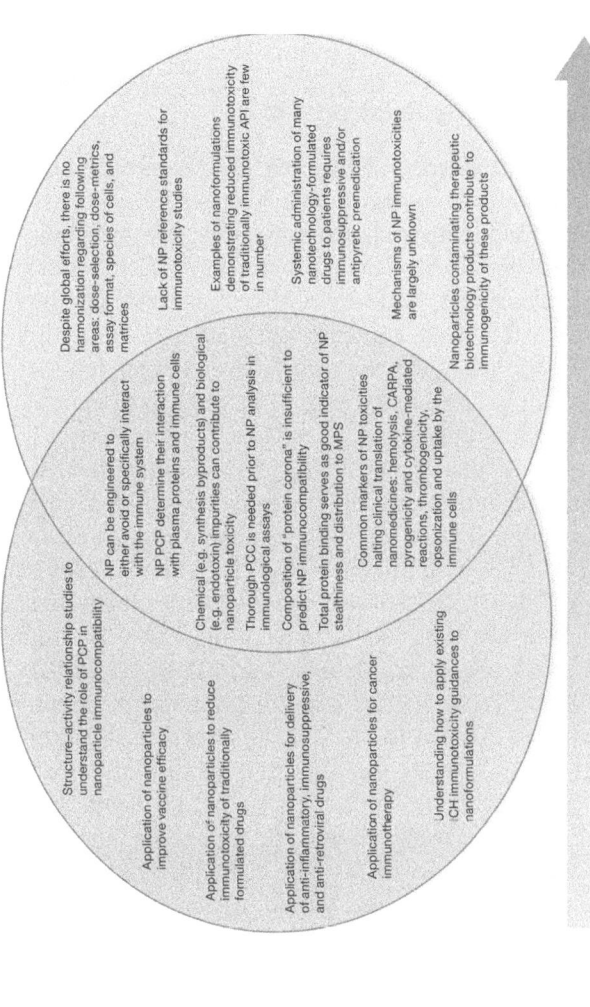

Figure 5.2 Achievements, disappointments, and lessons learned from the characterization of engineered nanomaterials over the past decade. This diagram outlines achievements (left circle) and disappointments (right circle) based on the studies dedicated to investigating nanoparticle immunotoxicity over the past decade. The overlapping area shows the lessons learned from these studies. API, active pharmaceutical ingredient; NP, nanoparticles; PCP, physicochemical properties; CARPA, complement activation-related pseudoallergy; ICH, International Conference on Harmonization.

Herein, we focus on the most prominent pieces of the nanoparticle-immune system puzzle, discussing what worked, what didn't, and what has been learned over the past 15 years of research on nanomaterials engineered for biomedical applications. A summary of achievements, disappointments, and lessons learned is presented in Fig. 5.2, and is further discussed below.

5.2 Achievements

5.2.1 Structure–Activity Relationship

The physicochemical properties of nanoparticles determine their interactions with proteins in biological matrices (e.g., blood plasma and alveolar fluid) and with the immune cells. The structure–activity relationships between the most prominent physicochemical properties of nanoparticles and their effects on the immune system that lead to the most common types of immunotoxicity are summarized in Fig. 5.3. Below, we review several examples.

Nanoparticles with cationic surfaces, or those that carry cationic ligands, interact with biological membranes electrostatically. This leads to cellular damage, which triggers hemolysis, platelet activation, and aggregation, and to the induction of leukocyte procoagulant activity (PCA) and disseminated intravascular coagulation (DIC) (Greish et al., 2012; Jones et al., 2012a; Jones et al., 2012b; Ziemba et al., 2012). For example, cationic dendrimers of different architecture and size (generation five [G5] and generation four [G4] poly (propylene imine) [PPI] dendrimers [Bhadra et al., 2005; Agashe et al., 2006], G4 polyamidoamine [PAMAM] dendrimers [Bhadra et al., 2003; Asthana et al., 2005], generation three [G3] PAMAM and G3 PPI dendrimers [Malik et al., 2000], as well as G4 poly-L-lysine [PLL] dendrimers [Agrawal et al., 2007]) were shown to be hemolytic both in vitro and in vivo. The in vitro percent hemolysis varied from 14 to 86% in whole blood from human donors and various animal species, and was dependent on the density of the surface groups. Likewise, cationic PAMAM dendrimers, but not their anionic and neutral counterparts, altered key platelet functions and perturbed plasma coagulation, which culminated with DIC (Greish et al., 2012; Jones et al., 2012a; Jones et al., 2012b).

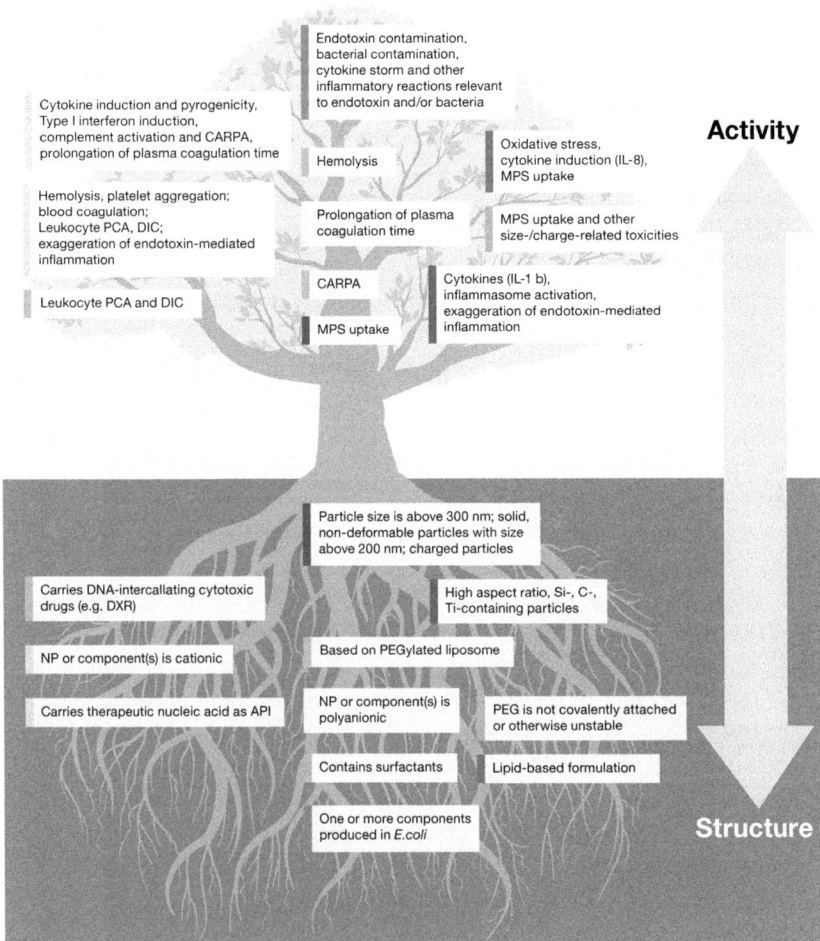

Figure 5.3 Structure–activity relationship summary. Shown are the structure–activity relationships between nanoparticles and their effects on the immune system. Each block listed in the bottom (structure) part of the figure is color-coded. To find what toxicity is related to the given structure block, please find the block in the top (activity) part of the figure marked with the color matching that of the structure block. PCA, procoagulant activity; DIC, disseminated intravascular coagulation; CARPA, complement activation-related pseudoallergy; MPS, mononuclear phagocytic system; IL, interleukin; PEG, polyethylene glycol; NP, nanoparticle; DXR, doxorubicin; API, active pharmaceutical ingredient; DNA, deoxyribonucleic acid.

The particle size, surface charge, and conformation of the polymer coating are important determinants of particle clearance by the mononuclear phagocytic system (MPS) in that smaller particles (100–200 nm) with unprotected surfaces and surfaces coated with a hydrophilic polymer in a "mushroom" configuration are primarily cleared by Kupffer cells in the liver; larger particles are eliminated by red pulp macrophages in the spleen. The addition of a hydrophilic polymer coating in a "brush" configuration protects particles from immune recognition, while increasing the particle size above 300 nm provides no protection, regardless of the polymer conformation (Gbadamosi et al., 2002). Exposure to high aspect ratio particles (e.g., carbon nanotubes, titanium nanobelts, cellulose nanofibers), as well as certain metallic particles (e.g., Si), results in inflammasome activation and the induction of proinflammatory cytokine interleukin (IL)-1β. These particles, as well as certain cationic and carbon-based particles, can exaggerate endotoxin-mediated inflammation (Baron et al., 2015).

The immunotoxicity of a nanoparticle is also influenced by the therapeutic payload it carries. For example, the induction of cytokines and type I interferons, the inflammatory reaction, the prolongation of plasma coagulation time, and complement activation are common dose-limiting toxicities of therapeutic nucleic acids (Levin, 1999). These toxicities are also commonly observed with nanoformulated nucleic acids, and this limits their translation into clinical use (Dobrovolskaia and McNeil, 2015a; Dobrovolskaia and McNeil, 2015b). Cytotoxic DNA-intercalating drugs used to treat cancer (e.g., doxorubicin, daunorubicin, and vincristine) are known to induce PCA and DIC (Wheeler and Geczy, 1990; Swystun et al., 2009; Kim et al., 2011). Formulating these drugs using nanotechnology carriers may help in avoiding the toxicity. However, if overcoming these toxicities is not considered during the design and optimization of nanoformulated versions of these drugs, both PCA and DIC may not be resolved.

5.2.2 Application of Nanoparticles to Improve Vaccine Efficacy

The efficacy of nanoparticle-based vaccines depends on the interactions between the particles and the target cells, and is determined by the physicochemical properties of the particles (size, shape, and surface functionalities) because these properties play a key role in particle recognition by the antigen-presenting cells (APCs). The type of nanocarrier is generally selected based on the type of immune response desired from the vaccine. Nanoparticles have been shown to provide a wide variety of advantages over conventional adjuvants. They can improve the solubility of hydrophobic antigens; provide controlled and sustainable release of antigens, therefore reducing the number of required immunizations; target antigens to specific cells and tissues, thus reducing side effects; prevent antigen degradation and deliver multiple antigens concurrently (reviewed in [Xiang et al., 2008; Xiang et al., 2010]). As of today, a wide variety of engineered nanomaterials from different classes (polymeric, chitosanic, magnetic, latex, gold, silica, and polystyrene) have been used successfully as antigen carriers and vaccine adjuvants (Tighe et al., 1998; Pavelic et al., 2001; Weiss et al., 2002; Walsh et al., 2003; Fifis et al., 2004a; Fifis et al., 2004b; Minigo et al., 2007; Mottram et al., 2007).

Depending on their size, particles are internalized by APCs via different pathways, including both pino- and phagocytosis (O'Hagan et al., 2001; Fifis et al., 2004a). Moreover, macrophages utilize multiple routes to take up the same types of nanoparticles (Franca et al., 2011). Several studies reported that smaller particles (20–200 nm) elicit stronger immune responses than their larger counterparts (O'Hagan et al., 2001; Fifis et al., 2004a; Fifis et al., 2004b; Minigo et al., 2007; Mottram et al., 2007; Manolova et al., 2008). For example, Plebanski and her group conducted a series of studies to demonstrate that 40–50 nm polystyrene particles induce potent CD4+ and CD8+ T-cell responses and do so more efficiently than their larger (>500 nm) counterparts. In contrast, particles >500 nm in size were more active in inducing interferon (IFN)-γ and antibody responses (Fifis et al., 2004a; Fifis et al., 2004b; Minigo et al., 2007; Mottram et al., 2007). Several studies demonstrated that small (<100 nm) nanoparticles quickly travel to the draining lymph nodes (LNs) after intradermal injection and effectively target LN-resident dendritic cells (DCs), B-cells, and macrophages (Manolova et al., 2008; Reddy et al., 2007b). These data suggested that large particles depend on interactions with and uptake by tissue-resident APCs, while smaller particles utilize both cell-associated migration and lymphatic drainage, thus providing better antigen presentation (Manolova et al., 2008).

Manipulation of nanoparticle size, shape, surface chemistry, and charge is generally employed to maximize antigen delivery to DCs. For example, small (<100 nm) nanoparticles were shown to be taken up more efficiently by DCs, while large (1 μm) particles were preferentially internalized by macrophages (Fifis et al., 2004a). Several other studies have also reported that ~50 nm is the optimal nanoparticle size for uptake by DCs (Aoyama et al., 2003; Nakai et al., 2003; Wang et al., 2011).

Particle size was also reported as a primary factor in determining the immunostimulatory profiles of vaccine formulations in that smaller particles (~220 nm) were more potent in inducing IFN-α responses, while their larger counterparts (~1200 nm) induced tumor necrosis factor (TNF)-α (Rettig et al., 2010). This difference was attributed to the type of the cells that internalized these particles: plasmacytoid DCs engulfed smaller particles, while macrophages preferred larger particles. Moreover, particle size was also suggested to be a key factor in determining the type of immunity induced. For example 40 nm nanoparticles promoted Th1 and CD8+ T-cell responses, while 100 nm particles induced Th2 responses (Fifis et al., 2004a; Fifis et al., 2004b; Mottram et al., 2007).

Tuning particle zeta potential is another approach that has been explored in vaccine design. For example, positively charged particles demonstrate greater uptake by DCs (Thiele et al., 2003; Foged et al., 2005; Villanueva et al., 2009) and induction of DC maturation (Thiele et al., 2001; Jilek et al., 2004; Little et al., 2004; Jilek et al., 2005; Reddy et al., 2007a). Cationic poly(D,L-lactic-co-glycolic) (PLG) particles improved the delivery of DNA adsorbed on the particle surface to APCs and induced greater cytotoxic T-lymphocyte responses compared to plain DNA antigen (Singh et al., 2000). More comprehensive coverage of this subject is available elsewhere (Fesenkova, 2013; Xiang et al., 2013).

5.2.3 Application of Nanoparticles for Delivery of Antiretroviral, Immunosuppressive and Anti-Inflammatory Drugs

5.2.3.1 Antiretroviral

Antiretroviral drug delivery has many assorted challenges, some of which are being effectively overcome using nanotechnology-based carriers. In addition to improving the solubility of antiretroviral drugs, nanoparticles are considered as a means to improve drug delivery to tissues and cells serving as viral reservoirs. When antiretroviral drugs are administered using conventional routes and formulations, the concentrations of these drugs in the plasma are usually higher than the concentrations found in the lymphoid tissue, which serves as a major depot for the virus (Fletcher et al., 2014). HIV can replicate in the lymphatic tissue even when the viral load in the peripheral blood is low; therefore, the need to enhance drug delivery into lymphoid tissue is recognized by many as an effective way of targeting HIV both in systemic circulation and at its depot sites (Fletcher et al., 2014).

The physicochemical properties of nanoparticles that influence lymphatic delivery are: size, charge, molecular weight, lipophilicity, and surface ligands (surfactants, PEG, hyaluronic acid, biotin, peptides, antigens, and lectins) (Cho and Lee, 2014; Singh et al., 2014). Several antiretroviral drugs have been formulated using a wet-bead milling process and showed good stability and the desired tissue distribution. These examples include rilpivirine solid drug nanoparticles (Verloes et al., 2008; Baert et al., 2009) and a cabotegravir (S/GSK1265744) nanocrystal formulation (Spreen et al., 2013). Other approaches focusing on the use of nanoformulations for oral delivery of these drugs are in progress in order to address another problem related to poor patient adherence to antiretroviral therapy (Prinapori and Di Biagio, 2015). A more comprehensive review of this subject is available in a recent review by Liptrott et al. (2016).

5.2.3.2 Immunosuppressive and anti-inflammatory

Nanoparticles can be immunosuppressive per se or used to deliver immunosuppressive drugs. For example, inhaling carbon nanotubes was shown to suppress humoral immune response via a mechanism involving the production of transforming growth factor beta (TGF-β) by alveolar macrophages and subsequent prostaglandin production by spleenocytes leading to the systemic immunosuppression (Mitchell et al., 2009). Other examples of nanoparticles displaying immunosuppressive activities without bearing a therapeutic payload include the imaging agent Resovist, a single intravenous administration of which resulted in suppression of the antibody response to the model antigen (Shen et al., 2011). The water-soluble fullerene derivative polyhydroxy C60 was shown to inhibit type I hypersensitivity reactions to allergens, both in vitro and in vivo (Ryan et al., 2007). Likewise, allergen-loaded poly(D,L-lactic-co-glycolic) acid (PLGA) particles, chitosan, poly(lactic acid) (PLA), poly [methyl vinyl ether-co-maleic anhydride) nanoparticles, and dendrosomes were used to suppress type I and type II reactions to environmental and food allergens (Roy et al., 1999; Scholl et al., 2004; Balenga et al., 2006; Gomez et al., 2007; Gomez et al., 2008), while synthetic peptide dendrimers were reported to block experimental allergic encephalomyelitis (Wegmann et al., 2008).

Other studies have shown the benefits of using nanoparticles for the targeted delivery of immunosuppressive and anti-inflammatory drugs, and to prevent the undesirable immunosuppression of small-molecule drugs (Stinchcombe et al., 2007). For example, using PLGA nanoparticles formulated to deliver glucocorticoids to inflamed joints in the mouse model of arthritis resulted in complete remission of the inflammatory response. The improved efficacy was due to the targeted and controlled release of steroids from the nanocarrier (Higaki et al., 2005). Liposomes loaded with clodronate were used to specifically eliminate macrophages in a swine model to protect animals from endotoxin-mediated lung injury (Gaca et al., 2003). Liposomal and polymeric nanoparticle reformulation of cyclosporine was reported to reduce off-target side effects (e.g., nephrotoxicity) (Freise et al., 1994; Italia et al., 2007). Tacrolimus delivery using lipid nanoparticles resulted in improved skin penetration and tissue deposition, as well as a reduction in side effects (Pople and Singh, 2012). Polylactide nanoparticles were used for ex vivo delivery of cyclosporine A into DCc (Azzi et al., 2010). Reinjection of these drug-loaded DCs into the footpads of mice improved drug delivery to the lymph nodes, where released cyclosporine suppressed T-cell proliferation (Azzi et al., 2010). Delivery of rapamycin by elastin-like polymeric nanoparticles resulted in reduced nephrotoxicity and injection site reactions, while demonstrating comparable efficacy (Shah et al., 2013).

The activity of liposomal formulations of glucocorticoids provides another example of how nanoparticles can alter a drug's tissue distribution so that it provides additional beneficial effects. In this example, the free drug affects T-lymphocytes, while its liposomal counterpart targets macrophages and induces an alternatively activated M2 phenotype, leading to the expression of anti-inflammatory cytokines and, consequently, reduced inflammation (Schweingruber et al., 2011). Nanotechnology is used not only to deliver single drugs, but also to co-deliver anti-inflammatory agents with different mechanisms of action. For example, dexamethasone-loaded PLGA nanoparticles can be combined with siRNA-targeting COX-2 to suppress inflammatory responses (Park et al., 2012). More examples illustrating the benefits of delivering immunosuppressive and anti-inflammatory drugs using nanoparticles have been recently reviewed elsewhere (Ilinskaya and Dobrovolskaia, 2014).

5.2.4 Application of Nanoparticles for Cancer Immunotherapy

Cancer immunotherapy is another rapidly growing branch of nanomedicine. Drugs used for cancer immunotherapy vary both in structure and by mechanism of action. For example, Iipilimumab (anti-CTLA4) directly interacts with and activates immune cells by removing co-inhibitory signal, while GVAX (GM-CSF tumor vaccine) improves tumor recognition by making the tumor more immunostimulatory (Ali and Lee, 2015; Lipson et al., 2015). A recent study by Fiering et al. demonstrated the use of iron oxide nanoparticles and an alternating magnetic field to induce local hyperthermia in melanoma. Interestingly, besides heat-mediated tumor ablation, this treatment resulted in a potent CD8+ T-cell–dependent response against the tumor, preventing the recurrence of tumor growth (Toraya-Brown et al., 2014). Another study demonstrated that tumor-associated myeloid-derived suppressor cells (MDSCs) characterized by high levels of oxidative reactions may be responsible for the degradation of chemotherapeutic drugs in the tumor environment, and that this degradation could be significantly reduced by drug loading onto functionalized carbon nanotubes (Seo et al., 2015). This example offers an effective way to prolong drug function in the tumor environment by using a nanodelivery approach. Other recent examples include using poly (ethyleneimine) nanoparticles for the delivery of antiPD-1 siRNA (Teo et al., 2015), branched polyethyleneimine-superparamagnetic iron oxide nanoparticles for the enhancement of Th1 cell polarization of DCs (Hoang et al., 2015), 6-thioguanine-loaded polymeric micelles for the depletion of MDSCs (Jeanbart et al., 2015), and a CD4-targeted, oil-in-water emulsion for the co-delivery of IL-2 and TGF-β (McHugh et al., 2015).

5.2.5 Understanding the Role of the Immune System in Nanoparticle Biodegradation

The therapeutic use of biodegradable nanoparticles is accompanied by fewer safety concerns than those of durable nanomaterials. Not surprisingly, the majority of currently marketed nanomedicines are represented by biodegradable, lipid-based materials (Etheridge et al., 2013). Macrophages are known to play an active role in internalizing and degrading these nanocarriers (Song et al., 2012).

The use of nonbiodegradable nanoparticles is often associated with concern regarding their bioaccumulation and long-term toxicity. In this context, recent studies demonstrating the unique role of activated neutrophils in the enzymatic digestion of durable nanoparticles are very encouraging. Specifically, activated neutrophils were reported to participate in the biodegradation of carbon nanotubes. Myeloperoxidase (MPO)-reactive intermediates and hypochlorous acid (HClO) generated by MPO are believed to contribute to the biodegradation process (Kagan et al., 2010a; Bianco et al., 2011; Shvedova et al., 2012), and the existence of this mechanism was confirmed in vivo using MPO-deficient mice (Shvedova et al., 2012). MPO was also found to have a role in the oxidative biodegradation of single-wall carbon nanotubes activated human neutrophils (Kagan et al., 2010a). Moreover, additional research suggests that peroxynitrite (ONOO$^-$)-driven oxidation, resulting from the enzymatic activities of NADPH-oxidase/inducible nitric oxide synthase (iNOS), functions as another carbon nanotube biodegradation pathway in activated macrophages (Vlasova et al., 2012; Bhattacharya et al., 2014; Farrera et al., 2014; Kagan et al., 2014). Recent data demonstrated that tumor-associated MDSCs expressing high levels of arginase, iNOS, NADPH oxidase, and MPO were also able to biodegrade carbon nanotubes via oxidative pathways. These MDSC properties were recently utilized for an interesting nanodelivery approach in which nitrogen-doped carbon nanotube cups (NCNCs) were loaded with therapeutic cargo (paclitaxel), sealed via conjugation with gold nanoparticles (GNPs), and opened by enzymatic oxidative processes within, or in the immediate proximity of, MDSCs. This mechanism was proposed to enhance antitumor immune responses by targeting and inactivating MDSCs using their own highly expressed oxidative enzymatic machinery (Zhao et al., 2015).

5.2.6 Application of Nanoparticles to Reducing the Immunotoxicity of Traditionally Formulated Drugs

In addition to their use as carriers of novel drugs, engineered nanomaterials are also increasingly being used to reformulate traditional drugs (low-molecular-weight drugs, therapeutic proteins, therapeutic antibodies, and nucleic acids). Reformulation of traditional drugs using nanotechnology has been shown to improve drug solubility and pharmacokinetics, as well as to reduce undesirable side effects. Below, we review several examples demonstrating how reformulation of traditional drugs using nanotechnology resulted in reduced immunotoxicity.

The traditional formulation of the cytotoxic oncology drug paclitaxel relies on the polyethoxylated castor oil excipient Cremophor-EL®, which is known to induce anaphylaxis in sensitive individuals. The anaphylactic reaction to Cremophor-EL is mediated by its ability to activate the complement system (Weiszhar et al., 2012). Due to this side effect, the Cremophor-EL formulation of paclitaxel (Taxol®) has to be administered via slow infusion, and after patient premedication with immunosuppressive drugs. In contrast to Taxol, the nanoalbumin formulation of paclitaxel (Abraxane®) is administered via push injection, and does not require premedication (Gradishar et al., 2005). Likewise, TNF-α failed in clinical trials because it induces systemic immunostimulation, resulting in fever and hypotension. However, TNF-α immobilized on the surface of PEGylated colloidal gold nanoparticles (Cyt6091) has successfully passed a Phase I trial (Libutti et al., 2010). Therapeutic protein immunogenicity, which leads to the formation of antibodies that neutralize the drug and, in some cases, endogenous proteins, is a common reason for the discontinuation of such drugs (Rosenberg et al., 2012). Liposomes were shown to reduce the immunogenicity of recombinant coagulation factor

FVIII and to protect encapsulated protein from the antibodies formed in response to the traditional formulation of this therapeutic protein (Ramani et al., 2008).

Other success stories demonstrating the reduction of immunotoxicity by reformulation of a traditional drug using nanotechnology-based carriers include the encapsulation of therapeutic antisense oligonucleotides into liposomes to prevent activation of the complement system (Klimuk et al., 2000) and to reduce cytokine-mediated toxicities (Yu et al., 2013), as well as the reformulation of the small-molecule oncology drug 5-fluorouracil, using chitosan nanoparticles to decrease its hematotoxicity (Giacalone et al., 2013; Cheng et al., 2014).

5.2.7 Understanding the Applicability of the Existing Regulatory Framework for Immunotoxicity Analysis of Nanoformulations

Several nanotechnology-formulated drugs have already reached the market. Examples include solid lipid nanoparticles (e.g., Leunesse®), liposomes (e.g., Doxil®), protein-based nanoparticles (e.g., Abraxane), as well as nanocrystals (e.g., Emend®). Regulatory experience gained from working with these formulations, along with data from preclinical reports, has helped to develop a better understanding of how to apply the regulatory framework established and used for small-molecule and macromolecular therapeutics to complex nanotechnology formulations (Tyner and Sadrieh, 2011; Bancos et al., 2013; Cruz et al., 2013; Tyner et al., 2015). The properties shared by nanotechnology-based carriers that have already reached the market are biodegradability (e.g., emulsions and liposomes) and the ability to quickly dissolve in the body (e.g., nanocrystals). Durable nanoparticles (e.g., metals and metal colloids) are expected to accumulate in the body and, as such, raise immunological safety concerns. These concerns stem from the notion that durable materials tend to distribute to the MPS (Sadauskas et al., 2009; Di Gioacchino et al., 2011; Moon et al., 2011; Umbreit et al., 2012). Accumulation of these materials in the MPS may affect the normal function of this system, and therefore this concern should not be ignored.

The FDA applies the same regulatory framework established for all small-molecule and macromolecular therapeutics to the regulation of complex nanotechnology products (Bancos et al., 2013). The manufacture, characterization, and non-clinical safety evaluation of combination products containing nanotechnology platforms have also been covered in several recent regulatory documents, which recommend conducting studies relevant to each component of the complex formulation (CDER, 2006; CDER, 2008). For example, the International Conference on Harmonization (ICH) S8 and S6 guidelines are consulted to estimate the immunotoxicity of small-molecule and macromolecular components, respectively, of any nanotechnology-formulated combination product (CDER, 2001; Bancos et al., 2013).

5.3 Disappointments

5.3.1 Harmonization of Testing Is Still in Progress

Ten years ago, international standards development organizations ASTM International and ISO established the E56 and TC229 committees, respectively, to lead the development of standard test methods for nanomaterials. This effort resulted in the development of three standardized test methods—ASTM E56-2524-08(2013), ASTM E56-2525-08(2013), and ISO 29701:2010—for the analysis of the hemolytic properties of nanoparticles, the effects of nanoparticulate materials on the formation of mouse colony-forming unit granulocyte/macrophage colonies, and of potential endotoxin contamination in nanomaterials, respectively (http://www.astm.org/COMMIT/SUBCOMMIT/E5603.htm; http://www.iso.org/iso/iso_catalogue/catalogue_tc/catalogue_tc_browse.htm?commid=381983). While these harmonized methods are important to support immunotoxicity studies, they are obviously insufficient to address the broad spectrum of end-points that are indicative of nanoparticle immunotoxicity. Despite global efforts, there is no harmonization in the following areas: dose selection, dose metrics, assay format, cell species, and matrices. Recently, an integrated approach was proposed to estimate relevant in vitro doses of tested nanomaterials in terms of total mass, surface area, and particle number (Cohen et al., 2014). While the results of this study are encouraging, adaptation of this approach by other laboratories and harmonization efforts are needed.

5.3.2 Lack of Nanoparticle Reference Standards for Immunotoxicity Studies

Reference standards are well-characterized materials with known properties. These materials can be used to validate toxicology protocols and ensure the quality of measurements specific to a given protocol measuring given end points. It is generally acknowledged that the lack of nanoparticle reference standards limits the validation of instruments, protocols, and materials used to assess exposure to nanomaterials and understand their biocompatibility. While many types of nanomaterials have been linked to certain types of immunotoxicity (e.g., cationic dendrimers are thrombogenic; PEGylated liposomes induce anaphylaxis), there are no standard reference materials to use for these and other types of immunotoxicity studies. Stefaniak et al. conducted a literature review and identified 25 nanomaterials that were considered to be candidate nanoparticle reference materials by standards development organizations worldwide (Stefaniak et al., 2013). Interestingly, this study found a limited consensus regarding the types of candidate nanoparticles between various organizations involved in the development of reference materials: the U.S. National Metrology Institute, the U.S. National Institute of Standards and Technology, the REFNANO project funded by the UK government, the Organization for Economic Cooperation and Development, and several NanoImpactNet projects (NanoImpactNet, NanoSustain, and NanoValid) funded by the European Commission (Stefaniak et al., 2013).

In the absence of consensus on and availability of relevant, well-characterized reference materials, immunotoxicity studies are conducted using traditional positive controls, such as lipopolysaccharide for cytokine induction, phytohemagglutinin for leukocyte proliferation, Triton X-100 for hemolysis, and cobra venom factor for complement activation (Dobrovolskaia, 2015). Preclinical studies often rely on nanomedicines with known clinical immunotoxicities (e.g., Doxil or Taxol for complement activation and anaphylaxis tests) (Dobrovolskaia, 2015). Despite the thorough characterization of their physicochemical properties, the main limitation in using these products as reference materials is their expense and limited accessibility to the nanotechnology research community.

5.3.3 Examples of Nanoformulations Demonstrating Reduced Immunotoxicity of Traditionally Immunotoxic APIs Are Fewer Than Expected

Despite clear examples demonstrating the potential of nanotechnology for reducing the immunotoxicity of traditional active pharmaceutical ingredients (APIs), the systemic administration of many nanoformulations to patients still requires immunosuppressive and/or antipyretic premedication. For example, the complexing of therapeutic nucleic acids with a nanocarrier is commonly based on electrostatic interactions. As such, the lipid and polymeric carriers used to formulate therapeutic nucleic acids tend to be cationic. As discussed earlier, cationic particles are known to be cytotoxic to a variety of immune cells, induce cytokine secretion, exaggerate endotoxin-mediated toxicities, activate complement, bind plasma proteins, trigger pro-coagulant activity, and may also affect protein conformation and function (Pantic, 2011; Boraschi et al., 2012; Dobrovolskaia and McNeil, 2013a). These data raise safety concerns. Indeed, several studies have demonstrated that base modifications were insufficient to reduce the immunostimulatory activity of certain nucleic acids (e.g., siRNA). Furthermore, lipid nanocarriers were shown to contribute to the drug's immunostimulation (Abrams et al., 2010). Examples of immunostimulatory lipids used to prepare nanocarriers for therapeutic nucleic acids include 2-(4-[(3β)-cholest-5-en-3-yloxy]butoxy)-*N*,*N*-di-methyl-3-[(9Z,12Z)-octadeca-9,12-dien-1-yloxy]propan-1-amine (CLinDMA) and PEG-dimyristoylglycerol (Abrams et al., 2010), protamine (Li et al., 1999), trimethyl ammonium propane-cholesterol (Kim et al., 2007), and 1,3-dioleoyl-3-trimethylammonium propane (DOTAP) (Li et al., 1999; Vasievich et al., 2011). Clinical studies investigating the safety of such formulations are often designed to prevent adverse reactions by premedicating patients with immunosuppressive cocktails containing immuno-suppressive agents (e.g dexamethasone), antipyretic agents (e.g acetaminophen), histamine H1 receptor blockers (e.g., diphenhydramine), and histamine H2 receptor blockers (e.g., ranitidine) (Coelho et al., 2013). It is clear that the advantages of tuning the physicochemical properties of nanoparticles for the purpose of reducing drug immunotoxicity have not yet been fully explored. The continuing evolution of the framework for evaluating nanoparticle immunotoxicity is the likely reason for this observation.

5.3.4 The Mechanisms of Nanoparticle Immunotoxicity Are Largely Unknown

The majority of studies reported in the past decade were focused on understanding the structure–activity relationship between the physicochemical properties nanoparticles and immunotoxicity. Uncovering the mechanisms of nanoparticle immunotoxicity is still a work in progress. Several interesting mechanisms have been recently described. For example, the inhibition of phosphoinositol-3-kinase (PI3K) by cationic PAMAM dendrimers has been suggested to contribute to the exaggeration of lipopolysaccharide (LPS)-induced leukocyte PCA in human peripheral blood mononuclear cells by these particles (Ilinskaya et al., 2014). The activation of mitogen-activated protein kinase (MAPK) p38 by gold nanoparticles functionalized with α-lipoyl-ω-methoxy poly (ethylene glycol) was proposed as a mechanism for exaggeration of LPS-triggered nitric oxide and IL-6 secretion in murine macrophages (Liu et al., 2012). The binding of colloidal gold nanoparticles to high-mobility group B-1 was implicated in the attenuation of nuclear factor-kappaB (NF-kB) signaling, the phosphorylation of JNK, and the secretion of TNF-α triggered by TLR9 activation by CpG oligonucleotides (Tsai et al., 2012).

The induction of oxidative and nitrosative stress by zinc oxide nanoparticles was described as a mechanism through which these particles activate redox-sensitive NF-κB and MAPK signaling pathways, leading to an inflammatory response in human monocytes (Senapati et al., 2015). The oxidative stress induced by the nanoemulsion Cremophor-EL was also suggested to trigger IL-8 production by human monocytes via a mechanism that bypasses gene expression (Ilinskaya et al., 2015).

It is also important to note that the effects of nanomaterials on immune cells may result in both suppression of the immune effector cells and activation of the immune regulatory (immunosuppressive) cells. Therefore, any discussion of nanoparticle immunotoxicity should consider specific immune cell subsets. For instance, airborne carbon nanotubes have recently been reported to induce rapid accumulation of pulmonary MDSCs in mice, which was associated with the accelerated growth of lung carcinomas in vivo (Shvedova et al., 2013). Further analysis of the mechanism revealed that carbon nanotubes may presensitize MDSCs to produce the immunosuppressive cytokine TGF-β, which contributes to the observed immunosuppression and, as a consequence, tumor growth (Shvedova et al., 2013). Thus, both uncovering the immunomodulatory properties of nanomaterials and understanding the molecular and cellular mechanisms of their activities are important for the clinical translation of these materials and potential minimization of any undesirable immunoreactivity.

Further research is clearly needed to uncover additional mechanisms and to link them to the physicochemical properties of nanoparticles.

5.3.5 Accidental Nanoparticulate Contaminants Contribute to the Immunogenicity of Therapeutic Proteins

The observation that particulate materials between 0.1 and 10 μm in size can contaminate recombinant protein therapeutics and contribute to their immunogenicity has generated increasing levels of concern (Carpenter et al., 2010). The mechanism by which this occurs has not been fully investigated; however, several factors, such as nanoparticle-triggered protein aggregation (Mire-Sluis et al., 2011; Van Beers et al., 2012); the adsorption of proteins on the particle surface, leading to the formation of highly immunogenic repeated-protein structures (Mire-Sluis et al., 2011; Van Beers et al., 2012); and the exaggeration of inflammatory responses triggered by trace amounts of endotoxin (Dobrovolskaia et al., 2016) may play a role. For example, tungsten microparticles originating from the tungsten pins used in the manufacturing process were shown to induce protein aggregation and increase the immunogenicity of a recombinant protein product (Liu et al., 2010).

Another study suggested that hydrophobic metal, glass, and polystyrene particulates adsorb proteins on the particle surface and contribute to protein aggregation and immunogenicity (Van Beers et al., 2012). Several other materials were named among common particulates found to contaminate therapeutic proteins, including cellulose and glass fibers, silicon oil, rubber, stainless steel, fluoropolymers, and plastics (Carpenter et al., 2010; Liu et al., 2010; Van Beers et al., 2012). Cellulose fibers were shown to exaggerate the production of endotoxin-induced pro-inflammatory cytokines (Dobrovolskaia et al., 2016). Gowns and other materials used in clean rooms during the manufacturing of therapeutic proteins—closures, filling pumps, containers, vial stoppers, etc.—serve as sources of these particulate contaminants (Carpenter et al., 2010). These data led regulatory agencies to require the detection and characterization of particulate materials that can contaminate therapeutic proteins; however, there is no general agreement as to what methods should be used (Carpenter et al., 2010). The major challenge in understanding the mechanism of therapeutic protein immunogenicity in the presence of contaminating particles is the limited quantities of these particulate materials. Even when a particulate contaminant is detected and isolated from the protein product, the quantity of the isolate is usually insufficient to conduct a follow-up mechanistic study (Carpenter et al., 2010). It is obvious that more research is needed to address this important issue.

5.4 Lessons Learned

5.4.1 Nanoparticles' Physicochemical Properties Are the Keys to Determining Particle Interaction with the Immune System

It is now well established that nanoparticles can be engineered to either avoid or specifically interact with the immune system. The tuning of nanoparticles to attain desirable attributes can be achieved by manipulating the physicochemical properties (size, charge, hydrophobicity, shape) of the particle that determine its interaction with plasma proteins and immune cells. This subject has been extensively reviewed elsewhere (Smith et al., 2013). We have also discussed many examples underlining this point in the achievement section of this paper.

5.4.2 Chemical (e.g., Synthesis Byproducts) and Biological (e.g., Endotoxin) Impurities Can Contribute to Nanoparticle Toxicity

It is very important to distinguish nanoparticle-mediated immunotoxicities from those triggered by chemical and biological impurities. The presence of traces of CTAB used as a stabilizing agent in the synthesis of gold nanorods was implicated in the cytotoxicity of these particles (Leonov et al., 2008). Iron and nickel used to catalyze reactions involved in the synthesis of carbon nanotubes were shown to trigger inflammatory reactions in response to nanotube exposure (Madani et al., 2013). Bacterial endotoxin is a common biological impurity affecting over 30% of preclinical-grade nanomaterials (Crist et al., 2013; Dobrovolskaia and McNeil, 2013b). If not properly identified in and eliminated from nanoformulations, endotoxin can confound the results of both nanoparticle immunotoxicity and efficacy studies. The elimination of endotoxin from nanoparticles was shown to reduce their immunotoxicity (Vallhov et al., 2006). Moreover, some nanomaterials, while not inflammatory themselves, were able to potentiate endotoxin-mediated inflammation. Silica- and carbon-based nanomaterials, as well as some metal oxides, have been shown to exaggerate endotoxin-mediated inflammation in the lungs (Inoue et al., 2006; Inoue et al., 2007; Shi et al., 2010; Inoue, 2011; Inoue and Takano, 2011), while cationic PAMAM dendrimers were reported to exaggerate endotoxin-induced leukocyte PCA (Dobrovolskaia et al., 2012; Ilinskaya et al., 2014). Strategies for endotoxin detection have been discussed elsewhere (Dobrovolskaia, 2015).

5.4.3 Thorough Physicochemical Characterization Is Needed Prior to Nanoparticle Analysis in Immunological Assays

Since the physicochemical properties of nanoparticles are the keys to determining particle interaction with the immune system, thorough particle characterization is needed prior to immunotoxicity studies (Clogston and Patri, 2013). Several examples have shown that instability of particle surface coatings results in inflammatory reactions, yet partial loss of the surface coating is undetectable by dynamic light scattering and zeta potential analysis, traditionally used for particle characterization (Clogston and Patri, 2013; Crist et al., 2013). Certain processes and reagents commonly involved in nanoparticle research (e.g., sterilization procedures and inhibitors used for signal transduction studies) may also affect nanoparticle integrity (Zheng et al., 2011; Dobrovolskaia et al., 2016). Missing such details may lead to misinterpretation of the study results and faulty conclusions. Another important lesson has been learned from using nanoparticles for drug delivery: drug conjugation to a nanotechnology-based carrier may change the drug's original properties. For example, celastrol conjugated to a dendrimer carrier retained its ability to suppress LPS-induced nitric oxide release, but lost its ability to inhibit production of pro-inflammatory cytokines (Boridy et al., 2012).

5.4.4 Total Protein Binding Serves as Good Indicator of Nanoparticle Stealthiness and Its Distribution to the MPS

Proteins bind to a nanoparticle surface instantaneously upon entry of the particle into the bloodstream. Some of these proteins stay on the surface as long as the particles circulate in the bloodstream, while others dissociate from the particle surface or get replaced by proteins with a higher binding affinity. Protein binding was shown to affect nanoparticle hydrodynamic size and charge (Dobrovolskaia et al., 2009b), and was also suggested to influence the way cells and tissues interact with and process the particles, ultimately guiding cellular uptake, clearance route, and tissue distribution (Goppert and Muller, 2005; Michaelis et al., 2006; Nagayama et al., 2007; Zensi et al., 2009). It is now well established that total protein binding can serve as an indicator of particle "stealthiness." Stealthy particles tend to stay in circulation longer. In contrast, particles with proteins bound to their surface are cleared by the cells of the MPS.

5.4.5 Composition of Protein Corona Is Insufficient to Predict Nanoparticle Immunocompatibility

In contrast to total protein binding, the composition of the protein corona has less predictive value. Complement and fibrinogen are abundant plasma proteins, and they have been reported as part of the so-called "protein corona" for many engineered nanomaterials. However, the presence of these proteins on the particle surface per se does not mean that the function of these proteins will be activated (Salvador-Morales et al., 2006; Salvador-Morales et al., 2008; Deng et al., 2009; Deng et al., 2011; Deng et al., 2012a; Deng et al., 2012b). The particle concentration needed to deplete these proteins to a level that would affect their function must be very high. Achieving such concentration in vitro is possible for some, but not all, nanoparticles. Moreover, protein levels in vivo vary from donor to donor, and even within the same donor, due to homeostasis. Therefore, the identification of the composition of a nanoparticle's protein corona cannot reliably serve as a predictor of particle immunotoxicity (Salvador-Morales et al., 2006; Salvador-Morales et al., 2008; Deng et al., 2009; Deng et al., 2011; Deng et al., 2012a; Deng et al., 2012b; Dobrovolskaia et al., 2014).

5.4.6 Common Markers of Nanoparticle Immunotoxicities Halting Clinical Translation of Nanomedicines

The toxicities that represent the most common safety concerns and reasons for nanoparticle failure in the preclinical stage include erythrocyte damage, thrombogenicity (platelet aggregation, plasma coagulation, DIC, and leukocyte PCA), cytokine-mediated inflammation and cytokine storming, pyrogenicity (mainly due to bacterial endotoxin contamination), and anaphylaxis and other complement activation-mediated reactions, as well as recognition and uptake by the cells of the MPS (Dobrovolskaia, 2015). As such, screening for these toxicities early in preclinical development helps in eliminating potentially toxic candidates. Figure 5.3 can be consulted as a guide to selecting the most appropriate screening method based on the immunotoxicity endpoint, which can be suggested by nanoparticle physicochemical properties.

5.5 Conclusions

The past decade has been full of exciting discoveries: unraveling trends in nanoparticle immunocompatibility, understanding the role of particle characterization, and finding therapeutic applications for a variety of nanocarriers. Since the mechanisms of nanoparticle immunotoxicity are not completely understood, the next decade of research should focus on identifying mechanisms and mapping them to the physicochemical properties of nanoparticles.

Disclosures and Conflict of Interest

This chapter was originally published as: Dobrovolskaia, M. A., Shurin, M., and Shvedova, A. A. Current understanding of interactions between nanoparticles and the immune system. *Toxicol. Appl. Pharmacol.* **299**, 78–89 (2015); DOI: 10.1016/j.taap.2015.12.022, and appears here, with edits, by kind permission of the publisher, Elsevier Inc. The content and conclusions of this publication are those of the authors and do not necessarily reflect the views or policies of the Department of Health and Human Services, National Institute for Occupational Safety and Health, Center for Disease Control, nor does the mention of trade names, commercial products, or organizations imply endorsement by the U.S. government. M.D.'s work has been funded with federal funds from the National Cancer Institute, National Institutes of Health, under contract HHSN261200800001E. The authors declare that they have no conflict of interest. No writing assistance was utilized in the production of this chapter and the authors have received no payment for its preparation.

Corresponding Author

Dr. Marina A. Dobrovolskaia
Nanotechnology Characterization Laboratory
Cancer Research Technology Program
Leidos Biomedical Research Inc.
Frederick National Laboratory for Cancer Research
NCI at Frederick, 1050 Boyles Street, Frederick MD 21702, USA
E-mail: marina@mail.nih.gov

About the Authors

Marina A. Dobrovolskaia is a senior principal scientist and head of the Immunology Section at the Nanotechnology Characterization Laboratory (NCL) in Frederick, Maryland, USA. She joined the NCL in February 2005 to establish immunology assay cascade and build a preclinical framework to conduct an immunological safety assessment of nanoparticle platforms and nanotechnology-formulated drugs and imaging agents. She currently directs characterization related to nanomaterials' interactions with components of the immune system. Dr. Dobrovolskaia's team develops, validates, and qualifies the performance of in vitro and in vivo assays to support the preclinical characterization of nanoparticles, and to monitor nanoparticle purity from biological contaminants such as bacteria, yeast, mold, and endotoxin, as well as conducting basic research to advance translational science and improve understanding of the mechanisms of nanoparticle immunotoxicity. Dr. Dobrovolskaia is a member of several working groups on Nanomedicine, Oligonucleotide Safety, and Endotoxin Detection. She has published more than 45 peer-reviewed papers pertaining to nanomaterial interactions with the immune system and prepared and edited 2 editions of the *Handbook of Immunological Properties of Engineered Nanomaterials*, which received international recognition. Dr. Dobrovolskaia is an invited speaker at numerous national and international nanotechnology-related conferences. She has served a four-year term as a special associate editor for *Nanomedicine: Nanotechnology, Biology and Medicine* (Elsevier) and is currently an editorial board member of the *Journal of Nanotoxicology and Nanomedicine* (IGI Global). Before joining the NCL, Dr. Dobrovolskaia worked as a research scientist in a GLP laboratory at PPD Development, Inc., in Richmond, Virgina, where she was responsible for the design, development, and validation of bioanalytical ligand-binding assays to support pharmacokinetic and toxicity studies in a variety of drug development projects. She received her MS degree from the Kazan State University in Russia, PhD from the N.N. Blokhin Cancer Research Center of the Russian Academy of Medical Sciences in Moscow, Russia, and MBA degree from Hood College in Frederick, Maryland, and completed two postdoctoral trainings at the National Cancer Institute in Frederick, Maryland, and the University of Maryland in Baltimore. She is also a member of Project Management Institute and a certified Project Management Professional. Her areas of research expertise include cell signaling, innate immunity, immunotoxicity of complex drug formulations, bioanalytical methodology, and endotoxin detection and quantification.

Michael R. Shurin is professor of Pathology and Immunology and associate director of Clinical Immunopathology at the University of Pittsburgh Medical Center, USA. He graduated from 2nd Medical School in Moscow and obtained his PhD degree in immunopharmacology. He was additionally trained at the University of Pittsburgh. His research focuses on the tumor microenvironment and its intrinsic and extrinsic regulation by different agents, including environmental factors.

Anna Shvedova's specialty interests include mechanism(s) of chronic skin and lung diseases caused by industrial chemicals, vapors, and particulates. Her studies have been focused on focus mechanisms of free radical reactions and inflammatory responses, molecular mechanisms of cell damage and apoptosis, oxidant injury, and cancer causing of occupational disease and injury, and on contributing to the development of biomarkers and valid strategies of intervention and prevention. Dr. Shvedova's lab is investigating the toxicology of nanomaterials and mechanism(s) of adverse effects of nanoscale products, development of novel cellular and animal models, and development of biomarkers of exposures. She has published over 140 scientific papers and book chapters in the toxicology field, co-editor of the first and second editions of *Adverse Effects of Engineered Nanomaterials: Exposure, Toxicology, and Impact on Human Health*, published by Elsevier. She was the founder and first president of the Dermal Toxicology Specialty Section of the Society of Toxicology (SOT). She has been serving as a scientific advisor of several activities focused on the dissemination of nanotechnologies to the public. She is a member of advisory board committees for US Army, Air Force, and NASA projects and NRC/Working Environment/Nanosafety, DK-2100 Copenhagen, Denmark. She is WP4/PI of the NANOMMUNE and partner of the NANOSOLUTION EU FW7 program projects. Dr. Shvedova has been honored with a SOT Public Communication Award in 2001; Alice Hamilton NIOSH Award for the paper of the year in Occupational Safety and Health in 2006 and 2009; Bullard-Sherwood Award: Research for Practice in 2011; and Women in Toxicology SOT Award in 2007. She is a board member of the working group on Skin Notation at the National Institute of Occupational Safety and Health/CDC, associate editor of *Toxicology and Applied Pharmacology*, and an editorial board member for *Regulatory Toxicology*.

References

Abrams, M. T., Koser, M. L., Seitzer, J., Williams, S. C., DiPietro, M. A., Wang, W., Shaw, A. W., Mao, X., Jadhav, V., Davide, J. P., Burke, P. A., Sachs, A. B., Stirdivant, S. M., Sepp-Lorenzino, L. (2010). Evaluation of efficacy, biodistribution, and inflammation for a potent siRNA nanoparticle: Effect of dexamethasone co-treatment. *Mol. Ther.*, **18**, 171–180.

Agashe, H. B., Dutta, T., Garg, M., Jain, N. K. (2006). Investigations on the toxicological profile of functionalized fifth-generation poly(propylene imine) dendrimer. *J. Pharm. Pharmacol.*, **58**, 1491–1498.

Agrawal, P., Gupta, U., Jain, N. K. (2007). Glycoconjugated peptide dendrimers-based nanoparticulate system for the delivery of chloroquine phosphate. *Biomaterials*, **28**, 3349–3359.

Ali, S., Lee, S. K. (2015). Ipilimumab therapy for melanoma: A mimic of leptomeningeal metastases. *AJNR Am. J. Neuroradiol.*, **36**, E69–E70.

Aoyama, Y., Kanamori, T., Nakai, T., Sasaki, T., Horiuchi, S., Sando, S., Niidome, T. (2003). Artificial viruses and their application to gene delivery. Size-controlled gene coating with glycocluster nanoparticles. *J. Am. Chem. Soc.*, **125**, 3455–3457.

Asthana, A., Chauhan, A. S., Diwan, P. V., Jain, N. K. (2005). Poly(amidoamine) (PAMAM) dendritic nanostructures for controlled site-specific delivery of acidic anti-inflammatory active ingredient. *AAPS PharmSciTech*, **6**, E536–E542.

Azzi, J., Tang, L., Moore, R., Tong, R., El Haddad, N., Akiyoshi, T., Mfarrej, B., Yang, S., Jurewicz, M., Ichimura, T., Lindeman, N., Cheng, J., Abdi,

R. (2010). Polylactide-cyclosporin A nanoparticles for targeted immunosuppression. *FASEB J.*, **24**, 3927–3938.

Baert, L., Van't Klooster, G., Dries, W., Francois, M., Wouters, A., Basstanie, E., Iterbeke, K., Stappers, F., Stevens, P., Schueller, L., Van Remoortere, P., Kraus, G., Wigerinck, P., Rosier, J. (2009). Development of a long-acting injectable formulation with nanoparticles of rilpivirine (TMC278) for HIV treatment. *Eur. J. Pharm. Biopharm.*, **72**, 502–508.

Balenga, N. A., Zahedifard, F., Weiss, R., Sarbolouki, M. N., Thalhamer, J., Rafati, S. (2006). Protective efficiency of dendrosomes as novel nano-sized adjuvants for DNA vaccination against birch pollen allergy. *J. Biotechnol.*, **124**, 602–614.

Bancos, S., Tyner, K. M., Weaver, J. L. (2013). Immunotoxicity testing of drug-nanoparticle conjugates: Regulatory considerations. In: Dobrovolskaia, M. A., McNeil, S. E., eds. *Handbook of Immunological Properties of Engineered Nanomaterials*, World Scientific Publishing, Singapore, pp. 671–685.

Baron, L., Gombault, A., Fanny, M., Villeret, B., Savigny, F., Guillou, N., Panek, C., Le Bert, M., Lagente, V., Rassendren, F., Riteau, N., Couillin, I. (2015). The NLRP3 inflammasome is activated by nanoparticles through ATP, ADP and adenosine. *Cell Death Dis.*, **6**, e1629.

Benning, P. J., Poirier, D. M., Ohno, T. R., Chen, Y., Jost, M. B., Stepniak, F., Kroll, G. H., Weaver, J. H., Fure, J., Smalley, R. E. (1992). C60 and C70 fullerenes and potassium fullerides. *Phys. Rev. B Condens. Matter*, **45**, 6899–6913.

Bhadra, D., Bhadra, S., Jain, S., Jain, N. K. (2003). A PEGylated dendritic nanoparticulate carrier of fluorouracil. *Int. J. Pharm.*, **257**, 111–124.

Bhadra, D., Yadav, A. K., Bhadra, S., Jain, N. K. (2005). Glycodendrimeric nanoparticulate carriers of primaquine phosphate for liver targeting. *Int. J. Pharm.*, **295**, 221–233.

Bhattacharya, K., Sacchetti, C., El-Sayed, R., Fornara, A., Kotchey, G. P., Gaugler, J. A., Star, A., Bottini, M., Fadeel, B. (2014). Enzymatic "stripping" and degradation of PEGylated carbon nanotubes. *Nanoscale*, **6**, 14686–14690.

Bianco, A., Kostarelos, K., Prato, M. (2011). Making carbon nanotubes biocompatible and biodegradable. *Chem. Commun. (Camb)*, **47**, 10182–10188.

Boraschi, D., Costantino, L., Italiani, P. (2012). Interaction of nanoparticles with immunocompetent cells: Nanosafety considerations. *Nanomedicine (London)*, **7**, 121–131.

Boridy, S., Soliman, G. M., Maysinger, D. (2012). Modulation of inflammatory signaling and cytokine release from microglia by celastrol incorporated into dendrimer nanocarriers. *Nanomedicine (London)*, **7**, 1149–1165.

Carpenter, J., Cherney, B., Lubinecki, A., Ma, S., Marszal, E., Mire-Sluis, A., Nikolai, T., Novak, J., Ragheb, J., Simak, J. (2010). Meeting report on protein particles and immunogenicity of therapeutic proteins: Filling in the gaps in risk evaluation and mitigation. *Biologicals*, **38**, 602–611.

CDER (2002). Guidance for industry: Immunotoxicology evaluation of investigational new drugs, Food and Drug Administration.

CDER (2006). Guidance for industry: Non-clinical safety evaluation of drug or biologic combinations, U.S. Department of Health and Human Services, Food and Drug Administration.

CDER (2015). Guidance for industry and review staff: Nonclinical safety evaluation of reformulated drug products and products intended for administration by alternative route, Food and Drug Administration.

Cheng, M., Chen, H., Wang, Y., Xu, H., He, B., Han, J., Zhang, Z. (2014). Optimized synthesis of glycyrrhetinic acid-modified chitosan 5-fluorouracil nanoparticles and their characteristics. *Int. J. Nanomed.*, **9**, 695–710.

Cho, H. Y., Lee, Y. B. (2014). Nano-sized drug delivery systems for lymphatic delivery. *J. Nanosci. Nanotechnol.*, **14**, 868–880.

Clogston, J. D., Patri, A. K. (2013). Importance of physicochemical characterization prior to immunological studies. In: Dobrovolskaia, M. A., McNeil, S. E., eds. *Handbook of Immunological Properties of Engineered Nanomaterials*, World Scientific Publishing, Singapore.

Coelho, T., Adams, D., Silva, A., Lozeron, P., Hawkins, P. N., Mant, T., Perez, J., Chiesa, J., Warrington, S., Tranter, E., Munisamy, M., Falzone, R., Harrop, J., Cehelsky, J., Bettencourt, B. R., Geissler, M., Butler, J. S., Sehgal, A., Meyers, R. E., Chen, Q., Borland, T., Hutabarat, R. M., Clausen, V. A., Alvarez, R., Fitzgerald, K., Gamba-Vitalo, C., Nochur, S. V., Vaishnaw, A. K., Sah, D. W., Gollob, J. A., Suhr, O. B. (2013). Safety and efficacy of RNAi therapy for transthyretin amyloidosis. *N. Engl. J. Med.*, **369**, 819–829.

Cohen, J. M., Teeguarden, J. G., Demokritou, P. (2014). An integrated approach for the in vitro dosimetry of engineered nanomaterials. *Part. Fibre Toxicol.*, **11**, 20.

Crist, R. M., Grossman, J. H., Patri, A. K., Stern, S. T., Dobrovolskaia, M. A., Adiseshaiah, P. P., Clogston, J. D., McNeil, S. E. (2013). Common Pitfalls

In Nanotechnology: Lessons Learned From NCI's Nanotechnology Characterization Laboratory. *Integr. Biol. (Camb)*, **5**, 66–73.

Cruz, C. N., Tyner, K. M., Velazquez, L., Hyams, K. C., Jacobs, A., Shaw, A. B., Jiang, W., Lionberger, R., Hinderling, P., Kong, Y., Brown, P. C., Ghosh, T., Strasinger, C., Suarez-Sharp, S., Henry, D., Van Uitert, M., Sadrieh, N., Morefield, E. (2013). CDER risk assessment exercise to evaluate potential risks from the use of nanomaterials in drug products. *AAPS J.*, **15**, 623–628.

Deng, Z. J., Liang, M., Monteiro, M., Toth, I., Minchin, R. F. (2011). Nanoparticle-induced unfolding of fibrinogen promotes Mac-1 receptor activation and inflammation. *Nat. Nanotechnol.*, **6**, 39–44.

Deng, Z. J., Liang, M., Toth, I., Monteiro, M., Minchin, R. F. (2012a). Plasma protein binding of positively and negatively charged polymer-coated gold nanoparticles elicits different biological responses. *Nanotoxicology*, **7**, 314–322.

Deng, Z. J., Liang, M., Toth, I., Monteiro, M. J., Minchin, R. F. (2012b). Molecular interaction of poly(acrylic acid) gold nanoparticles with human fibrinogen. *ACS Nano*, **6**, 8962–8969.

Deng, Z. J., Mortimer, G., Schiller, T., Musumeci, A., Martin, D., Minchin, R. F. (2009). Differential plasma protein binding to metal oxide nanoparticles. *Nanotechnology*, **20**, 455101.

Di Gioacchino, M., Petrarca, C., Lazzarin, F., Di Giampaolo, L., Sabbioni, E., Boscolo, P., Mariani-Costantini, R., Bernardini, G. (2011). Immunotoxicity of nanoparticles. *Int. J. Immunopathol. Pharmacol.*, **24**, 65S–71S.

Dobrovolskaia, M. A. (2015). Pre-clinical immunotoxicity studies of nanotechnology-formulated drugs: Challenges, considerations and strategy. *J. Control. Release*, **220**, 571–583.

Dobrovolskaia, M. A., Germolec, D. R., Weaver, J. L. (2009a). Evaluation of nanoparticle immunotoxicity. *Nat. Nanotechnol.*, **4**, 411–414.

Dobrovolskaia, M. A., McNeil, S. E. (2007). Immunological properties of engineered nanomaterials. *Nat. Nanotechnol.*, **2**, 469–478.

Dobrovolskaia, M. A., McNeil, S. E. (2013a). Immunological properties of engineered nanomaterials: An introduction. In: Dobrovolskaia, M. A., McNeil, S. E., eds. *Handbook of Immunological Properties of Engineered Nanomaterials*, World Scientific Publishing, Singapore, pp. 1–25.

Dobrovolskaia, M. A., McNeil, S. E. (2013b). Nanoparticles and endotoxin. In: Dobrovolskaia, M. A., McNeil, S. E., eds. *Handbook of Immunological Properties of Engineered Nanomaterials*, World Scientific Publishing, Singapore, pp. 77–115.

Dobrovolskaia, M. A., McNeil, S. E. (2015a). Immunological and hematological toxicities challenging clinical translation of nucleic acid-based therapeutics. *Expert Opin. Biol. Ther.*, **15**, 1023–1048.

Dobrovolskaia, M. A., McNeil, S. E. (2015b). Strategy for selecting nanotechnology carriers to overcome immunological and hematological toxicities challenging clinical translation of nucleic acid-based therapeutics. *Expert Opin. Drug Deliv.*, **12**, 1163–1175.

Dobrovolskaia, M. A., McNeil, S. E. (2016). Nanoparticles and endotoxin. In: Dobrovolskaia, M. A., McNeil, S. E., eds. *Handbook of Immunological Properties of Engineered Nanomaperials 1*, World Scientific Publishing, Singapore, pp. 143–186.

Dobrovolskaia, M. A., Neun, B. W., Man, S., Ye, X., Hansen, M., Patri, A. K., Crist, R. M., McNeil, S. E. (2014). Protein corona composition does not accurately predict hematocompatibility of colloidal gold nanoparticles. *Nanomedicine*, **10**, 1453–1463.

Dobrovolskaia, M. A., Patri, A. K., Potter, T. M., Rodriguez, J. C., Hall, J. B., McNeil, S. E. (2012). Dendrimer-induced leukocyte procoagulant activity depends on particle size and surface charge. *Nanomedicine (London)*, **7**, 245–256.

Dobrovolskaia, M. A., Patri, A. K., Zheng, J., Clogston, J. D., Ayub, N., Aggarwal, P., Neun, B. W., Hall, J. B., McNeil, S. E. (2009b). Interaction of colloidal gold nanoparticles with human blood: Effects on particle size and analysis of plasma protein binding profiles. *Nanomedicine*, **5**, 106–117.

Etheridge, M. L., Campbell, S. A., Erdman, A. G., Haynes, C. L., Wolf, S. M., McCullough, J. (2013). The big picture on nanomedicine: The state of investigational and approved nanomedicine products. *Nanomedicine*, **9**, 1–14.

Farrera, C., Bhattacharya, K., Lazzaretto, B., Andon, F. T., Hultenby, K., Kotchey, G. P., Star, A., Fadeel, B. (2014). Extracellular entrapment and degradation of single-walled carbon nanotubes. *Nanoscale*, **6**, 6974–6983.

Fesenkova, V. (2013). Nanoparticles and dendritic cells. In: Dobrovolskaia, M. A., McNeil, S. E., eds. *Handbook of Immunological Properties of Engineered Nanomaterials*, World Scientific Publishing, Singapore.

Fifis, T., Gamvrellis, A., Crimeen-Irwin, B., Pietersz, G. A., Li, J., Mottram, P. L., McKenzie, I. F., Plebanski, M. (2004a). Size-dependent immunogenicity: Therapeutic and protective properties of nano-vaccines against tumors. *J. Immunol.*, **173**, 3148–3154.

Fifis, T., Mottram, P., Bogdanoska, V., Hanley, J., Plebanski, M. (2004b). Short peptide sequences containing MHC class I and/or class

II epitopes linked to nano-beads induce strong immunity and inhibition of growth of antigen-specific tumour challenge in mice. *Vaccine*, **23**, 258–266.

Fletcher, C. V., Staskus, K., Wietgrefe, S. W., Rothenberger, M., Reilly, C., Chipman, J. G., Beilman, G. J., Khoruts, A., Thorkelson, A., Schmidt, T. E., Anderson, J., Perkey, K., Stevenson, M., Perelson, A. S., Douek, D. C., Haase, A. T., Schacker, T. W. (2014). Persistent HIV-1 replication is associated with lower antiretroviral drug concentrations in lymphatic tissues. *Proc. Natl. Acad. Sci. U. S. A.*, **111**, 2307–2312.

Foged, C., Brodin, B., Frokjaer, S., Sundblad, A. (2005). Particle size and surface charge affect particle uptake by human dendritic cells in an in vitro model. *Int. J. Pharm.*, **298**, 315–322.

Franca, A., Aggarwal, P., Barsov, E. V., Kozlov, S. V., Dobrovolskaia, M. A., Gonzalez-Fernandez, A. (2011). Macrophage scavenger receptor a mediates the uptake of gold colloids by macrophages in vitro. *Nanomedicine (London)*, **6**, 1175–1188.

Freise, C. E., Liu, T., Hong, K., Osorio, R. W., Papahadjopoulos, D., Ferrell, L., Ascher, N. L., Roberts, J. P. (1994). The increased efficacy and decreased nephrotoxicity of a cyclosporine liposome. *Transplantation*, **57**, 928–932.

Gaca, J. G., Palestrant, D., Lukes, D. J., Olausson, M., Parker, W., Davis Jr., R. D. (2003). Prevention of acute lung injury in swine: Depletion of pulmonary intravascular macrophages using liposomal clodronate. *J. Surg. Res.*, **112**, 19–25.

Gbadamosi, J. K., Hunter, A. C., Moghimi, S. M. (2002). PEGylation of microspheres generates a heterogeneous population of particles with differential surface characteristics and biological performance. *FEBS Lett.*, **532**, 338–344.

Giacalone, G., Bochot, A., Fattal, E., Hillaireau, H. (2013). Drug-induced nanocarrier assembly as a strategy for the cellular delivery of nucleotides and nucleotide analogues. *Biomacromolecules*, **14**, 737–742.

Gomez, S., Gamazo, C., Roman, B. S., Ferrer, M., Sanz, M. L., Irache, J. M. (2007). Gantrez AN nanoparticles as an adjuvant for oral immunotherapy with allergens. *Vaccine*, **25**, 5263–5271.

Gomez, S., Gamazo, C., San Roman, B., Ferrer, M., Sanz, M. L., Espuelas, S., Irache, J. M. (2008). Allergen immunotherapy with nanoparticles containing lipopolysaccharide from Brucella ovis. *Eur. J. Pharm. Biopharm.*, **70**, 711–717.

Gonzalez-Aramundiz, J. V., Cordeiro, A. S., Csaba, N., de la Fuente, M., Alonso, M. J. (2012). Nanovaccines: Nanocarriers for antigen delivery. *Biol. Aujourdhui*, **206**, 249–261.

Goppert, T. M., Muller, R. H. (2005). Polysorbate-stabilized solid lipid nanoparticles as colloidal carriers for intravenous targeting of drugs to the brain: Comparison of plasma protein adsorption patterns. *J. Drug Target.*, **13**, 179–187.

Gradishar, W. J., Tjulandin, S., Davidson, N., Shaw, H., Desai, N., Bhar, P., Hawkins, M., O'Shaughnessy, J. (2005). Phase III trial of nanoparticle albumin-bound paclitaxel compared with polyethylated castor oil-based paclitaxel in women with breast cancer. *J. Clin. Oncol.*, **23**, 7794–7803.

Gregoriadis, G., Wills, E. J., Swain, C. P., Tavill, A. S. (1974). Drug-carrier potential of liposomes in cancer chemotherapy. *Lancet,* **1**, 1313–1316.

Greish, K., Thiagarajan, G., Herd, H., Price, R., Bauer, H., Hubbard, D., Burckle, A., Sadekar, S., Yu, T., Anwar, A., Ray, A., Ghandehari, H. (2012). Size and surface charge significantly influence the toxicity of silica and dendritic nanoparticles. *Nanotoxicology,* **6**, 713–723.

Higaki, M., Ishihara, T., Izumo, N., Takatsu, M., Mizushima, Y. (2005). Treatment of experimental arthritis with poly(D, L-lactic/glycolic acid) nanoparticles encapsulating betamethasone sodium phosphate. *Ann. Rheum. Dis.*, **64**, 1132–1136.

Hoang, M. D., Lee, H. J., Jung, S. H., Choi, N. R., Vo, M. C., Nguyen-Pham, T. N., Kim, H. J., Park, I. K., Lee, J. J. (2015). Branched polyethylenimine-superparamagnetic iron oxide nanoparticles (bPEI-SPIONs) improve the immunogenicity of tumor antigens and enhance Th1 polarization of dendritic cells. *J. Immunol. Res.*, **2015**, 706379.

Ilinskaya, A. N., Clogston, J. D., McNeil, S. E., Dobrovolskaia, M. A. (2015). Induction of oxidative stress by *Taxol®* vehicle cremophor-EL triggers production of interleukin-8 by peripheral blood mononuclear cells through the mechanism not requiring de novo synthesis of mRNA. *Nanomedicine*, **11**, 1925–1938.

Ilinskaya, A. N., Dobrovolskaia, M. A. (2014). Immunosuppressive and anti-inflammatory properties of engineered nanomaterials. *Br. J. Pharmacol.*, **171**, 3988–4000.

Ilinskaya, A. N., Man, S., Patri, A. K., Clogston, J. D., Crist, R. M., Cachau, R.E., McNeil, S. E., Dobrovolskaia, M. A. (2014). Inhibition of phosphoinositol 3 kinase contributes to nanoparticle-mediated exaggeration of endotoxin-induced leukocyte procoagulant activity. *Nanomedicine (London)*, **9**, 1311–1326.

Inoue, K. (2011). Promoting effects of nanoparticles/materials on sensitive lung inflammatory diseases. *Environ. Health Prev. Med.*, **16**, 139–143.

Inoue, K., Takano, H. (2011). Aggravating impact of nanoparticles on immune-mediated pulmonary inflammation. *ScientificWorldJournal*, **11**, 382–390.

Inoue, K., Takano, H., Yanagisawa, R., Hirano, S., Kobayashi, T., Fujitani, Y., Shimada, A., Yoshikawa, T. (2007). Effects of inhaled nanoparticles on acute lung injury induced by lipopolysaccharide in mice. *Toxicology*, **238**, 99–110.

Inoue, K., Takano, H., Yanagisawa, R., Hirano, S., Sakurai, M., Shimada, A., Yoshikawa, T. (2006). Effects of airway exposure to nanoparticles on lung inflammation induced by bacterial endotoxin in mice. *Environ. Health Perspect.*, **114**, 1325–1330.

Italia, J. L., Bhatt, D. K., Bhardwaj, V., Tikoo, K., Kumar, M. N. (2007). PLGA nanoparticles for oral delivery of cyclosporine: Nephrotoxicity and pharmacokinetic studies in comparison to sandimmune neoral. *J. Control. Release*, **119**, 197–206.

Jeanbart, L., Kourtis, I. C., van der Vlies, A. J., Swartz, M. A., Hubbell, J. A. (2015). 6-Thioguanine-loaded polymeric micelles deplete myeloid-derived suppressor cells and enhance the efficacy of T cell immunotherapy in tumor-bearing mice. *Cancer Immunol. Immunother.*, **64**, 1033–1046.

Jilek, S., Merkle, H. P., Walter, E. (2005). DNA-loaded biodegradable microparticles as vaccine delivery systems and their interaction with dendritic cells. *Adv. Drug Deliv. Rev.*, **57**, 377–390.

Jilek, S., Ulrich, M., Merkle, H. P., Walter, E. (2004). Composition and surface charge of DNA-loaded microparticles determine maturation and cytokine secretion in human dendritic cells. *Pharm. Res.*, **21**, 1240–1247.

Jones, C. F., Campbell, R. A., Brooks, A. E., Assemi, S., Tadjiki, S., Thiagarajan, G., Mulcock, C., Weyrich, A. S., Brooks, B. D., Ghandehari, H., Grainger, D. W. (2012a). Cationic PAMAM dendrimers aggressively initiate blood clot formation. *ACS Nano*, **6**, 9900–9910.

Jones, C. F., Campbell, R. A., Franks, Z., Gibson, C. C., Thiagarajan, G., Vieira-de-Abreu, A., Sukavaneshvar, S., Mohammad, S. F., Li, D. Y., Ghandehari, H., Weyrich, A. S., Brooks, B. D., Grainger, D. W. (2012b). Cationic PAMAM dendrimers disrupt key platelet functions. *Mol. Pharm.*, **9**, 1599–1611.

Kagan, V. E., Konduru, N. V., Feng, W., Allen, B. L., Conroy, J., Volkov, Y., Vlasova, I. I., Belikova, N. A., Yanamala, N., Kapralov, A., Tyurina, Y. Y., Shi, J., Kisin, E. R., Murray, A. R., Franks, J., Stolz, D., Gou, P., Klein-

Seetharaman, J., Fadeel, B., Star, A., Shvedova, A. A. (2010a). Carbon nanotubes degraded by neutrophil myeloperoxidase induce less pulmonary inflammation. *Nat. Nanotechnol.*, **5**, 354–359.

Kagan, V. E., Kapralov, A. A., St Croix, C. M., Watkins, S. C., Kisin, E. R., Kotchey, G. P., Balasubramanian, K., Vlasova, I. I., Yu, J., Kim, K., Seo, W., Mallampalli, R. K., Star, A., Shvedova, A. A. (2014). Lung macrophages "digest" carbon nanotubes using a superoxide/peroxynitrite oxidative pathway. *ACS Nano,* **8**, 5610–5621.

Kagan, V. E., Shi, J., Feng, W., Shvedova, A. A., Fadeel, B. (2010b). Fantastic voyage and opportunities of engineered nanomaterials: What are the potential risks of occupational exposures? *J. Occup. Environ. Med.*, **52**, 943–946.

Kim, J. Y., Choung, S., Lee, E. J., Kim, Y. J., Choi, Y. C. (2007). Immune activation by siRNA/liposome complexes in mice is sequence-independent: Lack of a role for Toll-like receptor 3 signaling. *Mol. Cell*, **24**, 247–254.

Kim, S. H., Lim, K. M., Noh, J. Y., Kim, K., Kang, S., Chang, Y. K., Shin, S., Chung, J. H. (2011). Doxorubicin-induced platelet procoagulant activities: An important clue for chemotherapy-associated thrombosis. *Toxicol. Sci.*, **124**, 215–224.

Klimuk, S. K., Semple, S. C., Nahirney, P. N., Mullen, M. C., Bennett, C. F., Scherrer, P., Hope, M. J. (2000). Enhanced anti-inflammatory activity of a liposomal intercellular adhesion molecule-1 antisense oligodeoxynucleotide in an acute model of contact hypersensitivity. *J. Pharmacol. Exp. Ther.*, **292**, 480–488.

Leonov, A. P., Zheng, J., Clogston, J. D., Stern, S. T., Patri, A. K., Wei, A. (2008). Detoxification of gold nanorods by treatment with polystyrenesulfonate. *ACS Nano*, **2**, 2481–2488.

Levin, A. A. (1999). A review of the issues in the pharmacokinetics and toxicology of phosphorothioate antisense oligonucleotides. *Biochim. Biophys. Acta,* **1489**, 69–84.

Li, S., Wu, S. P., Whitmore, M., Loeffert, E. J., Wang, L., Watkins, S. C., Pitt, B. R., Huang, L. (1999). Effect of immune response on gene transfer to the lung via systemic administration of cationic lipidic vectors. *Am. J. Phys.*, **276**, L796–L804.

Libutti, S. K., Paciotti, G. F., Byrnes, A. A., Alexander Jr., H. R., Gannon, W. E., Walker, M., Seidel, G. D., Yuldasheva, N., Tamarkin, L. (2010). Phase I and pharmacokinetic studies of CYT-6091, a novel PEGylated colloidal gold-rhTNF nanomedicine. *Clin. Cancer Res.*, **16**, 6139–6149.

Lipson, E. J., Sharfman, W. H., Chen, S., McMiller, T. L., Pritchard, T. S., Salas, J. T., Sartorius-Mergenthaler, S., Freed, I., Ravi, S., Wang, H.,

Luber, B., Sproul, J. D., Taube, J. M., Pardoll, D. M., Topalian, S. L. (2015). Safety and immunologic correlates of melanoma GVAX, a GM-CSF secreting allogeneic melanoma cell vaccine administered in the adjuvant setting. *J. Transl. Med.*, **13**, 214.

Liptrott, N., Curley, P., Tatham, L. M., Owen, A. (2016). Opportunities and challenges in nanotechnology-enabled antiretroviral delivery. In: Dobrovolskaia, M. A., McNeil, S. E., eds. *Handbook of Immunological Properties of Engineered Nanomaterials* 3, World Scientific Publishing, Singapore, pp. 205–240.

Little, S. R., Lynn, D. M., Ge, Q., Anderson, D. G., Puram, S. V., Chen, J., Eisen, H. N., Langer, R. (2004). Poly-beta amino ester-containing microparticles enhance the activity of nonviral genetic vaccines. *Proc. Natl. Acad. Sci. U. S. A.*, **101**, 9534–9539.

Liu, Z., Li, W., Wang, F., Sun, C., Wang, L., Wang, J., Sun, F. (2012). Enhancement of lipopolysaccharide-induced nitric oxide and interleukin-6 production by PEGylated gold nanoparticles in RAW264.7 cells. *Nanoscale*, **4**, 7135–7142.

Liu, W., Swift, R., Torraca, G., Nashed-Samuel, Y., Wen, Z. Q., Jiang, Y., Vance, A., Mire-Sluis, A., Freund, E., Davis, J., Narhi, L. (2010). Root cause analysis of tungsten-induced protein aggregation in pre-filled syringes. *PDA J. Pharm. Sci. Technol.*, **64**, 11–19.

Luebke, R. (2012). Immunotoxicant screening and prioritization in the twenty-first century. *Toxicol. Pathol.*, **40**, 294–299.

Madani, S. Y., Mandel, A., Seifalian, A. M. (2013). A concise review of carbon nanotube's toxicology. *Nano Rev.*, **4**, 21521.

Malik, N., Wiwattanapatapee, R., Klopsch, R., Lorenz, K., Frey, H., Weener, J. W., Meijer, E. W., Paulus, W., Duncan, R. (2000). Dendrimers: Relationship between structure and biocompatibility in vitro, and preliminary studies on the biodistribution of 125I-labelled polyamidoamine dendrimers in vivo. *J. Control. Release*, **65**, 133–148.

Mallipeddi, R., Rohan, L. C. (2010). Progress in antiretroviral drug delivery using nanotechnology. *Int. J. Nanomed.*, **5**, 533–547.

Manolova, V., Flace, A., Bauer, M., Schwarz, K., Saudan, P., Bachmann, M. F. (2008). Nanoparticles target distinct dendritic cell populations according to their size. *Eur. J. Immunol.*, **38**, 1404–1413.

Marx, V. (2008). Poised to branch out. *Nat. Biotechnol.*, **26**, 729–732.

McHugh, M. D., Park, J., Uhrich, R., Gao, W., Horwitz, D. A., Fahmy, T. M. (2015). Paracrine co-delivery of TGF-beta and IL-2 using CD4-targeted nanoparticles for induction and maintenance of regulatory T cells. *Biomaterials*, **59**, 172–181.

Michaelis, K., Hoffmann, M. M., Dreis, S., Herbert, E., Alyautdin, R. N., Michaelis, M., Kreuter, J., Langer, K. (2006). Covalent linkage of apolipoprotein e to albumin nanoparticles strongly enhances drug transport into the brain. *J. Pharmacol. Exp. Ther.*, **317**, 1246–1253.

Minigo, G., Scholzen, A., Tang, C. K., Hanley, J. C., Kalkanidis, M., Pietersz, G. A., Apostolopoulos, V., Plebanski, M. (2007). Poly-L-lysine-coated nanoparticles: A potent delivery system to enhance DNA vaccine efficacy. *Vaccine*, **25**, 1316–1327.

Mire-Sluis, A., Cherney, B., Madsen, R., Polozova, A., Rosenberg, A., Smith, H., Arora, T., Narhi, L. (2011). Analysis and immunogenic potential of aggregates and particles: A practical approach, part 2. *BioProcess Int.*, **9**, 38–43.

Mitchell, L. A., Lauer, F. T., Burchiel, S. W., McDonald, J. D. (2009). Mechanisms for how inhaled multiwalled carbon nanotubes suppress systemic immune function in mice. *Nat. Nanotechnol.*, **4**, 451–456.

Moon, E. Y., Yi, G. H., Kang, J. S., Lim, J. S., Kim, H. M., Pyo, S. (2011). An increase in mouse tumor growth by an in vivo immunomodulating effect of titanium dioxide nanoparticles. *J. Immunotoxicol.*, **8**, 56–67.

Mottram, P., Leong, D., Crimeen-Irwin, B., Gloster, S., Xiang, S. D., Meanger, J., Ghildyal, R., Vardaxis, N., Plebanski, M. (2007). Type 1 and type 2 immunity following vaccination is influenced by nanoparticle size: Formulation of a model vaccine for respiratory syncytial virus. *Mol. Pharm.*, **4**, 73–84.

Nagayama, S., Ogawara, K., Minato, K., Fukuoka, Y., Takakura, Y., Hashida, M., Higaki, K., Kimura, T. (2007). Fetuin mediates hepatic uptake of negatively charged nanoparticles via scavenger receptor. *Int. J. Pharm.*, **329**, 192–198.

Nakai, T., Kanamori, T., Sando, S., Aoyama, Y. (2003). Remarkably size-regulated cell invasion by artificial viruses. Saccharide-dependent self-aggregation of glycoviruses and its consequences in glycoviral gene delivery. *J. Am. Chem. Soc.*, **125**, 8465–8475.

Newkome, G. R., Mishra, A., Moorefield, C. N. (2002). Improved synthesis of an ethereal tetraamine core for dendrimer construction. *J. Organomet. Chem.*, **67**, 3957–3960.

O'Hagan, D. T., MacKichan, M. L., Singh, M. (2001). Recent developments in adjuvants for vaccines against infectious diseases. *Biomol. Eng.*, **18**, 69–85.

Pantic, I. (2011). Nanoparticles and modulation of immune responses. *Sci. Prog.*, **94**, 97–107.

Park, J. S., Yang, H. N., Jeon, S. Y., Woo, D. G., Kim, M. S., Park, K. H. (2012). The use of anti-COX2 siRNA coated onto PLGA nanoparticles loading dexamethasone in the treatment of rheumatoid arthritis. *Biomaterials,* **33**, 8600–8612.

Pavelic, K., Hadzija, M., Bedrica, L., Pavelic, J., Dikic, I., Katic, M., Kralj, M., Bosnar, M. H., Kapitanovic, S., Poljak-Blazi, M., Krizanac, S., Stojkovic, R., Jurin, M., Subotic, B., Colic, M. (2001). Natural zeolite clinoptilolite: New adjuvant in anticancer therapy. *J. Mol. Med. (Berl),* **78**, 708–720.

Pople, P. V., Singh, K. K. (2012). Targeting tacrolimus to deeper layers of skin with improved safety for treatment of atopic *dermatitis—Part II*: In vivo assessment of dermatopharmacokinetics, biodistribution and efficacy. *Int. J. Pharm.,* **434**, 70–79.

Prinapori, R., Di Biagio, A. (2015). Efficacy, safety, and patient acceptability of elvitegravir/ cobicistat/emtricitabine/tenofovir in the treatment of HIV/AIDS. *Patient Prefer Adherence,* **9**, 1213–1218.

Ramani, K., Miclea, R. D., Purohit, V. S., Mager, D. E., Straubinger, R. M., Balu-Iyer, S. V. (2008). Phosphatidylserine containing liposomes reduce immunogenicity of recombinant human factor VIII (rFVIII) in a murine model of hemophilia A. *J. Pharm. Sci.,* **97**, 1386–1398.

Ramirez, A. P., Haddon, R. C., Zhou, O., Fleming, R. M., Zhang, J., McClure, S. M., Smalley, R. E. (1994). Magnetic susceptibility of molecular carbon: Nanotubes and fullerite. *Science,* **265**, 84–86.

Reddy, S. T., van der Vlies, A. J., Simeoni, E., Angeli, V., Randolph, G. J., O'Neil, C. P., Lee, L. K., Swartz, M. A., Hubbell, J. A. (2007a). Exploiting lymphatic transport and complement activation in nanoparticle vaccines. *Nat. Biotechnol.,* **25**, 1159–1164.

Reddy, S. T., van der Vlies, A. J., Simeoni, E., Angeli, V., Randolph, G. J., O'Neil, C. P., Lee, L. K., Swartz, M. A., Hubbell, J. A. (2007b). Exploiting lymphatic transport and complement activation in nanoparticle vaccines. *Nat. Biotechnol.,* **25**, 1159–1164.

Rettig, L., Haen, S. P., Bittermann, A. G., von Boehmer, L., Curioni, A., Krämer, S. D., Knuth, A., Pascolo, S. (2010). Particle size and activation threshold: A new dimension of danger signaling. *Blood,* **115**, 4533–4541.

Rosenberg, A. S., Verthelyi, D., Cherney, B. W. (2012). Managing uncertainty: A perspective on risk pertaining to product quality attributes as they bear on immunogenicity of therapeutic proteins. *J. Pharm. Sci.,* **101**, 3560–3567.

Roy, K., Mao, H. Q., Huang, S. K., Leong, K. W. (1999). Oral gene delivery with chitosan–DNA nanoparticles generates immunologic protection in a murine model of peanut allergy. *Nat. Med.,* **5**, 387–391.

Ryan, J. J., Bateman, H. R., Stover, A., Gomez, G., Norton, S. K., Zhao, W., Schwartz, L. B., Lenk, R., Kepley, C. L. (2007). Fullerene nanomaterials inhibit the allergic response. *J. Immunol.*, **179**, 665–672.

Sadauskas, E., Danscher, G., Stoltenberg, M., Vogel, U., Larsen, A., Wallin, H. (2009). Protracted elimination of gold nanoparticles from mouse liver. *Nanomedicine,* **5**, 162–169.

Salvador-Morales, C., Basiuk, E. V., Basiuk, V. A., Green, M. L., Sim, R. B. (2008). Effects of covalent functionalization on the biocompatibility characteristics of multi-walled carbon nanotubes. *J. Nanosci. Nanotechnol.*, **8**, 2347–2356.

Salvador-Morales, C., Flahaut, E., Sim, E., Sloan, J., Green, M. L., Sim, R. B. (2006). Complement activation and protein adsorption by carbon nanotubes. *Mol. Immunol.*, **43**, 193–201.

Scholl, I., Weissenbock, A., Forster-Waldl, E., Untersmayr, E., Walter, F., Willheim, M., Boltz-Nitulescu, G., Scheiner, O., Gabor, F., Jensen-Jarolim, E. (2004). Allergen-loaded biodegradable poly(D,L-lactic-co-glycolic) acid nanoparticles down-regulate an ongoing Th2 response in the BALB/c mouse model. *Clin. Exp. Allergy,* **34**, 315–321.

Schweingruber, N., Haine, A., Tiede, K., Karabinskaya, A., van den Brandt, J., Wust, S., Metselaar, J. M., Gold, R., Tuckermann, J. P., Reichardt, H. M., Luhder, F. (2011). Liposomal encapsulation of glucocorticoids alters their mode of action in the treatment of experimental autoimmune encephalomyelitis. *J. Immunol.*, **187**, 4310–4318.

Senapati, V. A., Kumar, A., Gupta, G. S., Pandey, A. K., Dhawan, A. (2015). ZnO nanoparticles induced inflammatory response and genotoxicity in human blood cells: A mechanistic approach. *Food Chem. Toxicol.*, **85**, 61–70.

Seo, W., Kapralov, A. A., Shurin, G. V., Shurin, M. R., Kagan, V. E., Star, A. (2015). Payload drug vs. nanocarrier biodegradation by myeloperoxidase- and peroxynitrite-mediated oxidations: Pharmacokinetic implications. *Nanoscale,* **7**, 8689–8694.

Shah, M., Edman, M. C., Janga, S. R., Shi, P., Dhandhukia, J., Liu, S., Louie, S. G., Rodgers, K., Mackay, J. A., Hamm-Alvarez, S. F. (2013). A rapamycin-binding protein polymer nanoparticle shows potent therapeutic activity in suppressing autoimmune dacryoadenitis in a mouse model of Sjögren's syndrome. *J. Control. Release,* **171**, 269–279.

Shen, C. C., Wang, C. C., Liao, M. H., Jan, T. R. (2011). A single exposure to iron oxide nanoparticles attenuates antigen-specific antibody production and T-cell reactivity in ovalbumin-sensitized BALB/c mice. *Int. J. Nanomed.*, **6**, 1229–1235.

Shi, Y., Yadav, S., Wang, F., Wang, H. (2010). Endotoxin promotes adverse effects of amorphous silica nanoparticles on lung epithelial cells in vitro. *J. Toxic. Environ. Health A*, **73**, 748–756.

Shvedova, A. A., Kapralov, A. A., Feng, W. H., Kisin, E. R., Murray, A. R., Mercer, R. R., St Croix, C. M., Lang, M. A., Watkins, S. C., Konduru, N. V., Allen, B. L., Conroy, J., Kotchey, G. P., Mohamed, B. M., Meade, A. D., Volkov, Y., Star, A., Fadeel, B., Kagan, V. E. (2012). Impaired clearance and enhanced pulmonary inflammatory/fibrotic response to carbon nanotubes in myeloperoxidase-deficient mice. *PLoS One*, **7**, e30923.

Shvedova, A. A., Tkach, A. V., Kisin, E. R., Khaliullin, T., Stanley, S., Gutkin, D. W., Star, A., Chen, Y., Shurin, G. V., Kagan, V. E., Shurin, M. R. (2013). Carbon nanotubes enhance metastatic growth of lung carcinoma via up-regulation of myeloid-derived suppressor cells. *Small*, **9**, 1691–1695.

Singh, M., Briones, M., Ott, G., O'Hagan, D. (2000). Cationic microparticles: A potent delivery system for DNA vaccines. *Proc. Natl. Acad. Sci. U. S. A.*, **97**, 811–816.

Singh, I., Swami, R., Khan, W., Sistla, R. (2014). Lymphatic system: A prospective area for advanced targeting of particulate drug carriers. *Expert Opin. Drug Deliv.*, **11**, 211–229.

Smith, D. M., Simon, J. K., Baker Jr., J. R. (2013). Applications of nanotechnology for immunology. *Nat. Rev. Immunol.*, **13**, 592–605.

Song, G., Wu, H., Yoshino, K., Zamboni, W. C. (2012). Factors affecting the pharmacokinetics and pharmacodynamics of liposomal drugs. *J. Lipid Res.*, **22**, 177–192.

Spreen, W. R., Margolis, D. A., Pottage Jr., J. C. (2013). Long-acting injectable antiretrovirals for HIV treatment and prevention. *Curr. Opin. HIV AIDS*, **8**, 565–571.

Stefaniak, A. B., Hackley, V. A., Roebben, G., Ehara, K., Hankin, S., Postek, M. T., Lynch, I., Fu, W. E., Linsinger, T. P., Thunemann, A. F. (2013). Nanoscale reference materials for environmental, health and safety measurements: Needs, gaps and opportunities. *Nanotoxicology*, **7**, 1325–1337.

Stinchcombe, T. E., Socinski, M. A., Walko, C. M., O'Neil, B. H., Collichio, F. A., Ivanova, A., Mu, H., Hawkins, M. J., Goldberg, R. M., Lindley, C., Claire Dees, E. (2007). Phase I and pharmacokinetic trial of carboplatin and albumin-bound paclitaxel, ABI-007 (abraxane) on three treatment schedules in patients with solid tumors. *Cancer Chemother. Pharmacol.*, **60**, 759–766.

Swystun, L. L., Shin, L. Y., Beaudin, S., Liaw, P. C. (2009). Chemotherapeutic agents doxorubicin and epirubicin induce a procoagulant phenotype on endothelial cells and blood monocytes. *J. Thromb. Haemost.*, **7**, 619–626.

Takagahara, T. (1987). Excitonic optical nonlinearity and exciton dynamics in semiconductor quantum dots. *Phys. Rev. B. Condens. Matter,* **36**, 9293–9296.

Teo, P. Y., Yang, C., Whilding, L. M., Parente-Pereira, A. C., Maher, J., George, A. J., Hedrick, J. L., Yang, Y. Y., Ghaem-Maghami, S. (2015). Ovarian cancer immunotherapy using PD-L1 siRNA targeted delivery from folic acid-functionalized polyethylenimine: Strategies to enhance T cell killing. *Adv. Healthc. Mater.*, **4**, 1180–1189.

Thiele, L., Merkle, H. P., Walter, E. (2003). Phagocytosis and phagosomal fate of surface-modified microparticles in dendritic cells and macrophages. *Pharm. Res.*, **20**, 221–228.

Thiele, L., Rothen-Rutishauser, B., Jilek, S., Wunderli-Allenspach, H., Merkle, H. P., Walter, E. (2001). Evaluation of particle uptake in human blood monocyte-derived cells in vitro. Does phagocytosis activity of dendritic cells measure up with macrophages? *J. Control. Release,* **76**, 59–71.

Tighe, H., Corr, M., Roman, M., Raz, E. (1998). Gene vaccination: Plasmid DNA is more than just a blueprint. *Immunol. Today,* **19**, 89–97.

Tomalia, D. A. (1991). Dendrimer research. *Science,* **252**, 1231.

Toraya-Brown, S., Sheen, M. R., Zhang, P., Chen, L., Baird, J. R., Demidenko, E., Turk, M. J., Hoopes, P. J., Conejo-Garcia, J. R., Fiering, S. (2014). Local hyperthermia treatment of tumors induces CD8(+) T cell-mediated resistance against distal and secondary tumors. *Nanomedicine,* **10**, 1273–1285.

Tran, T. H., Amiji, M. M. (2015). Targeted delivery systems for biological therapies of inflammatory diseases. *Expert Opin. Drug Deliv.,* **12**, 393–414.

Tsai, C. Y., Lu, S. L., Hu, C. W., Yeh, C. S., Lee, G. B., Lei, H. Y. (2012). Size-dependent attenuation of TLR9 signaling by gold nanoparticles in macrophages. *J. Immunol.*, **188**, 68–76.

Tyner, K., Sadrieh, N. (2011). Considerations when submitting nanotherapeutics to FDA/ CDER for regulatory review. *Methods Mol.Biol.*, **697**, 17–31.

Tyner, K. M., Zou, P., Yang, X., Zhang, H., Cruz, C. N., Lee, S. L. (2015). Product quality for nanomaterials: Current U.S. experience and perspective. *Wiley Interdiscip. Rev. Nanomed. Nanobiotechnol.*, **7**, 640–654.

Umbreit, T. H., Francke-Carroll, S., Weaver, J. L., Miller, T. J., Goering, P. L., Sadrieh, N., Stratmeyer, M. E. (2012). Tissue distribution and histopathological effects of titanium dioxide nanoparticles after intravenous or subcutaneous injection in mice. *J. Appl. Toxicol.*, **32**, 350–357.

Vallhov, H., Qin, J., Johansson, S. M., Ahlborg, N., Muhammed, M. A., Scheynius, A., Gabrielsson, S. (2006). The importance of an endotoxin-free environment during the production of nanoparticles used in medical applications. *Nano Lett.*, **6**, 1682–1686.

Van Beers, M. M., Gilli, F., Schellekens, H., Randolph, T. W., Jiskoot, W. (2012). Immunogenicity of recombinant human interferon beta interacting with particles of glass, metal, and polystyrene. *J. Pharm. Sci.*, **101**, 187–199.

Vasievich, E. A., Chen, W., Huang, L. (2011). Enantiospecific adjuvant activity of cationic lipid DOTAP in cancer vaccine. *Cancer Immunol. Immunother.*, **60**, 629–638.

Verloes, R., van't Klooster, G., Baert, L., van Velsen, F., Bouche, M.-P., Spittaels, K., Leempoels, J., Williams, P., Kraus, G., Wigerinck, P. (2008). TMC278 long acting—a parenteral nanosuspension formulation that provides sustained clinically relevant plasma Concentrations in HIV-negative volunteers. 17th International AIDS Conference Mexico City, Mexico City.

Villanueva, A., Cañete, M., Roca, A. G., Calero, M., Veintemillas-Verdaguer, S., Serna, C. J., Morales, M. D. P., Miranda, R. (2009). The influence of surface functionalization on the enhanced internalization of magnetic nanoparticles in cancer cells. *Nanotechnology*, **20**, 115103.

Vlasova, I. I., Vakhrusheva, T. V., Sokolov, A. V., Kostevich, V. A., Gusev, A. A., Gusev, S. A., Melnikova, V. I., Lobach, A. S. (2012). PEGylated single-walled carbon nanotubes activate neutrophils to increase production of hypochlorous acid, the oxidant capable of degrading nanotubes. *Toxicol. Appl. Pharmacol.*, **264**, 131–142.

Walsh, M. C., Banas, J. A., Mudzinski, S. P., Preissler, M. T., Graziano, R. F., Gosselin, E. J. (2003). A two-component modular approach for enhancing T-cell activation utilizing a unique anti-FcgammaRI-streptavidin construct and microspheres coated with biotinylated-antigen. *Biomol. Eng.*, **20**, 21–33.

Wang, J., Fu, L., Gu, F., Ma, Y. (2011). Notch1 is involved in migration and invasion of human breast cancer cells. *Oncol. Rep.*, **26**, 1295–1303.

Wegmann, K. W., Wagner, C. R., Whitham, R. H., Hinrichs, D. J. (2008). Synthetic peptide dendrimers block the development and expression of experimental allergic encephalomyelitis. *J. Immunol.*, **181**, 3301–3309.

Weiss, R., Scheiblhofer, S., Freund, J., Ferreira, F., Livey, I., Thalhamer, J. (2002). Gene gun bombardment with gold particles displays a particular Th2-promoting signal that over-rules the Th1-inducing effect of immunostimulatory CpG motifs in DNA vaccines. *Vaccine*, **20**, 3148–3154.

Weiszhar, Z., Czucz, J., Revesz, C., Rosivall, L., Szebeni, J., Rozsnyay, Z. (2012). Complement activation by polyethoxylated pharmaceutical surfactants: Cremophor-EL, tween-80 and tween-20. *Eur. J. Pharm. Sci.*, **45**, 492–498.

Wheeler, H. R., Geczy, C. L. (1990). Induction of macrophage procoagulant expression by cisplatin, daunorubicin and doxorubicin. *Int. J. Cancer*, **46**, 626–632.

Xiang, S. D., Scalzo-Inguanti, K., Minigo, G., Park, A., Hardy, C. L., Plebanski, M. (2008). Promising particle-based vaccines in cancer therapy. *Expert Rev. Vaccines*, **7**, 1103–1119.

Xiang, S. D., Fuchsberger, M., Karlson, T. D. L., Hardy, C. L., Selomulya, C., Plebanski, M. (2013). Nanoparticles, immunomodulation and vaccine delivery. In: Dobrovolskaia, M. A., McNeil, S. E., eds. *Handbook of Immunological Properties of Engineered Nanomaterials*, World Scientific Publishing, Singapore.

Xiang, S. D., Selomulya, C., Ho, J., Apostolopoulos, V., Plebanski, M. (2010). Delivery of DNA vaccines: An overview on the use of biodegradable polymeric and magnetic nanoparticles. *Wiley Interdiscip. Rev. Nanomed. Nanobiotechnol.*, **2**, 205–218.

Yu, B., Mao, Y., Bai, L. Y., Herman, S. E., Wang, X., Ramanunni, A., Jin, Y., Mo, X., Cheney, C., Chan, K. K., Jarjoura, D., Marcucci, G., Lee, R. J., Byrd, J. C., Lee, L. J., Muthusamy, N. (2013). Targeted nanoparticle delivery overcomes off-target immunostimulatory effects of oligonucleotides and improves therapeutic efficacy in chronic lymphocytic leukemia. *Blood*, **121**, 136–147.

Zaman, M., Good, M. F., Toth, I. (2013). Nanovaccines and their mode of action. *Methods*, **60**, 226–231.

Zensi, A., Begley, D., Pontikis, C., Legros, C., Mihoreanu, L., Wagner, S., Buchel, C., von Briesen, H., Kreuter, J. (2009). Albumin nanoparticles targeted with Apo E enter the CNS by transcytosis and are delivered to neurones. *J. Control. Release*, **137**, 78–86.

Zhao, Y., Burkert, S. C., Tang, Y., Sorescu, D. C., Kapralov, A. A., Shurin, G. V., Shurin, M. R., Kagan, V. E., Star, A. (2015). Nano-gold corking and enzymatic uncorking of carbon nanotube cups. *J. Am. Chem. Soc.*, **137**, 675–684.

Zheng, J., Clogston, J. D., Patri, A. K., Dobrovolskaia, M. A., McNeil, S. E. (2011). Sterilization of silver nanoparticles using standard gamma irradiation procedure affects particle integrity and biocompatibility. *J. Nanomed. Nanotechnol.*, **2011**, 001.

Ziemba, B., Halets, I., Shcharbin, D., Appelhans, D., Voit, B., Pieszynski, I., Bryszewska, M., Klajnert, B. (2012). Influence of fourth generation poly(propyleneimine) dendrimers on blood cells. *J. Biomed. Mater. Res. A*, **100**, 2870–2880.

Chapter 6

Auto-antibodies as Biomarkers for Disease Diagnosis

Angelika Lueking, Heike Göhler, and Peter Schulz-Knappe

Protagen AG, Dortmund, Germany

Keywords: antigen, auto-antibodies, autoimmune disease, biomarkers, companion diagnostics, diagnostic assay, immune response, microarrays, personalized medicine, recombinant proteins, risk assessment, UNIarray®

6.1 Introduction

New biomarkers with improved sensitivity and specificity are required to improve disease diagnosis and prognosis. Furthermore, to realize the concept of personalized medicine, new challenges of patient stratification and development of companion diagnostics need to be addressed. Consequently, biomarkers with diagnostic and prognostic value for cancer and autoimmune diseases will become more and more important. Auto-antibodies are a class of analytes that have attracted attention over the last years. They are directed against certain human proteins and induced by immune system activity in response to disease processes, e.g., in neurodegenerative diseases, cancer, or classical autoimmune diseases.

Here, we review the use of auto-antibody/antigen interactions for diagnostic assays and drug development strategies, which overcomes the technical problems and limitations of other proteomic markers found in the last decades. Different technology platforms are described enabling the discovery and validation of biomarkers as well as the development of diagnostic assays based on novel auto-antibody/antigen interactions.

6.2   Auto-antibodies as Biomarkers

Auto-antibodies are a class of biomarkers suitable for risk assessment, screening, prognosis, disease stratification, and therapy monitoring. Auto-antibodies, i.e., antibodies directed against certain human proteins, are induced by immune system activity in response to a disease process. Auto-antibody production reflects the immune response to a continuous remodeling of cells or tissues caused by protein turnover and chronic disease processes. In this context, the immune system fails to properly distinguish between self and nonself, and attacks its own cells and tissues. However, in so-called autoimmune diseases, the auto-antibodies present in blood are indicative for the clinical symptoms and the state of the disease. Prominent examples of autoimmune diseases are rheumatoid arthritis (RA), multiple sclerosis (MS), coeliac disease, diabetes mellitus type 1, systemic lupus erythematosus (SLE), Sjogren's syndrome, inflammatory bowel disease, and Hashimoto's thyroiditis.

Beyond classical autoimmune diseases, also in several cancer indications, the presence of auto-antibodies has been shown and was correlated to the disease state. Auto-antibodies against autologous tumor-associated antigens (TAAs) have been described (Anderson and LaBaer 2005). Most of them are altered, which renders them into an immunogenic form. They can be mutated (p53) (Soussi, 2000), overexpressed (NY-ESO-1) (Schubert et al., 2000), aberrantly degraded, or glycosylated (MUC-1) (von Mensdorff-Pouilly et al., 2000). Also, aberrant localization as described for cyclin B1 may provoke an immune response (Suzuki et al., 2005). It is speculated that the humoral response against TAAs is triggered by aberrant tumor cell death due to defective apoptosis or necrosis leading to the release of intracellular modified proteins with immunogenic potential. Tumor cell death also releases proteases that would generate cryptic self-epitopes.

Historically, the immune system was separated into two branches: humoral immunity and cellular immunity. The protective function of humoral immunization could be found in cell-free bodily fluids or serum and is mediated by secreted antibodies produced by activated B-lymphocytes. In contrast, the cell-mediated immunity does not involve antibodies but requires the activation of macrophages, natural killer cells, antigen-specific cytotoxic T-lymphocytes, and the release of different cytokines in response to an antigen. However, both systems are linked together by the activation of naïve B-cells in a T-cell-dependent manner. During T-cell-dependent activation, an antigen-presenting cell such as a macrophage or dendritic cell has digested the immunogenic antigen to peptides and presents this processed antigen to a helper T-cell (T_h-cell), which is then primed to this antigen. When a B-cell processes and presents the same antigen to the primed T_h-cell, the T-cell secretes several cytokines, which trigger the B-cell to proliferation and differentiation into plasma cells.

Up to now it has been quite difficult and time consuming to identify reliable serum markers, especially proteins and peptides, for the diagnosis of a certain disease. A particularly important reason is that such diagnostic markers are present in patient samples often only in minute and highly variable concentrations and have a limited stability. In contrast to this, the use of auto-antibodies as diagnostic markers has proven to be highly effective. Such antibodies can be detected by presenting their corresponding autoantigens in well-established assay formats, e.g., enzyme-linked immunosorbent assay, western blot, protein arrays, etc. A particular feature is their specific structure and high stability. They are present in serum or plasma in high concentrations and are not subjected to circadian rhythms or other short-term changes in physiological states. This means that sampling can occur any time because results are not influenced by the time of day of sampling or nutritional status. Due to their specificity and high affinity binding to their corresponding autoantigen, no enrichment or elaborated sample preparation is required. Enrichment occurs automatically during analysis by the binding of the auto-antibodies to the autoantigen. Potentially, auto-antibodies can also be used for the development of prognostic tests. Studies with SLE patients in the United States have shown that certain auto-antibodies could be detected as long as 10 years prior to the onset of the disease (Arbuckle et al., 2003). This highlights that assessing the immune response by measuring auto-antibodies is the most stable and efficient way of analyzing biomarkers for diagnostics. Several auto-antibodies are already established diagnostic markers (Table 6.1).

Table 6.1 Examples of established auto-antibody biomarkers and their corresponding antigens

Antigen	Disease	Antibody Scaffold
SS-A	systemic lupus erythematosus	IgG
Ro-52	systemic lupus erythematosus	IgG
CCP	rheumatoid arthritis	IgG
SS-B	systemic lupus erythematosus	IgG
Sm	systemic lupus erythematosus	IgG
Scl-70	systemic sclerosis	IgG
Jo-1	polymyositis	IgG
SLA	autoimmune liver disease	IgG
LKM1	autoimmune liver disease	IgG
AMA M2	autoimmune liver disease	IgG
Sp100	autoimmune liver disease	IgG
gp210	autoimmune liver disease	IgG
LBR	autoimmune liver disease	IgG

As indicated in Table 6.1, several disease-specific antibodies are known and used in established diagnostic test systems. The presence of many of these antibodies is associated with more than one autoimmune disease. For example, the detection of Ro-52 is typical for neonatal lupus erythematodes, Sjogren's syndrome, and SLE. This shows that the detection of just one antibody will be insufficient for the diagnosis of a disease. On the contrary, 116 different target antigens have been described in the literature for SLE patients (Sherer et al., 2004). Thus, the relatively small panel of target antigens/antibodies that are routinely measured does not probably assess the full heterogeneity of the disease. It is reasonable to assume that taking more disease-specific target antigens/antibodies into account will increase sensitivity and specificity of diagnosis.

Consequently, recent approaches measure a multitude of putative antibodies followed by data analysis using clustering or classification algorithms (Quintana et al., 2003; Li et al., 2007; Hueber et al., 2005). Due to cost effectiveness, the need for small sample volumes, and laborious procedures, the multiplex antigen array technology has become more and more accepted in this field (Robinson, 2006). This technology can be applied to improve diagnosis and used for the prediction of disease onset, classification of subjects into disease subgroups, as well as efficacy assessment of therapy regiments (Lueking and Cahill, 2006; Sharp and Utz, 2007).

6.3 Auto-antibodies for Companion Diagnostics Enabling Personalized Medicine

Auto-antibody signatures detected by blood screening can not only be useful for the diagnosis of a disease, but can also be used for patient stratification, i.e., divide patient populations into different groups, such as drug responders and nonresponders (Fig. 6.1). This is very desirable for the following reason: It is well established that for any disease, not all patients respond equally well to a standard drug treatment. On the contrary, in many cases a significant proportion of patients will either not respond at all or even show adverse effects as a result of the drug treatment. Ineffective drug treatments put an enormous cost burden on pharmaceutical companies during drug development and the health care insurance providers. Therefore, in recent years both the scientific community as well as regulatory agencies (Food and Drug Administration, European Medicines Agency) started recommending the development of diagnostic markers, assays, and tools to establish so-called companion diagnostics. They should enable a more targeted therapy to either specifically select the eligible patient population for a standard treatment or define a specific dosing scheme based on a specific molecular patient profile. This approach provided a rational basis of personalized medicine.

A prominent example is the monoclonal IgG1-λ antibody belimumab (Benlysta; GlaxoSmithKline, UK), which binds to the soluble human BLyS thereby inhibiting its biological activity. BLyS inhibits B-cell apoptosis and stimulates the differentiation of B-cells into immunoglobulin-producing plasma cells. In SLE, rheumatoid arthritis, and certain other autoimmune diseases, elevated levels of BLyS are believed to support the production of auto-antibodies, which may contribute to the destruction of healthy tissue. It has been shown in a phase II dose-ranging study that belimumab was only effective in patients with serologically active SLE patients (Wallace et al., 2009). Subgroup analysis revealed that the clinical end points were successfully reached in this large subgroup. Therefore, only seropositive patients were enrolled in the subsequent clinical phase III, in which the effectiveness of belimumab was proven. In 2010, the Food and Drug Administration approved Benlysta as the first novel SLE treatment in about 50 years. Sales are estimated to reach the $1 billion blockbuster threshold rather soon (according to Datamonitor Product Profile SLE, June 2011). The yearly treatment costs are in the range of $30,000 per patient, illustrating the high potential for any SLE drug, although a small percentage of SLE patients are responding to the drug.

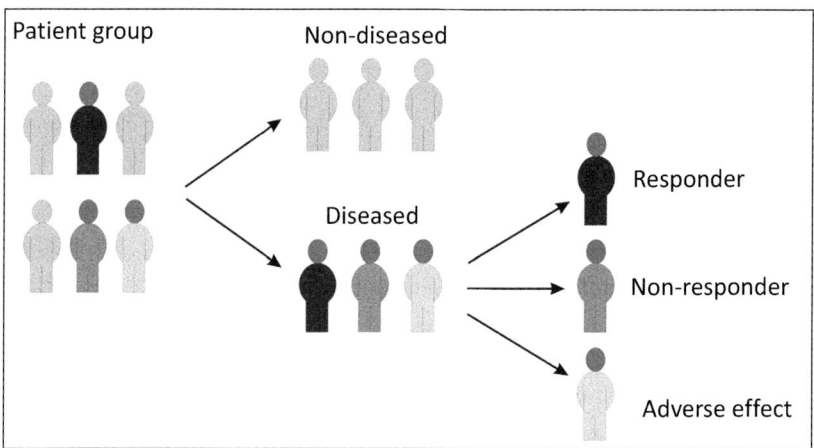

Figure 6.1 Use of biomarkers in diagnosis and personalized medicine. Diagnosis of a disease is based on the discriminative power of biomarker(s). Response or adverse effects of a patient to a drug can be predicted by biomarkers before or early during treatment.

Therefore, auto-antibodies in combination with their corresponding autoantigens have the potential to be used as companion diagnostics in this context, because they may enable classification of patients into different groups as indicated in Fig. 6.1.

6.4 Biomarker Discovery Strategies

The development of diagnostic markers comprises several phases from discovery to clinical assay development. Typically, the discovery process starts with high numbers of analytes tested against a low number of serum samples of cases and controls. Multiple rounds of verification and validation are carried out until the number of biomarkers is significantly decreased. Subsequently, four to ten biomarkers enter the phase of clinical assay development, whereas up to a few thousand samples are analyzed (Rifai et al, 2006).

Several techniques for autoantigen discovery are currently in use and encompass serological screening of cDNA expression libraries, phage display libraries, two-dimensional western blots, and different formats of protein arrays, such as planar or bead-based protein microarrays, peptide arrays, tissue arrays, and carbohydrates arrays.

The proteomics-based approach termed serological proteome analysis (SERPA9) combines two-dimensional electrophoresis, western blotting, and mass spectrometry (Klade et al., 2001). Proteins from tumor tissues or cell lines were separated by two-dimensional gel electrophoresis, transferred onto membranes, and incubated with serum samples from healthy people and patients. The auto-antibody signatures were compared and patient-associated protein spots were analyzed by mass spectrometry. Although the time-consuming construction of cDNA libraries is avoided and post-translational modifications are accessible for the screening approach, this technology has the major drawbacks of low reproducibility, low automation grade, and low sample throughput.

In the phage display approach, a cDNA library is constructed using tumor tissue, a cancer cell line, or short synthetic DNA sequences leading to peptides or proteins displayed on the phage surface. Auto-antibody screening in patient samples is done by a biopanning procedure involving succeeding rounds of immunoprecipitation and amplification/enrichment of auto-antibody-binding phages. However, the analysis of large numbers of patient and control samples as well as the quantitative analysis of the identified antigens or peptide epitopes requires further techniques such as protein microarrays.

The serological analysis of tumor antigens by recombinant cDNA expression cloning (SEREX) applies a cDNA expression library obtained from autologous tumor tissue (Sahin et al, 1995). By SEREX, several TAAs have been identified in various types of cancers, including lung, liver, breast, ovarian, prostate, and renal cancers (Tan et al., 2009). Proteins derived from aberrant transcripts highly specific for tumor activity can be detected. However, a general bias toward antigens that are highly expressed in the tumor tissue is found. Since often these libraries are cloned in a λ gt11 system, SEREX is time consuming, labor intensive, not amenable for automation, and therefore not suitable for the analysis of large patient numbers.

Protagen applied the UNIarray® technology platform, which enables a systematic approach to auto-antibody discovery in a unique way. The basis is founded in the availability of a large collection of recombinant human proteins. The company owns five tissue-specific recombinant human protein expression libraries (*Escherichia coli* expression, His-tag fusion proteins) enabling a high degree of automation in the downstream workflow. The largest library from human fetal brain represents >10,000 recombinant human protein expression products, i.e., potential autoantigens. Therefore, approximately 50% of the human genome can be currently accessed by the Protagen technology platform. More than 5,000 human-purified proteins are available for screening purposes. The interaction of auto-antibodies from patient samples with these potential autoantigens can be detected very fast with a high efficiency. Applying a hypothesis-free strategy, novel biomarker candidates can be identified and validated by miniaturized multiparameter assays such as planar or bead-based protein microarrays (Fig. 6.2). Using this approach, Protagen has identified (almost) exclusively novel, indication-specific sets of autoantigens in multiple sclerosis, rheumatoid arthritis, prostate cancer, and, in collaboration with academic partners, in Parkinson's disease, alopecia areata, dilated cardiomyopathy, and SLE (Lueking et al., 2005; Horn et al., 2006; Beyer et al., 2011; Massoner et al., 2011). In all indications studied so far, it became clear that for a precise diagnosis or differential diagnosis, multiple diagnostic marker panels will always be required.

An alternative strategy for the production of protein microarrays uses DNA as template immobilized onto a surface combined with an *in vitro* transcription and translation step. These approaches called nucleic-acid programmable protein array or DNA array to protein array (DAPA) (Sibani and LaBaer, 2011; He and Taussig, 2001) have the advantages that efforts for protein production can be circumvented and toxic proteins may be expressed *in vitro*. However, multiple process steps such as plasmid preparation, spotting of the DNA, and *in vitro* transcription and translation reactions are error prone and lead to considerable inter- and intra-batch differences, which may negatively affect the statistical and bioinformatical analysis.

Alternatively to recombinant proteins (derived from clonal expression and purification), the reverse phase protein microarray approach couples multidimensional liquid phase protein fractionation of localized and metastatic cancer tissue lysates to protein microarrays and subsequent antigen identification by mass spectrometry (Taylor et al., 2008). The analysis of the immunoreactive profile of these protein fractions is difficult, as each of the fractions consists of a number of different proteins and proteins may be represented by adjacent or different fractions. However, this platform enables the analysis of native antigens concerning post-translational modifications and presented linear and structural epitopes.

Tissue microarrays are not involved per se in biomarker discovery approaches but have a strong impact on the validation of discovered biomarkers. The analysis of miniaturized collections of arrayed tissues from pathologically evident tumor biopsies at the DNA, RNA, or protein level enables the linkage of molecular data with various tumor and patient data, such as clinicopathological information, survival, and treatment responses.

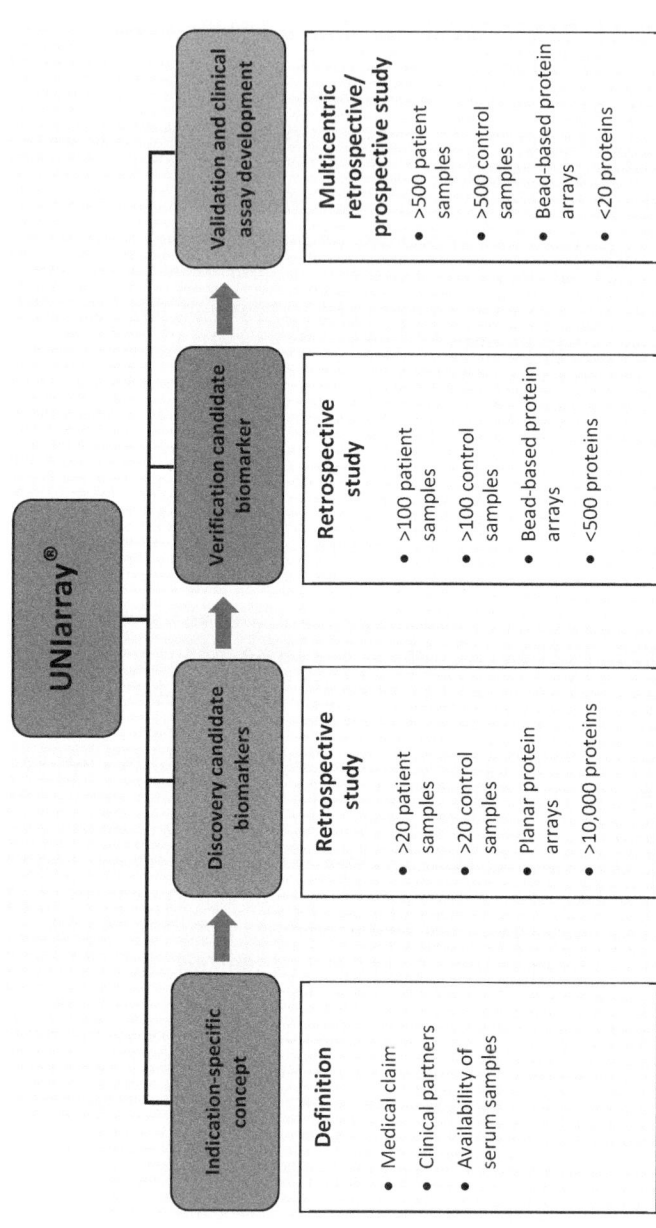

Figure 6.2 UNIarray® strategy shown as process chain for the discovery and validation of diagnostic relevant biomarkers based on antigen/auto-antibody interactions.

Peptide microarrays represent overlapping epitopes of short amino acid sequences from selected (predefined) antigens pointing to a hypothesis-driven approach. Recently this was extended to a hypothesis-free strategy by applying several thousand oligo-N-substituted glycines as unnatural synthetic molecules (so-called peptoids) to discover ligands that bind antibodies in the serum of patients with Alzheimer's disease (Reddy et al., 2011). Due to the synthetic unnatural design of the peptoids, it is expected that artificial "mimotopes" are represented, which may allow binding of antibodies directed against glycostructures or citrullinated antigens. However, as these peptoids represent a collection of artificial molecular shapes, a direct link to the native antigen is missing.

Interactions between proteins and carbohydrates are essential for various biological processes as the carbohydrates contained in glycoproteins, glycolipids, and proteoglycans are involved in recognition processes such as cell adhesion, migration, and signaling. To profile such interactions, carbohydrate microarrays containing polysaccharides, natural glycoconjugates, and mono- and oligosaccharides coupled to carrier molecules have been developed and used for the detection of serum auto-antibodies (Oyelaran and Gildersleeve, 2009). However, this technology is in its beginning and its impact on lead or therapeutic target discovery is, as yet, unclear.

Bead-based assays coupled with flow cytometry detection are a new and emerging technology platform in diagnostics allowing a high grade of multiplexing. The multiplex assay contains a set of different polystyrene beads, which can be differentiated by their different specific fluorescent color code. Each fluorescence-coded bead can be coupled to a specific target antigen following the quantitative determination of serum-contained auto-antibodies by flow cytometry detection. The auto-antibody–antigen interactions are measured with high accuracy and reproducibility at very high sensitivity.

6.5 Antigen/Auto-antibody Interactions as Biomarker Candidates

The autoimmune profile of the human covers a huge number of auto-antibodies, which display an enormous resource to identify novel marker candidates for diagnostic purposes. With access to protein collections covering the human proteome and to protein array technology, the systematic exploration of alterations in autoimmune profiles triggered by the onset or progression of diseases is feasible.

The first attempts using protein microarrays to characterize diagnostic relevant auto-antibodies were carried out in the field of autoimmune diseases. For rheumatoid arthritis, patterns of differential antigen recognition were found to be associated with a clinical subtype of rheumatoid arthritis. Autoreactivity directed against human cartilage gp39 and type II collagen was linked to less severe rheumatoid arthritis and against citrullinated epitopes to severe rheumatoid arthritis (Hueber et al., 2005). Using microarrays with a content of 70 autoantigens, Li and colleagues (Li et al., 2007) showed that serum samples of patients suffering from SLE and incomplete lupus erythematous displayed different autoimmune profiles.

Applying a strategy that uses thousands of antigens derived from a cDNA expression library to profile the humoral autoimmune repertoire, new putative autoantigens have been identified for different diseases such as dilated cardiomyopathy and alopecia areata (Lueking et al., 2005; Horn et al., 2006). Particularly, the two-step approach, including a discovery and a verification phase, resulted in the identification of eight antigen–auto-antibody interactions depicting a highly disease-specific autoimmune response by alopecia areata.

In contrast to proteomic studies, a hypothesis-driven approach was carried out by Quintana and colleagues (Quintana et al., 2008) to identify biomarker candidates for multiple sclerosis. Thereby, microarrays were produced, which contained 64 lipids and 268 protein fragments covering 40 proteins associated with the central nervous system (CNS) or the heat-shock response. Serum samples of different subtypes of multiple sclerosis such as relapse-remitting multiple sclerosis (RRMS), secondary progressive multiple sclerosis, and primary progressive multiple sclerosis were analyzed. All three subtypes were characterized by unique patterns of reactivity to CNS, whereas auto-antibodies against heat-shock proteins were only detected in RRMS serum samples. This suggested that auto-antibody signature links pathologic subtypes of multiple sclerosis and appears to reflect immune processes in the CNS.

Cerebrospinal fluid (CSF) of multiple sclerosis patients is characterized by the presence of immunoglobulin. The immunoglobulin is detected as oligoclonal bands in CSF and support current diagnosis. However, a common identity of OCB reactivity is not yet known and could be of great value to develop a diagnostic test. In a study by Beyer and colleagues (Beyer et al., 2011), samples of OCB-positive CSF of 20 patients with RRMS were compared to CSF of sex- and age-matched controls using the UNIarray® technology platform of Protagen AG. Interestingly, the functional annotation of the top 100 identified antigens results in a strong linkage to diabetes and the insulin signaling pathway. It has been reported that the prevalence of type 2 diabetes was higher in multiple sclerosis patients, perhaps caused by muscle degradation due to neuron degeneration or use of high dose methylprednisolone pulse (Hussein and Reddy, 2006). Several investigators have found some metabolic disorders linking both diseases, such as abnormalities in fat, calcium, and vitamin D metabolism.

In the last two decades, protein arrays have also been used to investigate the autoimmune profile of cancer patients and identify TAAs. Most of the TAAs have been identified in a discovery approach but have not been further validated to develop a diagnostic test (Casiano et al., 2006). For ovarian cancer, protein microarrays were employed to screen 30 serum samples of cancer patients and controls, respectively. Ninety-four antigens were identified that exhibit enhanced reactivity from serum samples of cancer patients relative to control serum samples (Hudson et al., 2007). For validation, specific antibodies against identified antigens were subjected to tissue microarray and antibodies against Lamin A/C, SSRP1, and RALBP1 were recognized to produce a robust signature of cancer. However, the three antigens were prevalent not only in ovarian cancer but also in tissues of many types of cancer and a subset of healthy tissue. Therefore, the diagnostic value of the candidate tissue markers remains elusive.

In the study of Anderson and colleagues (Anderson et al., 2011) three subsequent rounds of screening approaches, including discovery and verification phases, were carried out to identify biomarkers for the early detection of breast cancer. High-density custom protein arrays were used to analyze three cohorts of breast cancer case and control serum samples. In total, 285 serum samples were investigated and 28 TAAs were identified to react with auto-antibodies arising with the onset of breast cancer. Using these antigens, the discrimination of cases and controls achieved an area under the curve value of 0.756. The study design and validation of the identified TAAs indicate that these auto-antibody biomarkers have a high potential to enter the development of a clinical-grade assay.

In the study by Massoner and colleagues (Massoner et al., 2011), the autoimmune profile of prostate cancer patients was investigated. Using the UNIarray® strategy (Fig. 6.2), 160 serum samples of patients diagnosed with prostate cancer or benign diseases as well as healthy control individuals were investigated to identify autoimmune profile characteristics of each group. In the discovery screen, 408 proteins were identified as TAAs in serum samples of prostate cancer patients. In the validation screen, these 408 proteins were used to investigate whether the autoimmune profile of each group is discriminative and can be used by classification algorithms. After statistical analysis, 15 proteins were useful to yield an area under the curve value of 0.71, indicating a diverging autoimmune profile of prostate cancer and benign disease patients. Among these proteins, TTLL12 had been associated with prostate cancer, and six other proteins (RPIA, NOVA2, MAP2, HSPH1, RASSF7, and RBM15) had been described as relevant for other cancers. Thus auto-antibodies in serum samples of prostate cancer patients are directed against known cancer-associated proteins and novel proteins and they are stable among different patients.

6.6 Diagnostic Assays Based on Antigen/Auto-antibody Interactions

Following discovery, qualification, and verification, the successfully validated biomarker(s) is/are subjected to the development of a diagnostic assay. Currently in clinical laboratories, an increasing number of auto-antibodies are measured employing a broad spectrum of techniques and methods. The main techniques involving functionalized surfaces for multiplexed measurements are line-blot immunoassays, bead-based assays with flow cytometry detection, and antigen microarrays.

As an example for line immunoblot assays, the recomLine ANA/ENA (Mikrogen, Martinsried, Germany) allows the multiplex analysis of 14 antigens (RNP68, RNPA, RNPC, SmB, SmD, SSA60, SSA52, SSB, PO, PCNA, CEN-B, Scl70, Jo-1, and histones) in a single procedure. With the exception of histones, all antigens have a recombinant origin. Generally, line-blot assays are widely distributed and easy to use, but limited to a low number of analytes.

Currently, several companies supply commercial kits for the simultaneous measurement of different auto-antibodies by bead-based assays. ENA/ANA analysis based on this technology platform is supplied, for example, by Inova Diagnostics (San Diego, USA), Bio-Rad (Hercules, USA), BMD (Marne la Vallie, France), and Zeus Scientific (Raritan, USA).

Antigen microarrays may contain up to a thousand proteins immobilized to the surface resulting in a high multiplex grade. Companies supplying commercial kits based on protein microarrays include Randox (Belfast, United Kingdom) and Thermo Fisher Scientific (Phadia, Uppsala, Sweden). SQI Diagnostics (Toronto, Canada) combines the microtiter format with the protein microarray technology by spotting antigen panels into each well of a microtiter plate. Different in vitro diagnostic products for the diagnosis of autoimmune diseases such as rheumatoid arthritis, SLE, or celiac disease are commercially available. However, the number of antigens is in the range of tens to hundreds.

6.7 Conclusion

A large tool box is available for biomarker discovery. However, the reduction of a large panel of identified biomarkers to a small panel and transfer of these biomarker(s) to a clinical test remain as a future challenge. The access to disease-specific reference serum samples and a pipeline with defined benchmark data for validation will help to develop clinically relevant biomarkers.

The hunt for innovative diagnostics and prognostic biomarkers as well as biomarkers indicating drug failure before therapies are applied as gains of an increasing scientific and commercial importance. It is anticipated that companion diagnostics will have a widespread use in the pharmaceutical market as they promise to make drug development faster and clinical trials smaller. As a consequence, the time to market for a drug is shorter, peak sales are higher, and patent protection is longer. Also, existing drugs may more easily enter into new indications as shown for belimumab. As the use of such companion diagnostic assays can be integrated very early in clinical and even pre-clinical development, they can generate sales prior to achieving regulatory approval.

Disclosures and Conflict of Interest

This chapter was originally published as: Lueking, A., Göhler, H., Schulz-Knappe, P. (2014). Auto-antibodies as biomarkers for disease diagnosis. In Scholz, M., ed. *Biofunctional Surface Engineering*, chapter 10, pp. 233–249, and appears here, with edits, by kind permission of the publisher, Pan Stanford Publishing, Singapore. The authors declare that they have no conflict of interest and have no affiliations or financial involvement with any organization or entity discussed in this chapter. No writing assistance was utilized in the production of this chapter and the authors have received no payment for its preparation.

Corresponding Author

Dr. Angelika Lueking
Protagen AG
Otto-Hahn-Str. 15, 44227
Dortmund, Germany
angelika.lueking@protagen.com

References

Anderson, K. S. LaBaer, J. (2005). The sentinel within: Exploiting the immune system for cancer biomarkers. *J. Proteome Res.*, **4**(4), 1123–1133.

Anderson, K. S., Sibani, S., Wallstrom, G., Qiu, J., Mendoza, E. A., Raphael, J., et al. (2011). Protein microarray signature of autoantibody biomarkers for the early detection of breast cancer. *J. Proteome Res.*, **10**(1), 85–96.

Arbuckle, M. R., McClain, M. T., Rubertone, M. V., Scofield, R. H., Dennis, G. J., James, J. A., et al. (2003). Development of auto-antibodies before the clinical onset of systemic lupus erythematosus. *N. Engl. J. Med.*, **349**(16), 1526–1533.

Beyer, N. H., Lueking, A., Kowald, A., Frederiksen, J. L., Heegaard, N. H. (2011). Investigation of autoantibody profiles for cerebrospinal fluid biomarker discovery in patients with relapsing-remitting multiple sclerosis. *J. Neuroimmunol.*, **242**(1–2), 26–32.

Casiano, C. A., Mediavilla-Varela, M., Tan, E. M. (2006). Tumor-associated antigen arrays for the serological diagnosis of cancer. *Mol. Cell. Proteomics*, **5**(10), 1745–1759.

He, M., Taussig, M. J. (2001). Single step generation of protein arrays from DNA by cell-free expression and in situ immobilisation (PISA method). *Nucleic Acids Res.*, **29**(15), E73–73.

Horn, S., Lueking, A., Murphy, D., Staudt, A., Gutjahr, C., Schulte, K., et al. (2006). Profiling humoral autoimmune repertoire of dilated cardiomyopathy (DCM) patients and development of a disease-associated protein chip. *Proteomics,* **6**(2), 605–613.

Hudson, M. E., Pozdnyakova, I., Haines, K., Mor, G., Snyder, M. (2007). Identification of differentially expressed proteins in ovarian cancer using high-density protein microarrays. *Proc. Natl. Acad. Sci. U.S.A.*, **104**(44), 17494–17499.

Hueber, W., Kidd, B. A., Tomooka, B. H., Lee, B. J., Bruce, B., Fries, J. F., et al. (2005). Antigen microarray profiling of auto-antibodies in rheumatoid arthritis. *Arthritis Rheum.*, **52**(9), 2645–2655.

Hussein, W. I., Reddy, S. S. (2006). Prevalence of diabetes in patients with multiple sclerosis. *Diabetes Care,* **29**(8), 1984–1985.

Klade, C. S., Voss, T., Krystek, E., Ahorn, H., Zatloukal, K., Pummer, K., et al. (2001). Identification of tumor antigens in renal cell carcinoma by serological proteome analysis. *Proteomics*, **1**(7), 890–898.

Li, Q. Z., Zhou, J., Wandstrat, A. E., Carr-Johnson, F., Branch, V., Karp, D. R., et al. (2007). Protein array autoantibody profiles for insights into systemic lupus erythematosus and incomplete lupus syndromes. *Clin. Exp. Immunol.*, **147**(1), 60–70.

Lueking, A., Cahill, D. J. (2006). Protein biochips in the proteomic field. In: Hamacher, M., Marcus, K., Stühler, K., van Hall, A., Warscheid, B., Meyer, H. E., eds. *Proteomics in Drug Research*, Wiley-VCH Verlag GmbH & Co. KGaA, Weinheim, FRG.

Lueking, A., Huber, O., Wirths, C., Schulte, K., Stieler, K. M., Blume-Peytavi, U., et al. (2005). Profiling of alopecia areata autoantigens based on protein microarray technology. *Mol. Cell. Proteomics,* **4**(9), 1382–1390.

Massoner, P., Lueking, A., Goehler, H., Hopfner, A., Kowald, A., Kugler, K. G., et al. (2011). Serum auto-antibodies for discovery of prostate cancer specific biomarkers. *Prostate,* **72**(4), 427–436.

Oyelaran, O., Gildersleeve, J. C. (2009). Glycan arrays: Recent advances and future challenges. *Curr. Opin. Chem. Biol.,* **13**(4), 406–413.

Quintana, F. J., Farez, M. F., Viglietta, V., Iglesias, A. H., Merbl, Y., Izquierdo, G., et al. (2008). Antigen microarrays identify unique serum autoantibody signatures in clinical and pathological subtypes of multiple sclerosis. *Proc. Natl. Acad. Sci. U.S.A.,* **105**(48), 18889–18894.

Quintana, F. J., Getz, G., Hed, G., Domany, E., Cohen, I. R. (2003). Cluster analysis of human autoantibody reactivities in health and in type 1 diabetes mellitus: A bio-informatic approach to immune complexity. *J. Autoimmun.,* **21**(1), 65–75.

Reddy, M. M., Wilson, R., Wilson, J., Connell, S., Gocke, A., Hynan, L., et al. (2011). Identification of candidate IgG biomarkers for Alzheimer's disease via combinatorial library screening. *Cell,* **144**(1), 132–142.

Rifai, N., Gillette, M. A., Carr, S. A. (2006). Protein biomarker discovery and validation: The long and uncertain path to clinical utility. *Nat. Biotechnol.,* **24**(8), 971–983.

Robinson, W. H. (2006). Antigen arrays for antibody profiling. *Curr. Opin. Chem. Biol.,* **10**(1), 67–72.

Sahin, U., Tureci, O., Schmitt, H., Cochlovius, B., Johannes, T., Schmits, R., et al. (1995). Human neoplasms elicit multiple specific immune responses in the autologous host. *Proc. Natl. Acad. Sci. U.S.A.,* **92**(25), 11810–11813.

Schubert, U., Anton, L. C., Gibbs, J., Norbury, C. C., Yewdell, J. W., Bennink, J. R. (2000). Rapid degradation of a large fraction of newly synthesized proteins by proteasomes. *Nature,* **404**(6779), 770–774.

Sharp, V., Utz, P. J. (2007). Technology insight: Can autoantibody profiling improve clinical practice? *Nat. Clin. Pract. Rheumatol.,* **3**(2), 96–103.

Sherer, Y., Gorstein, A., Fritzler, M. J., Shoenfeld, Y. (2004). Autoantibody explosion in systemic lupus erythematosus: More than 100 different antibodies found in SLE patients. *Semin. Arthritis Rheum.,* **34**(2), 501–537.

Sibani, S., LaBaer, J. (2011). Immunoprofiling using NAPPA protein microarrays. *Methods Mol. Biol.*, **723**, 149–161.

Soussi, T. (2000). p53 antibodies in the sera of patients with various types of cancer: A review. *Cancer Res.*, **60**(7), 1777–1788.

Suzuki, H., Graziano, D. F., McKolanis, J., Finn, O. J. (2005). T-cell-dependent antibody responses against aberrantly expressed cyclin B1 protein in patients with cancer and premalignant disease. *Clin. Cancer Res.*, **11**(4), 1521–1526.

Tan, H. T., Low, J., Lim, S. G., Chung, M. C. (2009). Serum auto-antibodies as biomarkers for early cancer detection. *FEBS J.*, **276**(23), 6880–6904.

Taylor, B. S., Pal, M., Yu, J., Laxman, B., Kalyana-Sundaram, S., Zhao, R., et al. (2008). Humoral response profiling reveals pathways to prostate cancer progression. *Mol. Cell. Proteomics*, **7**(3), 600–611.

von Mensdorff-Pouilly, S., Petrakou, E., Kenemans, P., van Uffelen, K., Verstraeten, A. A., Snijdewint, F. G., et al. (2000). Reactivity of natural and induced human antibodies to MUC1 mucin with MUC1 peptides and n-acetylgalactosamine (GalNAc) peptides. *Int. J. Cancer*, **86**(5), 702–712.

Wallace, D. J., Stohl, W., Furie, R. A., Lisse, J. R., McKay, J. D., Merrill, J. T., et al. (2009). A phase II, randomized, double-blind, placebo-controlled, dose-ranging study of belimumab in patients with active systemic lupus erythematosus. *Arthritis Rheum.*, **61**(9), 1168–1178.

Chapter 7

The Accelerated Blood Clearance Phenomenon of PEGylated Nanocarriers

Amr S. Abu Lila, PhD,[a,b,c] and Tatsuhiro Ishida, PhD[a]

[a]*Department of Pharmacokinetics and Biopharmaceutics, Subdivision of Biopharmaceutical Sciences, Institute of Biomedical Sciences, Tokushima University, Japan*
[b]*Department of Pharmaceutics and Industrial Pharmacy, Faculty of Pharmacy, Zagazig University, Egypt*
[c]*Department of Pharmaceutics, College of Pharmacy, Hail University, Saudi Arabia*

Keywords: accelerated blood clearance (ABC) phenomenon, anti-PEG IgM, complement activation, B cell anergy, cytotoxic agents, effectuation phase, hypersensitivity reactions, induction phase, lipid dose, metronomic chemotherapy, nucleic acids, opsonins, particle size, PCBylation, polyethylene glycol (PEG), PEG-chain length, PEG-surface density, PEG terminal group, PEGylation, polyglycerol, repeated administration, splenic B cells, surface charge, third dose, time interval, T-cell independent antigens, Toll-like receptors

7.1 Introduction

PEGylation refers to the conjugation of polyethylene glycol (PEG), and it has been extensively applied in the pharmaceutical industry, particularly in the field of drug delivery, in order to improve the pharmacokinetic behavior of PEGylated therapeutics

Immune Aspects of Biopharmaceuticals and Nanomedicines
Edited by Raj Bawa, János Szebeni, Thomas J. Webster, and Gerald F. Audette
Copyright © 2018 Pan Stanford Publishing Pte. Ltd.
ISBN 978-981-4774-52-9 (Hardcover), 978-0-203-73153-6 (eBook)
www.panstanford.com

[1]. The crucial role of PEGylation derives mainly from the ability of PEG, a hydrophilic polymer, to attract water molecules, which results in a significant increase in the hydrodynamic size, and thereby, attenuates a rapid renal clearance of PEGylated products [2]. In addition, the steric stabilization that is imparted by the formation of a hydration zone around a PEGylated substance protects the PEGylated product against enzymatic degradation and against the surface binding of certain serum proteins (opsonins) that interact with the immune system [3, 4]. Consequently, PEGylation efficiently evades the recognition of PEGylated products by the mononuclear phagocyte system (MPS), which is an obstacle that has hindered the therapeutic efficacy of many non-PEGylated products.

However, despite the immunogenicity of PEG, there has been little attention paid to PEG in the past few decades until recently, when an increasing number of reports emphasized the potential negative side effects of PEG in drug delivery systems and/or protein therapeutics [5–8]. Several research groups have shown that anti-PEG antibodies, produced in response to a first dose of PEGylated liposome, are associated with the rapid clearance of subsequent doses of PEGylated liposomes, which is referred to as the accelerated blood clearance (ABC) phenomenon [9–11]. Surprisingly, the ABC phenomenon has not been restricted only to PEGylated liposomes. Remarkably, this phenomenon has been recognized upon repeated administration of other PEGylated nanocarriers, including nanoparticles [12], micelles [13], and microemulsions [14] and even with PEGylated proteins [15]. In addition, several animal studies and clinical observations concerning the use of Doxil® (a PEGylated liposomal formulation of the cytotoxic agent doxorubicin (DXR)) have shown that PEGylated liposomes can activate the complement system and potentially induce hypersensitivity reactions [7, 16, 17]. Consequently, since PEGylation of nanocarriers/protein therapeutics is currently considered to be one of the most favorable approaches in enhancing the *in vivo* fate of these products due to the reduction in their immunogenicity, the existence of anti-PEG antibodies, as well as the implications for the induction of the ABC phenomenon, is of clinical concern and requires thorough investigation.

7.2 Mechanism of ABC Phenomenon

Since the first observations of pharmacokinetic irregularities (the ABC phenomenon) upon repeated administration of PEGylated liposomes by Dams et al. [8], many reports have focused on explaining the underlying mechanism of the ABC phenomenon [10, 18–20]. Laverman et al. [10] identified two phases of the ABC phenomenon: the induction phase, in which the biological system is "primed" by the first injection of PEGylated liposomes, and, the effectuation phase, following the second, or subsequent, administration in which the PEGylated liposomes are rapidly opsonized and cleared from systemic circulation, by the macrophages of the MPS.

We previously reported that in rats and mice, the ABC phenomenon is mediated by a soluble serum factor, mainly IgM (anti-PEG IgM) and that the extent of the induction of the ABC phenomenon is strongly correlated with the level of anti-PEG IgM produced in response to a single injection of PEGylated liposomes [21–23]. Further studies have emphasized the vital role of the spleen in the production of anti-PEG IgM [24]. Splenectomy was found to significantly attenuate the production of anti-PEG IgM and consequently alleviate the induction of the ABC phenomenon following the administration of PEGylated liposomes [25]. In addition, although the follicle region in the spleen is the main compartment for B cells, recent studies have shown that PEGylated liposomes trigger its immune response via stimulation of the marginal zone B cells (MZ B cells) [5]. We [5] have emphasized the crucial contribution of MZ B cells over the follicular cells in the spleen, in anti-PEG IgM production, and in the recognition of PEGylated nanoparticles. Pre-treatment with cyclophosphamide has significantly suppressed the anti-PEG IgM production via the depletion of IgM-high cells in the MZ, particularly MZ B cells. These results provide strong evidence that the splenic MZ B cells are responsible for anti-PEG IgM production.

Furthermore, Semple et al. [26] reported that the ABC phenomenon was observed in BALB/c nu/nu (T cell-deficient) mice, but not in BALB/c SCID (T and B cells-deficient) mice. That study also showed that the extent of anti-PEG IgM antibody production following stimulation with PEGylated liposomes was substantially increased in the BALB/c nu/nu mice, but not in the BALB/c SCID mice. We also reported a similar finding [27]. These results provide strong evidence that the T cell-independent B cell response plays a crucial role in the induction of the ABC phenomenon.

T-cell independent (TI) antigens (or Thymus independent antigens) are known to activate B cells and to be responsible for the production of IgM antibodies in the early stages of immunization [28]. TI antigens generally fall into two classes (TI-1 or TI-2) and activate B cells by two different mechanisms. TI-1 antigens, at a lower concentration, are capable of specifically activating B cells and thus eliciting a specific antibody response. At higher concentrations, however, TI-1 antigens act as potent B-cell (mature and immature) mitogens, resulting in non-specific, polyclonal activation of B cells, and thereby, polyclonal IgM production [29, 30]. In contrast, TI-2 antigens only activate mature B cells and can induce an immunological response by extensively cross-linking the cell surface immunoglobulins of B cells, resulting in massive secretion of neutralizing antibodies, including IgM and IgG [28–30].

Roffler and colleagues [31, 32] have emphasized that anti-PEG IgM produced selectively following immunization with PEGylated β-glucuronide recognizes the repeating $-O-CH_2-CH_2-$ subunits of PEG. This provides the evidence that a PEG polymer in PEGylated nanoparticles/materials acts as a TI-2 antigen and that the repeating subunit may act as the immunogenic epitope of PEG and a binding site for the derived anti-PEG IgM. Based on the aforementioned data, we therefore envisage the following mechanism to explain the ABC phenomenon (Fig. 7.1): Once the PEGylated liposomes reach the spleen after the first injection, they bind/cross-link to surface immunoglobulins on reactive B-cells in the splenic marginal zone and consequently trigger the production of anti-PEG IgM antibody in a T-cell independent manner. Upon subsequent administration of PEGylated liposomes, the secreted IgM, in response to the first dose, selectively attaches to the PEG on these liposomes, and subsequently activates the complement system resulting in opsonization of the liposomes by C3 fragments and an enhanced uptake of the liposomes by the Kupffer cells in the liver.

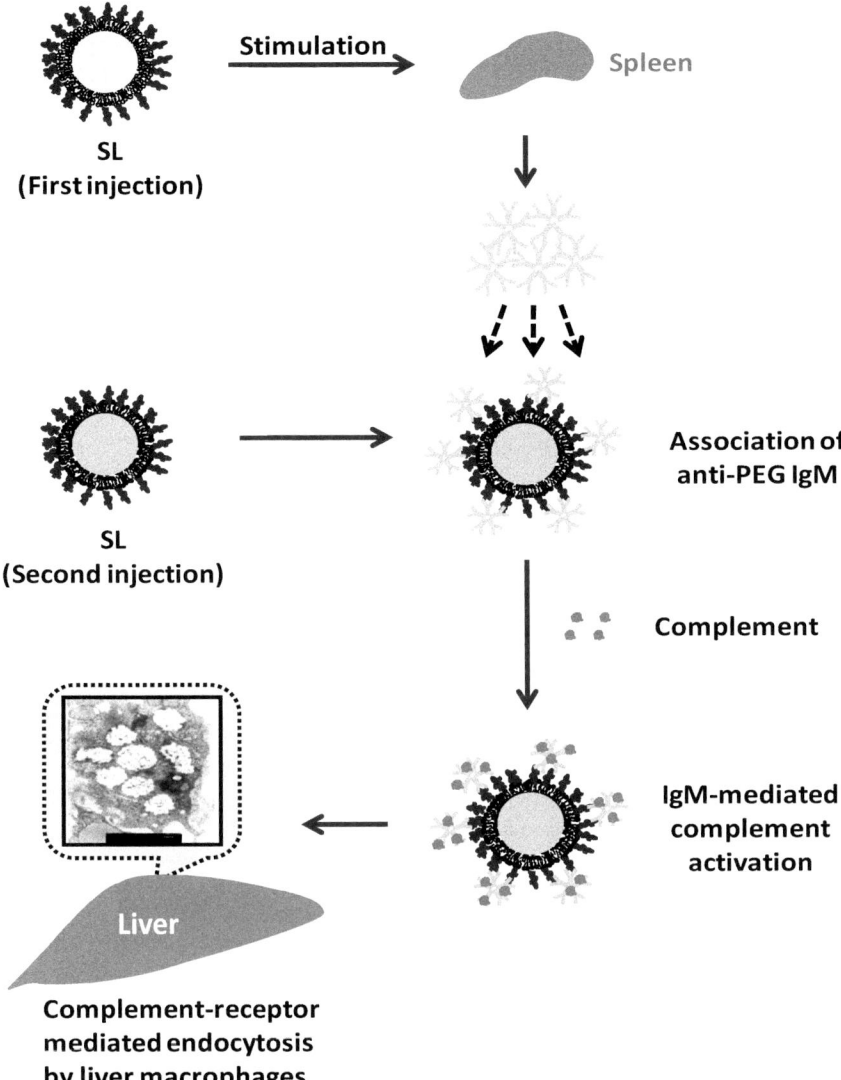

Figure 7.1 Sequence of events leading from anti-PEG IgM induction to the accelerated blood clearance of PEGylated liposomes.

Despite the fact that the spleen and anti-PEG IgM are involved in the ABC phenomenon, we [25] have reported that splenectomy failed to completely reverse the rapid clearance and increased hepatic accumulation of PEGylated liposomes to control values.

Therefore, we cannot exclude the involvement of another factor(s) and/or tissue(s) in the ABC phenomenon. A recent report by Deng and coworkers [33] revealed the occurrence of the ABC phenomenon following the subcutaneous administration of PEGylated solid-lipid nanoparticles (SLNs). They speculated that PEGylated SLNs drained from the subcutaneous injection site into the lymphatic network were phagocytosed in the lymph nodes, and they initiated the immune response with the production of anti-PEG IgM. Collectively, the immune response after repeated injections of PEGylated liposomes seems to be a complicated process, for which further studies must be conducted in order to clearly elucidate all the contributing factors and/or tissues.

7.3 Correlation Between Complement Activation and ABC Phenomenon

To date, with respect to the ABC phenomenon, researchers have focused mainly on the induction of anti-PEG IgM that is produced in response to a first dose of PEGylated nanoparticles in order to predict the magnitude of the ABC phenomenon. IgM antibodies alone, however, are unable to directly promote phagocytosis because IgM is not an opsonizing antibody due to the absence of Fc receptors for IgM on the surface of macrophages [34]. Alternatively, the binding of IgM can trigger the opsonization of complement components by complement activation via the classic pathway that subsequently promotes phagocytosis by Kupffer cells bearing complement receptors [34, 35].

Several studies have established the role of the complement system in the accelerated clearance of PEGylated liposomes from systemic circulation [20, 36, 37]. Dams et al. [8] showed that rats transfused with serum from rats pre-treated with PEGylated liposomes rapidly cleared the dose when they were treated with PEGylated liposomes for the first time. In addition, this accelerated clearance of the PEGylated liposomes was abolished by pre-heating the serum at 56°C for 30 min before transfusion, the temperature at which the complement is inactivated. In the same context, by using a single-pass liver perfusion technique, we showed that a first dose of PEGylated liposomes did not enhance the intrinsic phagocytic activity of the Kupffer cells. On the other hand, the serum obtained from rats pre-treated with PEGylated liposomes enhanced the hepatic uptake of a test dose of PEGylated liposomes. Furthermore, this serum-dependent uptake of a test dose by the liver was completely abolished by pre-treatment of the serum at 56°C for 30 min, which inhibited the complement activity [16]. These results strongly clarify the major role of anti-PEG IgM-mediated complement activation in initiating the accelerated blood clearance of PEGylated liposomes upon multiple administrations.

In a series of our studies [16, 23, 24], we reported that considerable IgM binding and complement consumption upon incubation with PEGylated liposomes has been observed only in sera from rats displaying a rapid clearance of the second dose. Yang et al. [37] also determined the residual complement activity in rat serum that was pre-treated with PEGylated liposomes. They demonstrated that a considerable portion of the complement was consumed when incubated with PEGylated liposomes in a phospholipid dose-dependent manner; the lower the injected dose, the greater the complement activation. Taken together, the presence of serum anti-PEG IgM and the subsequent binding of anti-PEG IgM to the PEGylated products are not necessarily related directly to the enhanced clearance of the products. Instead, it appears that subsequent complement activation following anti-PEG IgM binding is the most important rate-limiting step in dictating the *in vivo* fate of PEGylated products.

In addition to the crucial role of the complement system in the ABC phenomenon, several studies have verified that PEGylated products can activate the complement system after a first dose resulting in the induction of a PEGylated liposome-related hypersensitivity syndrome called C activation-related pseudoallergy (CARPA) [38–41]. Szebeni et al. [42] described how complement activation is responsible for infusion reactions observed in up to 25% of patients treated with Doxil®. Chanan-Khan et al. [43] also demonstrated that Doxil® therapy activated complement in the majority of patients and induced moderate to severe hypersensitivity reactions in about 50% of cancer patients infused with Doxil® for the first time. Serum complement terminal complex (SC5b-9) levels were elevated relative to the baseline in 80% of the patients who developed hypersensitivity reactions, suggesting that complement activation plays a casual role in hypersensitivity reactions caused by Doxil®. However, it remains unclear whether complement activation in response to a first dose is able to induce phagocytosis besides the generation of a hypersensitivity reaction.

7.4 Factors That Affect the Magnitude of the ABC Phenomenon

Several reports have emphasized the contribution of different factors on the incidence and magnitude of the ABC phenomenon. These factors include the physicochemical properties (particle size, surface charge, PEG-chain length, and degree of PEGylation), dosing intervals, the lipid dose used, and the nature of the encapsulated payload of the initially injected liposome.

7.4.1 Effect of Particle Size and Surface Charge

We [44] investigated the surface charge and liposomal size in the first injection for its effect on the induction of the ABC phenomenon. PEGylated liposomes with three different surface charges (+13.5, −46.15, and −1.51 mV) and three different sizes (100, 400, and 800 nm) were prepared. All the prepared liposomes, given as a first dose, affected the blood concentration and hepatic accumulation of subsequently injected $mPEG_{2000}$-liposomes in a comparable manner. Consequently, those results revealed that the charge and size of liposomes in the first dose were not critical for the induction of the phenomenon. On the contrary, Koide et al. [13] have indicated that the repeated injections of PEGylated polymeric micelles caused the ABC phenomenon in a size-dependent manner. They demonstrated that the pre-treatment of mice with 50.2 nm polymeric micelles significantly triggered the enhanced clearance and preferential hepatic accumulation of a subsequently injected test dose of PEGylated liposomes. On the other hand, smaller polymeric micelles (9.7 and 31.5 nm) failed to induce the accelerated clearance of a test-dose of PEGylated liposomes. They assumed that small polymeric micelles (31.5 nm or less) could escape the recognition by immune cells, while larger particles would be easily recognized by immune cells, presumably in the spleen, and thus would activate the immune system.

7.4.2 Effect of PEG-Surface Density and PEG-Chain Length

Despite the fact that the surface modification of colloidal particles with PEG markedly reduced their recognition by the cells of MPS and conferred them with long circulating characteristics when administered intravenously, many studies have reported that PEGylated nanocarriers might induce a T-cell-independent anti-PEG immune response and that PEG might act as an antigenic epitope [23, 45]. Consequently, the extent of the ABC phenomenon is thought to be related to both PEG-lipid concentration and PEG chain length [12, 46–48]. We have reported that liposomal surface modification with 5 mol% mPEG$_{2000}$-DSPE induced a maximal ABC effect in rats. Either decreasing or increasing the PEG density at the liposome surface beyond 5 mol%, however, attenuated the accelerated clearance of a subsequently injected dose rather than inducing it. We anticipated that mPEG$_{2000}$-DSPE at lower concentrations (<5 mol%) would fail to activate the splenic B cells, while excessively higher concentrations (>5 mol%) would result in a decrease in splenic B cells reactivity [47].

Contrary to our findings, Li et al. [49] reported that despite the similar anti-PEG IgM levels produced in response to the intravenous administration of liposomes modified with either 3% or 9% PEG, the latter formulation induced a more severe ABC phenomenon compared with that of the 3% PEG formulation. Li et al. emphasized that the 9% PEG formulation showed a much higher affinity to anti-PEG IgM and thus was easily recognized by the anti-PEG IgM produced in response to the first dose, compared with the 3% PEG formulation. Zhao et al. [46] also reported that solid lipid nanoparticles (SLNs) containing 10 mol% PEG produced a higher elimination rate and more splenohepatic uptake of the second dose than 5 mol% PEG-SLNs. They assumed that the magnitude of the ABC phenomenon was strongly correlated to the different circulation times of the initial dose imparted by the different PEG surface densities. Regarding 5 mol% PEG-SLNs, when the circulation time was only slightly prolonged, this allowed inefficient contact of the PEGylated SLNs with splenic B cells and resulted in the secretion of tiny amounts of anti-PEG IgM. On the other hand, 10 mol% PEG-SLNs exhibited a more extended circulation time, which allowed a more efficient degree of interaction with splenic B cells and a subsequent induction of larger amounts of anti-PEG IgM. This led to a more rapid clearance by comparison with the 5 mol% PEG-SLNs.

In addition to PEG surface density, several studies have indicated that the PEG chain length/molecular weight is also a crucial factor for the induction of the ABC phenomenon. We [44] reported that the injection of mPEG$_{5000}$-liposomes led to a significantly lower hepatic accumulation of subsequently injected liposomes relative to the injection of mPEG$_{2000}$-liposomes. On the contrary, in another study we [9] showed that PEG$_{30000}$-bovine serum albumin (PEG$_{30000}$-BSA) induced a higher anti-PEG IgM response and greater complement activation than PEG$_{2000}$-BSA.

7.4.3 Effect of PEG Terminal Group

Despite the increasing number of reports emphasizing the effect of PEG-surface density and PEG-chain length on the incidence/magnitude of the ABC phenomenon, little is known about the effect that PEG-DSPE terminal groups exert on the induction of the ABC phenomenon. Wang et al. [50] have recently investigated the fate of intravenously injected PEG-coated emulsions (DE) that were prepared using PEG terminated with a hydroxyl (OH), methoxy (OCH$_3$), carboxy (COOH), amino (NH$_2$), and thiol (SH) groups in rats. They demonstrated that DE-OCH$_3$ showed the longest circulation time after a single intravenous injection, followed by DE-SH and DE-COOH, while, DE-OH and DE-NH$_2$ showed the shortest half-lives. In addition, after multiple administrations, they correlated the incidence/magnitude of the ABC phenomenon to the circulation time of initially injected PEG-coated emulsions. They revealed that DE-SH triggered the fastest clearance of the subsequent dose of DE-OCH$_3$, while DE-NH$_2$ and DE-OH barely affected the clearance of the second dose. The unexpected increased immunogenicity of DE-SH over DE-OCH$_3$ was attributed to the presence of a thiol group, which presumably stimulated the proliferation and differentiation of

B cells either by inducing the synthesis of the cell membrane and cytosolic proteins or by reacting with follicular dendritic cells.

7.4.4 Time Interval

Previous reports on PEGylated nanoparticles have confirmed that the time interval between subsequent administration is a key factor in eliciting the ABC phenomenon [6, 8, 51]. These reports strongly correlated the production pattern of anti-PEG IgM to the incidence/magnitude of the ABC phenomenon. Production of anti-PEG IgM occurs in a wave pattern, meaning that after the first injection of PEGylated liposomes, anti-PEG IgM titers increase at day 3, peak at day 5, and then gradually decrease until they are undetectable by day 28 [9, 52].

Goins et al. [53] have previously reported unchanged pharmacokinetics of PEGylated liposomes following their sequential administration to rabbits in 6-week intervals. Oussoren and Storm [51] also reported similar pharmacokinetics for four doses of PEGylated liposomes in rats when they were administered in 24 or 48 h intervals. By contrast, in a series of our studies we showed that a prior administration of PEGylated liposomes in mice and rats triggered the rapid clearance and the preferential hepatic accumulation of a subsequently injected dose when administered within a time interval of 4 to 7 days (Fig. 7.2) [6, 19, 54].

Recently, Saadati et al. [55] investigated the pharmacokinetic behavior of PEGylated polymeric nanoparticles administered in rats at various time intervals after the first injection (3, 5, 7, 14, and 28 days), the ABC phenomenon was most pronounced at the time interval of 7 days between the first and the second injections, but the phenomenon was not so apparent at a time intervals of 3 or 14 days. On the other hand, when the time interval was extended to 28 days, the plasma concentration-time profile of the second dose was comparable to that of the control. Furthermore, Li et al. [56] have also demonstrated that the ABC phenomenon could only be induced in beagles following the intravenous administration of PEGylated liposomes when a second dose was administered within a 3-week time interval. A second dose administered 4 weeks after the first dose of PEGylated liposomes was not rapidly cleared from blood circulation. That study verified that the ABC phenomenon was mediated by the innate immune system via the production of anti-PEG IgM rather than by the adaptive immune system, because no immune memory was established.

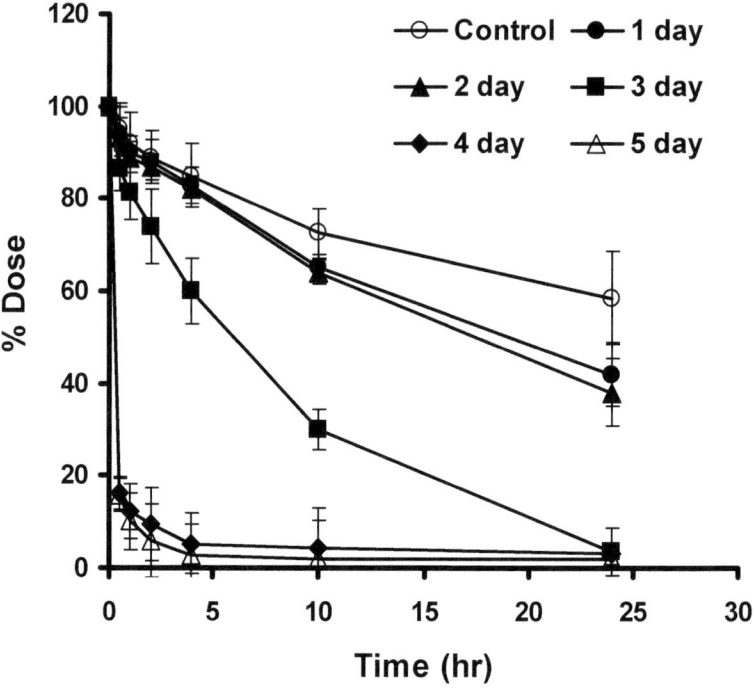

Figure 7.2 Accelerated blood clearance of a second dose of PEGylated liposomes. Rats were pre-treated with PEGylated liposomes (0.001 μmol phospholipids/kg). Then, a second dose of radio-labeled PEGylated liposomes (5 μmol phospholipids/kg) was injected at different time intervals. Each value represents the mean ± S.D. (n = 3). Modified from the author's original work [54].

7.4.5 Effect of Lipid Dose

Many reports have emphasized the critical role that a lipid dose of PEGylated nanocarriers plays on the induction of anti-PEG IgM and the subsequent accelerated clearance of PEGylated liposomes [19, 57]. In our earlier studies [19], we verified the strong inverse relationship between the lipid dose of the first injected dose of PEGylated liposomes and the magnitude of the ABC phenomenon within a range of 0.001–5 μmol phospholipid/kg.

The ABC phenomenon was significant when low phospholipid doses (0.001–0.1 μmol phospholipid/kg) of PEGylated liposomes were intravenously administered as the first dose. On the other hand, higher phospholipid doses (>5 μmol phospholipid/kg) significantly attenuated the incidence of the ABC phenomenon. These results clearly indicate that optimal PEGylated liposomes dosing (i.e., optimal amount of TI-2 antigens) is a prerequisite for the optimal priming of the immune system and the subsequent induction of the antibody response. It is postulated that at a low dose of PEGylated liposomes, the extent of B-cell receptor cross-linking might efficiently activate the cells and promote the production of anti-PEG IgM. On the other hand, when a high dose is first administered, the density of TI-2 antigens is too high, which causes splenic MZ B cells to induce an immune tolerance, which is referred to as anergy. In addition, continuous antigen binding and subsequent receptor signaling are known to be essential for the maintenance of anergy with B cells [58]. Therefore, a prolongation of the circulation time of PEGylated liposomes via the administration of a higher dose is assumed to extend the exposure of splenic MZ B cells to PEGylated liposomes, and thereby, to contribute to the induction of anergy. These observations agree with the clinical findings that emphasize the absence of the ABC phenomenon in patients treated with a clinically recommended dose of Doxil® (≥15 μmol phospholipid/kg) [57, 59].

    Metronomic chemotherapy, the frequent and continuous use of sequentially low-dose chemotherapeutics, has been advocated recently as a novel chemotherapeutic regimen with unique advantages in attenuating the multi-drug resistance of tumor cells [60, 61]. Nevertheless, based on the aforementioned data, the clinical application of PEGylated nanocarrier-based metronomic chemotherapy is assumed to be substantially compromised by the induction of the ABC phenomenon. Yang et al. [37] investigated the anti-tumor efficacy of epirubicin-containing PEGylated liposomes in S180 tumor-bearing mice following sequential low-dose injections. They demonstrated that the treatment efficacy disappeared after the sixth day of repeated injections, accompanied by an increased level of anti-PEG IgM and decreased residual complement activity in mouse serum. They speculated that the anti-PEG IgM-mediated accelerated clearance of subsequently injected doses of PEGylated epirubicin liposomes was strong evidence for an impairment of the therapeutic efficacy.

7.4.6 Effect of Third Dose

Many reports have emphasized the role of a first dose of PEGylated nanocarriers on priming the immune system against a sequentially administered second dose with the subsequent induction of the ABC phenomenon. Nevertheless, studies into the effect of an injection of more than two doses on the ABC phenomenon have been scarce. Dams et al. [8] demonstrated that the ABC phenomenon was less pronounced after a third or fourth dose in rats receiving weekly injections of empty PEGylated liposomes. We [62] also demonstrated that upon sequential administrations of DXR-containing PEGylated liposomes in Beagle dogs, the ABC phenomenon was pronounced after the second dose, but not after a third dose. We speculated that the toxic effect imparted by the large fraction of the second dose of DXR-containing liposomes accumulated in the spleen B cells would impair the production of anti-PEG IgM and thus result in a gradual recovery of pharmacokinetics following the administration of a third dose. Recently, Sadaati et al. [55] reported that, despite the elevated serum levels of anti-PEG IgM following the sequential administration of etoposide-loaded polymeric PEGylated nanoparticles, the pharmacokinetic behavior of a third dose was similar to that of the first dose. They postulated that the saturation of the MPS cells by the second dose of PEGylated nanoparticles might hinder the efficient uptake of a third dose, thereby resulting in a prolonged circulation time for the third dose.

7.4.7 Nature of Encapsulated Payload

7.4.7.1 Effect of encapsulating cytotoxic agents

Despite the fact that induction of the immune response upon repeated administration of PEGylated nanocarriers represents a tremendous impediment to clinical applications, it is amazing that the ABC phenomenon has not hindered the clinical application of PEGylated products containing cytotoxic drugs.

Laverman et al. [10] showed that repeated injections of Caelyx® (a commercial product of PEGylated DXR liposomes) did not induce the ABC phenomenon in a murine model, while empty PEGylated liposomes did. La-Beck et al. [59] demonstrated that the clearance of Doxil® does not change over a dose range of 10 and 60 mg/m^2. We also reported that encapsulation of DXR within liposomes significantly reduced the production of anti-PEG IgM and thus abrogated the immune response against PEGylated liposomes [20]. We postulated that DXR released from liposomes accumulating in the spleen inhibits B cell proliferation and/or kills proliferating B cells, and consequently impairs the production of PEG-specific IgM resulting in an abrogation of the accelerated clearance of a second dose of PEGylated liposomes. This postulation was supported by our finding that free DXR did not abrogate the induction of the ABC phenomenon by empty PEGylated liposomes [62]. Furthermore, we reported that repeated administration of oxaliplatin (l-OHP)-containing PEG-coated liposomes did not induce a significant anti-PEG IgM response, indicating that l-OHP encapsulated in PEG-coated liposomes was efficient in abrogating the ABC phenomenon [63]. Cui et al. [64] also reported that PEGylated mitoxantrone liposomes did not induce the ABC phenomenon. These results promoted the general assumption that the ABC phenomenon could occur only when empty or non-cytotoxic drug-containing PEGylated liposomes are repeatedly injected and thus lack the clinical implications upon repeated administration of PEGylated nanocarriers containing cytotoxic anti-neoplastic agents.

Nevertheless, Deng and coworkers [49, 65] recently showed that, unlike DXR and mitoxantrone, repeated injections of PEGylated liposomal topotecan could still induce a strong ABC phenomenon in Wistar rats, Beagle dogs, and mice. They attributed the induction of the ABC phenomenon, on the one hand, to the poor retention and rapid release of the encapsulated drug from liposomes with the formation of empty vesicles or vesicles containing a limited amount of the drug. On the other hand, they assumed that topotecan, a topoisomerase I inhibitor, is a cell-cycle phase-specific drug that can only inhibit the population of B cells in the splenic marginal zone occupying the S phase of the cell cycle. Accordingly, when compared with cell-cycle non-specific anticancer drugs, the toxic effect of topotecan on B cells may be potentially attenuated, with the result that a first injection of topotecan liposomes can still induce a strong ABC phenomenon for a second dose. In the same context, Saadati et al. [55] have evaluated the efficacy of etoposide, a cytotoxic drug, encapsulated within PEGylated polymeric nanoparticles on the induction of ABC in rats. They emphasized that drug incorporation did not affect the systemic clearance of subsequently injected doses. They suggested that etoposide, like topotecan, is a cell-cycle-specific drug that only affects the B cells in the G2/M phase. Thus, drug-loaded nanoparticles failed to inhibit the entire B cell population in the splenic marginal zone, which was confirmed by the production of a high level of IgM, and thereby, the induction of the ABC phenomenon.

Surprisingly, in a recent study, we [62] found that empty PEGylated liposomes as well as DXR containing PEGylated liposomes (Doxil®) caused anti-PEG IgM production, and thereby, induced a rapid clearance of subsequent doses in Beagle dogs in a Doxil® inverse-dose dependent manner. In that study, a lower dose (less than 2 mg DXR/m^2) induced the phenomenon, while a higher dose (20 mg DXR/m^2) did not. It was assumed that the limited amount of DXR released from tiny amounts of Doxil® might activate these immune cells rather than impairing the function of B lymphocytes, and in turn induce the production of anti-PEG IgM, which would finally elicit an enhanced clearance of subsequently injected Doxil®. On the other hand, at a high dose of Doxil®, DXR released in the spleen inhibited the proliferation of splenic B cells, thus impairing anti-PEG IgM production. In a similar manner, we further [52] reported that a lower dose of l-OHP-containing PEGylated liposomes (0.023 µg l-OHP/kg), rather than higher doses (2.3–2300 µg l-OHP/kg), retains the potency to induce a potent anti-PEG IgM response.

Collectively, the nature of the encapsulated drug, the concentration of drugs in the splenic marginal zone, and the time required for drugs to exert their effect on the B cells in the splenic marginal zone may all have marked effects on B cell proliferation, accordingly leading to different degrees of the ABC phenomenon.

7.4.7.2 Effect of encapsulating nucleic acid

Gene therapy represents a promising therapeutic modality for the treatment of genetic and acquired disorders. As an alternative to viral vectors, cationic liposomes have been introduced as promising carriers with low toxicity profiles and well-controlled gene delivery [66–68]. However, due to a low degree of *in vivo* transfection efficiency, multiple administrations are required in order to achieve the desirable therapeutic outcome. Hence, the ABC phenomenon encountered upon repeated administration of PEGylated liposomes may represent a potential barrier against the efficacy and/or safety of PEGylated cationic liposome, particularly when they carry immune-stimulating nucleic acids such as plasmid DNA (pDNA) or small interfering RNA (siRNA).

Semple et al. [26] verified the induction of a severe immune response in mice receiving repeated administrations of PEGylated liposomes carrying oligonucleotides (ODN), pDNA or RNA ribozyme, resulting in accelerated blood clearance and an increase in the mortality rate. Judge et al. [69] also reported that PEGylated lipid nanoparticles encapsulating pDNA triggered the abundant production of anti-PEG IgM, compared with empty PEGylated lipid nanoparticles, resulting in a severe reduction in gene expression relating to the encapsulated pDNA in tumor tissue following a second injection. A similar phenomenon was also observed after repeated injections of PEG-modified pDNA-lipoplexes [70]. Encapsulation of pDNA containing CpG motifs in PEG-coated cationic liposomes was found to further facilitate the induction of anti-PEG IgM production, resulting in a rapid clearance of a subsequently injected dose of PEG-coated lipoplex from blood circulation. Furthermore, the use of CpG-free pDNA instead of pDNA containing CpG motifs has significantly attenuated the immunogenic response against PEGylated cationic liposomes [70]. Given that the CpG motif in pDNA is a representative TLR9 agonist [71, 72], stimulation of the TLR9 signaling pathway by the CpG-motif in pDNA in immune competent cells is assumed to play a crucial role in the activation of innate immune systems with the subsequent induction of potent antibody responses to PEG-coated pDNA-lipoplexes.

In a recent study, we [73] investigated the contribution of a Toll-like receptor (TLR), exclusively TLR9, to the enhancement of anti-PEG IgM production following intravenous injection of pDNA-containing PEG-coated lipoplex (PDCL) in mice specifically lacking either myeloid differentiation primary-response protein 88 (MyD88), which is essential for most TLR signaling (except for TLR3), or TLR9, which recognizes CpG motifs in DNA [73]. PDCL injected into either MyD88 knock out (KO) or TLR9 KO mice failed to trigger a remarkable anti-PEG IgM response, compared with the wild type. This observation was attributed to the suppressed induction of cytokines such as IL-6, TNFα and IFNs in both MyD88 and TLR9 KO mice, which are known to participate in the induction of IgM production [74]. These results emphasized the crucial role of the MyD88/TLR9 signaling pathway in the induction of the antibody response, anti-PEG IgM production, following the injection of pDNA-containing PEG-coated lipoplex.

In the case of siRNA, we investigated the effect of siRNA encapsulation within PEGylated lipid nanoparticles (known as PEGylated wrapsome, PEG-WS) as well as the effect of the siRNA sequence on anti-PEG IgM production and promotion of the ABC phenomenon [75]. A PEGylated wrapsome (PEG-WS) was prepared by wrapping a core-lipoplex consisting of DC-6-14/POPC (50/50 molar ratio) with a neutral lipid layer of POPC containing a PEGylated lipid; PEG_{2000}-DSPE. Encapsulation of siRNA within PEG-WS was found to significantly attenuate the adjuvant effect of siRNA on the production of anti-PEG IgM, compared with siRNA in conventional PEGylated lipoplexes. In addition, chemical modification, such as 2'-O-methylation, of siRNA efficiently attenuated the production of anti-PEG IgM following the injection of PEGylated siRNA-lipoplex, presumably via rendering the siRNA less activator for TLR7 [76]. However, the exact contribution of TLR7 activation to the enhanced production of anti-PEG IgM in response to PEGylated siRNA-lipoplex was not completely elucidated.

Recently, by employing Toll-like receptor 7 knock out (TLR7 KO) mice and TLR7-deficient splenic B cell reconstituted SCID mice, we investigated how PEGylated siRNA-lipoplex activates the innate immune system through TLR7 and consequently enhances anti-PEG IgM production [77]. We found that the anti-PEG IgM production levels were much higher in wild type mice than in TLR7 KO mice, revealing the contribution of TLR7 activation in the robust production of anti-PEG IgM with a subsequent induction of the rapid clearance of subsequently injected PEGylated liposomes. In addition, we discriminated the contribution of direct activation of splenic B cells via intrinsic TLR7 from that of indirect activation of B cells by type 1 interferons secreted via the activation of TLR7 on plasmacytoid dendritic cells (pDCs) on the induction of anti-PEG IgM in response to PEGylated siRNA-lipoplex [79]. We further revealed that the induction of an anti-PEG IgM response was completely abolished in SCID mice reconstituted with the TLR7-deficient B cells, while the SCID mice reconstituted with wild type splenic B cells did produce an anti-PEG IgM response. These results confirm that direct activation of B cell-intrinsic TLR7, rather than activation of non B cell-extrinsic TLR7, plays a predominant role in robust anti-PEG IgM production.

7.4.8 Effects of Route and Manner of Administration

Recently, many reports have emphasized the contribution of the manner/route of drug administration to the magnitude of the ABC phenomenon. Li et al. [49] reported that, in Beagle dogs, the administration of PEGylated liposomes via slow intravenous infusion, with a total lipid dose of 12.5 µmol phospholipids/kg, instead of an intravenous bolus injection, with a total lipid dose of 50 µmol phospholipids/kg, was liable to result in the ABC phenomenon. They demonstrated that the rate of liposomal presentation to the systemic circulation would dictate the incidence and/or magnitude of the ABC phenomenon. They assumed that, following a slow intravenous infusion, the anti-PEG IgM levels produced in response to the first dose would be in excess of the PEGylated liposomes, and would thus facilitate the binding of antibodies to liposomal surface with a subsequent rapid clearance of liposomes from circulation. On the other hand, if a large dose of liposomes were rapidly injected into an animal, the IgM levels might not be sufficient to trigger the accelerated clearance of such a high dose of liposomes, which could subsequently lead to significant alterations in the pharmacokinetic profiles.

Interestingly, Zhao et al. [33] investigated the effect of different anatomical sites of injection on the pharmacokinetic behavior of subsequent intravenous (i.v.) injections of PEGylated SLNs. They demonstrated that after subcutaneous (s.c.) administration, the clearance of subsequent i.v. injections of PEGylated SLNs were accelerated. Moreover, by comparison with the initial i.v. injection, the subcutaneous route seemed to be more prone to induce a markedly accelerated clearance of the test dose. In a similar manner, Wang et al. [50] recently verified the occurrence of the ABC phenomenon following the repeated s.c. administration of PEG-coated emulsions. They emphasized the contribution of regional lymph nodes in the maintenance of an effective immune response against PEG-coated emulsions. These results clearly suggest a contribution of the manner/route of drug administration to the incidence of the ABC phenomenon following the sequential administration of PEGylated products.

7.4.9 Effect of Structure of Nanocarriers

Despite the fact that the ABC phenomenon is well recognized with PEGylated liposomes, many reports have revealed the elicitation of this phenomenon by pre-treatment with other PEGylated nanocarriers such as polymeric micelles [13], nanoparticles [12], microemulsions [14], and even PEGylated proteins [15]. Furthermore, several reports have shown that a single injection of PEG polymer does not induce the production of anti-PEG IgM [5]. This observation suggests that the induction of an immunogenic response only occurs against PEG conjugates in a haptenic manner; meaning that PEG itself cannot independently stimulate a sufficient immune response except after being conjugated [11, 78]. These results show that the colloidal structure of nanocarriers could substantially participate in the induction of the ABC phenomenon.

Koide et al. [13] demonstrated that pre-treatment of mice with empty polyethylene glycol-poly(β-benzyl L-aspartate) (PEG-PBLA) polymeric micelles significantly triggered the accelerated clearance of subsequently injected PEGylated liposomes. On the contrary, Ma et al. [79] reported that a weekly administration of gadolinium (Gd)-containing PEG-poly(L-lysine)-based polymeric micelles (Gd-micelles) did not cause the accelerated clearance of a second dose. In that study, the ABC phenomenon of the second dose of the Gd-containing PEGylated liposome (Gd-liposome) was induced by the first dose of both the Gd-liposome and the empty PEGylated liposome, but not by the first injection of the Gd-micelle. Such contentious results might be ascribed to the differences in the structures of the nanocarriers. From a structural perspective, the Gd-micelle has no hydrophobic portion. By contrast, the PEG-PBLA micelle, like PEGylated liposomes, consists of both a hydrophilic part, PEG, and a hydrophobic part, PBLA. Consequently, those researchers suggested that the presence of a hydrophobic core in the nanocarrier is a key factor influencing the induction of the ABC phenomenon.

In the same context, Shiraishi et al. [80] recently investigated the effect of the colloidal structure of PEGylated polymeric micelles on the induction of the ABC phenomenon. They employed 2 types of PEG-containing polymeric micelles: a micelle with a hydrophilic inner core (PEGP(Lys-DOTA-Gd)), and a micelle with a hydrophobic inner core (PEG-PBLA). They found that micelles with a hydrophilic inner core (PEGP(Lys-DOTA-Gd)) induced neither an anti-PEG IgM response nor the ABC phenomenon. By contrast, micelles with a hydrophobic inner core (PEG-PBLA) induced a pronounced ABC phenomenon. These results clearly confirmed the assumption that, despite the presence of a PEG outer shell, the presence of a hydrophobic core is a prerequisite for the induction of the ABC phenomenon.

Interestingly, Hara et al. [81] recently reported the occurrence of the ABC phenomenon following multiple administrations of lactosomes, despite the lack of a PEG moiety. Lactosome is a polymeric micelle composed of a hydrophilic poly-sarcosine block and a hydrophobic poly (L-lactic acid) block. They speculated that the hydrophilic polymer chain surrounding the hydrophobic core might act as a TI antigen, and thus, could induce an immune reaction against lactosomes upon repeated administration. In a further study, they [82] demonstrated that increasing the local density of the hydrophilic poly (sarcosine) chains around the hydrophobic core (surface density ≥ 0.3 chain/nm^2) significantly prevented the polymeric micelle from interacting with B-cell receptors, and thereby, abrogated the induction of the ABC phenomenon upon repeated administration.

7.5 Strategies to Abrogate/Attenuate Induction of the ABC Phenomenon

PEGylation is extensively applied in the pharmaceutical industry, particularly for drug delivery systems (DDS) in order to improve their *in vivo* fate following systemic administration. However, the induction of the ABC phenomenon upon repeated administration of PEGylated nanocarriers poses a tremendous challenge to their clinical application. Accordingly, many researchers are currently extending great efforts to find applicable strategies to abrogate/attenuate the induction of the ABC phenomenon. In this section, we will focus on some of these strategies.

7.5.1 Manipulation of Physicochemical Properties of PEGylated Nanocarriers

Manipulation of the physicochemical properties of PEGylated nanocarriers, such as particle size, has been used to efficiently alleviate induction of the ABC phenomenon. Koide et al. [13] demonstrated that decreasing the size of polymeric micelles to less than 30 nm, could efficiently abrogate the induction of the ABC phenomenon. In the same context, Zhang et al. [83] demonstrated the impact of the size of PEGylated uricases on triggering the ABC phenomenon in the treatment of hyperuricemic disorders [84]. Those researchers speculated that 40–60 nm is the lower size limit that can trigger the ABC phenomenon. Therefore, they revealed that removal of the uricase aggregates could successfully alleviate the ABC phenomenon for 5 kDa mPEG-modified uricase. These results show that controlling the size of the PEGylated products may be a viable means to abrogate the induction of the ABC phenomenon. However, preparing liposomal formulations with a particle size lower than 30 nm would be technically difficult and could potentially compromise the efficiency of encapsulating the payload.

7.5.2 Modification of PEG Moiety

Among the different strategies employed for abrogating/attenuating the immunogenicity of PEGylated products upon repeated administration, modification of the PEG moiety has received a great deal of attention. This strategy is aimed at conserving the potential of the PEG moiety for enhancing the circulation characteristics of the modified nanocarrier while imparting a PEG moiety less recognizable by the cells of the immune system. Ambegia et al. [85] and Webb et al. [86] have demonstrated that the *in vivo* fate and/or immunogenic response against liposomes could be adjusted by using a dissociable PEG-lipid with a different alkyl chain length. They revealed that surface modification of liposomes with easily dissociable PEG-lipids with shorter alkyl chains could efficiently attenuate the ABC phenomenon, compared with liposomes modified with conventional PEG-lipids, such as mPEG-DSPE and mPEG-CHOL. In a similar manner, Judge et al. [69] and Semple et al. [26] employed PEGylated lipids with smaller C14 lipid anchors in the preparation of PEGylated liposomes to alleviate the immunogenicity caused by repeated administration. However, this lipid exchange can result in defects in the membranes of the liposomes, which could induce a premature release of the encapsulated payload.

Another attempt has focused on the use of cleavable PEG-lipid derivatives via manipulating the linkage between lipids and the PEG moiety. Xu et al. [87] showed how the surface modification of liposomes with cleavable PEG-lipid derivatives, namely PEG-CHMC and PEG-CHEMS, which are linked via a single ester bond, could lessen/eliminate the occurrence of the ABC phenomenon in Wistar rats. They emphasized that the prompt cleavage of the ester bond in either PEG-CHMC or PEG-CHEMS by an esterase enzyme in blood circulation not only abrogated/attenuated the ABC phenomenon, but it also did not induce membrane defects in the liposomes. However, the capacity of PEG-CHMC or PEG-CHEMS with a single ester bond is still insufficient to provide liposomes with long-circulating characteristics, compared with non-cleavable PEGylated liposomes. Chen et al. [88] designed a double-smart mPEG-Hz-CHEMS-modified liposome in an attempt to reduce the occurrence of the ABC phenomenon. They employed a novel cleavable PEG-lipid derivative, mPEG-Hz-CHEMS, wherein the PEG moiety was linked to cholesterol by 2 ester bonds and 1 pH-sensitive hydrazone (Fig. 7.3). They speculated that the chemical bond (ester bonds) would gradually be cleaved in serum by the esterase enzyme, and in the low pH environment of a tumor tissue, the PEG moiety would be detached from the liposome surface allowing liposomal internalization/uptake by the tumor cells. This strategy is expected to evade the occurrence of the ABC phenomenon while not compromising the *in vivo* fate of the administered PEGylated liposomes.

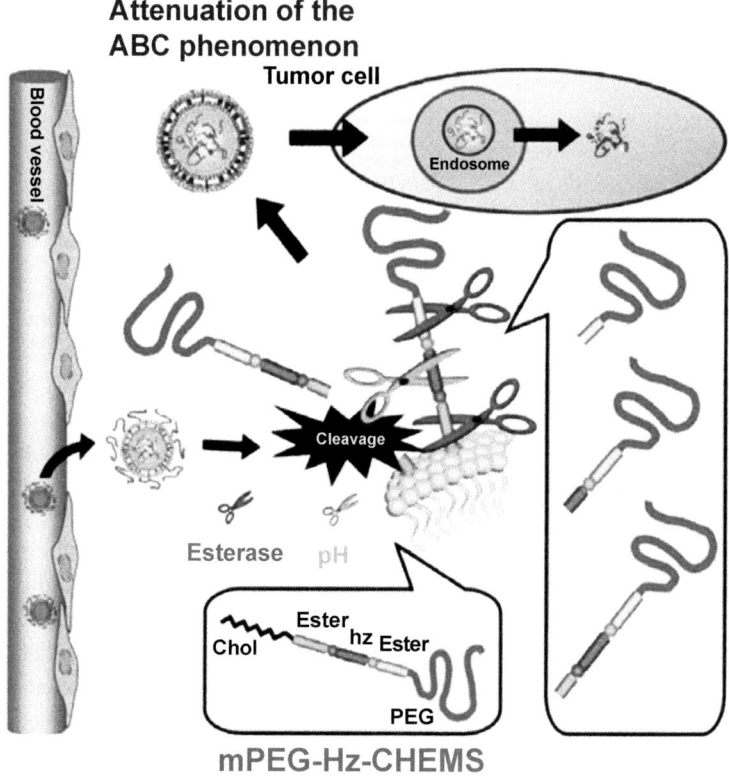

Figure 7.3 The strategy for double-cleavable smart mPEG-Hz-CHEMS-modified liposomes. Modified from the author's original work [88].

7.5.3 Use of Alternative Polymers

Although PEG represents the gold standard for the steric stabilization of nanocarriers, several attempts have used alternative polymers to alleviate/abrogate the incidence of the ABC phenomenon, while conserving the long-circulating characteristics of the modified nanocarrier. Polymers such as poly[*N*-(2-hydroxypropyl)methacrylamide)] [89], poly(hydroxyethyl L-glutamine), poly(hydroxyl ethyl L-asparagine) [90], poly(*N*-vinyl-2-pyrrolidone) (PVP) [91], and L-amino-acid-based biodegradable polymer-lipid conjugates [92] have been tested for their potential in abrogating the immune response towards an administered nanocarrier. Despite the fact that these polymers induced little, if any, immunogenic response, to date none has proven superior to PEG.

Recently, we employed a polyglycerol (PG)-derived lipid, as an alternative to PEG, to attenuate the immune response towards injected liposomes [93, 94]. We demonstrated that surface modification with PG elicits neither an anti-polymer immune response nor the ABC phenomenon upon repeated administration, resulting in enhanced accumulation of the prepared liposomes in tumor tissue. We postulated that the hydroxymethyl side group in the repeating $-(O\text{-}CH_2\text{-}CH(CH_2OH))_n-$ subunit of PG sterically hinders the interaction and effective binding to surface immunoglobulins on reactive splenic B-cells, and thus, prevents the direct stimulation of splenic B cells (Fig. 7.4). This could afford the attenuation of anti-PG IgM production, which may be responsible for eliciting the ABC phenomenon [93].

In a subsequent study [94], we confirmed that surface modification of lipoplex with PG efficiently attenuated the ABC phenomenon that is encountered upon repeated administrations of nucleic acids containing PEG-coated liposomes. We demonstrated that the absence of an immunogenic response against PG-coated pDNA-lipoplex enabled the efficient accumulation of the second dose in the tumor tissue of a tumor-bearing mouse model in a manner similar to the first dose. On the contrary, PEG-coated pDNA-lipoplex significantly elicited a strong immune response, which hindered the efficient intra-tumoral accumulation of a subsequently injected second dose.

Recently, Li et al. [95] developed a zwitterionic polymer poly(carboxy-betaine) (PCB) as an alternative to PEG. They demonstrated that cationic liposomes modified with DSPE-PCB$_{20}$ polymer showed a higher efficiency of siRNA encapsulation, cellular uptake and endosomal/lysosomal escape abilities, which in turn led to a significant release of siRNA into the cytoplasm. Furthermore, the PCBylated lipoplexes efficiently evaded the ABC phenomenon, which extended the plasma circulation time and resulted in a preferential intra-tumor accumulation of siRNA following multiple administrations.

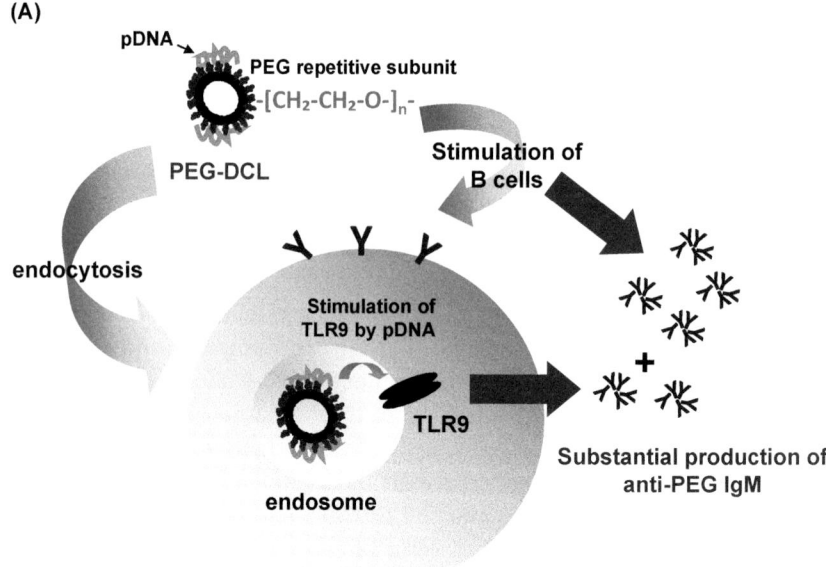

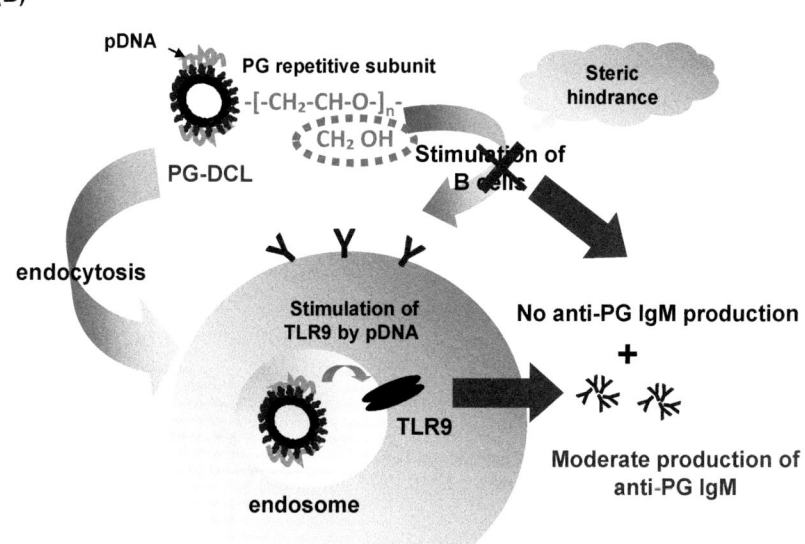

Figure 7.4 The possible mechanism underlying the differences in anti-polymer IgM production in response to (A) PEG-coated pDNA-lipoplex (PEG-DCL) or (B) PG-coated pDNA-lipoplex (PG-DCL). Modified from the author's original work [94].

7.5.4 Changing Administration Regimen

Many reports have emphasized the strong correlation between anti-PEG IgM production and the occurrence of the ABC phenomenon. Because anti-PEG IgM production occurs in a wave pattern following the first injection of PEGylated formulation [12, 22], a prolongation of the time interval between injections might be considered a simple means for reducing the magnitude of the ABC phenomenon [36]. However, this strategy might interfere with the optimal administration regimen for the encapsulated drugs, thereby limiting their therapeutic efficacy.

7.6 Clinical Implications of ABC Phenomenon

Polyethylene glycol (PEG) is recognized as a bio-inert, non-toxic and non-immunogenic material. These characteristics of PEG have led to its FDA approval as a vehicle or a modified material in foods, cosmetics, and pharmaceuticals. In addition, PEGylation, surface modification of nanocarriers and/or proteins with PEG, represents a milestone breakthrough in the field of drug delivery systems. PEGylation offers long-circulating characteristics to the modified nanocarriers, along with reduced recognition by the cells of MPS [3, 4]. In fact, PEGylated products such as peginteron® (a PEGylated interferon used in the treatment of hepatitis C virus) and Doxil® benefit from these characteristics and have been approved for use in clinical settings.

Notably, the first report revealing the immunogenicity of PEG was published in 1983 by Richter and Akerblom [96], who correlated the enhanced clearance of a second dose of PEGylated ovalbumin in mice to an elevated level of anti-PEG IgM antibodies. One year later, the authors emphasized the induction of anti-PEG IgM production in humans in response to treatment with PEGylated honeybee venom and PEGylated ragweed extract [78]. Despite the elevated levels of anti-PEG IgM in the serum of the patients in that study, no changes in response rates were observed. These findings contributed to a large extent in establishing the perception that PEG is non immunogenic.

Nevertheless, the immunogenicity of PEG has recently gained increased attention, particularly after PEGinesatide, an erythropoietic agent, was withdrawn from the market due to severe hypersensitivity reactions with even fatal consequences, which were presumably related to PEG [97, 98]. Furthermore, besides reports declaring that the immunogenicity of PEG might influence the safety of PEGylated products, several studies have implied that the ABC phenomenon of PEGylated products, mediated via the induction of anti-PEG IgM, could substantially attenuate the therapeutic potential of PEGylated products [8, 10, 21, 63]. Consequently, such immunogenicity of PEG-coated nanocarriers might represent a potential impediment against their development and use in clinical settings since the establishment of an antibody response can severely compromise both the safety and efficacy of an associated drug payload such as DNA, RNA or proteins.

7.7 Conclusion

PEGylation is being increasingly employed in the pharmaceutical field because of its promising properties for prolonging half-lives *in vivo* via increases in the hydrodynamic size and shielding against immunogenic epitopes by means of steric hindrance against the immune system. However, the immunogenicity encountered upon the administration of PEGylated nanocarriers/proteins seems to hamper the complete deployment of PEG. Anti-PEG IgM antibodies induced by a first dose of PEGylated nanocarriers such as liposomes are known to enhance the clearance of a subsequent dose injected a few days later, which is known as the "accelerated blood clearance" phenomenon. Furthermore, the ability of anti-PEG IgM, produced in response to a first dose of PEGylated nanocarriers, to induce hypersensitivity reactions has been confirmed clinically. As a result, the FDA has recently updated their guidelines to screen for anti-PEG antibodies during clinical trials of PEGylated therapeutics [99].

Disclosures and Conflict of Interest

The authors wish to thank Dr. J. L. McDonald for his helpful advice in writing the manuscript. This study was supported, in part, by a Grant-in-Aid for Scientific Research (B) 15H04639, the Ministry of Education, Culture, Sports, Science and Technology, Japan. The authors declare that they have no conflicts of interest. No writing assistance was utilized in the production of this chapter and the authors have received no payment for its preparation.

Corresponding Author

Dr. Tatsuhiro Ishida
Department of Pharmacokinetics and Biopharmaceutics
Subdivision of Biopharmaceutical Sciences
Institute of Biomedical Sciences
Tokushima University, 1-78-1, Sho-machi
Tokushima 770-8505, Japan
Email: ishida@tokushima-u.ac.jp

About the Authors

Amr S. Abu Lila received his master's degree in pharmaceutics from the Faculty of Pharmacy, Zagazig University, Egypt, in 2005 and his PhD in pharmaceutical life sciences from the Graduate School of Pharmaceutical Sciences, Tokushima University, Japan, in 2010. From 2011 to 2013, he was a JSPS postdoctoral fellow in Japan at the Graduate School of Pharmaceutical Sciences, Tokushima University. Currently, he is an associate professor in the Faculty of Pharmacy, Zagazig University. He has published more than 20 papers in peer-reviewed scientific journals, 4 review articles, and 3 book chapters. His current research focuses on anticancer therapeutics using liposomal drug delivery systems for the selective targeting of either chemotherapeutic agents or genes to tumor tissues.

Tatsuhiro Ishida graduated from Tokushima University, Japan, in 1993 and then received his master's degree in 1995 and his PhD in 1998 from the Faculty of Pharmaceutical Sciences, Tokushima University. From 1998 to 2000, he was a postdoctoral fellow at the University of Alberta (Pharmacology) Canada. In 2000, he became a lecturer in Pharmaceutical Sciences at Tokushima University and was promoted to associate professor in 2003. He has been a full professor there since 2013. He has published more than 100 peer-reviewed papers, more than 10 review articles, and 5 book chapters. He has given more than 30 presentations as an invited speaker at international conferences. He is interested in immunological responses against PEGylated therapeutics and the mechanisms behind the response.

References

1. Verhoef, J. J., Anchordoquy, T. J. (2013). Questioning the use of PEGylation for drug delivery. *Drug Deliv. Transl. Res.*, **3**, 499–503.
2. Harris, J. M., Chess, R. B. (2003). Effect of pegylation on pharmaceuticals. *Nat. Rev. Drug Discov.*, **2**, 214–221.
3. Pisal, D. S., Kosloski, M. P., Balu-Iyer, S. V. (2010). Delivery of therapeutic proteins. *J. Pharm. Sci.*, **99**, 2557–2575.
4. Papahadjopoulos, D., Allen, T. M., Gabizon, A., Mayhew, E., Matthay, K., Huang, S. K., Lee, K. D., Woodle, M. C., Lasic, D. D., Redemann, C., et al. (1991). Sterically stabilized liposomes: Improvements in pharmacokinetics and antitumor therapeutic efficacy. *Proc. Natl. Acad. Sci. U. S. A.*, **88**, 11460–11464.
5. Shimizu, T., Ishida, T., Kiwada, H. (2013). Transport of PEGylated liposomes from the splenic marginal zone to the follicle in the induction phase of the accelerated blood clearance phenomenon. *Immunobiology*, **218**, 725–732.
6. Ishida, T., Maeda, R., Ichihara, M., Irimura, K., Kiwada, H. (2003). Accelerated clearance of PEGylated liposomes in rats after repeated injections. *J. Control. Release*, **88**, 35–42.
7. Hamad, I., Hunter, A. C., Szebeni, J., Moghimi, S. M. (2008). Poly(ethylene glycol)s generate complement activation products in human serum through increased alternative pathway turnover and a MASP-2-dependent process. *Mol. Immunol.*, **46**, 225–232.
8. Dams, E. T., Laverman, P., Oyen, W. J., Storm, G., Scherphof, G. L., van Der Meer, J. W., Corstens, F. H., Boerman, O. C. (2000). Accelerated blood clearance and altered biodistribution of repeated injections of sterically stabilized liposomes. *J. Pharmacol. Exp. Ther.*, **292**, 1071–1079.
9. Shimizu, T., Ichihara, M., Yoshioka, Y., Ishida, T., Nakagawa, S., Kiwada, H. (2012). Intravenous administration of polyethylene glycol-coated (PEGylated) proteins and PEGylated adenovirus elicits an anti-PEG immunoglobulin M response. *Biol. Pharm. Bull.*, **35**, 1336–1342.

10. Laverman, P., Carstens, M. G., Boerman, O. C., Dams, E. T., Oyen, W. J., van Rooijen, N., Corstens, F. H., Storm, G. (2001). Factors affecting the accelerated blood clearance of polyethylene glycol-liposomes upon repeated injection. *J. Pharmacol. Exp. Ther.*, **298**, 607–612.

11. Ishida, T., Kiwada, H. (2013). Anti-polyethyleneglycol antibody response to PEGylated substances. *Biol. Pharm. Bull.*, **36**, 889–891.

12. Ishihara, T., Takeda, M., Sakamoto, H., Kimoto, A., Kobayashi, C., Takasaki, N., Yuki, K., Tanaka, K., Takenaga, M., Igarashi, R., Maeda, T., Yamakawa, N., Okamoto, Y., Otsuka, M., Ishida, T., Kiwada, H., Mizushima, Y., Mizushima, T. (2009). Accelerated blood clearance phenomenon upon repeated injection of PEG-modified PLA-nanoparticles. *Pharm. Res.*, **26**, 2270–2279.

13. Koide, H., Asai, T., Hatanaka, K., Urakami, T., Ishii, T., Kenjo, E., Nishihara, M., Yokoyama, M., Ishida, T., Kiwada, H., Oku, N. (2008). Particle size-dependent triggering of accelerated blood clearance phenomenon. *Int. J. Pharm.*, **362**, 197–200.

14. Joshi, M., Pathak, S., Sharma, S., Patravale, V. (2008). Solid microemulsion preconcentrate (NanOsorb) of artemether for effective treatment of malaria. *Int. J. Pharm.*, **362**, 172–178.

15. Lu, W., Wan, J., She, Z., Jiang, X. (2007). Brain delivery property and accelerated blood clearance of cationic albumin conjugated pegylated nanoparticle. *J. Control. Release*, **118**, 38–53.

16. Ishida, T., Kashima, S., Kiwada, H. (2008). The contribution of phagocytic activity of liver macrophages to the accelerated blood clearance (ABC) phenomenon of PEGylated liposomes in rats. *J. Control. Release*, **126**, 162–165.

17. Szebeni, J., Muggia, F., Gabizon, A., Barenholz, Y. (2011). Activation of complement by therapeutic liposomes and other lipid excipient-based therapeutic products: prediction and prevention. *Adv. Drug Deliv. Rev.*, **63**, 1020–1030.

18. Verhoef, J. J., Carpenter, J. F., Anchordoquy, T. J., Schellekens, H. (2014). Potential induction of anti-PEG antibodies and complement activation toward PEGylated therapeutics. *Drug Discov. Today*, **19**, 1945–1952.

19. Ishida, T., Harada, M., Wang, X. Y., Ichihara, M., Irimura, K., Kiwada, H. (2005). Accelerated blood clearance of PEGylated liposomes following preceding liposome injection: Effects of lipid dose and PEG surface-density and chain length of the first-dose liposomes. *J. Control. Release*, **105**, 305–317.

20. Ishida, T., Atobe, K., Wang, X., Kiwada, H. (2006). Accelerated blood clearance of PEGylated liposomes upon repeated injections: effect of doxorubicin-encapsulation and high-dose first injection. *J. Control. Release*, **115**, 251–258.
21. Ishida, T., Maeda, R., Ichihara, M., Mukai, Y., Motoki, Y., Manabe, Y., Irimura, K., Kiwada, H. (2002). The accelerated clearance on repeated injection of pegylated liposomes in rats: laboratory and histopathological study. *Cell. Mol. Biol. Lett.*, **7**, 286.
22. Ishida, T., Wang, X., Shimizu, T., Nawata, K., Kiwada, H. (2007). PEGylated liposomes elicit an anti-PEG IgM response in a T cell-independent manner. *J. Control. Release*, **122**, 349–355.
23. Wang, X., Ishida, T., Kiwada, H. (2007). Anti-PEG IgM elicited by injection of liposomes is involved in the enhanced blood clearance of a subsequent dose of PEGylated liposomes. *J. Control. Release*, **119**, 236–244.
24. Ishida, T., Ichihara, M., Wang, X., Kiwada, H. (2006). Spleen plays an important role in the induction of accelerated blood clearance of PEGylated liposomes. *J. Control. Release*, **115**, 243–250.
25. Ichihara, M., Shimizu, T., Imoto, A., Hashiguchi, Y., Uehara, Y., Ishida, T., Kiwada, H. (2010). Anti-PEG IgM response against PEGylated liposomes in mice and rats. *Pharmaceutics*, **3**, 1–11.
26. Semple, S. C., Harasym, T. O., Clow, K. A., Ansell, S. M., Klimuk, S. K., Hope, M. J. (2005). Immunogenicity and rapid blood clearance of liposomes containing polyethylene glycol-lipid conjugates and nucleic acid. *J. Pharmacol. Exp. Ther.*, **312**, 1020–1026.
27. Ichihara, H., Zako, K., Komizu, Y., Goto, K., Ueoka, R. (2011). Therapeutic effects of hybrid liposomes composed of phosphatidylcholine and docosahexaenoic acid on the hepatic metastasis of colon carcinoma along with apoptosis in vivo. *Biol. Pharm. Bull.*, **34**, 901–905.
28. Vinuesa, C. G., Chang, P. P. (2013). Innate B cell helpers reveal novel types of antibody responses. *Nat. Immunol.*, **14**, 119–126.
29. Mond, J. J., Lees, A., Snapper, C. M. (1995). T cell-independent antigens type 2. *Annu. Rev. Immunol.*, **13**, 655–692.
30. Mond, J. J., Vos, Q., Lees, A., Snapper, C. M. (1995). T cell independent antigens. *Curr. Opin. Immunol.*, **7**, 349–354.
31. Cheng, T. L., Chen, B. M., Chern, J. W., Wu, M. F., Roffler, S. R. (2000). Efficient clearance of poly(ethylene glycol)-modified immunoenzyme with anti-PEG monoclonal antibody for prodrug cancer therapy. *Bioconjug. Chem.*, **11**, 258–266.

32. Cheng, T. L., Wu, P. Y., Wu, M. F., Chern, J. W., Roffler, S. R. (1999). Accelerated clearance of polyethylene glycol-modified proteins by anti-polyethylene glycol IgM. *Bioconjug. Chem.*, **10**, 520–528.
33. Zhao, Y., Wang, C., Wang, L., Yang, Q., Tang, W., She, Z., Deng, Y. (2012). A frustrating problem: accelerated blood clearance of PEGylated solid lipid nanoparticles following subcutaneous injection in rats. *Eur. J. Pharm. Biopharm.*, **81**, 506–513.
34. Janeway, A. Jr. (2001). How the immune system works to protect the host from infection: a personal view. *Proc. Natl. Acad. Sci. U. S. A.*, **98**, 7461–7468.
35. Helmy, K. Y., Katschke, K. J. Jr., Gorgani, N. N., Kljavin, N. M., Elliott, J. M., Diehl, L., Scales, S. J., Ghilardi, N., van Lookeren Campagne, M. (2006). CRIg: a macrophage complement receptor required for phagocytosis of circulating pathogens. *Cell*, **124**, 915–927.
36. Taguchi, K., Ogaki, S., Watanabe, H., Kadowaki, D., Sakai, H., Kobayashi, K., Horinouchi, H., Maruyama, T., Otagiri, M. (2011). Fluid resuscitation with hemoglobin vesicles prevents Escherichia coli growth via complement activation in a hemorrhagic shock rat model. *J. Pharmacol. Exp. Ther.*, **337**, 201–208.
37. Yang, Q., Ma, Y., Zhao, Y., She, Z., Wang, L., Li, J., Wang, C., Deng, Y. (2013). Accelerated drug release and clearance of PEGylated epirubicin liposomes following repeated injections: a new challenge for sequential low-dose chemotherapy. *Int. J. Nanomed.*, **8**, 1257–1268.
38. Szebeni, J. (2014). Complement activation-related pseudoallergy: a stress reaction in blood triggered by nanomedicines and biologicals. *Mol. Immunol.*, **61**, 163–173.
39. Szebeni, J., Bedocs, P., Rozsnyay, Z., Weiszhar, Z., Urbanics, R., Rosivall, L., Cohen, R., Garbuzenko, O., Bathori, G., Toth, M., Bunger, R., Barenholz, Y. (2012). Liposome-induced complement activation and related cardiopulmonary distress in pigs: factors promoting reactogenicity of Doxil and AmBisome. *Nanomedicine*, **8**, 176–184.
40. Pieters, R., Hunger, S. P., Boos, J., Rizzari, C., Silverman, L., Baruchel, A., Goekbuget, N., Schrappe, M., Pui, C. H. (2011). L-asparaginase treatment in acute lymphoblastic leukemia: a focus on Erwinia asparaginase. *Cancer*, **117**, 238–249.
41. Garratty, G. (2004). Progress in modulating the RBC membrane to produce transfusable universal/stealth donor RBCs. *Transfus. Med. Rev.*, **18**, 245–256.
42. Szebeni, J., Baranyi, L., Savay, S., Bodo, M., Morse, D. S., Basta, M., Stahl, G. L., Bunger, R., Alving, C. R. (2000). Liposome-induced pulmonary

hypertension: properties and mechanism of a complement-mediated pseudoallergic reaction. *Am. J. Physiol. Heart Circ. Physiol.*, **279**, H1319–1328.

43. Chanan-Khan, A., Szebeni, J., Savay, S., Liebes, L., Rafique, N. M., Alving, C. R., Muggia, F. M. (2003). Complement activation following first exposure to pegylated liposomal doxorubicin (Doxil): possible role in hypersensitivity reactions. *Ann. Oncol.*, **14**, 1430–1437.

44. Wang, X. Y., Ishida, T., Ichihara, M., Kiwada, H. (2005). Influence of the physicochemical properties of liposomes on the accelerated blood clearance phenomenon in rats. *J. Control. Release*, **104**, 91–102.

45. Koide, H., Asai, T., Hatanaka, K., Akai, S., Ishii, T., Kenjo, E., Ishida, T., Kiwada, H., Tsukada, H., Oku, N. (2010). T cell-independent B cell response is responsible for ABC phenomenon induced by repeated injection of PEGylated liposomes. *Int. J. Pharm.*, **392**, 218–223.

46. Zhao, Y., Wang, L., Yan, M., Ma, Y., Zang, G., She, Z., Deng, Y. (2012). Repeated injection of PEGylated solid lipid nanoparticles induces accelerated blood clearance in mice and beagles. *Int. J. Nanomed.*, **7**, 2891–2900.

47. Ishida, T., Kiwada, H. (2008). Accelerated blood clearance (ABC) phenomenon upon repeated injection of PEGylated liposomes. *Int. J. Pharm.*, **354**, 56–62.

48. Bradley, A. J., Test, S. T., Murad, K. L., Mitsuyoshi, J., Scott, M. D. (2001). Interactions of IgM ABO antibodies and complement with methoxy-PEG-modified human RBCs. *Transfusion*, **41**, 1225–1233.

49. Li, C., Cao, J., Wang, Y., Zhao, X., Deng, C., Wei, N., Yang, J., Cui, J. (2012). Accelerated blood clearance of pegylated liposomal topotecan: influence of polyethylene glycol grafting density and animal species. *J. Pharm. Sci.*, **101**, 3864–3876.

50. Wang, C., Cheng, X., Sui, Y., Luo, X., Jiang, G., Wang, Y., Huang, Z., She, Z., Deng, Y. (2013). A noticeable phenomenon: thiol terminal PEG enhances the immunogenicity of PEGylated emulsions injected intravenously or subcutaneously into rats. *Eur. J. Pharm. Biopharm.*, **85**, 744–751.

51. Oussoren, C., Eling, W. M., Crommelin, D. J., Storm, G., Zuidema, J. (1998). The influence of the route of administration and liposome composition on the potential of liposomes to protect tissue against local toxicity of two antitumor drugs. *Biochim. Biophys. Acta*, **1369**, 159–172.

52. Nagao, A., Abu Lila, A. S., Ishida, T., Kiwada, H. (2013). Abrogation of the accelerated blood clearance phenomenon by SOXL regimen: promise for clinical application. *Int. J. Pharm.*, **441**, 395–401.

53. Goins, G. D., Yorio, N. C., Sanwo-Lewandowski, M. M., Brown, C. S. (1998). Life cycle experiments with Arabidopsis grown under red light-emitting diodes (LEDs). *Life Support Biosph. Sci.*, **5**, 143–149.
54. Ishida, T., Ichihara, M., Wang, X., Yamamoto, K., Kimura, J., Majima, E., Kiwada, H. (2006). Injection of PEGylated liposomes in rats elicits PEG-specific IgM, which is responsible for rapid elimination of a second dose of PEGylated liposomes. *J. Control. Release*, **112**, 15–25.
55. Saadati, R., Dadashzadeh, S., Abbasian, Z., Soleimanjahi, H. (2013). Accelerated blood clearance of PEGylated PLGA nanoparticles following repeated injections: effects of polymer dose, PEG coating, and encapsulated anticancer drug. *Pharm. Res.*, **30**, 985–995.
56. Li, C., Zhao, X., Wang, Y., Yang, H., Li, H., Tian, W., Yang, J., Cui, J. (2013). Prolongation of time interval between doses could eliminate accelerated blood clearance phenomenon induced by pegylated liposomal topotecan. *Int. J. Pharm.*, **443**, 17–25.
57. Lyass, O., Uziely, B., Ben-Yosef, R., Tzemach, D., Heshing, N. I., Lotem, M., Brufman, G., Gabizon, A. (2000). Correlation of toxicity with pharmacokinetics of pegylated liposomal doxorubicin (Doxil) in metastatic breast carcinoma. *Cancer*, **89**, 1037–1047.
58. Gauld, S. B., Benschop, R. J., Merrell, K. T., Cambier, J. C. (2005). Maintenance of B cell anergy requires constant antigen receptor occupancy and signaling. *Nat. Immunol.*, **6**, 1160–1167.
59. La-Beck, N. M., Zamboni, B. A., Gabizon, A., Schmeeda, H., Amantea, M., Gehrig, P. A., Zamboni, W. C. (2012). Factors affecting the pharmacokinetics of pegylated liposomal doxorubicin in patients. *Cancer Chemother. Pharmacol.*, **69**, 43–50.
60. Lien, K., Georgsdottir, S., Sivanathan, L., Chan, K., Emmenegger, U. (2013). Low-dose metronomic chemotherapy: a systematic literature analysis. *Eur. J. Cancer*, **49**, 3387–3395.
61. Scharovsky, O. G., Mainetti, L. E., Rozados, V. R. (2009). Metronomic chemotherapy: changing the paradigm that more is better. *Curr. Oncol.*, **16**, 7–15.
62. Suzuki, T., Ichihara, M., Hyodo, K., Yamamoto, E., Ishida, T., Kiwada, H., Ishihara, H., Kikuchi, H. (2012). Accelerated blood clearance of PEGylated liposomes containing doxorubicin upon repeated administration to dogs. *Int. J. Pharm.*, **436**, 636–643.
63. Abu Lila, A. S., Eldin, N. E., Ichihara, M., Ishida, T., Kiwada, H. (2012). Multiple administration of PEG-coated liposomal oxaliplatin enhances its therapeutic efficacy: a possible mechanism and the potential for clinical application. *Int. J. Pharm.*, **438**, 176–183.

64. Cui, J., Li, C., Wang, C., Li, Y., Zhang, L., Yang, H. (2008). Repeated injection of pegylated liposomal antitumour drugs induces the disappearance of the rapid distribution phase. *J. Pharm. Pharmacol.*, **60**, 1651–1657.

65. Ma, Y., Yang, Q., Wang, L., Zhou, X., Zhao, Y., Deng, Y. (2012). Repeated injections of PEGylated liposomal topotecan induces accelerated blood clearance phenomenon in rats. *Eur. J. Pharm. Sci.*, **45**, 539–545.

66. Li, S. D., Huang, L. (2006). Gene therapy progress and prospects: non-viral gene therapy by systemic delivery. *Gene Ther.*, **13**, 1313–1319.

67. Nakamura, K., Abu Lila, A. S., Matsunaga, M., Doi, Y., Ishida, T., Kiwada, H. (2011). A double-modulation strategy in cancer treatment with a chemotherapeutic agent and siRNA. *Mol. Ther.*, **19**, 2040–2047.

68. Gao, Y., Liu, X. L., Li, X. R. (2011). Research progress on siRNA delivery with nonviral carriers. *Int. J. Nanomed.*, **6**, 1017–1025.

69. Judge, A., McClintock, K., Phelps, J. R., Maclachlan, I. (2006). Hypersensitivity and loss of disease site targeting caused by antibody responses to PEGylated liposomes. *Mol. Ther.*, **13**, 328–337.

70. Tagami, T., Nakamura, K., Shimizu, T., Yamazaki, N., Ishida, T., Kiwada, H. (2010). CpG motifs in pDNA-sequences increase anti-PEG IgM production induced by PEG-coated pDNA-lipoplexes. *J. Control. Release*, **142**, 160–166.

71. Takeda, K., Akira, S. (2004). Microbial recognition by Toll-like receptors. *J. Dermatol. Sci.*, **34**, 73–82.

72. Hemmi, H., Takeuchi, O., Kawai, T., Kaisho, T., Sato, S., Sanjo, H., Matsumoto, M., Hoshino, K., Wagner, H., Takeda, K., Akira, S. (2000). A Toll-like receptor recognizes bacterial DNA. *Nature*, **408**, 740–745.

73. Hashimoto, Y., Uehara, Y., Abu Lila, A. S., Ishida, T., Kiwada, H. (2014). Activation of TLR9 by incorporated pDNA within PEG-coated lipoplex enhances anti-PEG IgM production. *Gene Ther.*, **21**, 593–598.

74. Hanten, J. A., Vasilakos, J. P., Riter, C. L., Neys, L., Lipson, K. E., Alkan, S. S., Birmachu, W. (2008). Comparison of human B cell activation by TLR7 and TLR9 agonists. *BMC Immunol.*, **9**, 39.

75. Tagami, T., Uehara, Y., Moriyoshi, N., Ishida, T., Kiwada, H. (2011). Anti-PEG IgM production by siRNA encapsulated in a PEGylated lipid nanocarrier is dependent on the sequence of the siRNA. *J. Control. Release*, **151**, 149–154.

76. Robbins, M., Judge, A., Liang, L., McClintock, K., Yaworski, E., MacLachlan, I. (2007). 2′-O-methyl-modified RNAs act as TLR7 antagonists. *Mol. Ther.*, **15**, 1663–1669.

77. Hashimoto, Y., Abu Lila, A. S., Shimizu, T., Ishida, T., Kiwada, H. (2014). B cell-intrinsic toll-like receptor 7 is responsible for the enhanced anti-PEG IgM production following injection of siRNA-containing PEGylated lipoplex in mice. *J. Control. Release*, **184**, 1–8.

78. Richter, A. W., Akerblom, E. (1984). Polyethylene glycol reactive antibodies in man: titer distribution in allergic patients treated with monomethoxy polyethylene glycol modified allergens or placebo, and in healthy blood donors. *Int. Arch. Allergy Appl. Immunol.*, **74**, 36–39.

79. Ma, H., Shiraishi, K., Minowa, T., Kawano, K., Yokoyama, M., Hattori, Y., Maitani, Y. (2010). Accelerated blood clearance was not induced for a gadolinium-containing PEG-poly(L-lysine)-based polymeric micelle in mice. *Pharm. Res.*, **27**, 296–302.

80. Shiraishi, K., Hamano, M., Ma, H., Kawano, K., Maitani, Y., Aoshi, T., Ishii, K. J., Yokoyama, M. (2013). Hydrophobic blocks of PEG-conjugates play a significant role in the accelerated blood clearance (ABC) phenomenon. *J. Control. Release*, **165**, 183–190.

81. Hara, E., Makino, A., Kurihara, K., Yamamoto, F., Ozeki, E., Kimura, S. (2012). Pharmacokinetic change of nanoparticulate formulation "Lactosome" on multiple administrations. *Int. Immunopharmacol.*, **14**, 261–266.

82. Hara, E., Ueda, M., Makino, A., Hara, I., Ozeki, E., Kimura, S. (2014). Factors influencing in vivo disposition of polymeric micelles on multiple administrations. *ACS Med. Chem. Lett.*, **5**, 873–877.

83. Zhang, C., Fan, K., Ma, X., Wei, D. (2012). Impact of large aggregated uricases and PEG diol on accelerated blood clearance of PEGylated canine uricase. *PLoS One*, **7**, e39659.

84. Kissel, P., Mauuary, G., Royer, R., Toussain, P. (1975). Letter: Treatment of malignant haemopathies and urate oxidase. *Lancet*, **1**, 229.

85. Ambegia, E., Ansell, S., Cullis, P., Heyes, J., Palmer, L., MacLachlan, I. (2005). Stabilized plasmid-lipid particles containing PEG-diacylglycerols exhibit extended circulation lifetimes and tumor selective gene expression. *Biochim. Biophys. Acta*, **1669**, 155–163.

86. Webb, M. S., Saxon, D., Wong, F. M., Lim, H. J., Wang, Z., Bally, M. B., Choi, L. S., Cullis, P. R., Mayer, L. D. (1998). Comparison of different hydrophobic anchors conjugated to poly(ethylene glycol): effects on the pharmacokinetics of liposomal vincristine. *Biochim. Biophys. Acta*, **1372**, 272–282.

87. Xu, H., Deng, Y., Chen, D., Hong, W., Lu, Y., Dong, X. (2008). Esterase-catalyzed dePEGylation of pH-sensitive vesicles modified with cleavable PEG-lipid derivatives. *J. Control. Release*, **130**, 238–245.

88. Chen, D., Liu, W., Shen, Y., Mu, H., Zhang, Y., Liang, R., Wang, A., Sun, K., Fu, F. (2011). Effects of a novel pH-sensitive liposome with cleavable esterase-catalyzed and pH-responsive double smart mPEG lipid derivative on ABC phenomenon. *Int. J. Nanomed.*, **6**, 2053–2061.

89. Whiteman, K. R., Subr, V., Ulbrich, K., Torchilin, V. P. (2001). Poly(Hpma)-coated liposomes demonstrate prolonged circulation in mice. *J. Liposome Res.*, **11**, 153–164.

90. Romberg, B., Oussoren, C., Snel, C. J., Carstens, M. G., Hennink, W. E., Storm, G. (2007). Pharmacokinetics of poly(hydroxyethyl-l-asparagine)-coated liposomes is superior over that of PEG-coated liposomes at low lipid dose and upon repeated administration. *Biochim. Biophys. Acta*, **1768**, 737–743.

91. Ishihara, T., Maeda, T., Sakamoto, H., Takasaki, N., Shigyo, M., Ishida, T., Kiwada, H., Mizushima, Y., Mizushima, T. (2010). Evasion of the accelerated blood clearance phenomenon by coating of nanoparticles with various hydrophilic polymers. *Biomacromolecules*, **11**, 2700–2706.

92. Metselaar, J. M., Bruin, P., de Boer, L. W., de Vringer, T., Snel, C., Oussoren, C., Wauben, M. H., Crommelin, D. J., Storm, G., Hennink, W. E. (2003). A novel family of L-amino acid-based biodegradable polymer-lipid conjugates for the development of long-circulating liposomes with effective drug-targeting capacity. *Bioconjug. Chem.*, **14**, 1156–1164.

93. Abu Lila, A. S., Nawata, K., Shimizu, T., Ishida, T., Kiwada, H. (2013). Use of polyglycerol (PG), instead of polyethylene glycol (PEG), prevents induction of the accelerated blood clearance phenomenon against long-circulating liposomes upon repeated administration. *Int. J. Pharm.*, **456**, 235–242.

94. Abu Lila, A. S., Uehara, Y., Ishida, T., Kiwada, H. (2014). Application of polyglycerol coating to plasmid DNA lipoplex for the evasion of the accelerated blood clearance phenomenon in nucleic acid delivery. *J. Pharm. Sci.*, **103**, 557–566.

95. Li, Y., Liu, R., Shi, Y., Zhang, Z., Zhang, X. (2015). Zwitterionic poly(carboxybetaine)-based cationic liposomes for effective delivery of small interfering RNA therapeutics without accelerated blood clearance phenomenon. *Theranostics*, **5**, 583–596.

96. Richter, A. W., Akerblom, E. (1983). Antibodies against polyethylene glycol produced in animals by immunization with monomethoxy polyethylene glycol modified proteins. *Int. Arch. Allergy Appl. Immunol.*, **70**, 124–131.

97. Mikhail, A. (2012). Profile of peginesatide and its potential for the treatment of anemia in adults with chronic kidney disease who are on dialysis. *J. Blood Med.*, **3**, 25–31.

98. Eckardt, K. U. (2013). Anaemia: The safety and efficacy of peginesatide in patients with CKD. *Nat. Rev. Nephrol.*, **9**, 192–193.

99. Guidance for Industry: Immunogenicity Assessment for Therapeutic Protein Products. Available at: http://www.fda.gov/downloads/drugs/guidancecomplianceregulatoryinformation/guidances/ucm338856 (accessed on March 2, 2017).

Chapter 8

Anti-PEG Immunity Against PEGylated Therapeutics

Amr S. Abu Lila, PhD,[a,b,c] and Tatsuhiro Ishida, PhD[a]

[a]*Department of Pharmacokinetics and Biopharmaceutics,
Subdivision of Biopharmaceutical Sciences, Institute of Biomedical Sciences,
Tokushima University, Japan*
[b]*Department of Pharmaceutics and Industrial Pharmacy,
Faculty of Pharmacy, Zagazig University, Egypt*
[c]*Department of Pharmaceutics, College of Pharmacy, Hail University, Saudi Arabia*

Keywords: accelerated blood clearance (ABC) phenomenon, anti-PEG IgM, anti-PEG IgG, complement activation, hypersensitivity reactions, PASylation, polyethylene glycol (PEG), PolyXen, proteins, XTEN, biological half-life, toxicity, mononuclear phagocyte system (MPS), PEGylation, hapten, adjuvant, immune system, liposomes, micelles, microemulsions, doxorubicin, Doxil®, uricase, allergy, PEG-modified bovine adenosine deaminase (PEG-ADA), PEG-INF-α, PEGylated phenylalanine ammonia lyase (PEG-PAL), PEGinesatide, chronic kidney disease (CKD), severe hypersensitivity reactions (HSR), PEGylated asparaginase (PEG-ASNase), PEGylated uricase (Pegloticase), C activation-related pseudoallergy (CARPA), class-2 thymus-independent (TI-2) antigen, methoxypoly(ethylene glycol) (mPEG), hydroxyl PEG (HO-PEG), immunogenicity, ovalbumin, PEGylated proteins, polyglycerol, poly(carboxybetaine), stealth polymers

8.1 Introduction

Polyethylene glycol (PEG) is one of the most widely used "stealth" polymers in the field of pharmaceutics for protein and drug

---

Immune Aspects of Biopharmaceuticals and Nanomedicines
Edited by Raj Bawa, János Szebeni, Thomas J. Webster, and Gerald F. Audette
Copyright © 2018 Pan Stanford Publishing Pte. Ltd.
ISBN 978-981-4774-52-9 (Hardcover), 978-0-203-73153-6 (eBook)
www.panstanford.com

carriers [1, 2]. PEGylation is the covalent coupling of polyethylene glycol (PEG) to peptides and proteins in order to significantly extend their biological half-life, reduce toxicity, and mask both humoral and cellular immunogenicity while sustaining therapeutic effectiveness [3, 4]. The covalent attachment of PEG to liposomes and colloidal nanocarriers can also suppress protein adsorption onto their surfaces, thereby rendering them invisible to the cells of the mononuclear phagocyte system (MPS), which extends their circulation time in the body and improves their overall therapeutic efficacy [5–7].

However, in contrast to the general perception that PEG is non-immunogenic, a growing body of literature has emerged claiming PEG to be immunogenic. Many reports have demonstrated that PEGylated substances can elicit antibody responses against PEG (anti-PEG) limiting the therapeutic efficacy and/or reducing the tolerance of PEGylated therapeutics [8, 9]. In addition, we and other research groups have demonstrated that anti-PEG antibodies are associated with the accelerated clearance of subsequently administered doses of PEGylated nanocarriers, which is referred to as "the accelerated blood clearance (ABC)" phenomenon [10–13].

Of interest, an emerging body of evidence is emphasizing the existence of naturally occurring anti-PEG antibodies in normal individuals who have never received PEGylated therapeutics systemically [14–16]. Armstrong et al. [14] revealed that these naturally occurring anti-PEG antibodies could prime the host immune system against administered PEGylated therapeutics, resulting in compromised therapeutic efficacy. Therefore, to claim PEG to be immunogenic and antigenic could be a disappointment, particularly with an increasing number of PEGylated products, such as PEGylated proteins and nanocarriers, anticipated to enter the market over the coming few years. Accordingly, a deep understanding of the prevalence and clinical implications of anti-PEG immunity is a prerequisite for the continual clinical application of PEGylated therapeutics.

8.2 PEG Immunogenicity in Animal Models

8.2.1 Anti-PEG Response to PEGylated Proteins

In animal models, PEG is generally believed to be immunologically inert, which may in part be due to the rapid renal clearance of the unconjugated polymer. Therefore, the immunogenicity of PEGylated substances was previously tested directly against a primary substance, rather than against covalently coupled PEG. However, a strong anti-PEG immunological response, manifested by the robust induction of antibodies to PEG, was observed upon conjugating this hydrophilic polymer to some proteins, particularly ovalbumin (OVA) [15] and uricase [17]. These findings potentially contradict the basic concept that PEG is a bio-inert, non-immunogenic substance. The first study demonstrating the induction of antibodies to PEG *in vivo* was conducted by Richter and Akerblom in 1983 [15], which was less than a decade after the introduction of protein PEGylation. They emphasized the induction of a strong anti-PEG immune response in rabbits following either intramuscular or subcutaneous injections of different PEGylated proteins in complete Freund's adjuvant. They showed that PEG, administered alone under similar conditions, elicited a weak and transitory immune response. Recently, we also confirmed the induction of an anti-PEG antibody response (mainly anti-PEG IgM) following a single intravenous injection of either PEGylated OVA or PEGylated bovine serum (BSA), similar to that observed with PEGylated liposomes, despite the fact that a single administration usually does not induce specific neutralizing antibodies to either OVA or BSA [8]. These results led to the assumption that the production of antibodies only occurs against PEG conjugates in a manner wherein PEG acts as a hapten. A hapten is a non-immunogenic small molecule that triggers an immune response only when conjugated to a large carrier such as a protein. This haptentic character of PEG is assumed to be dependent on the molecular weight and the immunogenicity of the conjugated protein, and on the presence of adjuvants [18–20].

Importantly, despite the increasing number of reports studying the anti-PEG antibody-mediated immunogenicity of PEGylated proteins in animals, only a few studies have verified the ability of the elicited anti-PEG antibodies to induce the rapid clearance of a subsequent dose of PEGylated protein products, such as a phenomenon observed upon sequential administration of PEGylated IFN β-1a in Rhesus monkeys [21]. This might be attributed, on the one hand, to the weak immunogenicity of the anchoring protein for PEG or, on the other hand, to the lack of a clear discrimination between anti-PEG antibodies and anti-anchoring protein antibodies.

8.2.2 Anti-PEG Response to PEGylated Nanocarriers

Despite the fact that PEGylation imparts the conjugated product with a prolonged circulation time *in vivo*, many reports have claimed that anti-PEG antibodies, produced in response to the first dose, may harmfully trigger the rapid systemic clearance of subsequently injected doses of PEGylated substances—the so-called "accelerated blood clearance (ABC)" phenomenon. In 1997, Moghimi and Gray [22] demonstrated that repeated administration of PEG-modified polystyrene particles to rats in 3–4-day intervals significantly triggered an accelerated clearance of the particles from systemic circulation by the cells of the MPS. Similarly, we and other research groups have revealed that "empty" PEGylated liposomes also elicit a strong immune response that results in an enhancement of systemic clearance with a corresponding preferential hepatic accumulation of subsequently administered doses of PEGylated liposomes in mice, rats, rabbits, dogs, and monkeys [23–25]. A mounting body of evidence has confirmed that anti-PEG antibodies, mainly anti-PEG IgM, elicited in response to PEGylated liposomes is likely to be responsible for such unexpected pharmacokinetic alterations. In our earlier study, we reported that the serum of pre-treated rats showed greater antibody adsorption onto PEGylated liposomes, compared with that of naïve animals, suggesting that anti-PEG antibodies were the prevailing serum factor responsible for the ABC phenomenon [26, 27]. Soon afterwards, we revealed that pre-treatment with "empty" PEGylated liposomes substantially triggered the production of anti-PEG antibodies, mainly anti-PEG IgM, which dictates the extent of the ABC phenomenon induction [23]. Other research groups [28–30] have also emphasized the correlation between the production levels of anti-PEG antibodies and the magnitude of the ABC phenomenon. The observed anti-PEG response has predominantly been that of IgM [23, 29, 31], although the development of anti-PEG IgG has also been reported [25, 32].

In further studies, we emphasized the vital role of anti-PEG IgM-mediated complement activation in the induction of the ABC phenomenon upon repeated administration of PEGylated liposomes. We demonstrated that serum from rats displaying an ABC response showed complement activation upon incubation with PEGylated liposomes [33]. In addition, by using a single-pass liver perfusion technique, we confirmed that heat-treatment of this serum alleviated the substantial hepatic clearance of PEGylated liposomes via the inhibition of the complement activity [34]. On the basis of our results [27, 31], we proposed the following tentative mechanism for the cause of this phenomenon. Anti-PEG IgM, which is produced in the spleen in response to the first dose, selectively binds to the PEG upon the second dose of liposomes injected several days later and subsequently activates the complement system, and, as a consequence, the liposomes are taken up by the Kupffer cells in the liver.

It is noteworthy that the induction of anti-PEG IgM and its implications on the pharmacokinetics of subsequently injected doses are not only restricted to PEGylated liposomes. Other classes of PEGylated nanocarriers, such as polymeric nanoparticles [35], micelles [36] and microemulsions [37], also elicit the response. Furthermore, different factors that include the lipid dose used, physicochemical properties (particle size, surface charge, PEG-chain length, and degree of PEGylation), the nature of the encapsulated payload of the initially injected liposome, and dosing intervals, can significantly affect the incidence and magnitude of the ABC phenomenon, as recently established in our previous publication [38].

8.3 PEG Immunogenicity in Humans

Despite the fact that PEGylation is used to introduce several therapeutic agents in both pre-clinical and clinical settings, including liposomal doxorubicin, uricase, interferon alpha (IFNα), and many other proteins, a growing body of literature clearly suggests that the induction of an anti-PEG antibody response is possible in humans [39–41]. However, in contrast to most animal studies, the anti-PEG antibody response in humans is more skewed toward IgG isotype antibodies [39, 42]. In addition, many reports have emphasized the natural existence of anti-PEG antibodies in the sera of normal donors despite the absence of treatment with PEGylated therapeutics [14]. Accordingly, both pre-existing and induced anti-PEG antibodies might represent a tremendous challenge to the future clinical application of PEGylated therapeutics [6, 42].

8.3.1 Pre-Existing Anti-PEG Antibodies in Normal Donors

In 1984, Richter and Akerblom first reported the pre-existence of anti-PEG antibodies, mostly anti-PEG IgM, in 0.2% of healthy normal subjects and in 3.3% of untreated allergy patients [19]. Almost 2 decades later, Armstrong et al. [14] reported the prevalence of higher anti-PEG antibodies titers, reaching up to 25%, in healthy blood donors. Such variations clearly reflect a significant increase in the prevalence of pre-existing anti-PEG antibodies in normal subjects over time. Both studies utilized passive hemagglutination of PEG-modified RBCs to detect PEG-specific antibodies, so the differences are unlikely to be caused by the method of detection. Since PEG has been classified by the FDA under the category of products that are Generally Recognized As Safe (GRAS), it has become commonly used in cosmetics, processed foods, and pharmaceuticals. In addition, PEG-containing surfactants, as well as PEG itself, are found in a vast majority of household and hygiene products (e.g., soap, shampoo, toothpaste, lotion, detergent). Therefore, it is easy to imagine that frequent exposure to PEG in daily use products could lead to the predictable formation of anti-PEG antibodies. However, the exact mechanism by which pre-existing anti-PEG antibodies are generated in individuals who have never received any previous treatment with PEGylated therapeutics has yet to be elucidated. Recently, Yang et al. [45] have postulated the following tentative mechanism: the human body is frequently subjected to certain conditions such as abrasions, ulcerations, and skin tears that may result in local inflammatory responses and recruitment of immune cells. Upon exposure to PEG in the daily use of common products, PEG is likely introduced to sites of inflammation and to come into close proximity to highly active immune cells, which in turn may be sufficient to drive the induction of anti-PEG antibodies. Accordingly, subsequent persistent exposure to PEGylated therapeutics may further induce a robust memory immune response to the polymer.

Besides the reports by Richter and Akerblom [19], as well as by Armstrong et al. [14], the prevalence of pre-existing anti-PEG antibodies has been further reported in both healthy donors and untreated controls during clinical trials. Tillmann and coworkers [40] observed the occurrence of relatively low levels of anti-PEG antibodies in healthy individuals (7–8%), whereas a high frequency of anti-PEG antibodies (up to 44%) was detected in patients with hepatitis C. However, these elevated levels of anti-PEG antibodies exerted no compromising effect on the therapeutic efficacy of PEGylated interferon used for the treatment of this chronic disease. Hershfield et al. [43] also reported the pre-existence of up to 19% of anti-PEG antibodies in the sera of control subjects prior to treatment with pegloticase, a PEGylated form of urate oxidase used in the treatment of refractory gout. As a result, the FDA has recently updated their guidelines to screen for anti-PEG antibodies during clinical trials of PEGylated therapeutics [44].

8.3.2 Impact of Anti-PEG Antibodies on the Efficacy of PEGylated Therapeutics

Despite the fact that the immunogenicity of PEG is well reported in animal models [8, 15, 17], early studies in humans on PEGylated therapeutics indicated the absence of an anti-PEG antibody response or a lack of clinical implications. In 1984, Richter et al. [19] first reported the induction of anti-PEG IgM production in humans during hyposensitization with methoxy-PEG-modified honeybee venom and PEGylated ragweed extract. In that study, however, the occurrence of anti-PEG antibodies did not appear to prime further immune responses, and thereby, no change in response rates was observed. Later on, in a clinical trial of PEG-modified bovine adenosine deaminase (PEG-ADA) in patients with adenosine deaminase deficiency, Chaffee et al. [45] reported the induction of an anti-PEG-ADA antibody response, mainly IgG, in 59% of treated patients. However, competitive ELISAs using ADA and different PEGylated proteins suggested that these antibodies were formed against ADA rather than the PEG moiety. Recently, Tillmann and colleagues [40] also evaluated the development, frequency and impact of anti-PEG antibodies in hepatitis C patients before and after treatment with PEG-INF-α. They revealed that despite the high prevalence of anti-PEG antibodies (44%) in the sera of patients with hepatitis C, which was significantly higher than the 6.9% in healthy controls, the presence of pre-existing anti-PEG antibodies did not appear to affect the efficacy of antiviral PEG-interferon therapy. The contribution of immune impairment and hepatic damage caused by HCV to the apparent lack of anti-PEG antibody effects were not elucidated. Collectively, these findings played an integral role in establishing the perception that anti-PEG antibody response has no clinical implications for the therapeutic efficacy of PEGylated therapeutics in humans.

Interestingly, unlike most clinical studies that have indicated the development of anti-PEG antibody responses in only a subset of patients, in a phase I dose-escalation study, Longo et al. [41] reported the development of anti-PEG antibodies in all phenylketonuria patients within 6 weeks following a subcutaneous injection of PEGylated phenylalanine ammonia lyase (PEG-PAL). However, neither pre-existing nor induced anti-PEG antibodies appeared to affect the therapeutic efficacy of a single dose of PEG-PAL. Notably, two patients from this study later suffered severe adverse reactions to intramuscular injections of medroxyprogesterone acetate, in which free PEG and polysorbate were included as excipients. However, the correlation between pre-dosing with PEG-PAL and the severe adverse reactions was not investigated.

In contrast to the aforementioned data, clinical studies have recently emphasized the contribution of the anti-PEG antibody response to the reduced therapeutic efficacy of PEGylated drugs and/or the development of severe side effects. PEGinesatide, an erythropoietic agent, was approved by the FDA in December 2011 for the treatment of anemia associated with chronic kidney disease (CKD) in adult patients on dialysis [46, 47]. However, soon after its approval, the drug was blamed for the induction of severe hypersensitivity reactions (HSR), which were fatal in 0.02% of patients within 30 min of the start of their first intravenous infusion leading to its withdrawal from the market in February 2013 [48]. Despite the fact that all HSR occurred upon the first exposure, and despite the lack of immunogenicity data, the contribution of cross-reactive antibodies in these patients cannot be overlooked.

Armstrong et al. [14] reported that when pediatric acute lymphoblastic leukemia patients were treated with PEGylated asparaginase (PEG-ASNase), 32% of them developed anti-PEG antibodies, which was mainly IgM. The presence of anti-PEG IgM was significantly correlated with the rapid clearance of PEG-ASNase, which in turn limited its therapeutic efficacy. Surprisingly, anti-PEG IgM was also detected in the sera of 13% of patients treated with unmodified ASNase, but no correlation was observed between serum ASNase activity and anti-PEG IgM levels.

The therapeutic efficacy of PEGylated uricase (Pegloticase), a drug used for the treatment of refractory gout, is also limited by the anti-PEG antibody response. In a phase III clinical trial, anti-PEG antibodies were detected in 89% of patients with refractory gout who had been treated with Pegloticase. This elevated level of anti-PEG antibodies triggered a rapid clearance of the drug, thereby compromising its therapeutic efficacy, particularly upon repeated administration [39]. In addition to rapid Pegloticase clearance, anti-PEG antibody-positive subjects also demonstrated an increased rate of infusion reactions [39, 43, 49], but the contribution of anti-PEG antibodies to adverse reactions to Pegloticase remains unclear.

8.3.3 Non-Antibody-Mediated Hypersensitivity Reactions

Besides the slow specific immunogenic response manifested in antibody formation against PEGylated liposomes, a growing body of literature has verified that the immune reaction against intravenously administered PEGylated liposomes may also involve rapid inflammatory-like hypersensitivity reactions, also known as infusion reactions, that do not involve IgE but arise as a consequence of activation of the complement C system [50, 51]. These anaphylactoid reactions can be distinguished within the Type I category of hypersensitivity reactions as "C activation-related pseudoallergy" (CARPA), a term that highlights the mechanism of the reaction [52–55] [38–41]. The first direct evidence for the causal relationship between C activation and hypersensitivity reactions (CARPA) to PEGylated liposomes was provided by Brouwers et al. [50], who reported three severe hypersensitivity reactions out of nine patients obtaining ^{99m}Tc-labeled pegylated liposomes for scintigraphic detection of bowel inflammation. Later on, extensive research by Szebeni et al. [51, 52, 54, 56] has shown that complement activation is responsible for infusion reactions observed in up to 25% of patients treated with Doxil®, which were non IgE-mediated. In addition, they have shown that these reactions can occur after a first administration, which implies that Doxil® can directly activate the complement system by the alternative pathway. Chanan-Khan et al. [57] also demonstrated that Doxil® therapy activated complement in the majority of patients and induced moderate to severe hypersensitivity reactions in 45% of cancer patients infused with Doxil® for the first time, regardless of sex or age of the patients. Importantly, Doxil® caused C activation, as manifested by the significant elevations of serum complement terminal complex (SC5b-9), in 21 out of 29 patients (72%) following infusion of the drug. Collectively, these data strongly suggested that C activation might be causal or a key contributing, but not rate-limiting factor in liposome-induced CARPA.

8.4 Properties of Anti-PEG Antibody Epitope

Many reports have highlighted the contribution of anti-PEG antibodies to the accelerated clearance of PEGylated therapeutics. However, the antigenic determinant of the PEG polymer remains a mystery. Based on its highly repetitive structure, PEG is acceptably classified as a class-2 thymus-independent (TI-2) antigen. TI-2 antigens are composed of identical repeating epitopes that act by cross-linking the cell-surface immunoglobulins of specific B cells leading to a significant and prolonged activation of B cells without co-stimulation by T cells [58]. In an earlier study, Richter and Akerblom [15] observed the inhibition of anti-PEG antibody precipitation with PEG_{300}, suggesting that the antigenic determinant of PEG may be a sequence of 6 to 7 $-CH_2-CH_2-O-$ units. Cheng and his colleagues [59] found that the monoclonal anti-PEG antibody (IgM) generated by immunization with PEGylated β-glucuronide recognizes the repeating sequence $-(O-CH_2-CH_2)_n-$ subunit (16 units) of PEG. Saifer et al. [60] later reported that triethylene glycol (MW 150–160) was bound by anti-PEG antibodies in direct and competitive ELISAs. These data suggest that the minimum epitope recognition by anti-PEG antibodies could range from three to four repeated oxyethylene units.

Nevertheless, in several studies, a single injection of PEG polymer alone did not elicit an anti-PEG response. On the other hand, PEG conjugated to proteins and/or nanocarriers will induce an anti-PEG immune response. Accordingly, the antigenic determinant for anti-PEG antibodies is assumed to occur at the linkage between PEG and other materials. Shiraishi et al. [29] observed cross-reactivity between anti-PEG IgMs elicited by hydrophobic PEGylated micelles and liposomes, but hydrophilic PEGylated micelles would induce neither an anti-PEG IgM response nor the ABC phenomenon. Accordingly, they proposed that the anti-PEG antibody epitope is the interphase between a hydrophobic core and conjugated PEG groups [30]. Furthermore, owing to the difference in the immune responses between free PEG and PEGylated therapeutics, PEG is projected to function as a hapten, which is a non-immunogenic small molecule that can elucidate an immune response only when conjugated to a larger carrier [6, 61].

8.5 Strategies to Avert Anti-PEG Antibody Responses

Based on the improved understanding of the prevalence and clinical implications of anti-PEG immunity in humans, many researchers are currently extending great effort to the development of applicable strategies to avert/attenuate the anti-PEG antibody responses.

8.5.1 Modification of the PEG Moiety

All currently approved PEGylated therapeutics contain methoxypoly(ethylene glycol) (mPEG). The use of mPEG in PEG conjugates of proteins and non-protein therapeutic agents has led to the recognition that the polymer components of such conjugates can induce anti-PEG antibodies that may accelerate the clearance and reduce the efficacy of the conjugates. These considerations have motivated the development of other forms of PEG. In a series of studies [62, 63], Sherman et al. assessed the role of the methoxy group in the immune responses to mPEG conjugates and the potential advantages of replacing mPEG with hydroxy PEG (HO-PEG) (Fig. 8.1). They employed rabbits immunized with either mPEG or HO-PEG conjugates of human serum albumin, human interferon-α, or porcine uricase as adjuvant emulsions [63]. They demonstrated that hydroxyl PEG (HO-PEG) represents a potentially less immunogenic and less antigenic alternative to mPEG. Accordingly, they suggested that using monofunctionally activated HO-PEG instead of mPEG in preparing conjugates for clinical use might decrease anti-PEG-mediated undesirable effects. However, since previous studies have revealed an avidity of both pre-existing and induced anti-PEG antibodies to the PEG backbone (repeated oxyethylene subunits), modifications of the terminal group of the PEG moiety might only have a moderate effect on abrogating the anti-PEG antibody response against PEGylated therapeutics.

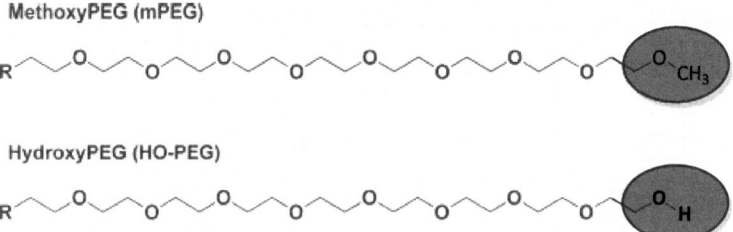

Figure 8.1 Partial structure of methoxyPEG and hydroxyPEG. R designates the rest of the polymer. The domains that may serve as epitopes for the formation of anti-PEG antibodies, namely, the distal terminal groups, are confined within the ovals.

8.5.2 Use of Alternative Polymers Instead of PEGylation

The use of alternative stealth polymers such as polyglycerol, Poly(carboxybetaine), XTEN peptide, PAS peptide, etc., have also received great attention. These polymers are less common in everyday household items and, therefore, may not represent the problem of pre-existing antibodies. Recently, Schellenberger et al. [64] developed an alternative technology based on new unstructured, hydrophilic polypeptides, called XTEN, which can be fused to the proteins or peptides imparting them with prolonged blood circulation. The XTEN technology has been shown to be safe and poorly immunogenic in many animal studies leading to their introduction in Phase I trials with two products: exenatide-XTEN fusion construct (VRS-859) for the treatment of type II diabetes and a growth hormone-XTEN fusion construct (VRS 317) for growth hormone deficiency.

Schlapschy et al. [65] developed conformationally disordered polypeptide chains with expanded hydrodynamic volume comprising the small residues proline, alanine, and serine (PAS). Because PAS showed stability in serum and lacked toxicity or immunogenicity in mice, PASylation technology is currently being adapted to typical biologics, such as the use of interferon and growth hormone to prolong circulation and boost bioactivity *in vivo*.

PolyXen® is an enabling platform technology for protein drug delivery. It uses the natural polymer polysialic acid (PSA) to prolong the drug's half-life and potentially improve the stability of therapeutic peptides and proteins [66]. PolyXen® technology has introduced two drug products into clinical trials: ErepoXen, an erythropoietin-PAS fusion construct, and SuliXen, a PAS-insulin.

It is noteworthy that antibodies against various natural and synthetic polymers have also been reported [67, 68], suggesting that such alternative polymers may also induce an immunogenic response upon repeated administration in humans.

8.6 Conclusion

In contrast to the general assumption that PEG is biologically inert, many reports have clearly underscored the potency and impact of anti-PEG immunity, which represents a particularly important concern in light of the increasing number of PEGylated therapeutic proteins and nanomedicines that are FDA-approved or are currently under clinical development. Therefore, patients should be pre-screened and monitored for anti-PEG antibodies prior to and throughout a course of treatment with PEGylated therapeutics. Furthermore, a deep understanding of the exact mechanism by which anti-PEG antibodies develop and are able to bind to the polymer is urgently needed, along with strategies to avert the challenges of PEG-specific immunity.

Disclosures and Conflict of Interest

The authors wish to thank Dr. J. L. McDonald for his helpful advice in writing the manuscript. This study was supported, in part, by a Grant-in-Aid for Scientific Research (B) 15H04639, the Ministry of Education, Culture, Sports, Science and Technology, Japan. The authors declare that they have no conflicts of interest. No writing assistance was utilized in the production of this chapter and the authors have received no payment for its preparation.

Corresponding Author

Dr. Tatsuhiro Ishida
Department of Pharmacokinetics and Biopharmaceutics
Subdivision of Biopharmaceutical Sciences
Institute of Biomedical Sciences
Tokushima University, 1-78-1, Sho-machi
Tokushima 770-8505, Japan
Email: ishida@tokushima-u.ac.jp

About the Authors

Amr S. Abu Lila received his master's degree in pharmaceutics from the Faculty of Pharmacy, Zagazig University, Egypt, in 2005 and his PhD in pharmaceutical life sciences from the Graduate School of Pharmaceutical Sciences, Tokushima University, Japan, in 2010. From 2011 to 2013, he was a JSPS postdoctoral fellow at the Graduate School of Pharmaceutical Sciences at Tokushima University in Japan. Currently, he is an associate professor in the Faculty of Pharmacy, Zagazig University. He has published more than 20 papers in peer-reviewed scientific journals, 4 review articles, and 3 book chapters. His current research focuses on anticancer therapeutics using liposomal drug delivery systems for the selective targeting of either chemotherapeutic agents or genes to tumor tissues.

Tatsuhiro Ishida graduated from Tokushima University, Japan, in 1993 and then received his master's degree in 1995 and his PhD in 1998 from the Faculty of Pharmaceutical Sciences, Tokushima University. From 1998 to 2000, he was a postdoctoral fellow at the University of Alberta (Pharmacology) Canada. In 2000, he became a lecturer in Pharmaceutical Sciences at Tokushima University and was promoted to associate professor in 2003. He has been a full professor there since 2013. He has published more than 100 peer-reviewed papers, more than 10 review articles, and 5 book chapters. He has given more than 30 presentations as an invited speaker at international conferences. He is interested in immunological responses against PEGylated therapeutics and the mechanisms behind the response.

References

1. Roberts, M. J., Bentley, M. D., Harris, J. M. (2002). Chemistry for peptide and protein PEGylation. *Adv. Drug Deliv. Rev.*, **54**, 459–476.
2. Zalipsky, S. (1995). Functionalized poly(ethylene glycol) for preparation of biologically relevant conjugates. *Bioconjug. Chem.*, **6**, 150–165.
3. Abuchowski, A., McCoy, J. R., Palczuk, N. C., van Es, T., Davis, F. F. (1977). Effect of covalent attachment of polyethylene glycol on immunogenicity and circulating life of bovine liver catalase. *J. Biol. Chem.*, **252**, 3582–3586.
4. Payne, R. W., Murphy, B. M., Manning, M. C. (2011). Product development issues for PEGylated proteins. *Pharm. Dev. Technol.*, **16**, 423–440.
5. Verhoef, J. J., Anchordoquy, T. J. (2013). Questioning the use of PEGylation for drug delivery. *Drug Deliv. Transl. Res.*, **3**, 499–503.
6. Garay, R. P., El-Gewely, R., Armstrong, J. K., Garratty, G., Richette, P. (2012). Antibodies against polyethylene glycol in healthy subjects and in patients treated with PEG-conjugated agents. *Expert Opin. Drug Deliv.*, **9**, 1319–1323.
7. Omata, D., Negishi, Y., Hagiwara, S., Yamamura, S., Endo-Takahashi, Y., Suzuki, R., Maruyama, K., Aramaki, Y. (2012). Enhanced gene delivery using Bubble liposomes and ultrasound for folate-PEG liposomes. *J. Drug Target*, **20**, 355–363.
8. Shimizu, T., Ichihara, M., Yoshioka, Y., Ishida, T., Nakagawa, S., Kiwada, H. (2012). Intravenous administration of polyethylene glycol-coated (PEGylated) proteins and PEGylated adenovirus elicits an anti-PEG immunoglobulin M response. *Biol. Pharm. Bull.*, **35**, 1336–1342.
9. Hashimoto, Y., Shimizu, T., Mima, Y., Abu Lila, A. S., Ishida, T., Kiwada, H. (2014). Generation, characterization and in vivo biological activity of two distinct monoclonal anti-PEG IgMs. *Toxicol. Appl. Pharmacol.*, **277**, 30–38.

10. Dams, E. T., Laverman, P., Oyen, W. J., Storm, G., Scherphof, G. L., van Der Meer, J. W., Corstens, F. H., Boerman, O. C. (2000). Accelerated blood clearance and altered biodistribution of repeated injections of sterically stabilized liposomes. *J. Pharmacol. Exp. Ther.*, **292**, 1071–1079.

11. Laverman, P., Carstens, M. G., Boerman, O. C., Dams, E. T., Oyen, W. J., van Rooijen, N., Corstens, F. H., Storm, G. (2001). Factors affecting the accelerated blood clearance of polyethylene glycol-liposomes upon repeated injection. *J. Pharmacol. Exp. Ther.*, **298**, 607–612.

12. Ishida, T., Maeda, R., Ichihara, M., Mukai, Y., Motoki, Y., Manabe, Y., Irimura, K., Kiwada, H. (2002). The accelerated clearance on repeated injection of pegylated liposomes in rats: laboratory and histopathological study. *Cell Mol. Biol. Lett.*, **7**, 286.

13. Nagao, A., Abu Lila, A. S., Ishida, T., Kiwada, H. (2013). Abrogation of the accelerated blood clearance phenomenon by SOXL regimen: promise for clinical application. *Int. J. Pharm.*, **441**, 395–401.

14. Armstrong, J. K., Hempel, G., Koling, S., Chan, L. S., Fisher, T., Meiselman, H. J., Garratty, G. (2007). Antibody against poly(ethylene glycol) adversely affects PEG-asparaginase therapy in acute lymphoblastic leukemia patients. *Cancer*, **110**, 103–111.

15. Richter, A. W., Akerblom, E. (1983). Antibodies against polyethylene glycol produced in animals by immunization with monomethoxy polyethylene glycol modified proteins. *Int. Arch. Allergy Appl. Immunol.*, **70**, 124–131.

16. Liu, Y., Reidler, H., Pan, J., Milunic, D., Qin, D., Chen, D., Vallejo, Y. R., Yin, R. (2011). A double antigen bridging immunogenicity ELISA for the detection of antibodies to polyethylene glycol polymers. *J. Pharmacol. Toxicol. Methods*, **64**, 238–245.

17. Tsuji, J., Hirose, K., Kasahara, E., Naitoh, M., Yamamoto, I. (1985). Studies on antigenicity of the polyethylene glycol (PEG)-modified uricas. *Int. J. Immunopharmacol.*, **7**, 725–730.

18. Ishida, T., Kiwada, H. (2013). Anti-polyethyleneglycol antibody response to PEGylated substances. *Biol. Pharm. Bull.*, **36**, 889–891.

19. Richter, A. W., Akerblom, E. (1984). Polyethylene glycol reactive antibodies in man: titer distribution in allergic patients treated with monomethoxy polyethylene glycol modified allergens or placebo, and in healthy blood donors. *Int. Arch. Allergy Appl. Immunol.*, **74**, 36–39.

20. Caliceti, P., Schiavon, O., Veronese, F. M. (2001). Immunological properties of uricase conjugated to neutral soluble polymers. *Bioconjug. Chem.*, **12**, 515–522.

21. Hu, X., Olivier, K., Polack, E., Crossman, M., Zokowski, K., Gronke, R. S., Parker, S., Li, Z., Nestorov, I., Baker, D. P., Clarke, J., Subramanyam, M. (2011). In vivo pharmacology and toxicology evaluation of polyethylene glycol-conjugated interferon beta-1a. *J. Pharmacol. Exp. Ther.*, **338**, 984–996.

22. Moghimi, S. M., Gray, T. (1997). A single dose of intravenously injected poloxamine-coated long-circulating particles triggers macrophage clearance of subsequent doses in rats. *Clin. Sci. (Lond.)*, **93**, 371–379.

23. Wang, X., Ishida, T., Kiwada, H. (2007). Anti-PEG IgM elicited by injection of liposomes is involved in the enhanced blood clearance of a subsequent dose of PEGylated liposomes. *J. Control. Release*, **119**, 236–244.

24. Semple, S. C., Harasym, T. O., Clow, K. A., Ansell, S. M., Klimuk, S. K., Hope, M. J. (2005). Immunogenicity and rapid blood clearance of liposomes containing polyethylene glycol-lipid conjugates and nucleic acid. *J. Pharmacol. Exp. Ther.*, **312**, 1020–1026.

25. Sroda, K., Rydlewski, J., Langner, M., Kozubek, A., Grzybek, M., Sikorski, A. F. (2005). Repeated injections of PEG-PE liposomes generate anti-PEG antibodies. *Cell Mol. Biol. Lett.*, **10**, 37–47.

26. Ishida, T., Harada, M., Wang, X. Y., Ichihara, M., Irimura, K., Kiwada, H. (2005). Accelerated blood clearance of PEGylated liposomes following preceding liposome injection: effects of lipid dose and PEG surface-density and chain length of the first-dose liposomes. *J. Control. Release*, **105**, 305–317.

27. Ishida, T., Ichihara, M., Wang, X., Kiwada, H. (2006). Spleen plays an important role in the induction of accelerated blood clearance of PEGylated liposomes. *J. Control. Release*, **115**, 243–250.

28. Judge, A., McClintock, K., Phelps, J. R., Maclachlan, I. (2006). Hypersensitivity and loss of disease site targeting caused by antibody responses to PEGylated liposomes. *Mol. Ther.*, **13**, 328–337.

29. Shiraishi, K., Hamano, M., Ma, H., Kawano, K., Maitani, Y., Aoshi, T., Ishii, K. J., Yokoyama, M. (2013). Hydrophobic blocks of PEG-conjugates play a significant role in the accelerated blood clearance (ABC) phenomenon. *J. Control. Release*, **165**, 183–190.

30. Kaminskas, L. M., McLeod, V. M., Porter, C. J., Boyd, B. J. (2011). Differences in colloidal structure of PEGylated nanomaterials dictate the likelihood of accelerated blood clearance. *J. Pharm. Sci.*, **100**, 5069–5077.

31. Ishida, T., Ichihara, M., Wang, X., Yamamoto, K., Kimura, J., Majima, E., Kiwada, H. (2006). Injection of PEGylated liposomes in rats elicits

PEG-specific IgM, which is responsible for rapid elimination of a second dose of PEGylated liposomes. *J. Control. Release*, **112**, 15–25.

32. Wunderlich, D. A., Macdougall, M., Mierz, D. V., Toth, J. G., Buckholz, T. M., Lumb, K. J., Vasavada, H. (2007). Generation and characterization of a monoclonal IgG antibody to polyethylene glycol. *Hybridoma (Larchmt)*, **26**, 168–172.

33. Ishida, T., Atobe, K., Wang, X., Kiwada, H. (2006). Accelerated blood clearance of PEGylated liposomes upon repeated injections: effect of doxorubicin-encapsulation and high-dose first injection. *J. Control. Release*, **115**, 251–258.

34. Ishida, T., Kashima, S., Kiwada, H. (2008). The contribution of phagocytic activity of liver macrophages to the accelerated blood clearance (ABC) phenomenon of PEGylated liposomes in rats. *J. Control. Release*, **126**, 162–165.

35. Saadati, R., Dadashzadeh, S., Abbasian, Z., Soleimanjahi, H. (2013). Accelerated blood clearance of PEGylated PLGA nanoparticles following repeated injections: effects of polymer dose, PEG coating, and encapsulated anticancer drug. *Pharm. Res.*, **30**, 985–995.

36. Koide, H., Asai, T., Hatanaka, K., Urakami, T., Ishii, T., Kenjo, E., Nishihara, M., Yokoyama, M., Ishida, T., Kiwada, H., Oku, N. (2008). Particle size-dependent triggering of accelerated blood clearance phenomenon. *Int. J. Pharm.*, **362**, 197–200.

37. Joshi, M., Pathak, S., Sharma, S., Patravale, V. (2008). Solid microemulsionpreconcentrate (NanOsorb) of artemether for effective treatment of malaria. *Int. J. Pharm.*, **362**, 172–178.

38. Abu Lila, A. S., Kiwada, H., Ishida, T. (2013). The accelerated blood clearance (ABC) phenomenon: clinical challenge and approaches to manage. *J. Control. Release*, **172**, 38–47.

39. Sundy, J. S., Ganson, N. J., Kelly, S. J., Scarlett, E. L., Rehrig, C. D., Huang, W., Hershfield, M. S. (2007). Pharmacokinetics and pharmacodynamics of intravenous PEGylated recombinant mammalian urate oxidase in patients with refractory gout. *Arthritis Rheum.*, **56**, 1021–1028.

40. Tillmann, B., Poulin-Charronnat, B. (2010). Auditory expectations for newly acquired structures. *Q. J. Exp. Psychol. (Hove)*, **63**, 1646–1664.

41. Longo, N., Harding, C. O., Burton, B. K., Grange, D. K., Vockley, J., Wasserstein, M., Rice, G. M., Dorenbaum, A., Neuenburg, J. K., Musson, D. G., Gu, Z., Sile, S. (2014). Single-dose, subcutaneous recombinant phenylalanine ammonia lyase conjugated with polyethylene glycol

in adult patients with phenylketonuria: an open-label, multicentre, phase 1 dose-escalation trial. *Lancet*, **384**, 37–44.

42. Armstrong, J. K. (2009). The occurrence, induction, specificity and potential effects of antibodies against poly(ethyleneglycol). In: Veronese, F. M., ed. *PEGylated Protein Drugs: Basic Science and Clinical Applications*, Birkhauser, Basel, pp. 147–168.

43. Hershfield, M. S., Ganson, N. J., Kelly, S. J., Scarlett, E. L., Jaggers, D. A., Sundy, J. S. (2014). Induced and pre-existing anti-polyethylene glycol antibody in a trial of every 3-week dosing of pegloticase for refractory gout, including in organ transplant recipients. *Arthritis Res. Ther.*, **16**, R63.

44. Guidance for Industry: Immunogenicity Assessment for Therapeutic Protein Products. Available at: http://www.fda.gov/downloads/drugs/guidancecomplianceregulatoryinformation/guidances/ucm338856.pdf (accessed on March 6, 2017).

45. Chaffee, S., Mary, A., Stiehm, E. R., Girault, D., Fischer, A., Hershfield, M. S. (1992). IgG antibody response to polyethylene glycol-modified adenosine deaminase in patients with adenosine deaminase deficiency. *J. Clin. Invest.*, **89**, 1643–1651.

46. Fishbane, S., Schiller, B., Locatelli, F., Covic, A. C., Provenzano, R., Wiecek, A., Levin, N. W., Kaplan, M., Macdougall, I. C., Francisco, C., Mayo, M. R., Polu, K. R., Duliege, A. M., Besarab, A. (2013). Peginesatide in patients with anemia undergoing hemodialysis. *N. Engl. J. Med.*, **368**, 307–319.

47. Macdougall, I. C., Provenzano, R., Sharma, A., Spinowitz, B. S., Schmidt, R. J., Pergola, P. E., Zabaneh, R. I., Tong-Starksen, S., Mayo, M. R., Tang, H., Polu, K. R., Duliege, A. M., Fishbane, S. (2013). Peginesatide for anemia in patients with chronic kidney disease not receiving dialysis. *N. Engl. J. Med.*, **368**, 320–332.

48. Eckardt, K. U. (2013). Anaemia: the safety and efficacy of peginesatide in patients with CKD. *Nat. Rev. Nephrol.*, **9**, 192–193.

49. Ganson, N. J., Kelly, S. J., Scarlett, E., Sundy, J. S., Hershfield, M. S. (2006). Control of hyperuricemia in subjects with refractory gout, and induction of antibody against poly(ethylene glycol) (PEG), in a phase I trial of subcutaneous PEGylated urate oxidase. *Arthritis Res. Ther.*, **8** (2006) R12.

50. Brouwers, A. H., De Jong, D. J., Dams, E. T., Oyen, W. J., Boerman, O. C., Laverman, P., Naber, T. H., Storm, G., Corstens, F. H. (2000). Tc-99m-PEG-liposomes for the evaluation of colitis in Crohn's disease. *J. Drug Target*, **8**, 225–233.

51. Szebeni, J. (2005). Complement activation-related pseudoallergy: a new class of drug-induced acute immune toxicity. *Toxicology*, **216**, 106–121.
52. Szebeni, J. (2014). Complement activation-related pseudoallergy: a stress reaction in blood triggered by nanomedicines and biologicals. *Mol. Immunol.*, **61**, 163–173.
53. Garratty, G. (2004). Progress in modulating the RBC membrane to produce transfusable universal/stealth donor RBCs. *Transfus. Med. Rev.*, **18**, 245–256.
54. Szebeni, J., Bedocs, P., Rozsnyay, Z., Weiszhar, Z., Urbanics, R., Rosivall, L., Cohen, R., Garbuzenko, O., Bathori, G., Toth, M., Bunger, R., Barenholz, Y. (2012). Liposome-induced complement activation and related cardiopulmonary distress in pigs: factors promoting reactogenicity of Doxil and AmBisome. *Nanomedicine*, **8**, 176–184.
55. Pieters, R., Hunger, S. P., Boos, J., Rizzari, C., Silverman, L., Baruchel, A., Goekbuget, N., Schrappe, M., Pui, C. H. (2011). L-asparaginase treatment in acute lymphoblastic leukemia: a focus on Erwiniaasparaginase. *Cancer*, **117**, 238–249.
56. Szebeni, J., Baranyi, L., Savay, S., Bodo, M., Morse, D. S., Basta, M., Stahl, G. L., Bunger, R., Alving, C. R. (2000). Liposome-induced pulmonary hypertension: properties and mechanism of a complement-mediated pseudoallergic reaction. *Am. J. Physiol. Heart Circ. Physiol.*, **279**, H1319–1328.
57. Chanan-Khan, A., Szebeni, J., Savay, S., Liebes, L., Rafique, N. M., Alving, C. R., Muggia, F. M. (2003). Complement activation following first exposure to pegylated liposomal doxorubicin (Doxil): possible role in hypersensitivity reactions. *Ann. Oncol.*, **14**, 1430–1437.
58. Cerutti, A., Cols, M., Puga, I. (2013). Marginal zone B cells: virtues of innate-like antibody-producing lymphocytes. *Nat. Rev. Immunol.*, **13**, 118–132.
59. Cheng, T. L., Wu, P. Y., Wu, M. F., Chern, J. W., Roffler, S. R. (1999). Accelerated clearance of polyethylene glycol-modified proteins by anti-polyethylene glycol IgM. *Bioconjug. Chem.*, **10**, 520–528.
60. Saifer, M. G., Williams, L. D., Sobczyk, M. A., Michaels, S. J., Sherman, M. R. (2014). Selectivity of binding of PEGs and PEG-like oligomers to anti-PEG antibodies induced by methoxyPEG-proteins. *Mol. Immunol.*, **57**, 236–246.
61. Verhoef, J. J., Carpenter, J. F., Anchordoquy, T. J., Schellekens, H. (2014). Potential induction of anti-PEG antibodies and complement

activation toward PEGylated therapeutics. *Drug Discov. Today*, **19**, 1945–1952.

62. Sherman, M. R., Saifer, M. G., Perez-Ruiz, F. (2008). PEG-uricase in the management of treatment-resistant gout and hyperuricemia. *Adv. Drug Deliv. Rev.*, **60**, 59–68.

63. Sherman, M. R., Williams, L. D., Sobczyk, M. A., Michaels, S. J., Saifer, M. G. (2012). Role of the methoxy group in immune responses to mPEG-protein conjugates. *Bioconjug. Chem.*, **23**, 485–499.

64. Schellenberger, V., Wang, C. W., Geething, N. C., Spink, B. J., Campbell, A., To, W., Scholle, M. D., Yin, Y., Yao, Y., Bogin, O., Cleland, J. L., Silverman, J., Stemmer, W. P. (2009). A recombinant polypeptide extends the in vivo half-life of peptides and proteins in a tunable manner. *Nat. Biotechnol.*, **27**, 1186–1190.

65. Schlapschy, M., Binder, U., Borger, C., Theobald, I., Wachinger, K., Kisling, S., Haller, D., Skerra, A. (2013). PASylation: a biological alternative to PEGylation for extending the plasma half-life of pharmaceutically active proteins. *Protein Eng. Des. Sel.*, **26**, 489–501.

66. Chen, C., Constantinou, A., Deonarain, M. (2011). Modulating antibody pharmacokinetics using hydrophilic polymers. *Expert Opin. Drug Deliv.*, **8**, 1221–1236.

67. Soshee, A., Zurcher, S., Spencer, N. D., Halperin, A., Nizak, C. (2014). General in vitro method to analyze the interactions of synthetic polymers with human antibody repertoires. *Biomacromolecules*, **15**, 113–121.

68. Specht, C., Schluter, B., Rolfing, M., Bruning, K., Pauels, H. G., Kolsch, E. (2003). Idiotype-specific CD4+CD25+ T suppressor cells prevent, by limiting antibody diversity, the occurrence of anti-dextran antibodies crossreacting with histone H3. *Eur. J. Immunol.*, **33**, 1242–1249.

Chapter 9

Complement Activation: Challenges to Nanomedicine Development

Dennis E. Hourcade, PhD,[a] Christine T. N. Pham, MD,[a] and Gregory M. Lanza, MD, PhD[b]

Divisions of Rheumatology[a] and Cardiology[b], Department of Medicine, Washington University School of Medicine, St. Louis, Missouri, USA

Keywords: alternative pathway, anaphylatoxins, atypical hemolytic uremic syndrome (aHUS), biomaterial, classical pathway, complement, complement activation, complement activation-related pseudoallergy (CARPA), complement cascade, dialysis membrane, hemolysis assay, infection, inflammation, lectin pathway, mannose-binding lectin (MBL), membrane attack complex (MAC), mononuclear phagocytic system (MPS), mouse model, nanomedicine, nanoparticle, nanotoxicities, PEGylated nanoparticle, pig model, pseudoallergic reactions, regulators of complement activation (RCA)

9.1 Introduction

Advances in materials science have made a variety of novel therapeutic devices possible. Prominent among them are nanoparticles that can target specific cells and tissues with imaging agents and drug payloads [1]. The physical nature of nanoparticles,

---

Immune Aspects of Biopharmaceuticals and Nanomedicines
Edited by Raj Bawa, János Szebeni, Thomas J. Webster, and Gerald F. Audette
Copyright © 2018 Pan Stanford Publishing Pte. Ltd.
ISBN 978-981-4774-52-9 (Hardcover), 978-0-203-73153-6 (eBook)
www.panstanford.com

in comparison to that of small molecules commonly employed as drugs or contrast media, raises additional safety issues: Unlike small molecules, nanoparticles are subject to the scrutiny of the immune system. Immune reactions arising from the interactions of nanoparticles with intended targets, blood-borne components or off-target tissues can result in unwanted consequences [2].

The complement system presents a rapid-acting, first-line defense of the intravascular space and other biological compartments from foreign invaders and facilitates the safe removal of apoptotic cells, immune complexes and cellular debris [3]. Complement (C) recognizes potential targets and marks them for clearance and/or lysis and initiates powerful inflammatory reactions. While fluid phase and membrane C regulators can mitigate limited C activation, robust systemic or local activation can overwhelm the regulatory process and pose dangerous health risks. This chapter summarizes the C response to biological targets and clarifies some of the important complement-related issues faced by nanotechnologists as they design next-generation therapeutics.

9.2 C Activation Pathways and Downstream Effectors

Complement activation is mediated by three major pathways (Fig. 9.1) [3, 4]. Each pathway responds to a different set of activators, ensuring that a wide range of dangerous agents are recognized: The classical pathway (CP) is triggered by antibody: antigen complexes, and the lectin pathway (LP) responds to specific carbohydrate moieties. The alternative pathway (AP) operates spontaneously at low levels and is further activated by a range of microbial surfaces. Each activation pathway results in the assembly of the C3 convertases, the central enzymes of the C cascade, which cleave the fluid phase protein C3, producing C3a and C3b. Nascent C3b can bind covalently to the target surface and mediate clearance by the mononuclear phagocytic system (MPS). Target-bound C3b can amplify C activity by providing an assembly point for additional AP convertases (the amplification loop). C3b can also form a complex with preformed C3 convertase to generate C5 convertase, a protease that cleaves C5 into C5a and C5b. C5b initiates the complement terminal pathway, culminating in the assembly of the membrane attack complex (MAC) in the plasma membrane. The MAC promotes cell lysis by disrupting membrane integrity. C3a and C5a, the anaphylatoxins, recruit immune cells that promote further inflammatory reactions [5, 6]. C3d, a C3 derivative, in complex with antigen, provides an adjuvant effect leading to the generation of higher antibody titer [7].

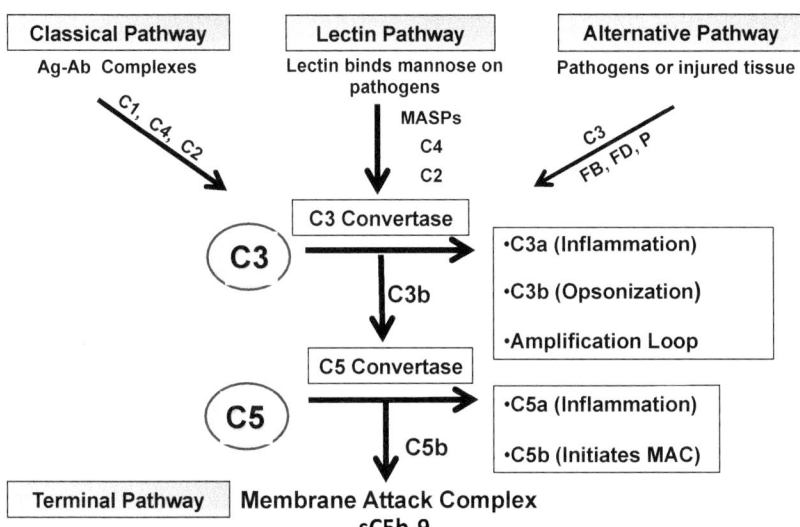

Figure 9.1 The complement cascade. Three activation pathways merge at the step of C3 activation (3, 4). The classical pathway is initiated by the binding of antibody to antigen and the lectin pathway by the binding of lectin to a carbohydrate. The alternative pathway turns over continuously and is further amplified in the presence of pathogens or injured tissue. Activation of the complement system leads to inflammation (release of anaphylatoxins C3a and C5a), opsonization (the coating of targets with C3b or C4b), and formation of the membrane attack complex (MAC). *Abbreviations*: MASP, mannose-binding lectin (MBL)-associated serine protease; FB, Factor B; FD, Factor D; P, properdin. Modified with permission from: Liszewski, M. K., Atkinson, J. P. (2016). Complement pathways. UpToDate, Post TW (Ed), UpToDate, Waltham, MA. For more information visit http://www.uptodate.com (accessed on October 6, 2016.) Copyright © 2016 UpToDate, Inc.

Complement activation can be rapid and robust: Antibody-initiated CP activation can mediate the deposition of 2×10^7 C3b molecules on a cellular target within 5 min [8]. The C cascade is normally held in check on host surfaces by the regulators of C activation (RCA) that inhibit convertase activity and MAC assembly [9]. These include fluid phase proteins Factor H (FH) and C4-binding protein (C4BP) as well as the membrane proteins CD46 (membrane cofactor protein, MCP), decay accelerating factor (DAF) and complement receptor 1 (CR1, CD35). These proteins inhibit C activation via two critical mechanisms: (1) They inhibit convertase activity directly by dissociating the convertase complexes (decay accelerating activity), and (2) they inhibit convertase formation by serving as cofactor for the cleavage of C3b and C4b by the fluid phase protease Factor I (FI) (cofactor activity). In addition, CD59, a membrane-bound non-RCA protein, protects host tissue by blocking the assembly of the MAC. These regulators, plus the intracellular pathways that respond to membrane perturbation [10], normally prevent the lysis of nucleated cells.

9.3 Role of C in Human Health and Disease

9.3.1 Complement Roles in Fighting Infection

The C system plays a critical role in the recognition and elimination of pathogenic microorganisms [11]. Individuals who lack C3, the central C protein, are at risk for infection by several major pathogens including *Streptococcus pneumonia*, *Neisseria meningitidis*, and *Haemophilus influenza* [12]. Complement activation directs the removal of invasive bacteria through two major mechanisms: (1) the clearance of opsonized bacteria by phagocytes and (2) the disruption of bacterial cell wall integrity by the MAC, leading to bacterial lysis.

Bacteria are divided into two taxonomic groups that differ markedly in their cell wall structures, the Gram-positive and Gram-negative bacteria [13]. In Gram-positive bacteria, the cytoplasmic membrane is immediately surrounded by a layer of peptidoglycan, a thin sheet of parallel glycan chains connected together with peptide crosslinks. The peptidoglycan sheet is in turn surrounded by a thick outer layer of acidic polysaccharides called teichoic acids. While this outer layer is subject to antibody (Ab)-dependent and Ab-independent CP activity, it is resistant to the assembly and function of the MAC. Therefore, the primary C-dependent host defense against Gram-positive pathogen such as *S. pneumonia* is via C3 opsonization followed by phagocytic clearance [14].

The cell envelope of Gram-negative bacteria contains two membranes, an inner cytoplasmic membrane and an outer membrane. Between them resides the peptidoglycan in a region known as the periplasm. The outer membrane contains a layer of lipopolysaccharides, or LPS, which extends into the cell exterior. The effect of C on Gram-negative bacteria is modulated by variations in these outer structures. Most Gram-negative bacteria can engage the CP and/or the AP and are subject to clearance by the MPS, but some, like *N. meningitidis,* the agent of meningococcal disease, are particularly vulnerable to the MAC. Those deficient in C alternative and terminal pathway components are highly prone to repeated infection by *N. meningitidis* [12].

9.3.2 Complement Regulation and Tissue Injury

Complement activity fluctuates between activation and regulation: When C recognizes a target, it ramps up its activity, clears and/or destroys the target, and then damps down to its basal level. Complement activity is generally limited to the target area, held in check by the RCA proteins. Complement regulatory failure can lead to serious consequences. Atypical hemolytic uremic syndrome (aHUS) is a severe kidney condition triggered by acute cell-based injury [15]. It is characterized by microangiopathic hemolytic anemia (red blood cells are fragmented as they flow through small blood vessels), thrombocytopenia (platelets are consumed in the microthrombi) and renal failure (fibrin-rich clots form in the renal glomerulus). aHUS is associated with an acute mortality rate of up to 25% and the development of permanent kidney damage requiring long-term dialysis in ~50% of patients. It occurs most often in individuals carrying a rare loss-of-function mutation in one of the AP regulatory proteins: FH, CD46 or FI. These individuals are haplo-insufficient (i.e., they are deficient in only one of two alleles), suggesting a tight and delicate balance between C activity and C regulation.

Complement activation has also long been implicated in immune complex-mediated diseases such as systemic lupus erythematosus (SLE) and various forms of vasculitides [16]. Although autoantibody-antigen-mediated tissue damage via the CP plays a major role in the pathogenesis of SLE, the AP and LP are increasingly recognized as important players in vasculitides [16], suggesting that the excessive activation of any of the C pathways may pose significant health risk.

9.4 Complement Activation by Biomaterials

Prior to the emergence of nanomedicine, numerous clinical procedures were in place that interfaced manufactured materials ("biomaterials") with biological tissues and/or fluids. Some of the outcomes have relevance to current nanomedicine design: Biomaterials generally lack the C regulatory capacity of cells and tissues and in many cases biomaterials have been shown to activate C [17]. During hemodialysis (HD), for example, patient blood exposed to a large surface area of dialysis membrane can result in profound C activation [18, 19]. The anaphylatoxins that are produced then engage receptors on neutrophils, mast cells, monocytes and other cells of myeloid origin, setting in motion multiple downstream events, including the release of potent pro-inflammatory cytokines and chemokines [5, 6]. The resulting chronic inflammation contributes to numerous adverse conditions, including increased risk for atherosclerosis and myocardial infarction [20], and strongly correlates with mortality in end-stage renal disease [21].

Although it is not fully understood how biomaterials trigger C activation, it is generally regarded to require the CP, with the AP serving to amplify the initial signal [22–24]. While the CP system utilizes specific antibody:antigen interactions and specialized pattern recognition mechanisms to identify native macromolecules, a biomaterial may activate the CP by a more complex route. For example, when a biomaterial surface such as polystyrene is exposed to the blood components it is immediately coated with a thin (~8 nm) film composed of a monolayer of plasma proteins that trigger C activation and convertase formation [22].

It has been proposed that interactions with artificial biomaterial surfaces (i.e., stents or implantable pumps) induce conformational changes or lead to the denaturation of plasma proteins. The composition and conformation of these altered adsorbed proteins, and not simply the biomaterial itself, trigger C [25]. Complement-dependent target recognition is mediated by conformational changes in its key recognition molecules (C1, C3, C4, antibody). It is possible that similar functional conformations are promoted when these molecules incorporate in the primary film or bind to the primary film on biomaterials.

9.5 Nanomedicine-Mediated C Activation

Some of the first indications that nanoparticles can activate C came from the clinical use of the anti-cancer drug Doxil®, a PEGylated liposomal formulation of doxorubicin (reviewed by Szebeni [26]). A fraction of clinical patients exhibited signs of cardiopulmonary distress that developed immediately after the start of infusion, including dyspnea, tachypnea, tachycardia, hypotension, chest pain, and back pain. Unlike IgE-mediated allergic reactions, the symptoms occurred within minutes upon first exposure to the drug, arguing against an adaptive (i.e., humoral) immune response. These non-IgE, "pseudoallergic" symptoms were also observed in rats and pigs used to investigate hemoglobin-containing liposomes as a potential red blood cell substitute [27, 28]. That C activation is the cause of these pseudoallergic reactions is supported by the close correlation of Doxil® dosage with severity of symptoms and evidence of C consumption, coinciding with a rise in the plasma concentration of C activation products [29]. Moreover, it was shown in pigs that C-specific inhibitors largely suppress the pseudoallergic symptoms further confirming that C activation plays a causal role in this response, also known as C activation-related pseudoallergy or CARPA [27, 30]. Since that time, the spectrum of reagents that cause pseudoallergic symptoms has been broadened to include micelle-solubilized drugs, certain antibodies, PEGylated proteins, contrast media and additional liposomal drugs [26].

9.5.1 Aspects of Nanomedicines That Activate/Modulate C

The C system employs several proteins to distinguish its biological targets. Many of these same recognition molecules and their respective activation pathways, including antibody and C1 (CP), mannose-binding lectin (MBL) (LP), and C3 (AP), have been shown to direct NP-dependent C activation. In general, the mechanisms that underlie NP recognition have not been fully clarified. It has been speculated that the polymers on the surface of liposomes, for example, may present pattern recognition receptors (PRRs) that serve as binding ligands for C protein [31]. Alternatively, as with other biomaterials, once introduced into a biological space nanoparticles are rapidly coated with a "corona" of adsorbed proteins, C proteins and antibodies among them. Fortuitous association of a recognition molecule to surface charge or chemical group could promote conformational changes that initiate the C cascade. Indeed, surface charge is known to play a critical role in the interaction of NPs with C: Nanoparticles bearing high surface charge, either positive or negative, strongly activate C, while those near neutrality have less activity. This can be an important issue because NPs are often designed with a positive or negative net overall surface charge in order to reduce self-aggregation. Other factors that modulate NP:C activity include particle size and curvature (reviewed by Szebeni [32]).

Much effort has been devoted to making "stealth" NPs that would travel "under the radar" of the immune system. NP coatings that can resist protein adsorption have been developed for this purpose. Many existing therapeutics are coated by covalently attached PEG chains, treatment that markedly extends NP circulation half-life. PEG effects on protein adsorption are dependent on numerous factors (reviewed by Salvador-Morales and Sim [33]). Briefly, at low surface chain density, PEG chains arrange in a "mushroom" conformation that provides low coverage of the underlying NP structure. At high surface chain density, PEG chains can take on a "brush" conformation that provides greater coverage [34]. The greater coverage afforded by the PEG brush conformation confers greater circulation lifetime, due to reduced uptake by the MPS, but it does not preclude C activation and may interfere with NP function. In principle, PEG terminal hydroxyl groups may present sites for covalent attachment to nascent C3b, leading to C activation in a non-Ab-mediated mechanism. Some investigations have introduced modified PEG chains or different polymers in their NP designs in order to reduce C activation. In some cases these alterations have reduced C activity [35, 36], while in other instances, these modifications induced a switch in the operating activation pathway [37]. Further, studies have shown the occurrence of anti-PEG antibodies, usually IgM, in patients without previous exposure to PEGylated therapeutics, that may trigger C activation [38]. The presence of these anti-PEG antibodies may be related an a priori exposure to PEG through daily commodities such as cosmetics and food products [39].

9.6 Current Methods to Measure Nanomedicine-Mediated C Activation

9.6.1 *In vitro* Assays

Use of Serum and Plasma. The activation of complement on a NP surface leads to the assembly of surface-bound C3 convertases that cleave C3 producing the C3a and C3b fragments. Most of the nascent C3b will diffuse into the fluid phase where it initiates the assembly of fluid phase convertases or is processed by FH/FI to soluble iC3b. A smaller fraction of nascent C3b will bind to the NP surface where it initiates the assembly of surface-bound convertases or is processed to surface-bound iC3b. These and other NP:C interactions have routinely been evaluated *in vitro* [40] by monitoring the generation of C activation products or the loss of C functional capacity when NPs are incubated in serum or plasma. Several important points must be kept in mind when designing these experiments (see the detailed discussion by Lachmann [41]). Among them:

(1) Human serum should be stored frozen at −80°C or lower and is best not thawed more than once.
(2) Serum depleted of one or more C components, produced in house or procured commercially, is often used to determine which C pathway is active. In these instances, the

recovery of full C activity must be demonstrated when the depleted serum is supplemented with the purified missing component(s).

(3) If plasma or whole blood is used, clotting should be inhibited with Hirudin or its derivatives, and not heparin, a C inhibitor.

(4) Mouse CP activity is extremely unstable and cannot be preserved with frozen serum. Best practice is to keep fresh mouse serum on ice to be used as soon as possible on the same day of collection.

(5) Mouse C proteins exhibit gender differences, with higher levels in males compared to females. Additionally, there are marked strain differences in C activity, C4 in particular.

ELISA. Many investigations have used commercially available ELISA kits to assay for the presence of C activation products following exposure to NPs *in vitro* [35, 37, 42–44]. In principle, C3a should be an excellent marker of *in vitro* NP-dependent C activity, since it is a highly amplified end-product of the cascade [8]. In practice, however, it is well established that *in vitro* incubation of serum or plasma produces a non-specific release of C3a due to the constant turnover of the AP. C3a generated *in vitro* is not cleared, as it is *in vivo*, and accumulates over the course of the reaction. The magnitude of the non-specific C3a release can be quite substantial.

During the course of our investigations of C activity associated with functionalized phospholipid nanoparticles, we observed spontaneous release of C3a within 15 min equivalent to the turnover of ~30% of all available C3 [45]. The NP-independent C3a "background" was indeed due to AP turnover during the incubation period. It was dependent on Factor B, an essential AP protein, and was not observed with heat-inactivated serum or with fresh serum. The non-specific C3a release resulted in a maximum signal that was 2–3 folds above background, thus greatly limiting assay sensitivity. Non-specific release assayed with the C4d ELISA was not observed, which was anticipated since C4 cleavage does not occur during spontaneous C3 turnover. Because C4d is not produced by NP-dependent AP activity, the use of C4d generation cannot provide a complete representation of C activity.

The terminal pathway provides an alternative ELISA target, soluble C5b-9 (sC5b-9, a nonfunctional byproduct of C5 convertase activity [46]), which has been regarded as more reliable than C3a as a marker of NP-dependent C activity. It is formed in the fluid phase by the association of the soluble C5b-7 complex with Vitronectin (Protein S) and subsequently with C8 and C9. However, in our experience the sensitivity of the sC5b-9 ELISA is also hampered by non-specific sC5b-9 production [45]. Importantly, two distinct examples of functionalized phospholipid NPs were studied that yielded little sC5b-9 production; yet they were highly C reactive using several other assays [45]. These data, which included C4d and C5a ELISA, PAGE Western blot analysis, hemolysis assays (see below), and mouse *in vivo* assays, indicated that in these two examples the CP was engaged, CP C3 convertases were assembled, much C3 was cleaved, but little C5 convertase activity occurred. These findings suggest that the functionalized phospholipid NP surfaces may not readily support C5 convertase assembly.

Hemolytic assay. Nanoparticle-dependent C activation can be evaluated indirectly by measuring the diminished C activity of NP-treated serum. The hemolysis assay has been used for decades to measure the C activity of human serum [47]. As such it has been of clinical importance in the recognition of C deficiencies. In practice, it is a titration procedure that defines the serum dilution that results in 50% lysis of Ab-sensitized sheep red blood cells (the CH50 metric) [47]. Several investigations have adapted the hemolysis assay to assess *in vitro* nanoparticle-dependent C activity [45, 48–50]. Hemolysis is dependent on the classical and terminal pathway proteins, including C1 through C9 (Fig. 9.1). The activation of C by NPs via any of the three activation C pathways will cause irreversible loss of C3, and C5 through C9, resulting in diminished residual hemolytic capacity. Since many of the C proteins that facilitate hemolysis are depleted by NP-dependent C activation, the pre-incubation of a serum sample with a C-reactive NP will diminish the subsequent hemolytic activity of that serum.

We developed a modified hemolysis assay to evaluate NP: C activation [45, 51]. The technique sensitivity was optimized to conform to parallel *in vivo* findings. A validation procedure with control particles and control reactions was established and new metric to quantify NP-dependent C activation was introduced. The non-specific background signal was relatively low, possibly because many of the critical CP proteins (C1, C2, C4) are not depleted by spontaneous AP turnover. The resulting assay was more sensitive in our experience than the ELISAs [45]. The hemolysis assay, however, had some disadvantages:

(1) Instability of the Ab-sensitized sheep erythrocytes necessitates use within 2 weeks following preparation.
(2) The C sensitivity of Ab-sensitized sheep erythrocytes varies from batch to batch and diminishes with prolonged storage.

While these factors were addressed with proper controls, they may still present challenges for normalizing and standardizing inter-lab assays.

PAGE/Western blot. Following the incubation of NPs with serum or plasma, NPs and/or supernatant fractions can also be examined by PAGE and Western blot for the presence of C3 and its activation fragments, especially iC3b [40, 51]. This method can be used to examine reaction kinetics or it can be adapted to animal models. The PAGE/Western blot approach is not generally conducive to precise quantitation and so the data obtained by this method are of a qualitative nature. Moreover, the PAGE/Western blot analysis of the supernatant fraction is also limited by fluid phase C3 fragments produced by spontaneous NP-independent AP activation (see above). The method is attractive, however, because the C3 fragments are identified by both Ab specificity and by relative migration rate. Moreover, concurrent analysis of the NP and the supernatant fractions derived from the same reactions can provide additional information.

Further obstacles to in vitro assays. In vitro C assays have well-understood intrinsic shortcomings, as discussed above. Some NPs, however, may be incompatible with certain assays, a situation that may be difficult to recognize. For example, if C3a is adsorbed to a specific NP, those assays that quantify C3a in the fluid phase will underestimate C activity. Conversely, NPs may support C activation but resist events further downstream [45, 50]. Beyond early screening, the complete assessment of NP:C interactions for those nanomedicine agents considered for clinical translation is best addressed with a panel of independent assay methods. Moreover, correlation with *in vivo* models (below) may be especially important in determining the extent of NP-mediated C activation.

9.6.2 *In vivo* Assays

9.6.2.1 Species selection for *in vivo* evaluation of NP-mediated C activation

When considering the use of animals to evaluate nanomedicine agents, the strengths and weaknesses of preclinical models must be considered in depth. Although the gross organ features of large- and small-animal models may reasonably resemble the human body, underlying differences in microscopic anatomy and function can have profound and unanticipated influence on the data acquired and its interpretation. Beyond the concerns of small-molecule medicines, which primarily exert their toxicity based on the biodistribution, metabolism, and biochemical activity of a drug and its metabolites, nanoparticles also include a physical aspect that differs among animal species and must be considered in the context of healthy and disease states within each species.

For example, in the mid-1970s, nuclear radiologists appreciated and reported that lung uptake during liver scanning with ^{99m}Tc sulfur colloid was substantially increased in patients with liver disease from a nominal 1–2% background level [52]. Under physiological conditions guinea pigs and toads had minimal pulmonary uptake of the nuclear colloid, but after substantial dosing with estrogens, markedly increased vascular lung uptake was appreciated. This alternative redistribution of nuclear signal was attributed to the emergence of pulmonary intravascular macrophages (PIMs). The authors suggested that the estrogenic compounds stimulated the reticuloendothelial system and induced monocyte proliferation and redistribution into the lung along with the particle clearance function of the liver. Many years later, other investigators performed biliary ligations to induce severe liver cholestasis in rats and showed that the decrease in liver particle clearance functionality was compensated for by the emergence of PIMs, which they suggested also led to increased sensitivity to endotoxin [53, 54]. PIMs are not normally present in human and rat pulmonary vasculature, but research dating into the 1980s has shown that the species of the order *Artiodactyla* (even-toed ungulates) naturally and abundantly harbor PIMs. These species utilize the lung as a primary clearance organ for bacteria and particles rather than the liver and spleen. In 1986, the biodistribution of radiolabeled colloidal gold and magnetic iron oxide nanoparticles in rats and calves was studied following intravenous injection. In rats, the particles were cleared from circulation predominantly by the liver, as expected, but in calves, the particle localized predominantly into the lungs. Extending the experiment to goats revealed similar findings. Electron microscopic studies localized the iron oxide particles in the calf lung to PIMs. From these initial data, the authors hypothesized that ruminants, unlike rodents and man, cleared particles from circulation predominantly through a PIM-mediated process as opposed to the spleen and liver MPS. The presence of PIMs and their confirmation and characterization as a clearance mechanism in ruminants were further extended to sheep and horses [55–63].

During the early 1990s, particles were under development to address compelling unmet medical needs. One such particle was Albunex®, the first successful gas-filled microbubble capable of intact pulmonary transit. Albunex® and subsequent microbubbles revolutionized cardiac ultrasound imaging by improving the definition of the endocardial blood border, which facilitated more sensitive and specific assessments of myocardial wall motion than had been previously possible [64]. In monkeys and rabbits, Albunex® provides cardiac acoustic enhancement with negligible side effects. In pigs, however, exposure to microbubbles markedly induced acute pulmonary hypertension. Pulmonary artery pressures increased in 22 s from a baseline of 17 mmHg to 42 mmHg with corresponding declines in blood pressure. The investigators suggested that the acute pulmonary hypertension in pigs was mediated by thromboxane A_2 (TXA_2) released from PIMs.

Similar work in swine involved *Oxygent*™, a fluorocarbon emulsion that was studied as an artificial blood substitute [65]. In anesthetized and conscious pigs given *Oxygent*™, mean pulmonary pressures increased acutely to 38 mmHg with associated skin flushing. The acute increase in pulmonary hypertension was attributed to TXA_2, whereas the "niacin-like" flushing was suggested to result from the release prostaglandins. Both are vasoactive arachidonic acid derivatives from cyclooxygenase metabolism, capable of eliciting pulmonary artery and venous constriction. Nonsteroidal anti-inflammatory drugs, such as ibuprofen and high-dose dexamethasone, were shown to blunt this acute pulmonary response, while anti-histamines, typically used for allergic and anaphylactoid immune responses, were ineffective [65]. In contrast to the muscular left ventricle, which can withstand increases in systemic blood pressure of 20 to 40 mmHg without secondary adverse effects on the cardiac stroke volume, similar acute increases in pulmonary hypertension can lead to a rapid failure of the thin-walled right ventricle [66]. Such extreme hemodynamic fluctuation may activate compensatory stress mechanisms that potentially induce a cytokine storm.

The pig model of CARPA is highly sensitive and remarkably reproducible [67, 68]. Although pigs and rats both respond to reactogenic liposomes with massive hemodynamic changes, data suggest that rats are 2–3 orders of magnitude less sensitive to liposomal-induced pseudoallergy than pigs. In addition, the changes in pulmonary vascular pressures in response to certain liposomes, such as AmBisome (liposomal amphotericin-B), remain remarkably similar in pigs [67, 68]. Soluble C receptor type 1 (sCR1), a known inhibitor of C, can completely block while anti-C5 antibody partially inhibits the pseudoallergic response, strongly suggesting that C activation is at least partially responsible for the observed acute and dramatic hemodynamic changes. The exquisite sensitivity of pigs to reactogenic liposomes points to a potential involvement of PIMs, which are present in large number in pigs [69]. These firmly attached, residential pulmonary cells express PRRs that may recognize repeating patterns found in PEG and polymers on the surface of liposomes and other particles, leading to a rapid release of inflammatory mediators, including TXA_2 [70, 71].

Although rodents are more similar to humans with regard to the influence of PIMs on *in vivo* nanoparticle assessments, they differ substantially in systemic particle clearance, which can influence the dosimetry of pharmacokinetic and safety assessments. Bulte et al. [72] recently reported the transit of large (250 nm) paramagnetic perfluorocarbon nanoparticles rapidly into the bile and gut in rats, which he imaged beginning 5 min post intravenous injection as an MRI cholangiogram (Fig. 9.1). While the MR imaging of this bioelimination route for intact NP had never been previously envisioned, the scientific appreciation of this excretion phenomena in rodents occurred decades ago. In 1958, Hampton [73] reported that colloidal particles (8% HgS in 500 μL or 25% ThO_2, Thorotrast) injected intravenously localized to the liver and spleen with transmission electron microscopy (TEM). Thorotrast particles were found in Kupffer cells, hepatocytes, and the biliary tree. Particles phagocytosed by Kupffer cells were retained indefinitely, but particles endocytosed by hepatocytes were transported rapidly into the biliary system as seen by Bulte et al. [72] by MRI. To eliminate the potential influence of systemic protein coating or corona on the handling of the particles within the liver, biliary retrograde infusions of Thorotrast colloidal particles were employed. TEM revealed that the biliary infused particles transited in reverse through hepatocytes into the space of Disse. Although some of the Thorotrast particles sequestered into Kupffer cells, much passed directly into the circulatory system. Similar subsequent evidence of NP excretion into rodent bile and feces includes silica particles from 50 to 200 nm [74], citrate-coated silver particles (~8 nm) [75], and iron oxide core high-density lipoproteins (~10 nm) [76]. In contrast to rodents, Juhlin reported in 1960 [77] that fluorescent spherical hydrophilic particles of methyl methacrylate injected intravenously in rabbits did not transit effectively into the bile, similar to the human result. Intravenous injection of particles (20 to 110 nm) produced a minimal biliary concentration and those over 60–140 nm were not appreciated in the bile. Particles in the range of 200–800 nm in diameter were not found in the bile despite a ~10-fold increase in systemic dose.

Rodents are frequently used to characterize the pharmacokinetics, pharmacodynamics, biodistribution, and bioelimination of nanomedicines for basic science and IND regulatory purposes. While rodent models offer a wealth of research opportunity, the data achieved with their use in NP-mediated responses must be viewed with an appreciation for the potential impact that rapid biliary clearance of particles may contribute. Rapid biliary excretion in rodents lowers the toxicity burden of particles that would otherwise be retained, leaving a better impression of biosafety than exists. Moreover, allometric scaling [78] to project NP pharmacokinetic, pharmacodynamics, and toxicity data from rodents to man may be compromised, since the concept of shape scaling implies that the physiological pathways are conserved. Indeed, all research with NP in preclinical animal models must be considered with regard to the potential influences of PIM density in the pulmonary vasculature and the rapid biliary clearance of large NP from the circulation into the gut, neither of which is prevalent in man.

9.6.2.2 NP:C interactions in the mouse model

The mouse is an often-studied animal model of C activation and regulation and there are several examples in which the mouse model system has been instrumental in elucidating the involvement of C in disease pathogenesis. While mice have not been shown to be good models of human C-mediated pseudoallergic reactions (see above), they still provide valuable insights into NP:C events, including NP-dependent C activation, NP opsonization, and NP uptake by the MPS, that may trigger profound physiological responses in humans. Recent investigations have shown remarkable similarities at these levels between the mouse and the human C response [45, 51, 79].

There are several advantages of using mouse models as investigative tools to elucidate NP:C effects. Soluble C activation products generated *in vivo* (i.e., iC3b, C3a) can be detected ex vivo by ELISA or PAGE/Western Blot without the non-specific background and potential adsorption issues that can complicate the *in vitro* assays [45, 51]. In some cases, NPs and phagocytic cells can be recovered from treated animals and NP opsonization and uptake can be examined *ex vivo* [79]. Numerous C-deficient mouse lines and neutralizing antibodies are available to facilitate investigations that focus on the C proteins that underlie NP:C interactions. On that note, mouse strains lacking B cells (the source of Ab) have been used to elucidate the contribution of natural Ab to NP-mediated C activation [51]. The parallels between mouse and human C, while considerable, do have their limits: Of particular note, the mouse lacks the classical pathway/lectin pathway C5 convertase [80]; failure to detect NP-dependent sC5b-9 and C5 cleavage products in mouse may not be generalized to other species. Also, the repertoire of mouse C regulators differs from that of humans although their regulatory mechanisms, decay acceleration activity, and cofactor activity are similar [81].

9.6.2.3 Individual differences

While NP:C interactions are dependent on the physical parameters of the NP, they also vary in intensity with the individual recipient. Chanan-Khan et al. [29], in their study of hypersensitivity reactions (HSR) to Doxil® treatment, demonstrated a wide variation in plasma sC5b-9 levels in cancer patients that correlated to HSR intensity. Variations in NP:C activity have also been observed *in vitro* [82, 83]. Using a hemolysis assay of NP:C interactions in a relatively small number of sera derived from healthy donors, we found some individual sera reacted much more vigorously to specific NPs than other serum samples. Moreover, significant variations in NP-dependent C activity between individuals of different race, gender, and age were observed [45].

The basis for NP:C variations are not generally understood, but several factors would likely play a part. Individual Ab repertoire would be expected to play an important role. There is also well-established variation in the potency of the C activation pathways and C regulators: Serum levels of the highly polymorphic LP recognition protein MBL can vary over 100-fold within the general population [84]. There are common and rare loss-of-function FH, FI, and CD46 genetic variants and gain-of-function FB and C3 genetic variants that would result in greater NP-dependent C activation and/or tissue vulnerability. An individual's state of health must be taken into account when the consequences of NP:C activity are considered. Chronic inflammation associated with NP:C interactions could exacerbate numerous pathologic conditions [5] and recent studies have indicated that C anaphylatoxins may promote tumorigenesis [85].

9.7 Conclusions

Unlike small molecules, NPs are subject to the scrutiny of the immune system. The activation of C, a major defender of the biological spaces, results in rapid target opsonization and uptake by the MPS, assembly of lytic factors, and the release of inflammatory mediators that can result in detrimental consequences. The mechanisms that underlie the activation of C by nanomaterials are not well defined and depend on both the NP structure and individual factors. *In vitro* and *in vivo* assays of NP:C interactions are valuable investigative and diagnostic tools, but all have drawbacks. When considering the use of animals to evaluate nanomedicine agents, the strengths and weaknesses of preclinical models must be considered in depth. Major challenges include the design of next-generation "stealth" NPs that can avoid C interactions and the development of strategies to safely identify individuals vulnerable to atypical NP:C activity.

Disclosures and Conflict of Interest

Dr. Hourcade would like to thank Dr. John Atkinson for his insightful comments and helpful critiques and Ms. Madonna Bogacki for excellent administrative assistance. This work was partially supported by NIH R01 AI051436 and NIH/FDA U01 NS073457. The authors have no conflict of interest to disclose. The authors have received no payment for the preparation of this chapter.

Corresponding Author

Dr. Dennis E. Hourcade
Division of Rheumatology
Washington University School of Medicine
660 South Euclid Avenue, Campus Box 8045
St. Louis, MO 63110, USA
Email: dhourcade@wustl.edu

About the Authors

Dennis Hourcade is professor of medicine at Washington University School of Medicine (WUSM) in the Division of Rheumatology, and director of the Protein Purification and Production Facility of the WUSM Rheumatic Diseases Core Center. He earned a BS in chemistry from the Massachusetts Institute of Technology and a PhD in biochemistry and molecular biology at Harvard University. His PhD dissertation was directed to the mechanisms that underlie DNA replication. Dr. Hourcade joined Dr. John Atkinson's HHMI lab at WUSM in 1986 to apply the techniques and mindset of the molecular biologist to the complement system. His investigations focused on the structure/function and evolution of the immune adherence receptor, CR1, and the genetic region that encodes the regulators of complement activation, the RCA cluster. In 1993, Dr. Hourcade joined the WUSM faculty and began long-term NIH-supported investigations of complement activation, in particular the assembly, regulation, and function of the alternative pathway convertases and the role of complement in autoimmune and inflammatory diseases. In 2009, he began collaborations with

Drs. Pham and Lanza that addressed the activation of complement by nanoparticles. They developed *in vitro* and *in vivo* models and used them to examine the complement activity of a large panel of functionalized phospholipid nanoparticles, and with collaborators at WUSM and the Pacific Northwest National Laboratory and the support of the NIH and FDA, they pioneered the bioinformatics analysis of NP:C interactions.

Christine Pham is associate professor of medicine at Washington University School of Medicine, Division of Rheumatology. She is a practicing rheumatologist and physician scientist with significant expertise in animal models of immune- and autoimmune-mediated inflammatory conditions with active NIH- and VA-sponsored research programs at Washington University and John Cochran VAMC in Saint Louis. She earned her MD at the University of Florida graduating as valedictorian of her class and pursued postgraduate residency training in internal medicine at Barnes Hospital in Saint Louis and clinical fellowship in Rheumatology at Washington University. Under the mentorship of Dr. Timothy Ley, Division of Bone Marrow Transplant at Washington University, Dr. Pham developed her own research program focusing on the role of innate immunity and proteases in immune/autoimmune diseases, with a special emphasis on inflammatory arthritides. She has collaborated extensively with Dr. Lanza to pioneer work in nanotherapeutics for the treatment of inflammatory joint diseases. Together with Dr. Hourcade, she developed the mouse model to test for nanoparticle-mediated complement activation. Dr. Pham previously served on the editorial board of the journal *Nanomedicine: Nanotechnology, Biology and Medicine* (2010 to 2015) and as a regular member of the NIH Nano Study Section (2009 to 2014).

Gregory Lanza is a professor of medicine and a board-certified cardiologist, with adjunctive appointments in biomedical engineering and biological sciences. He is the Oliver M. Langenberg Distinguished Professor of the Science and Practice of Medicine. He is the director of the Consortium for Translation Research in Advanced Imaging and Nanomedicine (C-TRAIN) and a member of the prestigious Siteman Cancer Center at Washington University in St. Louis. Dr. Lanza is an established NIH principle investigator with over 250 peer-reviewed published manuscripts, 30 US issued patents, and over 140 invited presentations since the turn of the century. He is a standing member of NIH grant review groups, a scientific advisor to corporations, government, academic programs and respected journals. Dr. Lanza received a BA from Colby College in Waterville, Maine, and a MS and PhD from the University of Georgia in 1981. He joined Monsanto Company in 1981, where as the preclinical research manager, he participated in the development of the first 14-day parenteral, controlled-release product for a 20,000 Kd protein (i.e., BST), now successfully marketed and sold as Posilac® by Elanco. He matriculated at Northwestern University Medical School in Chicago and received an MD degree in 1992. Dr. Lanza completed his Internal Medicine residency and Cardiology fellowship at Washington University (1994–1999). During his tenure at WU, he has co-invented and out-licensed numerous nanomedicine-based technologies for imaging and drug delivery. He co-founded Kereos, Inc, which brought two $\alpha v \beta 3$-targeted MR diagnostic agents to the clinic under US and ex-US INDs for cancer and atherosclerosis applications, helped to establish Washington University Center for Multiple Myeloma Nanotechnology program, founded Capella Imaging, LLC, developing an anti-fibrin nuclear agent to assist ventricular assist device management, and co-invented a new functional synthetic blood substitute, ErythroMer, now licensed to KaloCyte, LLC for development.

References

1. Petros, R. A., DeSimone, J. M. (2010). Strategies in the design of nanoparticles for therapeutic applications. *Nat. Rev. Drug Discov.*, **9**(8), 615–627.
2. Dobrovolskaia, M. A., McNeil, S. E. (2007). Immunological properties of engineered nanomaterials. *Nat. Nanotechnol.*, **2**(8), 469–478.
3. Ricklin, D., Hajishengallis, G., Yang, K., Lambris, J. D. (2010). Complement: A key system for immune surveillance and homeostasis. *Nat. Immunol.*, **11**(9), 785–797.
4. Volanakis, J. E. (1998). Overview of the complement system. In: Volanakis, J. E., Frank, M. M., eds. *The Human Complement System*

in *Health and Disease*, 10th ed., Marcel Dekker, Inc., New York, chapter 2, pp. 9–32.

5. Ricklin, D., Lambris, J. D. (2013). Complement in immune and inflammatory disorders: Pathophysiological mechanisms. *J. Immunol.*, **190**(8), 3831–3838.

6. Klos, A., Tenner, A. J., Johswich, K. O., Ager, R. R., Reis, E. S., Kohl, J. (2009). The role of the anaphylatoxins in health and disease. *Mol. Immunol.*, **46**(14), 2753–2766.

7. Fearon, D. T., Locksley, R. M. (1996). The instructive role of innate immunity in the acquired immune response. *Science*, **272**, 50–54.

8. Ollert, M. W., Kadlec, J. V., David, K., Petrella, E. C., Bredehorst, R., Vogel, C. W. (1994). Antibody-mediated complement activation on nucleated cells. A quantitative analysis of the individual reaction steps. *J. Immunol.*, **153**(5), 2213–2221.

9. Liszewski, M. K., Atkinson, J. P. (1998). Regulatory proteins of complement. In: Volanakis, J. E., Frank, M. M., eds. *The Human Complement System in Health and Disease*, 10th ed., Marcel Dekker, Inc., New York, chapter 7.

10. Triantafilou, K., Hughes, T. R., Triantafilou, M., Morgan, B. P. (2013). The complement membrane attack complex triggers intracellular Ca^{2+} fluxes leading to NLRP3 inflammasome activation. *J. Cell Sci.*, **126**(Pt 13), 2903–2913.

11. Walport, M. J. (2001). Complement. First of two parts. *N. Engl. J. Med.*, **344**(14), 1058–1066.

12. Figueroa, J. E., Densen, P. (1991). Infectious diseases associated with complement deficiencies. *Clin. Microbiol. Rev.*, **4**(3), 359–395.

13. Madigan, M. T., Martinko, J. M., Parker, J. (1997). Brock biology of microorganisms. In: Corey, P. F., ed. *Biology of Microorganisms*, Simon & Schuster/AViacom Company, Upper Saddle River, New Jersey.

14. Janoff, E. N., Rubins, J. B. (2000). *Invasive Pneumococcal Disease in the Immunocompromised Host*, Mary Ann Liebert, Inc, Larchmont, NY.

15. Liszewski, M. K., Atkinson, J. P. (2015). Complement regulators in human disease: Lessons from modern genetics. *J. Intern. Med.*, **277**(3), 294–305.

16. Chen, M., Daha, M. R., Kallenberg, C. G. (2010). The complement system in systemic autoimmune disease. *J. Autoimmun.*, **34**(3), J276–286.

17. Nilsson, B., Korsgren, O., Lambris, J. D., Ekdahl, K. N. (2010). Can cells and biomaterials in therapeutic medicine be shielded from innate immune recognition? *Trends Immunol.*, **31**(1), 32–38.
18. Ivanovich, P., Chenoweth, D. E., Schmidt, R., Klinkmann, H., Boxer, L. A., Jacob, H. S., Hammerschmidt, D. E. (1983). Symptoms and activation of granulocytes and complement with two dialysis membranes. *Kidney Int.*, **24**(6), 758–763.
19. Reis, E. S., DeAngelis, R. A., Chen, H., Resuello, R. R., Ricklin, D., Lambris, J. D. (2015). Therapeutic C3 inhibitor Cp40 abrogates complement activation induced by modern hemodialysis filters. *Immunobiology*, **220**(4), 476–482.
20. Galli, F. (2007). Protein damage and inflammation in uraemia and dialysis patients. *Nephrol. Dial. Transplant.*, **22**(Suppl 5), v20–36.
21. Stenvinkel, P., Alvestrand, A. (2002). Inflammation in end-stage renal disease: Sources, consequences, and therapy. *Semin. Dial.*, **15**(5), 329–337.
22. Andersson, J., Ekdahl, K. N., Lambris, J. D., Nilsson, B. (2005). Binding of C3 fragments on top of adsorbed plasma proteins during complement activation on a model biomaterial surface. *Biomaterials*, **26**(13), 1477–1485.
23. Tengvall, P., Askendal, A., Lundstrom, I. (1996). Complement activation by 3-mercapto-1,2-propanediol immobilized on gold surfaces. *Biomaterials*, **17**(10), 1001–1007.
24. Lhotta, K., Wurzner, R., Kronenberg, F., Oppermann, M., Konig, P. (1998). Rapid activation of the complement system by cuprophane depends on complement component C4. *Kidney Int.*, **53**(4), 1044–1051.
25. Ekdahl, K. N., Lambris, J. D., Elwing, H., Ricklin, D., Nilsson, P. H., Teramura, Y., Nicholls, I. A., Nilsson, B. (2011). Innate immunity activation on biomaterial surfaces: A mechanistic model and coping strategies. *Adv. Drug Deliv. Rev.*, **63**(12), 1042–1050.
26. Szebeni, J. (2014). Complement activation-related pseudoallergy: A stress reaction in blood triggered by nanomedicines and biologicals. *Mol. Immunol.*, **61**(2), 163–173.
27. Szebeni, J., Fontana, J. L., Wassef, N. M., Mongan, P. D., Morse, D. S., Dobbins, D. E., Stahl, G. L., Bunger, R., Alving, C. R. (1999). Hemodynamic changes induced by liposomes and liposome-encapsulated hemoglobin in pigs: A model for pseudoallergic cardiopulmonary reactions to liposomes. Role of complement and inhibition by soluble CR1 and anti-C5a antibody. *Circulation*, **99**(17), 2302–2309.

28. Szebeni, J., Wassef, N. M., Spielberg, H., Rudolph, A. S., Alving, C. R. (1994). Complement activation in rats by liposomes and liposome-encapsulated hemoglobin: Evidence for anti-lipid antibodies and alternative pathway activation. *Biochem. Biophys. Res. Commun.,* **205**(1), 255–263.

29. Chanan-Khan, A., Szebeni, J., Savay, S., Liebes, L., Rafique, N. M., Alving, C. R., Muggia, F. M. (2003). Complement activation following first exposure to pegylated liposomal doxorubicin (Doxil): Possible role in hypersensitivity reactions. *Ann. Oncol.,* **14**(9), 1430–1437.

30. Szebeni, J., (2005). Complement activation-related pseudoallergy: A new class of drug-induced acute immune toxicity. *Toxicology,* **216**(2–3), 106–121.

31. Andersen, A. J., Hashemi, S. H., Andresen, T. L., Hunter, A. C., Moghimi, S. M. (2009). Complement: Alive and kicking nanomedicines. *J. Biomed. Nanotechnol.,* **5**, 364–372.

32. Szebeni, J., Muggia, F., Gabizon, A., Barenholz, Y. (2011). Activation of complement by therapeutic liposomes and other lipid excipient-based therapeutic products: Prediction and prevention. *Adv. Drug Deliv. Rev.,* **63**(12), 1020–1030.

33. Salvador-Morales, C., Sim, R. B. (2013). Complement activation. In: Dobrovolskaia, M. A., McNeil, S. E., eds. *Frontiers in Nanobiomedical Research, Vol. 1: Handbook of Immunological Properties of Engineered Nanomaterials,* World Scientific Publishing, Hackensack, NJ, pp. 357–384.

34. Garbuzenko, O., Barenholz, Y., Priev, A. (2005). Effect of grafted PEG on liposome size and on compressibility and packing of lipid bilayer. *Chem. Phys. Lipids,* **135**(2), 117–129.

35. Salvador-Morales, C., Zhang, L., Langer, R., Farokhzad, O. C. (2009). Immunocompatibility properties of lipid-polymer hybrid nanoparticles with heterogeneous surface functional groups. *Biomaterials,* **30**(12), 2231–2240.

36. Moghimi, S. M., Hamad, I., Andresen, T. L., Jorgensen, K., Szebeni, J. (2006). Methylation of the phosphate oxygen moiety of phospholipid-methoxy(polyethylene glycol) conjugate prevents PEGylated liposome-mediated complement activation and anaphylatoxin production. *FASEB J.,* **20**(14), 2591–2593.

37. Hamad, I., Al-Hanbali, O., Hunter, A. C., Rutt, K. J., Andresen, T. L., Moghimi, S. M. (2010). Distinct polymer architecture mediates switching of complement activation pathways at the nanosphere-

serum interface: Implications for stealth nanoparticle engineering. *ACS Nano*, **4**(11), 6629–6638.

38. Garay, R. P., El-Gewely, R., Armstrong, J. K., Garratty, G., Richette, P. (2012). Antibodies against polyethylene glycol in healthy subjects and in patients treated with PEG-conjugated agents. *Expert Opin. Drug Deliv.*, **9**(11), 1319–1323.

39. Verhoef, J. J., Carpenter, J. F., Anchordoquy, T. J., Schellekens, H. (2014). Potential induction of anti-PEG antibodies and complement activation toward PEGylated therapeutics. *Drug Discov. Today*, **19**(12), 1945–1952.

40. Neun, B. W., Dobrovolskaia, M. A. (2011). Qualitative analysis of total complement activation by nanoparticles. *Methods Mol. Biol.*, **697**, 237–245.

41. Lachmann, P. J. (2010). Preparing serum for functional complement assays. *J. Immunol. Methods*, **352**(1–2), 195–197.

42. Hamad, I., Christy Hunter, A., Rutt, K. J., Liu, Z., Dai, H., Moein Moghimi, S. (2008). Complement activation by PEGylated single-walled carbon nanotubes is independent of C1q and alternative pathway turnover. *Mol. Immunol.*, **45**(14), 3797–3803.

43. Hamad, I., Hunter, A. C., Szebeni, J., Moghimi, S. M. (2008). Poly(ethylene glycol)s generate complement activation products in human serum through increased alternative pathway turnover and a MASP-2-dependent process. *Mol. Immunol.*, **46**(2), 225–232.

44. Thomas, S. N., van der Vlies, A. J., O'Neil, C. P., Reddy, S. T., Yu, S. S., Giorgio, T. D., Swartz, M. A., Hubbell, J. A. (2011). Engineering complement activation on polypropylene sulfide vaccine nano-particles. *Biomaterials*, **32**(8), 2194–2203.

45. Pham, C. T., Thomas, D. G., Beiser, J., Mitchell, L. M., Huang, J. L., Senpan, A., Hu, G., Gordon, M., Baker, N. A., Pan, D., Lanza, G. M., Hourcade, D. E. (2014). Application of a hemolysis assay for analysis of complement activation by perfluorocarbon nanoparticles. *Nanomed. Nanotechnol. Biol. Med.*, **10**(3), 651–660.

46. Plumb, M. E., Sodetz, J. M. (1998). Proteins of the membrane attack complex. In: Volanakis, J. E., Frank, M. M., eds. *The Human Complement System in Health and Disease*, 10th ed., Marcel Dekker, Inc., New York.

47. Whaley, K. (1985). *Measurement of Complement*. Churchill Livingstone, New York.

48. Meerasa, A., Huang, J. G., Gu, F. X. (2011). CH(50): A revisited hemolytic complement consumption assay for evaluation of nanoparticles

and blood plasma protein interaction. *Curr. Drug Deliv.*, **8**(3), 290–298.

49. Chonn, A., Cullis, P. R., Devine, D. V. (1991). The role of surface charge in the activation of the classical and alternative pathways of complement by liposomes. *J. Immunol.*, **146**(12), 4234–4241.

50. Pondman, K. M., Sobik, M., Nayak, A., Tsolaki, A. G., Jakel, A., Flahaut, E., Hampel, S., Ten Haken, B., Sim, R. B., Kishore, U. (2014). Complement activation by carbon nanotubes and its influence on the phagocytosis and cytokine response by macrophages. *Nanomed. Nanotechnol. Biol. Med.*, **10**(6), 1287–1299.

51. Pham, C. T., Mitchell, L. M., Huang, J. L., Lubniewski, C. M., Schall, O. F., Killgore, J. K., Pan, D., Wickline, S. A., Lanza, G. M., Hourcade, D. E. (2011). Variable antibody-dependent activation of complement by functionalized phospholipid nanoparticle surfaces. *J. Biol. Chem.*, **286**(1), 123–130.

52. Mikhael, M. A., Evens, R. G. (1975). Migration and embolization of macrophages to the lung: A possible mechanism for colloid uptake in the lung during liver scanning. *J. Nucl. Med.*, **16**(1), 22–27.

53. Chang, S. W., Ohara, N. (1994). Chronic biliary obstruction induces pulmonary intravascular phagocytosis and endotoxin sensitivity in rats. *J. Clin. Invest.*, **94**(5), 2009–2019.

54. Chang, S. W., Ohara, N. (1996). Pulmonary intravascular phagocytosis in liver disease. *Clin. Chest Med.*, **17**(1), 137–150.

55. Warner, A. E., Barry, B. E., Brain, J. D. (1986). Pulmonary intravascular macrophages in sheep. Morphology and function of a novel constituent of the mononuclear phagocyte system. *Lab. Invest.*, **55**(3), 276–288.

56. Warner, A. E., Brain, J. D. (1986). Intravascular pulmonary macrophages: A novel cell removes particles from blood. *Am. J. Physiol.*, **250**(4 Pt 2), R728–732.

57. Warner, A. E., Molina, R. M., Brain, J. D. (1987). Uptake of bloodborne bacteria by pulmonary intravascular macrophages and consequent inflammatory responses in sheep. *Am. Rev. Respir. Dis.*, **136**(3), 683–690.

58. Warner, A. E., Brain, J. D. (1990). The cell biology and pathogenic role of pulmonary intravascular macrophages. *Am. J. Physiol.*, **258**(2 Pt 1), L1–12.

59. DeCamp, M. M., Warner, A. E., Molina, R. M., Brain, J. D. (1992). Hepatic versus pulmonary uptake of particles injected into the portal circulation in sheep. Endotoxin escapes hepatic clearance causing pulmonary inflammation. *Am. Rev. Respir. Dis.*, **146**(1), 224–231.

60. Winkler, G. C. (1989). Review of the significance of pulmonary intravascular macrophages with respect to animal species and age. *Exp. Cell Biol.*, **57**(6), 281–286.
61. Longworth, K. E., Jarvis, K. A., Tyler, W. S., Steffey, E. P., Staub, N. C. (1994). Pulmonary intravascular macrophages in horses and ponies. *Am. J. Vet. Res.*, **55**(3), 382–388.
62. Longworth, K. E., Albertine, K. H., Staub, N. C. (1996). Ultrastructural quantification of pulmonary intravascular macrophages in newborn and 2-week-old lambs. *Anat. Rec.*, **246**(2), 238–244.
63. Longworth, K. E. (1997). The comparative biology of pulmonary intravascular macrophages. *Front. Biosci.*, **2**, d232–241.
64. Ostensen, J., Hede, R., Myreng, Y., Ege, T., Holtz, E. (1992). Intravenous injection of Albunex microspheres causes thromboxane mediated pulmonary hypertension in pigs, but not in monkeys or rabbits. *Acta Physiol. Scand.*, **144**(3), 307–315.
65. (1994). Abstracts from the XI Congress of the International Society for Artificial Cells, Blood Substitutes, and Immobilization Biotechnology (ISABI). July 24–27, 1994, Boston, Massachusetts. *Artif. Cells Blood Substit. Immobil. Biotechnol.*, **22**(5), A1–177.
66. Chin, K. M., Kim, N. H., Rubin, L. J. (2005). The right ventricle in pulmonary hypertension. *Coron. Artery Dis.*, **16**(1), 13–18.
67. Szebeni, J., Bedocs, P., Csukas, D., Rosivall, L., Bunger, R., Urbanics, R. (2012). A porcine model of complement-mediated infusion reactions to drug carrier nanosystems and other medicines. *Adv. Drug Deliv. Rev.*, **64**(15), 1706–1716.
68. Dezsi, L., Fulop, T., Meszaros, T., Szenasi, G., Urbanics, R., Vazsonyi, C., Orfi, E., Rosivall, L., Nemes, R., Kok, R. J., Metselaar, J. M., Storm, G., Szebeni, J. (2014). Features of complement activation-related pseudoallergy to liposomes with different surface charge and PEGylation: Comparison of the porcine and rat responses. *J. Control. Release*, **195**, 2–10.
69. Winkler, G. C. (1988). Pulmonary intravascular macrophages in domestic animal species: Review of structural and functional properties. *Am. J. Anat.*, **181**(3), 217–234.
70. Wassef, A., Janardhan, K., Pearce, J. W., Singh, B. (2004). Toll-like receptor 4 in normal and inflamed lungs and other organs of pig, dog and cattle. *Histol. Histopathol.*, **19**(4), 1201–1208.
71. Schneberger, D., Lewis, D., Caldwell, S., Singh, B. (2011). Expression of toll-like receptor 9 in lungs of pigs, dogs and cattle. *Int. J. Exp. Pathol.*, **92**(1), 1–7.

72. Bulte, J. W., Schmieder, A. H., Keupp, J., Caruthers, S. D., Wickline, S. A., Lanza, G. M. (2014). MR cholangiography demonstrates unsuspected rapid biliary clearance of nanoparticles in rodents: Implications for clinical translation. *Nanomed. Nanotechnol. Biol. Med.*, **10**(7), 1385–1388.

73. Hampton, J. C. (1958). An electron microscope study of the hepatic uptake and excretion of submicroscopic particles injected into the blood stream and into the bile duct. *Acta Anat. (Basel)*, **32**(3), 262–291.

74. Cho, M., Cho, W. S., Choi, M., Kim, S. J., Han, B. S., Kim, S. H., Kim, H. O., Sheen, Y. Y., Jeong, J. (2009). The impact of size on tissue distribution and elimination by single intravenous injection of silica nanoparticles. *Toxicol. Lett.*, **189**(3), 177–183.

75. Park, K., Park, E. J., Chun, I. K., Choi, K., Lee, S. H., Yoon, J., Lee, B. C. (2011). Bioavailability and toxicokinetics of citrate-coated silver nanoparticles in rats. *Arch. Pharm. Res.*, **34**(1), 153–158.

76. Skajaa, T., Cormode, D. P., Jarzyna, P. A., Delshad, A., Blachford, C., Barazza, A., Fisher, E. A., Gordon, R. E., Fayad, Z. A., Mulder, W. J. (2011). The biological properties of iron oxide core high-density lipoprotein in experimental atherosclerosis. *Biomaterials*, **32**(1), 206–213.

77. Juhlin, L. (1960). Excretion of intravenously injected solid particles in bile. *Acta Physiol. Scand.*, **49**, 224–230.

78. Huxley, J. (1972). *Problems of Relative Growth*, Dover, New York.

79. Inturi, S., Wang, G., Chen, F., Banda, N. K., Holers, V. M., Wu, L., Moghimi, S. M., Simberg, D. (2015). Modulatory role of surface coating of superparamagnetic iron oxide nanoworms in complement opsonization and leukocyte uptake. *ACS Nano*, **9**(11), 10758–10768.

80. Ebanks, R. O., Isenman, D. E. (1996). Mouse complement component C4 is devoid of classical pathway C5 convertase subunit activity. *Mol. Immunol.*, **33**(3), 297–309.

81. Molina, H., Wong, W., Kinoshita, T., Brenner, C., Foley, S., Holers, V. M. (1992). Distinct receptor and regulatory properties of recombinant mouse complement receptor 1 (CR1) and Crry, the two genetic homologues of human CR1. *J. Exp. Med.*, **175**, 121–129.

82. Szebeni, J., Muggia, F. M., Alving, C. R. (1998). Complement activation by Cremophor EL as a possible contributor to hypersensitivity to paclitaxel: An in vitro study. *J. Natl. Cancer Inst.*, **90**(4), 300–306.

83. Szebeni, J., Baranyi, B., Savay, S., Lutz, L. U., Jelezarova, E., Bunger, R., Alving, C. R. (2000). The role of complement activatin in hypersensitivity to pegylated liposomal doxorubicin. *J. Liposome Res.*, **10**, 347–361.
84. Garred, P., Larsen, F., Madsen, H. O., Koch, C. (2003). Mannose-binding lectin deficiency–revisited. *Mol. Immunol.*, **40**(2–4), 73–84.
85. Markiewski, M. M., DeAngelis, R. A., Benencia, F., Ricklin-Lichtsteiner, S. K., Koutoulaki, A., Gerard, C., Coukos, G., Lambris, J. D. (2008). Modulation of the antitumor immune response by complement. *Nat. Immunol.*, **9**(11), 1225–1235.

Chapter 10

Intravenous Immunoglobulin at the Borderline of Nanomedicines and Biologicals: Antithrombogenic Effect via Complement Attenuation

Milan Basta, MD, PhD

Biovisions, Inc, Potomac, Maryland, USA

Keywords: anaphylatoxins, antiphospholipid syndrome (APS), atherosclerosis, autoimmune diseases, C3a, C5a, C5b-9, complement, complement system, dermatomyositis, Fab fragment, Fc fragment, fibroblast growth factor 2 (FGF-2), hyperviscosity, immunodeficiency, immunomodulation, inflammation, interleukin (IL), intravenous immunoglobulin (IVIG), MAC, macrophage, matrix metalloproteinases (MMPs), myocardial infarction, plasminogen activator inhibitor (PAI-1), platelets, scavenging, sickle cell anemia, sickle cell disease (SCD), smooth muscle cells (SMC), thrombin-activated fibrinolysis inhibitor (TAFI), thrombo-embolic events, thrombogenesis, tissue-type plasminogen activator (tPA), tumor necrosis factor α (TNFα), vulnerable plaque

10.1 Introduction

Intravenous immunoglobulin (IVIG) is derived from a pool of plasma donors and consists of purified IgG molecules (>95%) and

---

Immune Aspects of Biopharmaceuticals and Nanomedicines
Edited by Raj Bawa, János Szebeni, Thomas J. Webster, and Gerald F. Audette
Copyright © 2018 Pan Stanford Publishing Pte. Ltd.
ISBN 978-981-4774-52-9 (Hardcover), 978-0-203-73153-6 (eBook)
www.panstanford.com

a vehicle that prevents immunoglobulin aggregation. It belongs to the group of therapeutic proteins referred to as biologicals, but based on the size of molecules (in the 8–15 nm), and the fact that its therapeutic application requires human processing (pooling, concentration, and filtration), by definition IVIG can be considered as a nanomedicine. It is a replacement therapy for patients with primary immunodeficiencies (PID) and is increasingly being used to treat (auto)immune conditions due to its immunomodulatory and anti-inflammatory properties. Its safety record is proven; almost all side effects are mild and are due to the rapid infusions of larger than recommended doses [1]. In a fraction (1%) of patients infused with IVIG, a temporary relationship with the occurrence of thromboembolic events is claimed to exist [2]. No evidence of a cause-and-effect relationship has been provided so far, while a significant body of evidence exists to support the opposite effect of IVIG: protection against thrombosis. In this chapter, the pathogenesis of three thromboembolic conditions—atherosclerosis, antiphospholipid syndrome and sickle cell anemia, characterized by tissue and organ damage induced by hypercoagulation and thrombus/emboli formation—is reviewed in light of their pathogenesis by complement-mediated inflammation and its modulation by IVIG. The data suggest that IVIG exerts protective, rather than an enhancing effect on coagulation/thrombus formation (Table 10.1).

Table 10.1 Thrombogenic activity/effects of complement fragments

C′ Fragment	Action	Cell Type	Effect
C5a	*Expression of MMP-1 and MMP-9	Macrophages	Extracellular matrix degradation and plaque destabilization
	**Synthesis and degranulation of PAI-1	Mast cells	Prothrombotic phenotype
	Chemotaxis	T-Ly and monocytes	Enhanced inflammation
	Production of IL-6, IL-1β, and TNFα	Leukocytes	Augmented inflammatory response

C' Fragment	Action	Cell Type	Effect
C5b-9	Production of TF and Secretion of vWF	Endothelial cells	Activation of coagulation and thrombin formation
	Expression of E-selectin, P-selectin, ICAM-1, VCAM-1	Endothelial cells	Proinflammatory phenotype
	Expression of L-selectin and integrins	Leukocytes	Adhesion to endothelial cells, initiation of inflammation
C3a	NLRP3 activation	Monocytes	Sustained inflammation
	Same effect as C5a, but in higher concentrations		

*Increases.
**Promotes.

10.2 Atherosclerosis

Atherosclerosis is the major risk factor associated with the onset of cardiovascular disorders such as stroke and myocardial infarction. It is recognized as a typical chronic inflammatory condition in which inflammation, triggered by complement activation, plays a fundamental role in the development and progression of atherosclerotic lesion formation, plaque rupture, and thrombosis [3].

10.2.1 Phases in Atherosclerotic Pathogenesis

(a) Endothelial cell dysfunction/activation is a crucial step in atherosclerotic pathogenesis that leads to proinflammatory, provasoconstrictive, and prothrombotic cell phenotype. It involves the following:

- Up-regulation of endothelial adhesion molecules such as E-selectin, P-selectin, ICAM-1, and VCAM-1
- Up-regulation of leukocyte adhesion molecules—L-selectin, integrins, and platelet-endothelial cell-adhesion molecule 1

- Increased endothelial permeability to lipoproteins and plasma proteins
- Leukocyte migration into the artery wall mediated by oxidized LDL, monocyte chemotactic protein 1 (MCP-1), IL-8, platelet-derived growth factor (PDGF), and macrophage colony-stimulating factor (M-CSF)

(b) Fatty streaks are formed by lipid-loaded monocytes and macrophages, so-called foam cells, along with T lymphocytes and smooth muscle cells (SMC) whose migration is stimulated by PDGF, fibroblast growth factor 2 (FGF-2), and TGFβ. T cell activation is the next step in this process, mediated by tumor necrosis factor α (TNFα), interleukin 2 (IL-2), and granulocyte-macrophage colony stimulating factor. Oxidized LDL, M-CSF, TNFα and IL-1 mediate foam cell formation, while platelet adherence and aggregation are induced by P-selectin, integrins, thromboxane A, and tissue factor (TF).

(c) Advanced complicated lesion

As fatty streaks advance, they form a fibrous cap that walls off the lesion from the lumen, representing a fibrous response to the injury. The cap covers a mixture of leukocytes, lipids and debris that together form a necrotic core due to apoptosis, necrosis and lipid accumulation. These lesions expand by continued leukocyte adhesion and entry. Macrophage accumulation is mediated by M-CSF, MCP-1 and oxidized LDLs. A fibrous cap is the result of PDGF, TGFβ, IL-1, and TNFα activation as well as a decrease in connective tissue degradation.

(d) Unstable fibrous plaques and their rupture

Rupture or ulceration of fibrous cap leads rapidly to thrombosis; thinning of the fibrous cap is induced by macrophages that secrete protelolytic enzymes such as metalloproteinases causing bleeding from the lumen of the artery, thrombus formation, and occlusion of the artery.

(e) Thrombosis

Exposure of subendothelial collagen and vWF to circulating platelets causes their adhesion, activation and aggregation; lesion-associated tissue factor binds to factor VII that triggers the coagulation cascade and thrombin generation. Further, thrombin promotes more platelet activation, cleavage of fibrinogen to form fibrin, activation of factor XIII (leading to the generation of platelet-rich cross-linked fibrin clot resistant to mechanical pressure and proteolytic degradation). Fibrin formation is countered by the fibrinolytic system involving tissue-type plasminogen activator (tPA) that cleaves plasminogen to form serine protease plasmin, which degrades fibrin to limit the extent of thrombus formation. Fibrinolytic inhibitors like plasminogen activator inhibitor (PAI-1) and thrombin-activated fibrinolysis inhibitor (TAFI) ensure that the fibrinolytic process is regulated and does not result in inappropriate dissolution of the thrombus [4].

10.2.2 Role of Complement in Atherosclerosis

Complement activation and subsequent generation of activated fragments, particularly C5a and C5b-9, influence many phases in the development and progression of atherosclerotic lesions, as well as thrombosis through the activation of platelets, promotion of fibrin formation, and impairment of fibrinolysis. The participation of the complement system in inflammation is consistent with the physiological role of the complement system that provides instant protection following the invasion of microorganisms. However, in the context of cardiovascular disorders, these same processes are detrimental and lead to the development of atherosclerosis, plaque rupture, and thrombosis [5].

Macrophages, which are localized in the rupture-prone shoulder regions of coronary plaques, are considered to play a major role in plaque destabilization and rupture through the production of matrix metalloproteinases (MMPs). The complement fragment C5a, present in human coronary lesions *in vivo* where it colocalizes with MMP-1 and MMP-9, increases MMP-1 and MMP-9 mRNA and antigenic levels in human macrophages *in vitro*. These effects were blocked by antibodies against the receptor C5aR/CD88 [6]. C5a exerts potent chemotactic and proinflammatory effects that mediate many phases of plaque formation such as up-regulation and endothelial expression of E, P selectins and ICAM-1 and VCAM-1 molecules, recruitment of monocytes and T lymphocytes and leukocyte synthesis of IL-1beta, IL-6 and TNFα to enhance an inflammatory response [7]. In addition, C3a and C5a promote mast cell synthesis and the release of PAI-1 leading to a prothrombotic phenotype [8]. When cultured and incubated with C5a, human monocyte-derived macrophages (MDM) and human plaque macrophages up-regulated PAI-1 via NF-kappaB activation, which could favor thrombus development and stabilization *in vivo* [9].

C5b-9 is present in atherosclerosis from the earliest to advanced lesions along with complement regulators such as Factor H, indicating active contribution to arterial inflammation [7]. C5b-9 induces endothelial cell TF expression and secretion of vWF that trigger the coagulation cascade, thrombin formation and platelet activation and adhesion. C5b-9 induces the formation of platelet microparticles from activated platelets and also induces the exposure of binding sites for factor VIIa and factor Va on platelet and microparticle membranes leading to tenase and prothrombinase complex formation promoting thrombin formation on the platelet surface. Platelets are capable of intrinsic production of complement components, suggesting a localized production of C5b-9. Complement activation on the surface of platelets generates C3a, C5a, and C5b-9, inducing platelet activation and alfa-granule release and aggregation. Altogether this brings a positive feedback loop for the activation of platelets and complement. C5b-9, therefore, stimulates platelet activation and aggregation, promotes thrombin production on the platelet surface, and favors fibrin formation, all of which indicates that complement activation is important for the consolidation of a forming thrombus. In addition, C5b-9 stimulates PAI-1 secretion by macrophages and mast cells, which inhibit fibrin degradation and enable the progression of the plaque. The activation of the alternative complement pathway gives rise to fibrin clots that are denser and lyse more slowly [10].

10.2.3 Use of IVIG in Animal Models of Atherosclerosis

Apolipoprotein E and low-density lipoprotein (LDL)-receptor-deficient mice develop accelerated atherosclerosis and atherosclerotic plaques that contain large numbers of T cells and macrophages [11]. Employing this model, several studies have been performed to demonstrate the efficacy of IVIG in the prevention and/or reduction of the atherosclerotic phenomena.

IVIG therapy was also effective when given in the later stages of atherosclerosis, 11 weeks after the onset of a high-fat diet containing 0.3% cholesterol. Fatty streak lesions were significantly reduced and their macrophage and CD4+ T cell accumulation suppressed [12]. It was determined that IVIG reduces atherosclerotic lesions via neutralizing/activated complement fragments by demonstrating reduced lesion area in IVIG-treated complement-sufficient mice, while the size of the lesions in C3-/- mice was not affected [13]. Pre-treatment with IVIG had an effect on intimal thickening (induced by placement of a periadventitial cuff over the right femoral artery) in a model of murine arterial injury. IVIG-treated mice (10 mg/mouse intraperitoneally for five consecutive days starting 1 day prior to cuff placement) showed significantly suppressed intimal thickening in comparison to control animals that received human serum albumin in the same dosage regimen and route of administration [15]. Considering the animal data as well as the potent anti-inflammatory effect of IVIG, it is highly likely that IVIG infused into atherosclerotic patients would exert an antithrombotic effect rather than induce thromboembolic (TE) events.

10.3 Antiphospholipid Syndrome

Antiphospholipid syndrome (APS) is an autoimmune, hypercoagulable state caused by antibodies to negatively charged phospholipids (aPL). APS is characterized by thrombosis in both arteries (myocardial infarction, stroke) and veins (deep vein thrombosis, pulmonary embolism) as well as pregnancy-related complications such as miscarriage, stillbirth, preterm delivery, or severe preeclampsia. In rare cases, APS leads to rapid organ failure due to generalized thrombosis; this is termed "catastrophic antiphospholipid syndrome" (CAPS) and is associated with a high mortality rate [15].

10.3.1 Role of Complement in APS

There is growing evidence obtained from both animal and human studies that the activation of the complement pathway may contribute to the pathogenesis of APS. Studies using models of induced thrombosis suggested that aPL antibodies against cardiolipin β2GPI required the presence of terminal complement components to induce thrombus formation, and mice deficient in C3 or C5 were found to be resistant to aPL-induced thrombosis. The infusion of human IgG antiphospholipid antibodies induced fetal loss in pregnant mice, an effect that was abrogated by the concurrent administration of a C3 convertase inhibitor [16]. Injection of mice with aPL antibodies induces significant adhesion of leukocytes to endothelial cells and an increase in thrombus size. Mice lacking complement components C3 and C5 were resistant to thrombosis and endothelial cell activation induced by aPL antibodies. Inhibition of C5 activation using anti-C5 monoclonal antibodies prevented the thrombogenic effect of aPL antibodies. These data suggest that complement activation mediates two important effector functions of aPL antibodies: induction of thrombosis and endothelial activation [17]. Complement component C5 (and particularly its cleavage product C5a) and neutrophils are identified as key mediators of fetal injury. Antibodies or peptides that block C5a-C5a receptor interactions prevent pregnancy complications [18]. The fact that F (ab')$_2$ fragments of APL antibodies do not mediate fetal injury and that C4-deficient mice are protected from fetal injury suggests that activation of the complement cascade is initiated via the classical pathway. Studies in factor B-deficient mice, however, indicate that the alternative pathway activation is required and amplifies complement activation [19]. Based on the aforementioned findings, it has been suggested that heparin prevents fetal loss in APS by inhibiting complement activation rather than by its anticoagulant effect. Treatment with heparin (unfractionated or low molecular weight) prevented complement activation *in vivo* and *in vitro* and protected mice from pregnancy complications induced by aPL antibodies. Other anticoagulants, such as hirudin and fondaparinux failed to prevent complement activation or pregnancy loss, suggesting that anticoagulation therapy alone is not sufficient to protect against miscarriage in APS [20]. Hypocomplementemia is common in patients with primary APS, reflecting complement activation and consumption, and was correlated with anticoagulant activity, suggesting that APL antibodies may activate monocytes and macrophages via anaphylatoxins produced in complement activation [21]. In *in vivo* experimental models, the thrombogenic activity of APL antibodies is associated with proinflammatory and procoagulant endothelial phenotypes. In addition, complement activation is required by aPL antibodies to display their thrombogenic activity in *in vivo* models. Blocking of complement activation as well as TNFα neutralization protect animals from aPL-induced fetal losses [19]. Studies using mice deficient in complement components and specific inhibitors to complement have demonstrated that activation of complement contributes to fetal loss, growth restriction and thrombosis. Inhibition of complement activation can prevent these complications. Use of a specific complement inhibitor to C5 has been used successfully in a patient with catastrophic APS undergoing renal transplantation [22].

10.3.2 Use of IVIG in Experimental APS

The animal model of APS is induced by passive transfer of anti-cardiolipin-$β_2$GP1 antibodies into naïve mice resulting in APS-characteristic findings such as fetal loss, impaired embryonic implantation and endothelial dysfunction, and thrombogenesis. The results of pre-clinical studies in which IVIG was used indicate decreased mortality, inhibition of thrombosis, reduced fetal resorption, improved pregnancy outcomes, and amelioration of endothelial cell inflammatory and thrombogenic phenotypes. There were no drug-related thromboembolic (TE) events.

IVIG treatment inhibited APL-induced endothelial cell activation and enhancement of thrombosis in mice passively infused with human APL-containing IgG, and this was associated with a decrease in APL levels. Similarly, IVIG lowered APL levels and inhibited thrombogenesis in mice immunized with *β*2GPI. Both recombinant and natural human Fc fragments of heterogeneous IgG against human anti-β2-GP1 antibodies were used as treatment in a mouse APS model. Both Fc fragments significantly decreased the levels of serum anti-β2-GP1 antibodies and exerted an anticoagulation effect by inhibiting thrombus formation and decreased mortality [23].

10.3.3 Use of IVIG in Human APS Studies

In human APS, IVIG has been used mainly for recurrent pregnancy loss and some other clinical manifestations of the syndrome.

In 20 studies involving 252 patients and over 3000 individual IVIG infusions, immunoglobulin therapy was not associated with thromboembolic complications. In addition, it was effective in preventing recurrent thrombosis when compared to conventional therapy. Other beneficial effects were the increase of live-birth rates, reduced antepartum complications (pre-eclampsia, fetal growth restriction, fetal distress due to placental insufficiency), and amelioration of autoimmune thrombocytopenia, hemolytic anemia and uretroplacental thrombosis.

A small cohort of patients with APS was treated with IVIG in addition to anticoagulant therapy. No arterial or venous thrombosis occurred in IVIG-treated patients, while in the control group that received conventional treatment, two patients developed acute cerebral ischemic attacks and one deep vein thrombosis [24]. In a pilot study, intravenous immune globulin did not improve pregnancy outcomes beyond those achieved with a heparin and low-dose aspirin treatment. Although not statistically significant, the findings of fewer cases of fetal growth restriction and neonatal intensive care unit admissions among the IVIG-treated pregnancies warrant a larger multicenter treatment trial [25]. Four out of five patients treated with IVIG at 400 mg/kg for 5 days monthly delivered healthy infants at term and one at 32 weeks with a diagnosis of fetal distress. The results suggest that immunoglobulin treatment may improve pregnancy outcomes beyond those observed with heparin and aspirin. Immunoglobulin treatment was not associated with major side effects [26]. A prospective, two-center trial study included 82 women with recurrent abortions of which 29 were treated with prednisone and low dose aspirin (LDA) in one center, and 53 received IVIG in the other center. Live-birth rates were the same between groups. Mean birth weight was higher in the IVIG group than in the prednisone plus LDA group. In the prednisone- plus LDA-treated patients, gestational hypertension and diabetes were significantly more frequent than in the IVIG-treated group (14 vs. 5% and 14 vs. 5%, respectively) [27].

Several case, small series, and large clinical studies in human aPL concluded that IVIG treatment significantly improved pregnancy outcome and complication rates. Neither preeclampsia nor fetal intrauterine growth retardation was observed. The immunoglobulin therapy was not associated with major side effects.

10.4 Sickle Cell Disease

Sickle cell disease (SCD) is an example of hyperviscosity and hypercoagulable conditions. A point mutation in hemoglobin gene comprises the molecular basis of anemia that ultimately, by involving multiple pathogenic factors, causes activation of endothelial cells and coagulation. The hallmark of the disease is thrombotic vaso-occlusion that can lead to multiple organ infarctions, including the brain, lung, spleen, kidney, and liver, caused by abnormally shaped erythrocytes captured by leukocytes adherent to endothelial cells [28]. Sickle cell vaso-occlusion is a complex multistep process likely involving interactions among sickle erythrocytes, leukocytes, and endothelial cells. Circulating activated monocytes have been detected in sickle cell patients. Activated monocytes secrete proinflammatory cytokines that can in turn induce endothelial cells to express ligands for sickle adhesion receptors, as well as tissue factor, thereby providing a link between sickle cell-mediated vascular occlusion and activation of blood coagulation. Markers of ongoing platelet activation, such as P-selectin expression on circulating platelets, increased plasma concentrations of platelet factor 4 and β-thromboglobulin, and increased numbers of circulating platelet microparticles, have been detected in patients with sickle cell anemia in both the absence and the presence of vaso-occlusive crises [29].

10.4.1 Complement in SCD

Previous studies documented complement activation in sickle cell disease patients and suggested that this contributes to an increased risk of infection. Alternative pathway activation is initiated by membrane phospholipid changes, which occur in sickled erythrocytes. Complement activation products in serial samples from sickle cell anemia patients were compared at baseline and during hospitalization for painful crisis to examine the correlation between complement activation and disease activity. Plasma concentrations of Bb, C4d, and C3a were measured as well as C3 bound to erythrocytes. Patients were subdivided into those with continuous pain and those with intermittent painful episodes. In patients with intermittent pain, there was little evidence of complement activation at baseline and increased plasma concentrations of Bb and C3a during painful crisis. Elevated C3a and C4d levels were observed in patients with continuous pain regardless of hospitalization status, suggesting a continuous underlying inflammatory process in these patients [30]. Complement activation is tightly regulated on the membrane of the normal erythrocyte; therefore, defective complement regulation by the sickle cell would be necessary for complement-dependent hemolysis to occur. A defect in the regulation of membrane attack complex (C5b-9) formation in sickle erythrocytes was detected, particularly in the densest cells. The defect is characterized by increased binding of C5b-7 and of C9 to denser sickle cells, which results in increased susceptibility of sickle cells to C5b-9-mediated lysis initiated by either C5b-6 or activated cobra venom factor. Among the densest sickle cells, irreversibly sickled cells are especially sensitive to reactive lysis. The similarity of this defect to that previously described in a patient with paroxysmal nocturnal hemoglobinuria suggests that complement-induced anemia could play a role in anemia associated with SCD [31].

Patients with sickle cell anemia and normal individuals were investigated with respect to plasma concentrations of the inflammatory markers lysozyme and myeloperoxidase and the complement activation marker C3d. The SCD patients showed significantly increased levels of myeloperoxidase and C3d, but not lysozyme, compared with the controls. The concentrations of myeloperoxidase and C3d in plasma showed a significant inverse correlation with the hemoglobin concentration. Myeloperoxidase and C3d showed a significant positive correlation. This suggests a role for the neutrophil and the complement system in the pathophysiology of sickle cell disease [32].

10.4.2 IVIG in Experimental Sickle Cell Disease

In sickle-cell mice, IVIG treatment reduced the number of adherent neutrophils and subsequently improved microcirculatory blood flow and survival of sickle cell animals. Furthermore, it was demonstrated that IVIG altered adhesion pathways and allowed an increase in neutrophil rolling as documented by digital fluorescence videomicroscopy [33]. Recent data using intravital microscopy in a sickle cell mouse model suggest that adherent leukocytes in postcapillary venules play a critical role in vaso-occlusion by capturing circulating sickle RBCs. In the course of studies to investigate the adhesion receptors mediating sickle RBC-WBC interactions, it was determined that control nonspecific immunoglobulin G (IgG) preparations displayed significant inhibitory activity. The effects of commercial IVIG preparations were studied and it was found that they inhibit RBC-WBC interactions in cremasteric venules in a dose-dependent manner. IVIG of at least 200 mg/kg dramatically reduced these interactions, even after TNF-α stimulation, and not only increased microcirculatory blood flow but also improved survival of sickle cell mice. These data raise the possibility that IVIG may have a beneficial effect on sickle cell-associated vaso-occlusion [34].

10.4.3 IVIG Use in Patients with SCD

IVIG has been used in humans to treat hyperhemolytic crises. The conclusion from 11 case and small series reports was that IVIG corrected anemia and resolved hemolysis without additional blood transfusions. There were no serious side effects of IVIG therapy in any of the studies. A clinical trial is currently under way to examine the effect of IVIG in patients with sickle cell pain crises.

10.5 Mechanism of IVIG Modulation of Vaso-Occlusive Disorders

Atherosclerosis, APS, and SCD have common pathogenic mechanisms and similar clinical manifestations underlined by endothelial cell proinflammatory and thrombogenic phenotypes. They are triggered by inappropriate (exaggerated) inflammation leading to acute thrombotic events in which the complement system plays a crucial role. Activated complement fragments such as C3a, C5a and the sublytic form of the terminal membrane complex C5b-9 mediate endothelial dysregulation, leukocyte adhesion, chemotaxis, synthesis of proinflammatory cytokines, platelet and coagulation activation, and induce synthesis of endothelial and leukocyte adhesion molecules (E and P selectin, VCAM-1, ICAM-1), proinflammatory cytokines (Il-1, Il-6 and TNF-α), and chemokines (MCP-1). In addition, these fragments mediate thrombogenesis at multiple levels by up-regulating endothelial cell tissue factor and von Willebrand factor with subsequent activation of the coagulation cascade, platelet activation and aggregation, formation of platelet microparticles, promotion of thrombin production on platelet and microparticle membranes and by exposing binding sites for factor VIIa and factor Va. Furthermore, C3a, C5a, and C5b-9 stimulate the production and secretion of PAI-1 by macrophages, mastocytes, and basophils. PAI-1 inhibits fibrin degradation and favors the progression of the thrombus.

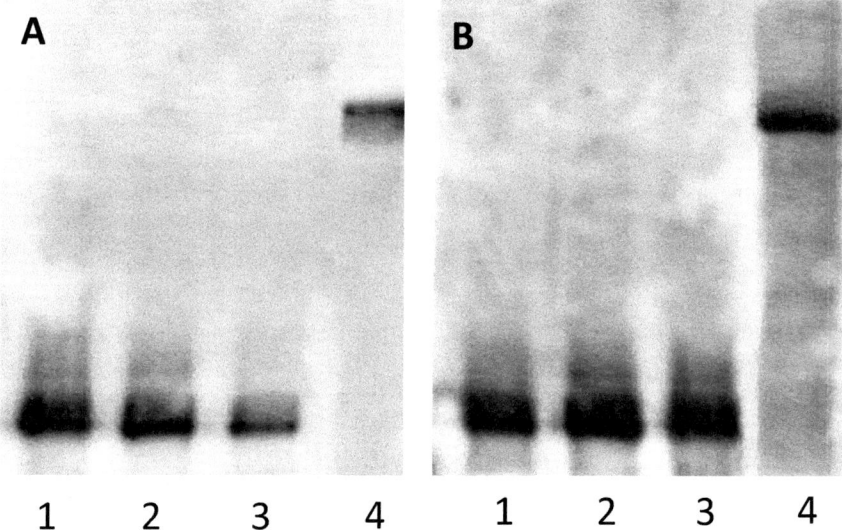

Figure 10.1 Evidence of covalent binding between anaphylatoxins C3a (Panel A) and C5a (Panel B) and F (ab')$_2$ immunoglobulin fragment. Purified C3a/C5a alone (Lanes 1 in both panels), C3a/C5a incubated with Fc fragments of immunoglobulin molecules (Lanes 2), C3a/C5a incubated with human albumin (Lanes 3), and C3a/C5a incubated with F (ab')$_2$ (Lanes 4) were subjected to SDS electrophoresis and subsequent Western blotting using corresponding monoclonal anti-anaphylatoxin antibodies. Both C3a and C5a bound covalently to F (ab')$_2$ fragments (as evidenced by the band shift to the molecular weight marker that represent the sum of the two molecules' weight) but not to Fc fragments or albumin.

IVIG exerts a potent anti-inflammatory and antithrombogenic effect due, in part, to its ability to scavenge potentially harmful complement fragments. In doing so, immunoglobulin molecules engage different structural regions—constant domain of the Fab fragment binds and neutralizes C3a and C5a ([35], Fig. 10.1), while the Fc region captures large fragments C3b and C4b, both of which are early precursors of the C5b-9 complex ([36], Fig. 10.2).

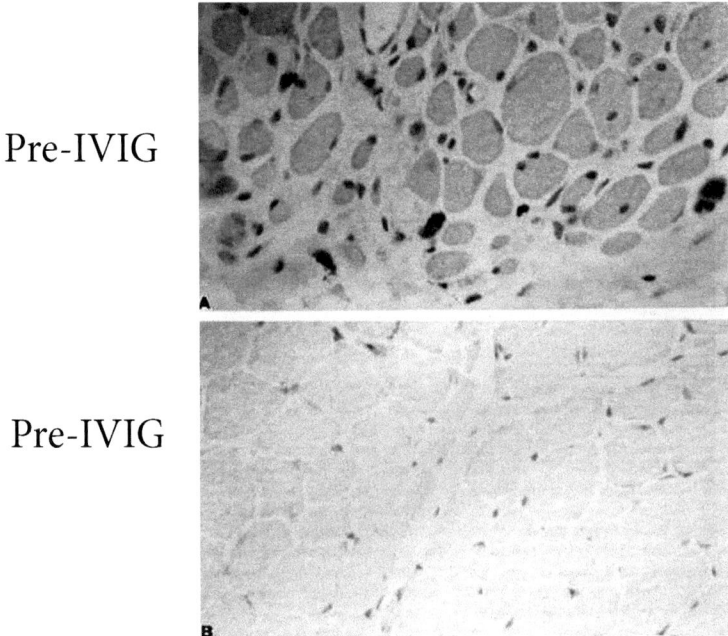

Figure 10.2 Staining for C3b-neo antigen in muscle biopsies of dermatomyositis (DM patients). Muscle biopsies from confirmed cases of DM were stained for C3b (neo) antigen before (Pre-IVIG) and 3 days after IVIG infusion (Post-IVIG). In baseline biopsies, C3b is seen between and within occasional muscle fibers. After IVIG treatment, no C3b staining is observed in corresponding biopsies, suggesting its removal from the circulation by infused immunoglobulin molecules.

10.6 Summary and Outlook

IVIG treatment of animals and humans with atherosclerosis, APS, and SCD was not associated with a high rate of thromboembolic events as would be expected if there was a cause-and-effect relationship between IVIG infusion and thrombogenesis. Animal models of these hypercoagulable and prothrombotic states would be perfect models in which to test the thrombotic potential of IVIG. Instead of confirming the IVIG-induced thromboembolic theory, the use of IVIG in these conditions proved the opposite effect—that IVIG is in fact antithrombogenic. Since anaphylatoxins C3a and C5a and the C5b-9 complex mediate most of the biological events underlying thrombogenesis (Table 10.1) it is plausible that their inhibition via immunoglobulin scavenging is the mechanism of anti-thrombogenic effect of IVIG. IVIG reduced vascular inflammation, improved blood flow, reduced plaque formation and progression, and improved clinical outcomes in atherosclerosis.

Disclosures and Conflict of Interest

Dr. Basta is the founder and president of BioVisions, Inc., Potomac, Maryland. The corporation focuses its scientific activities on pathogenic mechanisms (complement activation) shared by a wide range of diseases that represent major global health problems. No writing assistance was utilized in the production of this chapter and the author has received no payment for its preparation.

Corresponding Author

Dr. Milan Basta
Biovisions, Inc.
9012 Wandering Trail Dr
Potomac, MD 20854, USA
Email: Basta.Milan@gmail.com

About the Author

Milan Basta obtained his MD, MS, and PhD degrees from the University of Belgrade Medical School, Serbia. After his residency in pediatric surgery at the University Children's Hospital (Mother & Child Health Institute) in Belgrade, he pursued his postdoctoral training in immunology and complementology at several distinguished laboratories of the National Institutes of Health in Bethesda, Maryland. Over the ensuing years, he focused his research interest on the attenuation of complement-mediated immunopathology. Dr. Basta has achieved national and international recognition as a leader in the field of immunoglobulin-complement fragment interactions and the scavenging mechanism of the anti-inflammatory effect of high-dose intravenous immunoglobulin (IVIG). Dr. Basta has over 20 years of experience in complement research, specifically in the area of animal models of complement-mediated diseases and pre-clinical studies on the effectiveness of complement-attenuating therapeutics. He is the founder and president of BioVisions, Inc., located in Potomac, Maryland. The corporation focuses its scientific activities on pathogenic mechanisms (complement activation) shared by a wide range of diseases that represent major global health problems. The attenuation of complement may represent an effective therapeutic approach in many different, seemingly unrelated, pathologies. The company specializes in designing pre-clinical studies involving animal models of complement-mediated diseases for testing complement-modulating therapeutics. Its main strategy, both scientific and business, is the expansion of the use of existing plasma products (IVIG and C1-INH) to life-threatening and debilitating complement-mediated diseases.

References

1. Katz, U., Shoenfeld, Y. (2005). Review: Intravenous immunoglobulin therapy and thromboembolic complications. *Lupus*, **14**, 802–808.
2. Basta, M. (2014). Intravenous immunoglobulin-related thromboembolic events—an accusation that proves the opposite. *Clin. Exp. Immunol.*, **178**(Suppl 1), 153–155.
3. Ross, R. (1999). Atherosclerosis—an inflammatory disease. *N. Engl. J. Med.*, **340**, 115–126.
4. Furie, B., Furie, B. C. (2008). Mechanisms of thrombus formation. *N. Engl. J. Med.*, **359**, 938–949.
5. Torzewski, J., Bowyer, D. E., Waltenberger, J., Fitzsimmons, C. (1997). Processes in atherogenesis: Complement activation. *Atherosclerosis*, **132**, 131–138.

6. Speidl, W. S., Kastl, S. P., Hutter, R., Katsaros, K. M., Kaun, C., Bauriedel, G., Maurer, G., Huber, K., Badimon, J. J., Wojta, J. (2011). The complement component C5a is present in human coronary lesions *in vivo* and induces the expression of MMP-1 and MMP-9 in human macrophages *in vitro*. *FASEB J.*, **25**, 35–44.
7. Carter, A. M. (2012). Complement activation: An emerging player in the pathogenesis of cardiovascular disease. *Scientifica (Cairo)*, **2012**, 402783.
8. Wojta, J., Kaun, C., Zorn, G., Ghannadan, M., Hauswirth, A. W., Sperr, W. R., Fritsch, G., Printz, D., Binder, B. R., Schatzl, G., Zwirner, J., Maurer, G., Huber, K., Valent, P. (2002). C5a stimulates production of plasminogen activator inhibitor-1 in human mast cells and basophils. *Blood*, **100**, 517–523.
9. Kastl, S. P., Speidl, W. S., Kaun, C., Rega, G., Assadian, A., Weiss, T. W., Valent, P., Hagmueller, G. W., Maurer, G., Huber, K., Wojta, J. (2006). The complement component C5a induces the expression of plasminogen activator inhibitor-1 in human macrophages via NF-B activation. *J. Thromb. Haemost.*, **4**, 1790–1797.
10. Hertle, E., van Greevenbroek, M. J., Stehouwer, C. D. (2012). Complement C3: An emerging risk factor in cardiometabolic disease. *Diabetologia*, **55**, 881–884.
11. Persson, L., Borén, J., Nicoletti, A., Hansson, G. K., Pekna, M. (2005). Immunoglobulin treatment reduces atherosclerosis in apolipoprotein E-/- low-density lipoprotein receptor-/- mice via the complement system. *Clin. Exp. Immunol.*, **142**, 441–445.
12. Okabe, T. A., Kishimoto, C., Shimada, K., Murayama, T., Yokode, M., Kita, T. (2005). Effects of late administration of immunoglobulin on experimental atherosclerosis in apolipoprotein E-deficient mice. *Circ. J.*, **69**, 1543–1546.
13. Nicoletti, A., Kaveri, S., Caligiuri, G., Bariéty, J., Hansson, G. K. (1998). Immunoglobulin treatment reduces atherosclerosis in apo E knockout mice. *J. Clin. Invest.*, **102**, 910–918.
14. Keren, G., Keren, P., Barshack, I., Pri-Chen, S., George, J. (2001). The effect of intravenous immunoglobulins on intimal thickening in a mouse model of arterial injury. *Atherosclerosis*, **159**, 77–83.
15. Cervera, R., Bucciarelli, S., Plasín, M. A., Gómez-Puerta, J. A., Plaza, J., Pons-Estel, G., Shoenfeld, Y., Ingelmo, M., Espinos, G. (2009). Catastrophic antiphospholipid syndrome (CAPS): Descriptive analysis of a series of 280 patients from the "CAPS Registry". *J. Autoimmun.*, **32**, 240–245.

16. Samarkos, M., Mylona, E., Kapsimali, V. (2012). The role of complement in the antiphospholipid syndrome: A novel mechanism for pregnancy morbidity. *Semin. Arthritis Rheum.*, **42**, 66–69.
17. Pierangeli, S. S., Vega-Ostertag, M., Liu, X., Girardi, G. (2005). Complement activation: A novel pathogenic mechanism in the antiphospholipid syndrome. *Ann. N. Y. Acad. Sci.*, **1051**, 413–420.
18. Girardi, G., Berman, J., Redecha, P., Spruce, L., Thurman, J. M., Kraus, D., Hollmann, T. J., Casali, P., Caroll, M. C., Wetsel, R. A., Lambris, J. D., Holers, V. M., Salmon, J. E. (2003). Complement C5a receptors and neutrophils mediate fetal injury in the antiphospholipid syndrome. *J. Clin. Invest.*, **112**, 1644–1654.
19. Cavazzana, I., Manuela, N., Irene, C., Barbara, A., Sara, S., Orietta, B. M., Angela, T., Francesco, T., Luigi, M. P. (2007). Complement activation in anti-phospholipid syndrome: A clue for an inflammatory process? *J. Autoimmun.*, **28**, 160–164.
20. Girardi, G., Redecha, P., Salmon, J. E. (2004). Heparin prevents antiphospholipid antibody-induced fetal loss by inhibiting complement activation. *Nat. Med.*, **10**, 1222–1226.
21. Oku, K., Atsumi, T., Bohgaki, M., Amengual, O., Kataoka, H., Horita, T., Yasuda, S., Koike, T. (2009). Complement activation in patients with primary antiphospholipid syndrome. *Ann. Rheum. Dis.*, **68**, 1030–1035.
22. Lim, W. (2011). Complement and the antiphospholipid syndrome. *Curr. Opin. Hematol.*, **18**, 361–365.
23. Xie, W., Zhang, Y., Bu, C., Sun, S., Hu, S., Cai, G. (2011). Anti-coagulation effect of Fc fragment against anti-β2-GP1 antibodies in mouse models with APS. *Int Immunopharmacol.* **11**, 136–140.
24. Tenti, S., Guidelli, G. M., Bellisai, F., Galeazzi, M., Fioravanti, A. (2013). Long-term treatment of antiphospholipid syndrome with intravenous immunoglobulin in addition to conventional therapy. *Clin. Exp. Rheumatol.*, **31**, 877–882.
25. Branch, D. W., Peaceman, A. M., Druzin, M., Silver, R. K., El-Sayed, Y., Silver, R. M., Esplin, M. S., Spinnato, J., Harger, J. (2000). A multicenter, placebo-controlled pilot study of intravenous immune globulin treatment of antiphospholipid syndrome during pregnancy. The Pregnancy Loss Study Group. *Am. J. Obstet. Gynecol.*, **182**, 122–127.
26. Spinnato, J. A., Clark, A. L., Pierangeli, S. S., Harris, E. N. (1995). Intravenous immunoglobulin therapy for the antiphospholipid syndrome in pregnancy. *Am. J. Obstet. Gynecol.*, **172**, 690–694.

27. Vaquero, E., Lazzarin, N., Valensise, H., Menghini, S., Di Pierro, G., Cesa, F., Romanini, C. (2001). Pregnancy outcome in recurrent spontaneous abortion associated with antiphospholipid antibodies: A comparative study of intravenous immunoglobulin versus prednisone plus low-dose aspirin. *Am. J. Reprod. Immunol.*, **45**, 174–179.

28. Hoppe, C. C. (2014). Inflammatory mediators of endothelial injury in sickle cell disease. *Hematol. Oncol. Clin. North. Am.*, **28**, 265–286.

29. De Franceschi, L., Cappellini, M. D., Olivieri, O. (2011). Thrombosis and sickle cell disease. *Semin. Thromb. Hemost.*, **37**, 226–236.

30. Mold, C., Tamerius, J. D., Phillips, G. Jr. (1995). Complement activation during painful crisis in sickle cell anemia. *Clin. Immunol. Immunopathol.*, **76**, 314–320.

31. Test, S. T., Woolworth, V. S. (1994). Defective regulation of complement by the sickle erythrocyte: Evidence for a defect in control of membrane attack complex formation. *Blood*, **83**, 842–852.

32. Mohamed, A. O., Hashim, M. S., Nilsson, U. R., Venge, P. (1993). Increased *in vivo* activation of neutrophils and complement in sickle cell disease. *Am. J. Trop. Med. Hyg.*, **49**, 799–803.

33. Chang, J., Shi, P. A., Chiang, E. Y., Frenette, P. S. (2008). Intravenous immunoglobulins reverse acute vaso-occlusive crises in sickle cell mice through rapid inhibition of neutrophil adhesion. *Blood*, **111**, 915–923.

34. Turhan, A., Jenab, P., Bruhns, P., Ravetch, J. V., Coller, B. S., Frenette, P. S. (2004). Intravenous immune globulin prevents venular vaso-occlusion in sickle cell mice by inhibiting leukocyte adhesion and the interactions between sickle erythrocytes and adherent leukocytes. *Blood*, **103**, 2397–2400.

35. Basta, M., Van Goor, F., Luccioli, S., Billings, E. M., Vortmeyer, A. O., Baranyi, L., Szebeni, J., Alving, C. R., Carroll, M. C., Berkower, I., Stojilkovic, S. S., Metcalfe, D. D. (2003). F(ab)′2-mediated neutralization of C3a and C5a anaphylatoxins: A novel effector function of immunoglobulins. *Nat. Med.*, **9**, 431–438.

36. Basta, M., Dalakas, M. C. (1994). High-dose intravenous immunoglobulin exerts its beneficial effect in patients with dermatomyositis by blocking endomysial deposition of activated complement fragments. *J. Clin. Invest.*, **94**, 1729–1735.

Chapter 11

Lessons Learned from the Porcine CARPA Model: Constant and Variable Responses to Different Nanomedicines and Administration Protocols

Rudolf Urbanics, MD, PhD,[a,b] Péter Bedőcs, MD, PhD,[c] and János Szebeni, MD, PhD[a,b]

[a]*Nanomedicine Research and Education Center, Semmelweis University, Budapest, Hungary*
[b]*SeroScience Ltd., Budapest, Hungary*
[c]*Uniformed Services University of the Health Sciences, Bethesda, Maryland, USA*

Keywords: anaphylatoxins, anaphylaxis, animal models, arrhythmia blood cells, CARPA, complement, hemodynamic changes, hypersensitivity reactions, hypertension, hypotension, inflammatory mediators, nanomedicines, pigs, porcine model, pseudoallergy, tachyphylaxis, stress reaction, adverse drug reactions, drug toxicity, allergy, innate immunity, C3a, C5a, liposomes, nanoparticles

11.1 Introduction: CARPA as an Immune-Mediated Stress Reaction in Blood Triggered by Nanomedicines

Nanotechnology has achieved remarkable success in improving the therapeutic index of numerous drugs and agents by the use

---

Immune Aspects of Biopharmaceuticals and Nanomedicines
Edited by Raj Bawa, János Szebeni, Thomas J. Webster, and Gerald F. Audette
Copyright © 2018 Pan Stanford Publishing Pte. Ltd.
ISBN 978-981-4774-52-9 (Hardcover), 978-0-203-73153-6 (eBook)
www.panstanford.com

drug carrier nanosystems (nanocarriers) that carry and target the active pharmaceutical ingredient (API) to its site of action and/or control its ADME properties. However, the resulting increase of efficacy or decrease of toxicity is in some cases not without a price, as many of such nanodrugs or agents can cause adverse effects that the API alone would not do. We refer to them as reactogenic nanomedicines (RNMs). One of such adverse effects is complement (C) activation, which can lead to a syndrome called C activation-related pseudoallergy (CARPA). It is an acute and reversible immune reaction, also known as infusion reaction (or anaphylactic/anaphylactoid or idiosyncratic reaction), or non-immune allergy, whose symptoms, clinical significance, mechanism and many other properties were detailed previously in numerous reviews [1–8]. An apparently trivial observation 20 years ago that C activation underlies the major hemodynamic disturbance in rats following the infusion of liposomes [9], launched the progress of the CARPA concept, whose latest milestone is a broad claim that CARPA is a manifestation of a chemical stress on blood [8]. More explicitly, it is a universal defense process against the "threat" of nanomedicines, that can occur in any organism that has blood and C proteins therein, which falsely perceive the nanoparticles as pathogenic viruses. It is a stress reaction because just like in classical stress, an external harm ("pseudovirus") triggers a nonspecific, standard battery of physiological changes via a multiorgan "axis," namely the "immuno-cardiovascular" axis, which corresponds to the "hypothalamo-pituitary-adrenal" axis involved in classical stress [8]. Depending on the features of RNMs, their administration speed and the use of anti-allergic premedication, the prevalence of CARPA in man can reach 30–40% with mostly transient and mild symptoms. The reaction usually occurs unpredictably at the first use of the drug, as there is no known laboratory or any clinical test that could estimate the risk of a reaction. Because in rare cases CARPA may be severe, even fatal, the phenomenon is getting increasing regulatory attention. For example, the European Medicines Agency's latest guidance on generic liposome development recommends CARPA assays for preclinical safety testing [10]. However, it has not solidified to date which CARPA tests are appropriate, or best for regulatory purposes, which parameters need to be measured and under what conditions? In general, the use and utility of CARPA tests are in an early stage of scientific evaluation.

11.2 *In vitro* Tests for CARPA

In vitro testing of CARPA can be non-cellular and cellular. The non-cellular testing is based on the measurement of C activation by the RNM in serum, plasma or whole blood, using C split product ELISAs. There are many C split products that serve as activation marker and have commercially available pathway specific ELISAs, the best known being C3a, C5a, iC3b, C4d, Bb and SC5b-9. One issue in this regard is the inter-laboratory variation of test results, which problem has been addressed recently by efforts to produce international C standards [11–13]. The information that these in vitro assays provide on the risk of CARPA is limited, since they report only on the activity of afferent arm of CARPA, the extent of anaphylatoxin formation. The efferent arm, the body's response to anaphylatoxins, remains unknown. The most relevant cellular assays that can measure anaphylatoxin sensitivity are various basophil assays, which quantitate basophil leukocyte activation and/or secretion to model the mast cells' response to allergens or other RNMs [14–16]. We reported some preliminary, promising results with a basophil assay measuring CD203c upregulation as a predictor of liposomal CARPA [6], however, a study dedicated for the evaluation of the predictive value of the basophil test in pseudoallergy remains to be done.

11.3 Animal Models of Immune Toxicity: Which Is Good for CARPA Evaluation?

Animal models represent a major tool for the study of mechanisms in virtually all biomedical research. Hypersensitivity reactions result from a complex combination of genetic, environmental and temporal factors as well as complex interactions between the immune and other organ systems and the drug, making these reactions uniquely diverse. As delineated above, in vitro systems are unlikely able to mimic such complexity. There is a consensus in the field that HSRs can only be tested in whole animal models [17–19]. The critical question in this regard whether the standard, already accepted tests are appropriate for CARPA evaluation, or not.

Among the established animal models of immune toxicity, some can *a priori* be ruled out as CARPA test, based on their mode of operation. These include the mouse popliteal lymph node assay (PLNA) [20–22], the mouse ear-swelling test [23–25], the guinea pig (GP) maximization test (GPMT), the GP occluded patch test (Buehler's test) [26, 27] and the murine local lymph node assay (MLLA) [26]. In fact, the standard guinea pig assays were shown to be useless in predicting systemic hypersensitivity [28]. The measurement of blood lymphocyte counts and lymph node weight or assaying B, T or other immune cell proliferation or function evaluate long-term immune toxicity, and not CARPA, which is a short term toxicity. Taken together, to our best knowledge, none of the regulatory standard animal tests may be appropriate to assess CARPA.

11.4 Non-Standard Immunotoxicity Tests of CARPA in Different Animals

As pointed out, CARPA may be considered as a universal stress reaction, which implies that it may be present in most, if not all levels of mammalian evolution. Accordingly, there are a great number of reports in the literature on C activation and its consequences in various animals, including rats [9, 29–33], mice [33], dogs [34–37], rabbits [38, 39] nonhuman primates [40–49] and pigs [50–55]. Among these experimental systems, it is the pigs' response to C activator liposomes that best mimics the human infusion reactions to liposomes in terms of kinetics and spectrum of symptoms and the conditions of reaction induction (Table 11.1). For Doxil, for example, it was calculated that the drug dose that triggers CARPA in pigs corresponds to the dose that triggers infusion reactions in *hypersensitive* man [56]. These facts, taken together with the favorable ethical and financial aspects of working with pigs rather than dogs or primates, rationalize the use of pigs as a CARPA model. Thus, CARPA adds to the list of diseases that are successfully studied in pigs, as discussed below.

Table 11.1 Identities, similarities, and differences between CARPA in men and pigs

		Symptoms	
Comparison	Abnormalities	Human	Pig
Identical	Some cardiopulmonary and hemodynamic alterations	Difficulty of breathing, hypo- or hypertension, arrhythmia, tachycardia, bradycardia, edema	
	Body temperature	Fever	
	Blood cell changes	Leukopenia/leukocytosis, thrombocytopenia	
	Skin changes	Erythema, rash	
	Range of minimum reactogenic phospholipid dose in infusion or bolus	1–10 µg/kg/s	
Similar	Blood reaction markers	Rise of SC5b-9	Rise of TXB2
	Cardiopulmonary and hemodynamic alterations	Shortness of breath, fatigue, dizziness, fainting, swelling of the ankles, abdomen or legs, cyanosis, chest pain, passing out or dizziness	Rise of PAP, fall of SAP
	Time course	Symptoms start within 10 min after infusion and subside within 30–60 min	PAP rise between 3 and 10 min, returns to normal within 30–60 min
Different	Reaction frequency	<10%	>90%

11.5 The Use of Pigs as Disease Models

Pigs are widely used as large animal models in biomedical research [57–61], particularly in studies on cardiovascular diseases [62–67], trauma [68–71], sepsis [72–76], drug intoxications [54, 55], and, since 1999 [53], CARPA. As described in many previous experimental studies [50–53, 77] and a recent review [7], the porcine CARPA model represents a highly sensitive and reproducible model for the most serious, life threatening HSRs in man caused by RNM.

11.6 Technical Details of the Porcine CARPA Model

Figure 11.1 illustrates the setup and instruments used in the model. In brief, domestic pigs (usually 2–3 months old) or miniature pigs are initially sedated (with Calypsol/Xilazine) and, after tracheal intubation, anesthetized with isoflurane while breathing spontaneously. The animals thereafter undergo surgery to place multiple catheters into their circulation for the measurement of different hemodynamic parameters, administration of test drugs and blood sampling. Namely, a Swan-Ganz catheter is placed to the pulmonary artery wedge, for the measurement of pulmonary arterial pressure (PAP), and (optionally) central venous pressure (CVP) and cardiac output (CO). Additional catheters are placed into the femoral artery to record the systemic arterial pressure (SAP) and, (optionally), left ventricular end-diastolic pressure (LVEDP). The left femoral vein is canulated for blood sampling, and the external jugular vein for the administration of test articles and to maintain a slow drop infusion of saline (~3 mL/kg/h). For more sophisticated hemodynamic analysis to measure systemic vascular resistance (SVR) and pulmonary vascular resistance (PVR), additional catheters are placed and measurements and calculations are carried out. The hemodynamic, EKG and respiratory parameters are measured continuously, while blood cell counts, O_2 saturation, blood analytes (inflammatory and vasoactive mediators) and temperature are measured at predetermined times, usually in 10–20 min intervals. EKG leads I-III are placed at the standard Einthoven positions.

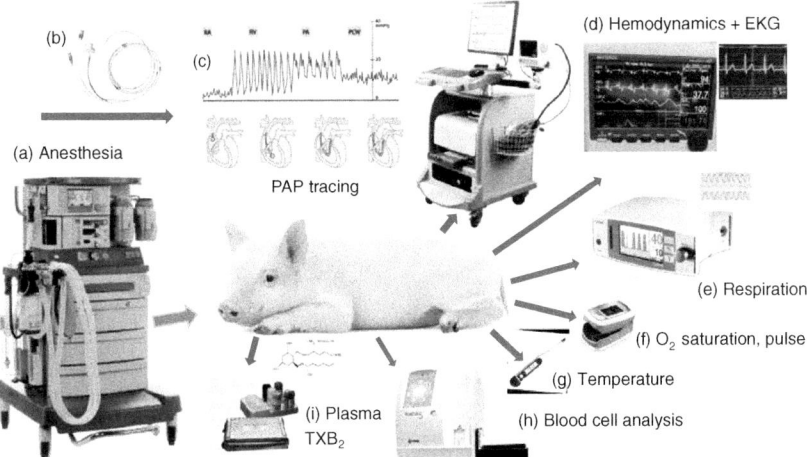

Figure 11.1 Instruments and endpoints measured in the porcine CARPA model. (a) anesthesia machine; (b) Swan-Ganz catheter; (c) blood pressure wave forms directing the passage of the tip of the Swan-Ganz catheter via the right atrium (RA), right ventricle (RV) and pulmonary artery (PA) until being wedged into the pulmonary capillary bed; (d) computerized multiple parameter hemodynamic monitoring system (1000 Hz sampling rate). From the continuous recording of SAP and PAP signals online averaging is performed and recorded, together with the heart rate, derived from SAP signal; (e) capnograph connected to the tracheal tube to measure respiratory rate, etCO$_2$ and inCO$_2$; (f) pulse oximeter (fixed on the tail) measures O$_2$ saturation in blood and pulse rate; (g) temperature is measured with a thermometer placed in the rectum; (h) veterinary hematology analyzer measuring all blood cell counts and WBC differential; (i) ELISA for measuring biomarkers of allergic/inflammatory reactions, e.g., TXB2, histamine, leukotrienes, adenosine, tryptase, PAF and C3 levels, etc.

11.7 The Symptom Tetrad

Figure 11.2 illustrates the four types of symptoms observed during CARPA in pigs: hemodynamic (a), hematological (b); laboratory (c) and skin (d) changes. Among the hemodynamic symptoms, the rise of PAP is the most prominent and reproducible measure of CARPA in the porcine model, which is invariably present with all RNMs. The transient, massive pulmonary hypertension is most likely due to the presence of pulmonary intravascular macrophages (PIM cells) in the lung of pigs, a theory based on

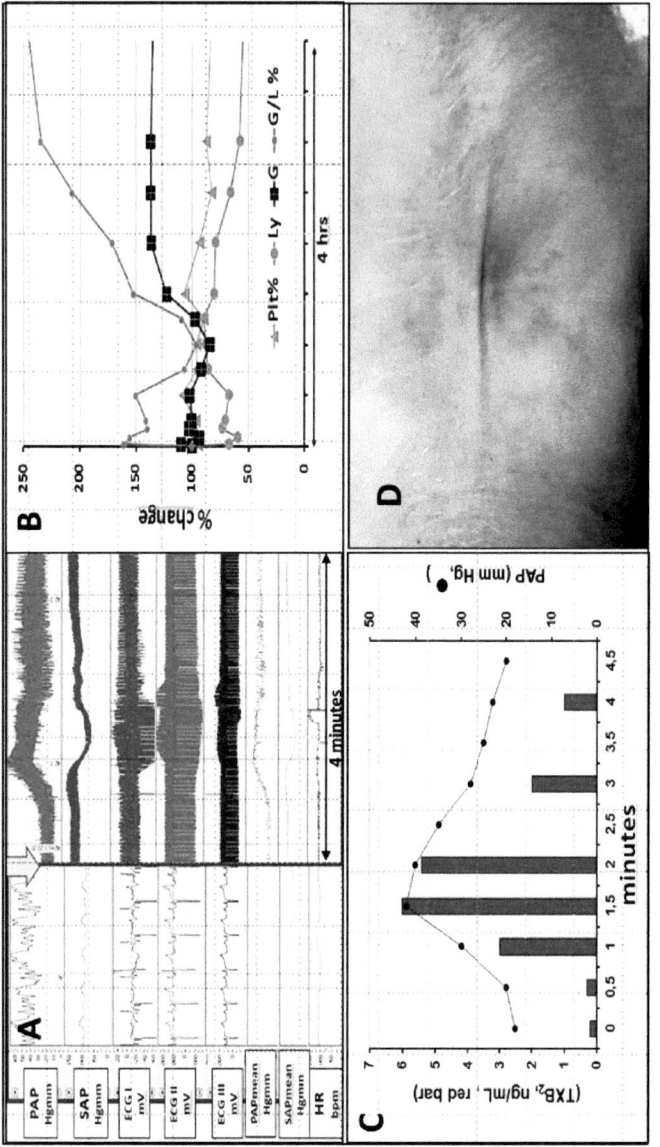

Figure 11.2 The symptom tetrad of CARPA in pigs. The hemodynamic (A) hematological (B), blood chemistry (C) and skin (D) changes are characteristic of porcine CARPA. Reproduced with permission, from [6, 7].

the speed (seconds) and prominence (maximal possible) of pulmonary changes in this species taken together with the known properties and functions of PIM cells [78]. Namely, PIM cells are directly exposed to blood and their function is to screen out from blood particulate pathogens. They can be activated both by anaphylatoxins and via particle binding to their surface receptors, and they respond to activation with massive secretion of vasoactive mediators [78]. These features are necessary and at the same time enough to explain the characteristic changes of PAP during CARPA in pigs.

The changes of SAP are more variable; it can drop, rise, display no change or undulate. The hematological changes typically include initial leukopenia followed by protracted leukocytosis and thrombocytopenia: among these, the leukopenic effect is the most frequent. Among the laboratory changes the rise of TXB2 is measured most often as it was found to show massive alterations in CARPA [53]. The exploration of changes of further allergy and inflammatory mediators (e.g., tryptase, leukotrienes, PAF, chemokines, cytokines) represents an unmet need in this field. Finally, the skin changes are rare and variable; they are seen only in case of very strong reactions.

11.8 Uniqueness of the Porcine CARPA Model

The uniqueness of the pig CARPA model lies in the identity, or close similarities of symptoms to the human CARPA reactions (Table 11.1) and the quantitative nature and reproducibility of measured endpoints, most significantly the rise of PAP. It should be stressed here again that the high reproducibility applies to the reactions to the same RNMs in different pigs, and NOT to the reactions to different RNMs.

11.9 Invariable Parameters of Porcine CARPA

As mentioned, the rise of PAP to i.v. bolus injection of certain RNMs is remarkably constant and reproducible if the dose and administration schedule is the same in all animals. This statement is demonstrated in Table 11.2, which summarizes the rise of PAP to 3 different RNMs in numerous independent experiments performed over 5 years.

Table 11.2 Inter-experimental variation of pulmonary hypertensive responses to (first) bolus administration of reactogenic liposomal nanomedicines and to zymosan

	1st PAP change (% of baseline ± S.E.M.)	n
Zymosan	368.9 ± 57.32	15
AmBisome	236.81 ± 100.91	7
Doxil	233.11 ± 91.79	12

Note: Data collected from n different experiments performed over years.

Yet another remarkable constancy of the model is the rise of PAP in the same animal, upon repetitive administration of a *non-tachyphylactic* RNM, such as large multilamellar liposomes (MLV) consisting of dimyristoyl phosphatidylcholine, dimyristoyl phosphatidylglycerol and cholesterol (50:5:45 mole ratios). Figure 11.3 illustrates this constancy: repeated injection of 5 mg lipid containing MLV in a pig at 30 to 60 min intervals eight times raised the PAP with negligible (6%) coefficient of variation [53].

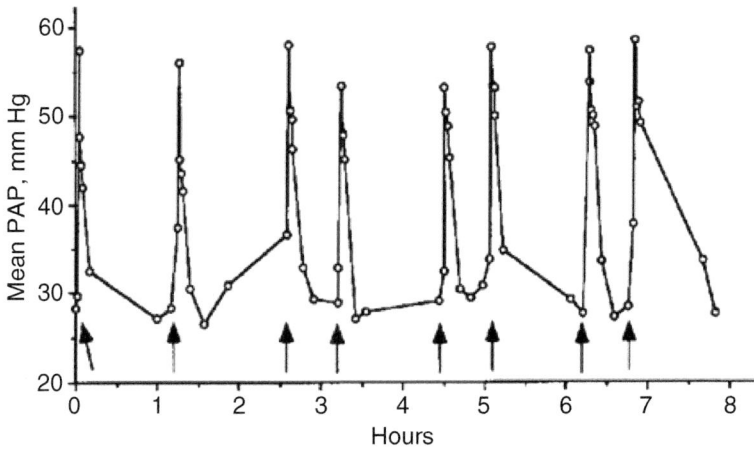

Figure 11.3 PAP responses to repetitive liposome injections. A pig was injected intravenously with 5 mg lipid-containing liposome boluses at indicated time points, and changes in PAP were recorded. Arrows indicate time points of injections. Reproduced from [53], with permission.

11.10 Variable Parameters of Porcine CARPA

11.10.1 Reaction Kinetics

The standard pulmonary reaction in porcine CARPA, observed immediately after bolus injection of RNMs consists of a rapid rise of PAP within 2–5 min after the bolus, a plateau reached in 5–15 min and decline to baseline or near baseline within 15–60 min (Fig. 11.2A and 11.5). However, there are unpublished examples for delayed start and/or protracted reactions as well, which we obtained with bolus administration of a surface conjugated liposomal nanomedicine (Fig. 11.4A) and upon stepwise infusion of a PEGylated liposomal drug candidate (Fig. 11.4B).

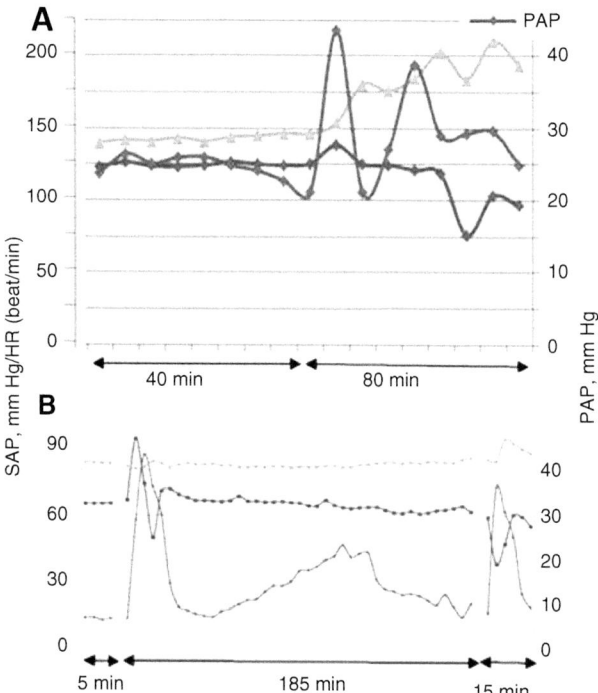

Figure 11.4 Examples of delayed and or protracted pulmonary hypertension caused by reactogenic nanomedicines. Excerpts from experiments performed in our laboratory, in collaboration with the suppliers of RNMs, which were conjugated (A) and PEGylated (B) small unilamellar liposomes (SUV). Blue, red and green indicate PAP, SAP and heart rate curves registered continuously. Unpublished data.

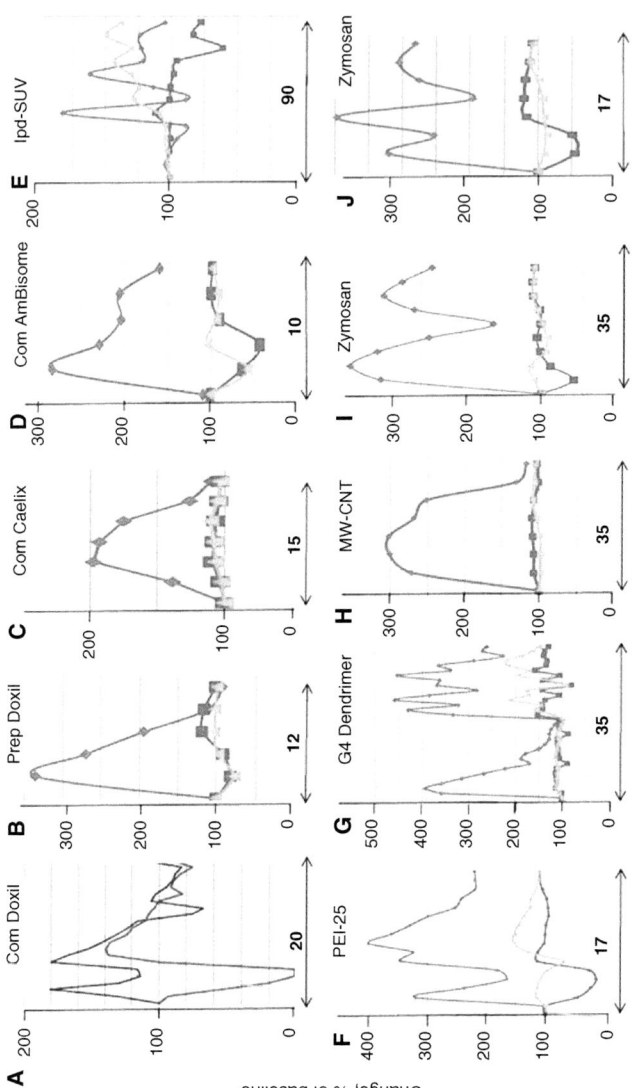

Figure 11.5 Variation of PAP and SAP waveforms. Panels were redrawn from different experiments wherein the CARPAgenic potential of nanoparticulate drugs or drug carriers were tested in pigs. Minutes indicate the times pan of reactions. Blue, red and green are PAP, SAP and heart rate curves respectively. Changes are shown in % of baseline. Abbreviations only here: com, commercial; prep, self-prepared; lpd, lipophilic prodrug-containing liposomes [79]; PEI25, 25 kD pegylated poly(ethylene imine) [80]; G4 dendrimer, 4th generation dendrimer; MWCNT, multiwall carbon nanotube. Most curves were reconstituted from experiments in refs. [7, 50, 51, 77]. The applied doses of different reaction triggers ranged in the 0.01–0.5 mg/kg range, for individual values see original references.

In the former experiment, testing the reactogenicity of a surface-conjugated SUV (Fig. 11.4A), the rise of PAP started only 40 min after bolus administration of the sample and the curve showed 3 progressively diminishing peaks over ~1.5 h. This observation remains to be understood, just as the late start of a protracted pulmonary hypertension observed during stepwise infusion of a PEGylated SUV drug candidate (Fig. 11.4B). After the initial, usual sharp and rapid reaction another reaction started at 30 min, the PAP peaked at 110 min and returned to normal after 180 min. These complex curves reflect complex interactions underlying the reactions, whose understanding will require a lot more work. It should be noted though that in clinical practice reactions can also start late, even hours after the infusion, and may also extend over hours.

11.10.2 Blood Pressure Wave Forms

Another feature of porcine CARPA is that the relative intensities and waveforms of PAP and SAP show substantial variation among different reaction inducers. Figure 11.5 shows a selection of different PAP/SAP/heart rate (HR) curves that were registered following bolus injection of different RNMs specified in the legend. The combination of different changes appears random at this time, as no known particle property can be associated with any particular time course or kinetic variables of PAP or SAP or HR. Nevertheless one observation can be explained: the initial splicing of PAP arises most likely from the coincidence of pulmonary hypertension with systemic hypotension, best discernible in Fig. 11.5A. The example shown for extended reaction is in Panel E, wherein a highly negative SUV formulation of an anticancer lipophilic (pro)drug was tested. The roller-coaster swings of PAP extended over 1.5 h, which may be attributed to metabolism or short-term physiological effects of the liposomes or the released drug leading to varying blood levels of different vasoactive mediators.

11.10.3 Tachyphylaxis

Yet another property of CARPA showing significant variation depending on the RNM is the presence/absence, and duration of tachyphylaxis, i.e. self-induced tolerance. The importance of tachyphylactic CARPA lies in the capability it offers for tolerance induction, i.e. to develop safe administration protocols for RNMs. In tachyphylactic CARPA the first dose of the RNM induces tolerance to the second or later administration of the same RNM, and if the first tolerogenic dose is made harmless, the hypersensitivity reactions (HSRs) can be entirely prevented. The utilization of tachyphylaxis for the prevention of HSR was first described for Doxil, wherein the HSR could be prevented by prior slow infusion with Doxebo [81] (Fig. 11.6A), a doxorubicin-free (placebo) Doxil that retains the tachyphylaxis-causing, self-tolerogenic effect of Doxil without causing major HSR [81]. However, not all RNMs can cause full tachyphylaxis, as it is a variable and structure-dependent property of RNMs; and it can also be absent (like in Fig. 11.3), partial (Fig. 11.6B) or biphasic (Fig. 11.6C).

To date, the absence of tachyphylaxis was observed with MLV (Fig. 11.3), zymosan (Fig. 11.6A), Ambisome [50], some PEI polymers [80] and, in general, highly charged nanoparticles (unpublished observations). Partial tachyphylaxis occurs more frequently with liposomes and it is a dose dependent phenomenon. The biphasic tachyphylaxis shown in Fig. 11.6C, wherein a section of nontachyphylactic responses switches to another nontachyphylatic section with smaller, but equal peaks, was observed for SAP after repetitive administration of the same dose of zymosan [7].

The mechanism of full or partial tachyphylaxis is not understood; based on its rapid appearance (minutes), it is most likely a passive phenomenon rather than an immune learning-based, cell-mediated active process. One theory, referred to as "double hit" hypothesis, points to allergy-mediating cells (mast cells, basophils, PIM cells) as anatomic sites of tachyphylaxis, and suggests that tachyphylactogenic RNMs trigger CARPA both via C activation and via direct binding to these allergy mediating cells, and the two signals together trigger these cells to release vasoactive mediators. If one of these mechanisms is not working for some reason at the time of repeated treatments, because a mediator gets consumed or a signal transduction pathway gets exhausted, tachyphylaxis will ensue [7, 78].

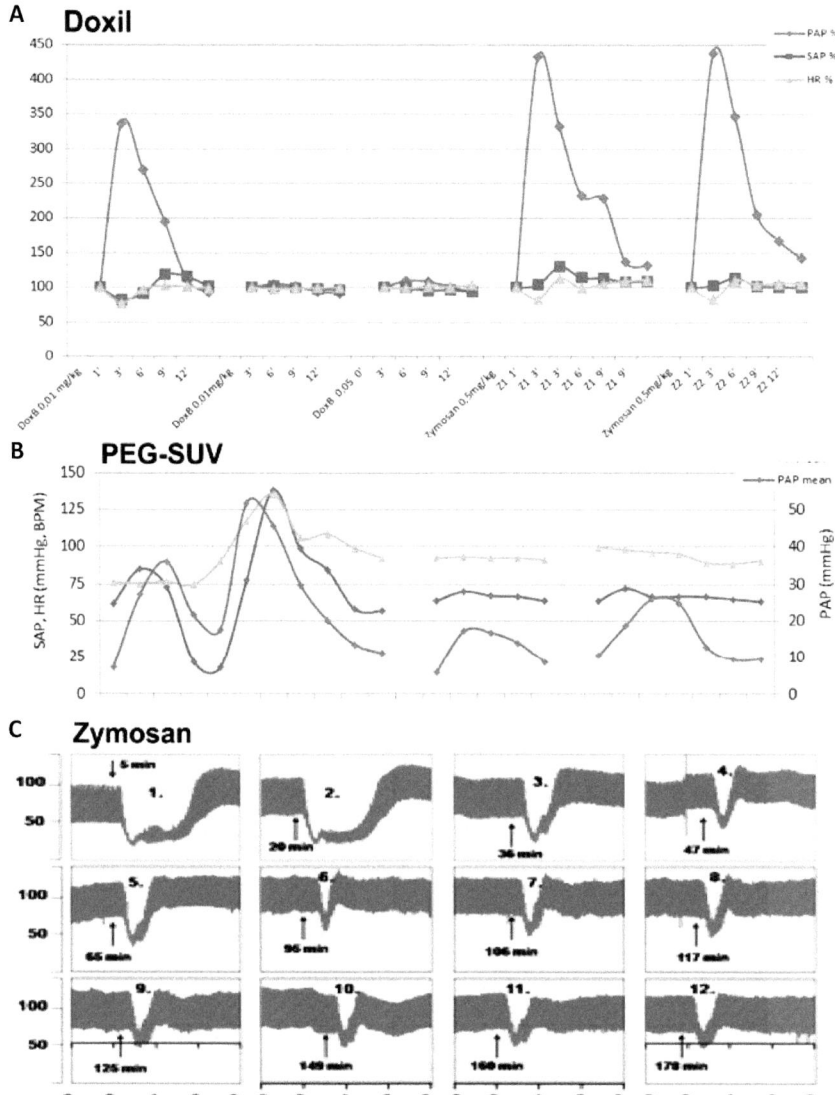

Figure 11.6 Full, partial and biphasic tachyphylaxis. (A) First 3 injections: repeated bolus administration of 0.01, 0.01 and 0.05 mg/kg Doxil, followed by 2 injections of 0.5 mg/kg zymosan. Figure shows the PAP (Blue), SAP (red) and HR (green) values as % of baseline. x-axis: minutes. Reproduced from [81]. (B) Repeated bolus administration of comparable doses of another PEGylated reactogenic liposome preparation, containing an anti-inflammatory steroid. Unpublished data. (C) Repeated bolus administration of 0.01 mg/kg zymosan in a pig, Y values are SAP, mm Hg. Reproduced from [7], with permission.

11.11 Summary and Future Directions

This review provided background information for placing the porcine CARPA model on the map of preclinical immunotoxicological tests for CARPA. It also focused, for the first time, on the constant and variable features of CARPA symptoms in pigs, particularly those seen in the hemodynamic response. The inter-animal and inter-experimental stability of PAP changes in response to a certain type of RNM, and switch to another type of response with another type of RNM represent remarkable, yet ill-understood features of the model that reflect the involvement of highly regulated, complex immunological and physiological processes.

Figure 11.7 illustrates this complexity by providing an overall scheme of molecular and cellular interactions in CARPA [8]. It shows the involvement of several organ systems, numerous cells and highly active biomolecules and mediators and multiple redundant pathways in both the afferent (triggering of allergy-mediating cells) and efferent arms (mediation of trigger signal to the effector cells) of CARPA.

Figure 11.8 provides further glimpses into the complex picture of CARPA pathogenesis, by highlighting the vicious cycle of hemodynamic and other organ derangements that may lead to the worst outcome of this immune toxicity; anaphylactic shock with circulatory collapse.

In summary, the constant and, at the same time highly variable symptoms of CARPA in pigs reflect the simplicity and at the same time the complexity of a "stress reaction in blood" [8], a novel immune phenomenon brought to light by the introduction of nanotechnology in pharmacotherapy. Further studies on the porcine CARPA model will hopefully reveal more details of the phenomenon and more understanding of its clinical use in the safety evaluation of novel nanomedicines.

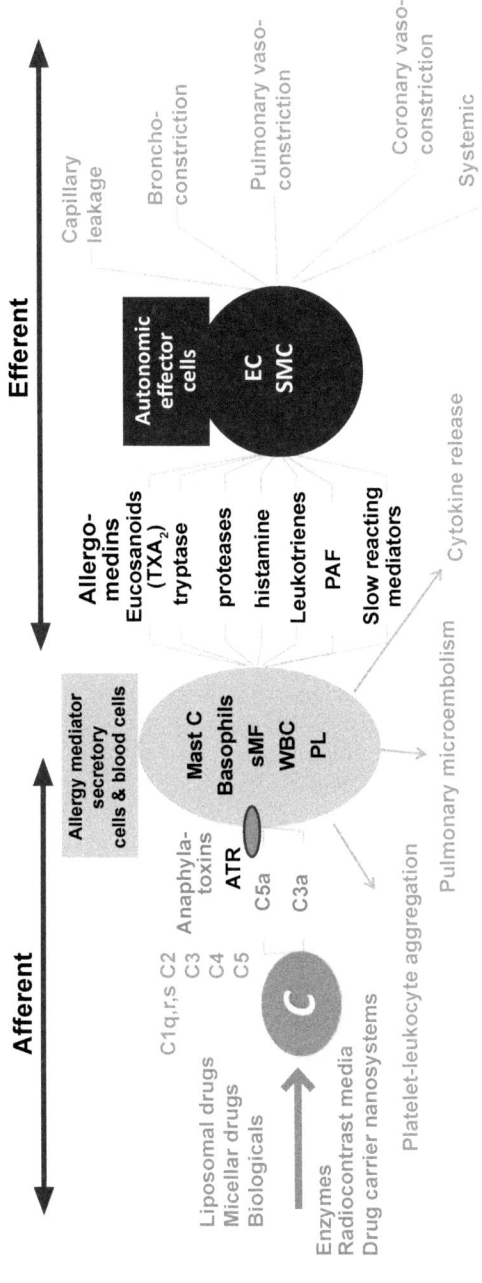

Figure 11.7 Afferent and efferent arms of CARPA. The hypothetical scheme illustrates the steps and interactions among a great number of cells and mediators involved in CARPA. AR, anaphylatoxin receptors; Mast C, mast cells; sMF, secretory macrophages; WBC, white blood cells; PL, platelets; EC, endothelial cells; SMC, smooth muscle cells. Different types of systems, cells, mediators and effects are color coded. Modified from [8], with permission.

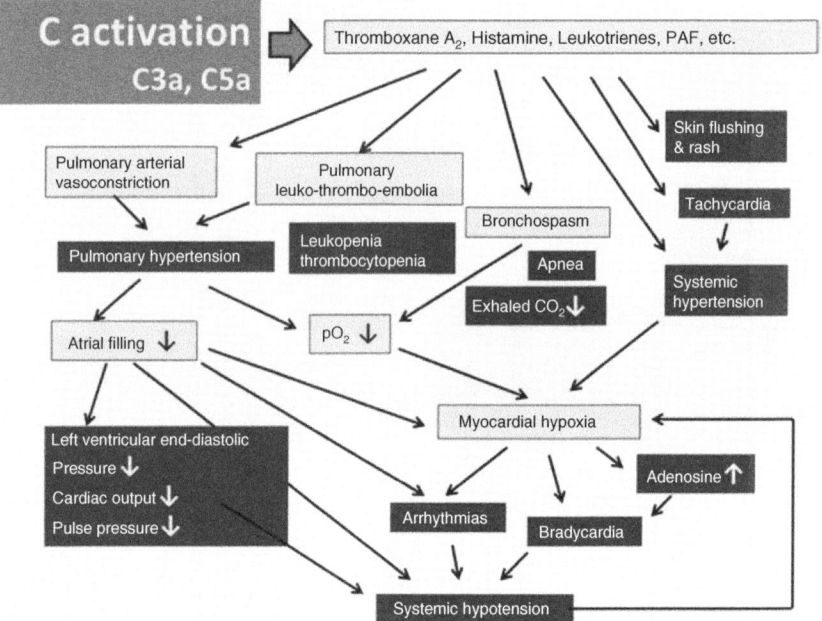

Figure 11.8 The vicious cycle of physiological changes in CARPA. The scheme illustrates the cause-effect relationships among different physiological changes that entail the activation of the CARPA cascade. The entries on blue and yellow background indicate clinical symptoms and underlying physiological changes, respectively. Reproduced from [8], with permission.

Disclosures and Conflict of Interest

This chapter was originally published as: Urbanics, R., Péter Bedőcs, P., Szebeni, J. Lessons learned from the porcine CARPA model: Constant and variable responses to different nanomedicines and administration protocols. *Eur. J. Nanomed.* **7**(3), 219–231 (2015); DOI 10.1515/ejnm-2015-0011, and appears here (with edits) by kind permission of the publisher, Walter de Gruyter GmbH, Berlin, Germany. J.S. and U.R. acknowledge the financial support to the Nanomedicine Research and Education Center at Semmelweis University from Gedeon Richter NyRT and EU FP7 projects No: 309820 (NanoAthero) and 310337 (CosmoPhos). Grants to SeroScience, Ltd, from EU FP7 projects No: 602923 (Thera-Glio) and 281035 (TransInt) are also acknowledged. The authors declare that they have no conflict of interest and have no affiliations or financial involvement with any organization or entity discussed in this chapter. No writing assistance was utilized in the production of this chapter and the authors have received no payment for its preparation.

Corresponding Author

Dr. János Szebeni
Nanomedicine Research and Education Center
Semmelweis University
1089 Nagyvárad tér 4, Budapest, Hungary
Email: jszebeni2@gmail.com

About the Authors

János Szebeni is director of the Nanomedicine Research and Education Center at Semmelweis University, Hungary. He is also the founder and CEO of a contract research SME "SeroScience," and full Professor of (immune) biology at Miskolc University, Hungary. He has held various guest professor and scientific positions in Hungary and abroad, mostly in the United States, where he lived for 22 years. His research on various themes in hematology, membrane biology, and immunology resulted more than 130 scientific papers and book chapters, 2 granted patents, and a book entitled *The Complement System: Novel Roles in Health and Disease* (Kluwer, 2004). Three fields stand out where he has been most active: artificial blood, liposomes, and the complement system. His original works led to the "CARPA" concept, which implies that complement activation underlies numerous drug-induced (pseudo)allergic (anaphylactoid) reactions.

Péter Bedőcs graduated cum laude as a doctor of medicine from Semmelweis University in 2004. He completed his PhD training at the university, focusing on pathophysiological effects and immunotoxicity of drug delivery systems utilizing nanotechnology. In collaboration with Semmelweis University, the Bay Zoltan Foundation in Hungary, and the Uniformed Services University of the Health Sciences (USUHS) in the United States, he participated in the development of an animal model for the testing of complement activation by intravenous therapeutic and diagnostic agents, which became an FDA-recommended preclinical test for nanopharmaceuticals. This also led to the establishment of an effective desensitization procedure to prevent immunotoxic reactions provoked by certain nanomedicines. Currently he works at the Defense and Veterans Center for Integrative Pain Management as a senior scientist and as an assistant professor in the Department of Military and Emergency Medicine at USUHS in Bethesda, Maryland, overseeing multiple clinical and animal studies in the area of patient safety in anesthesia, developing new effective modalities for pain management. Additionally, he continues to serve as a collaborator and consultant for the Malignant Hyperthermia Diagnostic Center at USUHS.

Rudolf Urbanics is head of the *in vivo* laboratory of Nanomedicine Research and Education Center of Semmelweis University, and SeroScience Ltd., an immunotoxicity CRO, since 2008 in Budapest, Hungary. He obtained his MD diploma and the PhD degree at Semmelweis Medical School, Budapest, Hungary. He had various research/collaboration positions at Max Planck Institute of System Physiology, at the University of Pennsylvania, Cerebrovascular Research Center, and at Pennsylvania Muscle Institute, working in the field of CNS regulation of blood flow/metabolism, ischemic/hypoxic disorders, stroke, and chronic neurodegenerative disease animal models. He was the deputy R&D director and head of the CNS Pharmacology Department at Biorex R&D Co. and worked at IVAX/Drug Research Institute Budapest as a leading researcher in safety and CNS pharmacology and later at the IVAX/Drug Research Institute, a subsidiary of TEVA as head of the *in vivo* Pharmacology Group. Currently, he is working with *in vivo* models of nanodrug-nanocarrier–induced, complement activation–related pseudoallergic reactions (CARPA), clarifying their immune-toxicological and safety hazards.

References

1. Szebeni, J. (2001). Complement activation-related pseudoallergy caused by liposomes, micellar carriers of intravenous drugs and radiocontrast agents. *Crit. Rev. Ther. Drug. Carr. Syst.*, **18**, 567–606.
2. Szebeni, J. (2004). Complement activation-related pseudoallergy: Mechanism of anaphylactoid reactions to drug carriers and radiocontrast agents. In: Szebeni, J., ed. *The Complement System: Novel Roles in Health and Disease*, Kluwer, Boston, pp. 399–440.
3. Szebeni, J. (2005). Complement activation-related pseudoallergy: A new class of drug-induced immune toxicity. *Toxicology*, **216**, 106–121.
4. Szebeni, J., Alving, C. R., Rosivall, L., Bunger, R., Baranyi, L., Bedocs, P., et al. (2007). Animal models of complement-mediated hypersensitivity reactions to liposomes and other lipid-based nanoparticles. *J. Liposome. Res.*, **17**, 107–117.
5. Szebeni, J., Muggia, F., Gabizon, A., Barenholz, Y. (2011). Activation of complement by therapeutic liposomes and other lipid excipient-based therapeutic products: Prediction and prevention. *Adv. Drug. Deliv. Rev.*, **63**, 1020–1030.
6. Szebeni, J. (2012). Hemocompatibility testing for nanomedicines and biologicals: Predictive assays for complement mediated infusion reactions. *Eur. J. Nanomed.*, **5**, 33–53.
7. Szebeni, J., Bedocs, P., Csukas, D., Rosivall, L., Bunger, R., Urbanics, R. (2012). A porcine model of complement-mediated infusion reactions to drug carrier nanosystems and other medicines. *Adv. Drug. Deliv. Rev.*, **64**, 1706–1716.
8. Szebeni, J. (2014). Complement activation-related pseudoallergy: A stress reaction in blood triggered by nanomedicines and biologicals. *Mol. Immunol.*, **61**, 163–173.
9. Szebeni, J., Wassef, N. M., Spielberg, H., Rudolph, A. S., Alving, C. R. (1994). Complement activation in rats by liposomes and liposome-encapsulated hemoglobin: Evidence for anti-lipid antibodies and alternative pathway activation. *Biochem. Biophys. Res. Commun.*, **205**, 255–263.
10. EMA-CHMP Reflection paper on the data requirements for intravenous liposomal products developed with reference to an innovator liposomal product. http://www.ema.europa.eu/docs/en_GB/document_library/Scientific_guideline/2013/03/WC500140351.pdf

11. Seelen, M. A., Roos, A., Wieslander, J., Mollnes, T. E., Sjoholm, A. G., Wurzner, R., et al. (2005). Functional analysis of the classical, alternative, and MBL pathways of the complement system: Standardization and validation of a simple ELISA. *J. Immunol. Methods*, **296**, 187–198.
12. Bergseth, G., Ludviksen, J. K., Kirschfink, M., Giclas, P. C., Nilsson, B., Mollnes, T. E. (2013). An international serum standard for application in assays to detect human complement activation products. *Mol. Immunol.*, **56**, 232–239.
13. Roos, A., Wieslander, J. (2014). Evaluation of complement function by ELISA. *Methods Mol. Biol.*, **1100**, 11–23.
14. Ocmant, A., Peignois, Y., Mulier, S., Hanssens, L., Michils, A., Schandené, L. (2007). Flow cytometry for basophil activation markers: The measurement of CD203c up-regulation is as reliable as CD63 expression in the diagnosis of cat allergy. *J. Immunol. Methods*, **320**, 40–48.
15. Boumiza, R., Debard, A. L., Monneret, G. (2005). The basophil activation test by flow cytometry: Recent developments in clinical studies, standardization and emerging perspectives. *Clin. Mol. Allergy*, **3**, 9.
16. Boumiza, R., Monneret, G., Forissier, M. F., Savoye, J., Gutowski, M. C., Powell, W. S., et al. (2003). Marked improvement of the basophil activation test by detecting CD203c instead of CD63. *Clin. Exp. Allergy*, **33**, 259–265.
17. Uetrecht, J. P. (1999). New concepts in immunology relevant to idiosyncratic drug reactions: The "danger hypothesis" and innate immune system. *Chem. Res. Toxicol.*, **12**, 387–395.
18. Choquet-Kastylevsky, G., Descotes, J. (1998). Value of animal models for predicting hypersensitivity reactions to medicinal products. *Toxicology*, **129**, 27–35.
19. Uetrecht, J. (2005). Role of animal models in the study of drug-induced hypersensitivity reactions. *AAPS J.*, **7**, E914–E921.
20. Lin, M., Sun, W., Wang, Y., Li, X., Jin, Y., Gong, W., et al. (2012). An intravenous exposure mouse model for prediction of potential drug-sensitization using reporter antigens popliteal lymph node assay. *J. Appl. Toxicol.*, **32**, 395–401.
21. Suda, A., Iwaki, Y., Kimura, M. (2000). Differentiation of responses to allergenic and irritant compounds in mouse popliteal lymph node assay. *J. Toxicol. Sci.*, **25**, 131–136.
22. Shinkai, K., Nakamura, K., Tsutsui, N., Kuninishi, Y., Iwaki, Y., Nishida, H., et al. (1999). Mouse popliteal lymph node assay for assessment

of allergic and autoimmunity-inducing potentials of low-molecular-weight drugs. *J. Toxicol. Sci.*, **24**, 95–102.

23. Gad, S. C. (1994). The mouse ear swelling test (MEST) in the 1990s. *Toxicology*, **93**, 33–46.
24. Cornacoff, J. B., House, R. V., Dean, J. H. (1992). Mouse ear swelling test (MEST). *Fundam. Appl. Toxicol.*, **19**, 157–158.
25. Gad, S. C. (1988). Standard mouse ear swelling test (MEST). *Fundam. Appl. Toxicol.*, **11**, 732–733.
26. Kimber, I., Hilton, J., Botham, P. A. (1990). Identification of contact allergens using the murine local lymph node assay: Comparisons with the Buehler occluded patch test in guinea pigs. *J. Appl. Toxicol.*, **10**, 173–180.
27. Robinson, M. K., Nusair, T. L., Fletcher, E. R., Ritz, H. L. (1990). A review of the Buehler guinea pig skin sensitization test and its use in a risk assessment process for human skin sensitization. *Toxicology*, **61**, 91–107.
28. Weaver, J. L., Staten, D., Swann, J., Armstrong, G., Bates, M., Hastings, K. L. (2003). Detection of systemic hypersensitivity to drugs using standard guinea pig assays. *Toxicology*, **193**, 203–217.
29. Rabinovici, R., Rudolph, A. S., Vernick, J., Feuerstein, G. (1994). Lyophilized liposome-encapsulated hemoglobin: Evaluation of hemodynamic, biochemical, and hematologic responses. *Crit. Care. Med.*, **22**, 480–485.
30. Rabinovici, R., Rudolph, A. S., Ligler, F. S., Smith, E. F., III, Feuerstein, G. (1992). Biological responses to exchange transfusion with liposome-encapsulated hemoglobin. *Circ. Shock*, **37**, 124–133.
31. Rabinovici, R., Rudolph, A. S., Yue, T.-L., Feuerstein, G. (1990). Biological responses to liposome-encapsulated hemoglobin (LEH) are improved by a PAF antagonist. *Circ. Shock*, **31**, 431–445.
32. Dézsi, L., Fülöp, T., Mészáros, T., Szénási, G., Urbanics, R., Vázsonyi, C., et al. (2014). Features of complement activation-related pseudoallergy to liposomes with different surface charge and PEGylation: Comparison of the porcine and rat responses. *J. Control. Release*, **195**, 2–10.
33. Dézsi, L., Rosivall, L., Hamar, P., Szebeni, J., Szénási, G. (2015). Rodent models of complement activation-related pseudoallergy: Inducers, symptoms, inhibitors and reaction mechanisms. *Eur. J. Nanomedicine*, **7**, 15–25.
34. Qiu, S., Liu, Z., Hou, L., Li, Y., Wang, J., Wang, H., et al. (2013). Complement activation associated with polysorbate 80 in beagle dogs. *Int. Immunopharmacol.*, **15**, 144–149.

35. Moghimi, S. M., Wibroe, P. P., Szebeni, J., Hunter, A. C. (2013). Surfactant-mediated complement activation in beagle dogs. *Int. Immunopharmacol.*, **17**, 33-34.
36. Summary, J. J., Dubick, M. A., Zaucha, G. M., Kilani, A. F., Korte, D. W., Wade, C. E. (1992). Acute and subacute toxicity of 7.5% hypertonic saline/6% dextran-70 (HSD) in dogs. 1. Serum immunoglobulin and complement responses. *J. Appl. Toxicol.*, **12**, 261-266.
37. Rossen, R. D., Michael, L. H., Kagiyama, A., Savage, H. E., Hanson, G., Reisberg, M. A., et al. (1988). Mechanism of complement activation after coronary artery occlusion: Evidence that myocardial ischemia in dogs causes release of constituents of myocardial subcellular origin that complex with human C1q in vivo. *Circ. Res.*, **62**, 572-584.
38. Awasthi, V. D., Garcia, D., Goins, B. A., Phillips, W. T. (2003). Circulation and biodistribution profiles of long-circulating PEG-liposomes of various sizes in rabbits. *Int. J. Pharm.*, **253**, 121-132.
39. Rudolph, A. S., Klipper, R. W., Goins, B., Phillips, W. T. (1991). In vivo biodistribution of a radiolabeled blood substitute: 99mTc-labeled liposome-encapsulated hemoglobin in an anesthetized rabbit. *Proc. Natl. Acad. Sci. U. S. A.*, **88**, 10976-10980.
40. Acierno, M. J., Labato, M. A., Stern, L. C., Mukherjee, J., Jakowski, R. M., Ross, L. A. (2006). Serum concentrations of the third component of complement in healthy dogs and dogs with protein-losing nephropathy. *Am. J. Vet. Res.*, **67**, 1105-1109.
41. Breathnach, R., Donahy, C., Jones, B. R., Bloomfield, F. J. (2006). Increased leukotriene B(4) production, complement C3 conversion and acid hydrolase enzyme concentrations in different leucocyte sub-populations of dogs with atopic dermatitis. *Vet. J.*, **171**, 106-113.
42. Aganezov, S. A., Shcherbak, I. G., Galebskaia, L. V., Solovtsova, I. M., Riumina, E. V. (1995). Activity of the complement system in blood serum of dogs after treatment of acute necrotic pancreatitis with boiled pancreatic juice. *Patol Fiziol Eksp Ter.*, (3), 18-20(in Russian, with English abstract).
43. Trail, P. A., Yang, T. J., Cameron, J. A. (1984). Increase in the haemolytic complement activity of dogs affected with cyclic haematopoiesis. *Vet. Immunol. Immunopathol.*, **7**, 359-368.
44. Muller-Peddinghaus, R., Schwartz-Porsche, D. (1983). Klinische Bedeutung der Serumkomplementierung bei Hunden [Clinical significance of serum complement in dogs]. *Zentralblatt fur Veterinarmedizin Reihe A*, **30**, 698-711.

45. Imai, T., Arii, H., Sato, T., Asakura, Y., Sakuraya, N., Sudo, I., et al. (1979). [Change in complement titers of dogs under endotoxemia]. *Masui. Jpn. J. Anesthesiol.*, **28**, 564–569.
46. Tawara, T., Hasegawa, K., Sugiura, Y., Harada, K., Miura, T., Hayashi, S., et al. (2008). Complement activation plays a key role in antibody-induced infusion toxicity in monkeys and rats. *J. Immunol.*, **180**, 2294–2298.
47. Henry, S. P., Beattie, G., Yeh, G., Chappel, A., Giclas, P., Mortari, A., et al. (2002). Complement activation is responsible for acute toxicities in rhesus monkeys treated with a phosphorothioate oligodeoxynucleotide. *Int. Immunopharmacol.*, **2**, 1657–1666.
48. Mugge, A., Heistad, D. D., Densen, P., Piegors, D. J., Armstrong, M. L., Padgett, R. C., et al. (1992). Activation of leukocytes with complement C5a is associated with prostanoid-dependent constriction of large arteries in atherosclerotic monkeys in vivo. *Atherosclerosis*, **95**, 211–222.
49. Hedin, H., Smedegard, G. (1979). Complement profiles in monkeys subjected to aggregate (immune complex) anaphylaxis, and following injection of soluble and particulate polysaccharides. *Int. Arch. Allergy Appl. Immunol.*, **60**, 286–294.
50. Szebeni, J., Bedőcs, P., Rozsnyay, Z., Weiszhár, Z., Urbanics, R., Rosivall, L., et al. (2012). Liposome-induced complement activation and related cardiopulmonary distress in pigs: Factors promoting reactogenicity of Doxil and AmBisome. *Nanomedicine NBM*, **8**, 176–184.
51. Szebeni, J., Baranyi, L., Savay, S., Bodo, M., Milosevits, J., Alving, C. R., et al. (2006). Complement activation-related cardiac anaphylaxis in pigs: Role of C5a anaphylatoxin and adenosine in liposome-induced abnormalities in ECG and heart function. *Am. J. Physiol. Heart Circul. Physiol.*, **290**, H1050–H1058.
52. Bodo, M., Szebeni, J., Baranyi, L., Savay, S., Pearce, F., Alving, C., et al. (2005). Rheoencephalographic evidence of complement activation-related cerebrovascular changes in pigs. *J. Cerebr. Blood Flow Met.*, **25**, S550.
53. Szebeni, J., Fontana, J. L., Wassef, N. M., Mongan, P. D., Morse, D. S., Dobbins, D. E., et al. (1999). Hemodynamic changes induced by liposomes and liposome-encapsulated hemoglobin in pigs: A model for pseudoallergic cardiopulmonary reactions to liposomes. Role of complement and inhibition by soluble CR1 and anti-C5a antibody. *Circulation*, **99**, 2302–2309.

54. Bedocs, P., Capacchione, J., Potts, L., Chugani, R., Weiszhar, Z., Szebeni, J., et al. (2014). Hypersensitivity reactions to intravenous lipid emulsion in swine: Relevance for lipid resuscitation studies. *Anesth. Analg.*, **119**, 1094–1101.
55. Buckenmaier, C. C., Capacchione, J., Mielke, A. R., Bina, S., Shields, C., Kwon, K. H., et al. (2012). The effect of lipid emulsion infusion on postmortem ropivacaine concentrations in swine: Endeavoring to comprehend a soldier's death. *Anesth. Analg.*, **114**, 894–900.
56. Szebeni, J., Baranyi, B., Savay, S., Lutz, L. U., Jelezarova, E., Bunger, R., et al. (2000). The role of complement activation in hypersensitivity to pegylated liposomal doxorubicin (Doxil®). *J. Liposome Res.*, **10**, 347–361.
57. Swindle, M. M., Makin, A., Herron, A. J., Clubb, F. J., Jr., Frazier, K. S. (2012). Swine as models in biomedical research and toxicology testing. *Vet. Pathol.*, **49**, 344–356.
58. Schook, L., Beattie, C., Beever, J., Donovan, S., Jamison, R., Zuckermann, F., et al. (2005). Swine in biomedical research: Creating the building blocks of animal models. *Anim. Biotechnol.*, **16**, 183–190.
59. Azria, M., Kiger, J. L. (1972). [Value of miniature swine in biomedical research]. *Therapie*, **27**, 723–732.
60. McClellan, R. O. (1968). Applications of swine in biomedical research. *Lab. Anim. Care.*, **18**, 120–126.
61. Bustad, L. K., McClellan, R. O. (1966). Swine in biomedical research. *Science*, **152**, 1526–1530.
62. Elmadhun, N. Y., Sabe, A. A., Robich, M. P., Chu, L. M., Lassaletta, A. D., Sellke, F. W. (2013). The pig as a valuable model for testing the effect of resveratrol to prevent cardiovascular disease. *Ann. N. Y. Acad. Sci.*, **1290**, 130–135.
63. Gabler, N. K., Osrowska, E., Imsic, M., Eagling, D. R., Jois, M., Tatham, B. G., et al. (2006). Dietary onion intake as part of a typical high fat diet improves indices of cardiovascular health using the mixed sex pig model. *Plant Food Hum. Nutr.*, **61**, 179–185.
64. Lanoye, L., Segers, P., Tchana-Sato, V., Rolin, S., Dogne, J. M., Ghuysen, A., et al. (2007). Cardiovascular haemodynamics and ventriculo-arterial coupling in an acute pig model of coronary ischaemia-reperfusion. *Exp. Physiol.*, **92**, 127–137.
65. Gross, D. R. (1997). Thromboembolic phenomena and the use of the pig as an appropriate animal model for research on cardiovascular devices. *Int. J. Artif. Organs.*, **20**, 195–203.

66. Fletcher, M. P., Stahl, G. L., Longhurst, J. C. (1990). In vivo and in vitro assessment of porcine neutrophil activation responses to chemoattractants: Flow cytometric evidence for the selective absence of formyl peptide receptors. *J. Leukocyte Biol.*, **47**, 355–365.
67. Suzuki, Y., Yeung, A. C., Ikeno, F. (2011). The representative porcine model for human cardiovascular disease. *J. Biomed. Biotechnol.*, **2011**, 195483.
68. Couret, D., de Bourmont, S., Prat, N., Cordier, P. Y., Soureau, J. B., Lambert, D., et al. (2013). A pig model for blunt chest trauma: No pulmonary edema in the early phase. *Am. J. Emerg. Med.*, **31**, 1220–1225.
69. Kieser, J. A., Weller, S., Swain, M. V., Neil Waddell, J., Das, R. (2013). Compressive rib fracture: Peri-mortem and post-mortem trauma patterns in a pig model. *Leg. Med. (Tokyo)*, **15**, 193–201.
70. Byard, R. W., Cains, G. E., Gilbert, J. D. (2007). Use of a pig model to demonstrate vulnerability of major neck vessels to inflicted trauma from common household items. *Am. J. Foren. Med. Path.*, **28**, 31–34.
71. Zierold, D., Perlstein, J., Weidman, E. R., Wiedeman, J. E. (2001). Penetrating trauma to the diaphragm: Natural history and ultrasonographic characteristics of untreated injury in a pig model. *Arch. Surg.*, **136**, 32–37.
72. Sauer, M., Altrichter, J., Mencke, T., Klohr, S., Thomsen, M., Kreutzer, H. J., et al. (2013). Role of different replacement fluids during extracorporeal treatment in a pig model of sepsis. *Ther. Apher. Dial.*, **17**, 84–92.
73. Sauer, M., Altrichter, J., Mencke, T., Klohr, S., Thomsen, M., Kreutzer, H. J., et al. (2012). Plasma separation by centrifugation and subsequent plasma filtration: Impact on survival in a pig model of sepsis. *Ther. Apher. Dial.*, **16**, 205–212.
74. Sauer, M., Altrichter, J., Kreutzer, H. J., Logters, T., Scholz, M., Noldge-Schomburg, G., et al. (2009). Extracorporeal cell therapy with granulocytes in a pig model of Gram-positive sepsis. *Crit. Care Med.*, **37**, 606–613.
75. Lobe, T. E., Woodall, D. L., Griffin, M. P. (1991). Early hemodynamic indicators of Gram-negative sepsis and shock in an infant pig model. *J. Pediatr. Surg.*, **26**, 1051–1057.
76. Dunn, D. L., Ferguson, R. M. (1982). Immunotherapy of Gram-negative bacterial sepsis: Enhanced survival in a guinea pig model by use of rabbit antiserum to *Escherichia coli* J5. *Surgery*, **92**, 212–219.

77. Szebeni, J., Baranyi, B., Savay, S., Bodo, M., Morse, D. S., Basta, M., et al. (2000). Liposome-induced pulmonary hypertension: Properties and mechanism of a complement-mediated pseudoallergic reaction. *Am. J. Physiol.*, **279**, H1319–H1328.
78. Csukas, D., Urbanics, R., Weber, G., Rosivall, L., Szebeni, J. (2015). Pulmonary intrvascular macrophages: Prime suspects as cellular mediators of porcine CARPA. *Eur. J. Nanomed.*, **7**, 27–36.
79. Kuznetsova, N. R., Sevrin, C., Lespineux, D., Bovin, N. V., Vodovozova, E. L., Mészáros, T., et al. (2012). Hemocompatibility of liposomes loaded with lipophilic prodrugs of methotrexate and melphalan in the lipid bilayer. *J. Control. Release*, **160**, 394–400.
80. Merkel, O. M., Urbanics, R., Bedőcs, P., Rozsnyay, Z., Rosivall, L., Toth, M., et al. (2011). In vitro and in vivo complement activation and related anaphylactic effects associated with polyethylenimine and polyethylenimine-grafted-poly (ethylene glycol) block copolymers. *Biomaterials*, **32**, 4936–4942.
81. Szebeni, J., Bedőcs, P., Urbanics, R., Bunger, R., Rosivall, L., Tóth, M., et al. (2012). Prevention of infusion reactions to PEGylated liposomal doxorubicin via tachyphylaxis induction by placebo vesicles: A porcine model. *J. Control. Release*, **160**, 382–387.

Chapter 12

Blood Cell Changes in Complement Activation-Related Pseudoallergy: Intertwining of Cellular and Humoral Interactions

Zsófia Patkó, MD, PhD,[a] and János Szebeni, MD, PhD, DSc[b,c]

[a]*Department of Nuclear Medicine, Borsod County University Hospital, Miskolc, Hungary*
[b]*Nanomedicine Research and Education Center, Institute of Pathophysiology, Semmelweis University School of Medicine, Budapest, Hungary*
[c]*SeroScience Ltd., Budapest, Hungary*

Keywords: anaphylactic shock, anaphylactoid reactions, anaphylatoxins, anaphylaxis, animal models, complement activation-related pseudoallergy (CARPA), hemodynamics, hypersensitivity reactions, liposomes, liposome-encapsulated hemoglobin, nanomedicines, nanoparticles, PEGylation, platelets, polyethylene glycol, white blood cells, adverse drug reactions, drug toxicity, allergy, innate immunity, C3a, C5a, macrophages, mast cells, vasoactive mediators, histamine, thromboxane, terminal complex, cytokines

12.1 Introduction

Anaphylatoxins, key mediators of complement activation-related pseudoallergy (CARPA), can activate white blood cells (WBCs) and

---

Immune Aspects of Biopharmaceuticals and Nanomedicines
Edited by Raj Bawa, János Szebeni, Thomas J. Webster, and Gerald F. Audette
Copyright © 2018 Pan Stanford Publishing Pte. Ltd.
ISBN 978-981-4774-52-9 (Hardcover), 978-0-203-73153-6 (eBook)
www.panstanford.com

platelets (PLs). This activation can lead to the binding of these cells to each other and also to capillary endothelial cells, entailing microthrombus formation and circulatory blockage mainly in the pulmonary and coronary microcirculation.

CARPA, as its name implies, is a non-Ig-E-mediated (pseudo-allergic) hypersensitivity reaction (HSR) that is triggered by C activation, or at least C activation plays a major contributing role. CARPA is best known in the context of nanotoxicity, since nanomedicines, i.e., particulate drugs and agents in the nanoscale range (10^{-9}–10^{-6} m) often cause such reactions. As reviewed earlier [1–8], and also discussed in other papers of this volume, the phenomenon represents an immune barrier to the clinical use of many promising nanomedicines. In essence, CARPA may be perceived as a chemical stress on blood that arises as a consequence of the similarity of nanomedicines to viruses, between which the immune system cannot make difference [8]. The entailing acute inflammatory reaction may have triple harm via (1) causing rapid clearance of the drug and, thus, reducing or eliminating its efficacy; (2) causing an acute illness in the host whose most severe manifestation is anaphylaxis; and (3) leading to immunogenicity, which turns drugs into vaccines and thus abolishes their therapeutic use [9, 10].

12.2 Blood Cell Changes in CARPA: Human and Animal Data

12.2.1 Human Studies

The best know HSRs whose pathomechanism is likely to involve CARPA are triggered by anticancer drugs administered in micellar solvents, like Taxol or Taxotere, liposomal drugs, like Doxil and AmBisome and radio-contrast agents, like iodine-containing contrast media. Clinical reports on hematological changes caused by the above drugs often list thrombocytopenia and leukopenia, but usually these changes represent cytotoxic, rather than the immune effects of reactogenic drugs. However, in the case of radiocontrast media, early association of leukopenia has been attributed to the immune reactivity of the contrast agent [11, 12].

12.2.2 Studies in Pigs

Figure 12.1a,b shows the platelet and white blood cell (WBC) changes observed in pigs that were treated with bolus injection of large multilamellar liposomes (MLV) [13]. It shows 10–30% drop of platelets and WBC counts within 30 min after injection in 6 of 8 animals. It is also seen that the individual variation of changes is substantial. Further studies in pigs analyzing blood cell changes during long-term infusion of PEGylated small unilamellar liposomes (SUV) showed initial leukopenia followed by leukocytosis, i.e., typical roller-coaster pattern of anaphylatoxin-induced WBC changes [14–16].

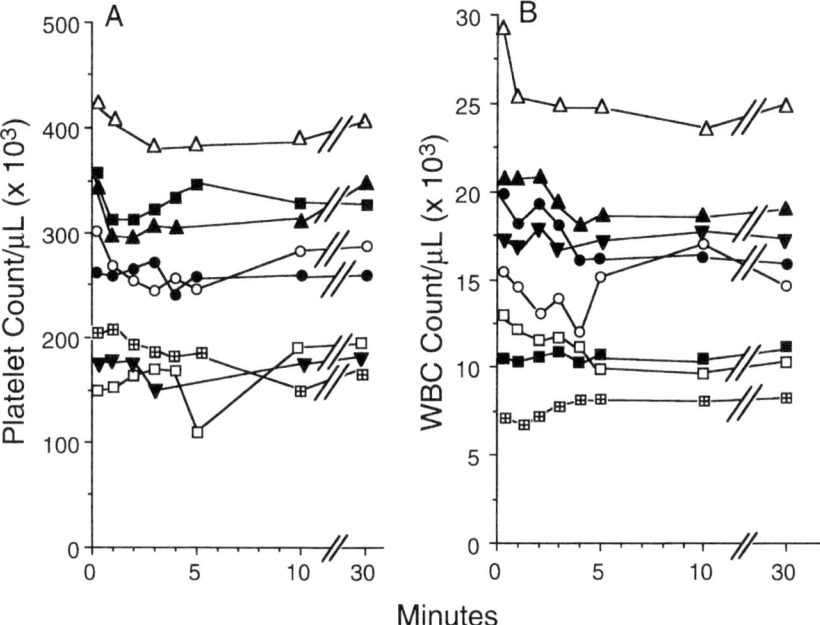

Figure 12.1 Liposome-induced changes in platelet (a) and white blood cell (WBC, b) counts in pigs. Cell counts were determined before injection of MLV and at different times thereafter, as indicated. Different symbols designate individual pigs. Reproduced from [13], with permission.

12.2.3 Studies in Rats

Liposomes and other C activators cause CARPA in rats, too, which phenomenon was first described and analyzed in detail by Rabinovici et al. in a series of studies starting in the late 1980s [17–19]. They injected liposome-encapsulated hemoglobin (LEH), a red cell substitute, along with appropriate controls, in conscious, normovolemic or exsanguinated rats and studied the hemodynamic and hematological consequences of treatment. LEH, as well as the carrier liposomes, caused thrombocytopenia, whose extent depended on the mode of LEH administration, namely, a 10% top load [17], 50% exchange transfusion [18] and 10% top load with lyophilized LEH [19] led to 60%, 40%, and 24% thrombocytopenia, respectively. Also in rats the above thrombocytopenia was associated with a rise, rather than drop of WBC count [17–19].

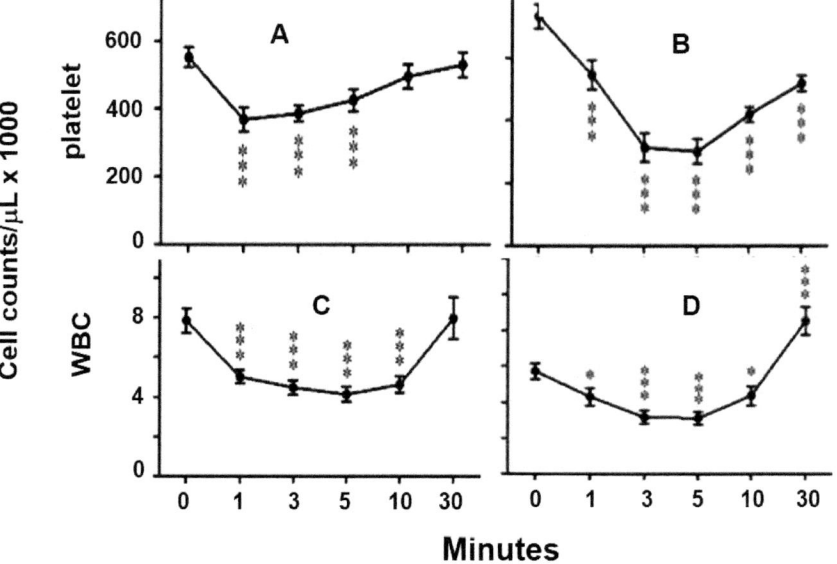

Figure 12.2 Hematologic effects of i.v. bolus administration of zymosan and AmBisome in rats. Values shown are mean ± SE (N = 8 rats in each group). *, **, ***: p < 0.05, 0.01, 0.001 vs. the time 0 value. The zymosan and AmBisome doses were 10 and 22 mg/kg, normalized to phospholipid amount in case of AmBisome. Modified from [20], with permission.

Consistent with the above reports 20 years ago, the rat studies in our laboratory using zymosan and liposomal amphotericin B (AmBisome) as reaction inducers, led to 30 and 60% drop in platelet counts between 1 and 3 min after i.v. injection of zymosan (Fig. 12.2a) and AmBisome (Fig. 12.2b), respectively. Also, we observed 50% leukopenia with zymosan (Fig. 12.2c) and 20% leukopenia with AmBisome (Fig. 12.2d), with the WBC counts reaching minimum around 4–5 min. In case of AmBisome, the WBC started to rise over baseline after 10 min (Fig. 12.2d), implying the start of reactive leukocytosis [20]. This study confirmed the huge difference between rats and pigs in sensitivity to reactogenic liposomes, as the reactogenic doses in these experiments were 100–1000-fold higher than that reported for pigs [20].

Regarding the destiny of platelets during thrombocytopenia in rats, Phillips et al. isolated and labeled rat platelets with ^{111}In, then re-infused the cells into the same animal [21]. Fifteen minutes later the animals were infused with a 10% top load of LEH, as well as carrier liposomes or free bovine hemoglobin (Hb). LEH, but not the controls, caused a transient 50% decrease in ^{111}In platelet activity 2–5 min post-infusion, which returned to baseline levels by 15 min. Tracing of labeled platelets with a gamma scintigraphic camera showed them to be sequestered in the lungs and liver [21].

12.3 Platelets and Their Role in CARPA

12.3.1 Platelets: The Mimosa of Blood Cells

Just like the plant mimosa, platelets are known to be very sensitive and fragile cells that rapidly respond to changes in their milieu. Besides their fundamental role in coagulation, they actively participate in inflammatory reactions and in the pathogenesis of prominent diseases, like hypertension [31–33], hypercholesterolemia with atherosclerosis [34], thrombosis, asthma [35, 36], complications of diabetes [37, 38] angina pectoris, myocardial infarction and stroke [39]. Other than injury, platelet activation can occur as a consequence of a variety of abnormal stimuli, e.g., shear stress, physical activity, major alterations in endocrine homeostasis. Figure 12.3 shows that the activation of platelets results in shape and surface changes that greatly expand the surface of these cells interacting with each other as well as with endothelium and circulating leukocytes. In addition, stimulated platelets release several soluble mediators, including PF4, PDGF, IL-1β, tissue factor and CD40L. The next sections will highlight details of platelet aggregation and its molecular and cellular interactions during CARPA.

 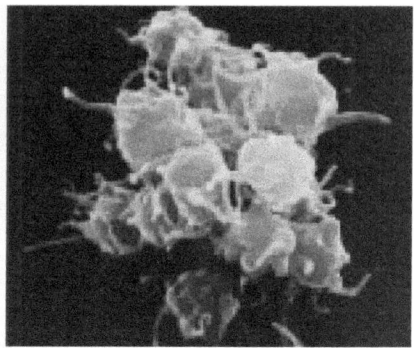

Resting platelets   **Activated platelets**

Figure 12.3 Electron micrographs of resting and activated platelets.

12.3.2 The Vicious Cycle of Platelet and Complement Activation

Resting platelets do not express anaphylatoxin receptors [28, 40], only activated platelets do so. When expressed, the binding of C3a and C5a to these receptors accelerate cell activation, thus linking inflammation to thrombosis [41–46]. Along this line, Patzelt et al. have shown that the expression of both anaphylatoxin-receptors on platelets strongly correlated with their state of activation, measured by the expression of activation markers P-selectin and SDF-1 [40]. On the other hand, activated platelets not only express anaphylatoxin receptors, but many other molecules that themselves activate C, and therefore accelerate C action. Thus, activated platelets have an intrinsic capacity to activate C, both the classical and alternative pathways. This has been shown when activated platelets were exposed to normal plasma or serum, and C activation was proportional to the extent of platelet activation and depended on the activator [41–45]. Platelets can activate C via different mechanisms, which have been proven for both adherent and activated platelets in suspension. In these processes, receptor for the globular head of C1q (gC1qR/p33), chondroitin sulfate and P-selectin were shown to play key roles [43]. In addition, the deposition of the terminal complex (C5b-9) on the platelet membrane also contributes the ensuing cellular changes, such as elevated surface P-selectin expression, transient membrane depolarization, granule secretion, translocation of phosphatidylserine to the outer membrane leaflet, enhanced platelet procoagulant activity and generation of platelet microparticles (PMP) [51–55].

Regarding the latter phenomenon, a unique feature of PMP is that they express 50- to 100-fold higher procoagulant activity compared to activated platelets [56]. They present glycoproteins Ib, IIb/IIIa, and P-selectin on their surface ready to interact with leukocytes and the vascular endothelium [57–59]. Furthermore, in a vicious cycle with C-induced PMP formation, PMP support *in situ* C activation via the classical and alternative pathways, leading to C3b and C5b-9 deposition [43]. These processes are held under control by C regulatory proteins (C1 inhibitor (C1-INH), CD55 and CD59 also expressed on PMP [43].

One example of a role of C in platelet activation is the thrombogenic action of certain bacteria, whereupon the binding of both C proteins and fibrinogen to Gp IIb/IIIa are needed for activation [60]. Another clinically relevant example of C activation by platelets is CARPA, wherein platelet activation is a key step in causing symptoms, most importantly pulmonary hypertension, which is due to microcirculatory blockade in the lung following WBC and platelet activation, aggregation, capillary margination and microthrombus formation.

12.3.3 The CARPA Cascade

Figure 12.4 shows the coupling of cellular and molecular interactions in CARPA, while Fig. 12.5 delineates the complex intertwining of different physiological processes leading to the typical symptoms of CARPA, specified in the red-shaded boxes of Fig. 12.5.

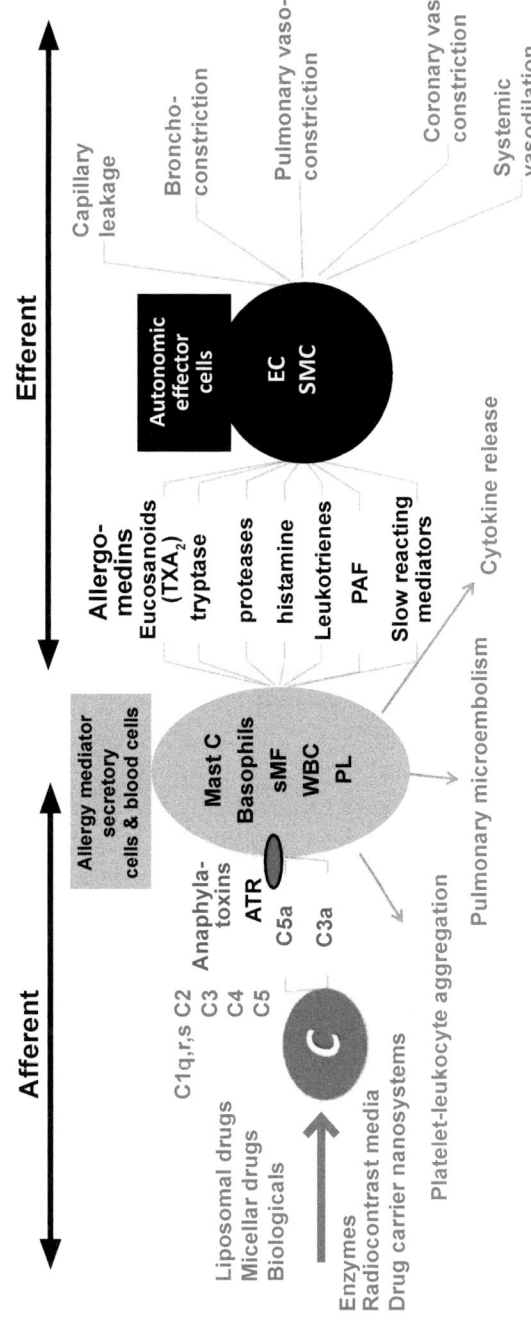

Figure 12.4 The CARPA cascade. The hypothetical scheme, sketched on the basis of collective information in refs. [2–4, 22–27], illustrates the steps and interactions among cells and reaction mediators. ATR, anaphylatoxin receptors; Mast C, mast cells; sMF, secretory macrophages; WBC, white blood cells; PL, platelets; EC, endothelial cells; SMC, smooth muscle cells. Reproduced from [8], with permission.

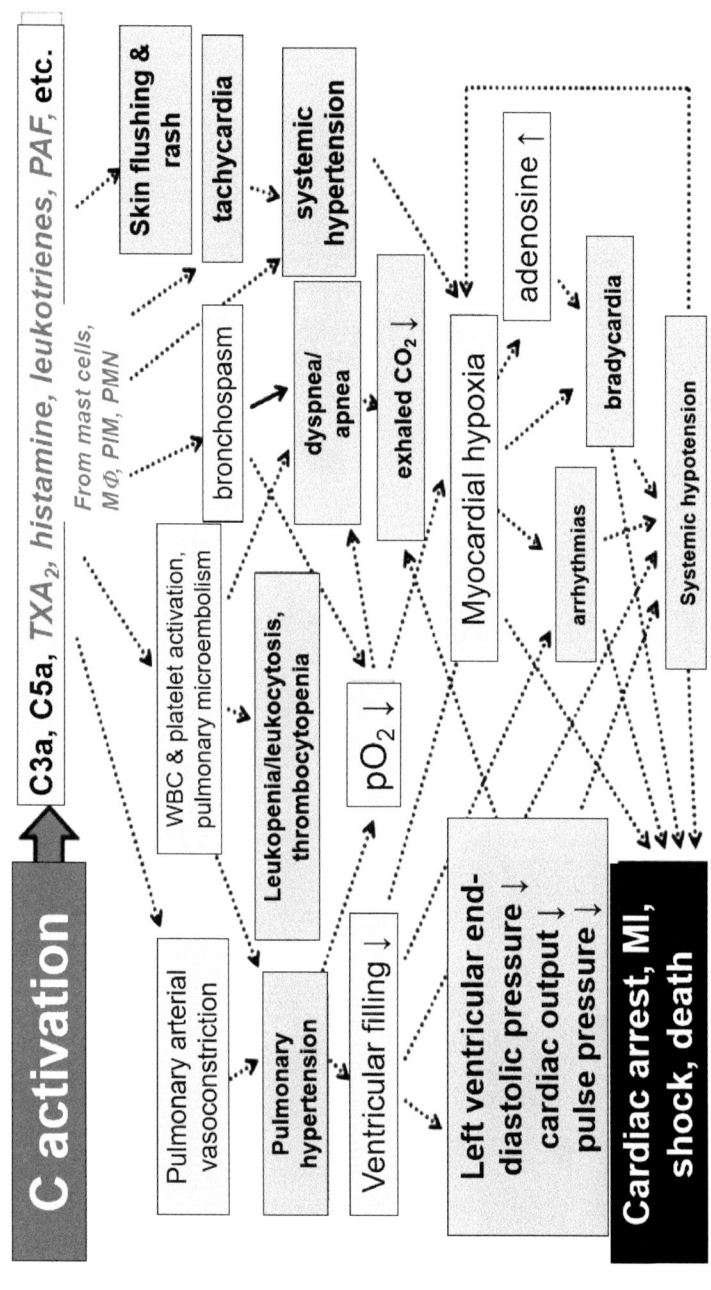

Figure 12.5 Inter-relations among the different abnormalities during CARPA, leading to clinical symptoms. Hypothetical scheme delineated on the basis of collective information in refs. [2–4, 22–27]. The physiological changes and clinical symptoms are boxed with yellow and blue background, respectively. Reproduced from [8] with permission.

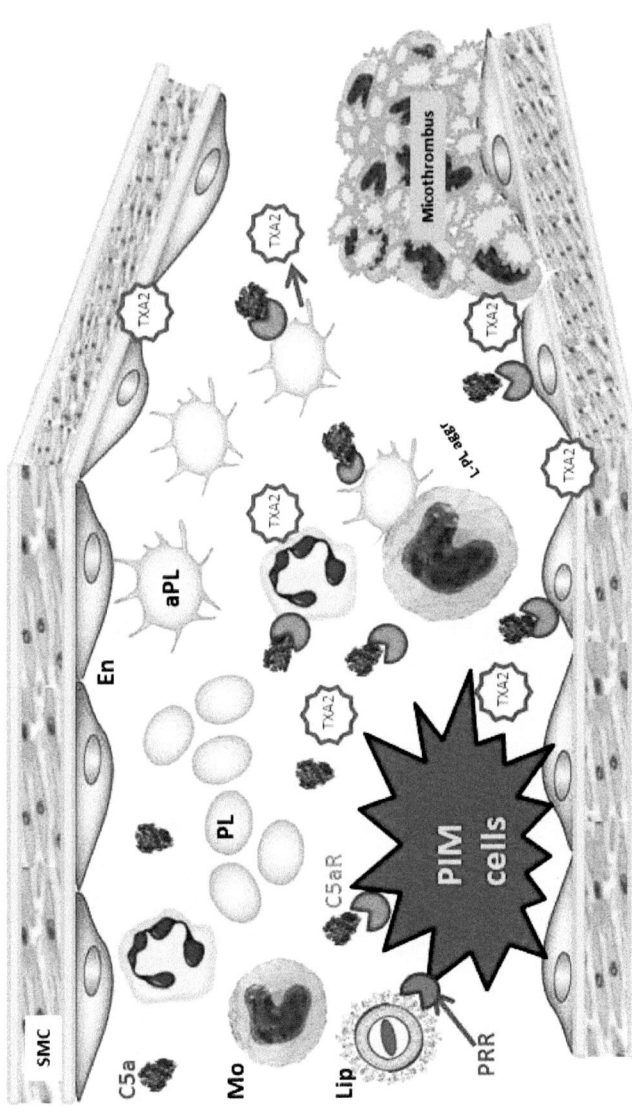

Figure 12.6 Molecular and cellular interactions underlying pulmonary hypertension during CARPA. The steps are described in the text. Abbreviations: SMC, smooth muscle cells; Mon, monocytes; PL, platelets; aPL. Activated platelets; C5aR, C5a receptor; L-PL aggr, leukocyte-platelet aggregates, PRR, patter recognition receptors, Lip, liposomes (Doxil).

These symptoms, in order of increasing severity, include headache, chills, fever, tachycardia, pruritus, rash, flushing, hypotension, hypertension, shortness of breath, apnea, facial swelling, chest pain, back pain, tightness in the chest and throat, arrhythmias, bradyarrhythmia, cyanosis, syncope, anaphylactic shock, myocardial infarction, cardiac arrest, death [2–4]. The blood cell changes, manifested in leukopenia followed by leukocytosis, thrombocytopenia, may include granulopenia and/or lymphopenia, and usually remain subclinical. In CARPA pathogenesis they are caused mainly by C activation-related anaphylatoxinemia, and they lead to dyspnea, due to blockage in the lung microcirculation [22–27]. Figure 12.6 illustrates the latter process, the cellular interactions leading to circulatory blockage in the pulmonary arterioles.

12.4 Zooming on Interactions, Receptors and Mediators in CARPA-Associated Blood Cell Changes

12.4.1 The Effects of Anaphylatoxins

The role of C3a and C5a in C-activation-related blood cell changes have been studied for a long time [22–27]. It was shown, among others, that C5a binds to monocytes/macrophages and neutrophils in a concentration-dependent, saturable process, which, in the case of neutrophils, leads to myeloperoxidase release, an index of degranulation, i.e., release of multiple vasoactive substances [28]. Nonmyeloid cells that express anaphylatoxin receptors include endothelial and epithelial cells and smooth muscle cells, responding to anaphylatoxin binding with increased permeability and contraction, respectively [22, 23]. There is, furthermore, ample direct experimental evidence for the causal role of anaphylatoxins in CARPA and associated blood cell changes, such as the inhibitory effect of an anti-porcine C5a antibody (GS1) [29] on liposome-induced porcine CARPA [13], and the prevention of LEH-induced thrombocytopenia in rats by decomplementing the animals [30]. The latter experiment showed that in C-depleted rats the platelet counts did not change following injection of LEH, while in control rats 40% thrombocytopenia developed within 4 min after treatment. This was followed by gradual return of platelet counts to normal levels over 60 min. The drop in circulating platelets was correlated with a rapid redistribution of ^{111}In-platelets from the circulation to the lungs and liver, whereas C-depleted rats showed no sequestration of PLs into these organs.

12.4.2 Leukocyte-Platelet Aggregation

Leukocytes and platelets are known to interact dynamically in a variety of physiological and pathophysiological processes. These interactions are initiated by activation of one of the participating cells and lead to aggregation and consequent microthrombus formation—as well as to further activation of the participating cells with consequent cytokine release.

The most potent initiator of heteroaggregate formation is platelet activation, which occurs in a variety of diseases from mild homeostatic changes to acute life-threatening events. Platelets, when activated, promptly bind to each other and circulating leukocytes. They are especially attracted to monocytes; therefore, monocyte-platelet aggregation has been identified as the most sensitive marker of platelet activation. Activated platelets then rapidly attach to the endothel or to circulating leukocytes, forming heteroaggregates. Similarly, to platelets, activated leukocytes express adhesive receptors that are involved in both leukocyte-platelet and leukocyte-endothel interactions. Accordingly, monocytes adhering to the endothelial surface were shown to have attached platelets as well on the surface. When coupled, platelets and leukocytes interact and further activate each other to have greater affinity to adhere to endothelial surfaces, especially in small vessels. Among many surface receptors key players in these interactions are platelet P-selectin and leukocyte Mac-1, securing relatively long attachment of these cells to each other and to the endothel, eliminating leukocytes and leukocyte-platelet aggregates from the circulation. These intercellular interactions also lead to the liberation of soluble factors, e.g., P-selectin and Mac-1 adhesion lead to the release of mediators that further upregulate their expression. Nevertheless, these processes are reversible in case the trigger stimulus disappears, and the participating cells soon re-enter the blood flow as normal cells. This explains the findings of animal and clinical data on transient leukopenia and thrombocytopenia as symptoms of adverse drug reactions.

12.4.3 Surface Receptor-Ligand Interactions in Leukocyte-Platelet Aggregation

As mentioned, activated platelets bind to leukocytes, above all to monocytes. As a consequence, elevated monocyte-platelet heteroaggregate levels were found in many conditions resulting platelet activation, e.g., high shear, endocrine disorders, diabetes, atherosclerosis, hypertension etc. Platelets attach to monocytes in a two-step process, mediated by selectins and integrins. As a first step, P-selectin is connected to the leukocyte P-selectin glycoprotein ligand (PSGL-1). This connection is then stabilized by the ligation of the leukocyte integrin Mac-1. On the platelet, GpIba has been shown to interact with Mac-1. Platelet integrin GpIIb/IIIa also contributes to the stabilization of the complex by Mac-1, when fibrinogen or other ligands, e.g., CD40L are present. This is very similar to the formation of platelet-platelet aggregates when GpIIb/IIIa receptors are connected via fibrinogen.

Figure 12.7 shows the receptors and ligands involved in platelet-monocyte interactions, as well as the inflammatory and thrombogenic mediators that liberate during their binding, namely, IL-1, CD40L from platelets and IL-8, MCP-1, TNF-α and PGE2 from monocytes. It should also be mentioned that the attachment of platelets to monocytes induces CD16 upregulation on CD14+ CD16− monocytes resulting in a phenotype change and a higher pro-inflammatory activity.

12.4.4 P-Selectin: A Linker between Platelet and C Activation

P-selectin is a well-known adhesive protein on platelets that binds to PSGL-1 on vascular endothelium and leukocytes as a consequence of platelet activation. It is stored in the alpha granules of platelets and becomes translocated to the surface membrane rapidly upon activation, therefore its expression is a classical marker of platelet activity. The molecular structure of P-selectin contains 9 C binding domains, one of which has been identified as a C3b binding site. Hence, when P selectin gets to the surface of platelets upon activation, a binding site for C3b gets exposed in order to clear the C3b generated during concurrent C activation. Interestingly, the pathway of platelet activation-related C activation turns from C1Q-mediated classical into alternative as the secretory products of a granules, C1 inhibitor and factor D—that cleaves factor B to its active form—starts to act [43, 46].

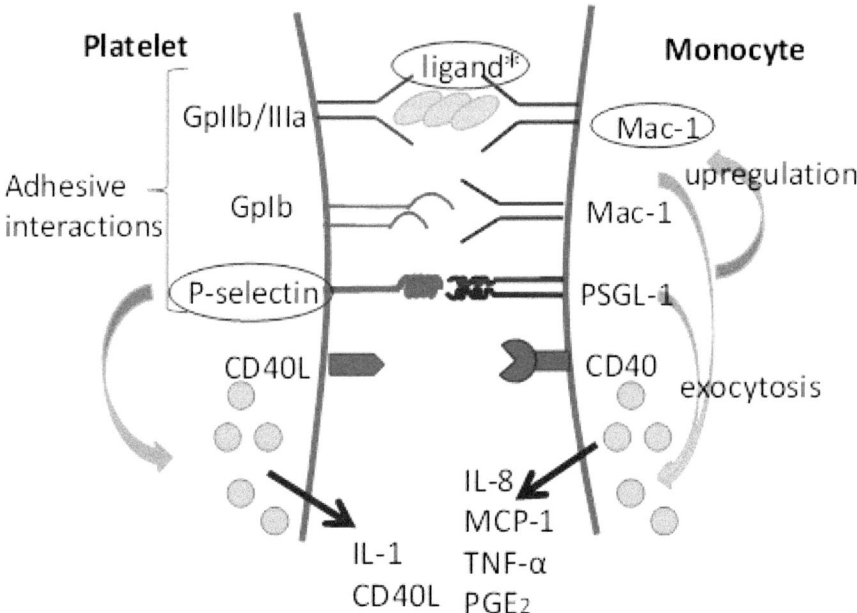

Figure 12.7 Platelet-monocyte adhesive interactions and inflammatory reactions following complement triggered activation. Platelet-monocyte heteroaggregation is initiated by the ligation of P-selectin to PSGL-1 on monocytes. As a result, integrin Mac-1 gets upregulated, and inflammatory mediators are released. Monocyte-platelet complex is then stabilized by Mac-1, which interacts with GpIb and GpIIb/IIIa on the platelet. Stabilization via GpIIb/IIIa and Mac-1 needs a soluble ligand present in the plasma or originating from the activated platelet such as fibrinogen or CD40L. CD40L also further increases inflammatory reactions of the monocyte.

12.4.5 The C1q Receptor

The receptor for the globular head of C1q (gC1qR) is expressed on the surface of adherent platelets and PMP. Collagen binding induces enhanced surface expression of the receptor, while other agonists fail to exhibit such function. The gC1qR molecule has been shown to interact not only with C1q but also with a variety of structurally distinct ligands, such as proteins of the blood coagulation system (HMW kininogen, factor XII). This molecule (gC1qR) also contributes to *in situ* classical pathway C activation [43, 47]. Chondroitin sulfate has been shown to be stored in platelet alpha granula and is exposed on the surface of platelets rapidly in response to a variety of agonists, including ADP, collagen, adrenalin and thrombin. As other glucosaminoglycans, chondroitin sulfate facilitates the binding of soluble proteins on the surface. C1q has been shown to bind to chondroitin sulfate in high amounts, while C activation was abolished in the absence of C1q. Following activation with thrombin receptor activating peptide (TRAP), chondroitin sulfate is expressed and C1q, C4, C3 and C9 are bound to the platelet surface [48]. As it has been demonstrated recently, C4BP and factor H also bind to platelet chondroitin sulfate-A. The chondroitin sulfate present in the alpha granules of platelets is fully sulfated [48]. In a clinical observation, injection of an over-sulfated form of chondroitin sulfate has been shown to cause fatal anaphylactoid reactions by activating both the C and the contact systems [49].

12.4.6 Complement Activation and Blood Cell Changes: The Whole Picture

Figure 12.8 attempts to give a comprehensive picture on C activation and cellular and molecular changes in blood, taking together all discussed facts and relationships. Thus, activation of surface adherent, as well as circulating platelets can be triggered by agonists or shear [44]. When platelets are activated, C1q, C3b, C4 and C5-9 are deposited on their surface. However, C activation is strictly regulated on platelets, therefore, binding of C elements on the platelet surface does not always result in activation. Platelet activation and consequent C activation may participate in several physiological and pathophysiological processes. Shear induces platelet activation and, in turn, C activation via the classical pathway. Activation of platelets by platelet agonists induces C activation mainly by the alternative pathway. This may result from the mechanism of platelet activation when it results in alpha granule release, which contains C1 inhibitor reducing the classical pathway activation. In this process, weak agonists such as ADP were found to induce less, and thrombin to induce more intense activation of the C cascade.

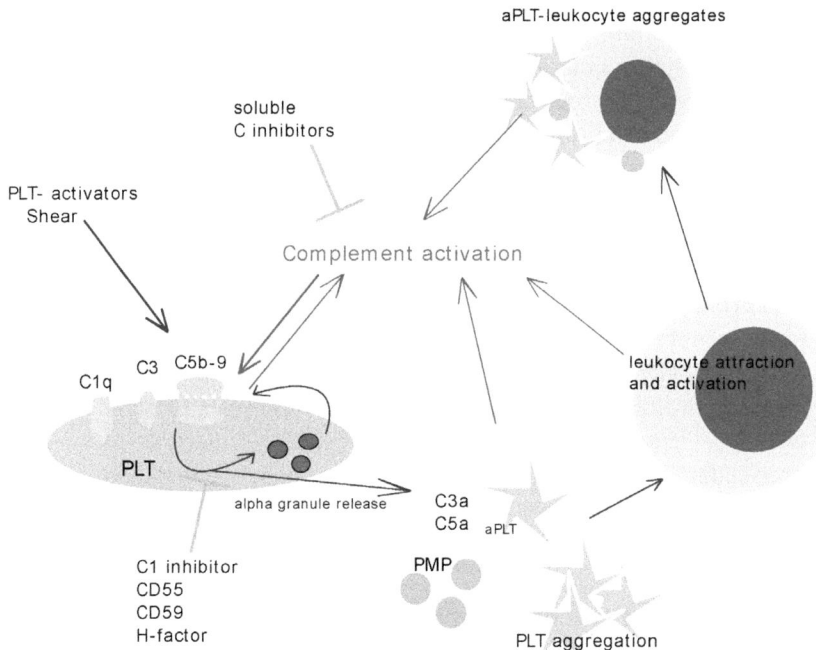

Figure 12.8 Promoters and inhibitors of a vicious cycle of platelet, WBC and C activation. Platelet activators and shear induce deposition of complement elements on platelet surface. Activation by C1q deposition, C3 or sublytic concentrations of C5b-9 lead to alpha granule secretion, release of PMP and activation of platelets. Secreted chemokines attract leukocytes to the site of activation. Activated platelets and PMP attach to leukocytes forming leukocyte-platelet aggregates. Both activated platelets and heteroaggregates further enhance complement activation, leading to an activation circle. However, this circle is limited by soluble and platelet surface complement inhibitors, e.g., CD55, CD59, C1-inhibitor or surface-bound factor H. PLT: platelet, aPLT: activated platelet, PMP: platelet microparticle.

As a consequence of platelet mediated C activation, activation of the clotting cascade can lead to the generation of C3a and C5a. Beside further activation of platelet functions in a positive feedback cycle, these anaphylatoxins invite leukocytes to the site of platelet activation and activate them [50]. This activation may result in the induction of inflammatory pathways as well as formation of heteroaggregates, which further enhances the functions of leukocytes and platelets. In the studies of Yin et al. [43], platelet mediated C activation was associated with generation of physiologically relevant levels of inflammatory peptides, C3a, C4a and C5a, which support recruitment of leukocytes and increase vascular permeability [46].

12.4.7 Complement and Monocyte-Platelet Aggregates

Despite data suggesting parallel functions of platelet-leukocyte interactions and C activation, the role of leukocyte-platelet aggregate formation in anaphylactoid reactions is not clear. Namely, pulmonary hypertension is one of the life-threatening side effects of excessive C activation. In anaphylactoid reactions, obstruction of pulmonary vessels by immune precipitates and aggregates of PMN leukocytes and platelets was observed. It also has been stated that (the second, protracted phase of) anaphylactoid reactions are less intense in leukopenic animals [64]. The hypotension observed during anaphylaxis was also more marked in normal than in leukopenic animals.

The administration of C1q inhibitor not only inhibited C, but also leukocyte and platelet activation in a pig to human transplantation model [65]. C1q INH also binds to E-and P-selectins and interferes with leukocyte adhesion. Leukocytes and platelets are activated also when pig kidney is perfused with human blood. In this experiment the C system was activated and significant leukocyte and platelet activation was observed. Administration of C1-INH/C1 inhibition significantly reduced leukocyte and platelet aggregation [65].

Figure 12.9 shows that C elements activate platelets and leukocytes, which is followed by the interaction of these cells. Activated platelets attach to leukocytes, and this interaction enhances prothrombotic and proinflammatory processes. Soluble mediators derived from this interaction, such as fibrinogen and CD40L stabilize heteroaggregates by ligation of the fibrinogen receptors of both cell types while leukocyte and platelet activation amplify C activation. These processes also lead to the liberation of proinflammatory and prothrombotic mediators and expression of adhesion molecules: P-selectin, GpIIb/IIIa, Mac-1-heteroaggregate formation.

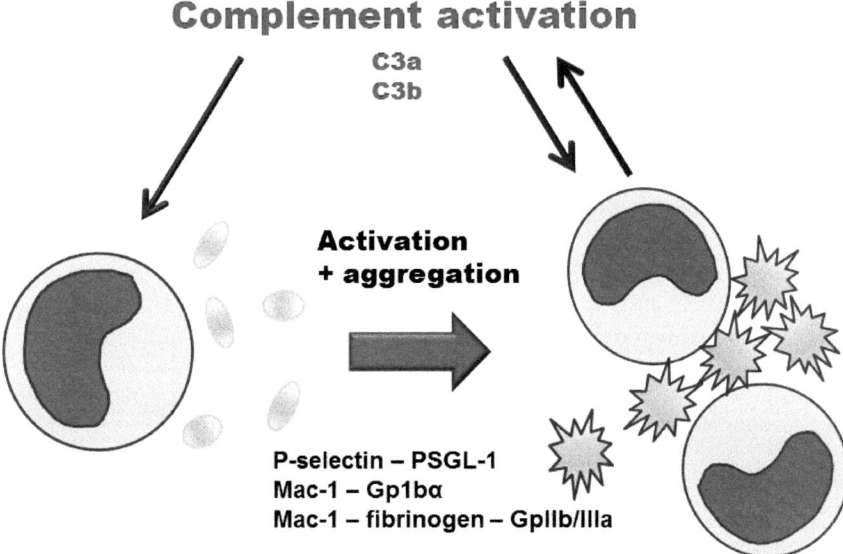

Figure 12.9 Scheme of interactions between leukocytes, platelets and C proteins. Activation of the C system has been associated with prelesional stages of atherosclerosis and the progression of the disease. Complement elements are deposited in atherosclerotic lesions [66]. In endothel C5b-9 promotes the secretion of vWF and elevated expression of P-selectin, E-selectin, intracellular adhesion molecules (ICAM), TF, MCP-1, IL-b, and IL-8 promoting leukocyte adhesion. Complement is also activated in ischemia-reperfusion injury [67].

12.4.8 Leukocyte and Platelet Counts in Anaphylaxis

In anaphylaxis leukocyte-platelet aggregates are rapidly cleared from circulation primarily because their entrapment in the pulmonary circulation [64]. Consequently, interruption of leukocyte-platelet interactions may have important therapeutic consequences to prevent complications of anaphylactoid reactions. One of these possibilities is the administration of GpIIb/IIIa inhibitor abciximab as it not only inhibits the fibrinogen receptor of platelets but also affects Mac-1, thereby (possibly) preventing the stabilization of leukocyte-platelet aggregates.

12.5 Conclusions

Nano-carriers represent a new, promising method for the treatment of serious, life-threatening diseases by delivering the drugs to specific cells. However, nano-sized drug-delivery systems also carry the possibility of serious, life-threatening side effects. Nanoparticles are in the size range of viruses; therefore, in contrast to chemical agents, they represent a challenge for the immune system. As a first-line of defense of the body, the C system may be activated when nano-materials get into the blood stream resulting in a non-IgE mediated immune-response: the C activation-related pseudoallergy (CARPA). In severe cases, the symptoms of CARPA mimic those observed in IgE mediated type I hypersensitive reactions, e.g., angioedema, bronchospasm, chest pain, dyspnea, choking, flush, fever, headache, hypotension, pulmonary hypertension etc.

The clinical relevance of CARPA lies in the rare, but severe, life-threatening reactions that cannot be predicted and may cause death in an occasional patient [1–8]. Such risk can be tolerated in case of life-saving (mostly anti-cancer) nanomedicines, but not in the case of diagnostic or non-life saving therapeutics. The CARPA concept has been developing over 20 years since the recognition of C activation as the underlying cause of hemodynamic side effects of liposome-encapsulated hemoglobin (LEH) [69]. It was suggested to represent a stress reaction in blood [8].

Disclosures and Conflict of Interest

This chapter was originally published as: Patkó, Z., Szebeni, J., Blood cell changes in complement activation-related pseudoallergy, *Eur. J. Nanomed.* **7**, 233244 (2015); DOI: https://doi.org/10.1515/ejnm-

2015-0021, and appears here, with edits, by kind permission of the publisher, Walter de Gruyter GmbH, Berlin, Germany. The authors declare that they have no conflict of interest and have no affiliations or financial involvement with any organization or entity discussed in this chapter. No writing assistance was utilized in the production of this chapter and the authors have received no payment for its preparation.

Corresponding Author

Dr. Zsófia Patkó
Department of Nuclear Diagnostics and Therapy
Borsod-Abaúj-Zemplén County Hospital
3526 Miskolc, Szentpeteri Kapu 72-76, Hungary
Email: patko.zsofia@gmail.com

About the Authors

Zsófia Patkó is working as a doctor at the Department of Nuclear Diagnostics and Therapy at Borsod County University Hospital, Miskolc, Hungary, and as assistant professor at the University of Miskolc. She graduated in 2004 at Semmelweis University, Budapest. During her PhD studies she has been doing research work at Albert Ludwigs University, Germany, and the Military Hospital Budapest, Hungary. She received her PhD in 2013 at Semmelweis University in the field of leukocyte-platelet interactions. She is interested in molecular and cellular imaging, and interactions of nanoparticles with the immune system and circulating blood elements.

János Szebeni is director of the Nanomedicine Research and Education Center at Semmelweis University. He is also the founder and CEO of a contract research SME "SeroScience," and full Professor of (immune) biology at Miskolc University, Hungary. He has held various guest professor and scientific positions in Hungary and abroad, mostly in the United States, where he lived for 22 years. His research on various themes in hematology, membrane biology, and immunology resulted more than 100 scientific papers (citations: >4550, H index: 35), 14 book chapters, 2 granted patents, and a book entitled *The Complement System: Novel Roles in Health and Disease* (Kluwer, 2004). Three fields stand out where he has been most active: artificial blood, liposomes, and the complement system. His original works led to the "CARPA" concept, which implies that complement activation underlies numerous drug-induced (pseudo) allergic (anaphylactoid) reactions.

References

1. Szebeni, J. (2001). Complement activation-related pseudoallergy caused by liposomes, micellar carriers of intravenous drugs and radiocontrast agents. *Crit. Rev. Ther. Drug Carr. Syst.*, **18**, 567–606.
2. Szebeni, J. (2004). Complement activation-related pseudoallergy: Mechanism of anaphylactoid reactions to drug carriers and radiocontrast agents. In: Szebeni, J., ed. *The Complement System: Novel Roles in Health and Disease*, Kluwer, Boston, pp. 399–440.
3. Szebeni, J. (2005). Complement activation-related pseudoallergy: A new class of drug-induced immune toxicity. *Toxicology*, **216**, 106–121.
4. Szebeni, J., Alving, C. R., Rosivall, L., Bünger, R., Baranyi, L., Bedöcs, P., Tóth, M., Barenholz, Y. (2007). Animal models of complement-mediated hypersensitivity reactions to liposomes and other lipid-based nanoparticles. *J. Liposome Res.*, **17**(2), 107–117.
5. Szebeni, J., Muggia, F., Gabizon, G., Barenholz, Y. (2011). Activation of complement by therapeutic liposomes and other lipid excipient-based therapeutic products: Prediction and prevention. *Adv. Drug Deliv. Rev.*, **63**, 1020–1030.
6. Szebeni, J. (2012). Hemocompatibility testing for nanomedicines and biologicals: Predictive assays for complement mediated infusion reactions. *Eur. J. Nanomed.*, **5**, 33–53.
7. Szebeni, J., Bedocs, P., Csukas, D., Rosivall, L., Bunger, R., Urbanics, R. (2012). A porcine model of complement-mediated infusion reactions to drug carrier nanosystems and other medicines. *Adv. Drug Deliv. Rev.*, **64**(15), 1706–1716.
8. Szebeni, J. (2014). Complement activation-related pseudoallergy: A stress reaction in blood triggered by nanomedicines and biologicals. *Mol. Immunol.*, **61**, 163–173.

9. Szebeni, J., Jiskoot, W. (2014). Immunological issues with nanomedicines: Immunogenicity, hypersensitivity, accelerated clearance and immune suppression. In: Torchillin, V., ed. *Handbook of Nanobiomedical Research*, World Scientific, Singapore, pp. 45–73.
10. Szebeni, J. (2015). Complement activation by nanomaterials. In: Fadeel, B., ed. *Handbook of Safety Assessment of Nanomaterials: From Toxicological Testing to Personalized Medicine*, Pan Stanford Publishing, Singapore, pp. 281–310.
11. Meth, M. J., Maibach, H. I. (2006). Current understanding of contrast media reactions and implications for clinical management. *Drug Saf.*, **29**, 133–141.
12. Kovoor, E. T., Morgan, G. (2010). Severe transient leukopenia following hysterosalpingography. *Gynecol. Obstet. Invest.*, **69**(3), 190–192.
13. Szebeni, J., Fontana, J. L., Wassef, N. M., Mongan, P. D., Morse, D. S., Dobbins, D. E., Stahl, G. L., Bunger, R., Alving, C. R. (1999). Hemodynamic changes induced by liposomes and liposome-encapsulated hemoglobin in pigs: A model for pseudoallergic cardiopulmonary reactions to liposomes. Role of complement and inhibition by soluble CR1 and anti-C5a antibody. *Circulation*, **99**(17), 2302–2309.
14. Hugli, T. E. (1981). The structural basis for anaphylatoxin and chemotactic functions of C3a, C4a, and C5a. *Crit. Rev. Immunol.*, **1**(4), 321–366.
15. Hugli, T. E., Marceau, F. (1985). Effects of the C5a anaphylatoxin and its relationship to cyclo-oxygenase metabolites in rabbit vascular strips. *Br. J. Pharmacol.*, **84**(3), 725–733.
16. Ember, J. A., Sanderson, S. D., Hugli, T. E., Morgan, E. L. (1994). Induction of interleukin-8 synthesis from monocytes by human C5a anaphylatoxin. *Am. J. Pathol.*, **144**(2), 393–403.
17. Rabinovici, R., Rudolph, A. S., Feuerstein, G. (1989). Characterization of hemodynamic, hematologic and biochemical responses to administration of liposome-encapsulated hemoglobin in the conscious, freely moving rat. *Circ. Shock*, **29**, 115–132.
18. Rabinovici, R., Rudolph, A. S., Ligler, F. S., Smith III, E. F., Feuerstein, G. (1992). Biological responses to exchange transfusion with liposome-encapsulated hemoglobin. *Circ. Shock*, **37**(2), 124–133.
19. Rabinovici, R., Rudolph, A. S., Vernick, J., Feuerstein, G. (1994). Lyophilized liposome-encapsulated hemoglobin: Evaluation of

hemodynamic, biochemical, and hematologic responses. *Crit. Care Med.*, **22**, 480–485.

20. Dézsi, L., Fülöp, T., Mészáros, T., Szénási, G., Urbanics, R., Vázsonyi, C., Őrfi, E., Rosivall, L., Nemes, R., Jan Kok, R., Metselaare, J. M., Storm, G., Szebeni, J. (2014). Features of complement activation-related pseudoallergy to liposomes with different surface charge and PEGylation: Comparison of the porcine and rat responses. *J. Control. Release*, **195**, 2–10.

21. Phillips, W. T., Klipper, R., Fresne, D., Rudolph, A. S., Javors, M., Goins, B. (1997). Platelet reactivity with liposome-encapsulated hemoglobin in the rat. *Exp. Hematol.*, **25**, 1347–1356.

22. Klos, A., Tenner, A. J., Johswich, K. O., Ager, R. R., Reis, E. S., Köhl, J. (2009). The role of the anaphylatoxins in health and disease. *Mol. Immunol.*, **46**, 2753–2766.

23. Klos, A., Wende, E., Wareham, K. J., Monk, P. N. (2013). International union of pharmacology. LXXXVII. Complement peptide C5a, C4a, and C3a receptors. *Pharmacol. Rev.*, **65**, 500–543.

24. Heideman, M., Hugli, T. E. (1984). Anaphylatoxin generation in multisystem organ failure. *J. Trauma*, **24**(12), 1038–1043.

25. Hugli, T. E. (1984). Structure and function of anaphylatoxins. *Springer Semin. Immunopathol.*, **7**, 193–219.

26. Marceau, F., Lundberg, C., Hugli, T. E. (1987). Effects of anaphylatoxins on circulation. *Immunopharmacology*, **14**, 67–84.

27. Mousli, M., Hugli, T. E., Landry, Y., Bronner, C. (1992). A mechanism of action for anaphylatoxin C3a stimulation of mast cells. *J. Immunol.*, **148**, 2456–2461.

28. Yancey, K. B., Lawley, T. J., Dersookian, M., Harvath, L. (1989). Analysis of the interaction of human C5a and C5a des Arg with human monocytes and neutrophils: Flow cytometric and chemotaxis studies. *J. Invest. Dermatol.*, **92**, 184–189.

29. Stahl, G. L., Behroozi, F., Morrissey, M. (1997). Monoclonal antibody, GS1, functionally inhibits porcine C5a induced neutrophil activation, but not C5b-9 formation. *FASEB J.*, **11**, A331.

30. Goins, B., Phillips, W. T., Klipper, R., Rudolph, A. S. (1997). Role of complement in rats injected with liposome-encapsulated hemoglobin. *J. Surg. Res.*, **68**, 99–105.

31. Nityanand, S., Pande, I., Bajpai, V. K., Singh, L., Chandra, M., Singh, B. N. (1993). Platelets in essential hypertension. *Thromb. Res.*, **72**(5), 447–454.

32. Nadar, S. K., Blann, A. D., Lip, G. Y. (2004). Plasma and platelet-derived vascular endothelial growth factor and angiopoietin-1 in hypertension: Effects of antihypertensive therapy. *J. Intern. Med.,* **256**(4), 331–337.

33. Nadar, S. K., Blann, A. D., Kamath, S., Beevers, D. G., Lip, G. Y. (2004). Platelet indexes in relation to target organ damage in high-risk hypertensive patients: A substudy of the Anglo-Scandinavian Cardiac Outcomes Trial (ASCOT). *J. Am. Coll. Cardiol.,* **44**(2), 415–422.

34. Broijersen, A., Hamsten, A., Eriksson, M., Angelin, B., Hjemdahl, P. (1998). Platelet activity *in vivo* in hyperlipoproteinemia–importance of combined hyperlipidemia. *Thromb. Haemost.,* **79**(2), 268–275.

35. Tian, J., Zhu, T., Liu, J., Guo, Z., Cao, X. (2015). Platelets promote allergic asthma through the expression of CD154. *Cell. Mol. Immunol.,* **12**(6), 700–707.

36. Cattaneo, M. (2015). The platelet P2 receptors in inflammation. *Hamostaseologie,* **35**, 262–266.

37. Jinchuan, Y., Zonggui, W., Jinming, C., Li, L., Xiantao, K. (2004). Upregulation of CD40–CD40 ligand system in patients with diabetes mellitus. *Clin. Chim. Acta,* **339**(1–2), 85–90.

38. Lupu, C., Manduteanu, I., Calb, M., Simionescu, N., Simionescu, M., Ionescu, M. (1993). Some major plasmalemma proteins of human diabetic platelets are involved in the enhanced platelet adhesion to cultured valvular endothelial cells. *Platelets,* **4**(2), 79–84.

39. Nadar, S. K., Lip, G. Y., Blann, A. D. (2004). Platelet morphology, soluble P selectin and platelet P-selectin in acute ischaemic stroke. The West Birmingham Stroke Project. *Thromb. Haemost.,* **92**(6), 1342–1348.

40. Patzelt, J., Mueller, K. A., Breuning, S., Karathanos, A., Schleicher, R., Seizer, P., Gawaz, M., Langer, H. F., Geisler, T. (2015). Expression of anaphylatoxin receptors on platelets in patients with coronary heart disease. *Atherosclerosis,* **238**, 289–295.

41. Peerschke, E. I., Yin, W., Grigg, S. E., Ghebrehiwet, B. (2006). Blood platelets activate the classical pathway of human complement. *J. Thromb. Haemost.,* **4**(9), 2035–2042.

42. Peerschke, E. I., Yin, W., Ghebrehiwet, B. (2008). Platelet mediated complement activation. *Adv. Exp. Med. Biol.,* **632**, 81–91.

43. Yin, W., Ghebrehiwet, B., Peerschke, E. I. (2008). Expression of complement components and inhibitors on platelet microparticles. *Platelets,* **19**(3), 225–233.

44. Peerschke, E. I., Yin, W., Ghebrehiwet, B. (2010). Complement activation on platelets: Implications for vascular inflammation and thrombosis. *Mol. Immunol.*, **47**(13), 2170–2175.
45. Peerschke, E. I., Andemariam, B., Yin, W., Bussel, J. B. (2010). Complement activation on platelets correlates with a decrease in circulating immature platelets in patients with immune thrombocytopenic purpura. *Br. J. Haematol.*, **148**(4), 638–645.
46. Del Conde, I., Cruz, M. A., Zhang, H., Lopez, J. A., Afshar-Kharghan, V. (2005). Platelet activation leads to activation and propagation of the complement system. *J. Exp. Med.*, **201**(6), 871–879.
47. Peerschke, E. I., Murphy, T. K., Ghebrehiwet, B. (2003). Activation-dependent surface expression of gC1qR/p33 on human blood platelets. *Thromb. Haemost.*, **89**(2), 331–339.
48. Hamad, O. A., Nilsson, P. H., Lasaosa, M., Ricklin, D., Lambris, J. D., Nilsson, B., Ekdahl, K. N. (2010). Contribution of chondroitin sulfate A to the binding of complement proteins to activated platelets. *PLoS One*, **5**(9), e12889.
49. Kishimoto, T. K., et al. (2008). Contaminated heparin associated with adverse clinical events and activation of the contact system. *N. Engl. J. Med.*, **358**(23), 2457–2467.
50. Fukuoka, Y., Hugli, T. E. (1988). Demonstration of a specific C3a receptor on guinea pig platelets. *J. Immunol.*, **140**(10), 3496–3501.
51. Wiedmer, T., Sims, P. J. (1985). Effect of complement proteins C5b-9 on blood platelets. Evidence for reversible depolarization of membrane potential. *J. Biol. Chem.*, **260**(13), 8014–8019.
52. Ando, B., Wiedmer, T., Hamilton, K. K., Sims, P. J. (1988). Complement proteins C5b-9 initiate secretion of platelet storage granules without increased binding of fibrinogen or von Willebrand factor to newly expressed cell surface GPIIb-IIIa. *J. Biol. Chem.*, **263**(24), 11907–11914.
53. Wiedmer, T., Esmon, C. T., Sims, P. J. (1986). Complement proteins C5b-9 stimulate procoagulant activity through platelet prothrombinase. *Blood*, **68**(4), 875–880.
54. Wiedmer, T., Esmon, C. T., Sims, P. J. (1986). On the mechanism by which complement proteins C5b-9 increase platelet prothrombinase activity. *J. Biol. Chem.*, **261**(31), 14587–14592.
55. Polley, M. J., Nachman, R. L. (1983). Human platelet activation by C3a and C3a des-arg. *J. Exp. Med.*, **158**(2), 603–615.

56. Sinauridze, E. I., Kireev, D. A., Popenko, N. Y., Pichugin, A. V., Panteleev, M. A., Krymskaya, O. V., Ataullakhanov, F. I. (2007). Platelet microparticle membranes have 50- to 100-fold higher specific procoagulant activity than activated platelets. *Thromb. Haemost.*, **97**(3), 425–434.

57. George, J. N., Pickett, E. B., Saucerman, S., McEver, R. P., Kunicki, T. J., Kieffer, N., Newman, P. J. (1986). Platelet surface glycoproteins. Studies on resting and activated platelets and platelet membrane microparticles in normal subjects, and observations in patients during adult respiratory distress syndrome and cardiac surgery. *J. Clin. Invest.*, **78**(2), 340–348.

58. Martinez, M. C., Tesse, A., Zobairi, F., Andriantsitohaina, R. (2005). Shed membrane microparticles from circulating and vascular cells in regulating vascular function. *Am. J. Physiol Heart Circ. Physiol.*, **288**(3), H1004– H1009.

59. Barry, O. P., Pratico, D., Savani, R. C., FitzGerald, G. A. (1998). Modulation of monocyte-endothelial cell interactions by platelet microparticles. *J. Clin. Invest.*, **102**(1), 136–144.

60. Miajlovic, H., Loughman, A., Brennan, M., Cox, D., Foster, T. J. (2007). Both complement- and fibrinogen-dependent mechanisms contribute to platelet aggregation mediated by Staphylococcus aureus clumping factor B. *Infect. Immun.*, **75**(7), 3335–3343.

61. Sims, P. J., Rollins, S. A., Wiedmer, T. (1989). Regulatory control of complement on blood platelets. Modulation of platelet procoagulant responses by a membrane inhibitor of the C5b-9 complex. *J. Biol. Chem.*, **264**(32), 19228–19235.

62. Mevorach, D., Mascarenhas, J. O., Gershov, D., Elkon, K. B. (1998). Complement-dependent clearance of apoptotic cells by human macrophages. *J. Exp. Med.*, **188**(12), 2313–2320.

63. Speidl, W. S., Exner, M., Amighi, J., Kastl, S. P., Zorn, G., Maurer, G., Wagner, O., Huber, K., Minar, E., Wojta, J., Schillinger, M. (2005). Complement component C5a predicts future cardiovascular events in patients with advanced atherosclerosis. *Eur. Heart J.*, **26**(21), 2294–2299.

64. Movat, H. Z., Uriuhara, T., Taichman, N. S., Rowsell, H. C., Mustard, J. F. (1968). The role of PMN-leucocyte lysosomes in tissue injury, inflammation and hypersensitivity. VI. The participation of the PMN-leucocyte and the blood platelet in systemic aggregate anaphylaxis. *Immunology*, **14**(5), 637–648.

65. Fiane, A. E., Videm, V., Johansen, H. T., Mellbye, O. J., Nielsen, E. W., Mollnes, T. E. (1999). C1-inhibitor attenuates hyperacute rejection and inhibits complement, leukocyte and platelet activation in an *ex vivo* pig-to-human perfusion model. *Immunopharmacology*, **42**(1–3), 231–243.
66. Seifert, P. S., Hansson, G. K. (1989). Complement receptors and regulatory proteins in human atherosclerotic lesions. *Arteriosclerosis*, **9**(6), 802–811.
67. Weiser, M. R., Williams, J. P., Moore, F. D., Jr., Kobzik, L., Ma, M., Hechtman, H. B., Carroll, M. C. (1996). Reperfusion injury of ischemic skeletal muscle is mediated by natural antibody and complement. *J. Exp. Med.*, **183**(5), 2343–2348.
68. van Gils, J. M., Zwaginga, J. J., Hordijk, P. L. (2009). Molecular and functional interactions among monocytes, platelets, and endothelial cells and their relevance for cardiovascular diseases. *J. Leukoc. Biol.*, **85**(2), 195–204.
69. Szebeni, J., Wassef, N. M., Spielberg, H., Rudolph, A. S., Alving, C. R. (1994). Complement activation in rats by liposomes and liposome-encapsulated hemoglobin: Evidence for anti-lipid antibodies and alternative pathway activation. *Biochem. Biophys. Res. Commun.*, **205**, 255–263.

Chapter 13

Rodent Models of Complement Activation-Related Pseudoallergy: Inducers, Symptoms, Inhibitors and Reaction Mechanisms

László Dézsi, PhD,[a,d] László Rosivall, MD, PhD,[b,c] Péter Hamar, MD, PhD,[b] János Szebeni, MD, PhD,[a,d,e] and Gábor Szénási, PhD[b]

[a]*Nanomedicine Research and Education Center,*
Semmelweis University, Budapest, Hungary
[b]*Department of Pathophysiology, Semmelweis University, Budapest, Hungary*
[c]*MTA-SE Pediatrics and Nephrology Research Group, Budapest, Hungary*
[d]*Seroscience Ltd., Budapest, Hungary*
[e]*Department of Nanobiotechnology and Regenerative Medicine,*
Faculty of Health, Miskolc University, Miskolc, Hungary

Keywords: anaphylaxis, anaphylactic shock, anaphylatoxins, anaphylaxis, arrhythmia, blood cells, CARPA, complement, hypersensitivity reactions, hypertension, inflammatory mediators, liposomes, mice, nanomedicines, pseudoallergy, rats, rodents, tachyphylaxis, thromboxane

13.1 Introduction

Complement activation-related pseudoallergy (CARPA) is a hypersensitivity reaction to the intravenous (i.v.) administration

---

Immune Aspects of Biopharmaceuticals and Nanomedicines
Edited by Raj Bawa, János Szebeni, Thomas J. Webster, and Gerald F. Audette
Copyright © 2018 Pan Stanford Publishing Pte. Ltd.
ISBN 978-981-4774-52-9 (Hardcover), 978-0-203-73153-6 (eBook)
www.panstanford.com

of nanoparticle-containing medicines (nanomedicines) in patients with occasionally serious consequences [1–4]. The details of CARPA and its mechanisms are described in other chapters of this Special Issue. New nanomedicines should be tested in animals prior to human administration in order to avoid toxicity, including CARPA. Currently the porcine model of CARPA is the best to predict potential safety concerns of nanomedicines as the pig is highly sensitive to i.v. treatment with liposomes and other nanoparticles, as the resulting physiological changes can be extrapolated to humans with high certainty [5]. However, new, less expensive models are also needed to advance the testing of nanomedicine-induced CARPA. Complement (C) activation can be easily tested *in vitro*, which is an inexpensive and rapid evaluation of potential safety risk [6, 7]. Such *in vitro* tests can be tailored to fulfill species specificity requirements. However, the problem with *in vitro*-based systems is that the effector arm of immune response, most importantly the cardiovascular system, is not present [8]. Therefore, reliable prediction of immunotoxicity requires a battery of tests which includes both *in vitro* and *in vivo* models [9]. Rodents are especially suitable for this purpose as a huge amount of information is available concerning the pathophysiology of C activation in rodent species. However, rodents seem to be much less sensitive to nanomedicines than pigs [10, 11]. The aim of the current review is to summarize the available information on CARPA in rodents to get an insight into various C activation mechanisms elicited by nanomedicines that may be present in various rodent species.

13.2 The CARPAgenic Effects of CVF, Zymosan and LPS in Rodents and Their Modulation with Complement Antagonists

Cobra venom factor (CVF) is a rapid activator of the complement system, and intravenous treatment of rodents with CVF is a model of acute respiratory distress syndrome, a severe illness due, in part, to C activation. Some of the physiological changes caused by CVF in the presence or absence of C inhibitors include the following observations. Pretreatment of anesthetized rats i.v. or p.o. with AcF-[OP(D-Cha)WR], a C5a receptor (C5aR) antagonist, markedly attenuated CVF-induced (4 IU/kg) drop in polymorphonuclear (PMN) cell count and the long-lasting hypotension, but it did not alter the transient increase in blood pressure. On the other hand, N(2)-[(2,2-diphenylethoxy)acetyl]-L-arginine, a C3a receptor (C3aR) antagonist, caused neutropenia on its own, which was similar to that caused by CVF. It also attenuated CVF-induced transient hypertension but did not alter hypotension. Inhibition of both C3 and C5 convertases by rosmarinic acid [12] inhibited all of the above responses. The increase in pulmonary vascular permeability was inhibited most by the C5aR antagonist. Rosmarinic acid was less effective, and the C3a receptor antagonist was the least effective in this respect. All three antagonists diminished the increases in plasma TNFα levels that peaked at 60 min after CVF administration [13].

A part of endotoxin (LPS) shock can be attributed to C activation as plasma levels of both C3a and C5a are markedly elevated in LPS-treated rats [14]. The underlying physiological changes include systemic hypotension and increased hematocrit, along with decreases in the leukocyte (PMN), monocyte, and platelet counts. Prior administration of a rat anti-C5a antibody failed to alter the hematologic changes and pulmonary edema caused by LPS, while the decrease in mean arterial pressure and the increase of hematocrit was partly prevented.

Zymosan is a ligand found on the surface of fungi, like yeast, and it is widely used to activate the alternative pathway of complement. It is a glucan with repeating glucose units connected by β-1,3-glycosidic linkages, which activates nuclear factor-κB (NF-κB) signaling in resident macrophages via toll-like receptors [15]. Damas et al. performed a series of studies to explore the hemodynamic, pulmonary and hematologic effects of zymosan in rats, and reported the following changes [16–21]. Intravenous treatment with zymosan reduced serum C hemolytic activity and caused leukopenia, thrombocytopenia, as well as decreased blood pressure and increased hematocrit as a result of extravasation of extracellular fluid in various vascular beds [16–21]. Most importantly, zymosan increased right ventricular systolic pressure and respiratory rate [18], which is also a key finding in pigs after liposome administration [22]. WEB 2086, a PAF antagonist, prevented the decreases in blood pressure and right ventricular systolic pressure, while indomethacin decreased the tachypnea and pulmonary hypertension but enhanced the drop in blood pressure and right ventricular systolic pressure. The vascular permeability change in the lung was abolished by indomethacin, and no plasma extravasation was found in rats made leukopenic by rabbit anti-neutrophil serum. On the other hand, WEB 2086, the antihistamine mepyramine, or the non-selective serotonin antagonist methysergide did not affect the vascular permeability response to zymosan in the lung. The zymosan-induced paw edema was prevented by pretreatment with the histamine H2 receptor antagonists, cimetidine and metiamide [19].

13.3 CARPA in Pregnancy

The potential C activation-related harmful effects of nanomedicines can be even more serious in pregnancy, as sustained C activation is suspected to contribute to the development of gestational complications and preeclampsia [23, 24]. Hypertension was induced in pregnant rats using the reduced uterine perfusion pressure (RUPP) model, and the animals were treated daily with the C5a receptor antagonist (C5aRA), PMX51 (acetyl-F-[Orn-P-(D-Cha)-WR]), the C3a receptor antagonist (C3aRA), and SB290157 (N2-[(2,2-diphenylethoxy) acetyl]-L-arginine) on gestational days 14–18. Both C3aRA and C5aRA partially reversed hypertension on gestational day 19, while only the C5aRA lowered tachycardia and attenuated the impaired endothelium-dependent relaxation in the mesenteric artery [25]. However, neither antagonist altered the decrease in plasma VEGF concentration or fetal retardation, but the C5aRA decreased the number of circulating neutrophils.

13.4 Characteristics of Liposome-Induced CARPA in Rats

The syndrome called later as CARPA was first demonstrated in conscious rats by Rabinovici et al. (1989) [26], who studied the hemodynamic, hematologic and blood chemistry effects of liposome-encapsulated hemoglobin (LEH), a potential red blood cell substitute [26]. The intravenous injection of LEH induced a transient, but relatively long lasting (<120 min) hypertension and tachycardia that was accompanied by increases in hematocrit and white blood cell count, while platelet count decreased. Plasma thromboxane B2 (TXB2, the stable metabolite of TXA2) levels increased in inverse correlation with platelet count. Injection of the hemoglobin-free liposome vehicle caused hypotension and tachycardia, increased hematocrit, white blood cell count and plasma TXB2 levels but decreased platelet count [26, 27]. In a subsequent study LEH was prepared using synthetic distearoyl phosphatidylcholine instead of hydrogenated soy lecithin. This change in formulation reduced the effects on heart rate and plasma TXB2 levels, while the administration of lyophilized LEH had no detectable hemodynamic, biochemical or hematologic effects [27, 28]. These results established that the size and compositions of liposomes are key modifiers of the hemodynamic and hematologic changes. The same authors also showed that pretreatment of rats with BN 50739, a platelet-activating factor (PAF) blocker prevented the LEH-induced CARPA, suggesting that PAF is a key mediator of CARPA in rats [29].

Treatment with LEH and, to a lesser extent, hemoglobin-free liposomes reduced plasma hemolytic C activity within a few minutes that came together with a reciprocal increase of plasma TXB2 levels in rats [30]. In an attempt to explore the mechanism of C activation, LEH was incubated in rat serum in the presence of EGTA/Mg^{++}, which inhibits C activation via the classical pathway, or the serum was preheated to 50°C, which inhibits C activation via the alternative pathway. Since heating alone prevented C consumption by LEH, it was concluded that LEH activated the alternative pathway [30]. Furthermore, administration of soluble C receptor type 1 (sCR1), or C depletion using cobra venom (CVF) factor prevented the LEH-induced increase in plasma TXB2 levels. These results established a causal relationship between LEH-induced C activation and the release of TXB2 [31]. In a later study it was revealed in conscious rats that treatment with liposome vesicles containing anionic phospholipid-methoxypolyethylene glycol (mPEG) conjugates decreased serum hemolytic C activity and increased plasma TXB2 levels, while the nonionic, methylated phospholipid-mPEG was free of such effects. Therefore, C activation was due to the zwitterionic phospholipid head-groups that should be avoided in order to produce safer vesicles for site-specific drug delivery [32].

Another research group packed contrast agents in liposomes of various lipid compositions in order to prevent glomerular filtration of the contrast agents, and thereby lengthening their circulation time. As to the high dose of contrast agents to be administered, the amount of liposomes was also high. Not surprisingly, 300 mg/kg i.v. hydrogenated soy phosphatidylcholine (HSPC) or 1,2-distearoyl-sn-glycero-3-phosphocholine (DSPC) induced drastic hypotension, decreased total peripheral resistance (TPR) and cardiac contractility [33]. On the other hand, soy phosphatidylcholine (SPC) or the addition of cholesterol to DSPC reduced the hemodynamic effects of liposomes at the same dose. However, C activation parameters were not followed. In a subsequent study, liposomes were injected i.v. in either maleic acid/NaOH, or Na_2-EDTA (pH 6.65), or Tris-HCl buffer (pH 7.33) and also contained 300 mg/kg iopromide at a 1 to 1 lipid-to-drug ratio. The acidic preparation induced hypotension, decreased TPR and cardiac contractility, while the buffered preparation was much less effective. Acetylsalicylic acid prevented the hemodynamic effects [34]. The authors concluded that the size, electric charge, and composition of liposomes were of major importance to elicit cardiovascular responses [34, 35].

In our laboratory we have applied the rat model of CARPA to investigate the immunological and hemodynamic responses to intravenous bolus injections of liposomes differing in surface properties. Systemic arterial blood pressure (SAP) and heart rate (HR) were continuously recorded in anesthetized male Wistar rats, and blood samples were taken to measure blood cell count and plasma TXB2 levels, as well as to determine total C activation using the classical C hemolytic (CH50) assay. The small unilamellar vesicles used in these studies, i.e., commercial Ambisome and a synthetic saturated PC (DPPC) and cholesterol-containing PEGylated liposome formulation wherein 2K-PEG is conjugated to cholesterol (Chol-PEG), had nearly an identical size and polydispersity, but had very different surface properties that represented two frequently applied surface modification. Namely, AmBisome is a surface conjugate-free, highly anionic (negatively charged) liposome, while Chol-PEG liposomes are neutral, surface-stabilized stealth vesicles (Table 13.1). To induce CARPA, zymosan was utilized for direct C activation, while AmBisome and 2K-PEG-Chol (Chol-PEG), as mentioned above, served as liposomal C activators.

Table 13.1 Characteristics of liposomes applied in the study by Dézsi et al. [11], reproduced with permission

Name	Character	Lipid composition	Mole ratios	Size, nm	PDI	Zeta potential, mV
AmBisome	Anionic, no PEG	HSPC/Chol/DSPG/Vit-E/Amph-B	49:23:18:0.3:9	98	0.12	−53.5
Chol-PEG	Neutral-PEGylated	DPPC/DSPE/Chol/2k-PEG-Chol	62:5:28:5	97	0.05	0.49

We have measured and compared the hemodynamic and hematologic effects of these liposomes in comparison with those caused by zymosan that served as a positive control. The effects of zymosan administration at 10 mg/kg i.v. are shown in Fig. 13.1. A gradual decrease in SAP by 40% after 10 min (Fig. 13.1A) could be observed, while HR (Fig. 13.1B) did not change. We have also seen significant leukopenia by 50% at 5 min that was restored by 30 min (Fig. 13.1C), which was associated with significant thrombocytopenia by 30% after 1 min (Fig. 13.1D). There was a severe reduction (by 60%) in hemolytic activity (Fig. 13.1E), while plasma TXB2 exhibited a significant, 4-fold rise (Fig. 13.1F).

AmBisome, calculated on its phospholipid (PL) content, was applied to rats at 22 mg PL/kg i.v. Administration of this lipid vesicle lead to a gradual decrease in SAP by 40% after 5 min (Fig. 13.2A), while no change in HR (Fig. 13.1B) was found. However, significant initial leukopenia by 50% was observed at 5 min, switching to leukocytosis by 10 min (Fig. 13.2C). This change paralleled the thrombocytopenia by 60% after 3–5 min (Fig. 13.2D). At this high dose, we have seen a reduction in hemolytic activity by 40% (Fig. 13.2E), however, plasma TXB2 rose only minimally (Fig. 13.2F). Except for somewhat different hematologic (Fig. 13.2C,D) and less TXB2 (Fig. 13.2F) changes, the effect of 22 mgPL/kg AmBisome was essentially identical to that seen with 10 mg/kg zymosan.

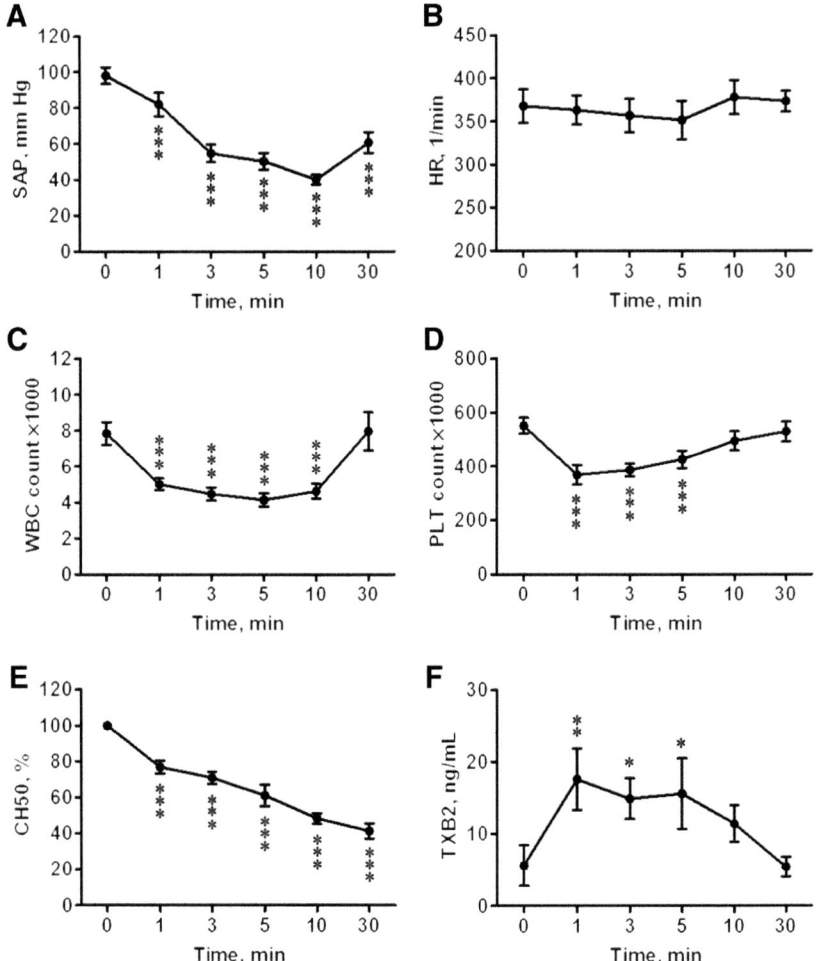

Figure 13.1 Pathophysiological changes in rats injected with 10 mg/kg i.v. zymosan. Dézsi et al. [11], reproduced with permission. Values shown are Mean±SE (n = 8). The curves were constructed from the 0, 1, 3, 5, 10 and 30 min readings of SAP and HR after injection, as well as of other parameters measured from blood samples taken at the same time points. *, **, ***: $p < 0.05, 0.01, 0.001$, respectively vs. the time 0 value.

Zymosan, being a well-known C activator, the practical identities of the measured physiological effects and C activation by AmBisome and zymosan provides strong support for C activation underlying the observed hemodynamic and hematologic changes.

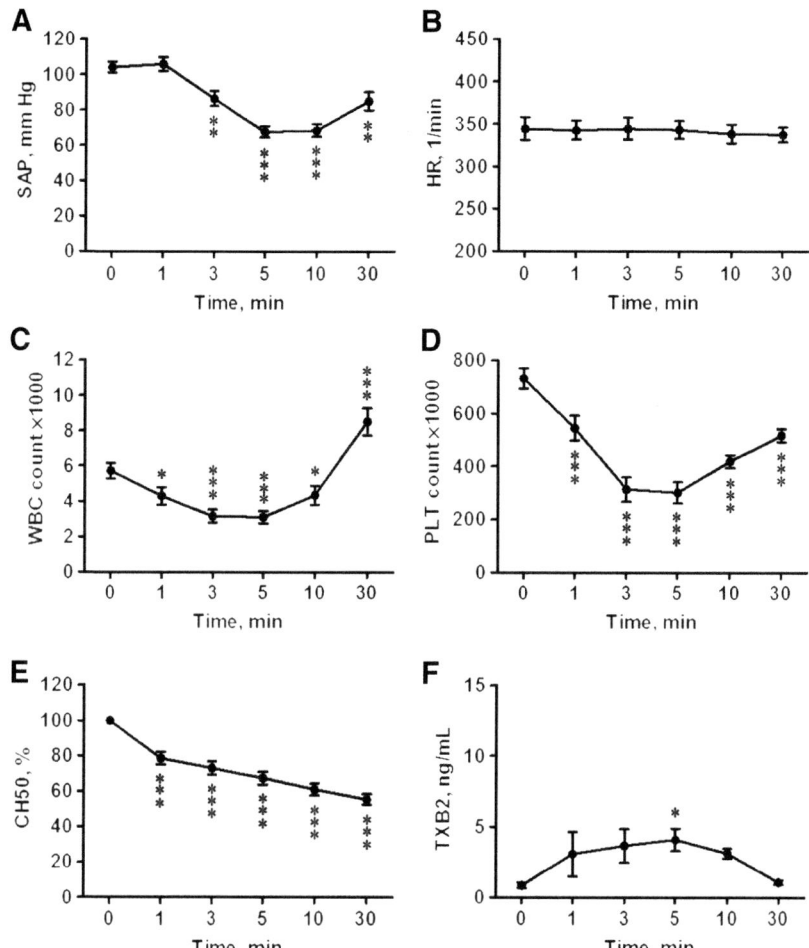

Figure 13.2 Pathophysiological changes in rats injected with 22 mg PL/kg i.v. AmBisome (n = 8). Dézsi et al. [11], reproduced with permission. Similar experiments to that of Fig. 13.1.

Finally, the efficacy of the two types of liposomes with different surface characteristics was compared. Chol-PEG liposomes at the dose of 60 mg PL/kg i.v. caused no changes in the measured parameters. Then, the effects of a 5-fold higher dose (300 mg PL/kg Chol-PEG) were tested that also resulted only in minor changes. There was a relatively small, although significant decrease in SAP (by 16%), while HR did not change. In parallel, we have found moderate leukopenia (but no leukocytosis), as well as thrombocytopenia (both 30%). A slight decrease in hemolytic activity and a small rise in plasma TXB2 could also be observed. Thus, Chol-PEG liposomes turned out to be substantially less effective C activators in rats compared with AmBisome or zymosan.

This study confirmed previous claims that rats are less sensitive to liposome-induced reactions than pigs. For example, the effective AmBisome dose to induce a similar drop in SAP was 2200-fold higher in rats than in pigs, since a 0.01 mgPL/kg i.v. was already effective in pigs. Another notion is that there could be huge differences between test agents of similar kind based on their physical characteristics. These figures provide strong evidence that the rat is not a sensitive model for immune toxicity screening or quantitative evaluation of the risk of CARPA. However, because the physiological changes in rats are essentially the same as those seen in pigs and humans, rats still provide a good model to study the reaction mechanisms of CARPA.

13.5 Effects of Complement Components C3a and C5a in the Guinea Pig

Complement C3a and C5a have distinct hemodynamic effects. Administration of porcine C5a or C5a des-Arg caused an immediate and short lived fall in blood pressure followed by a longer hypertensive response lasting for a few minutes in anesthetized guinea-pigs. Only the hypertensive effect was attenuated upon repeated administration, i.e., showed tachyphylaxis [36]. Bronchoconstriction followed the same time-course as hypotension, but the reaction was also tachyphylaxic [36]. The hypertensive effect was similarly, but only partly, reduced by pretreatment with histamine or alpha-adrenoceptor blockers, suggesting that C5a caused catecholamine liberation through the release of histamine. The second, hypertensive phase of the C5a effect is specific for the guinea pig as C5a causes mainly hypotension in other species. Administration of C5a des-Arg induced hypotension that lasted for more than 10 min, and was prevented by indomethacin pretreatment. Later studies extended the above findings by demonstrating that indomethacin, the thromboxane synthetase inhibitor U-63557A and the thromboxane receptor antagonist SQ 29,548 all attenuated, but the LTD4 antagonist L-649,923 failed to alter the C5a or C5a des-Arg-induced hypertension in anesthetized guinea-pigs. In summary, the tachyphylactic histamine and catecholamine release and the consequent hypertensive response induced by C5a was mediated by thromboxane. Contrary to the above findings, pyrilamine, a H1 antagonist did not alter, but phentolamine, an alpha-adrenergic antagonist inhibited the hypertensive effect. Treatment with C5a also decreased PMN and platelet counts. The C5a or C5a des-Arg-induced blood pressure rise was diminished after depleting the animals' platelets and white blood cells, while depleting the circulating PMN only had similar effects [37, 38].

The pulmonary response in anesthetized guinea-pigs to intravenous recombinant human C5a (rhC5a) was a reduction in dynamic lung compliance and an increase in pulmonary resistance. Similar to previous studies [39], bronchoconstriction followed the same time-course as hypotension, and it was also tachyphylaxic. Bronchoconstriction was not altered by pyrilamine despite an increase in plasma histamine levels. SQ 29,548, a selective thromboxane antagonist decreased the peak response only, while the superoxide dismutase and TXA2 inhibitor U-63557A altered the time course of the bronchoconstrictor response [39]. The time course and magnitude of bronchoconstriction was not affected by selective depletion of PMN, platelets or both. However, pyrilamine inhibited the bronchoconstriction after depletion of both circulating PMN and platelets [38]. Thus, similar to the hypertensive response, an increased pulmonary resistance due to bronchoconstriction was most likely mediated by the effect of thromboxane, while granulocytes and platelets were less important.

Similarly to C5a, treatment with the C3a-peptide (the last 21 amino acids of guinea pig C3a) caused a huge increase in pulmonary resistance and a decrease in dynamic lung compliance in guinea pigs [40]. The C3a-peptide also induced a transient systemic hypotension, followed by a longer hypertension lasting for about 5 min, and then blood pressure decreased for an additional 10–15 min. All these responses were absent in the C3a receptor-deficient (C3aR-) guinea pigs. Administration of recombinant human C5a (rhC5a) evoked almost the same responses to the C3a-peptide with similar magnitude and time course in normal and C3aR-guinea pigs. The ovalbumin-induced increase in pulmonary resistance and decrease in dynamic lung compliance was slightly delayed but was not attenuated in C3aR-vs. normal guinea pigs. On the other hand, the triphasic blood pressure response was altered in C3aR-guinea pigs in such a way that the hypertensive effect was aggravated and delayed and the late hypotensive effect was attenuated, suggesting a minor role of C3aR in the anaphylactic response in guinea pigs [40].

13.6 Effects of Complement Components C3a and C5a in the Rat

Treatment with C5a desArg (5 μg/rat) decreased mean arterial pressure and PMN, monocyte and platelet counts but did not alter the hematocrit in rats [14]. Administration of rhC5a induced an almost immediate hypotension lasting for more than 2 h, and resulted in a shorter, <30 min drop of circulating PMN cell counts in rats [41]. On the other hand, treatment with rhC3a caused a rapid, dose-dependent hypertension that lasted for a maximum of 5 min. The hypertensive effect was slightly potentiated by pretreatment with carboxypeptidase N inhibitor but was abolished by indomethacin. Administration of rhC3a elevated PMN cell counts with a delay of about 90 min at low doses, while at high doses it similarly elevated PMN cell counts that was preceded by a small neutropenia lasting for about 60 min, but this early response failed to reach the level of statistical significance. Pretreatment with carboxypeptidase N inhibitor abolished the delayed increase in PMN cell count that was caused by the low dose of rhC3a, but elicited an early and small neutropenia that was similar to that induced by the high dose of rhC3a [41].

One of two short peptide C5a agonists bound to C5aR on both PMN and macrophages while the other had affinity only for the macrophage C5aR. As a consequence, both C5a agonists decreased blood pressure of anaesthetized rats, while only the agonist with affinity for the granulocyte C5aR caused neutropenia [8]. These results gave an insight into the mechanism of the various effects of C activation, and raise the possibility of selectively altering various consequences of CARPA.

There are only a few studies on the effects of C in the pulmonary vasculature in rats. Therefore it is an important observation that the infusion of the 21-carboxy-terminal peptide of C3a (C3a57-77) caused an immediate pulmonary vasoconstriction lasting for 10 min in isolated, crystalloid buffer-perfused rat lungs. There was a parallel increase in lung-effluent TXB2 level, which was directly responsible for the pulmonary vasoconstriction as both indomethacin and the thromboxane synthetase inhibitors CGS 13080 and U63,357 inhibited the pulmonary arterial pressor response [42]. These results call the attention to the fact that the rat lung can be a direct target organ of CARPA despite the absence of pulmonary intravascular macrophages (PIM cells), which cells seem to have a pivotal role in making the lungs the primary responder organ in the highly CARPA sensitive pigs [11].

13.7 Effects of Complement Components C3a and C5a in the Rabbit, Hamster and Mouse

In anesthetized rabbits, treatment with C5a induced a systemic hypotension lasting for about 10 min, a fall in white blood cell count, and an increase in plasma histamine, PGE2, TXB2 and prostacyclin levels, while heart rate, cardiac contractility, hematocrit and platelet count did not change [43]. PMN cells almost fully disappeared from the blood, while lymphocyte cell counts decreased by about 50%. Central venous pressure increased in parallel with hypotension. All effects remained the same upon repeated administration of C5a. Pretreatment with indomethacin abolished the hemodynamic and prostaglandin responses but leukopenia reappeared. In contrast, the H1-receptor antagonist pyrilamine, and the thromboxane synthetase inhibitor dazoxibene failed to alter the hemodynamic responses, while the H2-receptor antagonist cimetidine attenuated the blood pressure drop and the elevations in plasma prostaglandin levels [43].

A detailed study evaluated leukopenia and subsequent leukocytosis activity of human C5a in a rabbits. Neutrophil, monocyte, eosinophil, and basophil counts all rapidly dropped upon intravenous treatment with human C5a, suggesting

C5a-activated leukocytes which become adherent, leading to sequestration, and depletion of cells from the circulation. However, C5a seemed to mobilize bone marrow causing a huge increase in monocyte, eosinophil, and basophil counts starting at 10–20 min after treatment. Indomethacin failed to alter the effects of C5a. Epinephrine, dexamethasone, lipopolysaccharide, and the prostanoid 15(S)-15-methyl PGF2α produced a different profile of leukocyte mobilization than that of C5a. It was hypothesized that C5a was directly responsible for the leukocitosis without the involvement of secondary mediators [44].

The hamster cheek pouch is a suitable model for studying the microvascular effects of C3a and C5a. Topical application of C3a (10 nM) caused local vasoconstriction, platelet aggregation, and increased vascular leakage of fluorescein-labeled dextran. Higher doses of C5a (20 or 100 nM) had the same effects and also resulted in accumulation of PMN. Pretreatment with mepyramine, a histamine H1 receptor blocker, partially inhibited the early phase (up to 5 min) of complement-induced extravasation, which was also partly due to the recruitment of PMN [45].

Liposome-induced CARPA was rarely evaluated in mice *in vivo* but the result from a number of excellent *in vitro* tests have been published, among others, a study by Banda et al., showed that iron-containing nanoparticles can activate complement in mouse serum via the lectin and alternative pathways [46]. A recent *in vivo* study has shown that intravenous treatment with a polyethoxylated castor oil-free, liposome-based paclitaxel formulation or paclitaxel-free liposomes caused hypersensitivity reactions after treatment, including shortness of breath and dyspnea. The paclitaxel formulation induced pulmonary edema, and increased serum sC5b-9 and lung histamine, i.e., caused C activation [47]. These results suggest that in mice also the lung seems to be the primary target organ similarly to pigs [22]. The role of zymosan causing plasma extravasation was widely investigated in the zymosan-induced peritoneal inflammation model in mice. Zymosan increased vascular permeability and caused peritoneal inflammation in Balb/c mice. The role of mast cells was found to be crucial in the zymosan-induced peritonitis, which was mediated via histamine receptors [48]. Other authors have found that leukotriene C4 synthase (LTC4S) mediates an increased vascular permeability [49]. These effects were lacking in mast cell-deficient WBB6F1 Balb/c mice and LTC4S$^{-/-}$ C57BL/6 mice [48, 49].

Although we consider the liposome-induced CARPA as a side-effect due to C activation, liposomes can activate PMN leukocytes that can lead to a therapeutic effect. In fact, empty liposomes were similarly effective to liposomal amphotericin B in alleviating invasive pulmonary aspergillosis in a non-neutropenic murine model [50].

Characteristic changes of *in vivo* observed parameters in various rodent models of CARPA are summarized in Table 13.2.

Table 13.2 Symptoms of CARPA in different rodent species

Symptoms	Rat	Mouse	Guinea pig	Rabbit
Rise of PAP (dyspnea)	+ [18]	+ [47]	+ [36, 39, 51]	+ [52]
Hypo/hypertension	+ [11]	+ [53]	+ [36]	+ [43, 54]
Hemoconcentration (rise of Hct)	+ [14, 26]	+ [53]	?	?
Leukopenia/leukocytosis	+ [11]	+ [55]	+ [38]	+ [43, 54]
Thrombocytopenia	+ [11]	+ [56]	+ [38]	+ [54]
Rise in TXB2	+ [11]	?	?	+ [43, 54]

Note: The presence of specified symptoms are shown with +mark, with corresponding references in brackets. ? means lack of information.

13.8 Conclusions

Although the effects of treatment with nanoparticle-containing carrier systems and medicines have not been fully explored in all rodent species, the collective results unequivocally prove that all major characteristics of CARPA are present in rodents including respiratory, hemodynamic and hematologic effects. The current information clearly shows that rodents represent appropriate models to study the reaction mechanisms of CARPA, as well as rodents are suitable for safety prediction of side effects of nanomedicines in humans, although at a much lower level of sensitivity.

Disclosures and Conflict of Interest

This chapter was originally published as: Dézsi, L., Rosivall, L., Hamar, P., Szebeni, J., Szénási, G. Rodent models of complement activation-related pseudoallergy: Inducers, symptoms, inhibitors and reaction mechanisms. *Eur. J. Nanomed.* 7, 1525 (2015); DOI: https://doi.org/10.1515/ejnm-2015-0002, and appears here (with edits) by kind permission of the publisher, Walter de Gruyter GmbH, Berlin, Germany. The authors declare that they have no conflict of interest and have no affiliations or financial involvement with any organization or entity discussed in this chapter. No writing assistance was utilized in the production of this chapter and the authors have received no payment for its preparation.

Corresponding Author

Dr. László Dézsi
Nanomedicine Research and Education Center
Semmelweis University
1089 Budapest, Nagyvárad tér 4, Hungary
Email: dezsi.laszlo@med.semmelweis-univ.hu

About the Authors

László Dézsi is an adjunct professor at the Nanomedicine Research and Education Center, Semmelweis University, Budapest, Hungary. He received his MSc degree in biology at Eötvös Loránd University, obtained his PhD and habilitation in physiology at Semmelweis University. He was a senior research associate at the 2nd Department of Physiology, Semmelweis University, and visiting scientist at Albert Ludwigs Universität, Freiburg, Germany and at University of Pennsylvania, Philadelphia, USA. He worked for Gedeon Richter Pharmaceutical Plc. for 13 years and was manager of Analgesic Research Laboratory of Richter and University of Pécs, Hungary. He was secretary of the Biomedical Engineering course at Budapest University of Technology and Economics and is now course director of "Cardiorespiratoric and neurophysiological measuring techniques" at Semmelweis.

His current field of research is nanomedicine studying complement activation related pseudoallergy (CARPA). He has published 51 original papers and 11 book chapters.

László Rosivall is a full professor, Széchenyi and Khwarizmi prizes laureate, head of the International Nephrology Research and Training Center, and PhD School of Basic Medical Sciences, and former head of the Department of Pathophysiology, Semmelweis University. He pioneered recognizing and characterizing the intrarenal renin-angiotensin system (RAS). Using nanotechnology he visualized the GFR *in vivo* and demonstrated special characteristics of the fenestration. He discovered a new, short loop feedback mechanism in the regulation of GFR. This unique JGA morphology and the high filtration volume in AA is one of the most striking recent observations of renal microcirculation and questions several basic renal physiological issues.

Peter Hamar is an associate professor in the Department of Pathophysiology, Semmelweis University. His major scientific interest is to understand the function and pathophysiologic role of small RNAs *in vivo* with the help of rodent models. A major roadblock to harnessing small RNSs for therapy is delivery to targeted cells, and to the appropriate intracellular compartment with nano-medicinal formulations. Therefore, he is studying different delivery methods. Besides being a principal investigator at Semmelweis, he intensely collaborates with the Immune Disease Institute at Harvard Medical School, Boston, USA. This collaboration was the first to harness RNA-interference for the kidney. He has co-authored 57 original papers.

János Szebeni is director of the Nanomedicine Research and Education Center at Semmelweis University. He is also the founder and CEO of a contract research SME "SeroScience," and full Professor of (immune) biology at Miskolc University, Hungary. He has held various

guest professor and scientific positions in Hungary and abroad, mostly in the United States, where he lived for 22 years. His research on various themes in hematology, membrane biology, and immunology resulted more than 130 scientific papers and book chapters, 2 granted patents, and a book entitled *The Complement System: Novel Roles in Health and Disease* (Kluwer, 2004). Three fields stand out where he has been most active: artificial blood, liposomes, and the complement system. His original works led to the "CARPA" concept, which implies that complement activation underlies numerous drug-induced (pseudo)allergic (anaphylactoid) reactions.

Gábor Szénási is a scientific adviser at the Department of Pathophysiology, Semmelweis University. He received his MSc degree from Eötvös Loránd University and PhD from the Hungarian Academy of Sciences. He was a senior research associate at the joint research group of 2nd Department of Internal Medicine, Semmelweis University and Hungarian Academy of Sciences, visiting research fellow at Baker Medical Research Institute, Melbourne, Australia, and laboratory head at EGIS Pharmaceutical Plc. for 18 years. His current research interests are in two main areas: pathophysiology of kidney fibrosis in chronic kidney disease and complement activation-related pseudoallergy (CARPA). He has published 85 peer-reviewed articles and book chapters and filed 28 patent applications.

References

1. Nath, P., Basher, A., Harada, M., Sarkar, S., Selim, S., Maude, R. J., et al. (2014). Immediate hypersensitivity reaction following liposomal amphotericin-B (AmBisome) infusion. *Trop. Doct.*, **44**, 241–242.
2. Burza, S., Sinha, P. K., Mahajan, R., Lima, M. A., Mitra, G., Verma, N., et al. (2014). Five-year field results and long-term effectiveness of 20 mg/kg liposomal amphotericin B (Ambisome) for visceral leishmaniasis in Bihar, India. *PLoS Negl. Trop. Dis.*, **8**, e2603.
3. Mukhtar, M., Aboud, M., Kheir, M., Bakhiet, S., Abdullah, N., Ali, A., et al. (2011). First report on Ambisome-associated allergic reaction

in two Sudanese leishmaniasis patients. *Am. J. Trop. Med. Hyg.,* **85**, 644–645.

4. Dummer, R., Quaglino, P., Becker, J. C., Hasan, B., Karrasch, M., Whittaker, S., et al. (2012). Prospective international multicenter phase II trial of intravenous pegylated liposomal doxorubicin monochemotherapy in patients with stage IIB, IVA, or IVB advanced mycosis fungoides: Final results from EORTC 21012. *J. Clin. Oncol.,* **30**, 4091–4097.

5. Szebeni, J., Bedocs, P., Csukas, D., Rosivall, L., Bunger, R., Urbanics, R. (2012). A porcine model of complement-mediated infusion reactions to drug carrier nanosystems and other medicines. *Adv. Drug Deliv. Rev.,* **64**, 1706–1716.

6. Neun, B. W., Dobrovolskaia, M. A. (2011). Qualitative analysis of total complement activation by nanoparticles. *Methods Mol. Biol.,* **697**, 237–245.

7. Gal, P., Dobo, J., Beinrohr, L., Pal, G., Zavodszky, P. (2013). Inhibition of the serine proteases of the complement system. *Adv. Exp. Med. Biol.,* **735**, 23–40.

8. Short, A. J., Paczkowski, N. J., Vogen, S. M., Sanderson, S. D., Taylor, S. M. (1999). Response-selective C5a agonists: Differential effects on neutropenia and hypotension in the rat. *Br. J. Pharmacol.,* **128**, 511–514.

9. Dobrovolskaia, M. A., McNeil, S. E. (2013). In vitro assays for monitoring nanoparticle interaction with components of the immune system. In: Dobrovolskaia, M. A., McNeil, S. E., eds. *Handbook of Immunological Properties of Engineered Nanomaterials,* World Scientific Publishing, Singapore, pp. 581–638.

10. Szebeni, J., Alving, C. R., Rosivall, L., Bunger, R., Baranyi, L., Bedocs, P., et al. (2007). Animal models of complement-mediated hypersensitivity reactions to liposomes and other lipid-based nanoparticles. *J. Liposome Res.,* **17**, 107–117.

11. Dezsi, L., Fulop, T., Meszaros, T., Szenasi, G., Urbanics, R., Vazsonyi, C., et al. (2014). Features of complement activation-related pseudoallergy to liposomes with different surface charge and PEGylation: Comparison of the porcine and rat responses. *J. Control. Release,* **195**, 2–10.

12. Gamaro, G. D., Suyenaga, E., Borsoi, M., Lermen, J., Pereira, P., Ardenghi, P. (2011). Effect of rosmarinic and caffeic acids on inflammatory and nociception process in rats. *ISRN Pharmacol.,* **2011**, 451682.

13. Proctor, L. M., Strachan, A. J., Woodruff, T. M., Mahadevan, I. B., Williams, H. M., Shiels, I. A., et al. (2006). Complement inhibitors selectively attenuate injury following administration of cobra venom factor to rats. *Int. Immunopharmacol.*, **6**, 1224–1232.
14. Smedegard, G., Cui, L. X., Hugli, T. E. (1989). Endotoxin-induced shock in the rat. A role for C5a. *Am. J. Pathol.*, **135**, 489–497.
15. Sato, M., Sano, H., Iwaki, D., Kudo, K., Konishi, M., Takahashi, H., et al. (2003). Direct binding of Toll-like receptor 2 to zymosan, and zymosan-induced NF-kappa B activation and TNF-alpha secretion are down-regulated by lung collectin surfactant protein A. *J. Immunol.*, **171**, 417–425.
16. Damas, J. (1991). Involvement of platelet-activating factor in the hypotensive response to zymosan in rats. *J. Lipid Mediators*, **3**, 333–344.
17. Damas, J., Bourdon, V., Remacle-Volon, G., Adam, A. (1990). Kinins and peritoneal exudates induced by carrageenin and zymosan in rats. *Br. J. Pharmacol.*, **101**, 418–422.
18. Damas, J., Lagneaux, D. (1991). Dissociation between the effects of zymosan on the systemic and pulmonary vessels of the rat. *Br. J. Pharmacol.*, **104**, 559–564.
19. Damas, J., Remacle-Volon, G. (1986). Mast cell amines and the oedema induced by zymosan and carrageenans in rats. *Eur. J. Pharmacol.*, **121**, 367–376.
20. Damas, J., Remacle-Volon, G., Bourdon, V. (1993). Platelet-activating factor and the vascular effects of zymosan in rats. *Eur. J. Pharmacol.*, **231**, 231–236.
21. Damas, J., Remacle-Volon, G., Nguyen, T. P. (1990). Inhibition by WEB 2086, a PAG-acether antagonist of oedema and peritonitis induced by zymosan in rats. *Arch. Int. Pharmacodyn. Ther.*, **306**, 161–169.
22. Szebeni, J., Bedocs, P., Rozsnyay, Z., Weiszhar, Z., Urbanics, R., Rosivall, L., et al. (2012). Liposome-induced complement activation and related cardiopulmonary distress in pigs: Factors promoting reactogenicity of Doxil and AmBisome. *Nanomedicine*, **8**, 176–184.
23. Hoffman, M. C., Rumer, K. K., Kramer, A., Lynch, A. M., Winn, V. D. (2014). Maternal and fetal alternative complement pathway activation in early severe preeclampsia. *Am. J. Reprod. Immunol.*, **71**, 55–60.
24. Burwick, R. M., Feinberg, B. B. (2013). Eculizumab for the treatment of preeclampsia/HELLP syndrome. *Placenta*, **34**, 201–203.

25. Lillegard, K. E., Loeks-Johnson, A. C., Opacich, J. W., Peterson, J. M., Bauer, A. J., Elmquist, B. J., et al. (2014). Differential effects of complement activation products c3a and c5a on cardiovascular function in hypertensive pregnant rats. *J. Pharmacol. Exp. Ther.*, **351**, 344–351.

26. Rabinovici, R., Rudolph, A. S., Feuerstein, G. (1989). Characterization of hemodynamic, hematologic, and biochemical responses to administration of liposome-encapsulated hemoglobin in the conscious, freely moving rat. *Circ. Shock*, **29**, 115–132.

27. Rabinovici, R., Rudolph, A. S., Feuerstein, G. (1990). Improved biological properties of synthetic distearoyl phosphatidyl choline-based liposome in the conscious rat. *Circ. Shock*, **30**, 207–219.

28. Rabinovici, R., Rudolph, A. S., Vernick, J., Feuerstein, G. (1994). Lyophilized liposome encapsulated hemoglobin: Evaluation of hemodynamic, biochemical, and hematologic responses. *Crit. Care Med.*, **22**, 480–485.

29. Rabinovici, R., Rudolph, A. S., Yue, T. L., Feuerstein, G. (1990). Biological responses to liposome-encapsulated hemoglobin (LEH) are improved by a PAF antagonist. *Circ. Shock*, **31**, 431–445.

30. Szebeni, J., Wassef, N. M., Spielberg, H., Rudolph, A. S., Alving, C. R. (1994). Complement activation in rats by liposomes and liposome-encapsulated hemoglobin: Evidence for anti-lipid antibodies and alternative pathway activation. *Biochem. Biophys. Res. Commun.*, **205**, 255–263.

31. Szebeni, J., Spielberg, H., Cliff, R. O., Wassef, N. M., Rudolph, A. S., Alving, C. R. (1997). Complement activation and thromboxane secretion by liposome-encapsulated hemoglobin in rats in vivo: Inhibition by soluble complement receptor type 1. *Artif. Cells Blood Substit. Immobil. Biotechnol.*, **25**, 347–355.

32. Moghimi, S. M., Hamad, I., Andresen, T. L., Jorgensen, K., Szebeni, J. (2006). Methylation of the phosphate oxygen moiety of phospholipid-methoxy(polyethylene glycol) conjugate prevents PEGylated liposome-mediated complement activation and anaphylatoxin production. *FASEB J.*, **20**, 2591–2593.

33. Muschick, P., Sachse, A., Leike, J., Wehrmann, D., Krause, W. (1995). Lipid dependent cardio-haemodynamic tolerability of liposomes in rats. *J. Liposome Res.*, **5**, 933–953.

34. Krause, W., Gerlach, S., Muschick, P. (2000). Prevention of the hemodynamic effects of iopromide-carrying liposomes in rats and pigs. *Invest. Radiol.*, **35**, 493–503.

35. Muschick, P., Wehrmann, D., Schuhmann-Giampieri, G., Krause, W. (1995). Cardiac and hemodynamic tolerability of iodinated contrast media in the anesthetized rat. *Invest. Radiol.*, **30**, 745–753.
36. Marceau, F., Lundberg, C., Hugli, T. E. (1987). Effects of the anaphylatoxins on circulation. *Immunopharmacology*, **14**, 67–84.
37. Fraser, D. G., Regal, J. F. (1990). C5a/C5ades-Arg-induced increase in blood pressure in the guinea pig: Role of thromboxane. *Immunopharmacology*, **19**, 59–68.
38. Fraser, D. G., Regal, J. F. (1991). Recombinant human C5a-induced bronchoconstriction in the guinea pig: Inhibition by an H1 antagonist after depletion of circulating granulocytes and platelets. *J. Pharmacol. Exp. Ther.*, **259**, 1213–1220.
39. Regal, J. F., Fraser, D. G. (1990). Recombinant human C5a-induced bronchoconstriction in the guinea-pig: A histamine independent mechanism. *Pulm. Pharmacol.*, **3**, 79–87.
40. Regal, J. F., Klos, A. (2000). Minor role of the C3a receptor in systemic anaphylaxis in the guinea pig. *Immunopharmacology*, **46**, 15–28.
41. Proctor, L. M., Moore, T. A., Monk, P. N., Sanderson, S. D., Taylor, S. M., Woodruff, T. M. (2009). Complement factors C3a and C5a have distinct hemodynamic effects in the rat. *Int. Immunopharmacol.*, **9**, 800–806.
42. Morganroth, M. L., Schoeneich, S. O., Till, G. O., Ward, P. A., Horvath, S. J., Glovsky, M. M. (1990). C3a57-77, a C-terminal peptide, causes thromboxane-dependent pulmonary vascular constriction in isolated perfused rat lungs. *Am. Rev. Respir. Dis.*, **141**, 296–300.
43. Lundberg, C., Marceau, F., Hugli, T. E. (1987). C5a-induced hemodynamic and hematologic changes in the rabbit. Role of cyclooxygenase products and polymorphonuclear leukocytes. *Am. J. Pathol.*, **128**, 471–483.
44. Kajita, T., Hugli, T. E. (1990). C5a-induced neutrophilia. A primary humoral mechanism for recruitment of neutrophils. *Am. J. Pathol.*, **137**, 467–477.
45. Bjork, J., Hugli, T. E., Smedegard, G. (1985). Microvascular effects of anaphylatoxins C3a and C5a. *J. Immunol.*, **134**, 1115–1119.
46. Banda, N. K., Mehta, G., Chao, Y., Wang, G., Inturi, S., Fossati-Jimack, L., et al. (2014). Mechanisms of complement activation by dextran-coated superparamagnetic iron oxide (SPIO) nanoworms in mouse versus human serum. *Part. Fibre Toxicol.*, **11**, 64.
47. Wang, H., Cheng, G., Du, Y., Ye, L., Chen, W., Zhang, L., et al. (2013). Hypersensitivity reaction studies of a polyethoxylated castor oil-free,

liposome-based alternative paclitaxel formulation. *Mol. Med. Rep.*, **7**, 947–952.

48. Kolaczkowska, E., Seljelid, R., Plytycz, B. (2001). Role of mast cells in zymosan-induced peritoneal inflammation in Balb/c and mast cell-deficient WBB6F1 mice. *J. Leukoc. Biol.*, **69**, 33–42.

49. Kanaoka, Y., Maekawa, A., Penrose, J. F., Austen, K. F., Lam, B. K. (2001). Attenuated zymosan-induced peritoneal vascular permeability and IgE-dependent passive cutaneous anaphylaxis in mice lacking leukotriene C4 synthase. *J. Biol. Chem.*, **276**, 22608–22613.

50. Lewis, R. E., Chamilos, G., Prince, R. A., Kontoyiannis, D. P. (2007). Pretreatment with empty liposomes attenuates the immunopathology of invasive pulmonary aspergillosis in corticosteroid-immunosuppressed mice. *Antimicrob. Agents Chemother.*, **51**, 1078–1081.

51. Regal, J. F., Fraser, D. G., Anderson, D. E., Solem, L. E. (1993). Enhancement of antigen-induced bronchoconstriction after intravascular complement activation with cobra venom factor. Reversal by granulocyte depletion. *J. Immunol.*, **150**, 3496–3505.

52. Heller, A., Kunz, M., Samakas, A., Haase, M., Kirschfink, M., Koch, T. (2000). The complement regulators C1 inhibitor and soluble complement receptor 1 attenuate acute lung injury in rabbits. *Shock*, **13**, 285–290.

53. Hsueh, W., Sun, X., Rioja, L. N., Gonzalez-Crussi, F. (1990). The role of the complement system in shock and tissue injury induced by tumour necrosis factor and endotoxin. *Immunology*, **70**, 309–314.

54. Bult, H., Herman, A. G., Laekeman, G. M., Rampart, M. (1985). Formation of prostanoids during intravascular complement activation in the rabbit. *Br. J. Pharmacol.*, **84**, 329–336.

55. Tousignant, J. D., Gates, A. L., Ingram, L. A., Johnson, C. L., Nietupski, J. B., Cheng, S. H., et al. (2000). Comprehensive analysis of the acute toxicities induced by systemic administration of cationic lipid: Plasmid DNA complexes in mice. *Hum. Gene Ther.*, **11**, 2493–2513.

56. Kiang, A., Hartman, Z. C., Everett, R. S., Serra, D., Jiang, H., Frank, M. M., et al. (2006). Multiple innate inflammatory responses induced after systemic adenovirus vector delivery depend on a functional complement system. *Mol. Ther.*, **14**, 588–598.

Chapter 14

Immune Reactions in the Delivery of RNA Interference-Based Therapeutics: Mechanisms and Opportunities

Kaushik Thanki, PhD,[a] Emily Falkenberg,[a] Monique Gangloff, PhD,[b] and Camilla Foged, PhD[a]

[a]*Department of Pharmacy, Faculty of Health and Medical Sciences, University of Copenhagen, Copenhagen, Denmark*
[b]*Department of Biochemistry, University of Cambridge, Cambridge, UK*

Keywords: small interfering RNA, RNA interference, nanomedicine, drug delivery, immunogenicity, toll-like receptors, TLR3, TLR4, TLR7, TLR8, TLR9, cationic lipids, bioconjugation, polyplexes, lipoplexes, immune reactions, TLR reporter cell lines, *in silico* tools, functional excipients, dendrimers, docking, TLR4 signalling pathways, gene silencing, MD2, pathogen-associated molecular patterns, stable nucleic acid lipid particles, lipid-polymer hybrid nanoparticles, nucleic acid recognition, lipid recognition

14.1 Background

RNA interference (RNAi) is an evolutionary conserved pathway by which double-stranded RNA (dsRNA) molecules mediate sequence-specific, post-transcriptional gene silencing in cells [1].

Immune Aspects of Biopharmaceuticals and Nanomedicines
Edited by Raj Bawa, János Szebeni, Thomas J. Webster, and Gerald F. Audette
Copyright © 2018 Pan Stanford Publishing Pte. Ltd.
ISBN 978-981-4774-52-9 (Hardcover), 978-0-203-73153-6 (eBook)
www.panstanford.com

Andrew Fire and Craig Mello discovered the pathway in their Nobel Prize-winning studies in *Caenorhabditis elegans* published in 1998 [2]. It has since been shown to be a crucial mechanism found in many eukaryotes, responsible for regulating gene expression in response to unfavourable environmental stimuli, and a part of the cellular innate defence against transposons and infection by RNA viruses [3]. The pathway has attracted immense research interest due to the therapeutic potential of harnessing it to manipulate expression of disease-associated genes, effectively offering a universal approach for treatment of genetic and viral diseases [4].

Various dsRNA molecules trigger RNAi, e.g., long dsRNAs, small interfering RNAs (siRNAs), short hairpin RNAs (shRNAs) and micro RNAs (miRNAs). The following will focus on exogenous siRNAs, as they circumvent limitations posed by other dsRNAs and thus currently hold the greatest promise as therapeutic agents. For example, in contrast to long dsRNAs, siRNAs do not elicit an interferon response [5]. Similarly, unlike miRNAs, siRNAs require perfect base pairing with mRNA to mediate target-specific gene silencing [6]. Exogenous siRNA molecules consist of a 19–21 base pair (bp) core duplex, followed by UU or TT 3′ dinucleotide overhangs on each strand, necessary for recognition by the RNAi machinery. The 5′ ends are phosphorylated, while the 3′ ends are hydroxylated (Fig. 14.1). The mechanism by which RNAi arrests *de novo* protein synthesis has been reviewed elsewhere [7, 8]. Briefly, siRNA targeting a specific gene is transfected into cells. In contrast to the longer dsRNA, siRNA is not cleaved by the endoribonuclease Dicer due to the smaller molecular size. Instead, synthetic siRNA is loaded directly into the RNA-induced silencing complex (RISC). The siRNA molecule is subsequently unwound, and the passenger strand is degraded, resulting in activation of RISC [9]. The remaining single RNA strand, usually the antisense strand, functions as a guide strand, which binds with perfect base complementarity to the target mRNA [7]. Binding activates argonaute 2 (AGO2) of the RISC complex, which in turn cleaves the mRNA phosphodiester backbone between nucleotides 10 and 11, relative to the 5′ end of the guide strand [10]. Cleavage of the mRNA induces its degradation by exonucleases, effectively preventing translation [5].

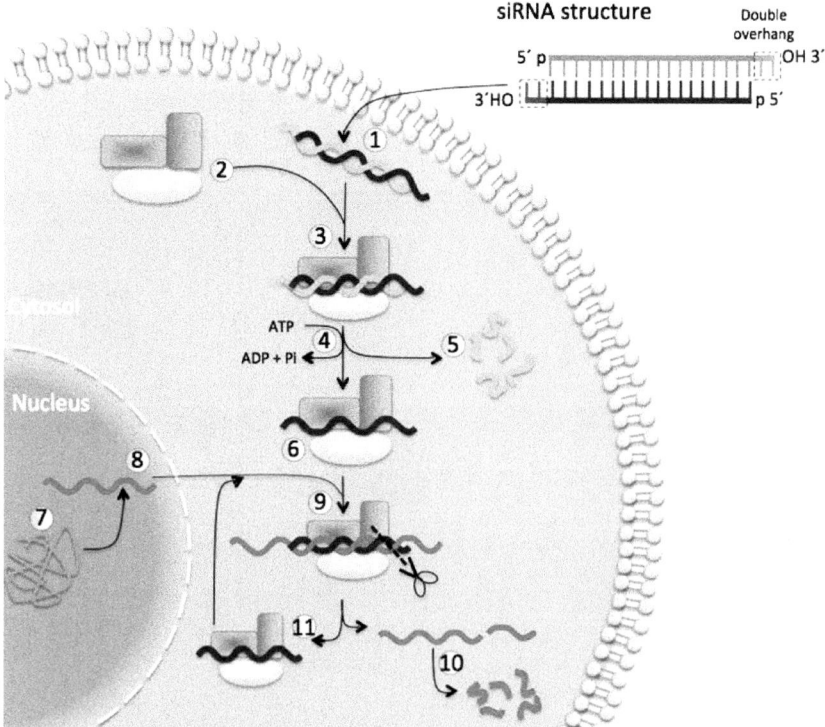

Figure 14.1 Gene silencing mechanism of siRNA. (1) siRNA is transfected into the target cell. (2) The RISC protein complex assembles, and (3) siRNA is loaded. (4) The siRNA is unwound into its constituent strands in an ATP-dependent process. (5) The passenger strand is degraded, and (6) the siRNA guide strand is bound to RISC, which is then activated. (7) The target gene in the nucleus is transcribed to mRNA, (8) which is exported to the cytosol. (9) The guide strand binds to the mRNA with perfect base complementarity, and (10) the mRNA is subsequently cleaved by AGO-2 and (10) degraded. (11) The activated RISC complex can bind to additional mRNA molecules, eventually amplifying the RNAi effect.

14.1.1 Hurdles to siRNA Delivery

Utility of siRNA therapeutics is entirely reliant on the ability to deliver siRNA molecules to their site of action, namely the cytosol of the target cells. The numerous delivery hurdles faced by siRNA can be broadly categorized into physical, chemical and biological barriers. The physicochemical properties of siRNA, e.g., the anionic phosphate backbone, hydrophilicity, and relatively large

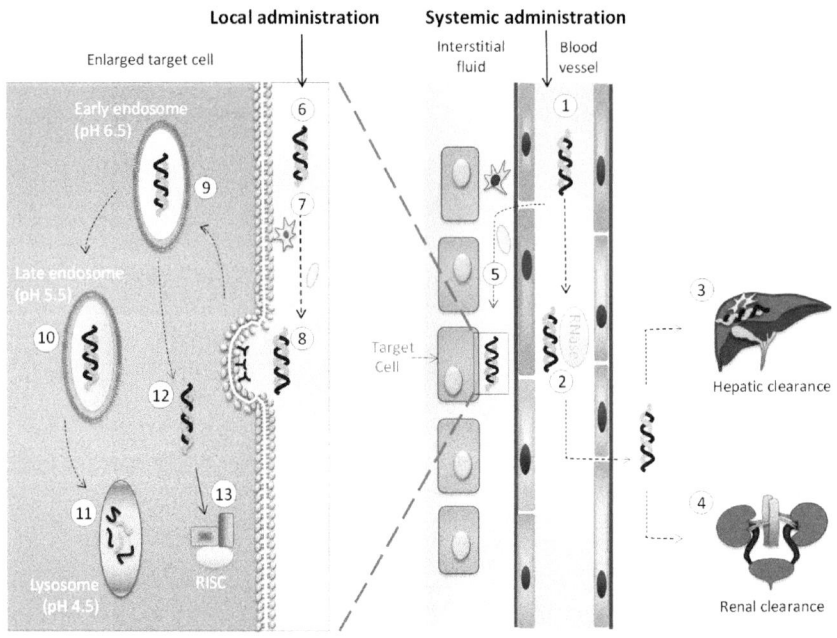

Figure 14.2 Biological barriers to siRNA delivery. (1) Systemically administered siRNA is injected into the blood circulation, where it is subjected to (2) endogenous nuclease degradation, (3) hepatic clearance and (4) renal clearance. Once in the liver, siRNA is phagocytosed by cells of the mononuclear phagocytic system. The siRNA remaining in the systemic circulation is passively transported to organs, (5) where siRNA may enter the interstitium, and undergoes further degradation by endonucleases and phagocytes during passive diffusion to the target cell. In contrast, (6) locally administered siRNA is delivered directly to the target tissue, (7) and is thus only subject to endonucleases and phagocytes located there. (8) Subsequently, siRNA may be internalized, primarily through endocytosis. (9) The resulting endocytic vesicles fuse with early endosomes, (10) which mature into late endosomes, (11) prior to fusing with lysosomes, where siRNA is degraded. To avoid degradation, (12) siRNA must undergo endosomal escape into the cytosol after which (13) siRNA is available for binding to the RNAi machinery.

molecular weight (approx. 13 kDa), advocate unfavourable cell uptake. In addition, biological barriers impede both extracellular and intracellular trafficking of siRNA (Fig. 14.2). The main extracellular challenges faced by systemically administered siRNA are (i) endogenous nuclease degradation, (ii) phagocytosis by the mononuclear phagocytic system (MPS), and (iii) renal clearance, collectively ensuing a plasma half-life of less than 10 min [11]. At the cellular level, siRNA has to overcome the hydrophobic cell membrane and enter the cytosol. If taken up *via* endocytosis, endosomal escape represents the major challenge to avoid lysosomal degradation [12].

By direct delivery of siRNA to the target by local administration, some of the extracellular hurdles encountered upon systemic delivery are bypassed, including the need to evade renal excretion and navigate the circulatory system, while simultaneously improving siRNA biodistribution to target cells. The large percentage of clinical trials employing local, rather than systemic delivery, is a reflection of this [13]. However, exploring systemic administration in the long term is essential to transfect difficult to reach cells, and realize the full potential of RNAi.

14.2 Overview of Approaches for Efficient Intracellular Delivery of siRNA

A number of highly diverse strategies have been explored to overcome siRNA delivery hurdles and translate RNAi therapy towards the clinic. The approaches can broadly be categorized into (i) chemical modification of siRNA, (ii) conjugation of siRNA with biomolecules, (iii) siRNA complexation, and (iv) encapsulation of siRNA in nanoparticles. At present, siRNA modification and siRNA encapsulation have been the most fruitful strategies in terms of reaching clinical trials, each with nine drug candidates in comparison to only two conjugation-based candidates [14]. The potentials and limitations of each approach are discussed in the following section, with specific focus on how delivery barriers are overcome and the associated side effects, drawing on the most widely applied examples of each.

14.2.1 Chemical Modification

The most straightforward delivery approach is chemical modification of siRNA molecules. Strategically positioned modifications can alter interactions between siRNA and cellular proteins, providing a means to overcome siRNA delivery barriers, including nuclease degradation and immunostimulation.

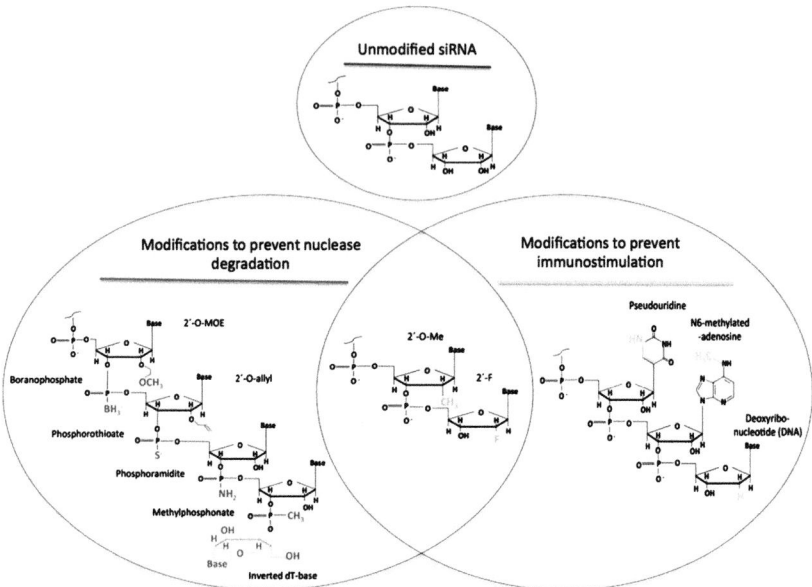

Figure 14.3 Examples of chemical modifications of siRNA molecules. The structure of unmodified siRNA is presented at the top. Modifications that reduce nuclease degradation and immunostimulation are presented to the left and right of the Venn diagram, respectively. The intersection represents modifications that simultaneously reduce nuclease degradation and immunostimulation.

Due to the presence of a hydroxyl group at the 2' position of each nucleotide, siRNA molecules are highly susceptible to RNase-catalysed hydrolysis. Early chemical modification analysis showed that the 2' hydroxyl groups are not involved in RISC binding and therefore not required to induce RNAi [15]. As a result, extensive modifications of the 2' position have been tested with the aim of minimizing RNase degradation, while maintaining siRNA activity. Two of the most commonly used modifications are addition of methyl or flouro groups (Fig. 14.3, left), which both decrease RNase recognition, and these modifications are well-tolerated throughout the duplex [15]. More bulky modifications, e.g., 2'-O-methoxy-ethyl-RNA (MOE) and 2'-O-allyl-RNA (Fig. 14.3) also reduce RNase degradation, but compromise RNAi potency, because these

these modifications disrupt the siRNA A-form helical structure, limiting their application to siRNA termini [15]. Additionally, while the 2′-O-Me modification is naturally occurring, the others are not, thus potential toxic side effects must be considered. Alternatively, the phosphate backbone can be modified. Replacement of a non-bridging oxygen with boron (boranophosphate), sulphur (phosphorothioate), nitrogen (phosphoramidite) or methyl groups (methylphosphonate) (Fig. 14.3), is compatible with RNAi activity, and hinders nuclease degradation [16]. However, these modifications can be cytotoxic and some, like boranophosphate, are challenging to synthesize. A final modification worth mentioning is end cap modification. Exonuclease attack occurs at the 3′-overhang of siRNA and progresses in the 3′→5′ direction. Addition of an inverted-dT base or other non-nucleotide groups to the 3′ end (Fig. 14.3), offers a strategy for protection against this exonuclease attack. While promising, such end cap modifications are immunostimulatory [17], and can only be added to the passenger strand termini without disturbing the RNAi process.

Chemical modification can also be harnessed to address immunostimulation, which is another key hurdle to RNAi therapy (Fig. 14.3, right). The innate immune system is activated when siRNA molecules are recognized by toll-like receptors (TLRs), e.g., TLR3, TLR7 and TLR8. The 2′-O-Me, 2′-F and phosphorothioate modifications discussed above are multifunctional in nature, because they also impede TLR recognition. Incorporation of as few as two 2′-O-Me nucleotides is sufficient to abrogate TLR recognition of an entire siRNA molecule [18]. Mechanistically, 2′ methyl groups act as competitive inhibitors of TLR7 thereby protecting them in *trans* [15]. Other modifications, which prevent immunostimulation in naturally occurring RNA molecules, can be incorporated in synthetic siRNAs to achieve a similar effect. These include incorporation of pseudouridine and N6-methylated-adenosine (Fig. 14.3), respectively, inhibiting binding to TLR7 and TLR3. Incorporation of N6-methyladenosine specifically evades immune stimulation by destabilizing the duplex structure, recognized by TLR3. However, pseudouridine and N6-methylated-adenosine modifications are utilized to a limited extent due to the success of other modifications, in particular 2′-O-Me [19]. Finally, 2′-deoxynucleotides (Fig. 14.3) have recently been reported to impede immune recognition, particularly dU and dT bases [20].

Unlike other modifications, this modification is exceptionally versatile with the possibility for incorporation in the entire passenger strand and the 5′ end of the guide strand without affecting RNAi potency [21].

Evidently, various modifications can be employed to reduce siRNA instability and immunostimulation. The discussed modifications have been found to increase siRNA half-life a thousand fold in human serum, and duplexes 90% modified with 2′-O-Me, 2′-F and DNA residues were shown to have no detectable effects on cytokine and interferon levels, while unmodified siRNA caused considerable immune induction [17]. However, the negative impact on RNAi potency and off-target effects associated with siRNA modifications limit their usefulness. A solution is to identify the most vulnerable positions in the duplex and modify only these specific positions.

14.2.2 Conjugation of siRNA to Biomolecules

An alternative strategy comprises conjugation of siRNA to a biomolecule. Depending on the specific type of conjugate, this approach may (i) enable target-specific delivery, (ii) improve serum stability and (iii) increase transfection efficiency (Table 14.1). Conjugation is primarily performed via cleavable linkages (e.g., acid-labile and reducible bonds) at the 3′ or 5′ termini of the sense strand, to facilitate intracellular release of siRNA and to ensure that the 5′ antisense terminus, essential for RNAi initiation, remains intact [22]. Below, focus will be on the first conjugate system to demonstrate *in vivo* efficacy, i.e., cholesterol–siRNA conjugates, and the more recent clinically tested conjugate system *N*-acetylgalactosamine (GalNAc)-siRNA conjugate.

Conjugation to cholesterol introduces hydrophobicity to the otherwise hydrophilic siRNA, enabling binding to serum albumin and lipoproteins, which improves siRNA pharmacokinetics. The siRNA half-life is extended from less than 10 min for the unconjugated siRNA to approx. 95 min for the conjugate, because ultrafiltration in the kidneys is reduced [23]. In addition, cellular uptake is enhanced via lipoprotein receptor-mediated endocytosis [24]. These two mechanisms have been shown to increase *in vivo* silencing by 73% [25]. However, repeated high doses (50 mg/kg) are required to effectuate knockdown, and the effect is restrained to lipoprotein-receptor expressing cells. Possible improvements involve identifying other lipophiles that interact more favourably with lipoproteins, or preassembling cholesterol–siRNA conjugates with lipoproteins prior to administration.

Table 14.1 Overview of selected siRNA conjugates and their associated therapeutic effects and drawbacks

Conjugates		Therapeutic effects and drawbacks
Small molecule based-	Lipophiles (Cholesterol, fatty acids and bile acids)	Improved serum stability Enhanced transfection efficiency Repeated high doses required
	Cell-penetrating peptides (CPPs)	Efficient cellular uptake May elicit immune responses and no improvement in serum stability
	Polyethylene glycol (PEG)	Increased serum stability Longer half-life due to increased hydrodynamic volume
Receptor ligand based-	Antibodies	Improved cell specificity and minimized off-target effects Form multimeric aggregates High-cost
	N-acetyl galactosamine (GalNAc)	Facilitated hepatocyte delivery and rapid cell uptake Improved stability profiles and long-lasting silencing Reduced toxicity and immunostimulation
	Aptamers	Cell-type specific delivery Low molecular weight facilitates interstitial transport Not all aptamers are efficiently internalized by cells

Recently, there has been a major shift away from lipophilic conjugates towards ligand conjugates that directly bind to cell-surface receptors [26]. An example is GalNAc conjugates that bind to hepatocyte asialoglycoprotein receptors, triggering clathrin-mediated endocytosis. Three GalNAc molecules are linked to one siRNA to increase receptor affinity. This type of siRNA conjugate has been shown to have extended half-life, rapid hepatocyte uptake and a wide therapeutic window, showing no immune activation at doses up to 300 mg/kg [14]. The most clinically advanced GalNAc–siRNA conjugate system is Alnylam's TTRsc, currently in phase III. However, the long-term safety and efficacy of siRNA–GalNAc conjugates remain to be determined.

14.2.3 Complexation of siRNA

Another, less clinically advanced delivery approach involves siRNA complexation with cationic lipids or polymers, forming lipoplexes or polyplexes, respectively. The complexes are formed by self-assembly *via* attractive electrostatic interactions between the anionic siRNA and the cationic molecules [27]. While showing improved transfection efficiency, siRNA complexes generally display low colloidal stability in physiological medium and significant toxicity, resulting in very few *in vivo* studies.

At present, a wide number of lipids have been tested for siRNA delivery. Lipoplexes promote cellular siRNA uptake by increasing hydrophobicity and facilitating endosomal escape [28]. However, most lipoplexes have failed in clinical trials due to excessive cationic charge, resulting in rapid clearance in first pass organs, hepatic elimination, immune activation and hence low transfection efficiency [29]. Ionizable and titratable lipids have been designed to overcome these limitations. At low pH, ionizable and titratable lipids are charged, enabling efficient complex formation. However, they are neutral at physiological pH, circumventing undesired side effects [29].

Polyplexes have also been extensively studied for siRNA delivery. Cationic polymers, including chitosan, polyethyleneimine (PEI) and poly-L-lysine (PLL), assemble with siRNA into polyplexes. They protect siRNA from enzymatic degradation, enhance transfection efficiency and facilitate endosomal escape, which may occur via the so-called proton sponge effect [28]. On the other hand, cationic polymers and linear PEI in particular, exhibit high cytotoxicity [30]. Employing hyperbranched polymers like polyamidoamine (PAMAM) and triazine to form dendriplexes provides an alternative. In comparison to polyplexes, dendrimers have been shown to form relatively more stable siRNA complexes with reduced cytotoxicity [31]. Modification, like PEGylation, is promising for shielding the cationic charge and mitigate cytotoxicity of polymer-based complexes [32]. While the solutions presented mask the underlying limitations of each system, it is clear that a better understanding of the delivery mechanisms, including internalization routes and immune activation is required to optimally design delivery systems [29].

14.2.4 Nanoparticle-Based Approaches

Nanoparticles, formed by encapsulation of siRNA, constitute another siRNA delivery approach, currently under extensive investigation. Nanoparticles encapsulating siRNA can be categorized into three groups; lipid-based, polymer-based and inorganic nanocarriers [33]. Generally, nanoparticles promote delivery by protecting siRNA from degradation, facilitating cell membrane permeation and enabling target-specific delivery through surface-bound ligands [34]. The following will focus on lipid- and polymer-based nanoparticles.

Lipid-based nanoparticles constitute the most widely used and clinically advanced delivery approach. These nanoparticles are usually composed of cationic lipid, with a neutral lipid, like cholesterol, to stabilize the particles and prevent excessive cationic charge [29]. While effective *in vitro*, these nanocarriers have drawbacks *in vivo* due to interaction with opsonins, resulting in rapid elimination from the bloodstream [26]. Innovative solutions to these problems have therefore been developed for systems, which have reached clinical trials. An example are stable nucleic acid-lipid particles (SNALPs), in which the cationic lipids have been optimized by altering their pK_a value; they are charged at the low pH in endosomes, but net neutral in the systemic circulation [35]. Likewise, biodegradable linkers have been employed to reduce cytotoxicity [26]. PEGylation of SNALPs reduces interaction with opsonins, which increases the circulation time. Collectively, this has contributed to significant improvements in siRNA delivery, with siRNA doses as low as 5 µg/kg capable of inducing *in vivo* gene silencing with concurrent low toxicity [36]. A recent report further suggested that SNALPs comprising novel lipid-like material [epoxide-derived lipidoid (C_{12-200})] enter cells via pinocytosis, bypass the endosomal route and thus evade immunostimulatory effects of loaded siRNAs [37].

Polymeric nanoparticles are less clinically developed. Initial investigations were performed using anionic biocompatible polymers, e.g., poly(lactic-co-glycolic acid) (PLGA) [26]. Since then, additional systems have been developed, with cyclodextrin-based nanoparticles being the most clinically advanced [14]. These nanoparticles are composed of polycationic cyclodextrin oligomer, as well as adamantane-PEG (AD-PEG), adamantane–PEG–transferrin (AD–PEG–Tf) and conjugated transferrin, to stabilize the cyclodextrin core, shield the cationic charge and improve cellular uptake by targeting the transferrin receptor CD71, respectively [38]. The documented efficacy of cyclodextrin polymer-based nanoparticles emphasises the benefits of multifunctional systems, and points towards prospective developments within siRNA delivery, involving these more complex systems to realize the full RNAi potential.

Yet another class of nanoparticles, referred as lipid-polymer hybrid nanoparticles (LPNs) have been explored for siRNA delivery [39]. Such an approach effectively mitigates the limitations of polymeric nanoparticles and liposomes [40]. The principle advantages associated with LPNs include (i) mechanical strength provided by the polymeric core, (ii) superior biocompatibility of the lipid coat, (iii) capability to load different types of active pharmacutical ingredients for co-delivery applications, (iv) efficient cellular uptake and membrane fusion characteristics, and (v) flexibility to develop tailor-made particles by attaching ligands for specific receptors [41].

14.3 Immune Reactions Elicited during RNAi Therapy

Over the course of evolution, the immune system has developed to recognize danger signals, i.e., pathogen-associated molecular patterns (PAMPs) and damage-associated molecular patterns (DAMPs), *via* pattern-recognition receptors (PRRs) [42–44]. Components of RNAi therapeutics, e.g., siRNA, cationic lipids and polymers, may also be recognized by PRRs, which are either membrane-bound, i.e., TLRs and C-type lectin receptors (CLRs), or cytoplasmic receptors *viz.* NOD-like receptors (NLRs) and RIG-I-like receptors (RLRs) [44]. The following section focuses on the immunological reactions that can occur during RNAi therapy.

14.3.1 Immune Reactions Stimulated by siRNA

Immunological responses may be elicited by siRNA under normal course of exposure. The innate immune system recognizes dsRNA and mediates a series of responses by induction of the secretion of interleukins (ILs) and interferons (IFNs) [45]. The interleukins, e.g., IL-1, IL-6, IL-12 and TNF-α, are secreted as first line of defence leading to strong inflammatory responses, which may be followed by activation of the adaptive immune system. The innate immune system comprises a variety of cell types. Mast cells residing in connective tissues and mucous membranes recognize danger signals and recruit other members of the innate immune system. Phagocytic cells of the innate immune system comprise macrophages, neutrophils and dendritic cells (DCs). Neutrophils, along with basophils and eosinophils, secrete a variety of toxic substances to combat the foreign invaders. These cell types recognize danger signals *via* PRRs. The adaptive immune system comprises specialized lymphocytes, i.e., B cells and T cells. These lymphocytes are initially primed by antigen-presenting cells (APCs), primarily DCs and macrophages, which link the innate and adaptive immune system. The adaptive immune system is highly specific and is characterized by having immunological memory. The B lymphocytes encountering known invaders secrete antigen-specific antibodies and in conjunction with T cells leads to their neutralization. Upon activation, CD8$^+$ T lymphocytes may proliferate and differentiate into cytotoxic T cells (CTLs), which in collaboration with phagocytic cells are responsible for clearance of pathogens. The release of IFN-α and IFN-β by stimulated plasmacytoid DCs imparts antiviral response, further recruiting additional CTLs, and a signalling cascade is initiated. The PRRs play a pivotal role in initiating such signalling cascades (Fig. 14.4). The siRNA is specifically recognized by the horseshoe-shaped ectodomain of TLR3 thereby resulting in formation of a stable homodimer mediated by C-terminal bringing the juxtamembrane region into close proximity for signal transduction [46]. The observations have been further corroborated by various studies in TLR3-deficient mice [47]. In addition, TLR7 recognizes ssRNA in a sequence-dependent manner. However, the *in vivo* stimulation of IFN-α secretion in TLR7-deficient mice was only possible when immunostimulatory RNA was complexed with cationic liposomes [48]. The results suggest a significant influence

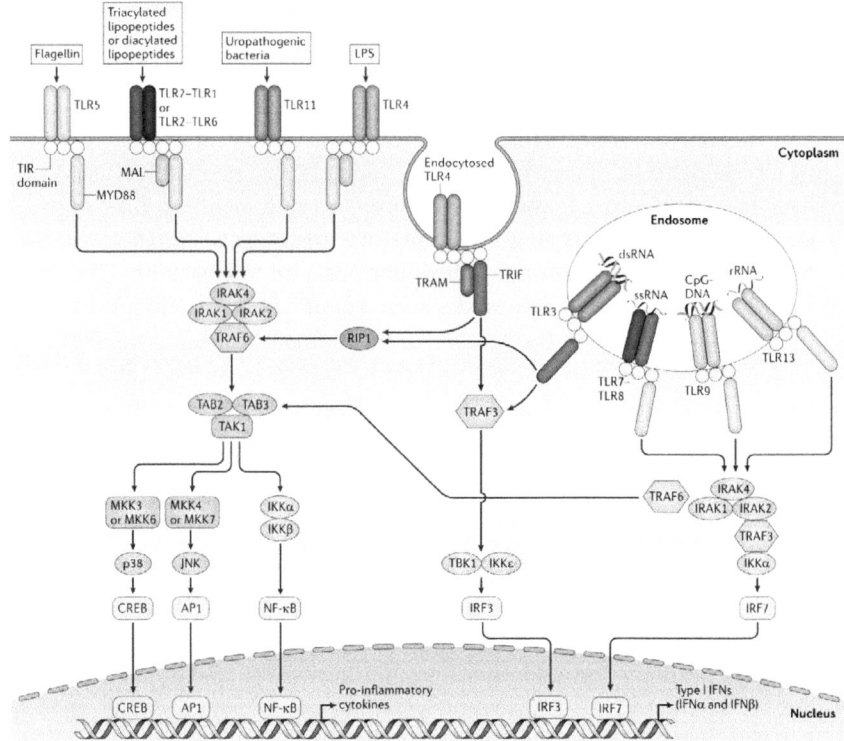

Figure 14.4 Mammalian TLR signalling pathways. TLR signalling is initiated by ligand-induced dimerization of receptors. Subsequently, the Toll–IL-1-resistance (TIR) domains of TLRs engage TIR domain-containing adaptor proteins [either myeloid differentiation primary-response protein 88 (MYD88) and MYD88-adaptor-like protein (MAL), or TIR domain-containing adaptor protein inducing IFNβ (TRIF) and TRIF-related adaptor molecule (TRAM)]. TLR4 translocates from the plasma membrane to the endosomes to switch signalling from MYD88 to TRIF. Engagement of the signalling adaptor molecules stimulates downstream signalling pathways that involve interactions between IL-1R-associated kinases (IRAKs) and the adaptor molecules TNF receptor-associated factors (TRAFs), and that lead to the activation of the mitogen-activated protein kinases (MAPKs) JUN N-terminal kinase (JNK) and p38, and to the activation of transcription factors. Two important families of transcription factors that are activated downstream of TLR signalling are nuclear factor-κB (NF-κB) and the interferon-regulatory factors (IRFs), but other transcription factors, such as cyclic AMP-responsive element-binding protein (CREB) and activator protein 1 (AP1), are also important. A major consequence of TLR signalling is the induction of pro-inflammatory cytokines, and in the case of the endosomal TLRs, the induction of type I interferon (IFN). dsRNA, double-stranded RNA;

IKK, inhibitor of NF-κB kinase; LPS, lipopolysaccharide; MKK, MAP kinase; RIP1, receptor-interacting protein 1; rRNA, ribosomal RNA; ssRNA, single-stranded RNA; TAB, TAK1-binding protein; TAK, TGFβ-activated kinase; TBK1, TANK-binding kinase 1. Reproduced with permission from [52].

of the delivery system on the overall immunological response towards therapeutic cargos. Furthermore, TLR8 can also recognize ssRNA in a sequence-specific manner with preference towards ssRNA containing GU-rich domains [49].

The cytoplasmic receptors additionally responsible for initiation of the signalling cascade include dsRNA-activated protein kinase (PKR) and retinoic acid-inducible gene-I (RIG-I). Approx. 11 bp of dsRNA have been shown to bind to the N-terminal of the dsRNA-binding domain (dsRBD) of PKR in a sequence-independent manner, resulting in autophosphorylation [50]. The activation of PKR leads to the phosphorylation of translation initiation factor, which subsequently inhibits translation of mRNA and release of IFNs. Interestingly, different structural motifs, i.e., conventional siRNA (containing two bp overhangs at the 3′ end of each strand) and blunt siRNA (no over hangs), did not show any discrepancy in the activation of PKR, being moderate in both cases [45]. On the other hand, RIG-I recognizes either ssRNA or dsRNA. However, the recognition is dependent on the sequence, motif length, and capping of the 5′ triphosphate group. An uncapped dsRNA of less than 300 bp (and as few as 10 bp), rich in poly-U/UC motif and with no overhangs, is perfectly recognized by RIG-I [51].

14.3.2 Immunogenic Potential of Functional Excipients Used in RNAi Therapy

The activation of PRRs initiates signalling cascades of immune reactions, and they recognize a series of molecular patterns (or structural motifs) of the pathogenic ligands. Hence, reasonable efforts have been made to understand the structure-activity relationship (SAR) of individual functional excipients, e.g., cationic lipids, as a function of their activation of the intracellular signalling pathways (*viz.* pro-inflammatory- and pro-apoptotic pathways) and subsequent toxicity (e.g., associated with membrane destabilizing effects) [53]. The cationic lipids *per se* can activate a series of inflammatory mediators, e.g., reactive oxygen species (ROS), cytokine secretion and induction of DC maturation, leading to initiation of signalling cascades. In addition, release of intracellular calcium ions is also triggered, further intensifying the deleterious effects of cationic lipids on homeostasis [54]. The SAR of cationic lipids suggests that activation of the signalling cascade and/or subsequent toxicity is dependent on (i) the type of hydrophilic headgroup, (ii) hydrophobic chains, and (iii) linker moieties [55]. Cationic phospholipids with saturated C_{12} or C_{14} fatty acid moieties showed higher DC activation, evident from CD80/86 upregulation, as compared to their longer counterparts with saturated C_{16} or C_{18} and unsaturated C_{18} fatty acid moieties [56].

A major concern attributed to the use of cationic lipids includes inhibition of protein kinase C leading to alterations in the normal function of the immune system. These effects of cationic lipids have been executed by series of either nuclear factor-κB (NF-κB)-dependent pathways [MYD88 pathway or Toll-interleukin receptor-domain-containing adapter-inducing interferon-β (TRIF) pathway] and/or the NF-κB independent pathway [TLR4 independent nuclear factor of activated T-cells (NFAT) pathway]. Irrespective of the signalling pathway, a series of kinases, e.g., mitogen activating protein kinases (MAPKs), c-Jun N-terminal kinases (JNKs), and extracellular signal regulated kinases (ERKs) are activated, along with proteases, e.g., caspases, ultimately resulting in inflammation and apoptosis (Fig. 14.4) [53].

PEI remains the most studied example of a cationic polymer used for RNAi and it is highly effective and versatile [57]. The principle toxicity concerns associated with PEI includes membrane disruption associated with pore formation, initiation of cell death pathways [58], complement activation [59] and induction of inflammatory responses [60]. The cationic polymers, dendrimers, and cyclodextrins all activate the complement system, thereby resulting in hypersensitivity reactions and immunological responses [58].

14.4 Recent Advancements in Predicting Immunological Complications

Recently, regulatory agencies have laid emphasis on the critical assessment of immunogenicity assays and associated strategies [61–64]. A series of guidelines and white papers have also been published on various aspects of bio-therapeutic products, including quality, safety and efficacy. Although these recommendations largely apply to the biotechnology-derived therapeutic proteins, the principles may also apply to RNAi-based drugs. Careful consideration of immunogenicity-related issues is necessary, as potential immune reactions cannot be foreseen until advanced stages of clinical development. In this context, yet another anticipated challenge includes careful selection of controls. A three-step approach is usually recommended, which comprises (i) screening assays, which are sensitive yet with high throughput capabilities, (ii) confirmatory assays, and (iii) bioassays [61]. A variety of screening assays are envisaged. Various aspects of these assays have been recently reviewed and are therefore not described further [65].

14.4.1 Reporter Cell Lines for Testing TLR Activity

During the past couple of decades, the knowledge has been expanded for design and development of PRR reporter cell lines. The primary uses of these cell lines include studies of (i) genetic functionalities, (ii) signalling pathways, (iii) screening of ligands for various types of PRRs, and (iv) up/down regulation of specific mediators, e.g., cytokines. Such cell-based reporter systems are obviously advantageous because they are quite simple, fast and provide reliable and reproducible information. In the current scenario, two different types of approaches are employed, *viz.* over-expressing cell lines and knock-out cell lines [66]. In the former case, the cells are engineered to express an inducible secreted reporter gene, which is subsequently measured as a function of response to any intended stimulus. In most cases, the cells lacking endogenous activity, yet equipped with most down-stream signalling components, are employed for such purposes, e.g., epithelial cells do not express TLRs and hence transfection with one or more TLRs can make them responsive to specific TLR agonist/antagonists, depending on the genetic manipulation. This approach has been demonstrated in series of cell lines, e.g., HEK293, HeLa, COS7, and CHO cell lines [67–70]. The knock-out cells lines, on the other hand, are usually helpful in assessing the interferon response to oligonucleotides. Ideally, the results of the screening experiments should be confirmed in assays using primary cells [66].

14.4.2 *In silico* Tools for Predicting Potential Immune Reactions

The growing number of known three-dimensional structures of PRRs bound to their activating ligands provides insight into the shape, size and charge requirements to mount an immune response. In this section, we address the use of computational tools to study the immunogenic potential of lipid-based siRNA delivery systems. More detailed information on the various structural aspects of PRRs and their molecular assembly can be found elsewhere [71].

14.4.2.1 Nucleic acid recognition

TLR3, located in endosomal compartments, recognizes dsRNA in a sequence-independent manner, ideally for oligonucleotides of minimum length 40–50 bp under acidic conditions at pH < 6.0. The ligand-binding domain of TLR3 contains 23 leucine-rich repeats (LRRs) that adopt a "horseshoe" shape [72, 73]. Upon dsRNA binding, receptor dimerization is triggered, which brings the C-terminal ends of the luminal domains in close proximity in a conformation that enables intracellular signalling, initiated by the recruitment of the adaptor protein TRIF *via* heterotypic TIR-TIR domain interactions [46]. Although dsRNAs as short as 21 bp have been shown to activate TLR3 [74], their binding mode differs from longer RNAs, and they are unable to contact both interaction sites. Nevertheless, they are sufficiently long to engage the C-terminal binding site from LRR19 to 21 [75]. It is currently unknown why such an organization is sufficiently stable to initiate signalling. It has been speculated that lateral clustering of neighbouring signalling units, possibly with the help of a bridging factor, might enhance their stability [76].

TLR7/8 possess larger ectodomains with 27 LRRs that adopt a donut-like ring structure in which the N-terminal contacts the C-terminal region. These receptors are sensors of ssRNA and can therefore only be activated by degradation products of siRNA in the endosomes under physiological conditions (Fig. 14.5). They detect specific nucleosides at the dimer interface in a binding site that involves LRR8 and LRR11-14 from one protomer, and LRR16-18 from the other one (Fig. 14.5). Recognition of ssRNAs takes place in a second location, encompassed by the concave surface of LRR1-5, LRR20' and part of the endoproteolytically processed Z-loop between LRR14 and 15. The sequence specificity of each receptor has recently been elucidated, and their crystal structure is known in complex with a variety of natural and synthetic ligands [77–79]. TLR7 is a sensor of G and U-containing ssRNA, as long as they are not in a terminal position. Hence, replacing the U with A reduces immunogenicity of siRNA [77].

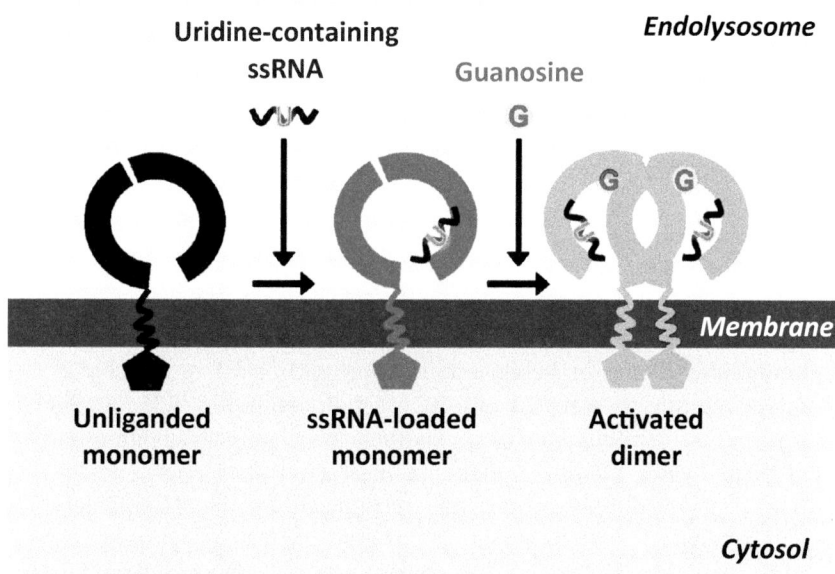

Figure 14.5 TLR7 is a sensor of G and U-containing ssRNA degradation products. Reproduced with permission from [77].

RIG-I preferentially recognizes two distinct RNA patterns: dsRNAs and ssRNA with 5' triphosphates in the cytoplasm to mount a type I IFN response [80, 81]. It was evident that further, ssRNA-5' triphosphate, but not 5' OH or a 5'-methyl guanosine cap, binds to the RIG-I repressor domain and promotes a conformational change that activates RIG-I signalling [82]. It was thought that self-RNAs like mRNAs are not recognized by RIG-I because of their caps, e.g., 7-methyl guanosine. However, it was recently shown that RIG-I cannot bind ssRNA, whereas dsRNA can be accommodated by RIG-I, regardless of such a modification [83]. Crystallographic analysis has shown that the m7G cap is accommodated by a conformational change in the helicase motif IVa [83]. Moreover, m7G capping works synergistically with 2′-O-Me to weaken RNA affinity by 200-fold and reduce ATPase activity. In contrast, dsRNA bearing a 5′ end nucleotide ribose cap abrogates signalling through a His residue at position 830 involved in cap discrimination. RIG-I can also fit 2-nucleotide 3′-overhangs such as those in siRNAs [84].

PKR is a stress-induced kinase and its N-terminal domain is a regulatory module that binds dsRNA. It is composed of two tandem binding motifs, which adopt the typical $\alpha\beta\beta\beta\alpha$ fold for dsRNA binding, referred to as dsRNA binding motif 1 and 2 (dsRBM1 and 2), respectively.

14.4.2.2 Lipid recognition

The most widely studied and promiscuous receptors for lipid recognition are probably TLR2 and TLR4. There is increasing evidence that PRRs have both canonical and atypical ligands. TLR2 senses cell-wall components, lipoteichoic acid and lipoproteins from gram-positive bacteria; lipoarabinomannan from mycobacteria; and zymosan from yeast [85]. Ligand recognition requires either homodimerization or heterodimerization with TLR1 or TLR6 via direct binding to their 20 LRR-long ectodomains [86, 87]. While lipopeptide binding has been characterized by crystallography, atypical ligand binding remains a mystery. It has been shown that triacylated lipopeptides are recognized by the TLR2-TLR1 heterodimer that accommodates the lipid chains in hydrophobic cavities within LRR9-12. TLR1 buries a single chain linked by an amide linkage almost perpendicular to the two ester-bound chains concealed in TLR2 [86]. In contrast, diacylated lipopeptides recruit TLR2-TLR6 heterodimers. The lack of hydrophobic cavity in TLR6 complements perfectly the lack of amide-bound lipid, while increased protein-protein interactions compensate the absence of ligand tether, compared to TLR1 heterodimers. TLR2, as well as NLRP3, have been implicated in cationic lipid signalling [88], but their mechanism of action is not fully resolved. Docking studies performed on a family of di-C18 cationic lipids suggest that they bind in a manner reminiscent of lipopeptides preferentially to TLR2-TLR1 heterodimers, despite only having two aliphatic chains [89].

TLR4 on the other hand has an extracellular region that is composed of 22 LRRs. MD-2 is the founding member of a class of small lipid-recognition proteins with an immunoglobulin-like fold and a deep hydrophobic cavity sandwiched between two beta-sheets [90]. MD-2 binds to the concave side of TLR4 contacting residues throughout LRR2 to 10. Lipopolysaccahride (LPS) binding to MD-2 induces the formation of an m-shaped receptor multimer composed of two copies of the TLR4–MD-2–LPS complex arranged symmetrically. The lipid A moiety of LPS bound within the hydrophobic core of MD-2 directly bridges the two TLR4 ectodomains. Five of the six lipid chains of *E. coli* LPS are buried deep inside the pocket, and the remaining chain is exposed to the surface of MD-2, forming a hydrophobic interaction with the conserved Phe residues of TLR4. A loop in MD-2 containing Phe 126 undergoes a conformational change that supports the hydrophobic interface by making further hydrophilic interactions with TLR4 at the dimer interface. LPS also contributes to TLR4 dimerization by forming ionic interactions with positively charged residues in TLR4 and MD-2. Di-C14 amidine interacts with critical residues involved in TLR4 activation that are not located at the LPS-binding site on MD-2 but are instead found at the TLR4 dimer interface [91]. It was previously proposed that nickel and cobalt ions trigger MD-2-dependent TLR4 dimerization and activation through chelation of His residues (H431, H456 and H458 on both TLR4 protomers) [92]. This implies that different anchoring points can lead to dimerization and foresees other possible ways of TLR4 activation that have still to be discovered. More importantly, the knowledge of the 3D structure of the receptor has guided the characterization of the atypical binding modes of both cationic lipid and ions, suggesting the potential adequacy of docking techniques for the latter.

14.4.2.3 Docking studies to assess PRR binding

Docking is used to predict the preferred orientation of molecules to one another. A given molecule (siRNA or transfection lipid) should not only bind to the PRR, it also has to favour its activated state in order to be immunogenic. Flexible docking and molecular dynamic simulations are computationally very heavy, while scoring functions are different from empirical binding energies. "What you see" is not always "what you get" [93]. While force fields have been successfully applied to crystallography, the deficiencies of the interatomic potentials often lead to incorrect modelling.

Despite their shortcomings, docking and molecular dynamic simulations should be assessed for knowledge-based design of improved RNAi therapeutics, and the predictions should be confirmed experimentally.

14.5 Conclusions and Future Perspectives

Undesired immunogenicity poses a major concern for the progression of RNAi therapeutics. Although the causes for immunogenicity are multifactorial, it may be possible to predict and/or precisely assess the contribution of the individual components. It may also be possible to circumvent such problems *via* knowledge-based design of materials based on SARs for ligands to the PRRs. The use of advanced technologies, e.g., systems biology, proteomics and genomics, is encouraged to simplify the physiological and pathophysiological complications and better understand the obtained data. A multidisciplinary approach is required to understand the complex mechanisms involved in immunological reactions. However, a collaborative effort should always be sought to provide feedback for improving biocompatibility of components of RNAi therapeutics, robustness of testing methodologies and/or precision of the predictive tools. A careful assessment of immunogenicity of RNAi-based drug candidates is required to improve safety and efficacy.

Disclosures and Conflict of Interest

No writing assistance was utilized in the production of this chapter and the authors have received no payment for its preparation. Nothing contained herein is to be considered as the rendering of legal advice.

Corresponding Author

Prof. Camilla Foged
Department of Pharmacy
University of Copenhagen, Building: 13-4-415b
Universitetsparken 2, DK-2100 Copenhagen, Denmark
Email: camilla.foged@sund.ku.dk

About the Authors

Kaushik Thanki is currently a post-doctoral fellow in the group of Prof. Camilla Foged at Department of Pharmacy, University of Copenhagen, Denmark. He is contributing as project leader to the Collaboration on the Optimization of Macromolecular Pharmaceutical Access to Cellular targets (COMPACT) project. He is a pharmacist by training and his current research is focused on addressing the pathophysiological conditions of chronic obstructive pulmonary disease (COPD) using formulation-enabled RNA interference therapy. During his short research career, he has collaborated closely with experts in the field of drug delivery from academia, industry and regulatory agencies. He gained in-depth experience in the formulation development required to ensure the quality, safety and efficacy of drug products.

Emily Falkenberg is a biomedical sciences bachelor student at Durham University, England. She is currently working as a research scholar student in the group of Prof. Camilla Foged at Department of Pharmacy, University of Copenhagen. Her work is focused on the synthesis and characterization of lead lipid structures that do not bind to TLR4 *in silico*. The lipids will be tested experimentally for *in vitro* transfection efficiency, toxicity and immunogenicity.

Monique Gangloff is a biochemist with a background in X-ray crystallography of nuclear receptors. She joined Prof. Nick Gay's laboratory in 2001 to work on toll receptors (Department of Biochemistry, University of Cambridge, England). Her current research interests focus on leucine-rich repeat kinase 2 (LRRK2), whose variants cause genetically inherited Parkinson's disease. In 2008, she has successfully used protein modelling and docking to define the mechanism of activation of TLR4 in collaboration with Prof. Clare Bryant (Department of Veterans Medicine), whose approach is to gain insight into innate immunity based on its species-specific

response to molecular patterns of microbial origins. More recently, she worked on the mode of action of non-canonical TLR ligands, such as cationic lipids developed for gene therapy and vaccine adjuvancy. This work is done in collaboration with Profs. Jean-Marie Ruysschaert (Structure et Fonction des Membranes Biologiques, Université Libre de Bruxelles) and Camilla Foged (Department of Pharmacy, University of Copenhagen).

Camilla Foged is head of the Vaccine Design and Delivery Group, Department of Pharmacy, University of Copenhagen, which comprises around 10 researchers. Her main research interest is advanced drug delivery aiming at designing new vaccine and nucleic acid delivery systems to improve therapy. The group is addressing drug delivery challenges using state-of-the-art technologies, and this has fostered innovative solutions and high-impact publications in drug delivery. The group's research goal is to improve disease prevention and treatment by designing nanoparticulate vaccine and nucleic acid formulations through an in-depth mechanistic knowledge of how the physicochemical properties of the particles affect their interaction with the environment, e.g., in formulation, *in vitro* and *in vivo*. Dr. Foged is widely recognized nationally and internationally in the field of drug delivery and has a well-established and extensive network of national and international collaborators in academia and industry. Her research endeavours have resulted in 3 patents applications and more than 90 international peer-reviewed publications.

References

1. Agrawal, N., Dasaradhi, P., Mohmmed, A., Malhotra, P., Bhatnagar, R. K., Mukherjee, S. K. (2003). RNA interference: Biology, mechanism, and applications. *Microbiol. Mol. Biol. Rev.*, **67**, 657–685.
2. Fire, A., Xu, S., Montgomery, M. K., Kostas, S. A., Driver, S. E., Mello, C. C. (1998). Potent and specific genetic interference by double-stranded RNA in Caenorhabditis elegans. *Nature*, **391**, 806–811.
3. Obbard, D. J., Gordon, K. H., Buck, A. H., Jiggins, F. M. (2009). The evolution of RNAi as a defence against viruses and transposable elements. *Philos. Trans. R. Soc. Lond. B Biol. Sci.*, **364**, 99–115.

4. Lorenzer, C., Dirin, M., Winkler, A. M., Baumann, V., Winkler, J. (2015). Going beyond the liver: Progress and challenges of targeted delivery of siRNA therapeutics. *J. Control. Release*, **203**, 1–15.

5. Chery, J. (2016). RNA therapeutics: RNAi and antisense mechanisms and clinical applications. Postdoc J., **4**, 35–50.

6. Kim, D., Rossi, J. (2008). RNAi mechanisms and applications. *Biotechniques*, **44**, 613–616.

7. Rana, T. M. (2007). Illuminating the silence: Understanding the structure and function of small RNAs. *Nat. Rev. Mol. Cell Biol.*, **8**, 23–36.

8. Sashital, D. G., Doudna, J. A. (2010). Structural insights into RNA interference. *Curr. Opin. Struct. Biol.*, **20**, 90–97.

9. Hannon, G. J. (2002). RNA interference. *Nature*, **418**, 244–251.

10. Elbashir, S. M., Lendeckel, W., Tuschl, T. (2001). RNA interference is mediated by 21- and 22-nucleotide RNAs. *Genes Dev.*, **15**, 188–200.

11. Wang, J., Lu, Z., Wientjes, M. G., Au, J. L. (2010). Delivery of siRNA therapeutics: Barriers and carriers. *AAPS J.*, **12**, 492–503.

12. Dominska, M., Dykxhoorn, D. M. (2010). Breaking down the barriers: siRNA delivery and endosome escape. *J. Cell Sci.*, **123**, 1183–1189.

13. Whitehead, K. A., Langer, R., Anderson, D. G. (2009). Knocking down barriers: Advances in siRNA delivery. *Nat. Rev. Drug Discov.*, **8**, 129–138.

14. Kanasty, R., Dorkin, J. R., Vegas, A., Anderson, D. (2013). Delivery materials for siRNA therapeutics. *Nat. Mater.*, **12**, 967–977.

15. Watts, J. K., Deleavey, G. F., Damha, M. J. (2008). Chemically modified siRNA: Tools and applications. *Drug Discov. Today*, **13**, 842–855.

16. Behlke, M. A. (2008). Chemical modification of siRNAs for in vivo use., *Oligonucleotides*, **18**, 305–320.

17. Morrissey, D. V., Lockridge, J. A., Shaw, L., Blanchard, K., Jensen, K., Breen, W., Hartsough, K., Machemer, L., Radka, S., Jadhav, V., Vaish, N., Zinnen, S., Vargeese, C., Bowman, K., Shaffer, C. S., Jeffs, L. B., Judge, A., MacLachlan, I., Polisky, B. (2005). Potent and persistent in vivo anti-HBV activity of chemically modified siRNAs. *Nat. Biotechnol.*, **23**, 1002–1007.

18. Judge, A. D., Bola, G., Lee, A. C., MacLachlan, I. (2006). Design of noninflammatory synthetic siRNA mediating potent gene silencing in vivo. *Mol. Ther.*, **13**, 494–505.

19. Bramsen, J. B., Kjems, J. (2012). Development of therapeutic-grade small interfering RNAs by chemical engineering. *Front. Genet.*, **3**, 154.
20. Eberle, F., Giessler, K., Deck, C., Heeg, K., Peter, M., Richert, C., Dalpke, A. H. (2008). Modifications in small interfering RNA that separate immunostimulation from RNA interference. *J. Immunol.*, **180**, 3229–3237.
21. Hogrefe, R. I., Lebedev, A. V., Zon, G., Pirollo, K. F., Rait, A., Zhou, Q., Yu, W., Chang, E. H. (2006). Chemically modified short interfering hybrids (siHYBRIDS): Nanoimmunoliposome delivery in vitro and in vivo for RNAi of HER-2. *Nucleosides Nucleotides Nucleic Acids*, **25**, 889–907.
22. Jeong, J. H., Mok, H., Oh, Y.-K., Park, T. G. (2008). siRNA conjugate delivery systems. *Bioconjug. Chem.*, **20**, 5–14.
23. De Paula, D., Bentley, M. V. L., Mahato, R. I. (2007). Hydrophobization and bioconjugation for enhanced siRNA delivery and targeting. *RNA*, **13**, 431–456.
24. Raouane, M., Desmaele, D., Urbinati, G., Massaad-Massade, L., Couvreur, P. (2012). Lipid conjugated oligonucleotides: A useful strategy for delivery. *Bioconjug. Chem.*, **23**, 1091–1104.
25. de Fougerolles, A., Vornlocher, H. P., Maraganore, J., Lieberman, J. (2007). Interfering with disease: A progress report on siRNA-based therapeutics. *Nat. Rev. Drug Discov.*, **6**, 443–453.
26. Juliano, R. L. (2016). The delivery of therapeutic oligonucleotides. *Nucleic Acids Res.*, **44**, 6518–6548.
27. Zhang, Y., Satterlee, A., Huang, L. (2012). In vivo gene delivery by nonviral vectors: Overcoming hurdles?. *Mol. Ther.*, **20**, 1298–1304.
28. Tseng, Y. C., Mozumdar, S., Huang, L. (2009). Lipid-based systemic delivery of siRNA. *Adv. Drug Deliv. Rev.*, **61**, 721–731.
29. Foged, C. (2012). siRNA delivery with lipid-based systems: Promises and pitfalls. *Curr. Top. Med. Chem.*, **12**, 97–107.
30. Kanasty, R. L., Whitehead, K. A., Vegas, A. J., Anderson, D. G. (2012). Action and reaction: The biological response to siRNA and its delivery vehicles. *Mol. Ther.*, **20**, 513–524.
31. Biswas, S., Torchilin, V. P. (2013). Dendrimers for siRNA delivery. *Pharmaceuticals*, **6**, 161–183.
32. Salcher, E. E., Wagner, E. (2010). Chemically programmed polymers for targeted DNA and siRNA transfection. *Top. Curr. Chem.*, **296**, 227–249.

33. Wan, C., Allen, T. M., Cullis, P. R. (2014). Lipid nanoparticle delivery systems for siRNA-based therapeutics. *Drug Deliv. Transl. Res.*, **4**, 74–83.
34. Ma, D. (2014). Enhancing endosomal escape for nanoparticle mediated siRNA delivery. *Nanoscale*, **6**, 6415–6425.
35. Semple, S. C., Akinc, A., Chen, et al. (2010). Rational design of cationic lipids for siRNA delivery. *Nat. Biotechnol.*, **28**, 172–U118.
36. Tam, Y. Y., Chen, S., Cullis, P. R. (2013). Advances in lipid nanoparticles for siRNA delivery. *Pharmaceutics*, **5**, 498–507.
37. Love, K. T., Mahon, K. P., Levins, C. G., Whitehead, K. A., Querbes, W., Dorkin, J. R., Qin, J., Cantley, W., Qin, L. L., Racie, T., Frank-Kamenetsky, M., Yip, K. N., Alvarez, R., Sah, D. W., de Fougerolles, A., Fitzgerald, K., Koteliansky, V., Akinc, A., Langer, R., Anderson, D. G. (2010). Lipid-like materials for low-dose, in vivo gene silencing. *Proc. Natl. Acad. Sci. U. S. A.*, **107**, 1864–1869.
38. Xu, C. F., Wang, J. (2015). Delivery systems for siRNA drug development in cancer therapy. *Asian J. Pharm. Sci.*, **10**, 1–12.
39. Colombo, S., Cun, D., Remaut, K., Bunker, M., Zhang, J., Martin-Bertelsen, B., Yaghmur, A., Braeckmans, K., Nielsen, H. M., Foged, C. (2015). Mechanistic profiling of the siRNA delivery dynamics of lipid-polymer hybrid nanoparticles. *J. Control. Release*, **201**, 22–31.
40. Zhang, L., Chan, J. M., Gu, F. X., Rhee, J. W., Wang, A. Z., Radovic-Moreno, A. F., Alexis, F., Langer, R., Farokhzad, O. C. (2008). Self-assembled lipid–polymer hybrid nanoparticles: A robust drug delivery platform. *ACS Nano*, **2**, 1696–1702.
41. Mandal, B., Bhattacharjee, H., Mittal, N., Sah, H., Balabathula, P., Thoma, L. A., Wood, G. C. (2013). Core-shell-type lipid-polymer hybrid nanoparticles as a drug delivery platform. *Nanomedicine*, **9**, 474–491.
42. Chen, G. Y., Nunez, G. (2010). Sterile inflammation: Sensing and reacting to damage. *Nat. Rev. Immunol.*, **10**, 826–837.
43. Kono, H., Rock, K. L. (2008). How dying cells alert the immune system to danger. *Nat. Rev. Immunol.*, **8**, 279–289.
44. Kumar, H., Kawai, T., Akira, S. (2011). Pathogen recognition by the innate immune system. *Int. Rev. Immunol.*, **30**, 16–34.
45. Whitehead, K. A., Dahlman, J. E., Langer, R. S., Anderson, D. G. (2011). Silencing or stimulation? siRNA delivery and the immune system. *Annu. Rev. Chem. Biomol. Eng.*, **2**, 77–96.

46. Botos, I., Liu, L., Wang, Y., Segal, D. M., Davies, D. R. (2009). The Toll-like receptor 3:dsRNA signaling complex. *Biochim. Biophys. Acta*, **1789**, 667–674.

47. Kawai, T., Akira, S. (2010). The role of pattern-recognition receptors in innate immunity: Update on Toll-like receptors. *Nat. Immunol.*, **11**, 373–384.

48. Hornung, V., Guenthner-Biller, M., Bourquin, C., Ablasser, A., Schlee, M., Uematsu, S., Noronha, A., Manoharan, M., Akira, S., de Fougerolles, A., Endres, S., Hartmann, G. (2005). Sequence-specific potent induction of IFN-alpha by short interfering RNA in plasmacytoid dendritic cells through TLR7. *Nat. Med.*, **11**, 263–270.

49. Heil, F., Hemmi, H., Hochrein, H., Ampenberger, F., Kirschning, C., Akira, S., Lipford, G., Wagner, H., Bauer, S. (2004). Species-specific recognition of single-stranded RNA via Toll-like receptor 7 and 8. *Science*, **303**, 1526–1529.

50. Zhang, Z., Weinschenk, T., Guo, K., Schluesener, H. J. (2006). siRNA binding proteins of microglial cells: PKR is an unanticipated ligand. *J. Cell Biochem.*, **97**, 1217–1229.

51. Kell, A. M., Gale, Jr., M. (2015). RIG-I in RNA virus recognition. *Virology*, **479-480**, 110–121.

52. O'Neill, L. A., Golenbock, D., Bowie, A. G. (2013). The history of Toll-like receptors–redefining innate immunity. *Nat. Rev. Immunol.*, **13**, 453–460.

53. Lonez, C., Vandenbranden, M., Ruysschaert, J. M. (2012). Cationic lipids activate intracellular signaling pathways. *Adv. Drug Deliv. Rev.*, **64**, 1749–1758.

54. Ouali, M., Ruysschaert, J. M., Lonez, C., Vandenbranden, M. (2007). Cationic lipids involved in gene transfer mobilize intracellular calcium. *Mol. Membr. Biol.*, **24**, 225–232.

55. Lv, H., Zhang, S., Wang, B., Cui, S., Yan, J. (2006). Toxicity of cationic lipids and cationic polymers in gene delivery. *J. Control. Release*, **114**, 100–109.

56. Vangasseri, D. P., Cui, Z., Chen, W., Hokey, D. A., Falo, Jr., L. D., Huang, L. (2006). Immunostimulation of dendritic cells by cationic liposomes. *Mol. Membr. Biol.*, **23**, 385–395.

57. Boussif, O., Lezoualc'h, F., Zanta, M. A., Mergny, M. D., Scherman, D., Demeneix, B., Behr, J. P. (1995). A versatile vector for gene and oligonucleotide transfer into cells in culture and in vivo: Polyethylenimine. *Proc. Natl. Acad. Sci. U. S. A.*, **92**, 7297–7301.

58. Ballarin-Gonzalez, B., Howard, K. A. (2012). Polycation-based nanoparticle delivery of RNAi therapeutics: Adverse effects and solutions. *Adv. Drug Deliv. Rev.*, **64**, 1717–1729.
59. Bartlett, D. W., Davis, M. E. (2007). Physicochemical and biological characterization of targeted, nucleic acid-containing nanoparticles. *Bioconjug. Chem.*, **18**, 456–468.
60. Beyerle, A., Braun, A., Banerjee, A., Ercal, N., Eickelberg, O., Kissel, T. H., Stoeger, T. (2011). Inflammatory responses to pulmonary application of PEI-based siRNA nanocarriers in mice. *Biomaterials*, **32**, 8694–8701.
61. FDA (2009). Guidance for Industry: Assay development for immunogenicity testing of therapeutic proteins. Available from: https://www.fda.gov/downloads/Drugs/GuidanceCompliance RegulatoryInformation/Guidances/UCM192750.pdf.
62. FDA (2014). Guidance for Industry: Immunogenicity assessment for therapeutic protein products. Available from: https://www.fda.gov/downloads/Drugs/GuidanceComplianceRegulatoryInformation/Guidances/UCM338856.pdf.
63. EMA (2012). Guideline on immunogenicity assessment of monoclonal antibodies intended for in vivo clinical use. Available from: http://www.ema.europa.eu/docs/en_GB/document_library/Scientific_guideline/2012/06/WC500128688.pdf.
64. ICH (2011). ICH harmonized tripartite guideline: Preclinical safety evaluation of biotechnology-derived pharmaceuticals S6(R1). Available from: http://www.ich.org/fileadmin/Public_Web_Site/ICH_Products/Guidelines/Safety/S6_R1/Step4/S6_R1_Guideline.pdf.
65. Wadhwa, M., Knezevic, I., Kang, H. N., Thorpe, R. (2015). Immunogenicity assessment of biotherapeutic products: An overview of assays and their utility. *Biologicals*, **43**, 298–306.
66. Dowling, J. K., Dellacasagrande, J. (2016). Toll-Like receptors: Ligands, cell-based models, and readouts for receptor action. *Methods Mol. Biol.*, **1390**, 3–27.
67. Latz, E., Visintin, A., Lien, E., Fitzgerald, K. A., Monks, B. G., Kurt-Jones, E. A., Golenbock, D. T., Espevik, T. (2002). Lipopolysaccharide rapidly traffics to and from the Golgi apparatus with the Toll-like receptor 4-MD-2-CD14 complex in a process that is distinct from the initiation of signal transduction. *J. Biol. Chem.*, **277**, 47834–47843.
68. Pridmore, A. C., Wyllie, D. H., Abdillahi, F., Steeghs, L., van der Ley, P., Dower, S. K., Read, R. C. (2001). A lipopolysaccharide-deficient

mutant of Neisseria meningitidis elicits attenuated cytokine release by human macrophages and signals via Toll-like receptor (TLR) 2 but not via TLR4/MD2. *J. Infect. Dis.*, **183**, 89–96.

69. Matsuguchi, T., Takagi, K., Musikacharoen, T., Yoshikai, Y. (2000). Gene expressions of lipopolysaccharide receptors, Toll-like receptors 2 and 4, are differently regulated in mouse T lymphocytes. *Blood*, **95**, 1378–1385.

70. Yoshimura, A., Lien, E., Ingalls, R. R., Tuomanen, E., Dziarski, R., Golenbock, D. (1999). Cutting edge: Recognition of Gram-positive bacterial cell wall components by the innate immune system occurs via Toll-like receptor 2. *J. Immunol.*, **163**, 1–5.

71. Gay, N. J., Symmons, M. F., Gangloff, M., Bryant, C. E. (2014). Assembly and localization of Toll-like receptor signalling complexes. *Nat. Rev. Immunol.*, **14**, 546–558.

72. Choe, J., Kelker, M. S., Wilson, I. A. (2005). Crystal structure of human Toll-like receptor 3 (TLR3) ectodomain. *Science*, **309**, 581–585.

73. Bell, J. K., Botos, I., Hall, P. R., Askins, J., Shiloach, J., Davies, D. R., Segal, D. M. (2006). The molecular structure of the TLR3 extracellular domain. *J. Endotoxin Res.*, **12**, 375–378.

74. Cho, W. G., Albuquerque, R. J., Kleinman, M. E., Tarallo, V., Greco, A., Nozaki, M., Green, M. G., Baffi, J. Z., Ambati, B. K., De Falco, M., Alexander, J. S., Brunetti, A., De Falco, S., Ambati, J. (2009). Small interfering RNA-induced TLR3 activation inhibits blood and lymphatic vessel growth. *Proc. Natl. Acad. Sci. U. S. A.*, **106**, 7137–7142.

75. Liu, L., Botos, I., Wang, Y., Leonard, J. N., Shiloach, J., Segal, D. M., Davies, D. R. (2008). Structural basis of Toll-like receptor 3 signaling with double-stranded RNA. *Science*, **320**, 379–381.

76. Luo, J., Obmolova, G., Malia, T. J., Wu, S. J., Duffy, K. E., Marion, J. D., Bell, J. K., Ge, P., Zhou, Z. H., Teplyakov, A., Zhao, Y., Lamb, R. J., Jordan, J. L., San Mateo, L. R., Sweet, R. W., Gilliland, G. L. (2012). Lateral clustering of TLR3:dsRNA signaling units revealed by TLR3ecd:3Fabs quaternary structure. *J. Mol. Biol.*, **421**, 112–124.

77. Zhang, Z., Ohto, U., Shibata, T., Krayukhina, E., Taoka, M., Yamauchi, Y., Tanji, H., Isobe, T., Uchiyama, S., Miyake, K., Shimizu, T. (2016). Structural analysis reveals that Toll-like receptor 7 is a dual receptor for guanosine and single-stranded RNA. *Immunity*, **45**, 737–748.

78. Tanji, H., Ohto, U., Shibata, T., Miyake, K., Shimizu, T. (2013). Structural reorganization of the Toll-like receptor 8 dimer induced by agonistic ligands. *Science*, **339**, 1426–1429.

79. Tanji, H., Ohto, U., Shibata, T., Taoka, M., Yamauchi, Y., Isobe, T., Miyake, K., Shimizu, T. (2015). Toll-like receptor 8 senses degradation products of single-stranded RNA. *Nat. Struct. Mol. Biol.*, **22**, 109–115.
80. Saito, T., Owen, D. M., Jiang, F., Marcotrigiano, J., Gale, Jr., M. (2008). Innate immunity induced by composition-dependent RIG-I recognition of hepatitis C virus RNA. *Nature*, **454**, 523–527.
81. Takahasi, K., Yoneyama, M., Nishihori, T., Hirai, R., Kumeta, H., Narita, R., Gale, Jr., M., Inagaki, F., Fujita, T. (2008). Nonself RNA-sensing mechanism of RIG-I helicase and activation of antiviral immune responses. *Mol. Cell.*, **29**, 428–440.
82. Hornung, V., Ellegast, J., Kim, S., Brzozka, K., Jung, A., Kato, H., Poeck, H., Akira, S., Conzelmann, K. K., Schlee, M., Endres, S., Hartmann, G. (2006). 5′-Triphosphate RNA is the ligand for RIG-I. *Science*, **314**, 994–997.
83. Devarkar, S. C., Wang, C., Miller, M. T., Ramanathan, A., Jiang, F., Khan, A. G., Patel, S. S., Marcotrigiano, J. (2016). Structural basis for m7G recognition and 2′-O-methyl discrimination in capped RNAs by the innate immune receptor RIG-I. *Proc. Natl. Acad. Sci. U. S. A.*, **113**, 596–601.
84. Ramanathan, A., Devarkar, S. C., Jiang, F., Miller, M. T., Khan, A. G., Marcotrigiano, J., Patel, S. S. (2016). The autoinhibitory CARD2-Hel2i Interface of RIG-I governs RNA selection. *Nucleic Acids Res.*, **44**, 896–909.
85. Gay, N. J., Gangloff, M. (2007). Structure and function of toll receptors and their ligands. *Annu. Rev. Biochem.*, **76**, 141–165.
86. Jin, M. S., Kim, S. E., Heo, J. Y., Lee, M. E., Kim, H. M., Paik, S. G., Lee, H. Y., Lee, J. O. (2007). Crystal structure of the TLR1-TLR2 heterodimer induced by binding of a tri-acylated lipopeptide. *Cell*, **130**, 1071–1082.
87. Kang, J. Y., Nan, X., Jin, M. S., Youn, S. J., Ryu, Y. H., Mah, S., Han, S. H., Lee, H., Paik, S. G., Lee, J. O. (2009). Recognition of lipopeptide patterns by Toll-like receptor 2-Toll-like receptor 6 heterodimer. *Immunity*, **31**, 873–884.
88. Lonez, C., Bessodes, M., Scherman, D., Vandenbranden, M., Escriou, V., Ruysschaert, J. M. (2014). Cationic lipid nanocarriers activate Toll-like receptor 2 and NLRP3 inflammasome pathways. *Nanomedicine*, **10**, 775–782.
89. Pizzuto, M., Gangloff, M., Scherman, D., Gay, N. J., Escriou, V., Ruysschaert, J. M., Lonez, C. (2017). Toll-like receptor 2 promiscuity is responsible for the immunostimulatory activity of nucleic acid nanocarriers. *J. Control. Release*, **247**, 182–193.

90. Inohara, N., Nunez, G. (2002). ML - a conserved domain involved in innate immunity and lipid metabolism. *Trends Biochem. Sci.*, **27**, 219–221.

91. Lonez, C., Irvine, K. L., Pizzuto, M., Schmidt, B. I., Gay, N. J., Ruysschaert, J. M., Gangloff, M., Bryant, C. E. (2015). Critical residues involved in Toll-like receptor 4 activation by cationic lipid nanocarriers are not located at the lipopolysaccharide-binding interface. *Cell. Mol. Life Sci.*, **72**, 3971–3982.

92. Schmidt, M., Raghavan, B., Muller, V., Vogl, T., Fejer, G., Tchaptchet, S., Keck, S., Kalis, C., Nielsen, P. J., Galanos, C., Roth, J., Skerra, A., Martin, S. F., Freudenberg, M. A., Goebeler, M. (2010). Crucial role for human Toll-like receptor 4 in the development of contact allergy to nickel. *Nat. Immunol.*, **11**, U814–U864.

93. Mobley, D. L., Dill, K. A. (2009). Binding of small-molecule ligands to proteins: "What you see" is not always "what you get". *Structure*, **17**, 489–498.

Chapter 15

Lipid Nanoparticle Induced Immunomodulatory Effects of siRNA

Ranjita Shegokar, PhD,[a] and Prabhat Mishra, PhD[b]

[a]*Free University of Berlin, Department of Pharmaceutics, Biopharmaceutics and NutriCosmetics, 12169 Berlin, Germany*
[b]*Pharmaceutics Division, Central Drug Research Institute, Lucknow, Uttar Pradesh, India*

Keywords: siRNA, shRNA, microRNA, nanomedicines, formulation issues, drug delivery systems, lipid nanoparticles, surface modification, targeting, clinical status, inflammation, cardiovascular therapeutics, antiviral agents, anticancer agents, liposomes, nanoparticles, toxicity, inhibition, medical sciences

15.1 Introduction

Gene therapy potentially targets the diseased site with limited risks [1]. Small RNAs (siRNA, shRNA, microRNA) play an entirely different role in the transcription and expression process for short-term nonviral nucleic acid delivery [2]. The field encompassing therapeutic applications of siRNA is versatile and includes use of siRNAs in therapeutics to the central nervous system (CNS), in inflammation or cardiovascular therapeutics and pain research. A lot of academic and industrial research is being done using siRNA as a therapeutic agent *in vitro* and *in vivo*. RNAi has also been rapidly adopted for the discovery and validation of gene function through the use of a sequence-specific short interfering RNA (siRNA). siRNA is a potent and specific inhibitor of gene expression and is being used as a new technology for drug target validation, studying functional genomics, and transgenic design and as a promising therapeutic agent for genetic diseases [3].

15.2 Discovery of siRNA

RNAi was first observed by plant biologists in the late 1980s [4], although their molecular structure remained unclear until the 1990s [4]. Later studies in the nematode *Caenorhabditis elegans* exhibited the gene-silencing mechanism [5]. The biology and mechanisms of actions of RNAi have been extensively reviewed thereafter [6–8]. siRNA can regulate gene expression by inhibiting the synthesis of the encoded protein in natural ways. Tabara et al. [9] concluded that transposon silencing is one of the natural functions of RNAi. RNAi techniques were quantitatively more efficient and more durable in cell culture compared with the antisense technology [10, 11]. One major advantage of RNAi over antisense oligonucleotides is that siRNA is based on a catalytic mechanism. Bertrand et al. [12, 13] concluded that siRNA is very efficient in inhibiting the synthesis of target proteins with improved specificity and reduced dose by 100- to 1000-fold.

15.2.1 Mechanism of Action

The mature siRNA shares partial complementary sequences in the 3′ UTR of a target mRNA. Cellular uptake of siRNA occurs via endocytosis. The mechanism of gene suppression was established by Fire et al. [14] using the 23mer to 25mer nucleotide sequence of RNA molecules (double-stranded siRNA) and the later synthesized single-stranded short hairpin RNA (shRNA) in 1998. This discovery was awarded the Nobel Prize for Medicine or Physiology in 2006.

The siRNA is generated by the cleavage of double-stranded RNA precursors of a cell by the RNAse III endonuclease Dicer

[15], which is complexed with the TAR-RNA-binding protein and avoids the contact of siRNA to the RNA-induced silencing complex (RISC) [16, 17]. The selectivity is based on the thermodynamic stabilities of the siRNA end [18, 19]. The end, which is less stable thermodynamic, helps in the unwinding of the 5′ end of the guide strand and this binds to Argonaute (Ago-2). The Ago domain, such as PAZ and MID, has specific functions in docking and anchoring the RNA, whereas PIWI has slicer activity [20]. Ago-2 recognizes and cleaves the messenger RNA molecule having perfect or near-perfect complementarity to the guide RNA. Partial complementarity between a siRNA and target mRNA (endogenous substrates for the RNAi machinery) may in some cases destabilize the transcripts if the binding mimics microRNA (miRNA) interactions with target sites. In the cytoplasm, the loop is removed by the RNAse III Dicer and only one of the two strands is loaded into RISC. The primary mechanism of action of miRNA is translation repression accompanied by message degradation (Fig. 15.1).

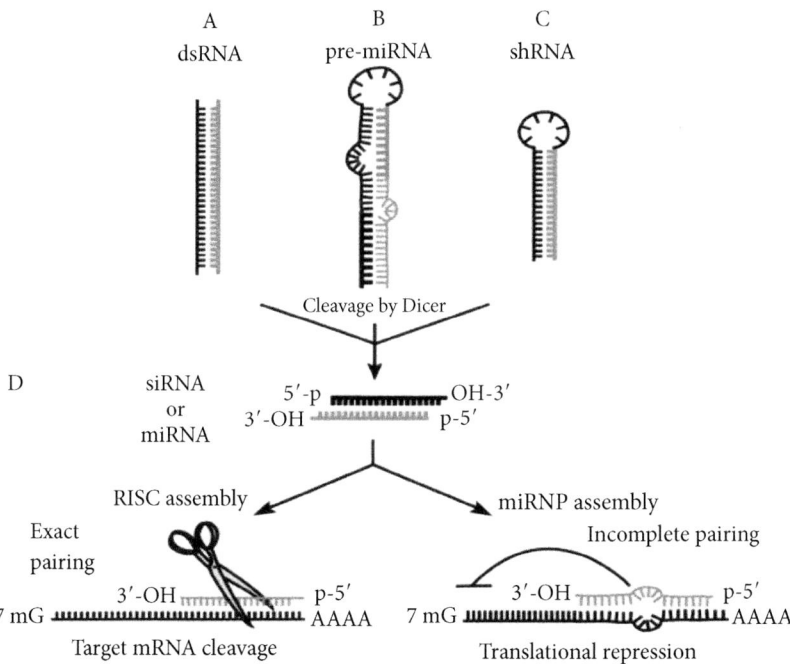

Figure 15.1 Sources for RNA-induced gene silencing include (A) dsRNAs, (B) miRNAs, (C) shRNAs, and (D) chemically synthesized siRNAs.

15.2.2 Advantages and Drawbacks

RNAi could provide an exciting new therapeutic modality for treating infections, cancer, neurodegenerative diseases, antiviral diseases (e.g., viral hepatitis and human immunodeficiency virus 1, HIV-1), Huntington's disease [21], hematological diseases, pain research and therapy, dominantly inherited genetic disorders, and many other illnesses [22]. However, the main hurdles are the poor pharmacokinetic properties of siRNA and major biological restrictions, such as off-target effects and interferon responses [23]. Also, problems such as low transfection efficiency and poor tissue penetration because of the polyanionic nature need attention [24]. Nonspecific immune stimulation *in vivo* has delayed their therapeutic applications. Still, the lack of an efficient delivery system to target and deliver the siRNA to the desired cells/target is the biggest limitation. Many groups are researching to find an optimal delivery tool that can be safely and repeatedly administered systemically and can deliver the siRNA specifically and efficiently to the target tissue. The majority of the academic, biotechnology, and pharmaceutical efforts are now focused on chemically synthesized small RNAs designed to manipulate miRNA expression. Various approaches such as lipid nanoparticles, polymeric nanoparticles, antibody conjugates, cyclodextrin nanoparticles [25], aptamer-siRNA conjugates [26], and chitosan nanoparticles as a delivery system for siRNA are being studied to overcome these issues.

15.2.3 Delivery of siRNA, Limitations, and Role of Biological Barriers

Avoidance of nonspecific uptake by the reticuloendothelial system (RES), especially the Kupffer cells in the liver and the macrophages in the spleen, dictates distribution. Even if the siRNA stays in the blood circulation long enough, extravasation of the formulation is required to reach the target tissue. Besides, all immunogenicity and other safety issues are still important concerns related to the use of viral vectors in human. However, viral delivery systems show toxicity and have relatively strong host responses resulting from the activation of the human immune system. Multiple nonviral-based delivery methods have been used *in vivo* for delivering siRNA, including hydrodynamic injection, cationic liposome encapsulation, cationic polymer complexes, and antibody-specific targeting delivery systems. Each of the current methods of gene delivery, whether viral or nonviral, has some limitations. Delivering siRNA *in vivo* mainly systematically to animal tissues is a complicated event and involves various physical, chemical, or biological issues as follows [27]:

(1) Stability of siRNA oligos in the extracellular and intracellular environments.
(2) Cell penetration: siRNA is anionic (40 negative phosphate charges on the siRNA backbone), hydrophilic, (7 nm in length, 13 kDa in weight) [28], and unable to enter cells by passive diffusion mechanisms.
(3) Rapid excretion: The siRNA (21-nucleotide double-stranded) is rapidly excreted through urine when administered systematically, even if siRNA molecules remain stable through chemical modifications.
(4) Degradation: siRNA oligos can be degraded by RNase activity within a short period in a serum environment.
(5) Low concentration at target: The nonspecific distributions of oligos throughout the body results in a significant decrease in the local concentration at the target site.
(6) Biological barriers: The siRNA oligos need to overcome tissue barriers in blood walls and endothelial wall to reach the site of action.
(7) Immune overstimulation: Might cause other adverse effects such as nonspecific events owing to the activation of innate immune responses [29, 30].

15.3 Strategies for Delivering siRNA

siRNA delivery can be classified into two main groups, viral and nonviral delivery systems (Fig. 15.2), and can be delivered by physical methods and/or by the introduction of a DNA plasmid in the nucleus of the target cell [31, 32]. In addition, the administration of precursor molecules or [33–35] delivering synthetic siRNA to the cytoplasm of the target cells is also possible.

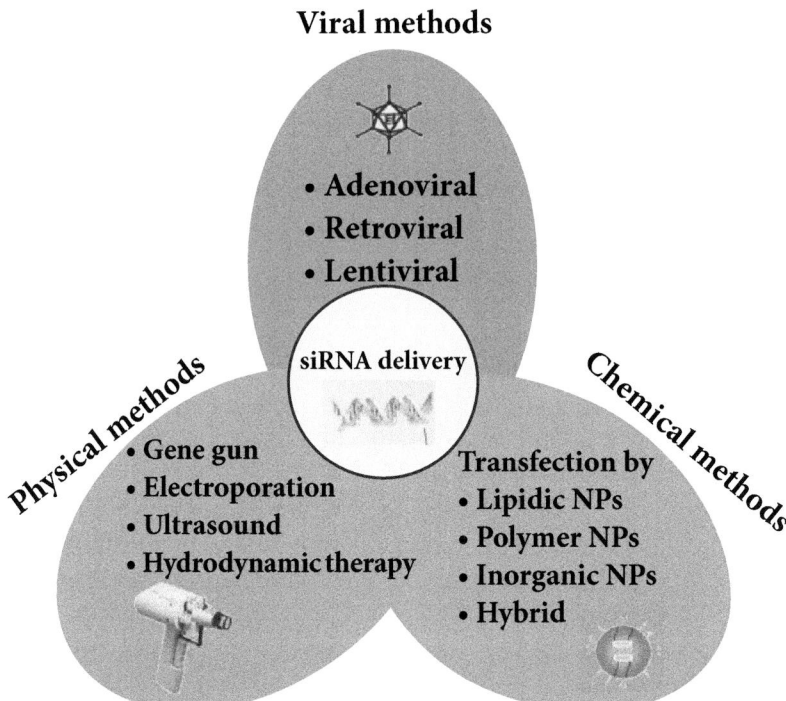

Figure 15.2 SiRNA delivery strategies *in vivo*. Drug delivery systems can solve stability (*in vivo* and *in vitro*), solubility issues and allow targeted or site-specific delivery.

Viral delivery systems are mostly used to deliver DNA plasmids or precursor molecules [36–38]. They exert high transduction efficacy, due to the inherent ability of viruses to transport genetic material into cells [39–41]. However, the potential of mutagenicity or oncogenesis, several host immune responses, and the high cost of production limit their application. For these reasons, different kinds of nonviral siRNA delivery systems have been explored [42, 43]. However, siRNAs require more effective delivery systems that allow siRNA stabilization, specific cell recognition, internalization and subcellular localization to the cytoplasm of target tissues and cells to exert it therapeutic effect [44]. Many nonviral carriers used to deliver DNA for gene therapy have been adopted for siRNA delivery. Cationic lipids and polymers are two major classes of nonviral delivery carriers that can form complexes with negatively charged siRNA.

15.4 Immune Response to siRNA Payload and Its Modulation

RNAi holds great promise to change the therapeutic modality in many diseases like cancer hepatitis, etc. However, naked siRNA is unstable in blood and serum because of the rapid degradation caused by endo- and exonucleases, with a very short half-life *in vivo*. In addition, if chemically un-modified, siRNA can induce immune responses and may lead to mostly unwanted off-target effects. It was observed that in vertebrates, an immune response can be induced by dsRNA, but that is part of the defense mechanism against viral infection. The dsRNA are generally sensed in the cytoplasmic and endosomal compartments by RNA-dependent kinases (PKRs) and induce an interferon (IFN) response that results in production of pro-inflammatory cytokines mainly by activating the NF-kappa-B dependent pathway. Although it was thought that short dsRNA-like siRNA would not generate immune responses, many studies report that siRNA can also produce robust immune responses. Other known mechanisms are helicase retinoid-acid-inducible gene I (RIG-I) and melanoma differentiation associated gene 5 (MDA-5), which may cause NF-kappa-B activation and IFN production. Both chemically synthesized siRNAs and siRNAs obtained from *in vitro* transcription with various lengths lead to the activation of PKR. IFN levels increase with an increase in length of dsRNA. TLRs can also recognize non-self RNA. Recognition of RNA by TLRs activates several signaling pathways, which leads to the activation of NF-κB and thus production of pro-inflammatory cytokines.

siRNAs can be optimized to avoid or reduce immune response generation. For example, GU-rich sequences should be minimized, length optimization and/or dose optimized and to perform an experimental screening of multiple siRNAs against the same target in order to identify the optimal candidate with minimal immune response. However, not just the naked siRNA but the delivery systems used for the systemic delivery of siRNAs also can initiate immune response. Cationic liposomes are commonly known to induce toxicity when injected in mice with effects ranging from blood clotting to immunostimulation. Cationic liposomes use the endosomal pathway to facilitate siRNA delivery, and as TLRs, RIG-I, and PKRs are expressed on the inner membrane of endosomes, they increase the exposure of siRNA to these receptors and thus increase the production of inflammatory mediators.

Immune modulation by the delivery system can be categorized in two ways: (1) inhibition of the immune response and (2) augmentation of the immune response. Inhibition of the immune response is commonly required in inflammatory disease where the down regulation of cytokine production is required, especially TNF α and interleukins, which lead to diseases such as collagen-induced arthritis, systemic lupus erythematous, LPS-induced inflammation, etc. Augmentation of the immune response is required when the general systemic antiviral effect is required. There, siRNA and its nanocarriers can elicit an IFN response that could affect viral infection as well as tumor growth (Fig. 15.3).

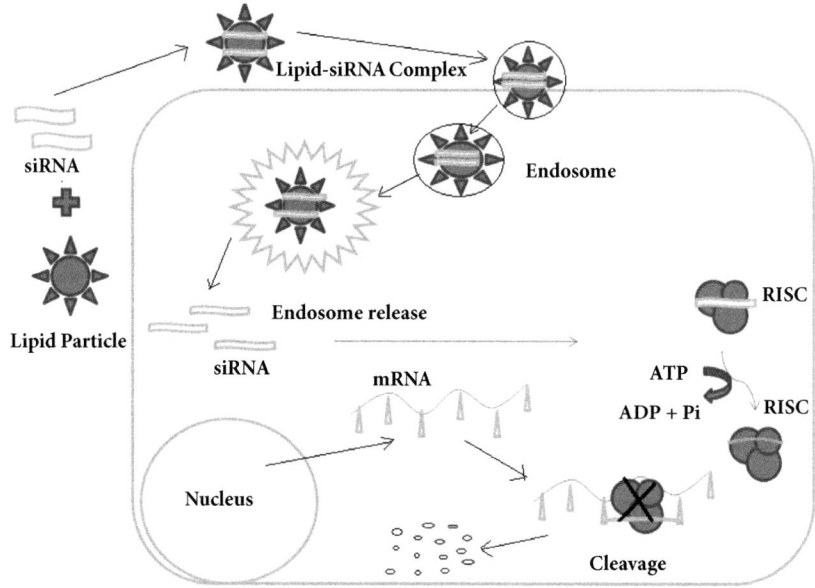

Figure 15.3 Mechanism of siRNA uptake using lipids as a surrogate vector.

15.4.1 Lipid-Based siRNA Delivery

Nanoparticles are appropriate delivery systems that facilitate siRNA transport in the cytosol through electrostatic interactions with negatively charged phospholipid bilayers or through specific targeting moieties. The siRNA complexes with the delivery (or transfection) reagent are often specifically referred to as lipoplexes, dendriplexes, or polyplexes depending on whether the vector used is a cationic lipid, dendrimer (branched like polymer structures), polymeric micelle, cationic cell–penetrating peptide, or polymer, respectively. A common mechanism is their net positive (cationic) charge that facilitates complex formation with the polyanionic nucleic acid and their interaction with a negatively charged cell membrane. Safety and efficient delivery of siRNA is prerequisite for the potential delivery of siRNA based therapy. Also, these systems avoid or reduce unfavorable immune response by siRNAs. The optimal systemic delivery systems for siRNA should be biocompatible, biodegradable, and nonimmunogenic. The systems should provide efficient delivery of siRNA into target cells or tissues with protection of the active double-stranded siRNA products from attack by serum nucleases. Next, the delivery systems must provide target tissue-specific distribution after systemic administration, avoiding rapid hepatic or renal clearance. Finally, after delivery into target cells via endocytosis, the systems should promote the endosomal release

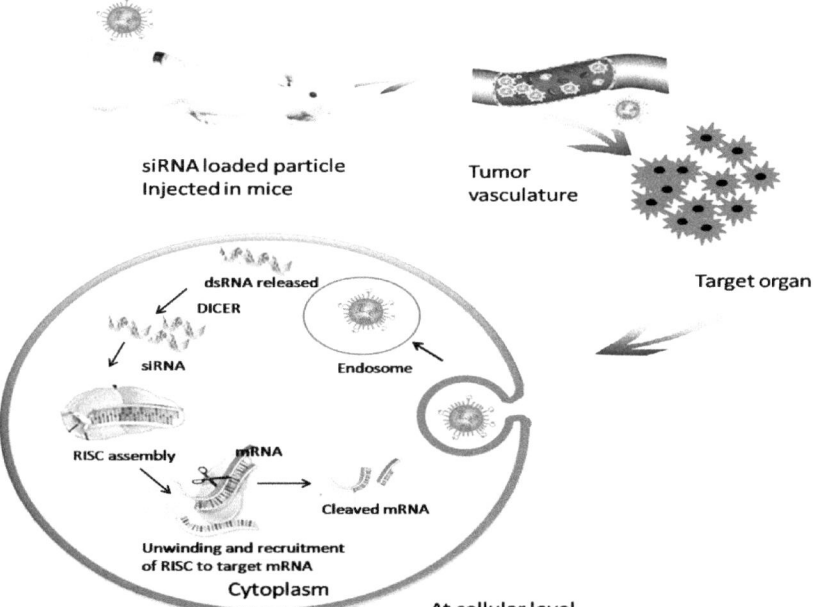

Figure 15.4 Journey of a siRNA loaded polymeric or lipid nanoparticle in extra and intracellular compartments.

of siRNA into the cytoplasm, allowing the interaction of siRNA with the endogenous RISC (Fig. 15.4). Other types of nonviral delivery strategies have been explored such as hydrodynamic injection [45, 46], particle bombardment [47] and electroporation [48, 49], microinjection to calcium co-precipitation and chemical methods. Cationic polymer, proteins and peptides, antibody or ligand-targeted conjugates mediated delivery of siRNA have been studied [50–55].

15.4.2 Liposomes

Encapsulation of siRNA in liposomes and other nanoparticles is one of the preferred approaches [56–59]. Cationic liposomes are one of the most attractive vehicles owing to their high transfection efficiency, favorable pharmacokinetic properties, and relatively low toxicity and immunogenicity. Moreover, cationic liposomes can protect siRNA from enzymatic degradation, and provide reduced renal clearance [60]. A clear understanding of liposome-delivery of siRNAs is still evolving, and more research is needed. The investigation of liposome delivery of siRNA *in vivo* is just in its infancy [61].

The i.v. administration of cationic liposomes in mice resulted in the accumulation of almost 60% of the dose of plasmid DNA in the liver [62]. In another study, lipid-mediated delivery of a siRNA against apolipoprotein B (apoB) was used to target apoB mRNA in the liver. No silencing activity was observed even at higher doses for the group without a formulation or chemical conjugation (>50 mg/kg) [63]. However, lipid-formulated siRNA showed more than 80% silencing of apoB mRNA and apoB-100 (single dose of 1 mg/kg dose) in a non-human primate [64] and mainly accumulated in the liver, spleen, and small intestine.

Encoded plasmid DNA (shRNA) encapsulated in multifunctional envelope nanodevice showed 96% inhibition of the marker luciferase gene [65, 66]. This effect was due to the detergent-like activity of the PEG phospholipid conjugate, which lyses liposomes at high concentrations. Electrostatic interaction between cationic and anionic lipids excludes the surface-bound water in the endosome membrane, promoting the formation of the inverted hexagonal (HII) phase [67], which not only destabilizes the endosome membrane but also promotes the de-assembly of the lipoplex. Cationic lipid (1,2-dioleoyl-3-trimethylammonium propane, DOTAP) destabilized showed knockdown of >75% of endogenous tissue necrosis factor (TNF-α) when given intraperitoneally [68]. The neutral lipid (L-alpha dioleoyl phosphatidyl ethanolamine, DOPE), destabilized the vesicular membranes of the endosome and lysosome and aided release of the lipoplex into the cytosol [68–70].

Endothelium-targeted cationic liposomes of siRNA, composed of β-L-arginyl-2,3-L-diaminopropionicacid-N-palmityl-N-oleylamide trihydrochloride (AtuFECT01), fusogenic 1,2-diphytanoyl-sn-glycero-3-phosphoethanolamine, and the PEGylated lipid N-(carbonylmethoxypolyethyleneglycol-2000)-1,2-distearoyl-sn-glycero-3-phospho-ethanol-amine sodium salt (DSPE-PEG) showed reduction of target protein levels in the vasculature of the heart, liver, and lung [71]. Sato et al. [72] studied the galactosylated liposomes-siRNA complex (75.3 nm, zeta potential 35.8 ± 1.8 mV) for gene silencing of endogenous hepatic gene expression. No liver toxicity for galactosylated liposomes/siRNA complex was observed [73–75].

Miyawaki-Shimizu et al. [76] prepared liposomes composed of 50% DDAB (dimethyl-dioctadecyl-ammoniumbromide) and 50% cholesterol (mol/mol) with a ratio of siRNA to liposome of 1:5 (w/v) in order to explore the role of caveoline-1 in regulating lung vascular permeability in male CD1 mice. The injected cationic liposome–siRNA complexes downregulated caveoline-1 expression in a concentration-dependent (0.4–1.3 mg/kg) and time-dependent manner (maximal suppression between 72 h and 144 h post-injection and full recovery at 168 h after siRNA administration). They also confirmed that caveolin-1 is essential for caveolar biogenesis and plays an important role as a negative regulator of interendothelial junction (IEJ) permeability *in vivo*. In 2006, Palliser et al. [77] reported topical vaginal preparations of siRNA-encapsulated cationic liposomes directed toward herpes simplex virus–specific genes.

15.4.3 Lipid Nanoparticles

Engineered cationic lipids can encapsulate negatively charged nucleic acids. Various parameters such as temperature, concentration, charge ratio, and lipid composition may affect the transfection efficiency of these lipid complexes, intracellular release of DNA from nucleic acid lipid complexes and translocation of the release of DNA to the nucleus.

Numerous cationic lipids generated by combinatorial synthesis have been screened for optimal siRNA delivery [78]. Kim et al. designed the delivery of siRNA using solid lipid nanoparticles (SLNs), which resulted in efficient target gene silencing and serum stability, with a minimal level of cytotoxicity [79].

The lipoplexes called stabilized nucleic acid particles (SNALP) were prepared by an ethanol dialysis method to deliver siRNA [2.5 mg/kg] systemically for silencing the apoB gene in mice and cynomolgus monkeys. SNALP were also prepared containing a diffusible poly(ethylene glycol)-lipid conjugate (PEG-lipid conjugate) that provides a neutral, hydrophilic coating to the particle's exterior [80–84]. Another sophisticated technique was described by Morrissey et al., who incorporated stabilized siRNA to target the hepatitis B virus into a specialized liposome to form a SNALP and administered it by intravenous injection into mice. The improved efficacy of siRNA-SNALP compared with free siRNA correlated with a longer half-life in plasma and the liver.

Zimmermann et al. showed sustained delivery of siRNA for more than 11 days to the liver with reduced apoB gene expression [64]. Heyes et al. synthesized various cationic lipids (1,2-distearyloxy-N,N-dimethyl-3-aminopropane (DSDMA), 1,2-dioleyloxy-N,N-dimethyl-3-aminopropane (DODMA), 1,2-dilinoleyloxy-N,N-dimethyl-3-aminopropane (DLinDMA), and 1,2-dilinolenyloxy-N,N-dimethyl-3-aminopropane (DLenDMA)) possessing 0, 1, 2, or 3 double bonds per alkyl chain, respectively, and determined its correlation between lipid saturation, fusogenicity, and efficiency of intracellular nucleic acid delivery [85].

Mevel et al. synthesized novel cationic lipids comprising of cholesteryl-moieties linked to guanidinium functional groups, and cationic lipids comprising of a dialkylglycylamide moiety conjugated with a polyamine or a guanidinium functional group. These cationic lipids were formulated into cationic liposomes with the neutral co-lipid dioleoyl-L-α-phosphatidylethanolamine (DOPE) [86–89] or with a recently reported neutral lipophosphoramidate derivative of histamine (MM27). They observed that liposomes prepared from the cationic lipid N',N'-dioctadecyl-N-4,8-diaza-10-aminodecanoylglycine amide (DODAG) and DOPE frequently mediated the highest levels of transfection *in vitro* in different cell lines studied (OVCAR-3, IGROV-1, and HeLa) both in the presence or absence of serum [90].

Folate conjugated to DSPE-PEG in combination with cholesteryl carboxy amidomethylene-N-hydroxylamine efficiently delivered siRNA into human nasopharyngeal KB cells [91]. Immunolipoplexes with monoclonal antibodies for insulin and ligands for transferrin receptor were found to be efficient for crossing the blood–brain barrier and siRNA delivery to the brain [92]. Herringson et al. [93] explored the potential of chelator lipid 3 (nitrilotriaceticacid)-ditetradecylamine (NTA3-DTDA) with neutral stealth liposomes to target siRNA to cells. They developed a novel method for incorporating siRNAs into lipoplexes utilizing helper lipids and the ionizable lipid 1,2-dioleoyl-3-dimethyl ammonium-propane (DODAP). This approach resulted in an efficient (50%) incorporation of siRNA into lipoplexes, which, when incorporated with Ni-NTA3-DTDA and engrafted with aHis-tagged form of murine CD4, can target siRNA to murine A20 B cells, *in vitro*.

Metal chelator lipid 3(nitrilotriaceticacid)-ditetradecylamine (NTA3-DTDA) liposomes were developed for engrafting molecules for targeting to specific cells. The incorporation of NTA3-DTDA into stealth liposomes containing an antigen and cytokines, allowed stable engraftment (by Ni^{2+}-chelating linkage) of His-tagged forms of B7.1, CD40, and ScFv and enabled targeting of the liposomes to T cells and dendritic cells *in vitro* and *in vivo* [94]. Two series of lipophilic siRNAs conjugated with derivatives of cholesterol, lithocholic acid, or lauric acid have been synthesized in order to improve the delivery of siRNA into human liver cells.

Khoury et al. [95] demonstrated cationic lipid 2-(3-[bis-(3-amino-propyl)-amino]-propylamino)-N-ditetradecylcarbamoylmethyl-acetamide (RPR209120) combined with DOPE can be efficient to deliver siRNAs designed to silence tumor necrosis factor α (TNF-α) in collagen-induced arthritis (CIA), which is among the most prominent cytokines in rheumatoid arthritis (RA) [96].

Flynn et al. [97] used lipofectamine to deliver IL12-p40siRNA for targeting the expression of IL12-p40 in a model of lipopolysaccharide (LPS)-induced inflammation. Significant reduction of an immune reaction in treated animals was because of reduced IL12 production in peritoneal macrophages. Fluorescein-labeled siRNAs were injected into adult mice for investigating cationic liposome mediated intravenous and intraperitoneal delivery to show that DOTAP containing liposomes can deliver siRNAs into various cell types and in contrast to mouse cells, siRNAs can activate the nonspecific pathway in human freshly isolated monocytes, resulting in TNF-α and IL-6 production [98]. A recent study has shown insufficient *in vivo* transfer to airway epithelial cells [99]. Following *in vivo* lung transfection using a Genzyme (http://www.genzyme.com) lipid (GL67) in mice, siRNAs were only visible in alveolar macrophages. SiRNAs targeted to β-galactosidase reduced β-gal mRNA levels in the airway epithelium of K18-lacZ mice by 30%. Addition of histone, poly-L-lysine, and protamine to some formulations of cationic lipids resulted in levels of delivery that were higher than either lipids or polymers alone. The combined formulations might also be less toxic [100].

Mirus Corporation has developed a lipopolyplex transfection reagent called TransIT-TKO® to deliver siRNA, which is composed of a charge-dense polycation and a cationic lipid. Charge-dense polymer form complexes with siRNA that are resistant to disassembly in physiological solutions, including serum. This allows complexes to be added directly to cell culture media that contains serum. Li, et al. [101] developed a lipid coated calcium phosphate (LCP) nanoparticle (NP) formulation for efficient delivery of siRNA to a xenograft tumor model by intravenous administration. Luciferase siRNA was used to evaluate the gene silencing effect in H-460 cells, which were stably transduced with a luciferase gene. The anisamide modified LCP NP silenced about 70% and 50% of luciferase activity for the tumor cells in culture and those grown in a xenograft model, respectively. The untargeted NP showed a very low silencing effect. SiRNA can be chemically modified by phosphodiester modification [102, 103] or two-sugar modification [63] or bioconjugated with lipids at one or both strands of siRNA. These modifications can protect siRNA from elimination and in activation but do not necessarily augment the efficacy of gene silencing *in vivo*.

Landen et al. [104] showed effective treatment with formulated anti-EphA2-siRNA, paclitaxel plus formulated control siRNA, and paclitaxel plus formulated anti-EphA2-siRNA (as combination therapy) twice weekly for 4 weeks. The combination therapy had the best results with a 86–91% reduction compared with a treatment with formulated control siRNA alone or a 67–82% reduction compared with nonspecific siRNA-DOPC and paclitaxel. No toxicities were observed during the study.

Zimmermann et al. treated cynomolgus monkeys (1 and 2.5 mg/kg SNALP-formulated siApoB) in a single injection and observed a clear and statistically significant dose-dependent gene silencing effect on cynomolgus liver ApoB mRNA. The mRNA levels were reduced by 68 ± 12% for the lower dose and about 90 ± 12% for the higher dose consistent across the liver for 11 days (reduction of 91 ± 15% for the higher dose at day 11). A significant reduction of Plasma ApoB, serum cholesterol, and LDL levels (low-density lipoprotein particle containing ApoB) was also observed in this period, with a maximal reduction of 62 ± 5.5% for cholesterol and 82 ± 7% for LDL at day 11 for the higher dose. In contrast, HDL levels (high-density lipoprotein particles without ApoB) showed no significant changes during this period. Targeted Cyclin D1 (CyD1) is an experimental model of intestinal inflammation to explore the role of leukocyte-expressed CyD1 in the pathogenesis of inflammation and the ability to use CyD1 as a target. These nanocarriers showed high cargo capacity and protected siRNA from degradation, allowed highly efficient intracellular delivery (due to the antibodies on their surface), and thus silence genes efficiently *in vivo* at low doses (2.5 mg/kg) with low off-target effects and toxicity [105].

Pirollo et al. prepared nano-immunoliposome complexes (scL complex, 100 nm) using the ethanol injection method and included MPB-DOPE (*N*-maleimido-phenylbutyrate-DOPE) at 5 molar percent of total lipids to allow conjugation of the TfRscFv and the peptide HoKC (a synthetic pH-sensitive histidylated oligolysine, designed to aid endosomal escape) [106]. Nanoparticles showed enhanced tumor delivery and specificity for both primary and metastatic cancers of various types, including prostate, pancreatic, and breast cancers. In another study [107], tumors induced in female thymic nude mice by s.c. inoculation with different cell types (human pancreatic (PANC-1) and breast (MDA-MB-435) cancer cells, human lung (H157) and colon carcinoma (H630) cells showed efficient silencing of the HER-2 gene (virtually elimination) and affected components in multiple signal-transduction pathways. Santel et al. formulated siRNA-lipoplexes (118 nm/46 mV) [71] upon i.v. administration at single dose showed predominant uptake of siRNA into endothelial cells in contrast to naked siRNA.

Lipidoid-siRNA formulations were developed by Akinc et al. Lipidoids are lipid-like molecules, synthesized by a one-step synthetic scheme (conjugate addition of alkyl-acrylates or alkyl-acrylamides to primary or secondary amines). They created a large library of lipidoids (over 1200 structurally diverse lipidoids) to find a new effective delivery system [78, 108]. Their optimized delivery system (LNP01, 50–60 nm) contains the lipidoid 98N12-5(1), cholesterol, and PEG-lipid (mPEG2000-C14) in a molar ratio of 42:48:10, and the formulation was stable for at least 5 months at 4, 25, and 37°C and increased the siRNA t1/2 in serum from about 15 min (unformulated, minimally chemically modified siRNA) to over 24 h (LNP01) *in vitro*. Biodistribution analysis showed that LNP01 can be administered repeatedly with equal *in vivo* efficacy over extended time periods as three cycles of bolus injection (5 mg/kg siRNA) against Factor VII, formulated with LNP01, which in C57BL/6 mice did not show any significant changes in silencing profiles compared with one bolus injection.

Liu, et al. studied the potential of novel siRNA delivery carrier composed of a cationic oligomer (PEI1200), a hydrophilic polymer (polyethylene glycol) and a biodegradable lipid-based cross-linking moiety, with an appropriate siRNA, targeting to CD133$^+$ cells to improve the efficacy of conventional chemotherapy. An MDR1-targeting siRNA (siMDR1) effectively reduced the expression of MDR1 in human colon CSCs (CD133$^+$-enriched cell population), resulting in significantly increasing the chemo-sensitivity to paclitaxel [109].

15.4.4 Surface-Modified siRNA Delivery

Recent success in using ligand-targeted complexes to deliver therapeutic siRNA to various tissues, receptors, and organs is very promising.

Chemical modification approaches

The stability of the siRNA molecule becomes a major issue and can be overcome by chemically modifying the basic RNA structure by changing the backbone of the RNA molecule. A chemical modification of siRNAs increases the stability by phosphorothioate addition to 39 ends and 29-O-methyl, 29-fluoro, and locked nucleic acid–type substitutions to the ribose backbone. Substitution of the 2′-OH group on the ribose ring with O-methyl group (2′-O-Me), a fluoro (2′-F) group, or a 2-methoxyethyl (2′-O-MOE) group resulted in prolonged half-lives (due to enhanced binding to serum proteins such as albumin) and RNAi stability of RNAi without affecting efficiency [110].

siRNA degradation in a cell culture system can be suppressed by linking a cholesterol molecule (phosphodiester (PO4) group with phosphothioate (PS)) to the 3′ end of the sense strand of a siRNA molecule [45]. The fatty acid conjugates with longer, saturated, alkyl chains, such as stearoyl and docosanyl can reduce target gene expression [111]. The modification of siRNA using 2,4-dinitrophenol (DNP) resulted in increased nuclease resistance as well as membrane permeability of siRNA [112]. Chemically modified siRNAs, such as sugar modifications (e.g., 2′-O methyl and 2′-deoxy-2′-fluoro (OMe/F)) or backbone modifications (e.g., phosphorothioate linkages), are often more stable against nucleases, enhance nuclease stability, and prolong siRNA half-life in serum without affecting its efficacy. Backbone modifications and bioconjugation with lipids and peptides are known to improve the stability and cellular uptake of siRNAs [113]. Strategies like enzyme active linkers, acid labile cross-linkers [114], pH-sensitive detergents [115], thermal sensitive liposomes [116], and reductive environment sensitive disulfide cross-linkers [117] have been explored to improve efficient self de-assembly of the nanoparticles.

Online siRNA design algorithms allow the identification of a target sequence, secondary structure, siRNA duplex end-stabilities, and minimization of sequence dependent off-target effects [118]. To predict off-target effects, AsiDesigner can be used. It is a freely accessible Web tool (http://sysbio.kribb.re.kr/AsiDesigner/) and provides stepwise off-target searching with BLAST and FASTA algorithms [119].

Surface modification of carriers systems [120]

Lipoplex can be decorated with different kinds of ligands such as antibodies, receptors, peptides, vitamins, oligonucleotides, and carbohydrates (Fig. 15.5) [121].

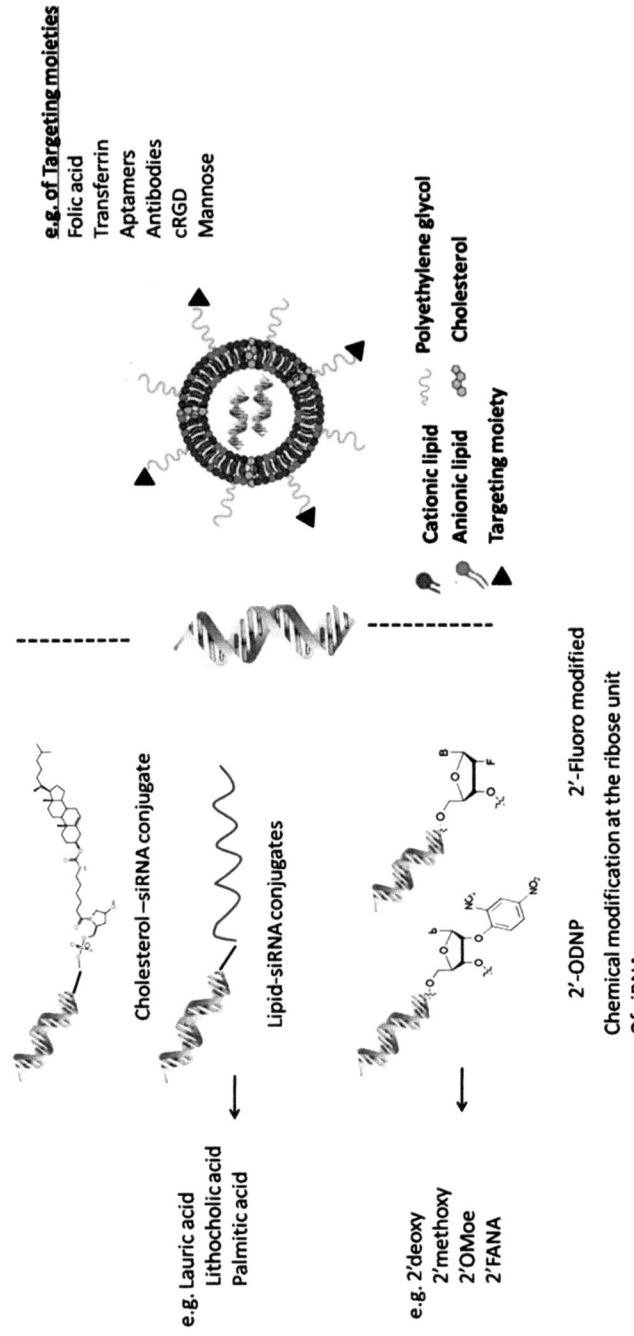

Figure 15.5 Surface modification options for delivery of siRNA and different ligand candidates available to modify surface properties offering required functionality.

An efficient systemic delivery of the siRNA to tumor and ocular neovasculature tissue in a herpes simplex virus (HSV) eye infection model was achieved through a Arg-Gly-Asp (RGD)-motif peptide ligand-targeted nanoparticle [122]. In another study, siRNA oligos of cationic lipid or polymer carriers containing the 5′-UGUGU-3′ motif induced a Toll-like receptor-mediated interferon response after either i.v. or i.p. administration [29]. Systemic and targeted delivery systems for siRNA have been studied using a protamine-antibody fusion protein [123]. siRNAs specific for human immunodeficiency virus (HIV)-1 capsid protein gag were complexed to a fusion protein composed of cationic protamine and HIV-1 envelope antibody. Aptamers have been suggested as targeting moieties for delivery systems. Lipofectamine 2000 [124] or cardiolipin analogs [125] are widely used for *in vitro* for a delivery of plasmid DNA or siRNA.

On the other hand, the surface charge can be masked by covering the vector with hydrophilic polymers, such as PEG, (poly)hydroxypropyl methacrylamide (pHPMA), and (poly)vinyl pyrrolidone (PVP) [126], which forms a dense hydrophilic network around the vector and limits the hydrophobic or electrostatic interactions with the extracellular medium. The result is a longer circulation time by avoiding MPS-uptake [127]. Mannose and mannose-related receptors are useful for macrophage or dendritic cells targeting. Transferrin or folate can be the best ligands to target cancer cells. Other ligands for targeting tumor cells are peptides with an arginine-glycine-aspartic acid (RGD) motif. Antibodies or its fragments can also be used as ligands [123]. Calando Pharmaceuticals (Pasadena, California) uses non-chemically modified siRNAs coupled to transferrin receptor targeting agents, composed of specialized cyclodextrin-based nanoparticles.

15.5 Clinical Status

Currently, several potential siRNA candidates are undergoing clinical trials for the treatment of macular degeneration, respiratory diseases, and cancers. The double-stranded RNA-based molecule, siRNA, has a high potential among biopharmaceutical therapeutics for gene silencing to treat diseases (Table 15.1). siRNA candidates entered clinical trials recently after their discovery.

Most siRNAs in clinical trials are administered by local delivery such as the intravitreal or intranasal routes. Several clinical trials are ongoing or planned for taking siRNA into the clinic for the treatment of important diseases such as macular degeneration, cancer, HIV and respiratory diseases. Currently many clinical trials are under way that use lipid nanoparticles for siRNA treatment of a variety of diseases and disorders. In addition, a number of siRNA drugs are in the pipeline of many pharmaceutical and biotechnology companies. An increase in new siRNA-based drugs, and a major revision of the pharmacopoeia, will soon be a reality [128]. siRNA (Sirna-027), Sirna Therapeutics, AGN211745, conducted their phase I trial for a chemically modified siRNA as a single intravitreal injection at doses up to 800 µm for the same therapeutic use of Cand5. The formulation was well tolerated by patients with improved visual acuity with no serious adverse events and has entered a phase II trial. Another AMD siRNA candidate, RTP-801i, blocks REDD-1 gene expression and has been investigated in a phase I clinical trial. Drug AGN211745 (Merck, Whitehouse Station, NJ) showed varied dose-dependent effects during phase I clinical trials to treat age-related macular degeneration (AMD) [129].

Table 15.1 Overview of ongoing and completed clinical trials with different types of siRNA

Type of siRNA	Formulation	Disease	Company name	Clinical status
Bevasiran-ib	Naked siRNA intravitreal injection	Age-related macular degeneration (AMD)	Acuity Pharmaceuticals and Opko Health Inc.	Completed, Phase II
AGN211745		AMD	Sirna Therapeutics, Inc	Phase I and II completed
RTP-801i	Naked siRNA intravitreal injection	AMD	Quark Pharmaceuticals	Phase I clinical trial ongoing
ALN-RSV01	Naked siRNA intranasal	Respiratory syncytial virus (RSV)	Alnylam Pharmaceuticals	Phase II completed

Type of siRNA	Formulation	Disease	Company name	Clinical status
ALN-VSP02	SNALP	Solid tumors, solid tumors with liver involvement	Alnylam Pharmaceuticals	Phase I completed
CALAA-01	Cyclodextrin nanoparticle, transferrin, PEG	Tumor	Calando pharmaceuticals	Entered Phase I trials but terminated
Atu027AG	Liposomes	Solid tumors	Silence Therapeutics GmbH	Phase I completed, Phase II ongoing
NUC B1000	Cationic lipid formulation	Hepatitis	Nucleonics	Phase I
TKM 080301	SNALP	NET and ACC, primary or secondary liver cancer	Tekmira Pharma	NET and ACC: Phase I recruiting, primary or secondary liver cancer Phase I completed
siRNA-EphA2-DOPC	Liposomes (neutral liposomes)	Solid tumors	M.D. Anderson Cancer Center	Phase I ongoing
siRNA tauRNAi	Liposome	Hepatocellular carcinoma	Marina Biotech (Bothell, WA, USA)	Preclinical stage
ALN-TTR02	LNP with ionizable cationic lipid DLinMC3-DMA	Transthyretin mediated amyloidosis (ATTR)	Alnylam Pharmaceuticals	Phase II completed
ALN-PCS	LNP with ionizable cationic lipid DLinMC3-DMA	Hypercholesterolemia	Alnylam Pharmaceuticals	Phase I completed

Recently Alnylam Pharmaceuticals completed two Phase I clinical trials for the first siRNA (ALN-RSV01) for targeting respiratory syncytial virus (RSV) in human volunteer subjects using a nasal spray. The formulation was found to be safe with mild adverse effects compared with placebos [24]. ALN-RSV01 is currently being investigated in a second phase II trial in naturally infected adult patients [130]. Nastech Pharmaceutical company Inc., (Seattle, http://www.nastech.com, now Marina Biotech) is currently developing siRNA as an influenza therapeutic and it is expected to be effective against all known human and avian influenza strains. Alnylam also is developing RNAi therapeutics for the pandemic flu, such as H5N1.

The pipeline is mainly dominated by anti-cancer siRNA candidates followed by macular degeneration drugs and equally for respiratory disorders. Alnylam Pharmaceuticals has developed ALN-VSP, which contains SNALPs as a delivery system (Tekmira's lipid-based nanoparticle technology) and has also entered phase I clinical trials. Nucleonics has begun a phase I human safety study for systemically administered siRNA (NUC B1000 cationic lipid formulation) designed to reduce the malignant effects of hepatitis.

15.6 Conclusions

RNAi therapeutics has quickly been established as a robust and effective gene silencing strategy and has several distinct advantages over traditional pharmaceutical drugs. The reason for the limited number of siRNA delivery systems in clinical trials is certainly the complexity of the approach. It is easier to administer naked or modified siRNA than to develop an efficient, reproducible, and safe delivery system, which has to undergo the same regulatory guidelines as the drug itself. The nature and extent of these gene expression changes are dependent on several factors, including delivery system variables such as its chemistry, lipid architecture, charge, and dose; the degree of saturation, nature of cationic lipids, fusogenicity, cellular uptake, gene silencing ability, and biological variables. Extensive investigations into the chemistry and physics of cationic lipids and their transfection efficiencies in a lipoplex formulation are equally important for transfection. The future expects further advances in the development of nontoxic liposomal and other forms of nanoparticle-based approaches as the vehicle allows relatively large amounts of siRNAs to be packaged and delivered at target site.

Disclosures and Conflict of Interest

The authors declare that they have no conflict of interest and have no affiliations or financial involvement with any organization or entity discussed in this chapter. This includes employment, consultancies, honoraria, grants, stock ownership or options, expert testimony, patents (received or pending), or royalties. No writing assistance was utilized in the production of this chapter and the authors have received no payment for its preparation.

Corresponding Author

Dr. Ranjita Shegokar
Freie Universität Berlin
Institute of Pharmacy
Department Pharmaceutics, Biopharmaceutics & NutriCosmetics
Kelchstraße 31, 12169 Berlin, Germany
Email: ranjita@arcslive.com

About the Authors

Ranjita Shegokar holds a PhD degree in pharmaceutical technology from SNDT University and has been a postdoctoral researcher at the Department of Pharmaceutics, Biopharmaceutics and NutriCosmetics at Free University of Berlin, Germany. Since last 8 years she is working in the pharmaceutical industry in various technical and R&D roles related to drug delivery systems. Currently, she serves as a Director of Pharma Business Unit at Capnomed GmbH, Germany. Dr. Shegokar has authored more than 50 research articles, 20 book chapters, and 150 research abstracts. She has edited multiple books in area of pharmaceutical drug delivery systems and challenges. She has filed several patent applications in area of active delivery. She is a recipient of several national and international awards for her research. Her areas of interest are oral solid dosage forms, dermal formulations, nanocrystals, lipid nanoparticles (SLNs/NLCs), nanoemulsions, and galenic formulations. https://www.researchgate.net/profile/Ranjita_Shegokar

Dr. Prabhat Ranjan Mishra is a principal scientist at the Pharmaceutics Division and associate professor at CSIR—Central Drug Research Institute, Lucknow, India. His current area of research is layer-by-layer technology–based nanoconstructs, nanocrystal technology, nanomaterials and nanodevices for health and infectious diseases, ligand/receptor-mediated drug targeting, novel drug delivery approaches for various parasitic diseases, and endocrine disorders, including osteoporosis.

References

1. Flotte, T. R. (2007). Gene therapy: The first two decades and the current state-of-the-art. *J. Cell Physiol.*, **213**(2), 301–305.
2. Zou, S., et al. (2010). Lipid-mediated delivery of RNA is more efficient than delivery of DNA in non-dividing cells. *Int. J. Pharm.*, **389**(1–2), 232–243.
3. Akhtar, S., Benter, I. (2007). Toxicogenomics of non-viral drug delivery systems for RNAi: Potential impact on siRNA-mediated gene silencing activity and specificity. *Adv. Drug Deliv. Rev.*, **59**(2–3), 164–182.
4. Bernstein, E., et al. (2001). Role for a bidentate ribonuclease in the initiation step of RNA interference. *Nature*, **409**(6818), 363–366.
5. Ahlquist, P. (2002). RNA-dependent RNA polymerases, viruses, and RNA silencing. *Science*, **296**(5571), 1270–1273.
6. Kim, D. H., Rossi, J. J. (2009). Overview of gene silencing by RNA interference. *Curr. Protoc. Nucleic Acid Chem.*, 36:16.1:16.1.1–16.1.10.
7. Kim, D., Rossi, J. (2008). RNAi mechanisms and applications. *Biotechniques*, **44**(5), 613–616.
8. de Fougerolles, A., et al. (2007). Interfering with disease: A progress report on siRNA-based therapeutics. *Nat. Rev. Drug Discov.*, **6**(6), 443–453.
9. Tabara, H., et al. (1999). The rde-1 gene, RNA interference, and transposon silencing in C. elegans. *Cell*, **99**(2), 123–132.

10. Hough, S. R., et al. (2003). Why RNAi makes sense. *Nat. Biotechnol.*, **21**(7), 731–732.
11. Lewis, D. L., et al. (2002). Efficient delivery of siRNA for inhibition of gene expression in postnatal mice. *Nat. Genet.*, **32**(1), 107–108.
12. Doudna, J. A., Cech, T. R. (2002). The chemical repertoire of natural ribozymes. *Nature*, **418**(6894), 222–228.
13. Bertrand, J. R., et al. (2002). Comparison of antisense oligonucleotides and siRNAs in cell culture and in vivo. *Biochem. Biophys. Res. Commun.*, **296**, 1000–1004.
14. Fire, A., et al. (1998). Potent and specific genetic interference by double-stranded RNA in Caenorhabditis elegans. *Nature*, **391**(6669), 806–811.
15. Zhang, H., et al. (2004). Single processing center models for human Dicer and bacterial RNase III. *Cell*, **118**(1), 57–68.
16. Zamore, P. D., et al. (2000). RNAi: Double-stranded RNA directs the ATP-dependent cleavage of mRNA at 21 to 23 nucleotide intervals. *Cell*, **101**(1), 25–33.
17. Elbashir, S. M., et al. (2001). Functional anatomy of siRNAs for mediating efficient RNAi in Drosophila melanogaster embryo lysate. *EMBO J.*, **20**(23), 6877–6888.
18. Schwarz, D. S., et al. (2003). Asymmetry in the assembly of the RNAi enzyme complex. *Cell*, 2003. **115**(2), 199–208.
19. Khvorova, A., Reynolds, A., Jayasena, S. D. (2003). Functional siRNAs and miRNAs exhibit strand bias. *Cell*, **115**(2), 209–216.
20. Kim, V. N., Han, J., Siomi, M. C. (2009). Biogenesis of small RNAs in animals. *Nat. Rev. Mol. Cell Biol.*, **10**(2), 126–139.
21. Gewirtz, A. M. (2007). RNA targeted therapeutics for hematologic malignancies. *Blood Cells Mol. Dis.*, **38**(2), 117–119.
22. Rohl, T., Kurreck, J. (2006). RNA interference in pain research. *J. Neurochem.*, **99**(2), 371–380.
23. Dorsett, Y., Tuschl, T. (2004). siRNAs: Applications in functional genomics and potential as therapeutics. *Nat. Rev. Drug Discov.*, **3**(4), 318–329.
24. Akhtar, S., Benter, I. (2007). Toxicogenomics of non-viral drug delivery systems for RNAi: Potential impact on siRNA-mediated gene silencing activity and specificity. *Adv. Drug Deliv. Rev.*, **59**(2–3), 164–182.
25. Hu-Lieskovan, S., et al. (2005). Sequence-specific knockdown of EWS-FLI1 by targeted, nonviral delivery of small interfering RNA

inhibits tumor growth in a murine model of metastatic Ewing's sarcoma. *Cancer Res.*, **65**(19), 8984–8992.

26. McNamara, J. O., et al. (2006). Cell type-specific delivery of siRNAs with aptamer-siRNA chimeras. *Nat. Biotechnol.*, **24**(8), 1005–1015.

27. Lu, P. Y., Woodle, M. C. (2005). Delivering siRNA in vivo for functional genomics and novel therapeutics. In: Appasani, K., ed. *RNA Interference Technology*, Cambridge University Press, London, pp. 303–317.

28. Akhtar, S., Benter, I. F. (2007). Nonviral delivery of synthetic siRNAs in vivo. *J. Clin. Invest.*, **117**(12), 3623–3632.

29. Judge, A. D., et al. (2005). Sequence-dependent stimulation of the mammalian innate immune response by synthetic siRNA. *Nat. Biotechnol.*, **23**(4), 457–462.

30. Hornung, V., et al. (2005). Sequence-specific potent induction of IFN-alpha by short interfering RNA in plasmacytoid dendritic cells through TLR7. *Nat. Med.*, **11**(3), 263–270.

31. Agrawal, N., et al. (2003). RNA interference: Biology, mechanism, and applications. *Microbiol. Mol. Biol. Rev.*, **67**(4), 657–685.

32. Fjose, A., et al. (2001). RNA interference: Mechanisms and applications. *Biotechnol. Annu. Rev.*, **7**, 31–57.

33. Aigner, A. (2007). Applications of RNA interference: Current state and prospects for siRNA-based strategies in vivo. *Appl. Microbiol. Biotechnol.*, **76**(1), 9–21.

34. Zhou, D., et al. (2006). RNA interference and potential applications. *Curr. Top. Med. Chem.*, **6**(9), 901–911.

35. Stanislawska, J., Olszewski, W. L. (2005). RNA interference–significance and applications. *Arch. Immunol. Ther. Exp.*, **53**(1), 39–46.

36. Wadhwa, R., et al. (2004). Know-how of RNA interference and its applications in research and therapy. *Mutat. Res.*, **567**(1), 71–84.

37. Hebert, C. G., Valdes, J. J., Bentley, W. E. (2008). Beyond silencing–engineering applications of RNA interference and antisense technology for altering cellular phenotype. *Curr. Opin. Biotechnol.*, **19**(5), 500–505.

38. Karkare, S., Daniel, S., Bhatnagar, D. (2004). RNA interference silencing the transcriptional message: Aspects and applications. *Appl. Biochem. Biotechnol.*, **119**(1), 1–12.

39. Kurreck, J. (2009). RNA interference: From basic research to therapeutic applications. *Angew. Chem. Int. Ed.*, **48**(8), 1378–1398.

40. Karagiannis, T. C., El-Osta, A. (2004). siRNAs: Mechanism of RNA interference, in vivo and potential clinical applications. *Cancer Biol. Ther.*, **3**(11), 1069–1074.

41. Aigner, A. (2006). Delivery systems for the direct application of siRNAs to induce RNA interference (RNAi) in vivo. *J. Biomed. Biotechnol.*, **2006**(4), 71659.

42. Suri, S. S., Fenniri, H., Singh, B. (2007). Nanotechnology-based drug delivery systems. *J. Occup. Med. Toxicol.*, **2**, 16.

43. Jiang, W., et al. (2007). Advances and challenges of nanotechnology-based drug delivery systems. *Expert Opin. Drug Deliv.*, **4**(6), 621–633.

44. Fattal, E., Bochot, A. (2008). State of the art and perspectives for the delivery of antisense oligonucleotides and siRNA by polymeric nanocarriers. *Int. J. Pharm.*, **364**(2), 237–248.

45. Liu, F., Song, Y., Liu, D. (1999). Hydrodynamics-based transfection in animals by systemic administration of plasmid DNA. *Gene Ther.*, **6**(7), 1258–1266.

46. Suda, T., Suda, K., Liu, D. (2008). Computer-assisted hydrodynamic gene delivery. *Mol. Ther.*, **16**(6), 1098–1104.

47. Belyantseva, I. A. (2009). Helios Gene Gun-mediated transfection of the inner ear sensory epithelium. *Methods Mol. Biol.*, **493**, 103–123.

48. Liu, F., Huang, L. (2002). Electric gene transfer to the liver following systemic administration of plasmid DNA. *Gene Ther.*, **9**(16), 1116–1119.

49. Liu, F., Huang, L. (2002). A syringe electrode device for simultaneous injection of DNA and electrotransfer. *Mol. Ther.*, **5**(3), 323–328.

50. Nguyen, D. N., et al. (2009). Drug delivery-mediated control of RNA immunostimulation. *Mol. Ther.*, **17**(9), 1555–1562.

51. Brown, W. L., et al. (2002). RNA bacteriophage capsid-mediated drug delivery and epitope presentation. *Intervirology*, **45**(4–6), 371–380.

52. Tao, W., et al. (2010). Noninvasive imaging of lipid nanoparticle-mediated systemic delivery of small-interfering RNA to the liver. *Mol. Ther.*, **18**(9), 1657–1666.

53. Baker, M. (2010). RNA interference: Homing in on delivery. *Nature*, **464**(7292), 1225–1228.

54. Sioud, M. (2010). RNA therapeutics: Function, design, and delivery. Preface. *Methods Mol. Biol.*, **629**, v–vii.

55. Popielarski, S. R., et al. (2005). A nanoparticle-based model delivery system to guide the rational design of gene delivery to the liver.

2. In vitro and in vivo uptake results. *Bioconjugate Chem.*, **16**(5), 1071–1080.

56. Kitamura, N., et al. (2009). Biodistribution of immunoliposome labeled with Tc-99m in tumor xenografted mice. *Ann. Nucl. Med.*, **23**(2), 149–153.

57. Park, J. W., Benz, C. C., Martin, F. J. (2004). Future directions of liposome- and immunoliposome-based cancer therapeutics. *Semin. Oncol.*, **31**(6 Suppl 13), 196–205.

58. Maruyama, K. (2002). PEG-immunoliposome. *Biosci. Rep.*, **22**(2), 251–266.

59. Maruyama, K., et al. (1990). Characterization of in vivo immunoliposome targeting to pulmonary endothelium. *J. Pharm. Sci.*, **79**(11), 978–984.

60. Lv, H., et al. (2006). Toxicity of cationic lipids and cationic polymers in gene delivery. *J. Control. Release*, **114**(1), 100–109.

61. Li, C. X., et al. (2006). Delivery of RNA interference. *Cell Cycle*, **5**, 2103–2109.

62. Kabanov, A. V. (1999). Taking polycation gene delivery systems from in vitro to in vivo. *Pharm. Sci. Technol. Today*, **2**(9), 365–372.

63. Soutschek, J., et al. (2004). Therapeutic silencing of an endogenous gene by systemic administration of modified siRNAs. *Nature*, **432**(7014), 173–178.

64. Zimmermann, T. S., et al. (2006). RNAi-mediated gene silencing in non-human primates. *Nature*, **441**(7089), 111–114.

65. Nakamura, Y., et al. (2007). Octaarginine-modified multifunctional envelope-type nano device for siRNA. *J. Control. Release*, **119**(3), 360–367.

66. Khalil, I. A., et al. (2007). Octaarginine-modified multifunctional envelope-type nanoparticles for gene delivery. *Gene Ther.*, **14**(8), 682–689.

67. Hafez, I. M., Maurer, N., Cullis, P. R. (2001). On the mechanism whereby cationic lipids promote intracellular delivery of polynucleic acids. *Gene Ther.*, **8**(15), 1188–1196.

68. Sorensen, D. R., Leirdal, M., Sioud, M. (2003). Gene silencing by systemic delivery of synthetic siRNAs in adult mice. *J. Mol. Biol.*, **327**(4), 761–766.

69. Farhood, H., et al. (1992). Effect of cationic cholesterol derivatives on gene transfer and protein kinase C activity. *Biochim. Biophys. Acta*, **1111**(2), 239–246.

70. Verma, U. N., et al. (2003). Small interfering RNAs directed against beta-catenin inhibit the in vitro and in vivo growth of colon cancer cells. *Clin. Cancer Res.*, **9**(4), 1291–1300.
71. Santel, A., et al. (2006). A novel siRNA-lipoplex technology for RNA interference in the mouse vascular endothelium. *Gene Ther.*, **13**(16), 1222–1234.
72. Sato, A., et al. (2007). Small interfering RNA delivery to the liver by intravenous administration of galactosylated cationic liposomes in mice. *Biomaterials*, **28**(7), 1434–1442.
73. Barreau, C., et al. (2006). Liposome-mediated RNA transfection should be used with caution. *RNA*, **12**(10), 1790–1793.
74. El Ouahabi, A., et al. (1996). Double long-chain amidine liposome-mediated self replicating RNA transfection. *FEBS Lett.*, **380**(1–2), 108–112.
75. Dwarki, V. J., Malone, R. W., Verma, I. M. (1993). Cationic liposome-mediated RNA transfection. *Methods Enzymol.*, **217**, 644–654.
76. Miyawaki-Shimizu, K., et al. (2006). siRNA-induced caveolin-1 knockdown in mice increases lung vascular permeability via the junctional pathway. *Am. J. Physiol. Lung Cell Mol. Physiol.*, **290**(2), L405–L413.
77. Palliser, D., et al. (2006). An siRNA-based microbicide protects mice from lethal herpes simplex virus 2 infection. *Nature*, **439**(7072), 89–94.
78. Akinc, A., et al. (2008). A combinatorial library of lipid-like materials for delivery of RNAi therapeutics. *Nat. Biotechnol.*, **26**(5), 561–569.
79. Kim, H. R., et al. (2008). Cationic solid lipid nanoparticles reconstituted from low density lipoprotein components for delivery of siRNA. *Mol. Pharmacol.*, **5**(4), 622–631.
80. Cullis, P. R. (2002). Stabilized plasmid-lipid particles for systemic gene therapy. *Cell. Mol. Biol. Lett.*, **7**(2), 226.
81. Fenske, D. B., MacLachlan, I., Cullis, P. R. (2002). Stabilized plasmid-lipid particles: A systemic gene therapy vector. *Methods Enzymol.*, **346**, 36–71.
82. Tam, P., et al. (2000). Stabilized plasmid-lipid particles for systemic gene therapy. *Gene Ther.*, **7**(21), 1867–1874.
83. Wheeler, J. J., et al. (1999). Stabilized plasmid-lipid particles: Construction and characterization. *Gene Ther.*, **6**(2), 271–281.
84. MacLachlan, I., Cullis, P., Graham, R. W. (1999). Progress towards a synthetic virus for systemic gene therapy. *Curr. Opin. Mol. Ther.*, **1**(2), 252–259.

85. Heyes, J., et al. (2005). Cationic lipid saturation influences intracellular delivery of encapsulated nucleic acids. *J. Control. Release*, **107**(2), 276–287.
86. Miller, A. D. (2008). Towards safe nanoparticle technologies for nucleic acid therapeutics. *Tumori*, **94**(2), 234–245.
87. Bhattacharya, S., Bajaj, A. (2009). Advances in gene delivery through molecular design of cationic lipids. *Chem. Commun. (Camb)*, **2009**(31), 4632–4656.
88. Martin, B., et al. (2005). The design of cationic lipids for gene delivery. *Curr. Pharm. Des.*, **11**(3), 375–394.
89. Ilies, M. A., Seitz, W. A., Balaban, A. T. (2002). Cationic lipids in gene delivery: Principles, vector design and therapeutical applications. *Curr. Pharm. Des.*, **8**(27), 2441–2473.
90. Boussif, O., et al. (2001). Enhanced in vitro and in vivo cationic lipid-mediated gene delivery with a fluorinated glycerophosphoethanolamine helper lipid. *J. Gene Med.*, **3**(2), 109–114.
91. Yoshizawa, T., et al. (2008). Folate-linked lipid-based nanoparticles for synthetic siRNA delivery in KB tumor xenografts. *Eur. J. Pharm. Biopharm.*, **70**(3), 718–725.
92. Boado, R. J. (2007). Blood-brain barrier transport of non-viral gene and RNAi therapeutics. *Pharm. Res.*, **24**(9), 1772–1787.
93. Herringson, T. P., Altin, J. G. (2009). Convenient targeting of stealth siRNA-lipoplexes to cells with chelator lipid-anchored molecules. *J. Control. Release*, **139**(3), 229–238.
94. van Broekhoven, C. L., et al. (2004). Targeting dendritic cells with antigen-containing liposomes: A highly effective procedure for induction of antitumor immunity and for tumor immunotherapy. *Cancer Res.*, **64**(12), 4357–4365.
95. Khoury, M., et al. (2006). Efficient new cationic liposome formulation for systemic delivery of small interfering RNA silencing tumor necrosis factor alpha in experimental arthritis. *Arthritis Rheum.*, **54**(6), 1867–1877.
96. Leng, Q., et al. (2005). Highly branched HK peptides are effective carriers of siRNA. *J. Gene Med.*, **7**(7), 977–986.
97. Flynn, M. A., et al. (2004). Efficient delivery of small interfering RNA for inhibition of IL-12p40 expression in vivo. *J. Inflamm. (Lond)*, **1**(1), 4.
98. Sioud, M., Sorensen, D. R. (2003). Cationic liposome-mediated delivery of siRNAs in adult mice. *Biochem. Biophys. Res. Commun.*, **312**(4), 1220–1225.

99. Griesenbach, U., et al. (2006). Inefficient cationic lipid-mediated siRNA and antisense oligonucleotide transfer to airway epithelial cells in vivo. *Respir. Res.*, **7**(1), 26.

100. Gao, X., Huang, L. (1996). Potentiation of cationic liposome-mediated gene delivery by polycations. *Biochemistry*, **35**(3), 1027–1036.

101. Li, J., et al. (2009). Biodegradable calcium phosphate nanoparticle with lipid coating for systemic siRNA delivery. *J. Control. Release*, **142**(3), 416–421.

102. Lendvai, G., et al. (2005). Biodistribution of 68Ga-labelled phosphodiester, phosphorothioate, and 2′-O-methyl phosphodiester oligonucleotides in normal rats. *Eur. J. Pharm. Sci.*, **26**(1), 26–38.

103. Braasch, D. A., et al. (2004). Biodistribution of phosphodiester and phosphorothioate siRNA. *Bioorg. Med. Chem. Lett.*, **14**(5), 1139–1143.

104. Landen, C. N., Jr., et al. (2005). Therapeutic EphA2 gene targeting in vivo using neutral liposomal small interfering RNA delivery. *Cancer Res.*, **65**(15), 6910–6918.

105. Peer, D., et al. (2008). Systemic leukocyte-directed siRNA delivery revealing cyclin D1 as an anti-inflammatory target. *Science*, **319**(5863), 627–630.

106. Pirollo, K. F., et al. (2006). Tumor-targeting nanoimmunoliposome complex for short interfering RNA delivery. *Hum. Gene Ther.*, **17**(1), 117–124.

107. Pirollo, K. F., et al. (2007). Materializing the potential of small interfering RNA via a tumor-targeting nanodelivery system. *Cancer Res.*, **67**(7), 2938–2943.

108. Akinc, A., et al. (2009). Development of lipidoid-siRNA formulations for systemic delivery to the liver. *Mol. Ther.*, **17**(5), 872–879.

109. Liu, C., et al. (2009). Novel biodegradable lipid nano complex for siRNA delivery significantly improving the chemosensitivity of human colon cancer stem cells to paclitaxel. *J. Control. Release*, **140**(3), 277–283.

110. Braasch, D. A., et al. (2003). RNA interference in mammalian cells by chemically-modified RNA. *Biochemistry*, **42**(26), 7967–7975.

111. Wolfrum, C., et al. (2007). Mechanisms and optimization of in vivo delivery of lipophilic siRNAs. *Nat. Biotechnol.*, **25**(10), 1149–1157.

112. Liao, H., Wang, J. H. (2005). Biomembrane-permeable and Ribonuclease-resistant siRNA with enhanced activity. *Oligonucleotides*, **15**(3), 196–205.

113. Moschos, S. A., Williams, A. E., Lindsay, M. A. (2007). Cell-penetrating-peptide-mediated siRNA lung delivery. *Biochem. Soc. Trans.*, **35**(Pt 4), 807–810.

114. Guo, X., Szoka, F. C., Jr. (2003). Chemical approaches to triggerable lipid vesicles for drug and gene delivery. *Acc. Chem. Res.*, **36**(5), 335–341.

115. Asokan, A., Cho, M. J. (2002). Exploitation of intracellular pH gradients in the cellular delivery of macromolecules. *J. Pharm. Sci.*, **91**(4), 903–913.

116. Needham, D., et al. (2000). A new temperature-sensitive liposome for use with mild hyperthermia: Characterization and testing in a human tumor xenograft model. *Cancer Res.*, **60**(5), 1197–1201.

117. Austin, C. D., et al. (2005). Oxidizing potential of endosomes and lysosomes limits intracellular cleavage of disulfide-based antibody-drug conjugates. *Proc. Natl. Acad. Sci. U.S.A.*, **102**(50), 17987–17992.

118. Kim, D. H., Rossi, J. J. (2007). Strategies for silencing human disease using RNA interference. *Nat. Rev. Genet.*, **8**(3), 173–184.

119. Park, Y. K., et al. (2008). AsiDesigner: Exon-based siRNA design server considering alternative splicing. *Nucleic Acids Res.*, **36**(Web Server issue), W97–103.

120. Watts, J. K., Deleavey, G. F., Damha, M. J. (2008). Chemically modified siRNA: Tools and applications. *Drug Discov. Today*, **13**(19–20), 842–855.

121. Rao, N. M. (2010). Cationic lipid-mediated nucleic acid delivery: Beyond being cationic. *Chem. Phys. Lipids*, **163**(3), 245–252.

122. Schiffelers, R. M., et al. (2004). Cancer siRNA therapy by tumor selective delivery with ligand-targeted sterically stabilized nanoparticle. *Nucleic Acids Res.*, **32**(19), e149.

123. Song, E., et al. (2005). Antibody mediated in vivo delivery of small interfering RNAs via cell-surface receptors. *Nat. Biotechnol.*, **23**(6), 709–717.

124. Dalby, B., et al. (2004). Advanced transfection with Lipofectamine 2000 reagent: Primary neurons, siRNA, and high-throughput applications. *Methods*, **33**(2), 95–103.

125. Chien, P. Y., et al. (2005). Novel cationic cardiolipin analogue-based liposome for efficient DNA and small interfering RNA delivery in vitro and in vivo. *Cancer Gene Ther.*, **12**(3), 321–328.

126. Ogris, M., Wagner, E. (2002). Targeting tumors with non-viral gene delivery systems. *Drug Discov. Today*, **7**(8), 479–485.
127. Gref, R., et al. (1994). Biodegradable long-circulating polymeric nanospheres. *Science*, **263**(5153), 1600–1603.
128. Gewirtz, A. M. (2007). On future's doorstep: RNA interference and the pharmacopeia of tomorrow. *J. Clin. Invest.*, **117**(12), 3612–3614.
129. Whelan, J. (2005). First clinical data on RNAi. *Drug Discov. Today*, **10**(15), 1014–1015.
130. DeVincenzo, J., et al. (2008). Evaluation of the safety, tolerability and pharmacokinetics of ALN-RSV01, a novel RNAi antiviral therapeutic directed against respiratory syncytial virus (RSV). *Antiviral Res.*, **77**(3), 225–231.

Chapter 16

Nanovaccines against Intracellular Pathogens Using *Coxiella burnetii* as a Model Organism

Erin J. van Schaik, PhD,[a] Anthony E. Gregory, PhD,[a] Gerald F. Audette, PhD,[b] and James E. Samuel, PhD[a]

[a]*Department of Microbial Pathogenesis and Immunology, College of Medicine, Texas A&M University, Texas, USA*
[b]*Department of Chemistry and Centre for Research on Biomolecular Interactions, York University, Ontario, Canada*

Keywords: adjuvant, antibody, CD4+ T cell, CD8+ T cell, cell-mediated immunity, Centers for Disease Control and Prevention (CDC), *Coxiella burnetii*, Coxiella-containing vacuole (CCV), delivery platform, gold nanoparticle, humoral immunity, immunity, infectious disease, intracellular pathogens, large cell variants (LCV), lipopolysaccharide (LPS), monodisperse particles, nanoparticle (NP), nanovaccine, pathogen-associated molecular patterns (PAMPs), Q fever, Qvax®, small cell variants (SCV), select agent, type IV secretion system (T4SS), uniform size, vaccine

16.1 Introduction

The breadth and applicability of nanoparticles (NPs) as vaccine delivery tools offer unique opportunities to researchers in

---

Immune Aspects of Biopharmaceuticals and Nanomedicines
Edited by Raj Bawa, János Szebeni, Thomas J. Webster, and Gerald F. Audette
Copyright © 2018 Pan Stanford Publishing Pte. Ltd.
ISBN 978-981-4774-52-9 (Hardcover), 978-0-203-73153-6 (eBook)
www.panstanford.com

developing safe and effective vaccines against intracellular pathogens. In this review, *Coxiella burnetii* has been highlighted as one such pathogen where this approach may be appropriate to eliminate the adverse effects associated with killed whole cell vaccination.

16.2 Using Nanomedicine to Tackle Intracellular Pathogens

Despite the success of vaccination efforts over the years and the advent of commercially available antibiotics in the 1930s, infectious diseases remain a persistent threat to public health. Tuberculosis, HIV/AIDS, and malaria alone account for more than 3.3 million deaths worldwide each year [1]. Some of the most deadly infectious diseases are caused by a group of pathogens that have a predominantly intracellular lifestyle, including a number of pathogens the Centers for Disease Control and Prevention (CDC) has designated Select Agents, such as *Burkholderia pseudomallei*, *Francisella tularensis*, *Brucella melitensis*, and *C. burnetii*. Intracellular pathogens can be bacterial, viral, fungal, or parasitic and share a common feature the ability to survive and replicate within a host cell. This often involves surviving in a range of conditions from harsh acidic environments such as a modified endosomes to the nutrient-rich cytosol [2, 3]. Surviving within difficult environments presents a major challenge, but in return, an intracellular lifestyle provides a pathogen with refuge from many antibiotics and components of the host immune system. Currently the most effective vaccines against many intracellular pathogens are either killed cells or attenuated mutants that induce a broad humoral and cell-mediated immune response. However, concerns regarding reactogenicity, wild-type reversion, and the use of these vaccines in immunocompromised individuals have driven the effort to alternate vaccine designs.

As the field of nanotechnology evolves, nanomedicine offers a new opportunity to overcome some of the unique challenges of combatting intracellular pathogens through the design and development of nanoparticles (NPs) to deliver therapeutic and/

or antigenic agents. This review uncovers some of the valuable properties attributed to NPs and how these properties can be manipulated to specifically target intracellular pathogens, using *C. burnetii* as an example. To frame this model organism, *C. burnetii* bacteriology, pathogenesis and components of protective immunity against Q fever will be reviewed followed by how nanovaccines could be used as a novel strategy to provide protection to this agent, and by adaptation of these principles, to other intracellular pathogens.

16.3   *C. burnetii* Bacteriology

C. burnetii is a Gram-negative, pleomorphic coccobacillus (0.2 to 0.4 µm wide, 0.4 to 1 µm long) due to a biphasic lifestyle with different morphologic variants like *Chlamydia* spp. [4]. The infectious particles, termed small cell variants (SCV), contain condensed chromatin and have thick cells walls (0.2–0.5 µm) [5, 6]. SCV are stable in the extracellular environment and highly resistant to heat, desiccation, osmotic shock, UV light and chemical disinfectants [7]. The metabolically active replicative form large cell variants (LCV) are more typical of Gram-negative bacteria and 0.5–1.0 µm in length [5, 6]. After intracellular internalization of SCV, a lag growth phase is observed which enables SCV to LCV morphogenesis [8].

16.4   Epidemiology and Clinical Manifestations

Q fever is a zoonotic disease with near worldwide distribution caused by *C. burnetii*. Although *C. burnetii* occupies a variety of animal reservoirs, including mammals, arthropods, and birds, there is a strong association with domestic ruminants, especially sheep and goats, leading to transmission to humans [9]. Infection in sheep and goats is usually asymptomatic in non-pregnant animals but can cause late-term abortion in pregnant animals with organisms in milk, urine, feces, and birthing fluids/products as principle sources of contamination [10]. Human infections are acquired primarily by inhalation of contaminated aerosols but can also occur from ingestion of contaminated milk [11]. *C. burnetii* has an extremely low infectious dose (<10 organisms) leading to disease. Low infectious dose, stability in the environment, and aerosol route transmission make it a potential bio-warfare agent and hence designation as a Category B Select Agent by the CDC. Exposure to contaminated aerosols can result in asymptomatic disease with sero-conversion in ~60% of cases or acute disease with symptoms occurring 10–26 days after exposure and resembling other flu-like febrile diseases with symptoms that include headache, fever, and chills and can progress to pneumonia with or without hepatitis [12]. Most patients resolve the infection and are then immune to re-infection with *C. burnetii*. Chronic disease develops in 1–5% of exposed individuals, which usually presents months to years after the initial exposure. Chronic disease has high lethality if untreated and effective treatment requires long courses of antibiotics (18 months to 4 years) [13]. The most frequent and serious presentation of chronic Q fever is endocarditis but other manifestations include chronic hepatitis, osteomyelitis, or chronic fatigue syndrome [14].

16.5 Virulence

The first virulence factor to be identified for *C. burnetii* was lipopolysaccharide (LPS), which is required for productive infection in animals but not for invasion and replication within host cells [15, 16]. During laboratory passage of *C. burnetii* in immuno-incompetent hosts such as tissue culture cells or embryonic eggs *C. burnetii* rapidly accumulate rough (phase II) variants characterized by a loss of O-antigen. Genetic events leading to this phenotype include large chromosomal deletions and single nucleotide polymorphisms in LPS biosynthetic pathway genes [17]. There are several reasons why LPS is considered a virulence factor for *C. burnetii*. Phase II *C. burnetii* are complement sensitive, while phase I (smooth) are resistant to the complement membrane attack complex [18]. Phase I LPS of *C. burnetii* sterically inhibits binding of antigen-specific antibodies on the bacterial surface [19]. Additionally, LPS is involved in additional forms of innate immune evasion that are discussed in subsequent sections of this chapter.

C. burnetii is unique among intracellular bacterial pathogens as its replicative niche (the Coxiella-containing vacuole, CCV) has properties of a phagolysosome, a typically microbicidal compartment [20]. Most intracellular pathogens unleash an arsenal of virulence factors that prevent fusion of their vacuole with lysosomes or escape the vacuolar compartment and replicate in the cytoplasm [21]. After binding to and passive internalization by macrophages, *C. burnetii* traffics through the default endocytic pathway to this phagolysosome [15, 22]. In the process, the CCV transits through the endosomal pathway by multiple fusion and fission events, initially acquiring markers EEA1 and Rab5, followed by acquisition of Rab7, LAMPI and II, and Cathepsin D, while acidifying [23]. The acidification of the vacuole is required to induce metabolic activity in *C. burnetii* and also required for secretion of virulence factors through the Type 4B Secretion System (T4BSS) [24, 25]. Metabolic activation coincides with transition of SVCs to LCVs and also expansion of the CCV; this expansion is dependent on bacterial protein expression, including the secretion of effectors into the cytosol [6, 26, 27]. Type 4 secretion systems are essential for effector release and are related to bacterial conjugative transfer and IncI plasmid conjugation systems [28]. The T4BSS of *C. burnetii* is closely related to the Dot/Icm system of *Legionella pneumophila* and several *L. pneumophila* Dot/Icm mutants can be complemented with *C. burnetii* homologues [29]. The *C. burnetii* Dot/Icm system is a major virulence factor and required for intracellular replication and animal infection [27, 30, 31]. Currently there are 143 reported *C. burnetii* T4BSS substrate proteins, many of which are involved in manipulating the host to create its replicative niche [32]. Of the 143 potential virulence determinants, only a handful have been characterized for their molecular functions. Interestingly, unlike its closest relative, *L. pneumophila* has many functionally redundant Dot/Icm effectors, such that mutants in only a few secreted effectors result in intracellular growth defective [31–33]. In contrast, *C. burnetii* effectors manipulate a number of host processes to remodel the CCV and prevent host cell death allowing for its intracellular lifestyle with many mutants being highly attenuated [34].

16.6 Immune Evasion Strategies

Pathogens are initially recognized by the innate immune system through pattern-recognition receptors (PRRs) on the surface of host immune cells, which recognize unique pathogen-associated molecular patterns (PAMPs). Bacterial pathogens use one of two strategies to evade the immune system, either by overt or by stealth attacks [35]. *C. burnetii* is a prime example of a stealth pathogen due to its ability to invade and replicate within macrophages with minimal stimulation of innate immune responses. Many of the mechanisms that *C. burnetii* uses to evade the innate immune system remain to be determined, but new research is shedding light on several of these strategies. The activation of immune responses is complicated by the fact that most laboratories use the LPS phase II variant (RSA439) since it can be used at BL2, yet this clone is distinct from virulent LPS phase I variants. For clarity in this review, immune evasion strategies will be restricted to a description of responses that occur after stimulation with virulent full-length LPS variants. Many studies have been performed using *in vitro* culture systems and determined that although *C. burnetii* is pro-inflammatory, the response is not as robust as against most pathogens. For example *C. burnetii* stimulates an atypical activation of macrophages after infection to an M2 state that results in lower secretion of several pro-inflammatory cytokines [36]. In addition, low levels of IL-1β are released after *C. burnetii* infection and recently a T4BSS effector, IcaA, was shown to prevent inflammasome activation, suggesting that *C. burnetii* inhibits this innate immune response [37, 38]. The most well-defined interactions of *C. burnetii* with the innate immune system that result in cytokine release are with toll-like receptor (TLR) 2 and TLR4; however, the role of these receptors during infection remains incompletely understood. TLR2 is a surface PRR on immune cells including macrophages that recognize lipoproteins on the surface of Gram-negative and Gram-positive bacteria [39]. Gram-negative bacteria are also recognized through their lipid A structure in LPS by TLR4 on the surface of innate immune cells, particularly dendritic cells (DCs) and macrophages [39]. Both TLR2 and TLR4 can induce an inflammatory response by signaling through MyD88, whereas TLR4 can also cause a type I interferon response through TRIF signaling [39].

Studies using purified lipid A from *C. burnetii* demonstrated that it is a dose-dependent antagonist of *E. coli* TLR4 LPS activation and these results are supported by the structural prediction of a tetra-acylated lipid A chain [40]. On the other hand, *C. burnetii* LPS is able to cause secretion of TNF-α in mouse peritoneal macrophages that is associated with expression of TLR4 [41]. Additionally TLR4 is involved in stimulation of antibody response as the kinetics of antibody production are delayed in TLR4-/- mice when compared to wild-type mice after infection with *C. burnetii* indicating that signaling occurs through TLR4 during infection [42]. Therefore, *C. burnetii* LPS may antagonize the effect of endotoxic LPS while still stimulating a response through TLR4. However, even though there is an association with TLR4, *C. burnetii* infected TRL4-/- mice are comparable in susceptibility to wild-type mice. In addition, polymorphisms in TLR4 in humans associated with more severe disease outcomes in multiple infections have no increased severity during acute Q fever [42–44]. The ability to antagonize the signaling of *E. coli* LPS through TLR4 suggests that even though *C. burnetii* does provide signaling through TLR4, the response does not mimic default signaling.

In vitro data suggest that *C. burnetii* LPS does not signal through TLR2 but that full-length LPS may shield TLR2 stimulation, while *in vivo* studies found that TLR2 is important for limiting bacterial growth in the lungs of mice infected with *C. burnetii* [43, 45]. In addition, peripheral blood mononuclear cells (PBMCs) from humans stimulated with *C. burnetii* secrete IL-1β and TNF-α in a TLR2-dependent manner [46]. However, once again polymorphisms in human TLR2 that are associated with more severe infections had no effect on acute Q fever [44]. In addition to signaling through TLR2 and TLR4, other TLRs are likely involved in the innate signaling as MyD88-deficient mice have higher bacterial loads then either TLR2- or TLR4-deficient mice [43]. Although animal experiments suggest that TLR2 may be important for controlling the infection, the phenotype is slight and therefore more research is required to completely determine what the outcomes are due to signaling or evasion through PRRs.

16.7 Qvax® and Correlates of Protective Immunity

Qvax® (CSL Biotherapies, Australia), a formalin-inactivated whole-cell phase I vaccine, is the only commercially available vaccine against *C. burnetii* and is only licensed for use in Australia. Qvax® provides long-lived protective immunity after a single vaccination [47]. A recent study demonstrated that long-term protection is conferred after vaccination lasting for at least 10 years in humans [48]. Potential vaccines must undergo serological analysis and skin-testing prior to vaccination as prior exposure to *C. burnetii* is associated with adverse side effects after vaccination. However, this strategy is costly and might not completely prevent adverse effects. Qvax® is not approved the use in the United States; FDA approval would be unlikely because of adverse reactions, which usually include sterile abscess at the site of injection, and occasional systemic responses [49]. The increase in the number of Q fever cases among war fighters renewed interest in the development of a safe and efficacious vaccine for use in the United States [50]. It is predicted that an FDA licensure of a new-generation vaccine would require immunity levels comparable to Qvax® without potential for adverse reactions in previously sensitize individuals. To develop such a vaccine, a comprehensive understanding of protective immunity and drivers of the adverse reaction to whole cells vaccine will be essential.

The components of immunity that confer protection have been dissected in mouse and guinea pig animal models of human Q fever. T cells have been demonstrated as essential for clearing infection in a number of studies. Andoh et al. (2007) found that three different mouse strains with T cell deficiencies were unable to control infection as they showed severe clinical disease and had large bacterial loads in their spleens [51]. Depletion of CD4+ or CD8+ T cells in mice or adoptive transfer of either into SCID mice was able to control pulmonary infection caused by *C. burnetii* [52]. The cell-mediated immunity induced by natural infection is thought to be dependent on the production of IFN-γ by T-cells [51]. The activation of infected *C. burnetii* macrophages with IFN-γ results in the inhibition of *C. burnetii* growth due to the production of reactive nitrogen and oxygen intermediates (RNI and ROI, respectively) [53]. In addition, IFN-γ-deficient mice are unable to control infection with *C. burnetii* [51]. On the other hand, vaccine-induced immunity in mice also causes a robust cell-mediated immune response that can be adoptively transferred and confer protection in naïve animals [54]. However, unlike the protection induced by nature infection, which is completely dependent on T cells responses vaccine-induced protection from formalin inactivated *C. burnetii* includes responses from B cells [51, 54].

Adoptive transfer of sera from vaccinated mice conferred a significant protection in immune-competent animals indicating that antibodies play a role in mediating protection [54]. The role of antibody in vaccine-induced protection has been documented in multiple studies over the years starting with work by Abinanti and Marmion (1957), which demonstrated that a mixture of antibodies and *C. burnetii* was not infectious in susceptible animals [55]. A second study found that IgM from a *C. burnetii*-sensitized individual was able to suppress the growth of *C. burnetii* in mouse spleens when mixed with the organism prior to infection [56]. Interestingly, although Ab-opsonized *C. burnetii* were able to stimulate maturation and cytokine production in dendritic cells (DCs) there was no decrease the ability of the opsonized *C. burnetii* to survive within phagocytes [57]. In addition, adoptive transfer of immune sera was just as protective in FcR-/- mice as wild-type mice indicating that antibody mediated protection does not require FcR receptor functions [57]. The complement system does not appear essential in antibody-mediated protection since passive transfer of immune sera to complement-deficient mice conferred the same level of protection as in wild-type mice [57]. Therefore, although antibody provides protection to *C. burnetii* infection, the mechanism of action appears unresolved. In conclusion, both humoral and cell-mediated immunity seem to contribute to the control *C. burnetii* infections where vaccine-induced antibodies are likely involved in prevention of clinical disease while cell-mediated immunity required for clearance.

16.8 Nanovaccines

The remarkable success of vaccines over the past century in limiting and, in some cases, eradicating numerous debilitating diseases can be attributed to the generation of acquired immunity. When an individual is exposed to a pathogen, an immune response consisting of humoral (antibodies) and cellular ($CD4^+$ and $CD8^+$ T cells) constituents is activated. Once triggered, antigen-specific B and T cell populations work to clear the pathogen using a variety of mechanisms. A fraction of these B and T cells then persist as memory cells to mount a faster, more efficient response should an individual encounter the same pathogen in the future [58, 59]. Using discrete components (subunits) of a pathogen, immune responses can be directed toward a specific antigen or a limited number of antigens. This type of response is predominantly humoral and often directed against a secreted gene product or a surface exposed antigen. This has been a successful vaccination development strategy in combating diseases such as tetanus, diphtheria, and anthrax whereby neutralizing antibodies raised against toxins secreted by these pathogens play an essential role in protective immunity [60–62]. Similarly, a key component of protection against *Streptococcus pneumoniae* and *Haemophilus influenzae* infections is the expansion of opsonizing antibodies raised against the respective capsular polysaccharides [63, 64]. In these instances where a pathogen has a predominantly extracellular lifestyle, generating a humoral response to a single antigen is particularly well suited since functional epitopes recognized by antibodies can inhibit essential functions. For intracellular pathogens, which are able to invade and replicate within host cells, the nature of protective immunity is less clearly defined and many of the vaccines designed against these pathogens were developed on an empirical basis using whole cell extracts from killed bacteria or attenuated strains (Table 16.1).

Intracellular pathogens are often equipped with a broad range of virulence mechanisms in order to survive within the inhospitable environments of the cell as describe above for *C. burnetii*. As such, a vaccine derived from a single antigen is less likely to be effective against intracellular pathogens. Moreover, the intracellular lifestyle of these pathogens means they can remain undetected by antibodies. While mounting a robust

Table 16.1 Summary of human diseases caused by intracellular pathogens alongside corresponding treatment and prophylactic options currently available

Pathogen	Disease	Residing cellular compartment	Treatment (duration)	Vaccine (type)
Bacteria				
Coxiella burnetii	Q fever	CCV (modified phagolysosome)	Acute: doxycycline (3 weeks) Chronic: doxycycline and hydroxychloroquine (18 months)	Qvax® (killed whole cell)—not approved outside of Australia
Mycobacterium tuberculosis	Tuberculosis	MPV (modified phagolysosome)	Isoniazid, rifampicin, ethambutol, and pyrazinamide (2 months) followed by isoniazid and rifampicin alone (4 months)	BCG (attenuated whole cell)— limited efficacy
Burkholderia pseudomallei	Melioidosis	Cytosol	Ceftazadine (2 weeks) followed by doxycycline (6 months)	None available
Salmonella enterica serovars Typhi	Typhoid fever	SCV (modified phagosome)	Ciprofloxacin (2 weeks)	Ty21a (attenuated whole cell) or Vi (subunit)
Listeria monocytogenes	Listeriosis	Cytosol	Ampicillin and gentamicin (2–6 weeks)	None available
Legionella pneumophila	Legionnaires disease	LCV (modified phagosome)	Azithromycin and ciprofloxacin (3 weeks)	None available
Virus				
HIV	AIDS	Endosome	Anti-retroviral therapy (indefinitely)	None available
Herpes simplex	Oral/genital herpes	Intranuclear replication compartments	Acyclovir, famciclovir, valacyclovir (2–3 weeks)	None available

(*Continued*)

Table 16.1 (*Continued*)

Pathogen	Disease	Residing cellular compartment	Treatment (duration)	Vaccine (type)
HCV	Hepatitis C	Endoplasmic reticulum	Ledipasvir and sofosbuvir (12 weeks)	None available
Fungus				
Aspergillus fumigatus	Aspergillosis	Phagosome	Itraconazole (24 weeks)	None available
Candida albicans	Candidiasis	Phagosome	Fluconazole (16 weeks)	None available
Parasite *Leishmania* **spp.**	Leishmaniasis	Parasitophorous vacuole	Sodium stibogluconate (3 weeks) and amphotericin B (8 weeks)	None available

antibody response against intracellular pathogens may play a role in controlling disease during the early stages of infection, it is generally believed that a strong cellular immune response is essential for protection against intracellular pathogens [65]. One of the difficulties in generating cellular immune responses is that soluble antigens do not readily activate this pathway for recognition by $CD8^+$ T cells. Antigen recognition by T cells occurs when an antigen-presenting cell (APC) encounters an antigen either within its cytosol, where it is degraded by the proteasome, or in the surrounding environment where it is engulfed and degraded via lysosomal fusion [66–69]. Processed antigens are then transported onto Major Histocompatibility Complex (MHC) I or II for presentation on the surface of APCs to $CD8^+$ and $CD4^+$ T cells, respectively [70–72]. $CD4^+$ T cells can be characterized further by their differentiation into functional subsets termed either T helper 1 (Th1) or T helper 2 (Th2) cells which is governed by the cytokine environment in which they mature [65, 73, 74]. Th1 cells are believed to play a vital role in driving protective cellular immunity against intracellular pathogens by inducing a pro-inflammatory response with the secretion of interleukin (IL)-2, interferon (IFN)-γ and tumor necrosis factor (TNF)-α [75, 76]. Additionally, a fraction of Th1 cells will persist as $CD4^+$ memory T cells, providing a quantitatively enhanced response upon secondary exposure [77]. Both $CD8^+$ and $CD4^+$ T cells have been shown to play essential roles in protective immunity against many intracellular pathogens, including *Mycobacterium tuberculosis*, *B. pseudomallei*, and *C. burnetii* [78–80].

In some instances, stimulatory adjuvants PAMPs such as unmethylated cytosine-guanine dinucleotide (CpG) found in bacterial DNA and monophosphoryl lipid A (MPLA) are co-administered with vaccines to potentiate the immune response toward a Th1 bias. These compounds have been characterized by their ability to activate co-stimulatory signals or intracellular signaling pathways. For MPLA, its major receptor in humans has been identified as TLR2 [81]. This triggers the release of pro-inflammatory cytokines via nuclear factor kappa-light-chain-enhancer of activated B cells (NF-κB), an important feature of the innate immune response [82]. Alternatively, CpG has been identified as binding to TLR9, which releases TNF-α and IL-12 through the expression of co-stimulatory molecules to induce a potent Th1 response [83, 84]. Cytokines themselves can also be used directly as an adjuvant, the most extensively studied being IL-2, IL-12 and IFNγ. However, problems with stability and a relatively high manufacturing cost mean that the use of cytokines for routine vaccination is unlikely [85, 86]. With a raised awareness of BSATs, as well as clinically relevant pathogens such as *M. tuberculosis* and *Salmonella enterica*, there is a clearly defined niche for the development of next generation vaccines that are not only well tolerated and cost effective but also able to provide robust protection against intracellular pathogens through the stimulation of cellular immunity.

Nanoparticles come in all shapes, sizes, and composites, including precious metals, synthetic polymers and viral expression systems. Among some of the most extensively studied within the scope of nanomedicine are gold NPs. This is due to their biocompatibility, low-toxicity, unique light-scattering properties, and pre-existing license for human use [87–89]. Furthermore, the size of gold NPs can be easily controlled to synthesize monodisperse nanoparticles by reducing chloroauric acid ($H(AuCl_4)$) with various reducing agents (Fig. 16.1)—a method first published by Turkevich et al. in 1952 [90]. The gold surface readily forms a strong covalent bond with thiol groups, allowing for surface modification with thiol-containing linker molecules or immunological stimulants. Generating monodisperse particles with a uniform shape is also essential for maintaining antigen loading consistency between batches [91]. NPs for use as vaccine delivery platforms have also been produced from various polymers, including poly(α-hydroxy acids), poly(amino acids) or polysaccharides [92–94]. In most cases, polymeric NPs are designed to encapsulate an antigen that can be used to extend the exposure period of antigens to host cells by slowing degradation rate of labile proteins [92, 95]. Polymer NPs are often biodegradable themselves, which is an attractive property given the concern over nanotoxicity in the environment [96–98]. As an alternative to synthetic NPs, virus-like particles (VLP) can be engineered using a baculovirus expression system to express recombinant antigenic proteins of choice [99]. Unlike viruses, VLPs assemble without any viral RNA, meaning they are non-replicating and non-infectious. Despite this, VLPs often elicit strong immune responses [100, 101]. VLPs expressing the human papillomavirus (HPV) L1 protein have been shown to produce strong humoral and cellular immune responses in both animal models and human clinical trials [102, 103].

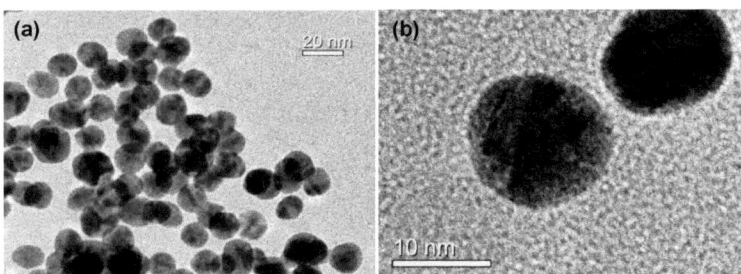

Figure 16.1 Transmission electron micrograph of 15 nm gold nanoparticles formed using a modified Turkevich method of citrate-reduced gold(III) chloride trihydrate. Image taken using a JEOL JEM-2010 transmission electron microscope at 100,000× magnification (a) and 400,000× magnification (b).

Antigen recognition by APCs is essential for generating antigen-specific B and T cell responses. In using NPs to deliver antigens, the efficacy of uptake into dendritic cells can be significantly increased compared with soluble antigen alone. This can be as high as a 30-fold increase in antigen uptake [104, 105].

For gold NPs, the size of particles alone is enough to influence the degree of cellular uptake. Researchers measuring intracellular gold content within HeLa cells after incubating with a range of NP sizes (14–100 nm) found a 50 nm particle was optimal for cellular uptake [106]. The shape and surface charge of a NP also play a pivotal role in the interactions between APCs and NPs. By modifying NPs with different poly(amino acids)/proteins, lipids, or polymers to adjust the surface charge, intracellular uptake is markedly increased in a variety of different cell types [107–111]. In general, cationic particles are more readily internalized due to the anionic nature of cell membranes. Additionally, spherical particles are more readily endocytosed than rod-shaped particles, which the investigators suggested could be due to the curvature of the NP reducing the number of available receptor sites for binding [106, 112]. In order to facilitate antigen uptake into APCs further, NPs functionalized with ligands complementary to APC receptors can specifically target these cell types to induce the desired immune response. Some of the most common targets for this purpose include mannose receptors, Fc receptors, CD11c-CD18 integrins and MHC receptors [113–118]. In doing so, the efficacy and tolerance of these vaccines is significantly improved by increasing their uptake and minimizing some of the less desirable, heterologous effects of vaccination.

The mechanisms by which antigens are taken up and subsequently processed by cells also influences the type of immune response induced [119, 120]. NPs made from either polymer or gold have been shown to enter various mammalian cell types via endosomal and non-endosomal pathways [121–124]. By entering cells via non-endosomal pathways, antigens are able to enter the proteasomal degradation pathway for the activation of $CD8^+$ T cells. A study based on pulmonary vaccination of C57Bl/6 mice with an antigen conjugated to 30 nm poly(propylene sulfide) (PPS) NPs demonstrated greater frequencies of antigen-specific $CD8^+$ T cells in the blood and spleen than soluble (unconjugated) antigen. This was also accompanied by a potent Th17 cytokine profile in $CD4^+$ T cells. Mice evaluated 50 days post vaccination were found to maintain higher levels of antigen-specific CD8+ effector memory T cells, providing long lasting cellular immunity [125]. In other cases, NPs may directly interact with components of the immune system to provoke an alternate response.

Using NPs synthesized from the copolymer poloaxmer, Reddy and colleagues were able to demonstrate the activation of complement though the alternative pathway. It was proposed that this occurs via a multi-step process resulting in terminal hydroxy groups binding to the exposed thioester of C3b [126]. As with many NP vaccine formulations, this is another example of how NPs are able to behave as a delivery platform while also providing an adjuvantal response during immunization.

Subunit vaccines offer the desirable benefits of generating an immune response toward a specific antigen without any of the adverse effects associated with killed whole cell or attenuated mutant vaccines. Unfortunately, these antigens are often poor immunogens and require an adjuvant to boost the immune response. In using NPs to deliver subunit vaccines, it is possible to increase the immunological response to an antigen due to an enhanced uptake by APCs and/or stimulation of innate immune pathways. Moreover, antigens encapsulated or decorated on NPs are presented to cells in a way that mimics their delivery during a natural infection and may facilitate a more appropriate immune response.

16.9 Why Are Nanoparticle Vaccines a Good Strategy for *C. burnetii*?

Studies into the correlates of protection from the formalin inactivated whole cell vaccine indicate that antibodies to LPS are an important component of protection against *C. burnetii* [54]. This, coupled with the fact that the lipid A molecule is a weak stimulator of TLR4, makes conjugation to nanoparticles a good strategy for an alternative vaccine [40]. Additionally, *C. burnetii* antigens could be packaged into degradable NPs for cytosolic degradation and MHCI presentation to CD8$^+$ T cells. This is a promising strategy since several CD8+ stimulating peptides have already been identified in *C. burnetii* as well as *in vivo* mouse infection studies indicating a crucial role for CD8$^+$ T cells in clearing infection [52, 127]. A strong Th1 response is also a key component of protection against *C. burnetii*; therefore, the identified CD4+ T cell stimulating *C. burnetii* antigens could be combined with NPs that are able to induce Th1 cytokines via exosome formation [128–130]. Since a single antigen is unlikely to confer protection, one strategy might be to conjugate several subunits onto the surface of NPs for presentation to APCs. This review has demonstrated that in using NPs to deliver vaccines, it is possible to tailor the immune response against an antigen so that an appropriate response, specific to the target pathogen, is generated.

Disclosures and Conflict of Interest

The authors declare that they have no conflict of interest and have no affiliations or financial involvement with any organization or entity discussed in this chapter. This includes employment, consultancies, honoraria, grants, stock ownership or options, expert testimony, patents (received or pending) or royalties. No writing assistance was utilized in the production of this chapter and the authors have received no payment for its preparation.

Corresponding Author

Dr. James E. Samuel
Department of Microbial Pathogenesis and Immunology
College of Medicine Texas A&M University
3107 Medical Research and Education Building
8447 State Hwy 47, Bryan, TX 77807, USA
Email: jsamuel@medicine.tamhsc.edu

About the Authors

Erin J. van Schaik is a research scientist in Prof. James E. Samuel's laboratory, which works on *C. burnetii* pathogenesis and vaccine development. She received her PhD from the University of Alberta, Canada, working on *Pseudomonas aeruginosa* in Dr. Randy Irvin's laboratory and then went on to do a postdoctoral fellowship in Dr. Don Woods' laboratory at the University of Calgary, Canada, working on *Burkholderia pseudomallei*.

Anthony E. Gregory is a postdoctoral research associate in Prof. James E. Samuel's laboratory, which works on *C. burnetii* vaccine development. He received his PhD from the University of Exeter, UK, in Dr. Richard Titball's laboratory, developing nanoparticle-based vaccines against *Burkholderia* spp., *Yersinia pestis*, and *Streptococcus pneumoniae*.

Gerald F. Audette is an associate professor of chemistry and a member of the Centre for Research on Biomolecular Interactions at York University, Canada. He conducted his doctoral training with Drs. Louis Delbaere and J. Wilson Quail, focusing on the elucidation of protein–carbohydrate interactions occurring during blood group recognition using high-resolution X-ray crystallography. His postdoctoral research was with Drs. Bart Hazes and Laura Frost at the University of Alberta, exploring the correlation between protein structure and biological activity of type IV pilins from *P. aeruginosa* and the type 4 secretion system from the conjugative F-plasmid of *E. coli*. His current research interests include structure/function studies of proteins involved in bacterial conjugation systems, as well as the type IV pilins and assembly systems from several bacterial pathogens, including *C. burnetii*. Dr. Audette is a coeditor of volumes 1–3 of Pan Stanford's Series on Clinical Nanomedicine and a subject editor of structural chemistry and crystallography for the journal *FACETS*.

James E. Samuel is a Regents Professor and head of the Department of Microbial Pathogenesis and Immunology and the Wofford Cain Endowed Chair of Infectious Disease in the College of Medicine at Texas A&M University, USA. He trained with Dr. Lou Mallavia working on *Coxiella burnetii* at Washington State University and then was postdoctoral fellow in Dr. Allison O'Brien's laboratory at Uniformed Services University of the Health Sciences in Bethesda, MD. He has worked for over 35 years with dual focus defining molecular pathogenic mechanisms and vaccine and diagnostic strategies against Q fever.

References

1. World Health Organization (2016). Global Health Observatory (DHO) Data. Available at: http://www.who.int/gho/en/ (accessed December 6, 2016).
2. Gruenberg, J., van der Goot, F. G. (2006). Mechanisms of pathogen entry through the endosomal compartments. *Nat. Rev. Mol. Cell Biol.*, **7**, 495–504.
3. Kumar, Y., Valdivia, R. H. (2009). Leading a sheltered life: Intracellular pathogens and maintenance of vacuolar compartments. *Cell Host Microbe*, **5**, 593–601.
4. McCaul, T. F., Williams, J. C. (1981). Developmental cycle of Coxiella burnetii: Structure and morphogenesis of vegetative and sporogenic differentiations. *J. Bacteriol.*, **147**, 1063–1076.
5. Sandoz, K. M., Sturdevant, D. E., Hansen, B., Heinzen, R. A. (2014). Developmental transitions of Coxiella burnetii grown in axenic media. *J. Microbiol. Methods*, **96**, 104–110.
6. Coleman, S. A., Fischer, E. R., Howe, D., Mead, D. J., Heinzen, R. A. (2004). Temporal analysis of Coxiella burnetii morphological differentiation. *J. Bacteriol.*, **186**, 7344–7352.
7. Williams, J. C. (1991). Infectivity, virulence, and pathogenicity of Coxiella burnetti for various hosts. In: *Q Fever: The Biology of Coxiella Burnetii*. CRC Press, Boca Raton, FL, pp. 21–71.
8. Sandoz, K. M., Popham, D. L., Beare, P. A., Sturdevant, D. E., Hansen, B., Nair, V., et al. (2016). Transcriptional profiling of *Coxiella burnetii* reveals extensive cell wall remodeling in the small cell variant developmental form. *PLoS One*, **11**, e0149957.
9. Maurin, M., Raoult, D. (1999). Q Fever. *Clin. Microbiol. Rev.*, **12**, 518–553.
10. van den Brom, R., van Engelen, E., Roest, H. I., van der Hoek, W., Vellema, P. (2015). Coxiella burnetii infections in sheep or goats: An opinionated review. *Vet. Microbiol.*, **181**, 119–129.
11. Raoult, D., Marrie, T., Mege, J. (2005). Natural history and pathophysiology of Q fever. *Lancet Infect. Dis.*, **5**, 219–226.
12. Marrie, T. J. (2010). Q fever pneumonia. *Infect. Dis. Clin. North Am.*, **24**, 27–41.
13. Calza, L., Attard, L., Manfredi, R., Chiodo, F. (2002). Doxycycline and chloroquine as treatment for chronic Q fever endocarditis. *J. Infect.*, **45**, 127–129.

14. Million, M., Thuny, F., Richet, H., Raoult, D. (2010). Long-term outcome of Q fever endocarditis: A 26-year personal survey. *Lancet Infect. Dis.*, **10**, 527–535.
15. Howe, D., Shannon, J. G., Winfree, S., Dorward, D. W., Heinzen, R. A. (2010). Coxiella burnetii phase I and II variants replicate with similar kinetics in degradative phagolysosome-like compartments of human macrophages. *Infect. Immun.*, **78**, 3465–3474.
16. Moos A, Hackstadt T. (1987). Comparative virulence of intra- and interstrain lipopolysaccharide variants of Coxiella burnetii in the guinea pig model. *Infect. Immun.*, **55**, 1144–1150.
17. Hoover, T. A., Culp, D. W., Vodkin, M. H., Williams, J. C., Thompson, H. A. (2002). Chromosomal DNA deletions explain phenotypic characteristics of two antigenic variants, phase II and RSA 514 (crazy), of the Coxiella burnetii nine mile strain. *Infect. Immun.*, **70**, 6726–6733.
18. Vishwanath, S., Hackstadt, T. (1988). Lipopolysaccharide phase variation determines the complement-mediated serum susceptibility of Coxiella burnetii. *Infect. Immun.*, **56**, 40-44.
19. Hackstadt, T. (1988). Steric hindrance of antibody binding to surface proteins of Coxiella burnetti by phase I lipopolysaccharide. *Infect. Immun.*, **56**, 802–807.
20. Levin, R., Grinstein, S., Canton, J. (2016). The life cycle of phagosomes: Formation, maturation, and resolution. *Immunol. Rev.*, **273**, 156–179.
21. Case, E. D. R., Samuel, J. E. (2016). Contrasting lifestyles within the host cell. *Microbiol. Spectr.*, **4**, doi:10.1128/microbiolspec.VMBF-0014-2015.
22. Baca, O. G., Klassen, D. A., Aragon, A. S. (1993). Entry of Coxiella burnetii into host cells. *Acta Virol.*, **37**, 143–155.
23. Kinchen, J. M., Ravichandran, K. S. (2008). Phagosome maturation: Going through the acid test. *Nat. Rev. Mol. Cell Biol.*, **9**, 781–795.
24. Omsland, A., Cockrell, D. C., Howe, D., Fischer, E. R., Virtaneva, K., Sturdevant, D. E., et al. (2009). Host cell-free growth of the Q fever bacterium Coxiella burnetii. *Proc. Natl. Acad. Sci. U. S. A.*, **106**, 4430–4434.
25. Newton, H. J., McDonough, J. A., Roy, C. R. (2013). Effector protein translocation by the *Coxiella burnetii* Dot/Icm type IV secretion system requires endocytic maturation of the pathogen-occupied vacuole. *PLoS One*, **8**, e54566.

26. Howe, D., Melnicáková J, Barák, I., Heinzen, R. A. (2003). Fusogenicity of the Coxiella burnetii parasitophorous vacuole. *Ann. N. Y. Acad. Sci.*, **990**, 556–562.

27. Carey, K. L., Newton, H. J., Lührmann, A., Roy, C. R. (2011). The *Coxiella burnetii* Dot/Icm system delivers a unique repertoire of type IV effectors into host cells and is required for intracellular replication. *PLoS Pathog.*, **7**, e1002056.

28. Nagai, H., Kubori, T. (2011). Type IVB secretion systems of Legionella and other gram-negative bacteria. *Front. Microbiol.*, **2**, 136.

29. Zamboni, D. S., McGrath, S., Rabinovitch, M., Roy, C. R. (2003). Coxiella burnetii express type IV secretion system proteins that function similarly to components of the Legionella pneumophila Dot/Icm system. *Mol. Microbiol.*, **49**, 965–976.

30. Beare, P. A., Gilk, S. D., Larson, C. L., Hill, J., Stead, C. M., Omsland, A., et al. (2011). Dot/Icm type IVB secretion system requirements for Coxiella burnetii growth in human macrophages. *mBio*, **2**, e00175-11.

31. Weber, M. M., Chen, C., Rowin, K., Mertens, K., Galvan, G., Zhi, H., et al. (2013). Identification of Coxiella burnetii type IV secretion substrates required for intracellular replication and coxiella-containing vacuole formation. *J. Bacteriol.*, **195**, 3914–3924.

32. Larson, C. L., Beare, P. A., Voth, D. E., Howe, D., Cockrell, D. C., Bastidas, R. J., et al. (2015). Coxiella burnetii effector proteins that localize to the parasitophorous vacuole membrane promote intracellular replication. *Infect. Immun.*, **83**, 661–670.

33. Martinez, E., Cantet, F., Fava, L., Norville, I., Bonazzi, M. (2014). Identification of OmpA, a *Coxiella burnetii* protein involved in host cell invasion, by multi-phenotypic high-content screening. *PLoS Pathog.*, **10**, e1004013.

34. van Schaik, E. J., Chen, C., Mertens, K., Weber, M. M., Samuel, J. E. (2013). Molecular pathogenesis of the obligate intracellular bacterium Coxiella burnetii. *Nat. Rev. Microbiol.*, **11**, 561–573.

35. Merrell, D. S., Falkow, S. (2004). Frontal and stealth attack strategies in microbial pathogenesis. *Nature,* **430**, 250–256.

36. Benoit, M., Barbarat, B., Bernard, A., Olive, D., Mege, J.-L. (2008). Coxiella burnetii, the agent of Q fever, stimulates an atypical M2 activation program in human macrophages. *Eur. J. Immunol.*, **38**, 1065–1070.

37. Graham, J. G., MacDonald, L. J., Hussain, S. K., Sharma, U. M., Kurten, R. C., Voth, D. E. (2013). Virulent Coxiella burnetii pathotypes productively

infect primary human alveolar macrophages. *Cell. Microbiol.*, **15**, 1012–1025.

38. Cunha, L. D., Ribeiro, J. M., Fernandes, T. D., Massis, L. M., Khoo, C. A., Moffatt, J. H., et al. (2015). Inhibition of inflammasome activation by Coxiella burnetii type IV secretion system effector IcaA. *Nat. Commun.*, **6**, 10205.

39. Kawai, T., Akira, S. (2010). The role of pattern-recognition receptors in innate immunity: Update on Toll-like receptors. *Nat. Immunol.*, **11**, 373–384.

40. Zamboni, D. S., Campos, M. A., Torrecilhas, A. C. T., Kiss, K., Samuel, J. E., Golenbock, D. T., et al. (2004). Stimulation of Toll-like receptor 2 by Coxiella burnetii is required for macrophage production of pro-inflammatory cytokines and resistance to infection. *J. Biol. Chem.*, **279**, 54405–54415.

41. Kubes, M., Kuzmová, Z., Gajdosová, E., Ihnatko, R., Mucha, V., Toman, R., et al. (2006). Induction of tumor necrosis factor alpha in murine macrophages with various strains of Coxiella burnetii and their lipopolysaccharides. *Acta Virol.*, **50**, 93–99.

42. Honstettre, A., Ghigo, E., Moynault, A., Capo, C., Toman, R., Akira, S., et al. (2004). Lipopolysaccharide from Coxiella burnetii is involved in bacterial phagocytosis, filamentous actin reorganization, and inflammatory responses through Toll-like receptor 4. *J. Immunol.*, **172**, 3695–3703.

43. Ramstead, A. G., Robison, A., Blackwell, A., Jerome, M., Freedman, B., Lubick, K. J., et al. (2016). Roles of TLR2, TLR4, and MyD88 during pulmonary Coxiella burnetii infection. *Infect. Immun.*, **84**, 940–949.

44. Everett, B., Cameron, B., Li, H., Vollmer-Conna, U., Davenport, T., Hickie, I., et al. (2007). Polymorphisms in Toll-like receptors-2 and -4 are not associated with disease manifestations in acute Q fever. *Genes Immun.*, **8**, 699–702.

45. Shannon, J. G., Howe, D., Heinzen, R. A. (2005). Virulent Coxiella burnetii does not activate human dendritic cells: Role of lipopolysaccharide as a shielding molecule. *Proc. Natl. Acad. Sci. U. S. A.*, **102**, 8722–8727.

46. Ammerdorffer, A., Schoffelen, T., Gresnigt, M., Oosting, M., den Brok, M., Abdollahi Roodsaz, S., et al. (2015). Recognition of Coxiella burnetii by Toll-like receptors and nucleotide-binding oligomerization domain-like receptors. *J. Infect. Dis.*, **211**, 978–987.

47. Marmion, B. P., Ormsbee, R. A., Kyrkou, M., Wright, J., Worswick, D. A., Izzo, A. A., et al. (1990). Vaccine prophylaxis of abattoir-associated

Q fever: Eight years' experience in Australian abattoirs. *Epidemiol. Infect.*, **104**, 275–287.

48. Kersh, G. J., Fitzpatrick, K. A., Self, J. S., Biggerstaff, B. J., Massung, R. F. (2013). long-term immune responses to Coxiella burnetii after vaccination. *Clin. Vaccine Immunol.*, **20**, 129–133.

49. Marmion, B. (2007). Q fever: The long journey to control by vaccination. *Med. J. Aust.*, **186**, 164–166.

50. Royal, J., Riddle, M. S., Mohareb, E., Monteville, M. R., Porter, C. K., Faix, D. J. (2013). Seroepidemiologic survey for Coxiella burnetii among US military personnel deployed to Southwest and Central Asia in 2005. *Am. J. Trop. Med. Hyg.*, **89**, 991–995.

51. Andoh, M., Zhang, G., Russell-Lodrigue, K. E., Shive, H. R., Weeks, B. R., Samuel, J. E. (2007). T cells are essential for bacterial clearance, and gamma interferon, tumor necrosis factor alpha, and B cells are crucial for disease development in Coxiella burnetii infection in mice. *Infect. Immun.*, **75**, 3245–3255.

52. Read, A. J., Erickson, S., Harmsen, A. G. (2010). Role of CD4(+) and CD8(+) T cells in clearance of primary pulmonary infection with Coxiella burnetii. *Infect. Immun.*, **78**, 3019–3026.

53. Brennan, R. E., Russell, K., Zhang, G., Samuel, J. E. (2004). Both inducible nitric oxide synthase and NADPH oxidase contribute to the control of virulent phase I Coxiella burnetii Infections. *Infect. Immun.*, **72**, 6666–6675.

54. Zhang, G., Russell-Lodrigue, K. E., Andoh, M., Zhang, Y., Hendrix, L. R., Samuel, J. E. (2007). Mechanisms of vaccine-induced protective immunity against Coxiella burnetii infection in BALB/c mice. *J. Immunol.*, **179**, 8372–8380.

55. Abinanti, F. R., Marmion, B. P. (1957). Protective or neutralizing antibody in Q fever. *Am. J. Hyg.*, **66**, 173–195.

56. Peacock, M. G., Fiset, P., Ormsbee, R. A., Wisseman, C. L. (1979). Antibody response in man following a small intradermal inoculation with Coxiella burnetii phase I vaccine. *Acta Virol.*, **23**, 73–81.

57. Shannon, J. G., Cockrell, D. C., Takahashi, K., Stahl, G. L., Heinzen, R. A. (2009). Antibody-mediated immunity to the obligate intracellular bacterial pathogen Coxiella burnetii is Fc receptor- and complement-independent. *BMC Immunol.*, **10**, 26.

58. Kaech, S. M., Wherry, E. J., Ahmed, R. (2002). Effector and memory T-cell differentiation: Implications for vaccine development. *Nat. Rev. Immunol.*, **2**, 251–262.

59. Kurosaki, T., Kometani, K., Ise, W. (2015). Memory B cells. *Nat. Rev. Immunol.*, **15**, 149–159.
60. Miyaji, E. N., Mazzantini, R. P., Dias, W. O., Nascimento, A. L., Marcovistz, R., Matos, D. S., et al. (2001). Induction of neutralizing antibodies against diphtheria toxin by priming with recombinant Mycobacterium bovis BCG expressing CRM197, a mutant diphtheria toxin. *Infect. Immun.*, **69**, 869–874.
61. Hanson, J. F., Taft, S. C., Weiss, A. A. (2006). Neutralizing antibodies and persistence of immunity following anthrax vaccination. *Clin. Vaccine Immunol.*, **13**, 208–213.
62. Boucher, P., Sato, H., Sato, Y., Locht, C. (1994). Neutralizing antibodies and immunoprotection against pertussis and tetanus obtained by use of a recombinant pertussis toxin-tetanus toxin fusion protein. *Infect. Immun.*, **62**, 449–456.
63. Amir, J., Scott, M. G., Nahm, M. H., Granoff, D.. M. (1990). Bactericidal and opsonic activity of IgG1 and IgG2 anticapsular antibodies to Haemophilus influenzae type b. *J. Infect. Dis.*, **162**, 163–171.
64. Bardardottir, E., Jonsson, S., Jonsdottir, I., Sigfusson, A., Valdimarsson, H. (1990). IgG subclass response and opsonization of Streptococcus pneumoniae after vaccination of healthy adults. *J. Infect. Dis.*, **162**, 482–488.
65. Seder, R. A., Hill, A. V. (2000). Vaccines against intracellular infections requiring cellular immunity. *Nature*, **406**, 793–798.
66. Hewitt, E. W. (2003). The MHC class I antigen presentation pathway: Strategies for viral immune evasion. *Immunology*, **110**, 163–169.
67. Schubert, U., Anton, L. C., Gibbs, J., Norbury, C. C., Yewdell, J. W., Bennink, J. R. (2000). Rapid degradation of a large fraction of newly synthesized proteins by proteasomes. *Nature*, **404**, 770–774.
68. Roche, P. A., Furuta, K. (2015). The ins and outs of MHC class II-mediated antigen processing and presentation. *Nat. Rev. Immunol.*, **15**, 203–216.
69. Michelet, X., Garg, S., Wolf, B. J., Tuli, A., Ricciardi-Castagnoli, P., Brenner, M. B. (2015). MHC class II presentation is controlled by the lysosomal small GTPase, Arl8b. *J. Immunol.*, **194**, 2079–2088.
70. Daniels, M. A., Devine, L., Miller, J. D., Moser, J. M., Lukacher, A. E., Altman, J. D., et al. (2001). CD8 binding to MHC class I molecules is influenced by T cell maturation and glycosylation. *Immunity*, **15**, 1051–1061.

71. Novak, E. J., Liu, A. W., Nepom, G. T., Kwok, W. W. (1999). MHC class II tetramers identify peptide-specific human CD4+ T cells proliferating in response to influenza A antigen. *J. Clin. Invest.*, **104**, R63-7.
72. Cresswell, P. (1994). Assembly, transport, and function of MHC class II molecules. *Annu. Rev. Immunol.*, **12**, 259–293.
73. Scott, P. (1993). Selective differentiation of CD4+ T helper cell subsets. *Curr. Opin. Immunol.*, **5**, 391–397.
74. Plaut, M., Pierce, J. H., Watson, C. J., Hanley-Hyde, J., Nordan, R. P., Paul, W. E. (1989). Mast cell lines produce lymphokines in response to cross-linkage of Fc epsilon RI or to calcium ionophores. *Nature*, **339**, 64–67.
75. Mosmann, T. R., Cherwinski, H., Bond, M. W., Giedlin, M. A., Coffman, R. L. (1986). Two types of murine helper T cell clone. I. Definition according to profiles of lymphokine activities and secreted proteins. *J. Immunol.*, **136**, 2348–2357.
76. Lighvani, A. A., Frucht, D. M., Jankovic, D., Yamane, H., Aliberti, J., Hissong, B. D., et al. (2001). T-bet is rapidly induced by interferon-γ in lymphoid and myeloid cells. *Proc. Natl. Acad. Sci. U. S. A.*, **98**, 15137–15142.
77. Ahmed, R., Gray, D. (1996). Immunological memory and protective immunity: Understanding their relation. *Science*, **272**, 54–60.
78. Canaday, D. H., Wilkinson, R. J., Li, Q., Harding, C. V., Silver, R. F., Boom, W. H. (2001). CD4+ and CD8+ T cells kill intracellular Mycobacterium tuberculosis by a perforin and Fas/Fas ligand-independent mechanism. *J. Immunol.*, **167**, 2734–2742.
79. Jenjaroen, K., Chumseng, S., Sumonwiriya, M., Ariyaprasert, P., Chantratita, N., Sunyakumthorn, P., et al. (2015). T-cell responses are associated with survival in acute melioidosis patients. *PLoS Negl. Trop. Dis.*, **9**, e0004152.
80. Read, A. J., Erickson, S., Harmsen, A. G. (2010). Role of CD4+ and CD8+ T cells in clearance of primary pulmonary infection with Coxiella burnetii. *Infect. Immun.*, **78**, 3019–3026.
81. Yang, R. B., Mark, M. R., Gray, A., Huang, A., Xie, M. H., Zhang, M., et al. (1998). Toll-like receptor-2 mediates lipopolysaccharide-induced cellular signalling. *Nature*, **395**, 284–288.
82. Kirschning, C. J., Wesche, H., Merrill Ayres, T., Rothe, M. (1998). Human Toll-like receptor 2 confers responsiveness to bacterial lipopolysaccharide. *J. Exp. Med.*, **188**, 2091–2097.
83. Rutz, M., Metzger, J., Gellert, T., Luppa, P., Lipford, G. B., Wagner, H., et al. (2004). Toll-like receptor 9 binds single-stranded CpG-DNA in

a sequence- and pH-dependent manner. *Eur. J. Immunol.*, **34**, 2541–2550.

84. Sparwasser, T., Koch, E.-S., Vabulas, R. M., Heeg, K., Lipford, G. B., Ellwart, J. W., et al. (1998). Bacterial DNA and immunostimulatory CpG oligonucleotides trigger maturation and activation of murine dendritic cells. *Eur. J. Immunol.*, **28**, 2045–2054.

85. Barouch, D. H., Truitt, D. M., Letvin, N. L. (2004). Expression kinetics of the interleukin-2/immunoglobulin (IL-2/Ig) plasmid cytokine adjuvant. *Vaccine*, **22**, 3092–3097.

86. Heath, A., Playfair, J. (1992). Cytokines as immunological adjuvants. *Vaccine*, **10**, 427–434.

87. Mukherjee, P., Bhattacharya, R., Wang, P., Wang, L., Basu, S., Nagy, J. A., et al. (2005). Antiangiogenic properties of gold nanoparticles. *Clin. Cancer Res.*, **11**, 3530–3534.

88. Connor, E. E., Mwamuka, J., Gole, A., Murphy, C. J., Wyatt, M. D. (2005). Gold nanoparticles are taken up by human cells but do not cause acute cytotoxicity. *Small*, **1**, 325–327.

89. Cheung, J. M., Scarsbrook, D., Klinkhoff, A. .V. (2012). Characterization of patients with arthritis referred for gold therapy in the era of biologics. *J. Rheumatol.*, **39**, 716–719.

90. Turkevich, J., Stevenson, P. C., Hillier, J. (1951). A study of the nucleation and growth processes in the synthesis of colloidal gold. *Discuss. Faraday Soc.*, **11**, 55–75.

91. Frens, G. (1973). Controlled nucleation for the regulation of the particle size in monodisperse gold suspensions. *Nature,* **241**, 20–22.

92. Li, X., Deng, X., Yuan, M., Xiong, C., Huang, Z., Zhang, Y., et al. (2000). In vitro degradation and release profiles of poly-DL-lactide-poly(ethylene glycol) microspheres with entrapped proteins. *J. Appl. Polym. Sci.*, **78**, 140–148.

93. Lu, J. M., Wang, X., Marin-Muller, C., Wang, H., Lin, P. H., Yao, Q., et al. (2009). Current advances in research and clinical applications of PLGA-based nanotechnology. *Expert Rev. Mol. Diagn.*, **9**, 325–341.

94. O'Donnell, P. B., McGinity, J. W. (1997). Preparation of microspheres by the solvent evaporation technique. *Adv. Drug Delivery Rev.*, **28**, 25–42.

95. Petersen, L. K., Phanse, Y., Ramer-Tait, A. E., Wannemuehler, M. J., Narasimhan, B. (2012). Amphiphilic polyanhydride nanoparticles stabilize Bacillus anthracis protective antigen. *Mol. Pharmaceutics*, **9**, 874–882.

96. Maurer-Jones, M. A., Gunsolus, I. L., Murphy, C. J., Haynes, C. L. (2013). Toxicity of engineered nanoparticles in the environment. *Anal. Chem.*, **85**, 3036–3049.
97. Sajid, M., Ilyas, M., Basheer, C., Tariq, M., Daud, M., Baig, N., et al. (2015). Impact of nanoparticles on human and environment: Review of toxicity factors, exposures, control strategies, and future prospects. *Environ. Sci. Pollut. Res.*, **22**, 4122–4143.
98. Matranga, V., Corsi, I. (2012). Toxic effects of engineered nanoparticles in the marine environment: Model organisms and molecular approaches. *Mar. Environ. Res.*, **76**, 32–40.
99. Liu, F., Wu, X., Li, L., Liu, Z., Wang, Z. (2013). Use of baculovirus expression system for generation of virus-like particles: Successes and challenges. *Protein Expression Purif.*, **90**, 104–116.
100. Zeltins, A. (2012). Construction and characterization of virus-like particles: A review. *Mol. Biotechnol.*, 1–16.
101. Roldao, A., Mellado, M. C., Castilho, L. R., Carrondo, M. J., Alves, P. M. (2010). Virus-like particles in vaccine development. *Expert Rev. Vaccines*, **9**, 1149–1176.
102. Koutsky, L. A., Ault, K. A., Wheeler, C. M., Brown, D. R., Barr, E., Alvarez, F. B., et al. (2002). A controlled trial of a human papillomavirus type 16 vaccine. *N. Engl. J. Med.*, **347**, 1645–1651.
103. Breitburd, F., Kirnbauer, R., Hubbert, N. L., Nonnenmacher, B., Trindinhdesmarquet, C., Orth, G., et al. (1995). Immunization with viruslike particles from cottontail rabbit papillomavirus (CRPV) can protect against experimental CRPV infection. *J. Virol.*, **69**, 3959–3963.
104. Akagi, T., Baba, M., Akashi, M. (2011). Biodegradable nanoparticles as vaccine adjuvants and delivery systems: Regulation of immune responses by nanoparticle-based vaccine. In: Kunugi, S., Yamaoka, T., eds. *Polymers in Nanomedicine*, Springer, Berlin Heidelberg, pp. 31–64.
105. Uto, T., Akagi, T., Toyama, M., Nishi, Y., Shima, F., Akashi, M., et al. (2011). Comparative activity of biodegradable nanoparticles with aluminum adjuvants: Antigen uptake by dendritic cells and induction of immune response in mice. *Immunol. Lett.*, **140**, 36–43.
106. Chithrani, B. D., Ghazani, A. A., Chan, W. C. W. (2006). Determining the size and shape dependence of gold nanoparticle uptake into mammalian cells. *Nano Lett.*, **6**, 662–668.
107. Merdan, T., Kopeček, J., Kissel, T. (2002). Prospects for cationic polymers in gene and oligonucleotide therapy against cancer. *Adv. Drug Delivery Rev.*, **54**, 715–758.

108. Foged, C., Brodin, B., Frokjaer, S., Sundblad, A. (2005). Particle size and surface charge affect particle uptake by human dendritic cells in an in vitro model. *Int. J. Pharm.*, **298**, 315–322.
109. Pedroso De Lima, M. C., Simões, S., Pires, P., Faneca, H., Düzgüneş, N. (2001). Cationic lipid-DNA complexes in gene delivery: From biophysics to biological applications. *Adv. Drug Delivery Rev.*, **47**, 277–294.
110. Boussif, O., Zanta, M. A., Behr, J. P. (1996). Optimized galenics improve in vitro gene transfer with cationic molecules up to 1000-fold. *Gene Ther.*, **3**, 1074–1080.
111. Zahr, A. S., Davis, C. A., Pishko, M. V. (2006). Macrophage uptake of core-shell nanoparticles surface modified with poly(ethylene glycol). *Langmuir*, **22**, 8178–8185.
112. Agarwal, R., Singh, V., Jurney, P., Shi, L., Sreenivasan, S. V., Roy, K. (2013). Mammalian cells preferentially internalize hydrogel nanodiscs over nanorods and use shape-specific uptake mechanisms. *Proc. Natl. Acad. Sci. U. S. A.*, **110**, 17247–17252.
113. Matsuo, H., Yoshimoto, N., Iijima, M., Niimi, T., Jung, J., Jeong, S. Y., et al. (2012). Engineered hepatitis B virus surface antigen L protein particles for in vivo active targeting of splenic dendritic cells. *Int. J. Nanomed.*, **7**, 3341–3350.
114. Reddy, S. T., Rehor, A., Schmoekel, H. G., Hubbell, J. A., Swartz, M. A. (2006). In vivo targeting of dendritic cells in lymph nodes with poly(propylene sulfide) nanoparticles. *J. Controlled Release*, **112**, 26–34.
115. Cho, N. H., Cheong, T. C., Min, J. H., Wu, J. H., Lee, S. J., Kim, D., et al. (2011). A multifunctional core-shell nanoparticle for dendritic cell-based cancer immunotherapy. *Nat. Nanotechnol.*, **6**, 675–682.
116. Carrillo-Conde, B., Song, E. H., Chavez-Santoscoy, A., Phanse, Y., Ramer-Tait, A. E., Pohl, N. L., et al. (2011). Mannose-functionalized "pathogen-like" polyanhydride nanoparticles target C-type lectin receptors on dendritic cells. *Mol. Pharm.*, **8**, 1877–1886.
117. Cruz, L. J., Tacken, P. J., Fokkink, R., Joosten, B., Stuart, M. C., Albericio, F., et al. (2010). Targeted PLGA nano-but not microparticles specifically deliver antigen to human dendritic cells via DC-SIGN in vitro. *J. Controlled Release*, **144**, 118–126.
118. Castro, F. V., Tutt, A. L., White, A. L., Teeling, J. L., James, S., French, R. R., et al. (2008). CD11c provides an effective immunotarget for the generation of both CD4 and CD8 T cell responses. *Eur. J. Immunol.*, **38**, 2263–2273.

119. Kamphorst, A. O., Guermonprez, P., Dudziak, D., Nussenzweig, M. C. (2010). Route of antigen uptake differentially impacts presentation by dendritic cells and activated monocytes. *J. Immunol.*, **185**, 3426–3435.

120. Joffre, O. P., Segura, E., Savina, A., Amigorena, S. (2012). Cross-presentation by dendritic cells. *Nat. Rev. Immunol.*, **12**, 557–569.

121. Iversen, T.-G., Skotland, T., Sandvig, K. (2011). Endocytosis and intracellular transport of nanoparticles: Present knowledge and need for future studies. *Nano Today*, **6**, 176–185.

122. Taylor, U., Klein, S., Petersen, S., Kues, W., Barcikowski, S., Rath, D. (2010). Nonendosomal cellular uptake of ligand-free, positively charged gold nanoparticles. *Cytometry A*, **77**, 439–446.

123. Sharma, G., Valenta, D. T., Altman, Y., Harvey, S., Xie, H., Mitragotri, S., et al. (2010). Polymer particle shape independently influences binding and internalization by macrophages. *J. Controlled Release*, **147**, 408–412.

124. Vercauteren, D., Vandenbroucke, R. E., Jones, A. T., Rejman, J., Demeester, J., De Smedt, S. C., et al. (2010). The use of inhibitors to study endocytic pathways of gene carriers: Optimization and pitfalls. *Mol. Ther.*, **18**, 561–569.

125. Nembrini, C., Stano, A., Dane, K. Y., Ballester, M., van der Vlies, A. J., Marsland, B. J., et al. (2011). Nanoparticle conjugation of antigen enhances cytotoxic T-cell responses in pulmonary vaccination. *Proc. Natl. Acad. Sci. U. S. A.*, **108**, E989–E997.

126. Reddy, S. T., van der Vlies, A. J., Simeoni, E., Angeli, V., Randolph, G. J., O'Neil, C. P., et al. (2007). Exploiting lymphatic transport and complement activation in nanoparticle vaccines. *Nat. Biotechnol.*, **25**, 1159–1164.

127. Xiong, X., Jiao, J., Gregory, A. E., Wang, P., Bi, Y., Wang, X., et al. (2016). Identification of *Coxiella burnetii* CD8+ epitopes and delivery by attenuated *Listeria monocytogenes* as a vaccine vector in a C57BL/6 mouse model. *J. Infect. Dis.*, doi: 10.1093/infdis/jiw470.

128. Xiong, X., Qi, Y., Jiao, J., Gong, W., Duan, C., Wen, B. (2014). Exploratory study on Th1 epitope-induced protective immunity against *Coxiella burnetii* infection. *PLoS One*, **9**, e87206.

129. Zhu, M., Tian, X., Song, X., Li, Y., Tian, Y., Zhao, Y., et al. (2012). Nanoparticle-induced exosomes target antigen-presenting cells to initiate Th1-type immune activation. *Small*, **8**, 2841–2848.

130. Chen, C., Dow, C., Wang, P., Sidney, J., Read, A., Harmsen, A., et al. (2011). Identification of CD4$^+$ T cell epitopes in *C. burnetii* antigens targeted by antibody responses. *PLoS One*, **6**, e17712.

Chapter 17

Immunogenicity Assessment for Therapeutic Protein Products[1]

U.S. Food and Drug Administration, Center for Drug Evaluation and Research, Silver Spring, Maryland, USA

Keywords: adverse events, aggregates, allergen, allergy, anaphylaxis, antibody, anti-drug antibody (ADA), antigen, bacterial, B-cells, biomarkers, CD28, chemical modifications, circulating immune complexes, clinical, clinically relevant, complement activation, C-reactive protein, cross-linking, cross-reactivity, cross-reactive immunologic material (CRIM), cryptic epitopes, cytokine, degradation, desensitization, disease dosing, efficacy, endogenous protein, enzyme, exposure, factor replacement therapy, formulation, genetic, glycosylation, granulocyte-macrophage colony-stimulating factor (GM-CSF), heterogeneous, high mobility group protein B1 (HMGB1), homogeneous, human leukocyte antigen (HLA), human serum albumin (HSA), hypersensitivity, IFN-beta, IgE, IgG, immune-deficient, immune responses, immune response modulating impurities (IRMI), immuno-

---

[1]This guidance document, courtesy of the Food and Drug Administration, has been prepared by the Division of Medical Policy Development, Office of Medical Policy, in the Center for Drug Evaluation and Research (CDER) in coordination with the Center for Biologics Evaluation and Research (CBER) at the FDA. It contains nonbinding recommendations. It represents the FDA's current thinking on this topic. It does not create or confer any rights for or on any person and does not operate to bind FDA or the public. You can use an alternative approach if the approach satisfies the requirements of the applicable statutes and regulations. If you want to discuss an alternative approach, contact the FDA staff responsible for implementing this guidance document.

Immune Aspects of Biopharmaceuticals and Nanomedicines
Edited by Raj Bawa, János Szebeni, Thomas J. Webster, and Gerald F. Audette
Copyright © 2018 Pan Stanford Publishing Pte. Ltd.
ISBN 978-981-4774-52-9 (Hardcover), 978-0-203-73153-6 (eBook)
www.panstanford.com

genicity, infusion reactions, interferon-alpha, interleukin-2, maternal, mitigation strategies, molecular structure, monitoring, monoclonal antibody, multispecific antibodies, neonatal, neutralizing antibody, non-acute immune response, particulate, pathophysiology, patient, patient history, polyethylene glycol (PEG), pegylation, pharmacokinetics, pharmacodynamics, polymorphism, posttranslational modifications, pregnancy, product custody, prophylactic, protein engineering, replacement therapy, risk-based approach, risk-benefit assessment, route of administration, safety, stability, T-cells, T-helper (Th) cells, therapeutic protein, titer, tolerance, vaccines, xenogenic

17.1 Introduction

This guidance document is intended to assist manufacturers and clinical investigators involved in the development of therapeutic protein products for human use.[2] In this document, the FDA outlines and recommends adoption of a risk-based approach to evaluating and mitigating immune responses to or adverse immunologically related responses associated with therapeutic protein products that affect their safety and efficacy. Any given approach to assessing and mitigating immunogenicity is determined on a case-by-case basis and should take into consideration the risk assessment we describe. For the purposes of this guidance document, immunogenicity is defined as the propensity of the therapeutic protein product to generate immune responses to itself and to related proteins or to induce immunologically related adverse clinical events.

This guidance document describes major clinical consequences of immune responses to therapeutic protein products and offers recommendations for risk mitigation in the clinical phase of development. It also describes product- and patient-specific factors

---

[2]See the draft guidance document for industry *Biosimilars: Questions and Answers Regarding Implementation of the Biologics Price Competition and Innovation Act of 2009* for FDA's interpretation of the category of "protein (except any chemically synthesized polypeptide)" in the amended definition of "biological product" in section 351(i) (1) of the Public Health Service Act (PHS Act). When final, this guidance document will represent the FDA's current thinking on this topic. The FDA updates guidances periodically. To make sure you have the most recent version of a guidance, check the FDA Drugs guidance Web page at http://www.fda.gov/RegulatoryInformation/Guidances/default.htm

that can affect the immunogenicity of therapeutic protein products. For each factor, recommendations are made for sponsors and investigators that may help them reduce the likelihood that these products will generate an immune response. Appendix A provides supplemental information on the diagnosis and pathophysiology of particular adverse consequences of immune responses to therapeutic protein products and brief discussions of the uses of animal studies and the conduct of comparative immunogenicity studies. Although this guidance document focuses on therapeutic protein products, the scientific principles may also apply to related products and biological entities, for example, peptides. Although this guidance document encompasses products used to modulate or modify the immune system, including those that are antigen specific, it does not cover products that are intended to induce a specific immune response to prevent or treat a disease or condition (such as vaccines to prevent infectious diseases) or to enhance the activity of other therapeutic interventions. This guidance document does not address assay development, which is covered in a separate guidance.[3]

FDA's guidance documents, including this guidance document, do not establish legally enforceable responsibilities. Instead, guidances describe the Agency's current thinking on a topic and should be viewed only as recommendations, unless specific regulatory or statutory requirements are cited. The use of the word *should* in Agency guidances means that something is suggested or recommended, but not required.

17.2   Background

Immune responses to therapeutic protein products may pose problems for both patient safety and product efficacy. Immunologically based adverse events, such as anaphylaxis, cytokine release syndrome, and cross-reactive neutralization of endogenous proteins mediating critical functions (see Appendix A.3), have caused sponsors to terminate the development of what otherwise may have been efficacious therapeutic protein

---

[3]See the draft guidance for industry *Assay Development for Immunogenicity Testing of Therapeutic Proteins*. When final, this guidance will reflect the FDA's current thinking on this topic.

products. Unwanted immune responses to therapeutic protein products may also neutralize their biological activities and result in adverse events not only by inhibiting the efficacy of the therapeutic protein product, but also by cross-reacting to an endogenous protein counterpart, leading to loss of its physiological function (e.g., neutralizing antibodies to therapeutic erythropoietin causes pure red cell aplasia by also neutralizing the endogenous protein) (Hermeling et al., 2004; Rosenberg and Worobec, 2004; Rosenberg and Worobec, 2005; Koren et al., 2008; Murphy, 2011). Because most of the adverse effects resulting from elicitation of an immune response to a therapeutic protein product appear to be mediated by humoral mechanisms, circulating antibodies to the therapeutic protein product has been the chief criterion for defining an immune response to this class of products.[4]

Both patient-related and product-related factors may affect immunogenicity of therapeutic protein products. These factors are critical elements in the immunogenicity risk assessment. Ideally, these factors should be taken into consideration in the early stages of therapeutic protein product development. Section 17.3 contains a detailed discussion of the nature of and the risk factors for immune responses to therapeutic protein products as well as possible mitigation strategies that may be employed.

17.3 Clinical Consequences

Treatment of patients with therapeutic protein products may result in immune responses of varying clinical relevance, ranging from antibody responses with no apparent clinical manifestations to life-threatening and catastrophic reactions. During therapeutic protein product development, elucidation of a specific underlying immunologic mechanism for immunologically related adverse events is encouraged, because this information can facilitate the development of strategies to help mitigate their risk (see Sections 17.3.2.1–17.3.2.5). The extent of information required

---

[4]IgG and IgE antibody responses are those most often associated with clinical adverse events, and their generation generally requires collaboration between antigen-specific T-helper cells and B-cells. Murphy, K. (2011). The humoral immune response. In: *Janeway's Immunobiology*, 8th ed., New York, Garland Science Publishing, pp. 367–408.

to perform a risk-benefit assessment will vary among individual products, depending on product origin and features, the immune responses of concern, the target disease indication, and the proposed patient population.

17.3.1 Consequences for Efficacy

Development of antibodies can limit product efficacy in patients treated with therapeutic protein products. Neutralizing antibodies can block the efficacy of therapeutic protein products by specifically targeting domains critical for efficacy. For example, antibodies binding to either the uptake or catalytic domain of a therapeutic enzyme may lead to loss of product efficacy. Loss of efficacy is problematic for all products, but is of utmost concern if the product is a lifesaving therapeutic. Neutralizing antibodies that cross-react with a nonredundant endogenous counterpart of the therapeutic protein product can also impact safety, as discussed in Section 17.3.2. Both neutralizing and non-neutralizing antibodies may alter the pharmacokinetics of the product by enhancing clearance (and thereby shortening serum half-life) or, conversely, by prolonging serum half-life and product activity. If present at high enough titer, a non-neutralizing antibody may mistarget the therapeutic protein product into the Fc receptor (FcR) bearing cells, thereby reducing or eliminating product efficacy (Brooks et al., 1998; Wang et al., 2008). Furthermore, although some antibody responses to therapeutic protein products may have no apparent effect on clinical safety or efficacy, they may promote the generation of neutralizing antibodies via the mechanism of epitope spreading of antibody responses (Disis et al., 2004; Hintermann et al., 2011). Pharmacodynamic biomarkers may be useful in the assessment of antibody-mediated interference with product activity, although correlation with clinical response is usually necessary to determine clinical relevance.

17.3.2 Consequences for Safety

The safety consequences of immunogenicity may vary widely and are often unpredictable in patients administered therapeutic protein products. Therefore, a high index of suspicion should be maintained for clinical events that may originate from such responses, even if the initial risk assessment suggests a lower risk of immunogenicity. The applicant should provide a rationale for the proposed immunogenicity testing paradigm, based on product- and patient-specific concerns. The following sections describe a few of the major safety concerns associated with immunogenicity:

17.3.2.1 Anaphylaxis

Anaphylaxis is a serious, acute allergic reaction characterized by certain clinical features. The definition currently accepted by the Agency relies on clinical diagnostic criteria and does not specify a particular immunologic mechanism (Sampson et al., 2006) (also see Appendix A.1). Historically, the definition of anaphylaxis has invoked the involvement of specific IgE antibodies. However, such a mechanistic definition may be problematic in the context of therapeutic protein product development and other clinical settings where it may not always be possible to identify a specific immunologic mechanism as the basis of an adverse event. To capture all potential adverse events of interest, the Agency recommends identifying all cases meeting the clinical diagnostic criteria of anaphylaxis, regardless of the presumed pathophysiology. Additional information, such as the assessment of serum histamine, serum tryptase, and complement components, following a reaction or the detection of product-specific IgE antibodies may help elucidate the pathophysiology of the anaphylactic response and thus guide control and mitigation strategies.

Furthermore, the presence of an anti-drug antibody (ADA) alone is not necessarily predictive of anaphylaxis or other hypersensitivity reactions. Correlation with clinical response is typically required to determine the clinical relevance of these antibodies. Determination of the underlying mechanism remains of interest, however, because anaphylaxis with confirmation of IgE involvement has certain prognostic implications for repeat exposure as well as for potential therapeutic options for mitigation.

17.3.2.2 Cytokine release syndrome

Cytokine release syndrome is a symptom complex caused by the rapid release of proinflammatory cytokines from target immune cells (Stebbings et al., 2007; Stebbings et al., 2013). Although cytokine release syndrome is not directly related to immunogenicity, the clinical presentation of cytokine release syndrome overlaps with anaphylaxis and other immunologically related adverse reactions. Distinguishing this symptom complex from these other types of adverse reactions is potentially useful for the purpose of risk mitigation. Although the underlying mechanisms may not be fully understood, in some cases the mechanism appears to relate to the cross-linking of activating cell surface expressed receptors, which are the targets of the therapeutic protein product (e.g., CD28 expressed on T-cells). A risk-based evaluation, focused on the mechanism of action of the therapeutic protein product as well as results of animal and *in vitro* evaluations should be performed to determine the need for collection of pre- and post-dose cytokine levels in the early phase of clinical development. In case of a clinical adverse event, such an evaluation may provide evidence to support the clinical diagnosis of cytokine release syndrome and help distinguish this entity from other acute drug reactions (e.g., anaphylaxis, see Appendix A.2).

17.3.2.3 "Infusion reactions"

Therapeutic protein products may elicit a range of acute effects, from symptomatic discomfort to sudden, fatal reactions that have often been grouped as "*infusion reactions*" in the past.

Although the term implies a certain temporal relationship, infusion reactions are otherwise not well defined and may encompass a wide range of clinical events, including anaphylaxis and other events that may not be directly related to antibody responses, such as cytokine release syndrome. In the absence of an agreed-upon definition for infusion reaction, the categorization of certain adverse events as infusion reactions without further detail is problematic and is not recommended. Sponsors are encouraged to use more-descriptive terminology when possible, noting the timing, duration, and specific signs and symptoms observed upon administration of a therapeutic protein product and to provide data from mechanistic studies which may facilitate a mitigation strategy.

17.3.2.4 Non-acute reactions

Anaphylaxis, cytokine release syndrome, and other acute reactions are temporally linked to administration of a therapeutic protein product. Delayed hypersensitivity (i.e., serum sickness) and immune responses secondary to immune complex formation typically have a subacute presentation. As a result, the association between administration of a therapeutic protein product and non-acute reactions may be more difficult to establish, and assessment of the underlying mechanism will likely require evaluation of circulating immune complexes and complement activation. Clinical signs may include delayed onset of fever, rash, arthralgia, myalgia, hematuria, proteinuria, serositis, central nervous system complications, and hemolytic anemia in the face of an ongoing antibody response to the therapeutic protein product (Hunley et al., 2004; Goto et al., 2009). When such a reaction is suspected, laboratory assessment for circulating immune complexes may help confirm the diagnosis. The necessity and details of a laboratory assessment will depend on the individual situation and should be discussed with the respective review division for the therapeutic protein product.

17.3.2.5 Cross-reactivity to endogenous proteins

ADA can have severe consequences if it cross-reacts to and inhibits a nonredundant endogenous counterpart of the therapeutic protein product or related proteins (Macdougall et al., 2012; Seidl et al., 2012). If the endogenous protein is redundant in biological function, inhibition of the therapeutic and endogenous proteins may not produce an obvious clinical syndrome until the system is stressed, because not all biological functions of an endogenous protein may be known or fully characterized (Stanley et al., 1994; Bukhari et al., 2011). Moreover, the long-term consequences of such antibodies may not be known. An additional potential consequence of cross-reactivity to an endogenous protein results from antibody responses to a therapeutic protein product that is a counterpart of an endogenous cell surface receptor or a counterpart of an endogenous cytokine that is membrane-expressed. Such antibodies may cross-reactively bind to the respective cell surface receptors or proteins, causing cytokine release or other manifestations of cellular activation.

For therapeutic protein product counterparts of endogenous proteins that are critical to normal fetal or neonatal development, neutralization of such endogenous proteins, resulting from antibodies to the therapeutic protein product that cross react to the endogenous counterpart, has the potential to negatively impact fetal or neonatal development when these immune responses are generated or boosted during pregnancy or breast feeding. As part of the risk evaluation, sponsors should consider the potential transmission of antibodies to the fetus by the placenta or to the developing neonate by human milk. Therefore, the risk of neutralizing antibody development following administration of such therapeutic protein products to women of childbearing potential should be strongly considered in light of the potential benefit. Moreover, the risk of neutralizing antibody development to endogenous proteins critical in growth and development beyond the neonatal period should be evaluated in studies in pediatric populations.

Although animal studies may provide useful information regarding the possible consequences of inhibition of an endogenous protein, particularly for endogenous proteins that are highly evolutionarily conserved, such studies are not considered to be predictive of the *likelihood* of an immune response to a therapeutic protein product in humans. Moreover, differences in the timing and extent of transplacental transfer of maternal antibodies may limit the utility of animal studies to assess *in utero* effects of cross-reactive antibodies to the endogenous counterpart of the therapeutic protein product.

17.4 Recommendations for Immunogenicity Risk Mitigation in the Clinical Phase of Development of Therapeutic Protein Products

Given the variety of factors that can affect immunogenicity, the risk assessment and the control and mitigation strategies will depend on the individual development program and should be considered at the earliest stage and at each subsequent stage of product development. The extent of immunogenicity safety information requiring premarketing and postmarketing will vary, depending on the potential severity of the consequences of such immune responses and the likelihood of their occurrence.

In terms of evaluating the clinical relevance of immune responses, the Agency has the following recommendations:

Development of assays for anti-drug antibody (ADA)

- Sponsors should develop and implement sensitive immunoassays commensurate with the overall product development program.[5] Concomitant assessment of levels of therapeutic protein product in the sample is recommended to assess the potential for the presence of the product to interfere with detection of an antibody in the assay.

Product-specific antibody sampling considerations

- Baseline samples for ADA testing should be collected, and the post-baseline sampling frequency and duration should reflect anticipated use of the product. More frequent sampling is appropriate during the initiation and early use of a new, chronically administered product; less frequent sampling may be appropriate after prolonged use. Repeat sampling should generally occur over periods of sufficient duration to determine whether these responses are persistent, neutralizing, and associated with clinical sequelae. Samples for antibody assessment should be drawn prior to administration of the therapeutic protein product.
- In addition to a prespecified sampling schedule, unscheduled sampling, triggered by suspected immunologically related adverse events, is necessary for establishing the clinical relevance of ADAs. Future sampling considerations for patients whose samples test positive for an antibody at the end of a study should be discussed with the respective review division for the therapeutic protein product. Informed consent should address the possibility for sampling beyond study termination.
- Banking of serum samples from clinical trials under appropriate storage conditions for future testing is always advisable.

---

[5]See the draft guidance for industry *Assay Development for Immunogenicity Testing of Therapeutic Proteins*, where assay development is covered in detail. When final, this guidance will reflect the Agency's current thinking on this topic. Guidance on appropriate assay development for immunogenicity testing is also available in the ICH guidances for industry *Q2A Text on Validation of Analytical Procedures* and *Q2B Validation of Analytical Procedures: Methodology*.

Dosing

- For first-in-human trials, a conservative approach in an appropriate medical setting with access to immediate supportive care in the event of a serious adverse event, such as anaphylaxis, should be taken. Staggered dosing among individual patients and dosing cohorts is appropriate. The trial design should include prespecified dose escalation criteria and adequate time intervals between dosing cohorts and, as appropriate for the pharmacokinetics and pharmacodynamics of the product, between individuals within a dosing cohort to assess toxicities prior to administration of subsequent doses or treatment of additional individuals. The need for such an approach will depend on individual circumstances.[6] Aside from first-in-human trials, there may be other situations where a similarly conservative approach is indicated, e.g., change in the route of administration, change in formulation, change in container closure system. As development progresses, dosing strategies and safety parameters can be modified based on clinical experience with the product and other products of the same class.
- Because it may be difficult to predict the incidence of product-specific antibodies in different clinical trial scenarios, dosing regimens in subsequent studies should be risk based and take into account the following: data from initial trials; the potential for and predicted effects of cross-reactivity to endogenous proteins; the severity of effects of neutralization of the therapeutic protein product (e.g., a lifesaving versus adjunctive treatment product); clinical parameters that impact immunogenicity in different patient populations; and the adequacy of proposed safety monitoring (Koren et al., 2008).
- Higher doses of therapeutic protein products do not uniformly overcome high titer and/or sustained or neutralizing antibody responses and may impact safety, e.g., may precipitate immune complex mediated disease or cause other toxicities. The appropriateness of such a dose

---

[6]See the guidance for industry, *Estimating the Maximum Safe Starting Dose in Initial Clinical Trials for Therapeutics in Adult Healthy Volunteers.*

escalation strategy will depend on the specific product, the magnitude of the antibody response, and the disease indication. A protocol defining specific safety monitoring evaluations and stopping rules should be developed prior to implementation of dose escalation to overcome an antibody response.

Adverse events

- The development of neutralizing antibody activity or the presence of sustained antibody titers may lead to loss of efficacy or an increased risk of an adverse reaction. In certain high-risk situations (e.g., assessment of a product with a nonredundant endogenous counterpart), real-time assessments for antibodies during a clinical trial may be recommended for safety reasons. Real-time assessments entail analyses of the samples as soon as possible after sampling, before banking of the samples, and prior to additional dosing. The need for such intensive monitoring will depend on individual circumstances.
- If clinically relevant immune responses are observed, sponsors are encouraged to study the underlying mechanism and identify any critical contributing factors. These investigations can facilitate development and adoption of potential control and mitigation strategies, including modification of product formulation and screening of higher-risk patients (see Section 17.5).
- In some cases, sponsors may choose to explore premedication, desensitization, or immune tolerance induction procedures as potential mitigation strategies. Given the risks associated with desensitization/immune tolerance induction procedures and the potential for premedication to mask early signs and symptoms of adverse events, the appropriateness of such procedures will depend on the nature of the specific indication, the target patient population, and the stage of development.

Comparative immunogenicity studies

- For all comparative immunogenicity studies (e.g., those comparing immunologically related adverse events, antibody incidence, titer, or neutralizing activity to product pre-

and post-manufacturing changes), a strong rationale and, when possible, prespecified criteria should be provided to justify what differences in incidence or severity of immune responses would constitute an unacceptable difference in product safety.[7] The same antibody assay should be used to enable valid comparisons (see Appendix A.6).

Postmarketing safety monitoring

- Robust postmarketing safety monitoring is an important component in ensuring the safety and effectiveness of therapeutic protein products. Because some aspects of postmarketing safety monitoring are product-specific, the FDA encourages sponsors to consult with the appropriate FDA review division to discuss the sponsor's proposed approach to postmarketing safety monitoring. Rare, but potentially serious, safety risks (e.g., immunologically related adverse events) may not be detected during preapproval clinical testing, because the size of the population exposed may not be large enough to assess rare events. In some cases, such risks may need to be evaluated through postmarketing surveillance or required studies or clinical trials.

17.5 Patient- and Product-Specific Factors That Affect Immunogenicity

17.5.1 Patient-Specific Factors That Affect Immunogenicity

Factors related to the target patient population may increase or decrease the potential for and the risk associated with an immune response. Therefore, caution is recommended when moving from one patient population to another, and a new risk assessment should be performed for each new patient population considered for treatment.

---

[7]For information on proposed biosimilar products, see the draft guidance for industry *Scientific Considerations in Demonstrating Biosimilarity to a Reference Product*. When final, this guidance will represent the FDA's current thinking on this topic.

17.5.1.1 Immunologic status and competence of the patient

Patients who are immune suppressed may be at lower risk of mounting immune responses to therapeutic protein products compared to healthy volunteers with intact immune responses. For example, 95 percent of immune-competent cancer patients generated neutralizing antibody to a granulocyte-macrophage colony-stimulating factor (GM-CSF) product; but only 10 percent of immune-compromised cancer patients did so in response to a GM-CSF product (Ragnhammar et al., 1994). Immune suppressive agents may diminish the immune response to therapeutic protein products. Thus, agents that kill antigen-activated lymphocytes and/or elicit activity of regulatory T-cells, such as methotrexate, have been shown to have a substantial effect on immunogenicity of co-administered monoclonal or other antibody products (Baert et al., 2003). In contrast to immune-deficient patients, patients with an activated immune system (e.g., patients with certain infections or autoimmune disease) may have augmented responses. Immune response generation may also be affected by patient age, particularly at the extremes of the age range (LeMaoult et al., 1997; PrabhuDas et al., 2011; Cuenca et al., 2013; Goronzy and Weyand, 2013). Particular caution with regard to immunogenicity and immune responses should be used in studies evaluating novel therapeutics in healthy volunteers (Li et al., 2001; Stebbings et al., 2007; Colombel et al., 2010; Garces et al., 2013).

Recommendation

In the development of therapeutic protein products, a rationale should be provided to support the selection of an appropriate study population, especially for first-in-human studies. The potential influence of concomitant medications on ADAs should be taken into consideration during all stages of clinical development.

17.5.1.2 Prior sensitization/history of allergy

Prior exposure to a therapeutic protein product or to a structurally similar protein may lead to pre-existing antibodies at baseline. This is a particular concern for patients receiving a replacement product, such as clotting factors or an enzyme replacement therapy, who may have antibodies to a previous product that could cross-react to an analogous product.

Sensitization to the excipients or process/product-related impurities of a therapeutic protein product may also predispose a patient to an adverse clinical consequence. For example, products produced from transgenic sources may contain allergenic foreign proteins, such as milk protein or protein from chicken eggs.

Because patient history may not capture all prior exposures that could generate a pre-existing antibody response or predict anaphylaxis, screening for pre-existing antibodies, e.g., inhibitors or neutralizing antibodies in factor replacement therapy, should be considered when appropriate.

Recommendation

Screening for a history of relevant allergies pertaining to the source material of the therapeutic protein product (e.g., produced in transgenic hen eggs versus mammalian cells) is recommended, and the appropriateness of additional clinical or laboratory tests prior to administration should be considered in light of the overall risk-benefit assessment.

17.5.1.3 Route of administration, dose, and frequency of administration

Route of administration can affect the risk of sensitization. In general, intradermal, subcutaneous, and inhalational routes of administration are associated with increased immunogenicity compared to intramuscular and intravenous (IV) routes. The IV route is generally considered to be the least likely to elicit an immune response. In conjunction with the route of administration, dose and frequency can also affect immunogenicity (Rosenberg and Worobec, 2004). For example, in certain circumstances, a lower dose administered intermittently may be more immunogenic than a larger dose administered without interruption. It should be noted that the effects of dose and frequency on ADA development may be affected by other factors, such as route of administration, product origin, and product-related factors that influence immunogenicity.

Recommendations

Immunogenicity should be considered when selecting an appropriate route of administration, especially for high-risk therapeutic protein products (e.g., therapeutic counterparts of nonredundant endogenous proteins) in first-in-human dosing.

Changes in the route of administration or dosing during product development may be associated with changes in the immunogenicity profile, and clinical safety data should be obtained to support such changes.

17.5.1.4 Genetic status

Genetic factors may modulate the immune response to a therapeutic protein product. In particular, some human leukocyte antigen (HLA) haplotypes may predispose patients to the development of undesirable antibody responses to specific products (Hoffmann et al., 2008). If both appropriate and feasible, HLA mapping studies may help define a subset of the patient population at increased risk. Moreover, genetic polymorphisms in cytokine genes may upregulate or downregulate immune responses (Donnelly et al., 2011).

Recommendation

Evaluation of genetic factors that may modulate the immune response to a therapeutic protein product is recommended in circumstances in which a subset of treated patients lose the clinical benefit of treatment or experience severe adverse events. For example, the subset of patients that generate neutralizing antibodies to IFN-beta products are more likely to possess distinct HLA haplotypes (Hoffmann et al., 2008). Thus, knowledge of the heightened susceptibility of patients with such HLA haplotypes may allow for measures to prevent such responses or for pursuit of other treatment options.

17.5.1.5 Status of immune tolerance to endogenous protein

Humans are not equally immunologically tolerant to all endogenous proteins. Thus, the robustness of immune tolerance to an endogenous protein affects the ease with which a therapeutic protein product counterpart of that endogenous protein can break such tolerance. Immunological tolerance in both protein-specific T- and B-cells depends on many factors, prominent among which is the abundance of the endogenous protein: immune tolerance is weaker for low-abundance and stronger for high-abundance proteins (Weigle, 1980; Goodnow, 1992; Haribhai et al., 2003).

The human immune system is not fully tolerant to low-abundance endogenous proteins, such as cytokines and growth factors, for which serum levels may be in the nanogram (ng)/milliter (mL) to picogram (pg)/mL range. This point is underscored by the presence of autoantibodies to cytokines and growth factors in healthy individuals, the development of antibodies to inflammatory cytokines, and the breaking of tolerance to endogenous proteins by the administration of exogenous recombinant therapeutic protein products (Hermeling et al., 2004; Rosenberg and Worobec, 2004; Rosenberg and Worobec, 2005; Koren et al., 2008).

When a human therapeutic protein product is intended as a replacement for an absent or deficient endogenous protein, patients with genetic mutations conferring a protein *knock out* phenotype may respond to the therapeutic protein product as to a foreign protein or neoantigen or may already be sensitized as a result of previous exposure to a similar therapeutic protein or related proteins from other sources. Such responses may abrogate the efficacy of the replacement therapy.

Recommendations

For a therapeutic protein product that is a counterpart of an endogenous protein, the robustness of immune tolerance to the endogenous protein should be investigated before initiating a clinical trial and such evaluation should consider the following as preeminent risks: if the clinical study is a first-in-human use, if the endogenous protein has a nonredundant physiological function, and if immune responses to the endogenous protein have been detected in the context of autoimmune diseases. Suggested evaluations include:

- Quantitating or gathering information on the level of the endogenous protein in serum in the steady state, as well as in conditions that may specifically elicit its production (Weigle, 1980).
- Assessing for or gathering information on the presence of pre-existing antibodies in healthy individuals and patient populations and on the frequency and role of such antibodies in autoimmune diseases (Bonfield et al., 2002; Hellmich et al., 2002).

- Evaluating immunogenicity, immune cell activation, inflammatory responses, and cytokine release in relevant animal studies to obtain insight and provide guidance for clinical safety assessments (Koren, 2002) (also see Appendix A.5).
- In patients requiring factor/enzyme replacement therapies, evaluation of patient tissue samples for the detection of endogenous protein or peptides (e.g., cross-reactive immunologic material (CRIM)), as well as for genetic mutations and HLA alleles (as appropriate), should be strongly considered to better predict the development of immune responses to the replacement therapy and to evaluate the need for tolerance induction mitigation strategies (Pandey et al., 2013b).
- Evaluating the extent of polymorphisms, including single nucleotide polymorphisms, when appropriate, in relevant patient populations to identify potential mismatches with the therapeutic protein product (Jefferis and Lefranc, 2009; Viel et al., 2009; Pandey et al., 2013a).

17.5.2 Product-Specific Factors That Affect Immunogenicity

Product-specific factors may increase or decrease the potential for and the risk associated with an immune response. Immunogenicity testing should be considered when changes are made to product-specific factors.

17.5.2.1 Product Origin (foreign or human)

Immune responses to nonhuman (i.e., foreign) proteins are expected and, as previously explained, may be anticipated for endogenous human proteins. Moreover, mismatches between the sequence of the endogenous protein of the patient and that of the therapeutic protein product caused by naturally occurring polymorphisms are one risk factor for the development of immune responses to the therapeutic protein product (Viel et al., 2009). However, the rapidity of development, the strength (titer), and the persistence of the response may depend on a number of factors, including the following: previous and ongoing environmental exposure and the mode of such exposure; the presence in the product of immunity-provoking factors, such as product aggregates and materials with adjuvant activity; and the product's inherent immunomodulatory activity (see Section 17.5.2.6). For example, environmental exposure to bacterial proteins from either commensal or pathogenic bacteria on skin or in the gut may be predisposed to the generation of immune responses when such bacterial proteins (either recombinantly or naturally derived) are used as therapeutics.

For proteins derived from natural sources, antibodies can develop not only to the desired therapeutic protein product, but also to other foreign protein components potentially present in the product. Furthermore, such foreign proteins may contain regions of homology to endogenous human proteins. The capacity of the foreign protein to break tolerance and induce antibody responses to the homologous human factor should be evaluated in the clinical trial. For example, during treatment with a bovine thrombin product, immune responses to bovine coagulation factor V, incidentally present in the product, led to the development of antibodies that cross-reacted to human factor V and resulted in life-threatening bleeding in some patients (Kessler and Ortel, 2009).

For monoclonal antibodies, product origin is an important factor that can influence immunogenicity. Although mouse antibodies have been shown to robustly elicit immune responses in humans as compared to chimeric, humanized and human monoclonal antibodies, it should be noted that chimeric, humanized and human monoclonal antibodies can also elicit a high rate of immunogenicity depending on the dosing regimen and patient population (Singh, 2011). In fact, some human antibodies developed using phage display may have significant ADA responses.

Moreover, novel structural formats, including fusion proteins, bispecific or multispecific antibodies (bivalent or tetravalent), single chain fragments, single domain antibodies, and specifically engineered antibodies with mutations in the constant or variable regions, may elicit immune responses, as such novel structures may create neoantigens or expose cryptic epitopes. In addition, site-specific mutations in constant regions may create novel *allotypes*, and the use of an *in vitro* affinity maturation process may result in novel *idiotypes*. An understanding of the increased immunogenicity associated with certain antibody products will require more complete characterization of the ADA response, such as identification of the target epitope(s) (Singh, 2011).

Recommendations

All therapeutic protein products should be evaluated for their content of and immune responses directed to incidental product components, including proteins and nonprotein components. A risk-based evaluation of potential immune responses to such process- and product-related impurities should be performed, and a testing program should be designed based on this evaluation. Foreign proteins intended for therapeutic use should be evaluated for molecular regions that bear strong homology to endogenous human proteins. When such homologies exist, assessment of antibodies to the homologous human protein should be made in addition to assessment of antibodies to the foreign therapeutic protein.

When developing assays to assess the immunogenicity of novel antibody-related products, appropriate controls should be incorporated into the assays to determine if the ADA response is directed against novel epitopes.

17.5.2.2 Primary Molecular Structure/Posttranslational Modifications

Primary sequence, higher-order structure, species origin, and molecular weight of therapeutic protein products are all important factors that may contribute to immunogenicity. Primary sequence analysis can reveal potentially immunogenic sequence differences in proteins that are otherwise relatively conserved between humans and animals. In some cases, nonhuman epitopes may elicit T-cell help or facilitate epitope spreading to generate an antibody response to the conserved human sequences (Dalum et al., 1997). Per Section 17.5.1.4, it is important to note that therapeutic protein products of human origin may elicit immune responses in subsets of patients with distinct HLA haplotypes as well as in patients whose endogenous protein amino acid sequence differs from that of the therapeutic protein product, even by single nucleotide polymorphisms.

Additional advanced analyses of a primary sequence are also likely to detect HLA class II binding epitopes in nonpolymorphic human proteins. Such epitopes may elicit and activate regulatory T-cells, which enforce self-tolerance, or, opposingly, could activate T-helper (Th) cells when immune tolerance to the endogenous protein is not robust (Barbosa and Celis, 2007; Tatarewicz et al., 2007; De Groot et al., 2008; Weber et al., 2009). However, if considered appropriate, engineering of changes to the primary sequence to eliminate immunogenic Th cell epitopes or addition of tolerogenic T-cell epitopes should be done cautiously, because these modifications may alter critical product quality attributes such as aggregation, deamidation, and oxidation and thus alter product stability and immunogenicity. Therefore, extensive evaluation and testing of critical product attributes should be performed following such changes. Primary sequence considerations are especially important in evaluation of the immunogenicity of fusion proteins, because the immune responses to neoantigens formed in the joining region may be elicited (Miller et al., 1999) and may then spread to conserved segments of the molecule. Fusion proteins consisting of a foreign protein and an endogenous protein are of particular concern because of the capacity of the foreign protein to elicit T-cell help for the generation of an antibody response to the endogenous protein partner. Similarly, bioengineered proteins involve the introduction of sequences not normally found in nature and may thus contain neo-epitopes. These epitopes have the potential to broadly elicit immune responses or may instead interact with HLA alleles found only in a subset of patients to induce immune responses (Kimchi-Sarfaty et al., 2013).

Chemical modifications of therapeutic protein products, such as oxidation, deamidation, aldehyde modification, and deimination, may elicit immune responses by, for example, modifying a primary sequence, causing aggregate formation, or altering antigen processing and presentation. Importantly, such changes may be well controlled during manufacture and storage, but may occur *in vivo* in the context of the relatively high pH of the *in vivo* environment or in inflammatory environments and cause loss of activity as well as elicitation of immune responses. Evaluation of therapeutic protein products in the context of the *in vivo* environments to which they are targeted can reveal susceptibility to such chemical modifications. (Huang et al., 2005; Demeule et al., 2006; Makrygiannakis et al., 2006). Susceptibility to chemical modifications of therapeutic protein products, and thus the possibility of the loss of activity or induction of immune responses *in vivo*, should prompt consideration of careful protein engineering.

Recommendations

Careful consideration should be given to the primary sequences chosen for the development of therapeutic protein products in general, and especially of therapeutic protein product counterparts of endogenous proteins in view of potential polymorphisms in endogenous proteins across human populations.

The ADA response to fusion molecules or engineered versions of therapeutic protein products should utilize assays that are able to assess reactivity to the whole molecule as well as to its distinct components. Immune responses directed to the intact protein product, but not reactive with either of the separate partner proteins, may be targeting novel epitopes in the fusion region.

Evaluation of therapeutic protein products in the *in vivo* milieu in which they function (e.g., in inflammatory environments or at physiologic pH) may reveal susceptibilities to modifications (e.g., aggregation and deamidation) that result in loss of efficacy or induction of immune responses. Such information may facilitate product engineering to enhance the stability of the product under such stress conditions. Sponsors should consider obtaining this information early in product design and development.

17.5.2.3 Quaternary structure: product aggregates and measurement of aggregates

Protein aggregates are defined as any self-associated protein species, with monomer defined as the smallest naturally occurring and/or functional subunit. Aggregates are further classified based on five characteristics: size, reversibility/dissociation, conformation, chemical modification, and morphology (Narhi et al., 2012). Aggregates ranging from dimer to visible particles that are hundreds of micrometers in size (Narhi et al., 2012) have been recognized for their potential to elicit immune responses to therapeutic protein products for over a half-century (Gamble, 1966). Mechanisms by which protein aggregates may elicit or enhance immune responses include the following: extensive cross-linking of B-cell receptors, causing efficient B-cell activation (Dintzis et al., 1989; Bachmann et al., 1993); and enhancing antigen uptake, processing, and presentation; and triggering immunostimulatory danger signals (Seong and Matzinger, 2004). Such mechanisms may enhance recruitment of the T-cell help needed for generation of high-affinity, isotype-switched IgG antibody, the antibody response most often associated with neutralization of product efficacy (Bachmann and Zinkernagel, 1997).

The potential clinical consequences of immune responses induced by protein aggregates may in large measure depend on the loss or preservation of native epitopes in the aggregate. Some antibodies generated by aggregates containing native proteins can bind to monomeric protein as well as to the aggregate, with the potential to inhibit or neutralize product activity. In contrast, some antibodies to denatured/degraded protein bind uniquely to the aggregated material, but not to native protein monomers, such as was the case with early preparations of human intravenous immune globulin (IVIG) (Barandun et al., 1962; Ellis and Henney, 1969). Responses to aggregates containing degraded epitopes have been shown to cause anaphylaxis, but do not inhibit or neutralize activity of the native protein (Ellis and Henney, 1969).

Critical information is lacking regarding the types and quantities of aggregates needed to generate immune responses for any given therapeutic protein product (Marszal and Fowler, 2012), although there is evidence that higher-molecular-weight aggregates and particles are more potent in eliciting such responses than lower-molecular-weight aggregates (Dintzis et al., 1989; Bachmann et al., 1993; Joubert et al., 2012). The aggregates formed and the quantities that efficiently elicit immune responses also may differ for different products and in different clinical scenarios. Furthermore, the use of any single method for the assessment of aggregates is not sufficient to provide a robust measure of protein aggregation. For example, the sole use of size exclusion chromatography may preclude detection of higher-molecular-weight aggregates that fail to traverse the column prefilter, yet may be the most crucial species in generating immune responses. Moreover, it has been recognized that subvisible particulates in the size range of 0.1–10 microns have a strong potential to be immunogenic, but are not precisely monitored by currently employed technologies (Berkowitz, 2006; Roda et al., 2009; Gross and Zeppezauer, 2010; Mahler and Jiskoot, 2012). These very large aggregates may contain thousands to millions of protein molecules and may be homogeneous or heterogeneous (e.g., protein molecules adhered to glass or metal particles).

Recommendations

It is critical for manufacturers of therapeutic protein products to minimize protein aggregation to the extent possible. Strategies to minimize aggregate formation should be developed as early as feasible in product development. This can be done by using an appropriate cell substrate, selecting manufacturing conditions that minimize aggregate formation, employing a robust purification scheme that removes aggregates to the greatest extent possible, and choosing a formulation (see Section 17.5.2.7) and container closure system (see Section 17.5.2.8) that minimize aggregation during storage. It is particularly important that product expiration dating take into account any increase in protein aggregates associated with protein denaturation or degradation during storage.

Methods that individually or in combination enhance detection of protein aggregates should be employed to characterize distinct species of aggregates in a product. Methods for measuring aggregation are constantly evolving and improving. Constant improvement and development of these methods should be considered in choosing one or more appropriate assays. Assays should be validated for use in routine lot release and stability evaluations, and several of them should be employed for comparability assessments. Animal studies may be useful in identifying aggregate species that have the potential to be immunogenic, although additional considerations (amount and types of aggregates, route of administration, etc.) may determine the extent to which such aggregate species pose clinical risk.

Assessment should be made of the range and levels of subvisible particles (2–10 microns) present in therapeutic protein products initially and over the course of the shelf life. Several methods are currently qualified to evaluate the content of subvisible particulates in this size range (Mahler and Jiskoot, 2012). As more methods become available, sponsors should strive to characterize particles in smaller (0.1–2 microns) size ranges. Sponsors should conduct a risk assessment of the impact of these particles on the clinical performance of the therapeutic protein product and develop control and mitigation strategies based on that assessment, when appropriate.

17.5.2.4 Glycosylation/pegylation

Glycosylation may strongly modulate immunogenicity of therapeutic protein products. Although foreign glycoforms such as mammalian xenogeneic sugars (Chung et al., 2008; Ghaderi et al., 2010), yeast mannans (Bretthauer and Castellino, 1999), or plant sugars (Gomord and Faye, 2004) may trigger vigorous innate and acquired immune responses, glycosylation of proteins with conserved mammalian sugars generally enhances product solubility and diminishes product aggregation and immunogenicity. Glycosylation indirectly alters protein immunogenicity by minimizing protein aggregation, as well as by shielding immunogenic protein epitopes from the immune system (Wei et al., 2003; Cole et al., 2004). Pegylation of therapeutic protein products has been found to diminish their immunogenicity via similar mechanisms (Inada et al., 1995; Harris et al., 2001), although immune responses to the polyethylene glycol (PEG) itself have been recognized and have caused loss of product efficacy and adverse safety consequences (Liu et al., 2011). Anti-PEG antibodies have also been found to be cross-reactive between pegylated products (Garay et al., 2012; Schellekens et al., 2013).

Recommendations

For proteins that are normally glycosylated, the use of a cell substrate production system and appropriate manufacturing methods that glycosylate the therapeutic protein product in a nonimmunogenic manner is recommended.

For pegylated therapeutic protein products, the ADA assay should be able to detect both the anti-protein antibodies and antibodies against the PEG moiety. The same principle may apply to modifications where the therapeutic protein products that are not pegylated but are modified with other high molecular weight entities, e.g., hydroxyethyl starch.

17.5.2.5 Impurities with adjuvant activity

Adjuvant activity can arise through multiple mechanisms, including the presence of microbial or host-cell-related impurities in therapeutic protein products (Verthelyi and Wang, 2010; Rhee et al., 2011; Eon-Duval et al., 2012; Kwissa et al., 2012). These innate immune response modulating impurities (IIRMIs), including lipopolysaccharide, β–glucan and flagellin, high-mobility group protein B1 (HMGB1), and nucleic acids, exert immune-enhancing activity by binding to and signaling through toll-like receptors or other pattern-recognition receptors present on B-cells, dendritic cells, and other antigen-presenting cell populations (Iwasaki and Medzhitov, 2010; Verthelyi and Wang, 2010). This signaling prompts maturation of antigen-presenting cells and/or serves to directly stimulate B-cell antibody production.

Recommendations

It is very important for manufacturers to minimize the types and amounts of such microbial or host-cell-related impurities in therapeutic protein products.

Assays to evaluate the types of IIRMIs present should be tailored to the relevant cell substrate. Because even trace levels of IIRMIs can modify the immunogenicity of a therapeutic protein product, the assays used to detect them should have sensitivities to assess levels that may lead to clinically relevant immune responses.

If biomarkers are used to detect and compare the presence of IIRMIs, they should be tailored to the IIRMIs that could be present in the product. Examples of biomarkers could include cytokine release and transcription factor activation from defined cell populations.

17.5.2.6 Immunomodulatory properties of the therapeutic protein product

The immunomodulatory activity of any given therapeutic protein product critically influences not only the immune response directed to itself but also immune responses directed to other co-administered therapeutic protein products, endogenous proteins, or even small drug molecules and may not be predictable. For example, interferon-alpha (Gogas et al., 2006; Tovey and Lallemand), interleukin-2 (Franzke et al., 1999), and GM-CSF (Hamilton, 2008) are not only relatively immunogenic of themselves but also are known to upregulate immune responses to endogenous proteins and to induce clinical autoimmunity. Immunosuppressive therapeutic proteins may globally downregulate immune responses, raising the possibility of serious infections. However, not all immunosuppressive therapeutic protein products suppress responses to themselves. For example, integrin and TNF monoclonal antibodies tend to be immunogenic. Thus, the immunogenicity of such therapeutic protein products should be evaluated empirically.

Recommendations

The immunomodulatory properties of therapeutic protein products, their effects on immune responses to themselves, and their capacity to induce autoimmunity should be monitored from the earliest stages of product development (Franzke et al., 1999; Gogas et al., 2006; Hamilton, 2008).

Vaccination using live attenuated organisms should be avoided when the therapeutic protein product is immunosuppressive. Updated vaccination status, compliant with local health care standards, is recommended for patients before administration of the therapeutic protein product.

17.5.2.7 Formulation

Formulation components are principally chosen for their ability to preserve the native conformation of the therapeutic protein in storage by preventing denaturation due to hydrophobic interactions, as well as by preventing chemical degradation, including truncation, oxidation, and deamidation (Cleland et al., 1993; Shire et al., 2004; Wakankar and Borchardt, 2006). Large protein excipients in the formulation, such as human serum albumin (HSA), may affect immunogenicity positively or negatively. Excipients such as HSA, although added for their ability to inhibit hydrophobic interactions, may coaggregate with the therapeutic protein or form protein adducts under suboptimal storage conditions (Braun and Alsenz, 1997). Polysorbate, a nonionic detergent, is the most commonly used alternative to HSA. The stability of both types of excipients (i.e., HSA and polysorbate) should be kept in mind for formulation purposes because they too are subject to modifications (e.g., oxidation), which may then pose a threat to the integrity of the therapeutic protein product.

Formulation may also affect immunogenicity of the product by altering the spectrum of leachables from the container closure system. Leachables from rubber stoppers have been shown to possess immune adjuvant activity, as shown in an animal experiment (Mueller et al., 2009). Organic compounds with immunologic activity as well as metals have been eluted from container closure materials by polysorbate-containing formulations, leading to increased oxidation and aggregation (Seidl et al., 2012).

Recommendations

Excipients should be evaluated for their potential to prevent denaturation and degradation of therapeutic protein products during storage. Interactions between excipients and therapeutic proteins should be carefully evaluated, especially in terms of co-aggregation or formation of protein-excipient adducts.

Excipient stability should be carefully considered when establishing product shelf life. Thorough analyses of leachables and extractables should be performed to evaluate the capacity of container closure materials to interact with and modify the therapeutic protein product. A risk assessment should be conducted, and control and risk mitigation strategies should be developed as appropriate.

17.5.2.8 Container closure considerations

Interactions between therapeutic protein products and the container closure may negatively affect product quality and immunogenicity. These interactions are more likely with prefilled syringes of therapeutic protein products. These syringes are composed of multiple surfaces and materials that interact with the therapeutic protein product over a prolonged time period and thus have the potential to alter product quality and immunogenicity.

The following are other container closure considerations pertinent to immunogenicity:

- Glass and air interfaces can denature proteins and cause aggregation in glass syringes and vials.
- Glass vials have been known to delaminate at higher pH and with citrate formulations, potentially creating protein-coated glass particles, which may enhance immunogenicity of the therapeutic protein product (Fradkin et al., 2011).
- Silicone oil-coated syringe components provide a chemical and structural environment on which proteins can denature and aggregate.
- Appropriate in-use stability studies should be performed to confirm that conditions needed to maintain product quality and prevent degradation are adequately defined.
- Leached materials from the container closure system may be a source of materials that enhance immunogenicity, either by chemically modifying the therapeutic protein product or by having direct immune adjuvant activity, including the following:
  o Organic compounds with immunomodulatory activity may be eluted from container closure materials by polysorbate-containing formulations: a leachable organic compound involved in vulcanization was found in a polysorbate formulated product when the stopper surfaces were not Teflon coated (Boven et al., 2005).
  o Metals that oxidize and aggregate therapeutic protein products or activate metalloproteinases have been found in various products contained in prefilled syringes or in vials. For example, tungsten oxide that leached from the syringe barrel was reported to cause protein aggregation (Bee et al., 2009); and leached metals from vial stoppers caused increased proteolysis of a therapeutic protein because of activation of a metalloproteinase that co-purified with the product.

Recommendations

Whenever possible, sponsors should obtain detailed information regarding a description of all raw materials used in the manufacture of the container closure systems for their products. Sponsors should conduct a comprehensive extractables and leachables laboratory assessment using multiple analytical techniques to assess the attributes of the container-closure system that could interact with and degrade protein therapeutic products.

Because the United States Pharmacopeia *elastomeric closures for injections* tests do not adequately characterize the impact of leachables in storage containers on therapeutic protein products under real-time storage conditions, leachables must be evaluated for each therapeutic protein product in the context of its storage container under real-time storage conditions.[8]

Testing for leachables should be performed on the product under stress conditions,[9] as well as under real-time storage conditions, because in some cases the amount of leachables increases dramatically over time and at elevated temperatures. Product compatibility testing should be performed to assess the effects of container closure system materials and all leachables on product quality.

17.5.2.9 Product custody

Products in their intended primary packaging container closure system should be tested for stability in protocols that include appropriate in-use conditions (e.g., light, temperature, and agitation) to identify conditions and practices that may cause product denaturation and degradation.

Given that most therapeutic protein products degrade on exposure to heat and light or with mechanical agitation, to ensure product quality, health care practitioners and patients should be educated regarding product storage, handling, and administration.

A secure supply chain is critical. Appropriate temperature-controlled transport and storage is of utmost importance in preserving product quality. For example, the storage of epoetin-α under inappropriate conditions by unauthorized vendors was associated with high levels of aggregates and antibody-mediated pure red cell aplasia (Fotiou et al., 2009).

---

[8] 21 CFR 600.11(b) and (h).
[9] See the ICH guidance for industry *Q1A(R2) Stability Testing Of New Drug Substances And Products*.

Recommendations

Patient educational materials (e.g., FDA-approved patient labeling providing instructions for use as required under 21 CFR 201.57 and 201.80) should explicitly identify appropriate storage and handling conditions of the product. Appropriate patient instruction by caregivers is vital to ensuring product quality and helping to minimize adverse impacts on product quality during product storage and handling. Appropriate temperature-controlled transport and storage should be ensured.

17.6 Conclusion

The consequences of immune responses to therapeutic protein products can range from no apparent effect to serious adverse events, including life-threatening complications such as anaphylaxis, neutralization of the effectiveness of lifesaving or highly effective therapies, or neutralization of endogenous proteins with nonredundant functions. Although immunogenicity risk factors pertaining to product quality attributes and patient/protocol factors are understood, immune responses to therapeutic protein products cannot be predicted based solely on characterization of these factors, but should be evaluated in the clinic. A risk-based approach, as delineated in this guidance document, provides investigators with the tools to develop novel protein therapeutics, evaluate the effect of manufacturing changes, and evaluate the potential need for tolerance-inducing protocols when severe consequences result from immunogenicity.

Appendix A

(1) **Diagnosis of Anaphylaxis**

The diagnosis of anaphylaxis is based on the following three clinical criteria, with anaphylaxis considered as highly likely when one of these criteria is fulfilled (Sampson et al., 2006):

(a) Acute onset of an illness (minutes to several hours) with involvement of the skin, mucosal tissue, or both (e.g.,

generalized hives, pruritus or flushing, swollen lips-tongue-uvula), **and at least one of the following:**
- Respiratory compromise (e.g., dyspnea, wheeze-bronchospasm, stridor, reduced peak expiratory flow, hypoxemia)
- Reduced blood pressure or associated symptoms of end-organ dysfunction (e.g., hypotonia [collapse], syncope, incontinence)

(b) Two or more of the following that occur rapidly after exposure *to a* **likely** *allergen for that patient* (minutes to several hours):
- Involvement of the skin-mucosal tissue (e.g., generalized hives, itch-flush, swollen lips-tongue-uvula)
- Respiratory compromise (e.g., dyspnea, wheeze-bronchospasm, stridor, reduced peak expiratory flow, hypoxemia)
- Reduced blood pressure or associated symptoms (e.g., hypotonia [collapse], syncope, incontinence)
- Persistent gastrointestinal symptoms (e.g., crampy abdominal pain, vomiting)

(c) Reduced blood pressure after exposure to a **known** allergen *for that patient* (minutes to several hours):
- Infants and children: low systolic blood pressure (age specific) or greater than 30-percent decrease in systolic blood pressure
- Adults: systolic blood pressure of less than 90 mm Hg or greater than 30-percent decrease from that person's baseline

Although none of the clinical criteria provide 100-percent sensitivity and specificity, it is believed that these criteria are likely to capture more than 95 percent of cases of anaphylaxis (Sampson et al., 2006).

Laboratory tests for evaluating anaphylaxis:

At present, there are no sensitive and specific laboratory tests to confirm the clinical diagnosis of anaphylaxis. Skin testing and *in vitro* diagnostic tests to determine the level of specific IgE antibodies directed against the therapeutic protein product, mediator release, or basophil activation may be useful for characterizing the underlying pathophysiology and may provide insight into potential mitigation strategies (Simons, 2010; Lee and Vadas, 2011).

However, the results of unvalidated tests should be interpreted with caution; and the clinical relevance of positive results from unvalidated tests may be uncertain during product development.

(2) **Cytokine Release Syndrome**

Monoclonal antibodies specific for cell surface receptors or for cell membrane expressed cytokines, as well as antibodies that develop in patients to therapeutic protein products that bind to cell surface receptors, have the potential to augment a product's intrinsic agonist activity and exacerbate infusion-related toxicities. *In vitro* assessments of the capacity of such therapeutic protein products to mediate cellular activation, including proliferation and cytokine release in human whole blood or peripheral blood mononuclear cells, are recommended. For products with the potential to incur a cytokine release syndrome, an initial starting dose below that obtained by traditional calculations and slower infusion rates, where applicable, may also be recommended (Duff, 2006). Pre- and post-admss

Data from both animal studies and *in vitro* assessments may provide information to guide development of therapeutic protein products with the potential to induce cytokine release. Although data from both animal studies and *in vitro* assessments may supplement each other, they generally are not fully predictive of the clinical occurrence or outcome. Therefore it is imperative that great caution is always exercised in the clinical development of products with the potential to mediate receptor cross-linking (see Sections 17.3.2.1 and 17.3.2.2). Although the traditional animal models used for toxicology testing (i.e., rat, mouse, dog, and cynomolgus monkey) rarely demonstrate overt toxicities related to lymphocyte activation and cytokine release, specific markers related to T-cell activation and cytokine release can be measured in routine toxicology studies, provided that the drug is pharmacologically active in the test species. These data may then be useful for predicting the potential for these agents to induce a cytokine release syndrome in the clinic or for evaluating the activity of second-generation agents that have been modified to reduce their level of T-cell activation. For example, cytokine production can be measured in blood samples obtained from treated animals during pharmacokinetic or general toxicology studies, provided that the amount of samples obtained does not compromise the health of the animals or the ability to evaluate the toxicology endpoints at the study termination point. When the evaluation of cytokine release is included in animal testing, the measurement of a cytokine panel that is as broad as possible and includes IL-2, IL-6, IFN-γ, and TNF-α, as well as other relevant cytokines indicative of cytokine release syndrome, is recommended. Such proposed animal studies should be discussed with FDA prior to initiation (Hsu et al., 1999; Norman et al., 2000). *In vitro* assessments of cellular activation, including proliferation and cytokine release in human whole blood or peripheral blood mononuclear cells, are important assessment tools that can help in overcoming the known limitations of animals in modeling activating stimuli in some T-cell subsets (Stebbings et al., 2007; Hellwig et al., 2008; Findlay et al., 2011; Romer et al., 2011; Stebbings et al., 2013). The impact of product cross-linking of cellular receptors should be considered in such studies. Signs of cellular activation *in vitro* should also be taken as an indication that the product has the potential to induce toxicities in the clinic, regardless of negative findings from animal studies.

(3) Non-Acute Immune Responses

Type III hypersensitivity responses, including those mediated by immune complexes and T-cells (delayed hypersensitivity responses in the older literature), are relatively rare with respect to therapeutic protein products; and a high degree of clinical suspicion is necessary for their diagnosis (Dharnidharka et al., 1998; Hunley et al., 2004; Gamarra et al., 2006; Goto et al., 2009). Signs and symptoms of immune complex deposition may include fever, rash, arthralgia, myalgia, hematuria, proteinuria, serositis, central nervous system complications, and hemolytic anemia. Immune complexes, composed of antibody and a therapeutic protein product, have been responsible for the development of glomerulonephritis and nephrotic syndrome in patients undergoing tolerance induction treatment (with factor IX and α-glucosidase) in the face of a high titer and sustained antibody response (Dharnidharka et al., 1998; Hunley et al., 2004). There have been case reports of immune complex disease with immune responses to monoclonal antibodies (Gamarra et al., 2006; Goto et al., 2009) and situations in which large doses of a monoclonal antibody targeting high levels of a circulating multivalent antigen may increase the likelihood of immune complex deposition (Gonzalez and Waxman, 2000).

If patients develop signs or symptoms suggestive of immune complex disease, appropriate laboratory assessments for circulating immune complexes and complement activation should be undertaken; and the administration of the therapeutic protein product should be suspended. In certain situations, development of tolerance induction therapies that eliminate the antibody response may be appropriate prior to further attempts at treatment.

(4) **Antibody Responses to Therapeutic Protein Products**

Antibodies to therapeutic protein products are classified as either neutralizing or non-neutralizing. Neutralizing antibodies bind to distinct functional domains of the therapeutic protein product and preclude their activity. For example, antibodies to therapeutic enzymes may bind to either the catalytic site, blocking catalysis of a substrate, or to the uptake domain, preventing uptake of the enzyme into the cell. In rare circumstances, a neutralizing antibody may act as a *carrier* and enhance the half-life of the product and prolong its therapeutic effect. As discussed in Section 17.3 of this guidance document, non-neutralizing antibodies bind to areas of the therapeutic protein product other than specific functional domains and may exhibit a range of effects on safety and efficacy—enhanced or delayed clearance of the therapeutic protein product, which may prompt consideration of dosing changes, induction of anaphylaxis, diminished efficacy of the product by causing uptake of the therapeutic protein product into FcR-expressing cells rather than the target cells, and facilitation of epitope spreading, allowing the emergence of neutralizing antibodies. However, they may have no apparent effect on either safety or efficacy.

The development of a neutralizing antibody is expected with administration of nonhuman proteins and in patients receiving factor/enzyme replacement therapies to whom such therapeutic protein products appear as foreign. However, a neutralizing antibody to a therapeutic protein product that cross reacts to an endogenous protein does not always arise in situations in which the endogenous factor is defective or absent by genetic mutation, as in the case of hemophilia A or lysosomal storage diseases. Neutralizing antibodies can develop in healthy individuals to some normal endogenous proteins because immune tolerance to some endogenous proteins is not robust and can be broken by a therapeutic protein homolog with sufficient provocation. For example, healthy volunteers treated with a thrombopoietin (TPO)-type protein mounted a neutralizing antibody response to the therapeutic, which cross-reactively neutralized endogenous TPO, inducing a prolonged state of thrombocytopenia in those formerly healthy individuals (Li et al., 2001). Thus, treatment with therapeutic counterparts of endogenous proteins serving a unique function or endogenous proteins present at low abundance should be undertaken with utmost caution. Neutralizing a antibody to a therapeutic protein product can also be catastrophic when it neutralizes the efficacy of a lifesaving therapeutic such as therapeutic enzymes for lysosomal storage disorders, and immune tolerance induction should be considered in such circumstances (Wang et al., 2008).

The loss of efficacy of mAbs in patients caused by immune responses to the mAb can be highly problematic, and the clinical consequences should not be minimized. Sponsors may consider development of immune tolerance induction regimens in such patients.

As discussed in Section 17.2.5 of the guidance document, if the endogenous protein is redundant in biological function (e.g., Type I interferons), neutralization of the therapeutic and endogenous protein may not appear to produce an obvious clinical syndrome. However, the more subtle effects of blocking endogenous factors, even though redundant in some functions, may not be apparent until the system is stressed, as not all biological functions of a factor may be known or fully characterized. Moreover, the effects of long-term persistence of a neutralizing antibody, as have been observed, for example, in a small percentage of patients with antibodies to IFN-β (Bellomi et al., 2003), would not be known from short-term follow-up and should be studied under the long term. Generally, for products given chronically, one year or more of immunogenicity data should be collected and evaluated unless a shorter duration can be scientifically justified. However, longer-term evaluation may be warranted depending on the frequency and severity of the consequences. In some cases, these studies may be done in the postmarket setting. Agreement with the Agency should be sought regarding the extent of data required before and after marketing.

In some circumstances, antibody responses, regardless of apparent clinical effect, should be serially followed until the levels return to baseline or an alternative approach is discussed with the Agency. Moreover, for patients in whom a therapeutic protein product appears to lose efficacy, regardless of the duration of the treatment course, it is important that an assessment be undertaken to determine whether the loss of efficacy is antibody mediated.

For patients who develop neutralizing antibodies or are considered at very high risk of developing neutralizing antibodies to a lifesaving therapeutic protein product (e.g., CRIM-negative patients with a deletion mutation for a critical enzyme who are given enzyme replacement therapy), consideration should be given to tolerance induction regimens in a prophylactic setting, before or concomitant with the onset of treatment (Wang et al., 2008; Mendelsohn et al., 2009; Messinger et al., 2012). Given the degree of immune suppression of such regimens, although far less than that of a therapeutic regimen to reverse an ongoing antibody response, careful safety monitoring should be undertaken throughout the duration of the protocol.

(5) Utility of Animal Studies

Immunogenicity assessments in animals are conducted to assist in the interpretation of animal study results (e.g., toxicology studies) and in the design of subsequent clinical and nonclinical studies (for additional information, see the ICH guidance for industry *S6(R1) Preclinical Safety Evaluation of Biotechnology-Derived Pharmaceuticals*).[10] They are generally limited in their ability to predict the *incidence* of human immune responses to a therapeutic protein product, but they may be useful in describing the *consequences* of antibody responses, particularly when an evolutionarily conserved, nonredundant endogenous protein is inhibited by cross-reactive antibodies generated to its therapeutic protein product counterpart. When available, animal models including hyperimmunized mice or gene knock out mice can be used to address potential consequences of inhibition of endogenous proteins. A special case is that of endogenous proteins that are vital to embryonic or fetal development whose elimination is embryonically lethal. In such situations, the use of conditional knock out mice may be useful for assessing potential consequences of neutralizing antibodies. As in human studies, consideration should be given to the potential transmission of antibodies to developing neonates by breast milk.

In contrast to proteins that mediate biologically unique functions, animal models are generally not useful for predicting consequences of immune responses to therapeutic protein products that are counterparts to endogenous proteins with redundant biological functions. Mice that are transgenic for genes encoding human proteins, humanized mice (i.e., immune-deficient mice with human immune systems), and mouse models of human diseases are increasingly being developed and may be considered for use in addressing multiple clinical issues, including immunogenicity.

In addition to appropriate animal studies, consideration should be given to *in vitro* and *in silico* analyses that may

---

[10]ICH guidance for industry *S6(R1) Preclinical Safety Evaluation of Biotechnology-Derived Pharmaceuticals* is available at http://www.fda.gov/downloads/Drugs/GuidanceComplianceRegulatoryInformation/Guidances/UCM074957.pdf

supplement animal studies to better or further elucidate risk for immunogenicity.

(6) Comparative Immunogenicity Studies

The need for and the extent and timing of clinical immunogenicity studies in the context of evaluating the effects of a manufacturing change will depend on such factors as the degree of analytical comparability between the product before and after the manufacturing change, findings from informative comparative animal studies, and the incidence and clinical consequences of immune responses to the product prior to the manufacturing change. For example, if the clinical consequence of an immune response is severe (e.g., when the product is a therapeutic counterpart of an endogenous protein with a critical, nonredundant biological function or is known to provoke anaphylaxis), more extensive immunogenicity assessments will likely be needed.[11]

References

Bachmann, M. F., Rohrer, U. H., et al. (1993). The influence of antigen organization on B cell responsiveness. *Science*, **262**(5138), 1448–1451.

Bachmann, M. F., Zinkernagel, R. M. (1997). Neutralizing antiviral B cell responses. *Annu. Rev. Immunol.*, **15**, 235–270.

Baert, F., Noman, M., et al. (2003). Influence of immunogenicity on the long-term efficacy of infliximab in Crohn's disease. *N. Engl. J. Med.*, **348**(7), 601–608.

Barandun, S., Kistler, P., et al. (1962). Intravenous administration of human gamma-globulin. *Vox Sang.*, **7**, 157–174.

Barbosa, M. D., Celis, E. (2007). Immunogenicity of protein therapeutics and the interplay between tolerance and antibody responses. *Drug Discov. Today*, **12**(15–16), 674–681.

Bee, J. S., Nelson, S. A., et al. (2009). Precipitation of a monoclonal antibody by soluble tungsten. *J. Pharm. Sci.*, **98**(9), 3290–3301.

---

[11]Guidance on development programs for biosimilar products is available in the draft guidance for industry *Scientific Considerations in Demonstrating Biosimilarity to a Reference Product*. When final, this guidance will reflect the FDA's current thinking on this topic.

Bellomi, F., Scagnolari, C., et al. (2003). Fate of neutralizing and binding antibodies to IFN beta in MS patients treated with IFN beta for 6 years. *J. Neurol. Sci.*, **215**(1–2), 3–8.

Berkowitz, S. A. (2006). Role of analytical ultracentrifugation in assessing the aggregation of protein biopharmaceuticals. *AAPS J.*, **8**(3), E590–605.

Bonfield, T. L., Russell, D., et al. (2002). Autoantibodies against granulocyte macrophage colony-stimulating factor are diagnostic for pulmonary alveolar proteinosis. *Am. J. Respir. Cell Mol. Biol.*, **27**(4), 481–486.

Boven, K., Stryker, S., et al. (2005). The increased incidence of pure red cell aplasia with an Eprex formulation in uncoated rubber stopper syringes. *Kidney Int.*, **67**(6), 2346–2353.

Braun, A., Alsenz, J. (1997). Development and use of enzyme-linked immunosorbent assays (ELISA) for the detection of protein aggregates in interferon-alpha (IFN-alpha) formulations. *Pharm. Res.*, **14**(10), 1394–1400.

Bretthauer, R. K., Castellino, F. J. (1999). Glycosylation of Pichia pastoris-derived proteins. *Biotechnol. Appl. Biochem.*, **30**(Pt 3), 193–200.

Brooks, D. A., Hopwood, J. J., et al. (1998). Immune response to enzyme replacement therapy: Clinical signs of hypersensitivity reactions and altered enzyme distribution in a high titre rat model. *Biochim. Biophys. Acta*, **1407**(2), 163–172.

Bukhari, N., Torres, L., et al. (2011). Axonal regrowth after spinal cord injury via chondroitinase and the tissue plasminogen activator (tPA)/plasmin system. *J. Neurosci.*, **31**(42), 14931–14943.

Chung, C. H., Mirakhur, B., et al. (2008). Cetuximab-induced anaphylaxis and IgE specific for galactose-alpha-1,3-galactose. *N. Engl. J. Med.*, **358**(11), 1109–1117.

Cleland, J. L., Powell, M. F., et al. (1993). The development of stable protein formulations: A close look at protein aggregation, deamidation, and oxidation. *Crit. Rev. Ther. Drug Carrier Syst.*, **10**(4), 307–377.

Cole, K. S., Steckbeck, J. D.. et al. (2004). Removal of N-linked glycosylation sites in the V1 region of simian immunodeficiency virus gp120 results in redirection of B-cell responses to V3. *J. Virol.*, **78**(3), 1525–1539.

Colombel, J. F., Sandborn, W. J., et al. (2010). Infliximab, azathioprine, or combination therapy for Crohn's disease. *N. Engl. J. Med.*, **362**(15), 1383–1395.

Cuenca, A. G., Wynn, J. L., et al. (2013). Role of innate immunity in neonatal infection. *Am. J. Perinatol.*, **30**(2), 105–112.

Dalum, I., Jensen, M. R., et al. (1997). Induction of cross-reactive antibodies against a self protein by immunization with a modified self protein containing a foreign T helper epitope. *Mol. Immunol.*, **34**(16–17), 1113–1120.

De Groot, A. S., Moise, L., et al. (2008). Activation of natural regulatory T cells by IgG Fc-derived peptide "Tregitopes". *Blood*, **112**(8), 3303–3311.

Demeule, B., Gurny, R., et al. (2006). Where disease pathogenesis meets protein formulation: Renal deposition of immunoglobulin aggregates. *Eur. J. Pharm. Biopharm.*, **62**(2), 121–130.

Dharnidharka, V. R., Takemoto, C., et al. (1998). Membranous glomerulonephritis and nephrosis post factor IX infusions in hemophilia B. *Pediatr. Nephrol.*, **12**(8), 654–657.

Dintzis, R. Z., Okajima, M., et al. (1989). The immunogenicity of soluble haptenated polymers is determined by molecular mass and hapten valence. *J. Immunol.*, **143**(4), 1239–1244.

Disis, M. L., Goodell, V., et al. (2004). Humoral epitope-spreading following immunization with a HER-2/neu peptide based vaccine in cancer patients. *J. Clin. Immunol.*, **24**(5), 571–578.

Donnelly, R. P., Dickensheets, H., et al. (2011). Interferon-lambda and therapy for chronic hepatitis C virus infection. *Trends Immunol.* **32**(9), 443–450.

Duff, E. G. O. P. O. C. T. C. P. G. W. (2006). Expert Group on Phase One Clinical Trials: Final Report. G. W. Duff.

Ellis, E. F., Henney, C. S. (1969). Adverse reactions following administration of human gamma globulin. *J. Allergy*, **43**(1), 45–54.

Eon-Duval, A., Broly, H., et al. (2012). Quality attributes of recombinant therapeutic proteins: An assessment of impact on safety and efficacy as part of a quality by design development approach. *Biotechnol. Prog.*, **28**(3), 608–622.

Findlay, L., Eastwood, D., et al. (2011). Comparison of novel methods for predicting the risk of pro-inflammatory clinical infusion reactions during monoclonal antibody therapy. *J. Immunol. Methods*, **371**(1–2), 134–142.

Fotiou, F., Aravind, S., et al. (2009). Impact of illegal trade on the quality of epoetin alfa in Thailand. *Clin. Ther.*, **31**(2), 336–346.

Fradkin, A. H., Carpenter, J. F., et al. (2011). Glass particles as an adjuvant: A model for adverse immunogenicity of therapeutic proteins. *J. Pharm. Sci.*, **100**(11), 4953–4964.

Franzke, A., Peest, D., et al. (1999). Autoimmunity resulting from cytokine treatment predicts long-term survival in patients with metastatic renal cell cancer. *J. Clin. Oncol.*, **17**(2), 529–533.

Gamarra, R. M., McGraw, S. D., et al. (2006). Serum sickness-like reactions in patients receiving intravenous infliximab. *J. Emerg. Med.*, **30**(1), 41–44.

Gamble, C. N. (1966). The role of soluble aggregates in the primary immune response of mice to human gamma globulin. *Int. Arch. Allergy Appl. Immunol.*, **30**(5), 446–455.

Garay, R. P., El-Gewely, R., et al. (2012). Antibodies against polyethylene glycol in healthy subjects and in patients treated with PEG-conjugated agents. *Expert Opin. Drug Deliv.*, **9**(11), 1319–1323.

Garces, S., Demengeot, J., et al. (2013). The immunogenicity of anti-TNF therapy in immune-mediated inflammatory diseases: A systematic review of the literature with a meta-analysis. *Ann. Rheum. Dis.*, **72**(12), 1947–1955.

Ghaderi, D., Taylor, R. E., et al. (2010). Implications of the presence of N-glycolylneuraminic acid in recombinant therapeutic glycoproteins. *Nat. Biotechnol.*, **28**(8), 863–867.

Gogas, H., Ioannovich, J., et al. (2006). Prognostic significance of autoimmunity during treatment of melanoma with interferon. *N. Engl. J. Med.*, **354**(7), 709–718.

Gomord, V., Faye, L. (2004). Posttranslational modification of therapeutic proteins in plants. *Curr. Opin. Plant Biol.*, **7**(2), 171–181.

Gonzalez, M. L., Waxman, F. J. (2000). Glomerular deposition of immune complexes made with IgG2a monoclonal antibodies. *J. Immunol.*, **164**(2), 1071–1077.

Goodnow, C. C. (1992). Transgenic mice and analysis of B-cell tolerance. *Annu. Rev. Immunol.*, **10**, 489–518.

Goronzy, J. J., Weyand, C. M. (2013). Understanding immunosenescence to improve responses to vaccines. *Nat. Immunol.*, **14**(5), 428–436.

Goto, S., Goto, H., et al. (2009). Serum sickness with an elevated level of human anti-chimeric antibody following treatment with rituximab in a child with chronic immune thrombocytopenic purpura. *Int. J. Hematol.*, **89**(3), 305–309.

Gross, P. C., Zeppezauer, M. (2010). Infrared spectroscopy for biopharmaceutical protein analysis. *J. Pharm. Biomed. Anal.*, **53**(1), 29–36.

Hamilton, J. A. (2008). Colony-stimulating factors in inflammation and autoimmunity. *Nat. Rev. Immunol.*, **8**(7), 533–544.

Haribhai, D., Engle, D., et al. (2003). A threshold for central T cell tolerance to an inducible serum protein. *J. Immunol.*, **170**(6), 3007–3014.

Harris, J. M., Martin, N. E., et al. (2001). Pegylation: A novel process for modifying pharmacokinetics. *Clin. Pharmacokinet.*, **40**(7), 539–551.

Hellmich, B., Csernok, E., et al. (2002). Autoantibodies against granulocyte colony-stimulating factor in Felty's syndrome and neutropenic systemic lupus erythematosus. *Arthritis Rheum.*, **46**(9), 2384–2391.

Hellwig, K., Schimrigk, S., et al. (2008). Allergic and nonallergic delayed infusion reactions during natalizumab therapy. *Arch. Neurol.*, **65**(5), 656–658.

Hermeling, S., Crommelin, D. J., et al. (2004). Structure-immunogenicity relationships of therapeutic proteins. *Pharm. Res.*, **21**(6), 897–903.

Hintermann, E., Holdener, M., et al. (2011). Epitope spreading of the anti-CYP2D6 antibody response in patients with autoimmune hepatitis and in the CYP2D6 mouse model. *J. Autoimmun.*, **37**(3), 242–253.

Hoffmann, S., Cepok, S., et al. (2008). HLA-DRB1*0401 and HLA-DRB1*0408 are strongly associated with the development of antibodies against interferon-beta therapy in multiple sclerosis. *Am. J. Hum. Genet.*, **83**(2), 219–227.

Hsu, D. H., Shi, J. D., et al. (1999). A humanized anti-CD3 antibody, HuM291, with low mitogenic activity, mediates complete and reversible T-cell depletion in chimpanzees. *Transplantation*, **68**(4), 545–554.

Huang, L., Lu, J., et al. (2005). In vivo deamidation characterization of monoclonal antibody by LC/MS/MS. *Anal. Chem.*, **77**(5), 1432–1439.

Hunley, T. E., Corzo, D., et al. (2004). Nephrotic syndrome complicating alpha-glucosidase replacement therapy for Pompe disease. *Pediatrics*, **114**(4), e532–535.

Inada, Y., Furukawa, M., et al. (1995). Biomedical and biotechnological applications of PEG- and PM-modified proteins. *Trends Biotechnol.*, **13**(3), 86–91.

Iwasaki, A., Medzhitov, R. (2010). Regulation of adaptive immunity by the innate immune system. *Science*, **327**(5963), 291–295.

Jefferis, R., Lefranc, M. P. (2009). Human immunoglobulin allotypes: Possible implications for immunogenicity. *MAbs*, **1**(4), 332–338.

Joubert, M. K., Hokom, M., et al. (2012). Highly aggregated antibody therapeutics can enhance the in vitro innate and late-stage T-cell immune responses. *J. Biol. Chem.*, **287**(30), 25266–25279.

Kessler, C. M., Ortel, T. L. (2009). Recent developments in topical thrombins. *Thromb. Haemost.*, **102**(1), 15–24.

Kimchi-Sarfaty, C., Schiller, T., et al. (2013). Building better drugs: Developing and regulating engineered therapeutic proteins. *Trends Pharmacol. Sci.*, **34**(10), 534–548.

Koren, E. (2002). From characterization of antibodies to prediction of immunogenicity. *Dev. Biol. (Basel)*, **109**, 87–95.

Koren, E., Smith, H. W., et al. (2008). Recommendations on risk-based strategies for detection and characterization of antibodies against biotechnology products. *J. Immunol. Methods*, **333**(1–2), 1–9.

Kwissa, M., Nakaya, H. I., et al. (2012). Distinct TLR adjuvants differentially stimulate systemic and local innate immune responses in nonhuman primates. *Blood,* **119**(9), 2044–2055.

Lee, J. K., Vadas, P. (2011). Anaphylaxis: Mechanisms and management. *Clin. Exp. Allergy*, **41**(7), 923–938.

LeMaoult, J., Szabo, P., et al. (1997). Effect of age on humoral immunity, selection of the B-cell repertoire and B-cell development. *Immunol. Rev.*, **160**, 115–126.

Li, J., Yang, C., et al. (2001). Thrombocytopenia caused by the development of antibodies to thrombopoietin. *Blood,* **98**(12), 3241–3248.

Liu, Y., Reidler, H., et al. (2011). A double antigen bridging immunogenicity ELISA for the detection of antibodies to polyethylene glycol polymers. *J. Pharmacol. Toxicol. Methods*, **64**(3), 238–245.

Macdougall, I. C., Roger, S. D., et al. (2012). Antibody-mediated pure red cell aplasia in chronic kidney disease patients receiving erythropoiesis-stimulating agents: New insights. *Kidney Int.*, **81**(8), 727–732.

Mahler, H. C., Jiskoot, W. (2012). *Analysis of Aggregates and Particles in Protein Pharmaceuticals*, John Wiley.

Makrygiannakis, D., af Klint, E., et al. (2006). Citrullination is an inflammation-dependent process. *Ann. Rheum. Dis.*, **65**(9), 1219–1222.

Marszal, E., Fowler, E. (2012). Workshop on predictive science of the immunogenicity aspects of particles in biopharmaceutical products. *J. Pharm. Sci.,* **101**(10), 3555–3559.

Mendelsohn, N. J., Messinger, Y. H., et al. (2009). Elimination of antibodies to recombinant enzyme in Pompe's disease. *N. Engl. J. Med.,* **360**(2), 194–195.

Messinger, Y. H., Mendelsohn, N. J., et al. (2012). Successful immune tolerance induction to enzyme replacement therapy in CRIM-negative infantile Pompe disease. *Genet. Med.,* **14**(1), 135–142.

Miller, L. L., Korn, E. L., et al. (1999). Abrogation of the hematological and biological activities of the interleukin-3/granulocyte-macrophage colony-stimulating factor fusion protein PIXY321 by neutralizing anti-PIXY321 antibodies in cancer patients receiving high-dose carboplatin. *Blood,* **93**(10), 3250–3258.

Mueller, R., Karle, A., et al. (2009). Evaluation of the immuno-stimulatory potential of stopper extractables and leachables by using dendritic cells as readout. *J. Pharm. Sci.,* **98**(10), 3548–3561.

Murphy, K. (2011). The humoral immune response. In: *Janeway's Immunobiology*, 8th ed., Garland Science Publishing, New York, pp. 367–408.

Narhi, L. O., Schmit, J., et al. (2012). Classification of protein aggregates. *J. Pharm. Sci.,* **101**(2), 493–498.

Norman, D. J., Vincenti, F., et al. (2000). Phase I trial of HuM291, a humanized anti-CD3 antibody, in patients receiving renal allografts from living donors. *Transplantation,* **70**(12), 1707–1712.

Pandey, G. S., Yanover, C., et al. (2013a). Polymorphisms in the F8 gene and MHC-II variants as risk factors for the development of inhibitory anti-factor VIII antibodies during the treatment of hemophilia A: A computational assessment. *PLoS Comput. Biol.,* **9**(5), e1003066.

Pandey, G. S., Yanover, C., et al. (2013b). Endogenous factor VIII synthesis from the intron 22-inverted F8 locus may modulate the immunogenicity of replacement therapy for hemophilia A. *Nat. Med.,* **19**(10), 1318–1324.

PrabhuDas, M., Adkins, B., et al. (2011). Challenges in infant immunity: Implications for responses to infection and vaccines. *Nat. Immunol.,* **12**(3), 189–194.

Ragnhammar, P., Friesen, H. J., et al. (1994). Induction of anti-recombinant human granulocyte-macrophage colony-stimulating factor (Escherichia coli-derived) antibodies and clinical effects in nonimmunocompromised patients. *Blood,* **84**(12), 4078–4087.

Rhee, E. G., Blattman, J. N., et al. (2011). Multiple innate immune pathways contribute to the immunogenicity of recombinant adenovirus vaccine vectors. *J. Virol.*, **85**(1), 315–323.

Roda, B., Zattoni, A., et al. (2009). Field-flow fractionation in bioanalysis: A review of recent trends. *Anal. Chim. Acta*, **635**(2), 132–143.

Romer, P. S., Berr, S., et al. (2011). Preculture of PBMCs at high cell density increases sensitivity of T-cell responses, revealing cytokine release by CD28 superagonist TGN1412. *Blood,* **118**(26), 6772–6782.

Rosenberg, A. S., Worobec, A. (2004). A risk-based approach to immunogenicity concerns of therapeutic protein products—Part 2: Considering host-specific and product-specific factors impacting immunogenicity. *Biopharm. Int.*, **17**, 34–42.

Rosenberg, A. S., Worobec, A. (2005). A risk-based approach to immunogenicity concerns of therapeutic protein products—Part 3: Effects of manufacturing changes in immunogenicity and the utility of animal immunogenicity studies. *Biopharm. Int.*, **18**, 32–36.

Sampson, H. A., Munoz-Furlong, A., et al. (2006). Second symposium on the definition and management of anaphylaxis: Summary report-- Second National Institute of Allergy and Infectious Disease/Food Allergy and Anaphylaxis Network symposium. *J. Allergy Clin. Immunol.,* **117**(2), 391–397.

Schellekens, H., Hennink, W. E., et al. (2013). The immunogenicity of polyethylene glycol: Facts and fiction. *Pharm. Res.*, **30**(7), 1729–1734.

Seidl, A., Hainzl, O., et al. (2012). Tungsten-induced denaturation and aggregation of epoetin alfa during primary packaging as a cause of immunogenicity. *Pharm. Res.*, **29**(6), 1454–1467.

Seong, S. Y.,. Matzinger, P. (2004). Hydrophobicity: An ancient damage-associated molecular pattern that initiates innate immune responses. *Nat. Rev. Immunol.*, **4**(6), 469–478.

Shire, S. J., Shahrokh, Z., et al. (2004). Challenges in the development of high protein concentration formulations. *J. Pharm. Sci.*, **93**(6), 1390–1402.

Simons, F. E. (2010). Anaphylaxis. *J. Allergy Clin. Immunol.*, **125**(2 Suppl 2), S161–181.

Singh, S. K. (2011). Impact of product-related factors on immunogenicity of biotherapeutics. *J. Pharm. Sci.,* **100**(2), 354–387.

Stanley, E., Lieschke, G. J., et al. (1994). Granulocyte/macrophage colony-stimulating factor-deficient mice show no major perturbation of hematopoiesis but develop a characteristic pulmonary pathology. *Proc. Natl. Acad. Sci. U. S. A.*, **91**(12), 5592–5596.

Stebbings, R., Eastwood, D., et al. (2013). After TGN1412: Recent developments in cytokine release assays. *J. Immunotoxicol.*, **10**(1), 75–82.

Stebbings, R., Findlay, L., et al. (2007). "Cytokine storm" in the phase I trial of monoclonal antibody TGN1412: Better understanding the causes to improve preclinical testing of immunotherapeutics. *J. Immunol.*, **179**(5), 3325–3331.

Tatarewicz, S. M., Wei, X., et al. (2007). Development of a maturing T-cell-mediated immune response in patients with idiopathic Parkinson's disease receiving r-metHuGDNF via continuous intraputaminal infusion. *J. Clin. Immunol.*, **27**(6), 620–627.

Tovey, M. G., Lallemand, C. (2010). Adjuvant activity of cytokines. *Methods Mol. Biol.*, **626**, 287–309.

Verthelyi, D., Wang, V. (2010). Trace levels of innate immune response modulating impurities (IIRMIs) synergize to break tolerance to therapeutic proteins. *PLoS One*, **5**(12), e15252.

Viel, K. R., Ameri, A., et al. (2009). Inhibitors of factor VIII in black patients with hemophilia. *N. Engl. J. Med.*, **360**(16), 1618–1627.

Wakankar, A. A., Borchardt, R. T. (2006). Formulation considerations for proteins susceptible to asparagine deamidation and aspartate isomerization. *J. Pharm. Sci.*, **95**(11), 2321–2336.

Wang, J., Lozier, J., et al. (2008). Neutralizing antibodies to therapeutic enzymes: Considerations for testing, prevention and treatment. *Nat. Biotechnol.*, **26**(8), 901–908.

Weber, C. A., Mehta, P. J., et al. (2009). T cell epitope: Friend or foe? Immunogenicity of biologics in context. *Adv. Drug Deliv. Rev.*, **61**(11), 965–976.

Wei, X., Decker, J. M., et al. (2003). Antibody neutralization and escape by HIV-1. *Nature*, **422**(6929), 307–312.

Weigle, W. O. (1980). Analysis of autoimmunity through experimental models of thyroiditis and allergic encephalomyelitis. *Adv. Immunol.*, **30**, 159–273.

Chapter 18

Assay Development and Validation for Immunogenicity Testing of Therapeutic Protein Products[1]

U.S. Food and Drug Administration, Center for Drug Evaluation and Research, Silver Spring, Maryland, USA

Keywords: adverse drug reactions, antibody(ies), anti-drug antibody (ADA), assay specificity, sensitivity, robustness, selectivity, performance, precision, drug tolerance, binding antibodies, neutralizing antibodies, cross reactivity, endogenous, counterpart protein, epitope mapping, hypersensitivity, immunoglobulins (Ig), IgG, IgM, IgE, intra-assay variability, isotyping, matrix interference, minimal required dilution (MRD), monoclonal antibodies, pharmacodynamics, pharmacokinetics, preexisitng (natural) antibodies, rheumatoid factor (RF), screening assay, neutralizing assay, confirmatory assay, system suitability control, therapeutic protein product, titer(ing), cut point, antibody-drug conjugates (ADCs), quality control (QC), validation

---

[1]This guidance document has been prepared by the Office of Medical Policy in the Center for Drug Evaluation and Research in cooperation with the Center for Biologics Evaluation and Research and the Center for Devices and Radiological Health at the Food and Drug Administration. This draft guidance document when finalized, will represent the current thinking of the Food and Drug Administration (FDA or Agency) on this topic. It does not establish any rights for any person and is not binding on the FDA or the public. An alternative approach may be employed if it satisfies the requirements of the applicable statutes and regulations. To discuss an alternative approach, contact the FDA office responsible for this guidance document as listed on the title page.

Immune Aspects of Biopharmaceuticals and Nanomedicines
Edited by Raj Bawa, János Szebeni, Thomas J. Webster, and Gerald F. Audette
Copyright © 2018 Pan Stanford Publishing Pte. Ltd.
ISBN 978-981-4774-52-9 (Hardcover), 978-0-203-73153-6 (eBook)
www.panstanford.com

18.1 Introduction

This guidance document provides recommendations to facilitate industry's development and validation of immune assays for assessment of the immunogenicity of therapeutic protein products during clinical trials. Specifically, this document includes guidance regarding the development and validation of screening assays, confirmatory assays, titering assays, and neutralization assays.[2,3] For the purposes of this guidance document, immunogenicity is defined as the propensity of the therapeutic protein product to generate immune responses to itself and to related proteins or to induce immunologically related adverse clinical events. The recommendations for assay development and validation provided in this document apply to assays for detection of anti-drug antibody(ies) (ADA).[4] This guidance document may also apply to some combination products on a case-by-case basis.[5]

This document does not discuss the product and patient risk factors that may contribute to immunogenicity.[6] This guidance document, including any discussions of terminology used in this

---

[2]This document specifically does not discuss the development or validation of anti-drug antibody(ies) (ADA) assays for animal studies; however, some concepts discussed are relevant to the design of ADA studies for nonclinical testing. Refer to the International Conference on Harmonisation (ICH) guidance for industry *S6(R1) Preclinical Safety Evaluation of Biotechnology-Derived Pharmaceuticals* for more information regarding immunogenicity assessments in animal toxicology studies. Also see the guidance for industry *Immunogenicity Assessment for Therapeutic Protein Products*, where the topic "Utility of Animal Studies" is covered in more detail. We update guidances periodically. To make sure you have the most recent version of a guidance, check the FDA guidance Web page at http://www.fda.gov/RegulatoryInformation/Guidances/default.htm.

[3]For information on clinical immunogenicity assessment of proposed biosimilar biological products, see the guidance for industry *Scientific Considerations in Demonstrating Biosimilarity to a Reference Product*.

[4]This guidance document does not pertain to immunogenicity assays for assessment of immune response to preventative and therapeutic vaccines for infectious disease indications.

[5]General information on combination products is available at http://www.fda.gov/%20CombinationProducts/default.htm

[6]See the guidance for industry *Immunogenicity Assessment for Therapeutic Protein Products*, where these topics are covered in more detail.

guidance document, does not apply to in vitro diagnostic products.[7] This guidance document revises the draft guidance for industry *Assay Development for Immunogenicity Testing of Therapeutic Proteins* issued in December 2009. The information in this guidance document has been reorganized for clarity and includes new information on titering and confirmatory assays.

In general, FDA's guidance documents do not establish legally enforceable responsibilities. Instead, guidances describe the Agency's current thinking on a topic and should be viewed only as recommendations, unless specific regulatory or statutory requirements are cited. The use of the word *should* in Agency guidances means that something is suggested or recommended, but not required.

18.2   Background

Patient immune responses to therapeutic protein products have the potential to affect product safety and efficacy.[8] The clinical effects of patient immune responses are highly variable, ranging from no effect at all to extremely harmful effects to patient health. Detection and analysis of ADA formation is a helpful tool in understanding potential patient immune responses. Information on immune responses observed during clinical trials, particularly the incidence of ADA induction and the implications of ADA responses for therapeutic protein product safety and efficacy, is crucial for any therapeutic protein product development program. Accordingly, such information, if applicable, should be included in the prescribing information as a subsection of the ADVERSE

---

[7]Per 21 CFR 809.3(a), "in vitro diagnostic products are those reagents, instruments, and systems intended for use in the diagnosis of disease or other conditions, including a determination of the state of health, in order to cure, mitigate, treat, or prevent disease or its sequelae. Such products are intended for use in the collection, preparation, and examination of specimens taken from the human body. These products are devices as defined in section 201(h) of the Federal Food, Drug, and Cosmetic Act (the act), and may also be biological products subject to section 351 of the Public Health Service Act.

[8]See the guidance for industry *Immunogenicity Assessment for Therapeutic Protein Products*.

REACTIONS section entitled "Immunogenicity." Therefore, the development of valid, sensitive, specific, and selective assays to measure ADA responses is a key aspect of therapeutic protein product development.

18.3 General Principles

The risk to patients of mounting an immune response to a therapeutic protein product will vary with the product. FDA recommends adoption of a risk-based approach to evaluating and mitigating immune responses to or immunologically related adverse clinical events associated with therapeutic protein products that affect their safety and efficacy.[9] Immune responses may have multiple effects, including neutralizing activity and the ability to induce hypersensitivity responses. Immunogenicity tests should be designed to detect ADA that could mediate unwanted biological or physiological consequences.

Screening assays, also known as binding antibody (BAb) assays, are used to detect all antibodies that bind to the therapeutic protein product. The specificity of BAb for the therapeutic protein product is established using confirmatory assays. ADA are further characterized using titering and neutralization assays. Titering assays are used to characterize the magnitude of the ADA response. It is important to characterize this magnitude with titering assays because the impact of ADA on safety and efficacy may correlate with ADA titer and persistence rather than incidence (Cohen and Rivera, 2010). Neutralization assays assess the ability of ADA to interfere with the therapeutic protein product-target interactions. Therefore, neutralizing antibodies (NAb) are a subset of BAb. It is important to characterize neutralizing activity of ADA with neutralization assays because the impact of ADA on safety and efficacy may correlate with NAb activity rather than ADA incidence (Calabresi, Giovannoni, et al., 2007; Goodin, Frohman, et al., 2007; Cohen and Rivera, 2010). Similarly, it may be important in some cases to establish NAb titers. Additional

---

[9]See the guidance for industry titled *Immunogenicity Assessment for Therapeutic Protein Products*.

characterization assays, such as isotyping, epitope mapping, and assessing cross-reactivity, e.g., to endogenous counterparts or to other products, may be useful.

The optimal time to design, develop, and validate ADA assays during therapeutic protein product development depends on the risk assessment of the product (Mire-Sluis, Barrett, et al., 2004; Gupta, Indelicato, et al., 2007; Shankar, Devanarayan, et al., 2008; Gupta, Devanarayan, et al., 2011). The sponsor should provide a rationale for the immunogenicity testing paradigm, preferably at the investigational new drug application (IND) stage, during phase 1. Because ADA assays are critical when immunogenicity poses a high clinical risk (e.g., assessment of a therapeutic protein product with a non-redundant endogenous counterpart) and real-time data concerning patient responses are needed, the sponsor should implement preliminary validated assays early, before and during phase 1, and obtain data in real time. Real-time assessments entail analyses of the samples as soon as possible after sampling, before banking of the samples, and prior to additional dosing when the dosing regimen allows. In lower risk situations, the sponsor may bank patient samples so they can be tested when suitable assays are available. FDA encourages sponsors to test samples during phase 1 and phase 2 studies using suitable assays. Samples derived from pivotal studies should be tested with fully validated assays. At the time of license application, the sponsor should provide data supporting full validation of the assays. Recommendations regarding the timing of ADA sample collection can be found in Section 18.7.1.[10]

Assays for detection of ADA facilitate understanding of the immunogenicity, safety, and efficacy of therapeutic protein products. However, the detection of ADA is dependent on key operating parameters of the assays (e.g., sensitivity, specificity),

---

[10]See the guidance for industry titled *Immunogenicity Assessment for Therapeutic Protein Products*, where immunogenicity risk assessment and mitigation considerations are covered in more detail. Guidance on appropriate assay development and validation for immunogenicity testing is also available in the ICH guidances for industry titled *Q2A Text on Validation of Analytical Procedures* and *Q2B Validation of Analytical Procedures: Methodology*.

which vary between assays.[11] Although information on ADA incidence is typically included in the prescribing information under an "Immunogenicity" subsection of the ADVERSE REACTIONS section, FDA cautions that comparison of ADA incidence among products, even for products that share sequence or structural homology, can be misleading. This is because detection of ADA formation is highly dependent on the sensitivity and specificity of the assay. Additionally, the observed incidence of ADA (including NAb) positivity in an assay may be influenced by factors such as method, sample handling, timing of sample collection, concomitant medications, and disease condition. Therefore, comparing immunogenicity rates among therapeutic protein products with structural homology for the same indication is unsound, even though fully validated assays are employed. When a true comparison of immunogenicity across different therapeutic protein products that have homology is needed, it should be obtained by conducting a head-to-head clinical study using a standardized assay under the same conditions that has equivalent sensitivity and specificity for both therapeutic protein products.[12]

The recommendations on assay development and validation provided in this guidance document are based on common issues encountered by the Agency upon review of immunogenicity submissions. Sponsors should contact FDA for any product-specific guidance. Isotyping and cross-reactivity assay designs should be discussed with FDA. Other publications may also be consulted for additional insight (see Mire-Sluis, Barrett, et al., 2004; Gupta, Indelicato, et al., 2007; Shankar, Devanarayan, et al., 2008; Gupta, Devanarayan, et al., 2011). In general, FDA recommends that sponsors develop assays that are optimized for sensitivity, specificity, selectivity, precision, reproducibility, and robustness (see Sections 18.4.3–18.4.7).

---

[11]See the United States Pharmacopeia (USP) General Chapter titled 1106 *Immunogenicity Assays–Design and Validation of Immunoassays to Detect Anti-Drug Antibodies* for a broader discussion of various assay types.

[12]For information on proposed biosimilar products, see the guidance for industry titled *Scientific Considerations in Demonstrating Biosimilarity to a Reference Product*.

18.4 Assay Design Elements

This section applies to all types of assays for detection of ADA, unless specified otherwise.

18.4.1 Testing Strategy

18.4.1.1 Multi-tiered testing approach

FDA recommends a multi-tiered ADA testing approach because of the size of some clinical trials and the necessity of testing patient samples at several time points. In this paradigm, a rapid, sensitive screening assay is initially used to assess clinical samples. The initial screening assay should be sensitive to low levels of low- and high-affinity ADA (see Section 18.5.1). Samples testing positive in the screening assay are then subjected to a confirmatory assay to demonstrate that ADA are specific for the therapeutic protein product. For example, a competition assay could confirm that antibody is specifically binding to the therapeutic protein product and that the positive finding in the screening assay is not a result of non-specific interactions of the test serum or detection reagent with other materials in the assay milieu such as plastic or other proteins.

Samples identified as positive in the confirmatory assay should be further characterized in other assays, such as titering and neutralization assays. In some cases, assays to detect cross-reactivity to other proteins with homology, such as the corresponding endogenous protein, may be needed. Further, tests to assess the isotype of the antibodies and their epitope specificity may also be recommended once samples containing antibodies are confirmed as positive.

18.4.1.2 Immunoglobulin isotypes

The initial screening assay should be able to detect all relevant immunoglobulin (Ig) isotypes. For non-mucosal routes of administration, and in the absence of anaphylaxis, the expected ADA isotypes are IgM and IgG. For mucosal routes of administration, IgA isotype ADA are also expected. Although FDA expects that all relevant isotypes be detected in screening assays, it is not necessary that the screening assay establish which isotypes are being detected. For example, assays using the bridging format may provide no information on which isotypes are being detected. Bridging assay format can theoretically detect antibodies of most isotypes, but may not detect IgG4 isotypes. In some circumstances the sponsor should develop assays that discriminate between antibody isotypes. For example, for therapeutic protein products where the risk for anaphylaxis is a concern, antigen-specific IgE assays should be developed. In addition, the generation of IgG4 antibodies has been associated with immune responses generated under conditions of chronic antigen exposure, such as with factor VIII treatment, and in erythropoietin-treated patients with pure red cell aplasia (Matsumoto, Shima, et al., 2001; Aalberse and Schuurman, 2002). Consequently, depending on the clinical concern, assessing for specific isotypes may be needed.

18.4.1.3 Epitope specificity

FDA recommends that the sponsor direct initial screening tests against the whole therapeutic protein product and, when relevant, its endogenous counterpart. For some therapeutic protein products, the sponsor may need to investigate the ADA to specific epitopes to which immune responses are specifically generated. For example, determination of epitope specificity is recommended for some fusion molecules because the region where the two molecules join may form a neoantigen, and immune responses to this region may arise. Because of epitope spreading, immune responses to other parts of the molecule may ensue, leading to the generation of antibodies to the therapeutic protein product or its endogenous counterpart (Prummer, 1997; Miller, Korn, et al., 1999; Disis, Goodell, et al., 2004; Thrasyvoulides, Liakata, et al., 2007; van der Woude, Rantapaa-Dahlqvist, et al., 2010; Hintermann, Holdener, et al., 2011). For these therapeutic protein products, FDA encourages sponsors to investigate the initiating event in the immune cascade. This knowledge may allow for modification to the protein to reduce its potential immunogenicity. Similarly, for therapeutic protein products with modifications, such as PEGylation, sponsors should develop assays to determine the specificity of ADA for the protein component as well as the modification to the therapeutic protein product. Also see, Sections 18.4.11.4 and 18.4.11.5.

18.4.2 Assay Cut Point

The cut point of the assay is the level of response of the assay that defines the sample response as positive or negative. Information specific to establishing the cut point for the respective assay types is provided in Sections 18.5 and 18.6. Establishing the appropriate cut point is critical to ensuring acceptable assay sensitivity.

The cut point of the assay can be influenced by a myriad of interfering factors, such as pre-existing antibodies, rheumatoid factor (RF), human anti-mouse antibodies, and the levels of product-related material or homologous proteins in the matrix. These factors should be considered early on in assay development when defining the cut point. Because samples from different target populations and disease states may have components that can cause the background signal from the assay to vary, different cut points may be needed for discrete populations being studied.

The cut point should be statistically determined using samples from treatment-naïve subjects.[13] By performing replicate assay runs with these samples, the variability of the assay can be estimated. During assay development, a small number of samples may be used to estimate the cut point. This may be done with as few as 5–10 samples from treatment-naïve subjects.

The specific approach employed to determine the cut point will depend on various factors. Specifically, because the cut point should identify any samples that produce a signal beyond that of the variability of the assay, the sponsor should consider the impact of statistically determined outlier values as well as true-positive samples when establishing the cut point. The sponsor should provide justification for the removal of any data points, along with the respective method used to determine their status as outliers. Positive values and samples may derive from non-specific serum factors or the presence of pre-existing antibodies in patient samples (Ross, Hansen, et al., 1990; Turano, Balsari, et al., 1992; Coutinho, Kazatchkine, et al., 1995; Caruso and Turano, 1997; van der Meide and Schellekens, 1997;

---

[13]Treatment-naïve subjects could be healthy individuals or a patient population not exposed to therapeutic protein product, depending on the stage of assay development or validation and on the availability of samples. Sponsors should provide justification for the appropriateness of the samples used.

Boes, 2000). Although pre-existing antibodies to a variety of endogenous proteins are present in healthy individuals, these can be much higher in some disease states. The sponsor should identify those samples with pre-existing antibodies, for example, through immunodepletion approaches, and remove them from the cut point analysis. If the presence of pre-existing antibodies is a confounding factor, it may be necessary to assign positive responses or a cut point based on the difference between individual patient results before and after exposure. It is possible to arrive at a reasonable value to define assay cut point through careful design consideration, such as utilizing the minimal required dilution (MRD) of the sample, removing statistical outliers from analyses, minimizing the impact of interfering factors, improving assay drug tolerance, and using an approach to account for pre-existing antibodies.

18.4.3 Sensitivity

18.4.3.1 Assay sensitivity

The sponsor should determine the sensitivity of the assay to have confidence when reporting immunogenicity rates. Assay sensitivity represents the lowest concentration at which the antibody preparation consistently produces either a positive result or readout equal to the cut point determined for that particular assay.[14] FDA recommends that screening and confirmatory ADA assays achieve a sensitivity of at least 100 nanograms per milliliter (ng/mL). Although traditionally FDA has recommended sensitivity of at least 250–500 ng/mL, recent data suggest that concentrations as low as 100 ng/mL may be associated with clinical events (Plotkin, 2010; Zhou, Hoofring, et al., 2013). However, it is understood that neutralization assays may not always achieve that level of sensitivity.

The assays should have sufficient sensitivity to enable detection of low levels of ADA before the amount of ADA reaches levels that can be associated with altered pharmacokinetic, pharmacodynamic, safety, or efficacy profiles. Because assessment of patient antibody levels will occur in the presence of biological matrix,

---

[14]See, USP General Chapter titled 1106 *Immunogenicity Assays—Design and Validation of Immunoassays to Detect Anti-Drug Antibodies* for a discussion on *Relative Sensitivity*.

testing of assay sensitivity should be performed with the relevant dilution of the same biological matrix (e.g., serum or plasma, with the same anticoagulant as the diluent, from the target population). The final sensitivity should be expressed as mass of antibody detectable/mL of undiluted matrix. Therefore, assay sensitivity should be reported after factoring in the MRD. Assay sensitivity should not be reported as titer. During development, sensitivity should be assessed using both individual as well as pooled samples from treatment-naïve subjects so that the suitability of the negative control can be established.

Assay sensitivity should be determined by testing serial dilutions of a positive control antibody of known concentration in pooled negative control matrix. The dilution series should be no greater than two- or threefold, and a minimum of five dilutions should be tested. Alternatively, sensitivity can be calculated by interpolating the linear portion of the dilution curve to the assay cut point. As noted previously, assay sensitivity should be reported in mass units per volume of undiluted matrix.

A purified preparation of antibodies specific to the therapeutic protein product should be used to determine the sensitivity of the assay so that assay sensitivity can be reported in mass units/mL of matrix. Antibodies used to assess sensitivity can take the form of affinity purified polyclonal preparations or monoclonal antibodies (mAb).

A low positive system suitability control containing a concentration of ADA slightly above the sensitivity of the assay should be used to ensure that the sensitivity of the assay is consistent across assay runs. The low positive system suitability control should be designed to fail in 1% of the runs (see Section 18.4.9.1).

18.4.3.2 Drug tolerance

Therapeutic protein product or the endogenous counterpart present in the serum may interfere with the sensitivity of the assay. Specifically, complexes formed between ADA and the therapeutic protein product, also called ADA-drug complexes, that prevent detection of ADA in the test format can form if product-related materials are present in the test sample. This is because ADA assays are generally designed to detect uncomplexed ADA. The assessment of assay sensitivity in the presence of the expected levels of interfering therapeutic protein product, also known as the assay's drug tolerance, is critical to understanding the suitability of the method for detecting ADA in dosed patients.[15] FDA recommends that the sponsor examine assay drug tolerance early in assay development. The sponsor may examine drug tolerance by deliberately adding different known amounts of purified ADA into individual ADA-negative control samples in the absence or presence of different quantities of the therapeutic protein product under consideration and determining quantitatively whether the therapeutic protein product interferes with ADA detection. Results obtained in the absence and presence of different quantities of the therapeutic protein product under consideration should be compared. There should be a relationship between the quantity of antibody and the amount of therapeutic protein product required for a specified degree of inhibition. Data from pharmacokinetic studies may be useful in establishing optimal sample collection times. Acid dissociation pretreatment or other approaches may be used to disrupt circulating ADA-drug complexes, which may lead to increased assay drug tolerance. Interference from the therapeutic protein product can be minimized if the sponsor collects patient samples at a time when the therapeutic protein product has decayed to a level where it does not interfere with assay results.

18.4.4 Specificity and Selectivity

Demonstrating assay specificity and selectivity is critical to the interpretation of immunogenicity assay results. Specificity refers to the ability of a method to detect ADA that bind the therapeutic protein product but not assay components such as surfaces or reagents. The assays should exclusively detect the target analyte, in this case the ADA.[16] The selectivity of an ADA assay is its ability to identify therapeutic protein product-specific ADA in a matrix such as serum or plasma that may contain potential interfering substances. Assay results may be affected by

---

[15]See, USP General Chapter titled 1106 *Immunogenicity Assays–Design and Validation of Immunoassays to Detect Anti-Drug Antibodies.*

[16]See, USP General Chapter titled 1106 *Immunogenicity Assays–Design and Validation of Immunoassays to Detect Anti-Drug Antibodies.*

interference from the matrix or from on-board therapeutic protein product.[17] Lack of assay specificity or selectivity can lead to false-positive results, which could obscure relationships between ADA response and clinical safety and efficacy measures. Demonstrating the specificity and selectivity of antibody responses to mAb, Fc-fusion protein, and Ig-fusion proteins poses particular challenges because of the high concentration of Ig in human serum. The sponsor should clearly demonstrate that the assay method specifically detects anti-mAb and not the mAb product itself, non-specific endogenous antibodies, or antibody reagents used in the assay. Similarly, for patient populations with a high incidence of RF, the sponsor should demonstrate that RF does not interfere with the detection method. Host cell proteins and other product-related impurities may interfere with demonstrating the assay specificity and selectivity as well.

A straightforward approach to addressing specificity and selectivity is to demonstrate that binding can be blocked by soluble or unlabeled purified therapeutic protein product. One approach is to incubate positive and negative control antibody samples with the purified therapeutic protein product or its components under consideration. Inhibition of signal in the presence of the relevant therapeutic protein product or its components demonstrates that the response is specific and selective. For responses to mAb products, inclusion of another mAb with the same Fc but different variable region can be critical. For responses to other proteins, an unrelated protein of similar size and charge can be used. If the assay is specific and selective for the protein in question, generally the addition of that protein in solution should reduce the response to background or the cut point, whereas the addition of an unrelated protein of similar size and charge should have no effect. Conversely, addition of the protein in question should have little effect on antibodies specific to an unrelated protein. Selectivity should further be evaluated by performing recovery studies, in which positive control antibodies are spiked into matrix at defined concentrations, and the positive control antibody signal is compared to that obtained from antibody spiked into assay buffer alone.

---

[17]See, USP General Chapter titled 1106 *Immunogenicity Assays–Design and Validation of Immunoassays to Detect Anti-Drug Antibodies.*

18.4.4.1 Matrix interference

An important consideration is how interference from the assay matrix, which is composed of the sample and the diluent, can affect assay performance. Components in the matrix other than therapeutic protein product can interfere with assay results. For example, different anticoagulants used during sample collection may have different effects in the assay, potentially affecting the assay sensitivity and linearity. Sponsors should evaluate different salt anticoagulant sample collection solutions for their effect on assay results.

Endogenous and exogenous components in serum or plasma may influence assay results, and it is usually necessary to dilute patient samples for testing to minimize such effects. The sponsor should examine the effect of such interferents by performing spike-and-recovery studies. The sponsor should define the dilution factor that will be used for preparation of patient samples before performing validation studies assessing potential interference of this matrix on assay results (see Section 18.4.4.2 on MRD).

Buffer components that are chemically related to the therapeutic protein product may also interfere in the assay. For example, polysorbate is chemically similar to polyethylene glycol (PEG) and therefore may interfere in the detection of anti-PEG antibodies. The chemical composition of the buffer should be carefully considered during assay development.

The sponsor may examine matrix interference by spiking different known amounts of purified ADA into the assay buffer in the absence or presence of different matrix components. Comparing the recovery of ADA in buffer alone with that in the matrix can provide input on the degree of interference from matrix components. Furthermore, such analysis may guide decisions on the MRD recommended for sample testing. In addition, the sponsor should examine other parameters affecting patient samples, such as hemolysis, lipemia, presence of bilirubin, and presence of concomitant medications that a patient population may be using. Samples that have very high antibody titers may need additional testing, such as with different dilutions of the competing product in the confirmatory assay, to ensure their identification.

18.4.4.2 Minimal required dilution

Matrix components can contribute to non-specific signal if undiluted, thereby obscuring positive results. Therefore, there is frequently a need to dilute patient samples to maintain a reasonable ability to detect ADA (sensitivity). Ideally, the MRD is the sample dilution that yields a signal close to that of the assay diluent and allows for the highest signal-to-noise ratio. MRD typically ranges from 1:5 to 1:100.

FDA recommends that the sponsor determine the MRD from a panel of appropriate number of samples from treatment-naïve subjects. Determination of MRD usually involves serially diluting treatment-naïve ADA-negative samples, as well as testing known amounts of purified antibody (at high, medium, and low concentrations) in serially diluted matrix in comparison to the same amount of antibody in buffer. This ensures a reasonable signal-to-noise ratio throughout the range of the assay. The MRD should be calculated using at least 10 individual serum samples; the appropriate number of samples will depend on various factors, including the variability of the individual samples.

Although the MRD ultimately selected by the sponsor will depend on the assay design and patient population, FDA recommends that dilutions not exceed 1:100. Higher dilution may result in the spurious identification of a negative response when patients may actually possess low levels of therapeutic protein product-specific antibodies, the occurrence of which can be related to significantly altered pharmacokinetics, pharmacodynamics, safety, or efficacy profiles. However, in some instances greater initial dilutions may be required, and the overall effect of such dilutions on assay sensitivity and immunogenicity risk assessment should be considered.

18.4.5 Precision

Precision is a measure of the variability in a series of measurements for the same material run in a method. Results should be reproducible within and between assay runs to assure adequate precision.[18] Demonstrating assay precision is critical to the assessment of ADA because assay variability is the basis for determining the cut points and ensuring that low positive samples are detected as positive. To provide reliable estimates, the sponsor should evaluate both intra-assay (repeatability) and inter-assay (intermediate precision) variability of assay responses.

18.4.6 Reproducibility

Reproducibility is an important consideration if an assay will be run by two or more independent laboratories during a study, and a sponsor should establish the comparability of the data produced by each laboratory.[19] In addition, the assays should have the same precision between different laboratories under the established assay operating conditions (for example, using the same instrument platform).

18.4.7 Robustness and Sample Stability

Assay robustness is an indication of the assay's reliability during normal usage[20] and is assessed by the capacity of the assay to remain unaffected by small but deliberate variations in method and instrument performance that would be expected under relevant, real-life circumstances in routine laboratory practice. For example, changes in temperature, incubation times, or buffer characteristics, such as pH and salt concentration, can all impact assay results. The complexity of bioassays makes them particularly susceptible to variations in assay conditions, and it is essential to

---

[18] For more information on precision, see the guidance for industry titled *Bioanalytical Method Validation*. Also see, USP General Chapter 1106 titled *Immunogenicity Assays–Design and Validation of Immunoassays to Detect Anti-Drug Antibodies*.

[19] For more information on reproducibility, see the guidance for industry *Bioanalytical Method Validation*. Also see, USP General Chapter 1106 titled *Immunogenicity Assays–Design and Validation of Immunoassays to Detect Anti-Drug Antibodies*, USP General Chapter 1225 titled *Validation of Compendial Procedures*, and the ICH guidance for industry *Q2B Validation of Analytical Procedures: Methodology*.

[20] For more information on robustness, see the ICH guidance for industry *Q2B Validation of Analytical Procedures: Methodology*. Also see, USP General Chapter 1106 titled *Immunogenicity Assays–Design and Validation of Immunoassays to Detect Anti-Drug Antibodies*.

evaluate and optimize parameters such as cell passage number, incubation times, and culture media components. The sponsor should examine robustness during the development phase, and if small changes in specific steps in the assay affect results, specific precautions should be taken to control their variability. FDA recommends storing patient samples in a manner that preserves antibody reactivity at the time of testing. FDA recommends that the sponsor avoid freeze-thaw cycles because freezing and thawing patient samples may also affect assay results. However, studies evaluating long-term stability of positive control antibodies may be useful.[21]

18.4.8 Selection of Format

A number of different assay formats and instrumentation are available that can be employed for detection of ADA. These include, but are not limited to, direct binding assays, bridging assays, and equilibrium binding assays. Each assay format has advantages and disadvantages, including rapidity of throughput, sensitivity, selectivity, dynamic range, ability to detect various Ig isotypes, ability to detect rapidly dissociating antibodies, and availability of reagents. One of the major differences between each of these assay formats is the number and vigor of washes, which can have an effect on assay sensitivity. All assays should be evaluated for their ability to detect rapidly dissociating antibodies such as IgM, which are common in early immune responses. Failure to detect such antibodies in early immune responses to therapeutic protein products may result in under-detection of true-positive antibody samples. Epitope exposure is also important to consider because binding to plastic or coupling to other agents, such as reporters (i.e., fluorochromes, enzymes, or biotin), can result in conformational changes of the antigen that can obscure, expose, modify, or destroy relevant antibody binding sites on the therapeutic protein product in question.

18.4.9 Selection of Reagents

Many components of the assays for ADA detection may be standard or obtained from commercial sources, for example,

---

[21]For more information on stability studies, see the FDA's guidance for industry titled *Bioanalytical Method Validation.*

commercially available reagents such as Protein A/G coated resins used in the depletion approach for confirmatory assays. Other components, however, including positive control antibodies, negative controls, and system suitability controls, may need to be generated specifically for the particular assay.

18.4.9.1 Development of positive control antibodies

Sponsors may use different or the same positive control antibodies to establish and monitor system suitability during routine assessment of assay performance, as well as to determine that the assay employed is fit for purpose. For system suitability controls, a positive control antibody, either mono- or polyclonal, used at concentrations adjusted to control the cut point and dynamic range levels, may be suitable.

Positive control antibodies frequently are generated by immunizing animals in the absence or presence of adjuvants. FDA recommends that positive control antibodies generated by immunizing animals be affinity purified using the therapeutic protein product. This approach enriches the polyclonal antibody preparation for ADA, which enables a more accurate interpretation of sensitivity assessment results. The selection of animal species when generating positive control antibodies should be carefully considered. For example, if an anti-human Ig reagent will be used as a secondary reagent to detect patient antibodies, the positive control antibodies and quality control (QC) samples should be detectable by that same reagent. When the positive control antibody is not detectable by that same reagent, an additional secondary reagent to detect the positive control antibody may be needed. In those cases, an additional positive control antibody for the secondary reagent used to detect human antibodies should be implemented to ensure that the reagent performs as expected. In some instances, the sponsor may be able to generate a positive control antibody from patient samples.[22] Although such antibodies can be very valuable, such samples are generally not available in early trials. Alternatively, individual mAb or panels of mAb may be used for positive control antibodies. Sponsors should discuss with FDA alternative

---

[22] Proper informed consent from patients is needed and should be planned ahead of time.

approaches to assay development and validation in the rare event that a sponsor is not able to generate a positive control antibody.

Ideally, the positive control antibody used to determine assay applicability for the purpose of the respective assay should reflect the anticipated immune response that will occur in humans. For therapeutic mAb, the sponsor should give special consideration to the selection of a positive control antibody for the assay. When animals are immunized with a chimeric, humanized, or human mAb to develop a positive control antibody, the humoral response may be against the human Fc and not the variable region of the molecule. Such positive control antibodies may not be relevant for the anticipated immune response in patients where the response is primarily directed to the antigen-binding regions.

Once a source of a positive control antibody has been identified, the sponsor should use that source to assess assay performance characteristics such as sensitivity, selectivity, specificity, and reproducibility. FDA recommends that sponsors generate and reserve positive control antibody solution for use as a quality or system suitability control. For assay development and validation, dilutions should be representative of a high, medium, and low value in the assay. This is needed even for qualitative assays to understand whether assay performance is acceptable across a broad range of antibody concentrations. Although high- and low-value QC samples should be used, medium-value QC samples for detection of ADA are generally not needed for monitoring system suitability during routine assessment of assay performance.

18.4.9.2 Development of negative controls

For negative control samples, it is recommended that when possible, the control population should have the same disease condition. The control samples should represent a similar gender, age, and concomitant medications so that the sample matrix is representative of the study population. Similarly, control samples should be collected and handled in the same manner as study samples with respect to, for example, type of anticoagulant used, sample volume, and sample preparation and storage, because these pre-analytical variables can impact the performance of control samples in the assay. It is frequently the case that such control samples are not available for use during development or pre-study validation exercises. In those situations, it is acceptable to use purchased samples or samples from healthy donors, but important parameters of assay performance such as cut point, sensitivity, and selectivity should be confirmed when samples from treatment-naïve subjects from the appropriate target population become available.

FDA recommends that the sponsor establish a negative control for validation studies and patient sample testing. In this regard, a pool of sera from an appropriate number of treatment-naïve subjects can serve as a useful negative control. Importantly, the value obtained for the negative control should be below but close to the cut point determined for the assay in the patient population being tested. Negative controls that yield values far below the mean value derived from individual serum samples used to establish the cut point may not be useful in ensuring proper assay performance.

18.4.9.3 Detection reagent consideration

The selection of a suitable detection reagent (i.e., reporter) depends on the assay format chosen. It is critical to minimize the non-specific signal from the detection reagent. The detection reagent chosen should have the adequate sensitivity required for the particular assay. These factors should be taken into consideration when deciding on the detection reagent.

18.4.9.4 Controlling non-specific binding

Every reagent, from the plastic of the microtiter plates to the developing agent, can affect assay sensitivity and non-specific binding. One of the most critical elements is the selection of the proper assay buffer and blocking reagents used to prevent non-specific binding to the solid surface. The sponsor should carefully consider the number and timing of wash steps as well as the detergents added to the assay buffer (i.e., blocking or wash buffer) to reduce background noise, but still maintain sensitivity. A variety of proteins can be used as blocking reagents to provide acceptable signal-to-noise ratio. However, these proteins may not all perform equivalently in specific immunoassays. For example, they may not bind well to all types of solid phases or may show unexpected cross-reactivity with the detecting reagent. Therefore, the sponsor may need to test several blocking agents to optimize assay performance. Moreover, including uncoated wells is insufficient to assess non-specific binding. Rather, determining the capacity of ADA to bind to an unrelated protein of similar size and charge that may be present in the sample may prove to be a better test of binding specificity.

18.4.10 Reporting Results for Qualitative and Semi-Quantitative Assays

Several approaches may be used to report positive antibody responses, and the appropriateness of the approach used should be evaluated on a case-by-case basis. The most common approach is qualitative, with patients reported as having a positive or negative antibody response.

For patients who are confirmed to be ADA positive, determining antibody levels can be informative because it allows for the stratified assessment of ADA levels and their impact on safety and efficacy. These relationships may not be elucidated unless ADA levels are determined. Positive antibody responses may be reported as a titer (e.g., the reciprocal of the highest dilution that gives a readout at or just above the cut point of the assay), when appropriate. The MRD should be factored in the calculations of titers and provided when reporting titers. Reporting levels of antibodies in terms of titers is appropriate and generally understood by the medical community. Values may also be zreported as amount of mass units of therapeutic protein product neutralized per volume serum with the caveat that these are arbitrary in vitro assay units and cannot be used to directly assess therapeutic protein product availability in vivo.

Unless the assay method used allows for independent determination of mass, antibody levels reported in mass units are generally not acceptable because they are based on interpolation of data from standard curves generated with a positive control antibody, and parallelism between the reference standard and test article cannot be assumed. Thus, FDA does not consider it necessary nor desirable for the sponsor to report patient antibody results in terms of mass units unless (1) the results are determined by quantitative means or (2) a universally accepted and accessible source of validated antibody is available as a control and parallelism between the dilution curves of the control antibody and patient samples has been demonstrated. Furthermore, even if parallelism is demonstrated, because the reference standard and test articles are likely to contain different populations of antibodies, the absolute mass units cannot be calculated. Therefore, FDA understands that the mass units reported are relative rather than absolute values.

18.4.11 Other Considerations for Assay Development

A myriad of factors can affect the assessment of antibody levels, such as patient sample variability, therapeutic protein product-dose response of the cells used to generate the standard curve in a cell-based neutralization bioassay, affinity and avidity of the ADA, and concentration of competing product in confirmatory assays. Accounting for such factors is important to understand and analyze assay variability and avoid errors. Common factors that should be considered include the following:

18.4.11.1 Pre-existing antibodies

A growing body of evidence in the medical literature suggests that B-cells and T-cells with specificity for a number of self-proteins exist naturally and may even be heightened in some disease states, such as in patients subjected to cytokine therapy or suffering from a variety of immunological or immunoinflammatory diseases (Coutinho, Kazatchkine, et al., 1995; van der Meide and Schellekens, 1997; Boes, 2000). For example, antibodies to interferon can be found in normal individuals (Ross, Hansen, et al., 1990; Turano, Balsari, et al., 1992; Caruso and Turano, 1997). Less surprisingly, subjects may have pre-existing antibodies to foreign antigens, such as bacterial products, most likely as a result of exposure to the organism or cross-reactivity. Pre-existing antibodies may have clinical effects and may affect the efficacy of the therapeutic protein product being tested. An alternative to the qualitative screening assay approach may be needed to assess the quantity and quality of ADA when pre-existing antibodies are present. For example, testing samples for an increase in ADA using a semi-quantitative assay type such as a titering assay (see Sections 18.5.3 and 18.6.4) can provide information on the impact of a therapeutic protein product on product immunogenicity that is not provided by a qualitative assay.

18.4.11.2 Rheumatoid Factor

Measuring immune responses to therapeutic protein products that possess Ig tails, such as mAb and Fc-fusion proteins, may be particularly difficult when RF is present in serum or plasma. RF is generally an IgM antibody that recognizes IgG, although other Ig specificities have been noted. Consequently, RF will bind Fc regions, making it appear that specific antibody to the therapeutic protein product exists. Several approaches for minimizing interference from RF have proven useful, including treatment with aspartame (Ramsland, Movafagh, et al., 1999) and careful optimization of reagent concentrations so as to reduce background binding. When examining immune responses to Fc-fusion proteins in clinical settings where RF is present, FDA recommends developing an assay specific for the non-Fc region of the proteins.

18.4.11.3 Monoclonal antibodies

Some special considerations pertain to the detection of antibodies against mAb. Animal-derived mAb, particularly those of rodent origin, are expected to be immunogenic with the immune response directed against the whole mAb molecule. In the early days of the therapeutic mAb industry, this was a key reason for the failure of clinical trials (Kuus-Reichel, Grauer, et al., 1994).

Technologies reducing the presence of non-human sequences in mAb, such as chimerization and humanization, have led to a dramatic reduction but not elimination of immunogenicity. In these cases, the immune responses are directed largely against the variable regions of the mAb (Harding, Stickler, et al., 2010; van Schouwenburg, Kruithof, et al., 2014). As immune responses against the variable regions of human mAb are anticipated, FDA does not expect that the use of human mAb will further reduce immunogenicity by a significant margin. The assays that can detect the reactivity against variable regions are considered more appropriate to evaluate the potential impact of antibodies against mAb-based therapeutics in patients. However, engineering of Fc portion (e.g., modification of the levels of afucosylation) in human antibodies may affect immunogenicity. Many of these concerns also pertain to Fc-fusion proteins containing a human Fc region.

18.4.11.4 Conjugated proteins

Because antibody-drug conjugates (ADCs) are antibodies conjugated with small molecule drugs, they represent a classic hapten-carrier molecule. Therefore, the immunogenicity assays should be able to measure the responses to all components of the ADC therapeutic protein product, including the antibody, linker-drug, and new epitopes that may result from conjugation. When ADCs need to be labeled for immunogenicity assays, the conjugation should be performed carefully because ADCs are already modified. The potential for increased hydrophobicity of the labeled molecules may cause aggregation, and therefore the stability and solubility of these capture reagents should be adequately characterized.

18.4.11.5 Products with multiple functional domains

Some proteins possess multiple domains that function in different ways to mediate clinical efficacy. An immune response to one domain may inhibit a specific function while leaving others intact. Examination of immune responses to therapeutic protein products with multiple functional domains may require development of multiple assays to measure immune responses to different domains of the molecules.

18.5 Assay Development

Information specific to development of respective assay types is provided in Sections 18.5.1–18.5.4 below. These sections supplement information relevant to all assay types provided in Section 18.4.

18.5.1 Development of Screening Assay

Based on the multi-tiered approach discussed previously in Section 18.4.1, the first assay to be employed for detection of ADA should be a highly sensitive screening assay that detects low- and high-affinity ADA. Approximately 10 individual samples may be used to estimate the cut point early in assay development; however, this may need to be adjusted when treatment-naïve samples from the target population become available. A low but defined false-positive rate is desirable for the initial screening assay because it maximizes detection of true positives. Subsequent assays can be employed to exclude false-positive results when determining the true incidence of immunogenicity.

18.5.2 Development of Confirmatory Assay

Because the screening assay is designed to broadly detect the presence of antibodies that bind product in serum samples with a defined false-positive rate, FDA recommends that the sponsor develop assays to confirm the binding of antibodies that are specific to the therapeutic protein product. Implementation of a suitable confirmatory assay is important to prevent data on ADA false-positive patients from confounding the analyses of the impact of ADA on safety and efficacy.

18.5.2.1 Selection of format for confirmatory assay

It is expected that the selected confirmatory assay will be at least as sensitive as the screening assay but have higher specificity and at least as good selectivity in order to identify any false-positive samples. The method and instrument platform selected may be similar to or different from those used for the screening assay. Frequently, both screening and confirmatory assays use the same method and instrument platform. In such cases, the sensitivity of each assay will need to be determined in mass units and confirmed using system suitability controls to ensure that the assay is sensitive to the presence of binding antibody. When using a binding competition assay, the concentration of competing product should be optimized to confirm the presence of antibodies throughout and above the range of the assay.

18.5.2.2 Cut point of confirmatory assay

If a competitive inhibition format is selected, a recommended approach to determining the cut point uses the data from the binding of antibody-negative treatment-naïve patient samples in the presence of the competitor, which is usually the therapeutic protein product. In this case, the amount of therapeutic protein product used to establish the cut point should be the same as the amount of therapeutic protein product that will be used as a competitive inhibitor in the assay. However, this approach may not be appropriate when dealing with samples where pre-existing antibodies are present in the treatment-naïve population. In those cases, the sponsor should exclude true positives from the cut point assessment. In rare cases when baseline negative samples are not available, sponsors may evaluate changes in titer or use an orthogonal method to confirm samples that screen positive.

18.5.3 Development of Titering Assay

18.5.3.1 Titer determination

Titers are defined as the maximal dilution where a sample gives a value above the screening cut point. Titers are often informative and can be linked to clinical impact of the ADA. Titering assays can be particularly informative when patients have pre-existing antibodies. Titering assays most often are performed using the same platform as the screening assay. Sera are tested in sequential dilutions. Alternatively, titer may be determined by extrapolating the dilution to the assay cut point using the linear portion of the dose response curve.

18.5.3.2 Cut point of titering assay

When patients have pre-existing ADA, treatment-boosted ADA responses may be identified by post-treatment increases in titer. A cut point for defining the treatment-emergent or boosted responses is needed. Frequently this cut point is determined as a titer that is two dilution steps greater than the pre-treatment titer, when twofold dilutions are used to determine the titer. If titer is established by extrapolating the dilution curve to the assay cut point, treatment-emergent responses may be determined using estimates of assay variability.

18.5.4 Development of Neutralization Assay

In vitro neutralization assays provide an indication of the potential of the ADA to inhibit the biological activity of the product. Such NAb can interfere with the clinical activity of a therapeutic protein product by preventing the product from reaching its target or by interfering with receptor-ligand interactions. The testing method selected to assess neutralizing potential for ADA-positive samples should be based on the mechanism of action of the therapeutic protein product.

18.5.4.1 Selection of format for neutralization assay

Two formats of assays have been used to measure NAb activity: cell-based bioassays and non-cell-based competitive ligand-binding assays. Selection of the appropriate assay format depends on various factors. These factors include, but are not limited to, the mechanism of action of the therapeutic protein product, its ability to reflect the in vivo situation most closely, and the selectivity, sensitivity, precision, and robustness of the assay. FDA recommends that neutralization assays use a cell-based bioassay format depending on the therapeutic protein product's mechanism of action because, frequently, cell-based bioassays more closely reflect the in vivo situation and therefore provide more relevant information than ligand-binding assays. Because the cell-based bioassays are often based on the product's potency, historically the format of these assays has been extremely variable. The choice and design of potency bioassays are generally based on a cell line's ability to respond to the product in question and the potency bioassay's relevance to the therapeutic protein product's mechanism of action.

The cellular responses measured in these bioassays are numerous and can include outcomes such as phosphorylation of intracellular substrates, calcium mobilization, proliferation, and cell death. In some cases, sponsors have developed cell lines to express relevant receptors or reporter constructs. When therapeutic protein products directly stimulate a cellular response, the direct effect of NAb on reducing bioactivity in the bioassay can be measured. When therapeutic protein products indirectly impact cellular activity; for example, by blocking a receptor-ligand interaction, the indirect effect of the NAb on restoring bioactivity in a bioassay can be measured. Generally, bioassays have significant variability and a limited dynamic range for their activity curves. Such problems can make development and validation of neutralization assays difficult.

There are cases when ligand-binding assay formats may be used. One such case is when sufficiently sensitive or selective cell-based bioassays cannot be developed. Another case is when the therapeutic protein product does not have a cell-based mechanism of action; for example, enzyme therapeutic protein products that target serum proteins. Ligand-binding assays may also be appropriate for therapeutic protein products that bind serum ligands, preventing them from interacting with their receptor. However, cell-based bioassays may still be more appropriate for such therapeutic protein products to demonstrate that ADA are inhibiting cellular activity. Sponsors should discuss using ligand-binding assays with FDA in such cases.

18.5.4.2 Activity Curve of Neutralization Assay

The sponsor should carefully consider the dose response curve (product concentration versus activity) before examining other elements of neutralization assay validation. Assays with a small dynamic range may not prove useful for determination of neutralizing activity. Generally, the neutralization assay will employ a single concentration of therapeutic protein product with a single dilution of antibody. Consequently, the sponsor should choose a therapeutic protein product concentration whose activity readout is sensitive to inhibition. If the assay is performed at concentrations near the plateau of the dose-response curve (marked "No" in Fig. 18.1, below), it may not be possible to discern samples with low amounts of NAb. FDA recommends that the neutralization assay be performed at therapeutic protein product concentrations that are on the linear range of the curve (marked "Yes" in Fig. 18.1). The assay should also give reproducible results.

The x-axis (Concentration) indicates a concentration of the therapeutic protein product, and the y-axis (Activity) indicates resultant activity; for example, the concentration of cytokine secretion of a cell line upon stimulation with the therapeutic protein product. The curve demonstrates a steep response to a therapeutic protein product that plateaus at approximately 300.

The "No" arrow indicates a concentration of a therapeutic protein product that would be inappropriate to use in a single dose neutralization assay because it would represent a range of concentrations where the activity induced by the therapeutic protein would be relatively insensitive to inhibition by NAb. The "Yes" arrow represents a range of concentrations on the linear part of the curve where the activity induced by the therapeutic protein product would be sensitive to neutralization by antibody.

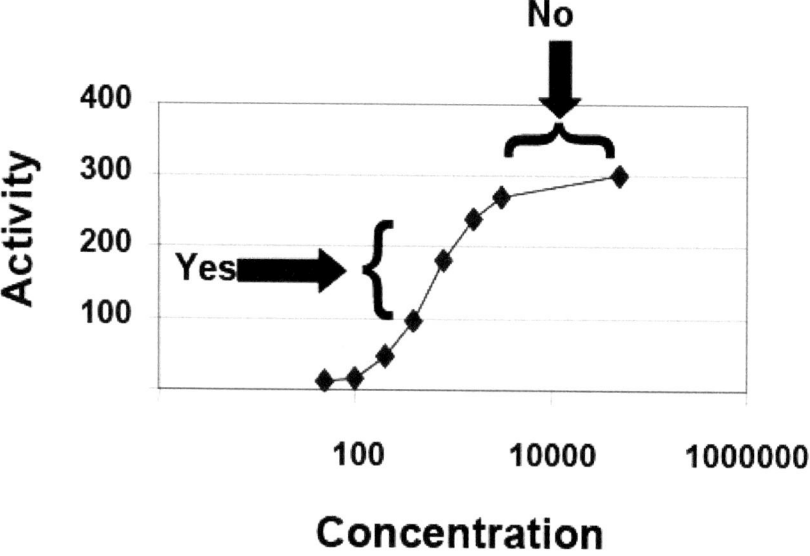

Figure 18.1 Activity curve for a representative therapeutic protein product.

18.5.4.3 Considerations for matrix interference for neutralization assay

The matrix can cause interference with neutralization assays, particularly as serum or plasma components may enhance or inhibit the activity of a therapeutic protein product in bioassays. For example, sera from patients with particular diseases may contain elevated levels of one or more cytokines that might serve to activate cells in the bioassay and obscure the presence of NAb by increasing the response to the original stimulatory factor or therapeutic protein product. Therefore, the sponsor should understand matrix effects in these assays. Approaches such as enriching for ADA from serum or plasma samples may be appropriate for these types of situations. However, this approach may result in the loss of NAb, and consequently will require careful examination and validation by the sponsor.

The concentration of therapeutic protein product employed in the neutralization assay has a critical impact on assay sensitivity. FDA recognizes that although the use of low concentrations of therapeutic protein product may lead to a neutralization assay that is more sensitive to inhibition by antibodies, very low concentrations of therapeutic protein product may result in poor precision of the assay. Also see Section 18.4.4.1 for general information on matrix interference.

18.5.4.4 Cut point of neutralization assay

Determination of assay cut point has historically posed a great challenge for neutralization assays. As with all assays, the cut point should be determined based on the assay variability established using samples from treatment-naïve subjects. If neutralization assays are performed on samples that tested positive in screening and confirmatory assays, a 1% false-positive rate is acceptable. If neutralization assays are used for screening, a 5% false-positive rate should be used (see Section 18.6.2.2). If the degree of sample variation makes it difficult to assess NAb activity, other approaches may be considered but should be discussed with FDA before implementation. Alternatively, exploring other assay formats that lead to less variability and provide a more accurate assignment of cut point may be necessary. Also see Section 18.4.2 for general information on assay cut point.

18.2.4.5 Additional considerations for neutralization assay

Because neutralization assays are most commonly performed only on samples that are confirmed to have antigen-specific ADA, confirmatory approaches are not usually necessary. However, because of the complexity of bioassays, confirmation of assay specificity may be useful in determining whether patients have mounted a true NAb response. The sponsor should consider the following approaches:
  (a) Unrelated inhibitory molecules may cause neutralizing activity, and sometimes it may be unclear whether the

observed neutralizing activity is caused by neutralizing antibodies or by other inhibitory molecules. Test results from baseline pre-exposure samples may be informative. When there is concern that there is non-specific inhibition, antibody depletion assays should be performed to evaluate whether the neutralizing activity is truly caused by ADA and not caused by other inhibitory molecules.

(b) Cell lines may be responsive to multiple stimuli other than the therapeutic protein product under study. In such cases, the presence of NAb can be examined in the presence of the therapeutic protein product, which should be blocked by a specific NAb response, versus alternative stimuli, which should not be blocked by a specific NAb response.

(c) Serum may contain components such as soluble receptors or endogenous product counterparts that may yield false results in the neutralization assay. In such instances, adding test serum or plasma samples directly to the bioassay in the absence of therapeutic protein product may be useful in understanding assay results.

18.6 Assay Validation

Assay validation is a process of demonstrating, by the use of specific laboratory investigations, that the performance characteristics of the ADA assay employed are suitable for its intended use.[23] The level of validation depends on the stage of product development and the risks of consequences of immunogenicity to patients associated with the therapeutic protein product. A partial validation involving assessments of assay sensitivity, specificity, and precision requirements with less emphasis on robustness, reproducibility, and stability may be adequate for the earlier stages of clinical development such as phase 1 and phase 2 studies. However, as a scientific matter, as stated in Section 18.6.1, fully validated assays should be used for pivotal and postmarketing studies.

---

[23]See, USP General Chapter 1106 titled *Immunogenicity Assays–Design and Validation of Immunoassays to Detect Anti-Drug Antibodies*. Also see, the guidance for industry *Bioanalytical Method Validation*, USP General Chapter 1225 titled *Validation of Compendial Procedures*, and the ICH guidance for industry *Q2(R1)* titled *Validation of Analytical Procedures: Text and Methodology*.

Information specific to validation of respective assay types is provided in Sections 18.6.2–18.6.5. These sections supplement information relevant to all assay types provided in Sections 18.4 and 18.6.1.

18.6.1 General Considerations for Assay Validation

Samples derived from pivotal studies should be tested with fully validated assays. At the time of license application, the sponsor should provide data supporting full validation of the assays. Validation includes all of the procedures that demonstrate that a particular assay used for quantitative measurement of ADA in a given sample is reliable and reproducible for the intended use. The fundamental parameters for validation include (1) cut point, (2) sensitivity, (3) specificity and selectivity, (4) precision, (5) reproducibility when relevant, and (6) robustness of some assay features and stability of reagents and control samples. The acceptability of clinical data generated by an assay corresponds directly to the criteria used to validate the assay.

Determination of cut point is a fundamental aspect of assay validation. If treatment-naïve samples from the appropriate patient population are not available for the pre-study validation exercise, alternative samples may be used. Frequently these are samples from commercial sources. When alternative samples are used to determine the cut point in the validation exercise, the cut point should be determined again once samples from the appropriate population (e.g. treatment-naïve patients) are available. The cut point validated using the appropriate samples should be used to determine whether samples are positive for ADA.

For validation of the fundamental assay parameters, FDA recommends, at the minimum, that inter-assay precision be evaluated on at least 3 different days with two analysts each preparing a minimum of six otherwise independent preparations of the same sample using the same instrument platform and model. Intra-assay precision should be evaluated with a minimum of six independent preparations of the same sample per plate independently prepared by the same analyst. In cases where intra-assay or inter-assay precision has a coefficient of variance (%CV) greater than 20%, sponsors should consider the need to refine the assay parameters to optimize the assay precision to the extent possible or provide justification to explain why higher %CV should be acceptable. Alternatively, in assays with low throughput (e.g., titer assay) when it may not be possible to run six independent preparations of the same sample on a plate, intra-assay precision should be evaluated with a minimum of three independent preparations of the same sample per plate and at least nine total independent preparations of the same samples. Samples should include negative controls and positive samples whose testing yields values in the low, medium, and high levels of the assay dynamic range. The sponsor should evaluate inter-instrument and inter-operator precision when relevant. Assays should have comparable precision between different operators under the same operating conditions.

When changes are made to a previously validated method, the sponsor should exercise judgment as to how much additional validation is needed. During the course of a typical product development program, a defined ADA assay may undergo modifications. Occasionally, samples may need to be re-tested with the optimized validated assay; therefore, provisions should be made to preserve sufficient sample volume under conditions that allow for re-testing until the assays have been completely validated and evaluated by the Agency.[24]

Critical method parameters, for example, incubation times and temperatures, should be validated to demonstrate that the assay performs as expected within predetermined ranges for these parameters. Generally, the low, middle, and high values of the allowed range are tested in the validation exercise.

Additional parameters may need to be validated depending on the method (or technology) and instrument platform used for the assay. For example, surface plasmon resonance assays should be validated for surface stability upon regeneration, and criteria should be set for baseline performance of the chip. The efficiency and stability of the labeled[25] reagents should be established. The sponsor should examine robustness during the

---

[24]See the guidance for industry titled *Bioanalytical Method Validation* for different types and levels of validation. Also see, USP General Chapter 1106 titled *Immunogenicity Assays – Design and Validation of Immunoassays to Detect Anti-Drug Antibodies*.

[25]A reagent is considered *labeled* if it is conjugated or fused to a moiety that will aid in its capture or visualization. For example, conjugation to biotin, streptavidin, or a fluorochrome. *Unlabeled* reagent is a reagent (for example, a drug) that is not *labeled*.

development phase, and if small changes in specific steps in the assay affect results, specific precautions should be taken to control their variability.

18.6.2 Validation of Screening Assay

18.6.2.1 Sensitivity of screening assay

All the general considerations for assay validation discussed previously apply to validation of screening assay. As noted earlier, the sensitivity is particularly important in the initial screening assay because these results dictate the further analysis of the sample.

18.3.2.2 Cut point of screening assay

The cut point should be determined statistically with a minimum of 50 samples tested on at least 3 different days by at least two analysts using suitable statistical methods. FDA recommends that the cut point for screening assays be determined by a 90% one-sided lower confidence interval for the 95th percentile of the negative control population (Shen, Dong, et al., 2015). This will assure at least a 5% false-positive rate with a 90% confidence level. This approach improves the probability of the assay identifying all patients who may develop antibodies. The statistical method used to determine the cut point should be based on the statistical distribution of the data. For example, the 95th percentile of the normal distribution is estimated by the mean plus 1.645 standard deviation. Other approaches may be used for estimating 95th percentile, including the use of median and median absolute deviation value instead of mean and standard deviation.

The mean response of negative control samples may be constant or may vary between assays, plates, or analysts. When the mean is constant, a cut point may be established during assay validation that can be applied to the assay in-study. This is frequently called a fixed cut point. When the mean varies between assays, plates, or analysts but the variance around the mean is constant, a normalization factor can be statistically determined and applied in-study. This is also known as a floating cut point. When both the mean and variance vary, a cut point must be established for each assay, plate, or analyst. This is known as a dynamic cut point. One drawback of the dynamic cut point is the need to have more replicates of the negative control in the assay. Dynamic cut points should not be used to compensate for deficient assay optimization.

18.6.3 Validation of Confirmatory Assay

Confirmatory assays should be fully validated in a manner similar to screening and neutralization assays because these assays raise some specific issues. As a scientific matter, the studies to validate the assay will depend on the assay format and instrumentation chosen. If these assays are based on competition for antigen binding[26] by the antibodies in patient samples and the measurement is loss of response, it is critical to identify the degree of inhibition or depletion that will be used to ascribe positivity to a sample. In the past, fixed percentages of binding reduction were used, but these numbers were often arbitrary and are unlikely to be relevant for all assays.

FDA recommends establishing a cut point based on the assessment of the binding changes observed in samples that are known to lack the antibodies when competing antigen is added. FDA also recommends that the sensitivity of the confirmatory assay be confirmed using a low concentration of the positive control antibody.

For the estimation of the confirmatory assay cut point, an 80% one-sided lower confidence interval for the 99th percentile is recommended. Because the purpose of this assay is to eliminate false-positive samples arising as a result of non-specific binding, it is adequate to use a 1% false-positive rate for the calculation of the confirmatory cut point. The use of tighter false-positive rates such as 0.1% is not recommended because it will lead to an increased risk of false-negative results. See Section 18.4.2 for general information on assay cut point.

If the confirmatory assay format is a competiton assay in which a competitor, usually unlabeled therapeutic protein

---

[26]*Competition for antigen binding* refers to a competition assay where the ability of antigen-specific antibodies to bind to either labeled or plate-bound antigen is inhibited by unlabeled or soluble antigen.

product,[27] will be added to the reaction mixture to inhibit ADA binding to the capture reagent for the cut point assay, the same concentration of unlabeled therapeutic protein product should be added to the samples when determining the confirmatory cut point.

18.6.4 Validation of Titering Assay

The principles of assay validation described in Section 18.6.1 apply in general to validation of titering assays. The cut point of the titration assay may be the same as or different from that of the screening assay. When the titering assay is not used for screening and the cut point is different than that of the screening assay, the validation of the separate titration method cut point can become necessary; for example, when the signal from the assay diluent or matrix causes higher results than the screening assay cut point because of a blocking effect of serum or if samples at a dilution higher than the MRD do not generate consistently negative results, i.e., when the screening cut point falls on the lower plateau of the positive-control dilution curve.[28]

18.6.5 Validation of Neutralization Assay

A minimum of 30 samples tested on at least 3 different days by at least two analysts should be used to determine the cut point, using suitable statistical methods.

FDA recognizes that not all ADA are neutralizing, and it can be difficult to identify positive control antibodies with neutralizing capacity. Further, if an affinity purified polyclonal positive control antibody preparation is used, it is likely that only a portion of the antibodies are neutralizing, which can make the assay appear less sensitive. Therefore, it is important to validate assay sensitivity.

Sponsors should validate assay specificity for cell-based neutralization bioassays. As mentioned, for cells that may be responsive to stimuli other than the specific therapeutic protein product, the ability to demonstrate that NAb only inhibit the

---

[27]See footnote 25.
[28]See, USP General Chapter 1106 titled *Immunogenicity Assays — Design and Validation of Immunoassays to Detect Anti-Drug Antibodies*.

response to therapeutic protein product and not the response to other stimuli is a good indication of assay specificity. In such studies, FDA recommends that the other stimuli be employed at a concentration that yields an outcome similar to that of the therapeutic protein product. The sponsor should also confirm the absence of alternative stimuli in patient serum (see Sections 18.4.3 and 18.4.4).

Cell-based neutralization bioassays frequently have reduced precision when compared to ligand-binding assays because biologic responses can be inherently more variable than carefully controlled binding studies. Consequently, the sponsor should perform more replicates for assessment of precision and assessment of patient responses than for the screening assay (see Section 18.4.5).

Additional parameters that should be validated are assay performance when cells at the low, middle, and high range of the allowed passage numbers, cell density, and cell viability are used (see Section 18.4.7).

18.7 Implementation of Assay Testing

18.7.1 Obtaining Patient Samples

FDA recommends that the sponsor obtain pre-exposure samples from all patients. Because there is the potential for pre-existing antibodies or confounding components in the matrix, understanding the degree of reactivity before treatment is essential. The sponsor should obtain subsequent samples, with the timing depending on the frequency of dosing. Optimally, samples taken 7 to 14 days after the first exposure can help elucidate an early IgM response. Samples taken at 4 to 6 weeks after the first exposure are generally optimal for determining IgG responses. For individuals receiving a single dose of therapeutic protein product, the above time frame may be adequate. However, for patients receiving a therapeutic protein product at multiple times during the trial, the sponsor should obtain samples at appropriate intervals throughout the trial and also obtain a sample approximately 30 days after the last exposure.

Obtaining samples at a time when there will be minimal interference from the therapeutic protein product present in the serum is essential. A sponsor should consider the therapeutic protein product's half-life to help determine appropriate times for sampling. This is especially important for mAb products because these products can have half-lives of several weeks or more; and depending on the dosing regimen, the therapeutic mAb itself could remain present in the serum for months. Under circumstances when testing for IgE is needed, the timing of sample collection should be discussed with FDA.

The level of therapeutic protein product that interferes with the assay, as determined by immune competition, may also help define meaningful time points for sampling. If therapeutic protein product-free samples cannot be obtained during the treatment phase of the trial, the sponsor should take additional samples after an appropriate washout period (e.g., five half-lives). Obtaining samples to test for meaningful antibody results can also be complicated if the therapeutic protein product in question is itself an immune suppressant. In such instances, the sponsor should obtain samples from patients who have undergone a washout period either because the treatment phase has ended or because the patient has dropped out of the study.

Samples to determine serum concentrations of therapeutic protein product should be obtained at the same time as immunogenicity samples. Testing such samples can provide information on whether the therapeutic protein product in the samples may be interfering with ADA testing and whether ADA may be altering the therapeutic protein product's pharmacokinetics.

18.7.2 Concurrent Positive and Negative Quality Controls

If the sponsor completes the proper validation work and makes the cut point determinations, the immunogenicity status of patients should be straightforward to determine. However, positive control or QC samples are critical and should be run concurrently with patient samples. We recommend that these samples span a level of positivity with QC samples having a known negative, low, and high reactivity in the assay. More important, the QC samples should be diluted in the matrix in which patient samples will be examined; for example, the same percent serum or plasma (specify salt anticoagulant used). In this way, the sponsor ensures that the assay is performing to its optimal degree of accuracy and that patient samples are correctly evaluated. For the low-positive QC sample, we recommend that a concentration be selected that, upon statistical analysis, would lead to the rejection of an assay run 1% of the time. Such an approach would ensure the appropriate sensitivity of the assay when performed on actual patient samples. The concentration of high-positive QC samples should be set to monitor prozone effects.[29]

FDA also recommends that these QC samples be obtained from humans or animals possessing antibodies that are detected by the secondary detecting reagent, to ensure that negative results that might be observed are truly caused by lack of antigen reactivity and not caused by failure of the secondary reagent. This issue is not a problem for antigen bridging assays because labeled antigen is used for detection.

18.7.3 Confirmation of Cut Point in the Target Population

Samples from different populations can have different background activity in ADA assays. Therefore, it is necessary to confirm that the cut point determined during assay validation is suitable for the population being studied. Similarly, if samples used to determine the cut point during assay validation were not obtained and handled in a manner that represents how samples will be obtained and handled in-study, the cut point should also be confirmed with appropriate samples in-study. A sufficient number of samples from the target population should be used, and justification for the number used should be provided. If sufficient numbers of samples are not available, agreement with the Agency should be sought for the number of samples to be used.

18.8 Documentation

The rationale and information for the immunogenicity testing paradigm should be provided in module 5.3.1.4 of the electronic

---

[29]Prozone effects (also referred to as hook effects) are a reduction in signal that may occur as a result of the presence of a high concentration of a particular analyte or antibody and may cause false-negative results.

common technical document (eCTD) on *Reports of Bioanalytical and Analytical Methods for Human Studies*.[30] The standard operating procedure of the respective assay being used should be provided to the FDA, together with the results of the validation studies and relevant assay development information for parameters that were not validated, such as the MRD, the stimulatory concentration of therapeutic protein product used in the NAb assay, and some robustness parameters that are critical for assay performance (see Section 18.7. Documentation in the draft guidance for industry titled *Bioanalytical Method Validation*).[31]

References

Aalberse, R. C., Schuurman, J. (2002). IgG4 breaking the rules. *Immunology*, **105**(1), 9–19.

Boes, M. (2000). Role of natural and immune IgM antibodies in immune responses. *Mol. Immunol.*, **37**(18), 1141–1149.

Calabresi, P. A., Giovannoni, G., et al. (2007). The incidence and significance of anti-natalizumab antibodies: Results from AFFIRM and SENTINEL. *Neurology*, **69**(14), 1391–1403.

Caruso, A., Turano, A. (1997). Natural antibodies to interferon-gamma. *Biotherapy*, **10**(1), 29–37.

Cohen, B. A., Rivera, V. M. (2010). PRISMS: The story of a pivotal clinical trial series in multiple sclerosis. *Curr. Med. Res. Opin.*, **26**(4), 827–838.

Coutinho, A., Kazatchkine, M. D., et al. (1995). Natural autoantibodies. *Curr. Opin. Immunol.*, **7**(6), 812–818.

---

[30]See the FDA Web site for further information on eCTD submissions. For more information about the agreed-upon common format for the preparation of a well-structured Efficacy section of the CTD for applications that will be submitted to regulatory authorities, see the ICH guidance for industry *M4E: The CTD — Efficacy*. For more information on how sponsors and applicants must organize the content they submit to the FDA electronically for all submission types under section 745A(a) of the FD&C Act, see the guidance for industry (and the technical specification documents it incorporates by reference) titled *Providing Regulatory Submissions in Electronic Format — Certain Human Pharmaceutical Product Applications and Related Submissions Using the eCTD Specifications*.

[31]When final, this guidance document (chapter) will represent the FDA's current thinking on this topic. To make sure you have the most recent version of any guidance, check the FDA guidance Web page at: http://www.fda.gov/RegulatoryInformation/Guidances/default.htm

Disis, M. L., Goodell, V., et al. (2004). Humoral epitope-spreading following immunization with a HER-2/neu peptide based vaccine in cancer patients. *J. Clin. Immunol.*, **24**(5), 571–578.

Goodin, D. S., Frohman, E. M., et al. (2007). Neutralizing antibodies to interferon beta: Assessment of their clinical and radiographic impact: An evidence report: Report of the Therapeutics and Technology Assessment Subcommittee of the American Academy of Neurology. *Neurology*, **68**(13), 977–984.

Gupta, S., Devanarayan, V., et al. (2011). Recommendations for the validation of cell-based assays used for the detection of neutralizing antibody immune responses elicited against biological therapeutics. *J. Pharm. Biomed. Anal.*, **55**(5), 878–888.

Gupta, S., Indelicato, S. R., et al. (2007). Recommendations for the design, optimization, and qualification of cell-based assays used for the detection of neutralizing antibody responses elicited to biological therapeutics. *J. Immunol. Methods*, **321**(1–2), 1–18.

Harding, F. A., Stickler, M. M., et al. (2010). The immunogenicity of humanized and fully human antibodies: Residual immunogenicity resides in the CDR regions. *MAbs*, **2**(3), 256–265.

Hintermann, E., Holdener, M., et al. (2011). Epitope spreading of the anti-CYP2D6 antibody response in patients with autoimmune hepatitis and in the CYP2D6 mouse model. *J. Autoimmun.*, **37**(3), 242–253.

Kuus-Reichel, K., Grauer, L. S., et al. (1994). Will immunogenicity limit the use, efficacy, and future development of therapeutic monoclonal antibodies?. *Clin. Diagn. Lab. Immunol.*, **1**(4), 365–372.

Matsumoto, T., Shima, M., et al. (2001). Immunological characterization of factor VIII autoantibodies in patients with acquired hemophilia A in the presence or absence of underlying disease. *Thromb. Res.*, **104**(6), 381–388.

Miller, L. L., Korn, E. L., et al. (1999). Abrogation of the hematological and biological activities of the interleukin-3/granulocyte-macrophage colony-stimulating factor fusion protein PIXY321 by neutralizing anti-PIXY321 antibodies in cancer patients receiving high-dose carboplatin. *Blood*, **93**(10), 3250–3258.

Mire-Sluis, A. R., Barrett, Y. C., et al. (2004). Recommendations for the design and optimization of immunoassays used in the detection of host antibodies against biotechnology products. *J. Immunol. Methods*, **289**(1–2), 1–16.

Plotkin, S. A. (2010). Correlates of protection induced by vaccination. *Clin. Vaccine Immunol.*, **17**(7), 1055–1065.

Prummer, O. (1997). Treatment-induced antibodies to interleukin-2. *Biotherapy*, **10**(1), 15–24.

Ramsland, P. A., Movafagh, B. F., et al. (1999). Interference of rheumatoid factor activity by aspartame, a dipeptide methyl ester. *J. Mol. Recognit.*, **12**(5), 249–257.

Ross, C., Hansen, M. B., et al. (1990). Autoantibodies to crude human leucocyte interferon (IFN), native human IFN, recombinant human IFN-alpha 2b and human IFN-gamma in healthy blood donors. *Clin. Exp. Immunol.*, **82**(1), 57–62.

Shankar, G., Devanarayan, V., et al. (2008). Recommendations for the validation of immunoassays used for detection of host antibodies against biotechnology products. *J. Pharm. Biomed. Anal.*, **48**(5), 1267–1281.

Shen, M., Dong, X., et al. (2015). Statistical evaluation of several methods for cut-point determination of immunogenicity screening assay. *J. Biopharm. Stat.*, **25**(2), 269–279.

Thrasyvoulides, A., Liakata, E., et al. (2007). Spreading of antibody reactivity to non-thyroid antigens during experimental immunization with human thyroglobulin. *Clin. Exp. Immunol.*, **147**(1), 120–127.

Turano, A., Balsari, A., et al. (1992). Natural human antibodies to gamma interferon interfere with the immunomodulating activity of the lymphokine. *Proc. Natl. Acad. Sci. U. S. A.*, **89**(10), 4447–4451.

van der Meide, P. H., Schellekens, H. (1997). Anti-cytokine autoantibodies: Epiphenomenon or critical modulators of cytokine action. *Biotherapy*, **10**(1), 39–48.

van der Woude, D., Rantapaa-Dahlqvist, S., et al. (2010). Epitope spreading of the anti-citrullinated protein antibody response occurs before disease onset and is associated with the disease course of early arthritis. *Ann. Rheum. Dis.*, **69**(8), 1554–1561.

van Schouwenburg, P. A., Kruithof, S., et al. (2014). Functional analysis of the anti-adalimumab response using patient-derived monoclonal antibodies. *J. Biol. Chem.*, **289**(50), 34482–34488.

Zhou, L., Hoofring, S. A., et al. (2013). Stratification of antibody-positive subjects by antibody level reveals an impact of immunogenicity on pharmacokinetics. *AAPS J*, **15**(1), 30–40.

Chapter 19

The "Sentinel": A Conceptual Nanomedical Strategy for the Enhancement of the Human Immune System

Frank J. Boehm[a] and Angelika Domschke, PhD[b]

[a]*NanoApps Medical, Inc., Vancouver, British Columbia, Canada*
[b]*Angelika Domschke Consulting LLC, Duluth, Georgia, USA*

Keywords: human immune system, innate immune system, adaptive immune system, immune enhancement, immune augmentation, immunotherapy, immunomodulation, nanomedicine, nanotherapeutics, nanotechnology, autonomous nanomedical devices, nanomedical device, nanoparticles, nanomaterial, engineered nanomaterials, nanomedical platform, visionary, conceptual, sentinel, Global Health Care Equivalency (GHCE), Vascular Cartographic Scanning Nanodevice (VCSN), molecular manufacturing, molecular disassembly array, medical paradigm shift, T cells, B cells, drug resistance, nanopropulsion, diapedesis

19.1 Introduction

The human immune system comprises an array of cells that function in a highly orchestrated manner toward the defense of the "self" from myriad pathogenic entities, which are matched by an extensive heterogeneity of immune cells. However, dangerous viruses and bacteria may evolve strategies for circumventing the immune system, by acquiring stealth capabilities that prevent their detection and destruction. Hence, they may proliferate unchecked at the host's peril. Further, there are ever increasing instances of antibiotic resistance that have the potential to manifest as severe epidemic and pandemic health crises on a global scale [1–3].

We propose that through the emergence of advanced nanomedical technologies over the next few decades, the intrinsic human immune system might be significantly augmented through the incorporation of what we refer to as a conceptual "sentinel" nanomedical platform. It might be possible that these sentinels, which would be exceptionally "trained," would continually scan for, identify, and rapidly eradicate virtually any invading bacterial or viral pathogen, organic and inorganic toxin, or other unidentified "non-self" entity in the human body.

As the result of this nanomedical capacity, individuals would rarely become ill and might be endowed with considerably extended longevity. Hence, beneficial sentinel capacities might serve as an important facet of a far more extensive worldwide vision for nanomedicine. When combined with nanotechnology (molecular manufacturing) and artificial intelligence (imbued within quantum computers), nanomedicine may give rise to a condition that the authors refer to as "Global Health Care Equivalency" (GHCE), where every individual on the planet would have equivalent access to advanced, efficacious, and cost effective nanomedical diagnostics and therapeutics no matter where they happen to reside, or under what conditions they live.

19.2 Brief Survey of Current Nanomedical Research: Toward Immune System Augmentation

Over the last decades, knowledge associated with the immensely complex functions of the immune system has increased tremendously. While we are still deciphering this enormously multifaceted system and its interactions, research has already commenced to make significant strides toward the enhancement of specific aspects of the immune system and the engineering of artificial immune responsive entities, where indeed this developmental progress is exponential. Here we focus on only a very few highlights that closely align with the topic of this chapter.

Current research with engineered nanoparticles might have the potential to lead to the beneficial augmentation of certain facets of the human immune system though various nanomedical strategies. Contingent on the class of nanomaterials and their specific dimensional, morphological, electronic, and chemical, attributes, they may either convey a hindrance or benefit to immune cells and the integrity of innate and adaptive immune function. Hence, stringent safety and efficacy protocols should be established toward the development of these prospective nanomedical therapies to ensure that they have being thoroughly vetted prior to their implementation for any form of immune modulation.

Toward this end, Smith et al. explored how certain nanomaterials (e.g., titanium dioxide, carbon nanotubes, and fullerenes) (Fig. 19.1) may impact the immune system. In their concluding remarks, they state: "Interactions between ENMs [engineered nanomaterials] and the immune system are unavoidable, and concerns about the possibility of ENM-mediated immunomodulation promote a growing need to evaluate the effects of these novel materials on the many facets of the immune system. Conversely, ENM toxicity that is mediated by one or more types of immune cells adds a further complication. However, by investigating the bidirectional ENM–immune cell interactions, the resulting toxic effects, and the mechanisms by which these effects occur, we can better characterize the hazards these materials pose in order to select—with knowledge and forethought—appropriate nanotherapeutics for use in medicinal applications" [4].

Zhao et al. demonstrated the use of non-toxic functionalized carbon nanotubes (fCNTs) (200–400 nm long) to enhance the uptake of CpG oligodeoxynucleotides by tumor-associated inflammatory phagocytic cells, which had the effect of activating them, in vitro and in vivo. Conversely, pristine CNTs and free CpG exhibited no anti-tumor effects. It was shown that a low-dose injection of CNT–CpG complexes in mice eliminated gliomas by triggering natural killer cells and CD8 cells. The researchers reported that this nanomedical strategy may result in the development of a "more robust anti-tumor response without inducing toxicity" [5].

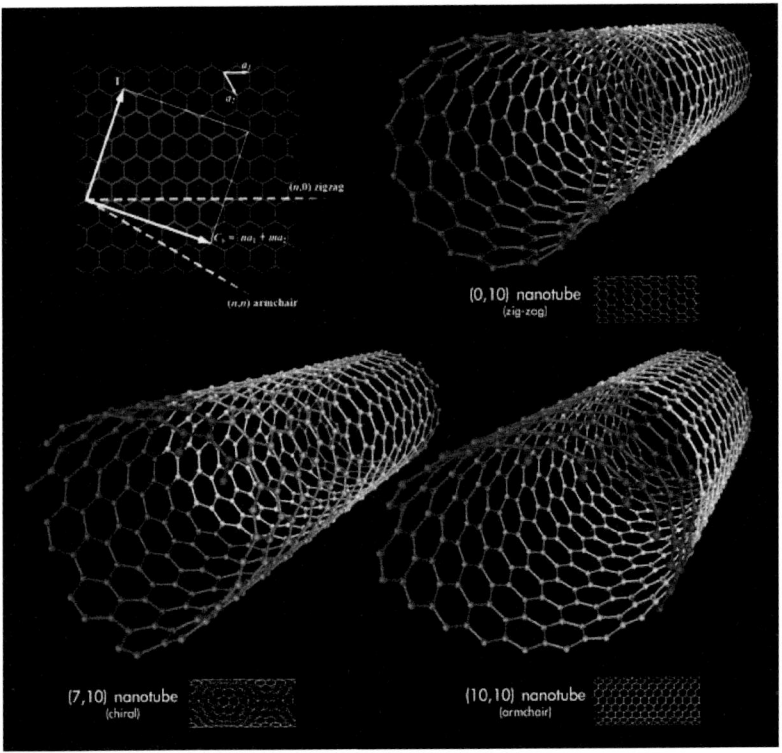

Figure 19.1 The three types of carbon nanotubes. Image credit: Michael Ströck, February 1, 2006, under Creative Commons License.

Luminescent porous silicon nanoparticles were investigated by Gu et al., who activated antigen-presenting cells with the intent of enhancing the immune system. These nanoparticles, which were host to agonistic monoclonal antibodies (FGK45) to the antigen-presenting cell CD40 receptor (co-stimulatory receptor and a necrosis factor receptor resident on dendritic cells, B cells, and macrophages), exhibited significantly augmented B cell activation. These nanoparticle stimulators elicited a cell response that corresponded to a 30–40 times higher concentration of free FGK45 monoclonal antibodies [6].

Spatz et al. recently described the development of specifically biofunctionalized gold nanoparticle-infused droplets of water-in-oil emulsions as 3D antigen-presenting cell analogs that could interact with T cells. Nanopatterning and microfluidics were employed to create these droplets, which were comprised of triblock PFPE–PEG–PFPE and diblock PFPE–PEG–Gold surfactants, combined at various concentration ratios toward obtaining stable emulsion droplets containing diverse gold NP densities. For this study, the gold nanoparticles were tethered to rhodamine B to assess efficacy; however, the aim would be to employ them as anchoring nodes for various bioactive molecules that have the capacity to custom-activate T cells. Earlier preclinical studies have shown that the "adoptive transfer of regulatory T cells can exhibit a marked beneficial impact on different autoimmune diseases" [7–9]. The researchers state: "The ability of T cells to exert forces in all three dimensions on the biomolecules held by the drop may also be important in evaluating the affinity and function of antigen receptors, ..." [10].

In an ultimate scenario, the T cells of a cancer patient might be extracted, trained with antigens that are specific to his/her particular cancer, and reintroduced into the patient to specifically target and eradicate the cancer [11].

19.3 Conceptual "Sentinel" Nanomedical Platform for the Significant Enhancement of the Human Immune System

19.3.1 Description

It is plausible that with the advent of advanced autonomous nanomedical devices, the human immune system might be significantly augmented via a conceptual "sentinel" class nanomedical platform. The sentinel (Fig. 19.2) platform would comprise autonomous approximately 1 micron in diameter nanomedical devices, which might be administered in the thousands or tens of thousands, and have the capacity to significantly enhance the innate and adaptive components of the human immune system to the degree that virtually any "non-self" pathogenic microorganism, organic/inorganic toxin, or unknown entity that an individual may be exposed to will be rapidly identified and preemptively eradicated prior to the onset of any negative physiological impact.

Figure 19.2 Artistic representation of "sentinel" class nanomedical device. Reproduced from Svidinenko, Y., Nanobotmodels Company. Figure obtained from *Nanomedical Device and Systems Design: Challenges, Possibilities, Visions* (2013) and appears here with kind permission of CRC Press/Taylor Francis Group.

19.3.2 Envisaged Capabilities

Nascent configurations of the sentinel platform are envisioned to operate in collaboration with the intrinsic human immune system, to provide a supportive role as an ultrasensitive "first responder" network to rapidly immobilize and molecularly tag any form of invasive entity for degradative processing. Further, if required, sentinel nanodevices would have the capacity to penetrate through tissues (via diapedesis) to track and eliminate intrusive species. In those instances where the identification of foreign entities is deemed inconclusive, a default protocol would be instituted, following a comprehensive evaluation, to ensure that these entities are completely eradicated through highly localized hyperthermia, cavitation, oxidation, or by nanomechanical disruption and disassembly. As sentinel nanodevices become more sophisticated via the integration of more deeply imbued advanced artificial intelligence (AI), they might serve (if deemed necessary for patients with severely compromised immune systems), as a robust (temporary or permanent) adjunct or alternative to the human immune system altogether [12].

19.3.2.1 Pathogen Eradication

It is conceivable that practically any known species of pathogenic microorganism; even unknown species thereof, might be effectively eliminated via the considerable nanomedical armamentarium that would be at the disposal of sentinel nanodevices. Cumulatively, they would serve as an insurmountable obstacle to the propagation of infectious agents, in that any attempted response-driven adaptation to circumvent sentinel capacities would be futile; these agents would still be identified as non-self and destroyed. Any pathogenic microorganism, be they virus, bacteria, fungi, mycobacteria, fungi, protozoa, or helminths (parasitic worms), would be eradicated via the infliction of irremediable damage to their structural integrity on multiple fronts, in conjunction with the subsequent degradation and destruction of their internal components.

The sentinel might be considered as micron scale Swiss Army knife that is replete, aside for other capabilities, with a selection of potent and readily deployable nanoscale pathogen destroying tools. These nanometric implements might include a retractable lance (to puncture and irreversibly disrupt the cell membranes of pathogenic agents), a gold-tipped cell-penetrating electrode to initiate highly localized hyperthermia via surface plasmon resonance, and a free radical generator for emitting guided and highly localized plumes of hydroxyl radicals and/or additional powerful oxidizing agents to incapacitate intrusive pathogens [12].

19.3.2.2 Toxin degradation

Toward the nullification of any threatening organic or inorganic toxins that may be introduced into the human body, the sentinel might house an onboard "molecular disassembly array," which would constitute a dynamic grid that is populated by thousands of diamondoid manipulator tooltips. These tooltips could enable the sentinel to immobilize and progressively proceed "through" recalcitrant toxins, and in effect, disassembling them layer by layer at the molecular level until they are deemed neutralized and non-toxic. "Optional atomic layer by atomic layer disassembly might be utilized to disable the deleterious effects of individual toxic organic/inorganic molecules" [12].

19.3.2.3 Negation of drug resistance

Significant benefits provided by sentinels may include the strong potential to relegate the existing vaccination paradigm to obsolescence. In addition to eradicating all non-self species from the human body, sentinels might also be programmed to quickly scrutinize, differentiate, and destroy any seemingly healthy cells that may have been commandeered by viruses to propagate stealth infection processes and replication. Moreover, in effect they might collaboratively function as a highly effectual stop against the proliferation of malaria (*Plasmodium falciparum*), HIV/AIDS, influenza, and "superbugs," which may evolve antibiotic resistance [13–15]. As a result, these capabilities may culminate in the establishment of a powerful "drug free" paradigm shift, through the utilization of (to reiterate from above), cell membrane disruption via highly localized oxidation, cavitation, hyperthermia, or nanomechanical degradation to eliminate any type of infectious agent and/or cellular abnormality within the human body, inclusive of metastatic cancer cells. To ensure destructive efficacy against recalcitrant pathogens or toxins, a molecular scale disassembly protocol might be engaged [12].

19.3.3 Sentinel Mode of Operation

Sentinel class nanomedical devices would be mass-uploaded with a complete set of comprehensive data to facilitate the identification of all known pathogens, chemicals, and toxins that pose threats, or potential threats, to human health. A percentage of the nanodevices could be programmed to constantly "patrol" the vasculature (down to capillary level, with the smallest capillary being ~3 microns in diameter) and lymphatic system for the presence of non-self entities, and all sentinels would have the ability, if necessary, to traverse tissues to capture and eradicate such entities. They would be endowed with quantum computational capabilities, which would work in conjunction with a range of on board nanosensor probes to specifically distinguish any suspect entities that they encounter. Sentinels would be fully autonomous in vivo by virtue of onboard propulsion and navigation, and would be able to harvest chemical energy (e.g., via glucose) from the local in vivo environment. However, for the most part, in order to conserve energy and resources, only a small portion of the full contingent of nanodevices would passively circulate through the vasculature and lymphatic system to serve as "scouts." The remaining population might power down to a hibernation/standby status, while residing unobtrusively within existing lymphoid organs until such time that they are activated, via a scout alert protocol, to rapidly and effectively deal with the presence of pathogenic or toxic intruders.

Sentinel propulsion might be enabled by oscillating piezoelectronic nanorods or fins, flagella-like appendages, or through other nanopropulsive strategies [16]. Navigation for the total population of sentinels would be facilitated by previously generated digital 3D maps of the entire vasculature and lymphatic system of a particular individual/patient via another class of nanomedical device referred to as the Vascular Cartographic Scanning Nanodevice (VCSN) [17]. These maps would also be uploaded by the sentinels en masse, and prior to their being ingressed into an individual/patient, a micron sized fiducial beacon would be deployed and self-affixed/anchored to a stable solid structure (preferably bone) where it would remain static and serve as an orientational reference node to facilitate precise navigation.

19.3.4 Advantages and Disadvantages

The most immediately evident and palpable advantageous aspect of the sentinel nanomedical platform in significantly augmenting the human immune system would be that individuals so imbued would no longer become ill, at least not from any infectious diseases, or through toxicity. They would not contract colds, influenza, or any other virally or bacterially propagated infectious diseases, including hepatitis, sexually transmitted diseases including HIV/AIDs, or other serious conditions such as tuberculosis, malaria, Ebola, Zika, etc. Additionally, cumulative arthrosclerosis associated plaque materials, cellular lipofuscin aggregates, as well as Alzheimer's related beta-amyloid plaques might be effectively eliminated via other dedicated classes of nanomedical devices.

When sufficiently advanced, the sentinel platform would negate the requirement of vaccines, as it would efficaciously eradicate virtually all known invasive viral and bacterial species, and would be programmed to destroy even unknown species thereof. Further, these individuals would be completely protected from any form of toxicity (save for perhaps radioactive poisoning, which may itself be somehow neutralized within the human body in the future) and poisoning from chemicals, or from poisonous insect, snake, or scorpion, etc., bites.

A secondary and not insignificant aspect might be that these capacities would undoubtedly have considerable positive impacts on human longevity, as the cumulatively degradative physiological and psychological tolls that typically attend various illnesses would be absent.

A potential disadvantage for the integration of the sentinel platform would center on the overall security of the system and its individual constituents. As with any sophisticated computationally based technology, it might be not be out of the realm of possibility that sufficiently focused, resourceful, and technically savvy perpetrators with nefarious intent might conceivably exploit an unforeseen vulnerability to undermine sentinel functionality. Further, despite stringent fabrication, safety, and testing protocols; with the deployment of tens of thousands of sentinels, there may be a risk of malfunction in individual units. On the topic of security, Boehm notes... "Multiple redundancies in programming and established failsafe protocols would negate the possibility that any "self" healthy cell or biological entity would even be approached by sentinels. Any deviation from established programming would result in immediate nanodevice shutdown and its subsequent retrieval for egress" [12].

19.3.5 Significant Paradigm Shifts across Medical and Social Domains

It is becoming ever clearer that our immune systems function as incredibly complex, yet extremely well orchestrated communities of autonomous entities that operate on multiple interactive levels, spanning molecular, cellular, tissue, organ, cognitive, and whole body, domains. Research derived from this knowledge is envisioned to create equally elegant and autonomously functioning systems that work in a synergistically orchestrated manner within the human body.

The discovery and creation of these autonomous systems will follow the trajectory of the rising trend across society for self-discovery and autonomous empowerment as relates to one's own health. This inclination is fortified via easy access to scientific knowledge, which bolsters the desire for making personal health decisions, while accepting individual responsibility for our health measures and overall wellbeing. The envisaged sentinel nanomedical platform constitutes but one projection of the most positive future outcomes of this movement, which the authors propose will culminate in a condition that the authors refer to as Global Health Care Equivalency (GHCE). A book titled: *Nanotechnology, Nanomedicine, and AI: Toward the Dream of Global Health Care Equivalency* (CRC Press/Taylor & Francis) is currently under preparation by the authors, which will articulate this conceptual platform, and explore how GHCE might initiate significant positive paradigm shifts across the medical and health care domains on a global scale.

Under GHCE, every individual on the planet would have equivalent access to advanced, efficacious, and cost effective nanomedical diagnostics and therapeutics, no matter where they happen to reside, or under what conditions they live. The authors elaborate further that:

> *The attainment of GHCE might serve to significantly reduce the perception of individuals in the developing world of being marginalized, at least in terms of health care, which may ultimately translate to conflict reduction. In the developed world, GHCE would serve to dramatically reduce health care expenditures across the board. Synergies between nanotechnology, nanomedicine, and AI may enable this vision on a global scale. Progress toward*

this goal will be incremental, with each successive wave of nanomedical technologies being more advanced than the previous wave.

The tipping point for GHCE will appear with the emergence of Molecular Manufacturing, a topic of the book in preparation by the authors and titled: *Molecular Manufacturing: Emergence of the Grand Equalizer* (CRC Press/Taylor & Francis). Molecular manufacturing [18] will make possible the cost effective fabrication of the sentinel, in addition to numerous other classes of advanced autonomous nanomedical devices that Boehm describes in his first book [19].

19.4 Conclusion

Over the last 40 years, incredible work has been achieved by dedicated and visionary scientists and entrepreneurs worldwide, toward deciphering the immensely complex immune system. Today we acknowledge that there indeed remain many unknowns, yet at the same time cumulative global efforts are propelling our knowledge exponentially forward at a very rapid pace, which has never before been experienced. It is hoped that this chapter might present a vision of a positive future with the advent of advanced nanomedical technologies that will beneficially change society.

Disclosures and Conflict of Interest

The opinions and perspectives here reflect the current views of the authors. The authors declare that they have no conflict of interest. No writing assistance was utilized in the production of this chapter and the authors have received no payment for its preparation.

Corresponding Author

Frank Boehm
1987 W14th Avenue
Vancouver, BC, Canada V6J 2K1
Email: frankboehm@nanoappsmedical.com

About the Authors

Frank Boehm serendipitously encountered the concept of nanotechnology on the internet in ~1996 and immediately become fascinated with its virtually limitless potential, particularly as it relates to the field of medicine. He passionately proceeded to autodidactically absorb knowledge on almost every aspect of nanotechnology and nanomedicine and began to evolve, and textually articulate, a number of advanced concepts and designs for various nanomedical components, devices, and systems. Concomitantly, he corresponded (and continues to) with myriad research scientists and thought leaders from across the globe, in the disciplines of nanotechnology and nanomedicine. In recognizing the immense potential of nanomedicine to impart positive paradigm shifts across the medicine domain (e.g., precisely targeted drug delivery, vascular and neurological plaque removal, non-invasive surgical procedures, physiological system and longevity enhancement, human space exploration) he was deeply motivated to write more extensively on the topic. In 2005, he garnered a publishing contract with CRC Press (Taylor & Francis), and over the ensuing eight years compiled a book manuscript (along with seven contributing authors) entitled: *Nanomedical Device and Systems Design, Challenges, Possibilities, Visions*. In parallel he formed the startup, NanoApps Medical, Inc., and serves as its CEO. The aim of this company is to investigate and develop advanced, innovative, and cost effective nanomedical diagnostic and therapeutic devices and systems for the benefit of individuals in both the developing and developed worlds.

Dr. Angelika Domschke is an accomplished, dynamic, and creative research and development leader with 20 years of experience, at the forefront of innovation in the design and implementation of novel materials and products for top international medical device companies and start-ups. She holds a PhD in polymer chemistry and has extensive expertise in program management, from concept to product launch. She has been awarded 26 patent families, which generated significant revenue encompassing medical devices, nanosensors, and implants, as well as materials test methods. As acting director and senior industry leader, she successfully implemented the uniquely product-driven master program in translational medicine for the City College of New York, as part of the educational portfolio of the City University, New York.

References

1. Martínez, J. L., Baquero, F. (2014). Emergence and spread of antibiotic resistance: Setting parameter space. *Ups. J. Med. Sci.*, **119**(2), 68–77.
2. Grundmann, H., Klugman, K. P., Walsh, T., Ramon-Pardo, P., Sigauque, B., Khan, W., Laxminarayan, R., Heddini, A., Stelling, J. (2011). A framework for global surveillance of antibiotic resistance. *Drug Resist. Updat.*, **14**(2), 79–87.
3. Hu, S. H., Yuan, S. X., Qu, H., Jiang, T., Zhou, Y. J., Wang, M. X., Ming, D. S. (2016). Antibiotic resistance mechanisms of Myroides sp. *J. Zhejiang Univ. Sci. B*, **17**(3), 188–199.
4. Smith, M. J., Brown, J. M., Zamboni, W. C., Walker, N. J. (2014). From immunotoxicity to nanotherapy: The effects of nanomaterials on the immune system. *Toxicol. Sci.*, **138**(2), 249–255.
5. Zhao, D., Alizadeh, D., Zhang, L., Liu, W., Farrukh, O., Manuel, E., Diamond, D. J., Badie, B. (2011). Carbon nanotubes enhance CpG uptake and potentiate antiglioma immunity. *Clin. Cancer Res.*, **17**(4), 771–782.
6. Gu, L., Ruff, L. E., Qin, Z., Corr, M., Hedrick, S. M., Sailor, M. J. (2012). Multivalent porous silicon nanoparticles enhance the immune activation potency of agonistic CD40 antibody. *Adv. Mater.*, **24**(29), 3981–3987.
7. Tang, Q., Henriksen, K. J., Bi, M., Finger, E. B., Szot, G., Ye, J., Masteller, E. L., McDevitt, H., Bonyhadi, M., Bluestone, J. A. (2004). In vitro-expanded antigen-specific regulatory T cells suppress autoimmune diabetes. *J. Exp. Med.*, **199**(11), 1455–1465.
8. Tarbell, K. V., Petit, L., Zuo, X., Toy, P., Luo, X., Mqadmi, A., Yang, H., Suthanthiran, M., Mojsov, S., Steinman, R. M. (2007). Dendritic cell-expanded, islet-specific CD4+ CD25+ CD62L+ regulatory T cells restore normoglycemia in diabetic NOD mice. *J. Exp. Med.*, **204**(1), 191–201.
9. Haile, L. A., von Wasielewski, R., Gamrekelashvili, J., Krüger, C., Bachmann, O., Westendorf, A. M., Buer, J., Liblau Manns, M. P.,

Korangy, F., Greten, T. F. (2008). Myeloid-derived suppressor cells in inflammatory bowel disease: A new immunoregulatory pathway. *Gastroenterology,* **135**(3), 871–881.

10. Platzman, I., Janiesch, J. W., Spatz, J. P. (2013). Synthesis of nanostructured and biofunctionalized water-in-oil droplets as tools for homing T cells. *J. Am. Chem. Soc.,* **135**(9), 3339–3342.

11. Bourzac, K. (2013). Training immune cells to combat disease immunology: Researchers trap immune cells in droplets of water in oil in hopes of reprogramming them. *Chemical & Engineering News.* Available at: http://cen.acs.org/articles/91/web/2013/03/Training-Immune-Cells-Combat-Disease.html (accessed on April 24, 2017).

12. Boehm, F. J. (2013). Nanomedicine in regenerative biosystems, human augmentation, and longevity. In: *Nanomedical Device and Systems Design: Challenges, Possibilities, Visions,* CRC Press, Boca Raton, FL, USA, chapter 17, 653–741.

13. Adegoke, A. A., Faleye, A. C., Singh, G., Stenström, T. A. (2016). Antibiotic resistant superbugs: Assessment of the interrelationship of occurrence in clinical settings and environmental niches. *Molecules,* **22**(1), 29.

14. McKendry, R. A. (2012). Nanomechanics of superbugs and superdrugs: New frontiers in nanomedicine. *Biochem. Soc. Trans.,* **40**(4), 603–608.

15. Colatrella, S., Clair, J. D. (2014). Adapt or perish: A relentless fight for survival: Designing superbugs out of the intensive care unit. *Crit. Care Nurs. Q,* **37**(3), 251–267.

16. Boehm, F. J. (2013). Design challenges and considerations for nanomedical in vivo aqueous propulsion, surface ambling, and navigation. In: *Nanomedical Device and Systems Design: Challenges, Possibilities, Visions*, CRC Press, Boca Raton, FL, USA, chapter 3, 71–170.

17. Boehm, F. J. (2013). Exemplar nanomedical vascular cartographic scanning nanodevice. In: *Nanomedical Device and Systems Design: Challenges, Possibilities, Visions*, CRC Press, Boca Raton, FL, USA, chapter 1, 3–15.

18. Boehm, F. J. and Domschke, A. (2013). Quandary: Are molecularly manufactured burgers imbued with the life force? *Institute for Ethics and Emerging Technologies.* Available at: http://ieet.org/index.php/IEET/more/boehm20160115.

19. Boehm, F. J. (2013). *Nanomedical Device and Systems Design: Challenges, Possibilities, Visions,* CRC Press, Boca Raton, FL, USA.

Chapter 20

Immunotherapy for Gliomas and Other Intracranial Malignancies

Mario Ganau, MD, PhD,[a,b] Gianfranco K. I. Ligarotti, MD,[b] Salvatore Chibbaro, MD, PhD,[c] and Andrea Soddu, PhD[a]

[a]*Brain and Mind Institute, University of Western Ontario, Canada*
[b]*Department of Neurosurgery, Queen Elizabeth Hospital Birmingham, UK*
[c]*Department of Neurosurgery, Hôpital Universitaire Strasbourg, France*

Keywords: nanoneurosurgery, central nervous system (CNS), World Health Organization (WHO), neuro-oncology, brain tumors, gliomas, high grade gliomas (HGG), metastases, intrinsic tolerance mechanisms, dendritic cells (DC), tumor infiltrating lymphocytes, intracranial lymphatic vessels, epidermal growth factor receptor (EGFR), toll-like receptors (TLR), biomedical engineering, immunotherapy, adoptive immunotherapy, cancer immunotherapy, chemotherapy, clinical trials, nanodrugs, checkpoint inhibitors, cancer vaccines, monoclonal antibodies, oncolytic virus therapies, adoptive cell therapies, stereotactic radiosurgery (SRS)

20.1 Regional Immunotherapy: A Rising Trend in Nanomedicine

Over the last 5 years, the scientific community witnessed a rising trend in nanomedicine: the regional administration of cell-based and viral vector-based therapies for solid malignancies (see Fig. 20.1). Multimodality therapies that combine regional

---

Immune Aspects of Biopharmaceuticals and Nanomedicines
Edited by Raj Bawa, János Szebeni, Thomas J. Webster, and Gerald F. Audette
Copyright © 2018 Pan Stanford Publishing Pte. Ltd.
ISBN 978-981-4774-52-9 (Hardcover), 978-0-203-73153-6 (eBook)
www.panstanford.com

immunotherapy with other local and systemic therapies are demonstrating continuous growth as the field of immunotherapy keeps on expanding [1–3].

Until recently, brain tumors were somehow not considered amenable for those treatments. For a long time, in fact, the most accepted theories suggested that the requirements for the initiation of immune responses within the central nervous system (CNS) were significantly more stringent than in other organs or compartments of the human body [4–9].

However, some groundbreaking discoveries, such as the identification of functional lymphatic vessels lining the dural sinuses, and the presence of tumor infiltrating lymphocytes in brain tumors, recently confirmed that anti-tumor responses are engendered also in response to malignant intracranial lesions [10–12]. As such, many surgical interventions initially developed for intrapleural and intraperitoneal delivery of the so-called "adoptive immunotherapy" were eventually optimized to enhance their local delivery also intracranially.

Although extremely promising, this is not a story of immediate success: Immunotherapy for CNS pathologies is still in its early stages, and before that it went through many failures, mostly due to a lack of specificity of these treatments (durable responses were noted only in a small proportion of treated patients) and the potential for serious toxicity [13, 14].

Noteworthy, unlike chemotherapy, which acts directly on the tumor, cancer immunotherapy exerts its effects on the immune system and demonstrates new kinetics that involve building a cellular immune response, followed by changes in tumor burden and, hopefully, in patient survival [15].

Thus, its introduction in the therapeutic armamentarium is bringing up new paradigms in neuro-immunology and neuro-oncology, including the identification of predictive biomarkers, the designing of tailored therapeutic regimens, and the reconsideration of established endpoints [16].

Following a brief introduction on the difficulties of conceiving immunotherapy as an additional treatment tool for brain tumors; this chapter will focus on the ongoing clinical trials and will highlight some success stories that are paving the way for a more widespread use of immunotherapies in the multimodality management of intracranial malignancies in the next 10 years.

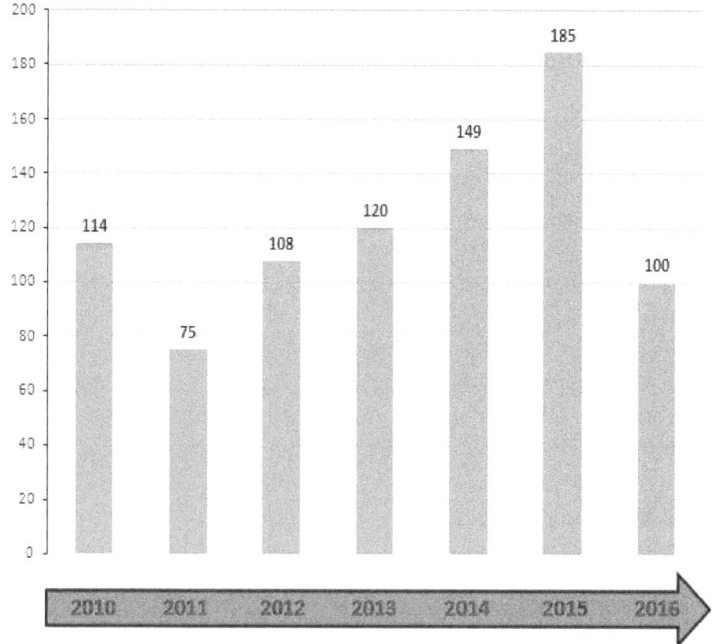

Figure 20.1 Number of published articles (y-axis) on "Immunotherapy for brain tumors" indexed on Pubmed, per year of publication (x-axis) (*Source: PubMed.gov*).

20.2 Primary and Secondary Brain Tumors

The most common primary tumors of the CNS are gliomas; in this group of astrocytic lineage, the most aggressive ones, namely high-grade gliomas (HGG), include class III and IV of the World Health Organization (WHO) Classification, such as anaplastic astrocytomas, glioblastomas, and gliosarcomas. HGG are associated with disproportionately high morbidity and mortality regardless the application of state of the art treatment strategies; in fact, because of their histological heterogeneity (35–40% of them have epigenetic modifications as the underlying mechanism driving malignancy) their treatment is extremely complex. As a result, the outcome remains poor, with a median survival of only 14.6 months [2, 7, 17].

Brain metastases are the most common intracranial malignancy and, despite advances in prevention and early diagnosis, their incidence has steadily risen over time. The cerebral blood flow represents 15% of the cardiac output, constantly, and primary tumors are known to escape local hypoxia by releasing in the bloodstream circulating tumor cells at an exponential rate. As such, it is no wonder that an estimated 25–45% of all cancers sooner or later will develop brain metastases, with lung and breast cancers showing a strong organotropism for the CNS [2, 3, 18]. Until recently, those lesions were considered as a homogenous condition, uniformly treated with whole brain radiotherapy alone or with surgical resection for large lesions and stereotactic radiosurgery for smaller lesions. Increasingly, specific systemic medical therapies are being used to treat brain metastases based on the primary site of disease, nonetheless, as of today; they still represent a devastating clinical reality, carrying an estimated survival time of less than one year in most of the cases [19].

Primary and secondary brain tumors overexpress several antigens, and it is therefore difficult to explain the reasons for a defective response of the immune system toward those lesions [20–23]. Actually, the delay between the initial stages of oncogenesis and the clinical presentation is certainly sustained by a wide variety of direct and indirect tolerance mechanisms to immune suppression adopted by those tumors [24–26].

20.3 Current Approaches to Immunotherapy for Brain Tumors

Briefly, there could be two ways of using immunotherapy for brain tumors: The first is the "active immunotherapy" designed to boost the patient's native immune response, and the second is "adoptive immunotherapy," where in vitro activated immune cells or specific molecules (e.g., antibodies) directly targeting tumor cells are injected [11]. As such, six broad categories of immunotherapies for brain tumors can be currently identified: adjuvant immunotherapies, checkpoint inhibitors, cancer vaccines, monoclonal antibodies, oncolytic virus therapies, and adoptive cell therapies.

- **Adjuvants** are substances that boost the immune response. They can be used alone or combined with other immunotherapies to prevent immunosuppression or sustain immune reaction to cancer cells. Examples of this class are agonists that target toll-like receptors (TLR), a family of pattern recognition receptors that function as primary sensors of the innate immune system to recognize microbial pathogens or cancers. TLR agonists can be used either alone or in combination with tumor antigens and showed promise in terms of both enhancing immune responses and eliciting anti-tumor activity [27–29].

- **Checkpoint inhibitors** target molecules involved in immunoregulatory pathways. Their aim, by blocking these inhibitory molecules, is to unleash or enhance innate anti-cancer immune responses. Following the observation of striking effects of drugs that target CTLA-4 or PD-1 against melanoma and other tumor entities, it was recognized that these drugs may also be active against metastatic tumor lesions in the brain. Their therapeutic activity against primary brain tumors is currently being investigated within clinical trials in both adult and pediatric populations [4, 30, 31].
- **Cancer vaccines** are designed to elicit an immune response against tumor-specific or tumor-associated antigens. A vast array of vaccine strategies (used alone on in combination with radiotherapy) advanced from preclinical studies to active clinical trials in patients with recurrent or newly diagnosed HGG, and brain metastases. Those vaccines are usually based on peptides, heat shock proteins, autologous tumor cells, and DC [32–35].
- **Monoclonal antibodies** (also known as monoclonal antibody drug conjugates) are designed to target specific antigens on tumors. An example is antibodies directed towards epidermal growth factor receptor (EGFR): Stable in the bloodstream, those drug conjugates release potent cytotoxic agents only once inside targeted cancer cells [36]. Anti-EGFR antibodies are currently evaluated for the treatment of patients with various solid tumors including HGG [37].
- **Oncolytic virus therapies** are based on modified virus strains that selectively target tumor cells inducing an immune response against the cancer. The production of virus progeny leads to secondary infection and spreading within the bulk tumor eventually destroying it, while

sparing normal cells. Selective replication of an oncolytic virus in tumor cells can be achieved by taking advantage of aberrations found in cancer cells, which include defective innate anti-viral and apoptotic response to viral infection. Thus, viral genes that are not essential for viral growth but are required for viral propagation in normal cells can be mutated or deleted to engender tumor selectivity [38]. Many viruses, including both DNA (i.e., herpes simplex virus) and RNA (i.e., poliovirus) viruses are currently studied as potential platforms for cancer therapeutics in HGG [39–41].

- **Adoptive cell therapies** are based on methods aimed at re-engineering immune cells to enhance their activity and improve the immune system's anti-cancer response. For instance, T cells obtained from glioma patients can be engineered ex vivo to express chimeric antigen receptors specific for glioma antigens (CAR T cells). The expansion and function of adoptively transferred CAR T cells can be potentiated by the lymphodepletive and tumoricidal effects of standard of care chemotherapy and radiotherapy, so that on reinfusion into the patient CAR T cells will carry out multiple antineoplastic activities all at once [42, 43].

20.4 Review of Ongoing Clinical Trials

Many years of extensive research led to the elucidation of the basic mechanisms behind immune surveillance and brain tumor evasion; because of this, a revolution spurred in neuro-oncology and dramatically influenced the way effective immunotherapeutic strategies are conceived and developed. The ongoing implementation of those immunotherapeutic concepts into the clinical routine has the potential to provide a powerful addition to current therapies against various brain tumors. Herein we will discuss the state of the art in term of clinical trials for immunotherapy on both primary and secondary brain tumors.

By the end of 2016, 87 national and international clinical trials on immunotherapies for gliomas were registered on the *ClinicalTrials.gov* portal of the U.S. National Institutes of Health. So far, only 29 studies have actually completed data collection and analysis; out of them, just in 3 cases final data have already been uploaded in the portal (NCT01280552, NCT00643097, and

NCT00323115) and the publications of the related analyses are expected in due time. From a review of the literature available to date on gliomas, it is possible to conclude that Phase I/II trials have assessed the efficacy of increasing immune activity mostly using vaccines made from cytotoxic T cells, autologous tumor cells, or dendritic cells (DC). Studies to decrease tumor immunoresistance have focused on cytokine modulation of known immunosuppressive factors in the tumor microenvironment. Several early studies reported a survival benefit only when different forms of immunotherapy were used simultaneously [44].

Due to the partial response of brain metastases to standard treatment options and the restricted therapeutic indications (i.e., poor performance status are not candidates for cytotoxic chemotherapy, surgery or stereotactic radiosurgery (SRS)), there is a strong clinical rationale for the use of targeted therapies, including immunotherapies, to widen the inclusion criteria [3]. A search conducted at the end of 2016 on the portal *ClinicalTrials.gov* retrieved 14 registered national and international clinical trials on immunotherapy for brain metastases. Among them, 9 are still ongoing and 3 did not start recruiting yet. Analyzing the characteristics of those trials (mostly in Phase I/II, only 3 reached Phase III), it is possible to confirm that the main attention of researchers is focused toward checkpoint inhibitors, adoptive cell therapies and cancer vaccines (especially DC ones); and interestingly one of those studies was exclusively focused on pediatric tumors. To date, only 2 trials on brain metastases completed their data collection (NCT00576537 in 2001; NCT01875601 in 2015) for primary outcomes, but none of them published the final results yet.

20.5 Neuro-Oncology and Immunotherapy: An Outlook for the Next 10 Years

A greater understanding of the mechanisms exploited by tumor cells to suppress the immune system and evade destruction has provided a wealth of potential treatment strategies, enabling the development of active immunotherapies that target specific components of the immune system. Based on the review of concluded and ongoing clinical trials, it is possible to affirm that exploiting the anti-cancer effect of the immune system with the use of vaccines, viral vectors, and more lately with immune checkpoint inhibitors and chimeric antigen receptor modification, seems to be a promising therapeutic strategy in a broad spectrum of CNS tumors. Unfortunately, most immunotherapeutic agents did not show sufficient activity in early trials, and whereas some were advanced to Phase III investigation, the results of randomized comparisons were poor or are still pending in most of the cases. Likely contributing factors include ineffective or marginally effective agents, the still incomplete understanding of human tumor immunology, and the absence of adequate tools for development, including criteria for refined trial endpoints. For instance, targeting of unique antigens restricted to the tumor itself is the most important parameter in advancing DC vaccines. In order to overcome intrinsic mechanisms of immune evasion observed in HGG, the future of DC-based therapy lies in a multi-antigenic vaccine approach. Successful targeting of multiple antigens will require a comprehensive understanding of all immunologically relevant oncological epitopes present in each tumor (for instance through nano-immuno-multiplexing recognition of relevant markers), thereby permitting a rational vaccine design [35, 45].

In conclusion, novel approaches based on immunotherapies are totally different when compared to conventional treatments, in terms of efficacy or toxicity; there is therefore a need for the development of new tools for conducting related laboratory and clinical researches. In particular, the durable responses observed with immunotherapeutic agents and their kinetics requires the definition of new efficacy endpoints in clinical trials. The introduction of immune-related response criteria and exploratory endpoints, such as landmark survival, and their adoption from regulatory authorities will be essential for better and faster reporting of efficacy in future clinical trials. In the next 10 years, most of the clinical trials currently ongoing on the use of immunotherapy for brain tumors will be completed and some molecules will hopefully advance beyond Phase III, as has already happened for other tumors (i.e., melanoma, lymphomas, etc.). The current perspective is that immunotherapy will prove to be a successful resource only if it will be considered together with other existing treatment tools to provide our patients with a truly personalized/tailored therapeutic approach.

Disclosures and Conflict of Interest

The opinions and perspectives here reflect the current views of the authors. The authors declare that they have no conflict of interest and have no affiliations or financial involvement with any organization or entity discussed in this chapter. No writing assistance was utilized in the production of this chapter and the authors have received no payment for its preparation. This chapter was written at the Brain and Mind Institute, University of Western Ontario, Canada, where Dr. Ganau has conducted joint projects in the priority area of nanotechnology.

Corresponding Author

Dr. Mario Ganau
Suite 2204—70 Temperance St,
MH5 0B1 Toronto, Canada
Email: mario.ganau@alumni.harvard.edu

About the Authors

Mario Ganau is a neurosurgeon and scientist interested in the application of breakthrough technologies in clinical practice. A former Global Clinical Scholar from Harvard Medical School, he holds a PhD in nanotechnology, a PhD in biomedical engineering, and a DU in neuropharmacology from the University of Trieste, University of Cagliari, and UPMC-Paris Sorbonne University, respectively.

Ganfranco K. I. Ligarotti is a major of the Italian Air Force with a robust experience in oncologic neurosurgery. He graduated from the University of Milan and has a longstanding research track witnessed by an ongoing involvement in several EU research projects and pinpointed by his participation in the XXII Antarctic Expedition with the Programma Nazionale di Ricerche in Antartide (PNRA).

Salvatore Chibbaro is a neurosurgeon specialized in the management of skull base tumors. He carries out clinical and research activities at Strasbourg University Hospital, France. Dr. Chibbaro holds master's degrees in microsurgery and neuro-oncology and a PhD in biotechnology from Université Paris VII, France.

Andrea Soddu is an assistant professor of physics at the University of Western Ontario, Canada, where he is also member of the Brain and Mind Institute. Prof. Soddu holds a PhD in particle physics from Virginia University and completed several postdocs in Taiwan, Israel, and Belgium.

References

1. Zeltsman, M., et al. (2016). Surgical immune interventions for solid malignancies. *Am. J. Surg.*, **212**(4), 682–690.
2. Ganau, M., et al. (2015). Radiosurgical options in neuro-oncology: A review on current tenets and future opportunities. Part II: Adjuvant radiobiological tools. *Tumori*, **101**(1), 57–63.
3. Ganau, M., et al. (2014). Radiosurgical options in neuro-oncology: A review on current tenets and future opportunities. Part I: Therapeutic strategies. *Tumori*, **100**(4), 459–465.
4. Gomez, G. G., Kruse, C. A. (2006) Mechanisms of malignant glioma immune resistance and sources of immunosuppression. *Gene Ther. Mol. Biol.*, **10**(A), 133–146.
5. Read, S. B., et al. (2003). Human alloreactive CTL interactions with gliomas and with those having upregulated HLA expression from exogenous IFN-gamma or IFN-gamma gene modification. *J. Interferon Cytokine Res.*, **23**(7), 379–393.
6. Hickey, W. F. (2001). Basic principles of immunological surveillance of the normal central nervous system. *Glia*, **36**(2), 118–124.
7. Vauleon, E., et al. (2010). Overview of cellular immunotherapy for patients with glioblastoma. *Clin. Dev. Immunol.*, **2010**, pii: 689171.
8. Tambuyzer, B. R., et al. (2009). Microglia: Gatekeepers of central nervous system immunology. *J. Leukocyte Biol.*, **85**(3), 352–370.

9. Calzascia, T., et al. (2005). Homing phenotypes of tumor-specific CD8 T cells are predetermined at the tumor site by crosspresenting APCs. *Immunity*, **22**(2), 175–184.

10. Quattrocchi, K. B., et al. (1999). Pilot study of local autologous tumor infiltrating lymphocytes for the treatment of recurrent malignant gliomas. *J. Neurooncol.*, **45**, 141–157.

11. Louveau, A., et al. (2015). Structural and functional features of central nervous system lymphatic vessels. *Nature,* **523**(7560), 337–341.

12. Tsugawa, T., et al. (2004). Sequential delivery of interferon-alpha gene and DCs to intracranial gliomas promotes an effective antitumor response. *Gene Ther.*, **11**(21), 1551–1558.

13. Roth, P., et al. (2016). Immunotherapy of brain cancer. *Oncol. Res. Treat.,* **39**(6), 326–334.

14. Ganau, M., et al. (2012). Challenging new targets for CNS-HIV infection. *Front. Neurol.,* **3**, 43.

15. Ganau, M. (2014). Tackling gliomas with nanoformulated antineoplastic drugs: Suitability of hyaluronic acid nanoparticles. *Clin. Transl. Oncol.,* **16**(2), 220–223.

16. Ganau, L., et al. (2015). Management of gliomas: Overview of the latest technological advancements and related behavioral drawbacks. *Behav. Neurol.,* **2015**, 862634.

17. Talacchi, A., et al. (2010). Surgical treatment of high-grade gliomas in motor areas. The impact of different supportive technologies: A 171-patient series. *J. Neurooncol.,* **100**(3), 417–426.

18. Irmisch, A., Huelsken, J. (2013). Metastasis: New insights into organ-specific extravasation and metastatic niches. *Exp. Cell Res.*, **319**(11), 1604–1610.

19. Sinha, R., et al. (2016). The evolving clinical management of cerebral metastases. *Eur. J. Surg. Oncol.*, doi: 10.1016/j.ejso.2016.10.006. [Epub ahead of print].

20. Saikali, S., et al. (2007). Expression of nine tumour antigens in a series of human glioblastoma multiforme: Interest of EGFRvIII, IL-13Ralpha2, gp100 and TRP-2 for immunotherapy. *J. Neurooncol.*, **81**(2), 139–148.

21. Okada, H., et al. (2009). Immunotherapeutic approaches for glioma. *Crit. Rev. Immunol.*, **29**(1), 1–42.

22. Lauterbach, H., et al. (2006). Adoptive immunotherapy induces CNS dendritic cell recruitment and antigen presentation during clearance of a persistent viral infection. *J. Exp. Med.*, **203**(8), 1963–1975.

23. Masson, F., et al. (2007). Brain microenvironment promotes the final functional maturation of tumor-specific effector CD8+ T cells. *J. Immunol.*, **179**(2), 845–853.
24. Akasaki, Y., et al. (2004). Induction of a CD4+ T regulatory type 1 response by cyclooxygenase-2-overexpressing glioma. *J. Immunol.*, **173**(7), 4352–4359.
25. Schiltz, P. M., et al. (2002). Effects of IFN-γ and interleukin-1β on major histocompatibility complex antigen and intercellular adhesion molecule-1 expression by 9L gliosarcoma: Relevance to its cytolysis by alloreactive cytotoxic T lymphocytes. *J. Int. Cyt. Res.*, **22**, 1209–1216.
26. Ahn, B. J., et al. (2013). Immune-checkpoint blockade and active immunotherapy for glioma. *Cancers (Basel)*, **5**(4), 1379–1412.
27. Gnjatic, S., et al. (2010). Toll-like receptor agonists: Are they good adjuvants? *Cancer J.*, **16**(4), 382–391.
28. Deng, S., et al. (2014). Recent advances in the role of toll-like receptors and TLR agonists in immunotherapy for human glioma. *Protein Cell*, **5**(12), 899–911.
29. Xiong, Z., Ohlfest, J. R. (2011). Topical imiquimod has therapeutic and immunomodulatory effects against intracranial tumors. *J. Immunother.*, **34**(3), 264–269.
30. Ring, E. K., et al. (2016). Checkpoint proteins in pediatric brain and extracranial solid tumors: Opportunities for immunotherapy. *Clin. Cancer Res.*, **2016**, 1829.
31. Spagnolo, F., et al. (2016). Survival of patients with metastatic melanoma and brain metastases in the era of MAP-kinase inhibitors and immunologic checkpoint blockade antibodies: A systematic review. *Cancer Treat. Rev.*, **45**, 38–45.
32. Sayegh, E. T., et al. (2014). Vaccine therapies for patients with glioblastoma. *J. Neurooncol.*, **119**(3), 531–546.
33. Garg, A. D., et al. (2016). Dendritic cell vaccines based on immunogenic cell death elicit danger signals and T cell-driven rejection of high-grade glioma. *Sci. Transl. Med.*, **8**(328), 328ra27.
34. Li, G., et al. (2015). Neurosurgery concepts: Key perspectives on dendritic cell vaccines, metastatic tumor treatment, and radiosurgery. *Surg. Neurol. Int.*, **6**, 6.
35. Batich, K. A., et al. (2015). Enhancing dendritic cell-based vaccination for highly aggressive glioblastoma. *Expert Opin. Biol. Ther.*, **15**(1), 79–94.

36. Harding, J., Burtness, B. (2005). Cetuximab: An epidermal growth factor receptor chemeric human-murine monoclonal antibody. *Drugs Today*, **41**(2), 107–127.
37. Reilly, E. B., et al. (2015). Characterization of ABT-806, a humanized tumor-specific anti-EGFR monoclonal antibody. *Mol. Cancer Ther.*, **14**(5), 1141–1151.
38. Ning, J., Wakimoto, H. (2014). Oncolytic herpes simplex virus-based strategies: Toward a breakthrough in glioblastoma therapy. *Front. Microbiol.*, **5**, 303.
39. Sosnovtseva, A. O., et al. (2016). Sensitivity of C6 glioma cells carrying the human poliovirus receptor to oncolytic polioviruses. *Bull. Exp. Biol. Med.*, **161**(6), 821–825.
40. Delwar, Z. M., et al. (2016). Tumour-specific triple-regulated oncolytic herpes virus to target glioma. *Oncotarget,* **7**(19), 28658–28669.
41. Vera, B., et al. (2016). Characterization of the antiglioma effect of the oncolytic adenovirus VCN-01. *PLoS One*, **11**(1), e0147211.
42. Sengupta, S., et al. (2016). Chimeric antigen receptors for treatment of glioblastoma: A practical review of challenges and ways to overcome them. *Cancer Gene Ther.*, doi: 10.1038/cgt.2016.46. [Epub ahead of print].
43. Riccione, K., et al. (2015). Generation of CAR T cells for adoptive therapy in the context of glioblastoma standard of care. *J. Vis. Exp.*, **96**, e52937.
44. Ruzevick, J., et al. (2012). Clinical trials with immunotherapy for high-grade glioma. *Neurosurg. Clin. N. Am.*, **23**(3), 459–470.
45. Ganau, M., et al. (2015). A DNA-based nano-immunoassay for the label-free detection of glial fibrillary acidic protein in multicell lysates. *Nanomedicine,* **11**(2), 293–300.

Chapter 21

Engineering Nanoparticles to Overcome Barriers to Immunotherapy

Randall Toy, PhD, and Krishnendu Roy, PhD

The Wallace H. Coulter Department of Biomedical Engineering, Georgia Institute of Technology and Emory University, Georgia, USA

Keywords: cancer immunotherapy, drug delivery, intracellular delivery, targeted nanoparticles, tissue permeation, vaccines, toll-like receptor (TLR), vascular endothelial growth factor (VEGF), CD20, chimeric antigen receptor T-cells (CAR T-cells), DNA and RNA immunotherapeutics, siRNA, RNA interference, biodistribution, particle replication in nonwetting templates (PRINT) technology, nanoparticle asymmetry, cytokines, inflammation, nanoparticle surface chemistry, corona effect, ligand density, nonspecific uptake, polymeric poly(lactic-*co*-glycolic acid) (PLGA), release kinetics

21.1 Introduction

Immunotherapy is a burgeoning field that holds promise for making an impact in the treatment of incurable disorders, for example, cancer, HIV, emerging infectious diseases, inflammatory diseases, and autoimmune disorders. A wide range of therapeutic modalities have been developed to regulate immunity, which include vaccines (e.g., melanoma gp-100), recombinant cytokines (e.g., GM-CSF, IL-7, IL-12), monoclonal antibodies (e.g., anti-CTLA4, anti-PD1), autologous T-cells, and small molecules designed for specific intracellular targets (e.g., IDO1 inhibitors, COX2 inhibitors, toll-like receptor [TLR] agonists) [1].

Over 10 therapeutic monoclonal antibodies have been approved for use in immuno-oncology, with targets that include B-lymphocyte antigen (CD20), receptor tyrosine protein kinase erbB-2 in breast cancers (HER2), vascular endothelial growth factor (VEGF), CD52, and CD33 [2].

New immunotherapies have also been successfully combined with existing therapeutic interventions. Co-delivery of immunotherapy with chemotherapy, B-Raf proto-oncogene inhibitors, and VEGF-directed therapy have all been shown to amplify antitumor responses [3]. Newly developed virus-like particles have demonstrated immunostimulatory capabilities which can be harnessed for immunotherapy for metastatic cancer [4]. In addition, the field of T-cell receptor engineering and the manufacturing of chimeric antigen receptor T-cells (CAR T-cells) have enabled improved immune recognition of tumor antigens [5].

Despite this wealth of new technologies, the efficacy and widespread adoption of immunotherapy has been limited. The major challenge lies in delivering an immunotherapy to a specific target without causing harm to healthy tissues or inducing a feedback pathway that counteracts the mechanism of the immunotherapy. Nonspecific delivery of proinflammatory cytokines and monoclonal antibody therapies has the potential to induce systemic toxicity. In a similar fashion, the adoptive transfer of cells potentially can induce autoimmunity at off-target sites [6]. The development of cancer immunotherapies is stifled by the widespread presence of immune tolerance at the tumor site. Low immunogenicity of tumor antigens, the proliferation of immunosuppressive cells (e.g., myeloid-derived suppressor cells, regulatory T-cells), and the increased production of immunosuppressive cytokines (e.g., IL-10, TGF-β) work together to limit the antitumor response elicited by immunotherapies [7]. Autoimmune diseases, conversely, have the opposite problem of inducing systemic immune suppression that renders patients susceptible to infectious disease [8]. The overarching question is, therefore, how do we deliver the optimal amount of immunotherapy to a specific site, with appropriate kinetics and dosing schedule, without inducing deleterious side effects that outweigh the benefits of the therapy?

Nanoparticle platforms may serve as a solution to these drug delivery problems that constrain immunotherapy. Because of their larger size in comparison to small molecule therapeutics, nanoparticles have unique transport properties and biodistribution behavior. Moreover, the physical properties of nanoparticles (i.e., size, shape, charge, ligand density, and charge) can be engineered to facilitate the tuning of biodistribution, site-specific targeting, immunogenicity, detectability by medical imaging, and therapeutic loading. Nanoparticles are capable of delivering immunomodulatory agents directly to the tumor microenvironment, inducing immune tolerance, and conjugating directly to adoptively transferred T-cells for regulation of priming [9–12]. In addition, nanoparticles have been formulated to deliver cancer vaccines to antigen-presenting cells. The enhanced delivery of antigens loaded onto nanoparticles as cancer vaccines is evident through decreased tumor proliferation in comparison to tumor treated with soluble antigens [13]. When formulated as hydrophobic, solid-in-oil dispersions, nanoparticle delivery can be enhanced through the hydrophobic, protective stratum corneum of the epidermis. This enables transcutaneous vaccine delivery, which can be internalized by dendritic cells that subsequently traffic to the lymphatic system via the lymph nodes [14].

Nanoparticle constructs also enable the intracellular delivery of DNA and RNA immunotherapeutics. Loading of nucleic acid therapeutics onto a nanoparticle significantly enhances its ability to travel to a target site and enter a cell. In addition to the widespread presence of RNA-degrading enzymes *in vivo*, the delivery of free RNA molecules is impeded by their negative charge. This charge limits their ability to travel across the cellular membrane, which is also negatively charged [15].

By the mechanism of silencing the production of inflammatory cytokines, nanoparticles loaded with siRNA have demonstrated potency against melanoma. Therapeutic efficacy is even further enhanced when the siRNA nanoparticles are delivered concurrently with nanoparticles that deliver tumor antigens and immune adjuvants, such as the CpG oligonucleotide [16, 17]. Concurrent delivery of pDNA antigen, CpG oligonucleotide, and siRNA targeted to IL-10 was able to enhance the Th1/Th2 cytokine ratio to favor an antitumor response [18]. Another nanoparticle vaccine consisting of an immune response modifier, imiquimod, and STAT3 siRNA boosts the expression of co-stimulatory molecules (CD86), increases the production of IL-2, and enhances cytolytic T-cell activity after delivery to dendritic cells [19]. To present a wider array of antigens to boost the antitumor response, tumor lysate vaccines have been developed. It appears to be advantageous to deliver tumor lysate on nanoparticles; lysate-loaded particles were able to stimulate dendritic cell migration, upregulate co-stimulatory and MHC expression, and slow tumor growth to a greater degree than tumor lysate in soluble form [20]. It should be noted that combinatorial nanoparticle therapies are not limited to the delivery of vaccines. Polyamine/lipid nanoparticles loaded with siRNA designed for several gene targets were designed for delivery to the vascular endothelium. The construct was able to effectively simultaneously silence Tie1, Tie2, VEGFR-2, VE-cadherin, and ICAM-2 specifically in lung endothelial cells *in vivo* [21]. Depending on the material property of the selected nanoparticle, biodistribution can also be monitored using medical imaging modalities (i.e., iron oxide nanoparticles for magnetic resonance imaging applications). This strategy was applied to monitor the efficacy of gene therapy to mitigate immune rejection of heart transplants in rats [22].

The versatility of nanoparticles suggests that they can easily elevate immunotherapy efficacy to another level, but in reality, their efficacy is limited by a set of unique drug delivery problems. Nanoparticle targeting may be slightly more specific than small molecule targeting, but serum protein opsonization usually leads to their accumulation in phagocytic cells. In addition, there are some tissue interfaces (e.g., the blood brain barrier) which are not conducive to nanoparticle penetration [23]. Nanoparticles do have the enhanced ability to accumulate by passive targeting into highly angiogenic tumors. Increased vascular permeability, which is due to rapid tumor angiogenesis, permits extravasation of nanoparticles into the tumor interstitium by a phenomenon known as the "Enhanced Permeation and Retention Effect." High interstitial pressures caused by the extravasation of proteins that stifle lymphatic flow, however, often impede the flow of nanoparticles into the tissue [24]. If a nanoparticle can successfully evade phagocytic clearance, it then faces the challenge of traveling to its targeted cellular compartment for its intended biological effect to be realized. For example, the delivery of immune adjuvants is often targeted to specific TLRs, retinoic acid-inducible gene 1 (RIG-I) like receptors, or nucleotide-binding oligomerization domain (NOD)-like receptors, which may be located on the cell membrane, on membrane-bound organelles, or in the cytoplasm [25, 26]. The delivery of DNA and interfering RNA requires localization to the nucleus or the cytoplasm, respectively [27]. Unfortunately, nanoparticles have a tendency to traffic through vesicles by the clathrin-mediated endocytosis, caveolae-dependent endocytosis, or micropinocytosis pathway. All of these pathways converge into the endolysosomes, where low pH deactivates nucleic acids [28]. Further engineering is required to deliver a RNA-loaded nanoparticle to the cytosol, where interaction with the RNA interference silencing complex can occur.

In this review, we will evaluate how nanoparticles can be engineered so they can overcome these obstacles and deliver immunotherapies more efficiently to their target sites (Fig. 21.1). First, we will discuss how the size, shape, and surface chemistry of a nanoparticle affects multiple biological processes. We will focus on how these nanoparticle design parameters influence cellular recognition and internalization, transport through the vasculature, biodistribution, and the elicited immune response. Then, we will discuss methods in which nanoparticles can be engineered to maximize immunotherapeutic efficacy. The engineering approaches discussed will include (a) targeting immunotherapeutic nanoparticles to specific tissues and cells, (b) using environment-sensitive biomaterials to optimize immunotherapy delivery from nanoparticles, (c) designing nanoparticles that are able to penetrate into deep tissue, (d) optimizing the intracellular delivery of nanoparticles for gene delivery and RNA interference, (e) designing nanoparticles to enhance vaccine delivery, and (f) designing nanoparticles to boost the antitumor immune response (Table 21.1).

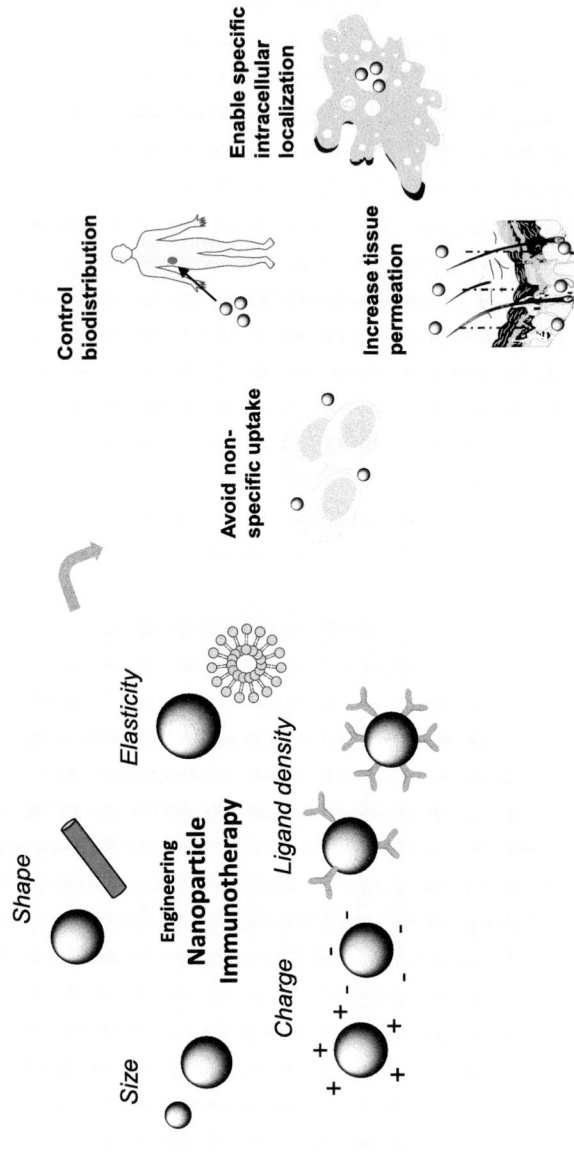

Figure 21.1 Engineering nanoparticle immunotherapy. Nanoparticles of unique size, shape, elasticity, charge, and ligand density can be formulated to enhance the delivery of immunotherapies. Through an understanding of the effect of each of these parameters on biotransport and immunogenicity, nanoparticles can be designed to control the biodistribution of immunotherapies, evade nonspecific uptake by phagocytic cells, increase tissue permeation, and enable specific localization to targeted cellular compartments.

Table 21.1 Strategies to enhance the efficacy of nanoparticle immunotherapy

Strategy	Nanoparticle design implementation	References
Reducing nonspecific uptake	Hydrophilic polymers (e.g., PEG)	29
	Engineering of shape	30
	Cellular hitchhiking	31, 32
	Engineering of nanoparticle rigidity	33
Enhancing tissue permeation	Tissue-penetrating peptides	34–39
	Chemical modification to increase permeation	40–42
Targeting to immune cells	Dendritic cell and macrophage targeting	43, 44
	T lymphocyte targeting	45, 46
	B lymphocyte targeting	47, 48
Targeted intracellular delivery	Cationic polymers	49, 50
	pH-sensitive biomaterials	51–53
	Virus-derived cell-penetrating peptides	54–56
	Direct cytosolic delivery via scavenger receptor	57, 58
Controlling release kinetics	Controlled release rate	59, 60
	Enzyme-triggered release	61, 62
	Photothermal triggered release	63–65

Application	Nanoparticle design implementation	References
Boosting nanoparticle vaccines	Manipulating antigen presentation	66–69
	Lymphoid organ targeting	70–75
Boosting the anti-tumor immune response	Stimulating immune activation	76–79
	Targeting the tumor microenvironment	80–83

21.2 Engineering Nanoparticles to Manipulate Transport and the Immune Response

The physical characteristics of a nanoparticle critically affect its *in vivo* transport and the immune response it triggers. Each design parameter has its own unique contributions to an immunotherapeutic nanoparticle's biotransport and toxicity profile. We highlight the effect of manipulating the design parameters of nanoparticle size, shape, charge, ligand density, and elasticity individually, but all parameters and their interplay must be considered when formulating a nanoparticle with efficient therapeutic delivery and low toxicity.

21.2.1 Nanoparticle Size

Nanoparticle size is a design parameter that can be tuned to enhance the targeted delivery and subsequent efficacy of nanoparticle immunotherapies. The size of a nanoparticle critically affects its pharmacokinetics, vascular transport, and cellular uptake. While small nanoparticles (<10 nm) have a tendency to be cleared in the kidneys, larger nanoparticles are more likely to be cleared by the liver and the spleen [84]. Nanoparticles that are greater than 200 nm in diameter are similar in size to fenestrations in the spleen, so particle elasticity also plays a role in the rate of splenic clearance for larger particles [85]. In addition to affecting the clearance rate, nanoparticle size affects how the particle's transport is mediated by blood flow. While the motion of smaller nanoparticles is primarily governed by diffusion, the motion of larger nanoparticles is governed by a combination of diffusion and convective flow [86]. The size effect on nanoparticle transport, therefore, has ramifications on how efficiently a nanoparticle can extravasate from a blood vessel and enter either a tumor or an inflammation site. When interstitial pressures are high, fast blood flow is required to direct larger nanoparticles deep into the tumor interstitial space. Therefore, large nanoparticles have a variable intratumoral distribution which depends on regional blood flow [87]. Whole body biodistribution of nanoparticles is also influenced by particle size. In a study comparing polystyrene spheres of diameters ranging from 0.1 to 10 μm, particle accumulation in the liver decreased as particle size increased.

At the same time, particle accumulation in the lungs increased as particle size increased [88]. This difference in uptake behavior highlights that the optimal particle size for cellular internalization is dependent on cell type. Moreover, size may be optimized to maximize the rate of uptake into cells. With HeLa cells, it was observed that the uptake of 50 nm nanoparticles was increased over the uptake of smaller 14 nm nanoparticles and larger 74 nm nanoparticles [89]. The effect of nanoparticle size on pharmacokinetics, transport, and internalization will manifest itself in downstream effects as well. To evaluate the effect of size on nanoparticle immunogenicity, micro- and nano-sized polylactide particles were formulated and loaded with pneumococcal antigens. It was subsequently found that the IgG responses to the particle vaccine depend on size [90]. An evaluation of nanoparticles manufactured by the Particle Replication in Nonwetting Templates (PRINT) technology, loaded with the ovalbumin (OVA) antigen, showed that 80 × 180 nm PEGylated nanoparticles elevated anti-OVA IgG titers significantly more than 1 µm PEGylated nanoparticles. Moreover, size is dominant over polyethylene glycol (PEG) linker length in influencing the humoral response [91].

21.2.2 Nanoparticle Shape

Particle shape is also a critical design parameter that influences how a nanoparticle immunotherapy moves while in the blood circulation, becomes internalized by cells, and stimulates an immune response. Initial nanoparticle formulations were produced primarily in spherical shapes, but recent advances in nanoparticle engineering have generated a wide portfolio of shapes that include rods, prisms, cubes, stars, and disks [92]. Nanoparticle asymmetry promotes particle tumbling toward the wall of blood vessels under flow (margination), which is caused by a nonuniform distribution of hydrodynamic forces acting on the particle. Asymmetry of nanoparticles also enhances nanoparticle penetration and distribution inside solid tissues and tumors. Inside tumor spheroids, nanodisks were observed to accumulate at higher amounts throughout a tumor spheroid than similarly sized nanorods. This difference in accumulation can be explained by the difference in interactions that particles of different shapes have with the cell membrane. As asymmetric nanoparticles approach the cell membrane at different contact angles, their rates of internalization will differ [93]. More specifically, this interaction angle will affect the energy required for particle internalization [94]. When an elongated particle's major axis lies tangential to a cell membrane, internalization is more difficult than if the particle's minor axis was aligned tangential to the cell membrane. In agreement with these findings, hydrogel nanorods with a higher aspect ratio were internalized by cancer cells, endothelial cells, and dendritic cells more slowly than hydrogel nanodiscs. This difference in uptake rate could be attributed to differences in adhesion forces between the cell membrane and particles of different shapes, the strain energy required for membrane deformation, and the effect of sedimentation or local particle concentration at the cell surface (Fig. 21.2) [95].

Like size, a nanoparticle's shape also influences the immunological response to the particle. An evaluation of spheres, cubes, and rods designed to stimulate antibody production against the West Nile Virus demonstrated that spheres and cubes induce the secretion of proinflammatory cytokines (TNF-α IL-6, IL-12, GM-CSF), while rods induce the secretion of inflammasome-related cytokines (IL-1 β, IL-18). Interestingly, the spheres and cubes were concurrently not internalized efficiently as rod-shaped particles. The aspect ratio of a rod-shaped particle also influences dendritic cell maturation and production; rods with a high aspect ratio induce significantly higher production of IL-1 β, IL-6, and IL-12 than rods with a shorter aspect ratio [96]. An evaluation of both nanoparticle size and shape suggest that surface area is a key parameter that influences the immune response [97]. This assertion is backed by the observation that cytokine production (TNF-α, IL-6) was also induced to different degrees by triangular, square, and pentagonal RNA nanoparticles in mouse macrophage cultures [98]. It has also been suggested that particle shape influences the Th1/Th2 polarization of the immune response. In a comparison of spherical and rod-shaped nanoparticles, it was found that spheres produced a Th1 biased response against OVA, while rods produced a Th2 biased response [99].

It is, therefore, important to consider that nanoparticle shapes may trigger different intracellular signaling pathways, which can lead to unique immune responses.

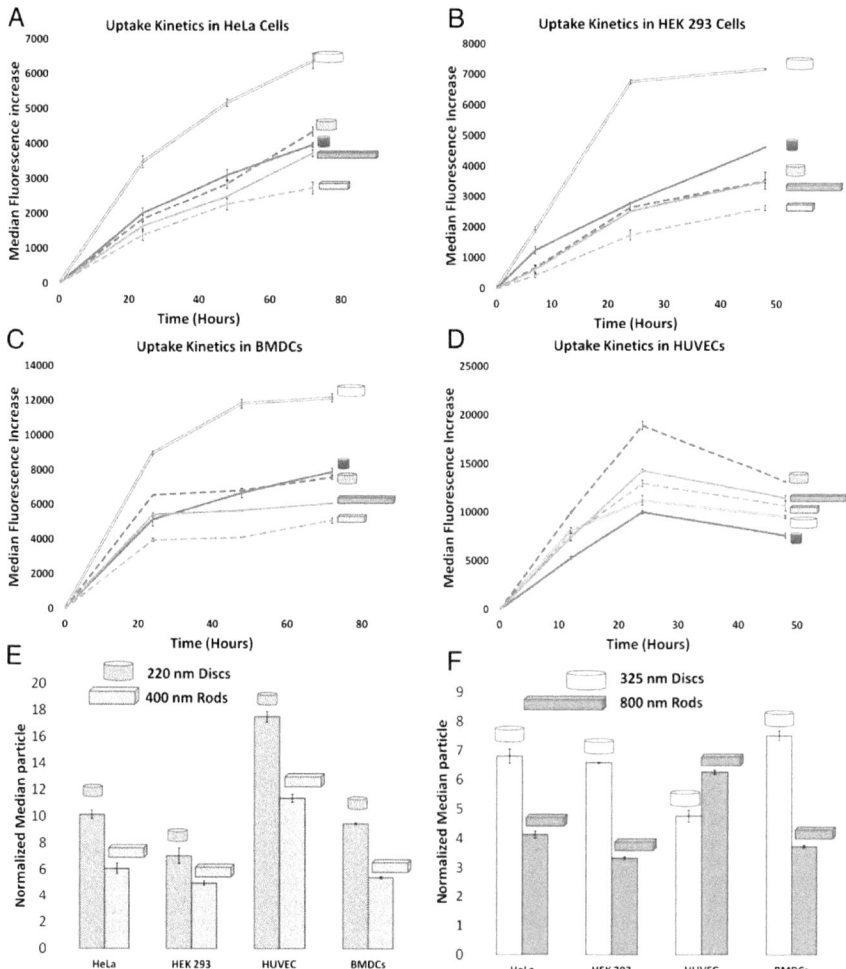

Figure 21.2 Shape affects the internalization of nanoparticles. Cellular-uptake kinetics of different shape-specific nanoparticles in various cell lines. (A) HeLa cells, (B) HEK 293 cells, (C) BMDCs, and (D) HUVEC cells. In A–D, red lines are for nanodiscs (hollow for 325 × 100 nm disks, dashed for 220 × 100 nm disks, and solid for 80 × 70 nm disks), and blue lines are for nanorods (dashed for 400 × 100 × 100 nm rods and solid for 800 × 100 × 100 nm rods). Error bars are SD with $n = 3$ for each data point. (E, F) Normalized median particle uptake per cell (indicates relative number of particles internalized by cells when normalized to 100 particles of 80 × 70 nm disks) at the maximum internalization time point (72 h for HeLa and BMDC, 48 h for HEKs, and 24 h for endothelial cells). Reproduced with permission from ref. 95. Copyright (2013) National Academy of Sciences.

21.2.3 Nanoparticle Charge

The tuning of nanoparticle surface chemistry has enabled a breadth of applications for nanoparticle immunotherapeutic delivery. Introduction of charge to nanoparticles enables the loading of moieties on the particle surface by electrostatic interactions. Linear polyethylenimine (lPEI) has been established as a cationic polymer which can complex with DNA and RNA and deliver its cargo into the cytoplasm [100]. PEI can be further modified to include degradable linkers and functional groups to enhance cellular uptake [101]. Through similar mechanisms, cationic lipid nanoparticles also can shuttle siRNA into the cytoplasm of cells [102].

The surface charge of a nanoparticle has the dual effect of enabling gene delivery and modulating the immune response. When positively charged, antigen-conjugated nanoparticles for pulmonary immunization stimulate antibody production, germinal B-cell expansion, CD41 T-cell activation, and expression of MHC II and coactivating receptors on T-cells [103]. When charge and hydrophilicity are varied in mesoporous silicon nanoparticles, downstream CD3, CD4, and CD8 proliferation are altered and leads to biasing toward a Th1 or Th2 response [104].

Application of positively charged or negatively charged nanoparticles to the skin revealed that positive charge could deepen skin penetration 2–6 fold [105].

A major issue with cationic nanoparticles has been acute systemic toxicity and nonspecific systemic immune-stimulation. It is widely accepted that almost all cationic particles stimulate acute inflammation, the exact mechanism of which are still being elucidated. A potential mechanism could be that cationically charged polymers used to synthesize nanoparticles can trigger pattern-recognition receptors in immune cells [106]. Several reports have indicated that the polysaccharide chitosan induces inflammasome response [107, 108]. This toxicity/immunogenicity issue hinders nanoparticle usage in the clinic and many promising formulations that showed encouraging results *in vitro* have failed to move past preclinical studies. Significant effort must be invested in understanding how charged nanoparticles (as well as uncharged ones) interact with the serum and extracellular molecules once injected and how the so called "corona effect" as well as the fundamental structure of the nanoparticle-material induces specific signaling pathways in immune cells, endothelial cells (which could be a major source of proinflammatory cytokines), and fibroblasts [109].

21.2.4 Nanoparticle Ligand Density

In addition to size, shape, and charge, the ligand density of a nanoparticle plays a critical role in its biodistribution, cellular uptake, cellular association, and immune response. Ligand conjugation to a nanoparticle can reduce off-target cytotoxicity. Simple functionalization of mesoporous silicon nanoparticles with amines can significantly reduce systemic toxicity and increases the maximum tolerated dose of particles [110]. In addition, polymers such as PEG have long been used to enhance particle hydrophilicity and reduce clearance by the reticuloendothelial system. Studies evaluating the effect of PEG density have also revealed that the degree of PEG surface loading changes the particle biodistribution between hepatocytes and Kupffer cells [111]. In addition to mediating the rate of off-target particle uptake, ligand density and distribution also influences the rate of internalization into targeted cells [112]. Internalization and externalization rates of a targeted receptor will affect the optimal ligand density that maximizes nanoparticle uptake, as was seen with folate-targeted liposomes [113]. Particle shape must also be taken into consideration when determining optimal ligand density. For a ligand with a fixed length and flexibility, a spherical nanoparticle will interact with a different number of receptors at an interface than an asymmetric nanoparticle. This was observed in a study comparing the association of symmetric and asymmetric ligand-decorated PRINT hydrogels with alveolar macrophages [114]. Not surprisingly, enhancement in the cell uptake of nanoparticles correlates to enhancement in therapeutic efficacy due to the nanoparticle. When the antibody surface density was increased on DEC-205 targeted nanoparticle dendritic cell vaccines, the rate of uptake and subsequent induction of CD36 expression on dendritic cells increased [115]. Ligand spacing and multiplicity have also been demonstrated to critically affect peptide-DNA complex induced TLR9 activation [116].

21.2.5 Nanoparticle Elasticity

Tuning the flexibility of a nanoparticle also affects antibody-mediated targeting, endocytosis, and phagocytosis. Softer particles have a prolonged blood circulation residence time and increased organ deposition. It is hypothesized that this is caused by the deformation of softer particles by macrophages into shapes which are more difficult to internalize [117]. Particle endocytosis, however, occurs more quickly with flexible nanoparticles. When the rates of endocytosis of HER2 targeted flexible liposomes and rigid mesoporous silica particles into HT29 colon cancer and SKBR3 cancer cells were compared, it was found that the liposomes underwent endocytosis more quickly than the silica particles [118]. These findings reflect the importance of considering particle design, cell type, and the mode of internalization when developing a nanoparticle immunotherapy for a given application.

21.3 Improving Nanoparticle Design to Enhance Immunotherapy Efficacy

21.3.1 Designing Nanoparticles to Control Where and When Immunotherapies Are Delivered

21.3.1.1 Reduction of nonspecific uptake

Designing a nanoparticle that avoids nonspecific uptake is as important as designing a nanoparticle which can be internalized efficiently by its target. Hydrophilic polymers, such as PEG, have long been used in nanoparticle formulations to reduce macrophage uptake and promote long circulation. Anchoring of anti-CD40 antibodies and CpG oligonucleotides to PEGylated lipid nanoparticles decreases the incidence of off-target side effects while maintaining therapeutic efficacy [29]. Engineering of nanoparticle shape will also aid nanoparticles to evade the macrophages of the reticuloendothelial system. For example, shaping mesoporous silica nanoparticles as long rods will slow their rate of excretion when compared to mesoporous silica nanoparticles shaped as short rods [30]. Another means to hide from circulating macrophages is a process known as "cellular hitchhiking," in which nanoparticles act as stowaways on the surface of nonimmunogenic cells. The cloaking of nanoparticles with the cell membranes of red blood cells (RBCs) has proven useful in Type II autoimmune diseases, where antibodies opsonize RBCs for phagocytosis. Cell membrane decorated nanoparticles can also serve as a sink for anti-RBC antibodies, which prevents phagocytosis and subsequent destruction of healthy RBCs by macrophages [31]. The incorporation of cell membrane components to polymeric poly(lactic-*co*-glycolic acid) (PLGA) particles provides the added benefit of the ability to load lipid adjuvants (e.g., monophosphoryl lipid A), to enhance the efficacy of tumor vaccines [119]. There has also been implementation of an elegant strategy that reduces nonspecific nanoparticle uptake through combination of all of the aforementioned approaches: active targeting, shape optimization, and cellular hitchhiking. Combination of these three approaches was able to significantly reduce off-target accumulation of nanoparticles targeted to ICAM-2 on the vascular endothelium in the lung [32]. Nanoparticle rigidity may be another parameter that can be altered to reduce nonspecific nanoparticle uptake. In a comparison of soft and rigid discoidal polymeric nanoparticles, it was found that softer nanoparticles were internalized by macrophages less quickly than their rigid counterparts [33].

21.3.1.2 Enhancing tissue penetration

As with nanoparticle chemotherapy, it is challenging to deliver nanoparticle immunotherapy deep into a tumor. The rapid rate of tumor development results in the generation of highly vascularized regions at the periphery of a tumor surrounding an avascular core [120]. Blockage of the lymphatic system with extravasated proteins causes high interstitial pressures, which also hinder the transport of nanoparticles into the tumor site. Therefore, it is a significant challenge to increase the distance in which nanoparticles travel from a blood vessel into the deep tissue space. To overcome this challenge, peptide and chemical modifications to the nanoparticle surface have been developed to improve the tissue penetration [34]. For example, polyarginines have been used to enhance the skin permeation of lipid nanoparticles. An added benefit of surface functionalization with polyarginines is that the modified particles increased retention in the dermis after administration [35]. Another such "tumor-penetrating peptide" is cyclic CRGDK/RGPD/EC, cyclized between the two cysteines with a disulfide bond iRGD, which binds to overexpressed αv integrins. After binding to the αv integrin, a proteolysis-induced structural change converts iRGD to a substrate for neuropilin-1 and neuropilin-2. Proteins of the neuropilin family facilitate angiogenesis after interaction with VEGF, of which overexpression is frequently a hallmark of tumor progression. Through this mechanism, the iRGD peptide is able to deliver coadministered dextran into peritoneal tumors and increase the therapeutic index of the chemotherapeutic doxorubicin [36]. Administration of iRGD-conjugated indocyanine liposomes to angiogenic endothelial cells also confirmed that iRGD enables nanoparticle permeation beyond the vascular endothelium (Fig. 21.3) [37]. When iRGD was conjugated to a microenvironment-responsive and multistage nanoparticle and administered to mice with 4T1 orthotopic breast tumors, increased nanoparticle permeation was observed throughout the tumor and was accompanied by a decrease in tumor burden [38]. Tumor-penetrating and membrane-translocating peptides have also been used to enhance the transport of nanocomplexes with siRNA to silence ID4, a prominent oncogene, in ovarian tumors. Conjugation of a tumor-penetrating peptide to the nanoparticle enabled deep localization away from the vascular endothelium and significantly enhanced siRNA delivery in comparison to naked nanoparticles [39]. Addition of these peptides to nanoparticle immunotherapies can potentially equalize distribution within both tissues and tumors. Cell-penetrating peptides have already been evaluated to enhance the delivery of vaccines to the mucosa. In an intranasal application, poly(*N*-vinylacetamide-*co*-acrylic acid) was modified with D-octaarginine to deliver OVA to the mucosa. The cell-penetrating peptide modified vaccine significantly elevated OVA-specific IgG titers in the sera of vaccinated mice [121].

Chemical modification of nanoparticles can also modify its permeation and cell-penetration properties. For instance, the addition of Pluronic F127 modified lipid vesicles to chitosan nanoparticles was found to enhance epithelial mucosa penetration. Furthermore, the addition of a polyethylene oxide corona could further improve transport of the nanoparticles throughout the mucosa [40]. Imidazole modification has also been shown to increase the tissue and mucosa permeation of chitosan nanoparticles, which provides an opportunity for the delivery of DNA vaccines and immunotherapeutic siRNA delivery [41]. Intravenous administration of imidazole-modified chitosan-siRNA nanoparticles led to a 49% knockdown in GAPDH protein expression in the lungs, while unmodified nanoparticles only induced 11% knockdown [42].

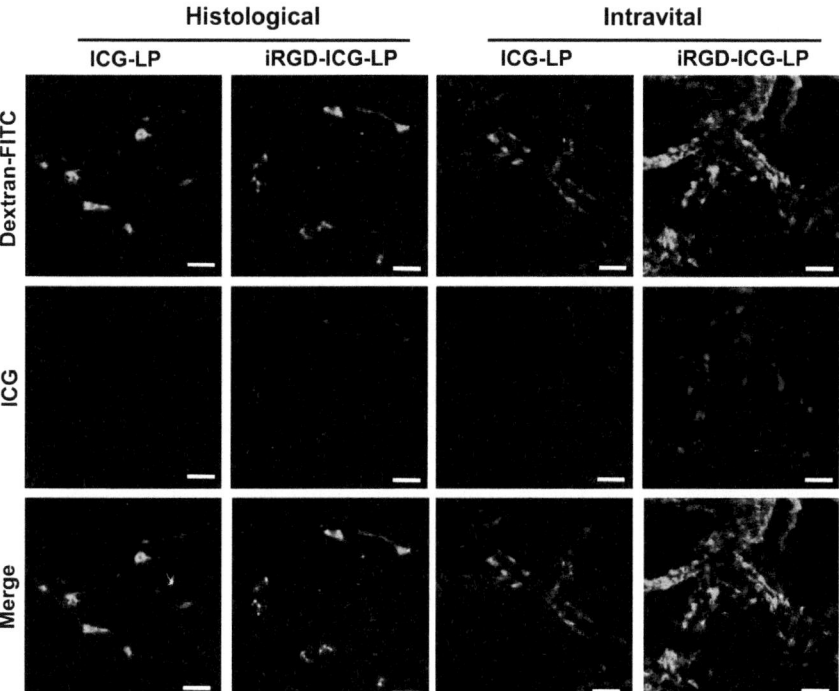

Figure 21.3 iRGD enhances the endothelial permeation of indocyanine-labeled liposomes. The binding and penetration of iRGD–ICG-LPs or ICG-LPs to angiogenic endothelial cells were assessed with intravital and histological examination. The tumor vascular images were captured at 10 min after injecting 40 kDa FITC-Dextran. The frozen sections were examined under a confocal microscope. Green represented the blood vessels labeled by FITC-Dextran and red represented ICG-loaded nanoparticles (Scale bar, 50 μm). Reproduced with permission from ref. 37. Copyright (2015), with permission from Elsevier.

21.3.1.3 Targeting nanoparticles to immune cells

As mentioned, one challenge that limits the efficacy of nanoparticle immunotherapy is determining how to control nanoparticle biodistribution. One common strategy to enhance nanoparticle immunotherapy is to attach a ligand or peptide that facilitates homing to a particular target. Monoclonal antibodies can also be conjugated to nanoparticles for the specific delivery of antigens and adjuvants to dendritic cells (anti-CD40) for tumor vaccination and TGF-β and IL-2 to induce regulatory T-cells (anti-CD4) for the downregulation of autoimmune disease [43, 44].

Nanoparticles targeted to dendritic cells using monoclonal antibodies against CD40, DEC-205 (C-type lectin), and CD11c (integrin receptor) were able to stimulate IL-12 production and co-stimulatory marker CD86 more effectively than nontargeted nanoparticles [122]. Nanoparticles for immunotherapy can also be targeted to dendritic cells and macrophages through functionalization with mannose, which binds to the mannose receptor (CD206). Mannose conjugation to nanoparticles coloaded with tumor vaccines and TLR adjuvants elevated interferon-γ levels in the spleen and slowed tumor proliferation in comparison to untargeted nanoparticles [45]. The success of this approach is not limited to mannose; a galactosylated cationic dextran has also been used to formulate nanoparticles that can deliver oligo-nucleotides to tumor-associated macrophages 2.5 times more efficiently than their ungalactosylated counterparts [46].

Cell targeting of nanoparticles is not limited to antigen presenting cells—methods have been also developed to target nanoparticles to B and T lymphocytes for immunotherapeutic delivery. Antibody fragments specific for cell surface antigens on adoptively transferred T-cells have been used for the targeted delivery of liposomes with IL-2 [123]. Dendrimers, which have the advantage of multivalent targeting capability, have been used to deliver small molecule payloads to B-cells [47]. Calcium phosphate nanoparticles loaded with protein antigen have also been developed to activate antigen-specific B-cells. The antigen-specific B-cells could internalize a large number of the nanoparticles, which resulted in increased expression of the early activation marker CD69, the increased expression of co-stimulatory marker CD86, and extensive cross-linking of the B-cell receptors [48].

21.3.1.4 Tailoring the intracellular delivery of nanoparticle immunotherapies

Once a nanoparticle immunotherapy reaches its target cell, it must be able to travel to the appropriate intracellular compartment for biological activity to occur. For the delivery of DNA plasmid vaccines, trafficking to the nucleus is required. For the application of delivering siRNA to inhibit checkpoint blockade, nanoparticles must be able to escape the endosomal pathway into the cytoplasm, where interaction with the RNA-induced silencing complex occurs. Initial attempts at facilitating the endosomal escape of nanoparticles involved polymeric modifications to create a cationic particle surface (e.g., polyethylenimine) [49]. It has been proposed that endosomal escape is then induced by a "proton sponge effect." The effect is due to the protonation of cationic polymers at low pH, which causes an influx of protons followed by an influx of chloride ions and water. Rapid water influx is then hypothesized to cause the lysosome to swell and eventually burst, which frees nanoparticles from the endolysosomal trafficking pathway [50]. More advanced strategies to facilitate nanoparticle escape into the cytosol rely on the lower pH in the milieu of the late endosomes and lysosomes. One antigen delivery system employs an antigen-loaded liposome modified with biodegradable polysaccharides that become fusogenic under acidic conditions. After fusion with the endosomal membrane, the particle can deliver antigens into the cytoplasm. These antigens can subsequently be presented to CD81 T-cells and induce cytotoxic T-lymphocytes [51]. In pH sensitive galactosyl dextran-retinal nanogels, it is hypothesized that the cleavage of hydrazone bonds at acidic pH both disassemble the nanogel and induce lysosomal rupture by the proton sponge effect [52]. Micelleplexes that disrupt the lysosomal membrane by two separate mechanisms have also been developed. The cationic micelleplex can become protonated at endosomal pH, which induces the proton sponge effect. In addition, the pH sensitive micelleplex was designed to release amphotericin that creates pores which further destabilize the endolysosomal membrane [53].

Another means of enhancing the intracellular delivery of nanoparticle immunotherapies is to incorporate cell-penetrating peptides on the particle surface. When attached to small molecules, cell-penetrating peptides can translocate with their cargo across the plasma membrane. Nanoparticles with cell-penetrating peptides conjugated to their surface will continue to enter cells through endocytic pathways; the peptides, however, enable the nanoparticle to penetrate the membrane of the endolysosomes to gain entry into the cytosol [54].

More advanced cell-penetrating peptides are derived from viral coat proteins, which enable viruses to escape the endosome and proceed to infect its host. Viral pH sensitive peptides, such as GALA or KALA, undergo a conformational change at low pH that promotes the destabilization and fusion of lipid membranes. It has been validated that these pH-sensitive, fusogenic peptides can be conjugated to cationic liposomes to boost transfection efficiency [55]. Conjugation of KALA to lipid nanoparticles also have demonstrated increased immunostimulatory abilities, measured by the upregulation of interferon-γ, IP-10, and IL-1β from bone marrow-derived dendritic cells, when compared to soluble CpG or nanoparticles without KALA (Fig. 21.4) [56]. GALA-modified lipid nanoparticles have also been used to deliver siRNA targeting SOCS1 (suppressor of cytokine signaling) in dendritic cells, which led to enhanced phosphorylation of STAT1 and the increased production of proinflammatory cytokines [124]. More elaborate nanoparticle designs enable targeted mitochondrial delivery through a mitochondria-fusogenic lipid envelope surrounded by an endosome-fusogenic lipid envelope. The exterior lipid envelope facilitates endosomal escape and a mitochondrial targeting signal enable fusion of the nanoparticle with the mitochondrial membrane, where cargo can be delivered [125].

Yet, even another strategy to enhance intracellular delivery is to bypass traditional mechanisms of nanoparticle internalization (clathrin-mediated endocytosis, caveolae mediated endocytosis, and micropinocytosis) altogether. These alternative internalization routes fully bypass the endolysosomal pathway without disrupting the intracellular vesicular compartments. One target pathway is the scavenger receptor BI (SR-BI), which can be targeted with high density lipoproteins (HDLs) [57].

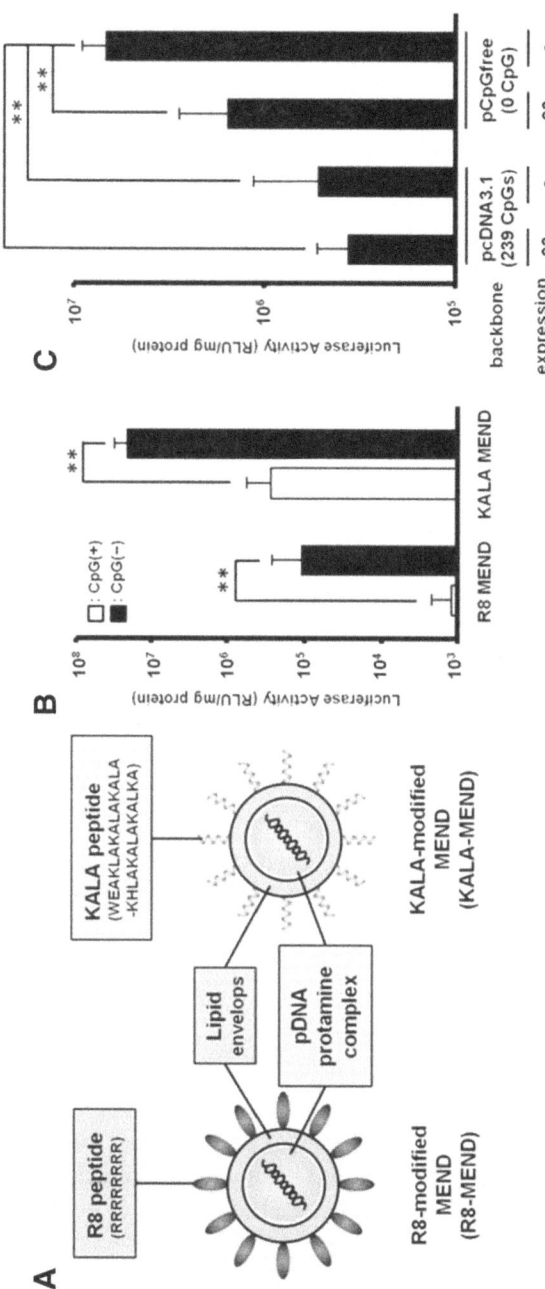

Figure 21.4 Viral peptides enhance the endosomal escape of nanoparticles. (A) Illustration of the R8-MEND (left) and KALA-MEND (right). (B) The R8-MEND and KALA-MEND encapsulating a conventional pDNA (pcDNA3.1-Luc; opened bar) or CpG-free pDNA (pCpGfree-Luc(0); closed bar) were transfected to BMDCs. Data were presented as the mean 6 SD of three independent experiments. Statistical differences were evaluated by one-way ANOVA, followed by Student's t test ($**p < 0.01$). (C) The transfection activity of KALA-MENDs encapsulating a pDNA with various set of backbone and inserts was also evaluated. Data were presented as the mean 6 SD of three independent experiments. Statistical analyses were performed by one-way ANOVA, followed by Bonferroni test. $**p < 0.01$ versus pCpGfree-Luc(0). Reproduced with permission from ref. 56.

Reconstituted HDL has been combined with cholesterol-siRNA to make siRNA nanocarriers that can bypass the endolysosomal pathway and reach the cytosol, where silencing can occur. The HDL-modified cholesterol-siRNA carriers could successfully downregulate VEGF expression in a MCF-7 breast tumor model that expressed the SR-BI receptor [58].

With all of these intracellular delivery strategies, it is important to consider the balance between efficient delivery and toxicity and immunogenicity. Cationic polymers, while effective at facilitating endosomal escape, also may induce cytotoxicity at high doses. Traditional cell penetrating peptides, which permeate the cell membrane, also will cause harm to cells at very high doses [126]. If new fusogenic viral peptides are conjugated to a nanoparticle, it is essential to evaluate if their systemic immunogenicity are outweighed by the increased therapeutic efficacy which is enabled by more site-specific intracellular delivery.

21.3.1.5 Controlling release kinetics of nanoparticle-based immunotherapies

So far, we have discussed ways to deliver nanoparticles to targeted locations and also ways to prevent delivery to undesired locations. In addition to where an immunotherapy is delivered, it is also important to consider when and for how long it is delivered. The intelligent design of nanoparticles for immunotherapy delivery can facilitate such precise control of release kinetics. This is especially important in immunotherapy applications for autoimmune disease, which requires prolonged maintenance of therapy. One approach to control kinetics is to use degradable nanoparticles, which can deliver immunotherapies with higher efficacy, at a slower rate, and with reduced toxicity. Biodegradable PLGA nanoparticles can deliver TLR adjuvants over several days, which have the advantage of increasing adjuvant uptake by dendritic cells and prolonging dendritic cell activation [59]. When nanoparticles are used to slowly deliver antigen over a long period of time, it is evident that the long-term memory response is enhanced [60].

Another approach to control when an immunotherapy is delivered is to incorporate functionalities that trigger drug release by internal or external mechanisms. We have already mentioned therapeutic approaches that rely on pH changes to facilitate site-specific intracellular delivery. The use of enzyme-sensitive linkers could also be incorporated into future nanoparticle designs for controlled release. For example, the incorporation of a matrix metalloprotease-2 responsive linker into a nanoparticle could facilitate the specific delivery of an immunotherapy in the tumor microenvironment [61]. In a similar fashion, incorporation of matrix metalloproteinase-9 sensitive lipopeptides into liposomes may facilitate the release of immunotherapy specifically at the location of a tumor [62]. External mechanisms to trigger immunotherapy release primarily rely on the use of near-infrared light to trigger chemical degradation or thermal ablation. For example, triggerable copper sulfide (CuS) nanoparticles have been formulated with a chitosan surface coating to deliver CpG oligonucleotide. Upon excitation with near-infrared light, the nanoparticles undergo disintegration and reassemble into polymer complexes exhibiting enhanced tumor retention. In addition, the new complexes can traffic efficiently to TLR-9-expressing-endosomes in plasmacytoid dendritic cells, which promotes innate immunity through the activation of natural killer cells. At the same time, photothermal ablation can perturb the tumor microenvironment and dislodge tumor antigens for recognition and ingestion by antigen presenting cells within the tumor stroma. The subsequent activation of antigen presenting cells will lead to cross-priming of tumor antigen-specific T-cells within the tumor draining lymph nodes [63]. Gold nanoparticle photothermal therapy has also been used for immunomodulation in tumors. It has been observed that photothermal therapy can promote a tumor-specific immune response in melanomas, which resulted in extensive proliferation of CD41 helper T-cells and CD81 cytotoxic T-cells [64]. Thermal ablation facilitated by gold nanoparticles also generates danger-associated molecular patterns, which activate inflammasome complexes that activates caspases to cleave precursors of proinflammatory cytokines [65].

21.3.2 Designing Nanoparticles to Fine-Tune the Immune Response in Vaccines and Tumor Immunotherapies

21.3.2.1 Nanoparticle vaccines

Nanoparticle vaccines have been established to significantly upregulate T-cell responses over equivalent soluble vaccines [66]. To maximize this effect, it is essential to consider both a nanoparticle vaccine's physical characteristics and material to induce a specific, desired immune response. For instance, the immune response elicited from a polymersome, which is a watery core particle, differs from the immune response elicited by a solid core nanoparticles composed of poly(propylene sulfide) (PPS). The polymersome encapsulates its antigen cargo, while antigen is conjugated to the surface of the PPS nanoparticle. It was found that antigen delivered in polymersomes tended to enhance CD4 responses, while antigen delivered by PPS nanoparticles tended to enhance CD8 responses. These differences are attributed to the difference in tendency to present antigen to MHC II or MHC I pathways [67]. Nanocarrier porosity also affects the manner in which antigen is encapsulated and how downstream immune responses are elicited. Three silica nanocarriers of different porosities each induced different levels of IgG and IgA antibody production [68]. Interestingly, antigen presentation is also more critical in dictating the cellular uptake of nanoparticle vaccines than the presence of targeting ligands, such as mannose [69].

Significant efforts have also been dedicated to the targeting of nanoparticle vaccines to the lymphoid organs. Transport to the lymph nodes is essential for antigen presentation leading to T-cell activation, which leads to cytotoxic T-cell responses and B-cell activation that stimulates the production of high affinity antibodies [70]. Nanoparticle size has been identified as a critical parameter that influences targeting to the lymph node. Evaluation of 20, 45, and 100 nm PPS nanoparticles administered intradermally demonstrated an inverse correlation between lymph node retention and nanoparticle size. Moreover, a targeting ligand was not necessary for significant accumulation of the 20 nm nanoparticles to the lymph nodes [71]. The same preference for smaller sized nanoparticles was observed after intravascular administration—30 nm polymeric micelles were found to extravasate and accumulate at metastatic lymph nodes more than 70 nm micelles of similar chemical composition [72].

As an alternative strategy, nanoparticles can also be preloaded into dendritic cells, which can home to the lymph nodes after injection. In this situation, the optimal nanoparticle size maximizes dendritic cell uptake without inducing toxicity. Delivery of immune adjuvants (e.g., CpG) on the nanoparticles can enhance antigen copresentation and upregulate costimulatory molecule expression at the lymph nodes [73].

To further enhance lymph node homing, nanoparticles can be coated with lipid membranes with incorporated ganglioside GM3. This facilitates interaction with Siglec1 on myeloid dendritic cells and macrophages, which is responsible for B-cell, CD81 T-cell, and iNKT priming and activation [74]. To evaluate the localization and biodistribution of nanoparticle vaccines, evaluation of trafficking to the lymph nodes can be conducted *in vivo* by magnetic resonance imaging (MRI) or single-photon emission computed tomography (SPECT) with iron oxide or radioisotope labeling, respectively [75].

21.3.2.2 Enhancement of the antitumor immune response

In addition to serving as tumor vaccines, nanoparticles favorably modulate the immune response through multiple means. One challenge with cancer immunotherapy is the presence of immune suppression, which mutes the immune response in the tumor microenvironment. An example of an immune suppression mechanism is checkpoint blockade, in which tumor cells upregulate ligands that bind to T-cell receptors (e.g., PD-1) that downregulate cytolytic activity. Nanoparticle immunotherapies have been designed to interfere with this immune suppression mechanism. In a melanoma model, polymeric nanoparticles loaded with cytotoxic T lymphocyte-associated molecule 4 (CTLA-4) siRNA have been able to increase the number of antitumor CD81 T-cells while simultaneously decrease the number of regulatory T-cells (T-regs) (Fig. 21.5) [76]. The delivery of immune adjuvants by PPS nanoparticles to dendritic cells in the draining lymph node also increases the CD8 to CD4 T-cell ratio, which also leads to slowed tumor growth in a melanoma model [77]. PPS nanoparticles functionalized with exposed hydroxyl groups also can engage the complement system, which is indicated by high C3a release from the antibody-antigen complex [78]. Hyaluronan nanoparticles have the unique property of initiating an innate immune response upon interaction with CD44, which is a tumor-specific marker in some forms of leukemia [79].

To further boost therapeutic efficacy, it is advantageous to develop means to target tumor environments with higher specificity. A powerful "next generation" targeted nanoparticle for immunotherapy applications is the aptamer, which consists of single-stranded RNA or DNA oligonucleotides that can form structures which have high affinity to their targets. An aptamer acting as an agonist to CD40, which enhances the immune response by promoting B-cell clonal expansion and germinal center formation, increased the median survival of mice with A20 B-cell lymphoma by approximately 10 days [80]. Another ssDNA aptamer has been designed to target the CD30 receptor, which trimerizes to activate cellular signaling that triggers cell apoptosis in anaplastic large cell lymphoma (Fig. 21.6) [81]. Aptamers have also been used against targets in the tumor stroma. The versatility of aptamer design can facilitate the pursuit of multiple targets; in a breast cancer study, it was shown that the targeting of both the VEGF receptor and the 4-1BB receptor, a costimulatory receptor that promotes the survival and activation of activated CD81 cells, provided survival benefit over treatment against each target individually [82]. Another innovative strategy exploits how the natural biodistribution of healthy lymphocytes mirrors the biodistribution of hard-to-reach tumors. By loading nanoparticles onto T-cells, chemotherapies with poor pharmacokinetics have successfully been delivered to disperse lymphomas with significantly higher therapeutic effect than soluble drug or free nanoparticles [83].

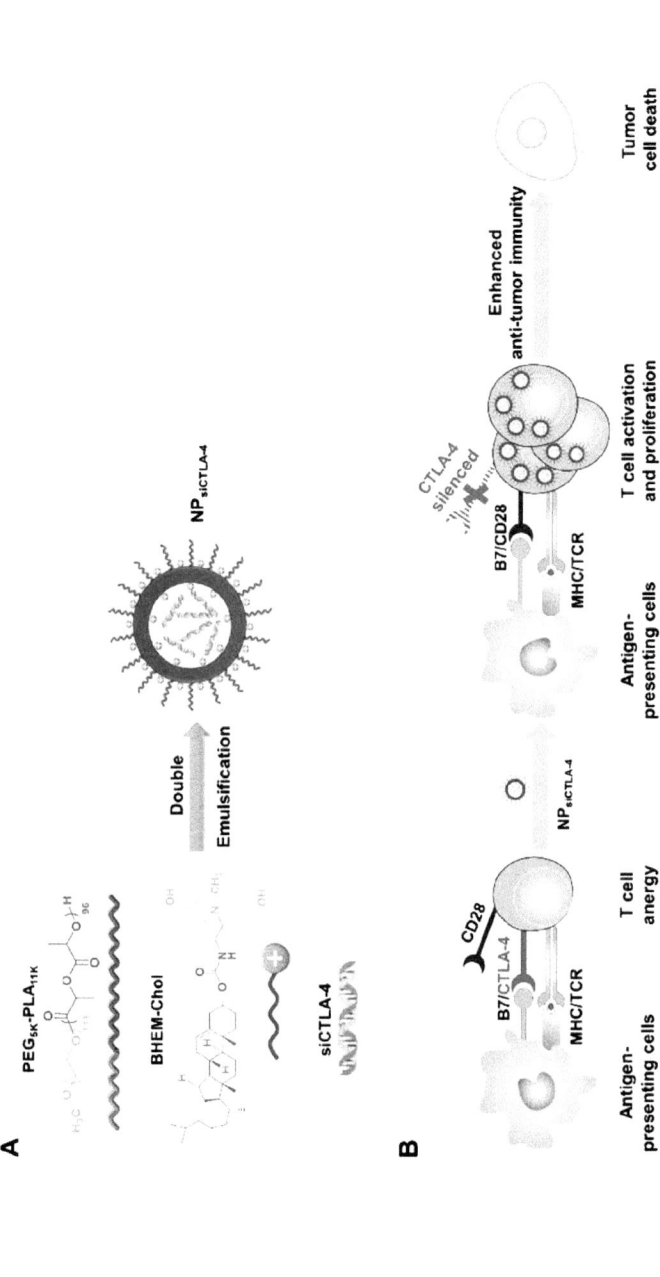

Figure 21.5 Nanoparticles to treat immune suppression. (A) Preparation of siCTLA-4-encapsulated nanoparticles (NP$_{siCTLA-4}$) with poly(ethylene glycol)-*block*-poly(D,L-lactide) and a cationic lipid BHEM-Chol by double emulsification. (B) Enhancing T-cell-mediated immune responses by blocking CTLA-4 using NP$_{siCTLA-4}$. CTLA-4 plays a strong inhibitory role in T-cell activation and proliferation, which significantly curbs T-cell-mediated tumor rejection. NPsiCTLA-4-mediated CTLA-4 knockdown enhanced the activation and proliferation of T-cells, which inhibited the overall growth of tumors. Reproduced with permission from ref. 76. Copyright (2016), with permission from Elsevier.

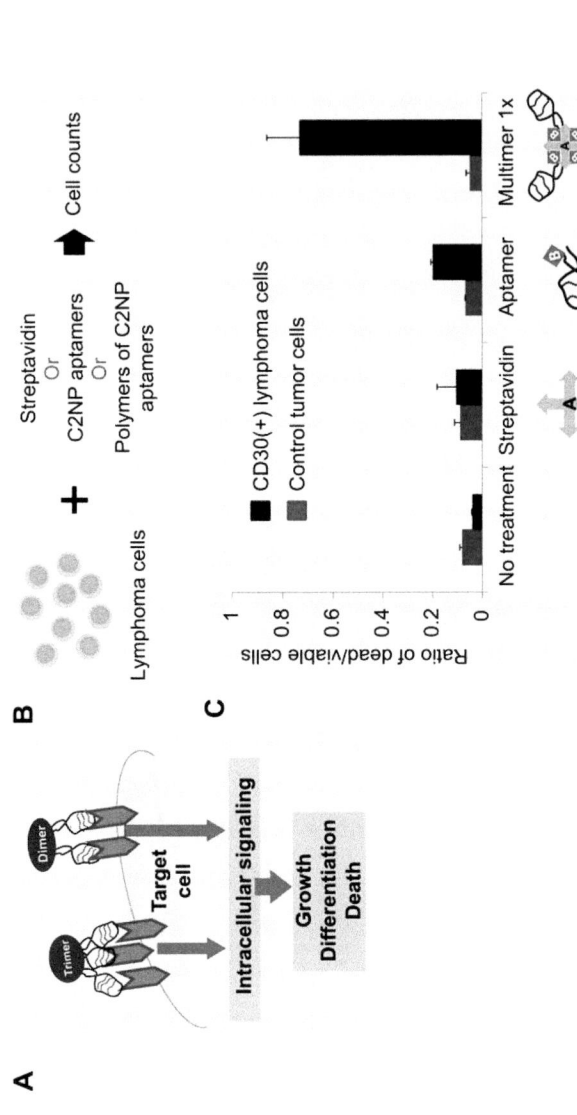

Figure 21.6 Targeted multivalent aptamers deliver biotherapy to anaplastic large cell lymphoma. (A) Schema showing receptor oligomerization inducing downstream signaling. CD30-associated signaling is activated by its ligand through trimerization of the receptor, leading to varied outcomes that range from apoptosis to proliferation. (B) CD30-positive and -negative cells were incubated without any treatment; in presence of control streptavidin, monomeric aptamer C2NP, and multimeric aptamer C2NP, for 72 h to detect aptamer-mediated CD30 signal transduction. A multivalent CD30 aptamer was made using biotinylated C2NP (33) with streptavidin (13). (C) The multivalent CD30 aptamer-induced signaling, resulting in a higher percentage of dead cells in CD30-positive ALCL (K299 cells), and had no effect on cell death in CD30-negative (HL60) cells. Reproduced with permission from ref. 81. Copyright (2013), with permission from Elsevier.

21.4 Conclusions

We return to the question posed at the beginning of the article, how do we deliver the optimal amount of immunotherapy to a specific site, with appropriate kinetics and dosing schedule, without inducing deleterious side effects that outweigh the benefits of the therapy? Nanoparticles are an appealing solution to this complicated drug delivery problem. They are versatile, engineered platforms that can be tailored to home to specific targets. By fine-tuning nanoparticle size, shape, and elasticity, nanoparticles can be guided to travel to a specific site. To decrease accumulation in off-target sites and minimize toxicity that is a result of nonspecific localization, surface modifications can be implemented to target nanoparticles to specific cells and prevent uptake by other cells. The formulation of nanoparticles with stimuli-sensitive materials (e.g., pH, enzymatic) or inside hydrogels and matrices enables immunotherapeutic release only when the particles have reached their target microenvironment. It is important to realize that a single design modification will likely not maximize site-specific delivery, minimize off-target accumulation, and optimize immunotherapy delivery timing, tissue penetration, and intracellular delivery. An ideal design will achieve a set of these goals and will be most effective when tailored specifically to its desired therapeutic use.

In the clinic, nanoparticles have been approved for the delivery of chemotherapy to cancer patients. Liposomal formulations loaded with chemotherapeutics such as doxorubicin, paclitaxel, and daunorubicin are recommended as neo-adjuvant therapies in ovarian cancer, Kaposi's sarcoma, and multiple myeloma. At the same time, nanoparticles are under clinical evaluation for the delivery of therapeutic siRNA, the detection of angiogenesis and micrometastatic lesions, and for the measurement of cancer biomarkers in the blood. Nanoparticle immunotherapies, however, have not traveled as far along the research and development pipeline. For clinical adoption to become a reality, it is essential to improve the design of nanoparticle immunotherapies so their targeting specificity is enhanced. In addition, it is important to continue boosting the efficacy of nanoparticle immunotherapies under development to justify their use over existing therapeutic regimens. One avenue of research will be to continue the evaluation of new combinations of immunotherapies. When developing and testing combinatorial immunotherapies, it is important to realize that a set of nanoparticles of different designs, not a single particle, may be necessary to maximize therapeutic effect. Mechanistic studies should be combined with *in silico* studies to understand how the immune response evolves over time with the treatment and the disease. At the same time, it is important to consider the immunogenicity of the nanoparticle delivery vehicle in combination with the immunogenicity of the therapeutics they carry.

It is also essential to grasp how a nanoparticle immunotherapy affects immune homeostasis. While design modifications to a nanoparticle may enable site-specific targeting and targeted delivery to specific cell compartments, these same modifications may disrupt cellular membranes, induce apoptosis, or interact with pattern recognition receptors and produce proinflammatory cytokines. The common approach to enhance cancer immunotherapy efficacy by reversing immune suppression could lead to increased T-cell activation or proinflammatory cytokine generation in healthy tissues. In addition, therapeutic delivery above the optimal dose could potentially activate immune suppression mechanisms that were previously dormant. Immunotherapies for autoimmune diseases face the opposite challenge. While immune suppression could favorably reduce inflammation in host tissues, appropriate dosing must be used to prevent increased susceptibility to infectious diseases. For these endeavors to be successful, it is necessary to understand how nanoparticle immunotherapies perturb the complex network of signaling pathways involved in immunity. The ability to intelligently engineer nanoparticle immunotherapies will allow us to effectively attack new targets as they are discovered. Using nanoparticles, long-term, synergistic, reactive immunotherapy regimens can be developed to keep diseases of immune dysregulation in check.

Disclosures and Conflict of Interest

This chapter was originally published as: Toy, R., and Roy, K. Engineering nanoparticles to overcome barriers to immunotherapy. *Bioengineering & Translational Medicine*, **1**, 47–62 (2016), DOI: 10.1002/btm2.10005, under the Creative Commons Attribution

License (https://creativecommons.org/licenses/by/3.0/) and appears here, with edits, by kind permission of the publisher, John Wiley & Sons. The authors acknowledge support from the Georgia Tech Foundation and the Robert A. Milton Chaired Professorship and wish to thank Pallab Prabhan for helpful discussions regarding the content of the manuscript. The authors declare that they have no conflict of interest and have no affiliations or financial involvement with any organization or entity discussed in this chapter. No writing assistance was utilized in the production of this chapter and the authors have received no payment for its preparation.

Corresponding Author

Dr. Krishnendu Roy
The Wallace H. Coulter Department of Biomedical Engineering
at Georgia Tech and Emory
The Parker H. Petit Institute for Bioengineering and Biosciences
Georgia Institute of Technology
EBB 3018, 950 Atlantic Dr. NW, Atlanta, GA 30318, USA
Email: krish.roy@gatech.edu

References

1. Adams, J. L., Smothers, J., Srinivasan, R., Hoos, A. (2015). Big opportunities for small molecules in immuno-oncology. *Nat. Rev. Drug Discov.*, **14**(9), 603–622.
2. Weiner, L. M., Surana, R., Wang, S. (2010). Monoclonal antibodies: Versatile platforms for cancer immunotherapy. *Nat. Rev. Immunol.*, **10**(5), 317–327.
3. Mahoney, K. M., Rennert, P. D., Freeman, G. J. (2015). Combination cancer immunotherapy and new immunomodulatory targets. *Nat. Rev. Drug Discov.*, **14**(8), 561–584.
4. Lizotte, P. H., Wen, A. M., Sheen, M. R., et al. (2016). In situ vaccination with cowpea mosaic virus nanoparticles suppresses metastatic cancer. *Nat. Nanotechnol.*, **11**(3), 295–303.
5. June, C. H., Maus, M. V., Plesa, G., et al. (2014). Engineered T cells for cancer therapy. *Cancer Immunol. Immunother.*, **63**(9), 969–975.
6. Caspi, R. R. (2008). Immunotherapy of autoimmunity and cancer: The penalty for success. *Nat. Rev. Immunol.*, **8**(12), 970–976.

7. Makkouk, A., Weiner, G. J. (2015). Cancer immunotherapy and breaking immune tolerance: New approaches to an old challenge. *Cancer Res.*, **75**(1), 5–10.
8. Nepom, G. T., St Clair, E. W., Turka, L. A. (2011). Challenges in the pursuit of immune tolerance. *Immunol. Rev.*, **241**(1), 49–62.
9. Goldberg, M. S. (2015). Immunoengineering: How nanotechnology can enhance cancer immunotherapy. *Cell*, **161**(2), 201–204.
10. Fan, Y., Moon, J. J. (2015). Nanoparticle drug delivery systems designed to improve cancer vaccines and immunotherapy. *Vaccines*, **3**(3), 662–685.
11. Almeida, J. P., Lin, A. Y., Figueroa, E. R., Foster, A. E., Drezek, R. A. (2015). In vivo gold nanoparticle delivery of peptide vaccine induces anti-tumor immune response in prophylactic and therapeutic tumor models. *Small*, **11**(12), 1453–1459.
12. Hunter, Z., McCarthy, D. P., Yap, W. T., et al. (2014). A biodegradable nanoparticle platform for the induction of antigen-specific immune tolerance for treatment of autoimmune disease. *ACS Nano*, **8**(3), 2148–2160.
13. Li, N., Peng, L. H., Chen, X., et al. (2014). Antigen-loaded nanocarriers enhance the migration of stimulated Langerhans cells to draining lymph nodes and induce effective transcutaneous immunization. *Nanomedicine*, **10**(1), 215–223.
14. Kitaoka, M., Naritomi, A., Hirakawa, Y., Kamiya, N., Goto, M. (2015). Transdermal immunization using solid-in-oil nanodispersion with CpG oligodeoxynucleotide adjuvants. *Pharm. Res.*, **32**(4), 1486–1492.
15. Resnier, P., Montier, T., Mathieu, V., Benoit, J. P., Passirani, C. (2013). A review of the current status of siRNA nanomedicines in the treatment of cancer. *Biomaterials*, **34**(27), 6429–6443.
16. Xu, Z., Wang, Y., Zhang, L., Huang, L. (2014). Nanoparticle-delivered transforming growth factor-beta siRNA enhances vaccination against advanced melanoma by modifying tumor microenvironment. *ACS Nano*, **8**(4), 3636–3645.
17. Luo, Z., Wang, C., Yi, H., et al. (2015). Nanovaccine loaded with poly I:C and STAT3 siRNA robustly elicits anti-tumor immune responses through modulating tumor-associated dendritic cells in vivo. *Biomaterials*, **38**, 50–60.
18. Pradhan, P., Qin, H., Leleux, J. A., et al. (2014). The effect of combined IL10 siRNA and CpG ODN as pathogen-mimicking microparticles on Th1/Th2 cytokine balance in dendritic cells and protective immunity against B cell lymphoma. *Biomaterials*, **35**(21), 5491–5504.

19. Heo, M. B., Lim, Y. T. (2014). Programmed nanoparticles for combined immunomodulation, antigen presentation and tracking of immunotherapeutic cells. *Biomaterials*, **35**(1), 590–600.
20. Liu, S. Y., Wei, W., Yue, H., et al. (2013). Nanoparticles-based multi-adjuvant whole cell tumor vaccine for cancer immunotherapy. *Biomaterials*, **34**(33), 8291–8300.
21. Dahlman, J. E., Barnes, C., Khan, O. F., et al. (2014). In vivo endothelial siRNA delivery using polymeric nanoparticles with low molecular weight. *Nat. Nanotechnol.*, **9**(8), 648–655.
22. Guo, Y., Chen, W., Wang, W., et al. (2012). Simultaneous diagnosis and gene therapy of immunorejection in rat allogeneic heart transplantation model using a T-cell-targeted theranostic nanosystem. *ACS Nano*, **6**(12), 10646–10657.
23. Wohlfart, S., Gelperina, S., Kreuter, J. (2012). Transport of drugs across the blood-brain barrier by nanoparticles. *J. Control. Release*, **161**(2), 264–273.
24. Ernsting, M. J., Murakami, M., Roy, A., Li, S. D. (2013). Factors controlling the pharmacokinetics, biodistribution and intratumoral penetration of nanoparticles. *J. Control. Release*, **172**(3), 782–794.
25. Kwissa, M., Nakaya, H. I., Oluoch, H., Pulendran, B. (2012). Distinct TLR adjuvants differentially stimulate systemic and local innate immune responses in nonhuman primates. *Blood*, **119**(9), 2044–2055.
26. Barber, G. N. (2011). Cytoplasmic DNA innate immune pathways. *Immunol. Rev.*, **243**(1), 99–108.
27. Gavrilov, K., Saltzman, W. M. (2012). Therapeutic siRNA: Principles, challenges, and strategies. *Yale J. Biol. Med.*, **85**(2), 187–200.
28. Sahay, G., Alakhova, D. Y., Kabanov, A. V. (2010). Endocytosis of nanomedicines. *J. Control. Release*, **145**(3), 182–195.
29. Kwong, B., Liu, H., Irvine, D. J. (2011). Induction of potent anti-tumor responses while eliminating systemic side effects via liposome-anchored combinatorial immunotherapy. *Biomaterials*, **32**(22), 5134–5147.
30. Huang, X., Li, L., Liu, T., et al. (2011). The shape effect of mesoporous silica nanoparticles on biodistribution, clearance, and biocompatibility in vivo. *ACS Nano*, **5**(7), 5390–5399.
31. Copp, J. A., Fang, R. H., Luk, B. T., et al. (2014). Clearance of pathological antibodies using biomimetic nanoparticles. *Proc. Natl. Acad. Sci. U. S. A.*, **111**(37), 13481–13486.

32. Anselmo, A. C., Kumar, S., Gupta, V., et al. (2015). Exploiting shape, cellular-hitchhiking and antibodies to target nanoparticles to lung endothelium: Synergy between physical, chemical and biological approaches. *Biomaterials*, **68**, 1–8.
33. Key, J., Palange, A. L., Gentile, F., et al. (2015). Soft discoidal polymeric nanoconstructs resist macrophage uptake and enhance vascular targeting in tumors. *ACS Nano*, **9**(12), 11628–11641.
34. Ruoslahti, E. (2012). Peptides as targeting elements and tissue penetration devices for nanoparticles. *Adv. Mater.*, **24**(28), 3747–3756.
35. Shah, P. P., Desai, P. R., Channer, D., Singh, M. (2012). Enhanced skin permeation using polyarginine modified nanostructured lipid carriers. *J. Control. Release*, **161**(3), 735–745.
36. Sugahara, K. N., Scodeller, P., Braun, G. B., et al. (2015). A tumor-penetrating peptide enhances circulation-independent targeting of peritoneal carcinomatosis. *J. Control. Release*, **212**, 59–69.
37. Yan, F., Wu, H., Liu, H., et al. (2015). Molecular imaging-guided photothermal/photodynamic therapy against tumor by iRGD-modified indocyanine green nanoparticles. *J. Control. Release*, **224**, 217–228.
38. Cun, X., Chen, J., Ruan, S., et al. (2015). A novel strategy through combining iRGD peptide with tumor-microenvironment-responsive and multistage nanoparticles for deep tumor penetration. *ACS Appl. Mater. Interfaces*, **7**(49), 27458–27466.
39. Ren, Y., Cheung, H. W., von Maltzhan, G., et al. (2012). Targeted tumor-penetrating siRNA nanocomplexes for credentialing the ovarian cancer oncogene ID4. *Sci. Transl. Med.*, **4**(147), 147ra112.
40. Li, X., Guo, S., Zhu, C., et al. (2013). Intestinal mucosa permeability following oral insulin delivery using core shell corona nano-lipoparticles. *Biomaterials*, **34**(37), 9678–9687.
41. Ghosn, B., van de Ven, A. L., Tam, J., et al. (2010). Efficient mucosal delivery of optical contrast agents using imidazole-modified chitosan. *J. Biomed. Opt.*, **15**(1), 015003.
42. Ghosn, B., Singh, A., Li, M., et al. (2010). Efficient gene silencing in lungs and liver using imidazole-modified chitosan as a nanocarrier for small interfering RNA. *Oligonucleotides*, **20**(3), 163–172.
43. Rosalia, R. A., Cruz, L. J., van Duikeren, S., et al. (2015). CD40-targeted dendritic cell delivery of PLGA-nanoparticle vaccines induce potent antitumor responses. *Biomaterials*, **40**, 88–97.

44. McHugh, M. D., Park, J., Uhrich, R., Gao, W., Horwitz, D. A., Fahmy, T. M. (2015). Paracrine co-delivery of TGF-beta and IL-2 using CD4-targeted nanoparticles for induction and maintenance of regulatory T cells. *Biomaterials*, **59**, 172–181.

45. Silva, J. M., Zupancic, E., Vandermeulen, G., et al. (2015). In vivo delivery of peptides and toll-like receptor ligands by mannose-functionalized polymeric nanoparticles induces prophylactic and therapeutic antitumor immune responses in a melanoma model. *J. Control. Release*, **198**, 91–103.

46. Huang, Z., Zhang, Z., Jiang, Y., et al. (2012). Targeted delivery of oligonucleotides into tumor-associated macrophages for cancer immunotherapy. *J. Control. Release*, **158**(2), 286–292.

47. Shah, N. D., Parekh, H. S., Steptoe, R. J. (2014). Asymmetric peptide dendrimers are effective linkers for antibody-mediated delivery of diverse payloads to b cells in vitro and in vivo. *Pharm. Res.*, **31**(11), 3150–3160.

48. Temchura, V. V., Kozlova, D., Sokolova, V., Uberla, K., Epple, M. (2014). Targeting and activation of antigen-specific B-cells by calcium phosphate nanoparticles loaded with protein antigen. *Biomaterials*, **35**(23), 6098–6105.

49. Kasturi, S. P., Sachaphibulkij, K., Roy, K. (2005). Covalent conjugation of polyethyleneimine on biodegradable microparticles for delivery of plasmid DNA vaccines. *Biomaterials*, **26**(32), 6375–6385.

50. Benjaminsen, R. V., Mattebjerg, M. A., Henriksen, J. R., Moghimi, S. M., Andresen, T. L. (2013). The possible "proton sponge" effect of polyethylenimine (PEI) does not include change in lysosomal pH. *Mol. Ther.*, **21**(1), 149–157.

51. Yuba, E., Kanda, Y., Yoshizaki, Y., et al. (2015). pH-sensitive polymer-liposome-based antigen delivery systems potentiated with interferon-gamma gene lipoplex for efficient cancer immunotherapy. *Biomaterials*, **67**, 214–224.

52. Wang, C., Li, P., Liu, L., et al. (2016). Self-adjuvanted nanovaccine for cancer immunotherapy: Role of lysosomal rupture-induced ROS in MHC class I antigen presentation. *Biomaterials*, **79**, 88–100.

53. Yu, H., Zou, Y., Wang, Y., et al. (2011). Overcoming endosomal barrier by amphotericin B-loaded dual pH-responsive PDMA-b-PDPA micelle-plexes for siRNA delivery. *ACS Nano*, **5**(11), 9246–9255.

54. Erazo-Oliveras, A., Muthukrishnan, N., Baker, R., Wang, T. Y., Pellois, J. P. (2012). Improving the endosomal escape of cell-penetrating

peptides and their cargos: Strategies and challenges. *Pharmaceuticals (Basel)*, **5**(11), 1177–1209.

55. Futaki, S., Masui, Y., Nakase, I., et al. (2005). Unique features of a pH-sensitive fusogenic peptide that improves the transfection efficiency of cationic liposomes. *J. Gene Med.*, **7**(11), 1450–1458.

56. Miura, N., Shaheen, S. M., Akita, H., Nakamura, T., Harashima, H. (2015). A KALA-modified lipid nanoparticle containing CpG-free plasmid DNA as a potential DNA vaccine carrier for antigen presentation and as an immune-stimulative adjuvant. *Nucleic Acids Res.*, **43**(3), 1317–1331.

57. Plebanek, M. P., Mutharasan, R. K., Volpert, O., Matov, A., Gatlin, J. C., Thaxton, C. S. (2015). Nanoparticle targeting and cholesterol flux through scavenger receptor type B-1 inhibits cellular exosome uptake. *Sci. Rep.*, **5**, 15724.

58. Ding, Y., Wang, Y., Zhou, J., et al. (2014). Direct cytosolic siRNA delivery by reconstituted high density lipoprotein for target-specific therapy of tumor angiogenesis. *Biomaterials*, **35**(25), 7214–7227.

59. Jewell, C. M., Lopez, S. C., Irvine, D. J. (2011). In situ engineering of the lymph node microenvironment via intranodal injection of adjuvant-releasing polymer particles. *Proc. Natl. Acad. Sci. U. S. A.*, **108**(38), 15745–15750.

60. Kanchan, V., Katare, Y. K., Panda, A. K. (2009). Memory antibody response from antigen loaded polymer particles and the effect of antigen release kinetics. *Biomaterials*, **30**(27), 4763–4776.

61. Zhu, L., Kate, P., Torchilin, V. P. (2012). Matrix metalloprotease 2-responsive multifunctional liposomal nanocarrier for enhanced tumor targeting. *ACS Nano*, **6**(4), 3491–3498.

62. Banerjee, J., Hanson, A. J., Gadam, B., et al. (2009). Release of liposomal contents by cell-secreted matrix metalloproteinase-9. *Bioconjug. Chem.*, **20**(7), 1332–1339.

63. Guo, L., Yan, D. D., Yang, D., et al. (2014). Combinatorial photothermal and immuno cancer therapy using chitosan-coated hollow copper sulfide nanoparticles. *ACS Nano*, **8**(6), 5670–5681.

64. Bear, A. S., Kennedy, L. C., Young, J. K., et al. (2013). Elimination of metastatic melanoma using gold nanoshell-enabled photothermal therapy and adoptive T cell transfer. *PLoS One*, **8**(7), e69073.

65. Nguyen, H. T., Tran, K. K., Sun, B., Shen, H. (2012). Activation of inflammasomes by tumor cell death mediated by gold nanoshells. *Biomaterials*, **33**(7), 2197–2205.

66. Li, A. V., Moon, J. J., Abraham, W., et al. (2013). Generation of effector memory T cell-based mucosal and systemic immunity with pulmonary nanoparticle vaccination. *Sci. Transl. Med.*, **5**(204), 204ra130.
67. Stano, A., Scott, E. A., Dane, K. Y., Swartz, M. A., Hubbell, J. A. (2013). Tunable T cell immunity towards a protein antigen using polymersomes vs. solid-core nanoparticles. *Biomaterials*, **34**(17), 4339–4346.
68. Wang, T., Jiang, H., Zhao, Q., Wang, S., Zou, M., Cheng, G. (2012). Enhanced mucosal and systemic immune responses obtained by porous silica nanoparticles used as an oral vaccine adjuvant: Effect of silica architecture on immunological properties. *Int. J. Pharm.*, **436**(1–2), 351–358.
69. Thomann-Harwood, L. J., Kaeuper, P., Rossi, N., Milona, P., Herrmann, B., McCullough, K. C. (2013). Nanogel vaccines targeting dendritic cells: Contributions of the surface decoration and vaccine cargo on cell targeting and activation. *J. Control. Release*, **166**(2), 95–105.
70. Andorko, J. I., Hess, K. L., Jewell, C. M. (2015). Harnessing biomaterials to engineer the lymph node microenvironment for immunity or tolerance. *AAPS J.*, **17**(2), 323–338.
71. Reddy, S. T., Rehor, A., Schmoekel, H. G., Hubbell, J. A., Swartz, M. A. (2006). In vivo targeting of dendritic cells in lymph nodes with poly(propylene sulfide) nanoparticles. *J. Control. Release*, **112**(1), 26–34.
72. Cabral, H., Makino, J., Matsumoto, Y., et al. (2015). Systemic targeting of lymph node metastasis through the blood vascular system by using size-controlled nanocarriers. *ACS Nano*, **9**(5), 4957–4967.
73. Zhou, Q., Zhang, Y., Du, J., et al. (2016). Different-sized gold nanoparticle activator/antigen increases dendritic cells accumulation in liver-draining lymph nodes and CD81 T cell responses. *ACS Nano*, **10**(2), 2678–2692.
74. Xu, F., Reiser, M., Yu, X., Gummuluru, S., Wetzler, L., Reinhard, B. M. (2016). Lipid-mediated targeting with membrane-wrapped nanoparticles in the presence of corona formation. *ACS Nano*, **10**(1), 1189–1200.
75. Cobaleda-Siles, M., Henriksen-Lacey, M., Ruiz de Angulo, A., et al. (2014). An iron oxide nanocarrier for dsRNA to target lymph nodes and strongly activate cells of the immune system. *Small*, **10**(24), 5054–5067.
76. Li, S. Y., Liu, Y., Xu, C. F., et al. (2016). Restoring antitumor functions of T cells via nanoparticle-mediated immune checkpoint modulation. *J. Control. Release*, **231**, 17–28.

77. Thomas, S. N., Vokali, E., Lund, A. W., Hubbell, J. A., Swartz, M. A. (2014). Targeting the tumor-draining lymph node with adjuvanted nanoparticles reshapes the antitumor immune response. *Biomaterials*, **35**(2), 814–824.

78. Thomas, S. N., van der Vlies, A. J., O'Neil, C. P., et al. (2011). Engineering complement activation on polypropylene sulfide vaccine nanoparticles. *Biomaterials*, **32**(8), 2194–2203.

79. Mizrahy, S., Raz, S. R., Hasgaard, M., et al. (2011). Hyaluronan-coated nanoparticles: The influence of the molecular weight on CD44-hyaluronan interactions and on the immune response. *J. Control. Release*, **156**(2), 231–238.

80. Soldevilla, M. M., Villanueva, H., Bendandi, M., Inoges, S., Lopez-Diaz de Cerio, A., Pastor, F. (2015). 2-Fluoro-RNA oligonucleotide CD40 targeted aptamers for the control of B lymphoma and bone-marrow aplasia. *Biomaterials*, **67**, 274–285.

81. Parekh, P., Kamble, S., Zhao, N., Zeng, Z., Portier, B. P., Zu, Y. (2013). Immunotherapy of CD30-expressing lymphoma using a highly stable ssDNA aptamer. *Biomaterials*, **34**(35), 8909–8917.

82. Schrand, B., Berezhnoy, A., Brenneman, R., et al. (2014). Targeting 4-1BB costimulation to the tumor stroma with bispecific aptamer conjugates enhances the therapeutic index of tumor immunotherapy. *Cancer Immunol. Res.*, **2**(9), 867–877.

83. Huang, B., Abraham, W. D., Zheng, Y., Bustamante Lopez, S. C., Luo, S. S., Irvine, D. J. (2015). Active targeting of chemotherapy to disseminated tumors using nanoparticle-carrying T cells. *Sci. Transl. Med.*, **7**(291), 291ra294.

84. Owens, D. E. III, Peppas, N. A. (2006). Opsonization, biodistribution, and pharmacokinetics of polymeric nanoparticles. *Int. J. Pharm.*, **307**(1), 93–102.

85. Petros, R. A., DeSimone, J. M. (2010). Strategies in the design of nanoparticles for therapeutic applications. *Nat. Rev. Drug Discov.*, **9**(8), 615–627.

86. Toy, R., Hayden, E., Shoup, C., Baskaran, H., Karathanasis, E. (2011). The effects of particle size, density and shape on margination of nanoparticles in microcirculation. *Nanotechnology*, **22**(11), 115101.

87. Toy, R., Hayden, E., Camann, A., et al. (2013). Multimodal in vivo imaging exposes the voyage of nanoparticles in tumor microcirculation. *ACS Nano*, **7**(4), 3118–3129.

88. Muro, S., Garnacho, C., Champion, J. A., et al. (2008). Control of endothelial targeting and intracellular delivery of therapeutic enzymes by modulating the size and shape of ICAM-1-targeted carriers. *Mol. Ther.*, **16**(8), 1450–1458.

89. Chithrani, B. D., Ghazani, A. A., Chan, W. C. (2006). Determining the size and shape dependence of gold nanoparticle uptake into mammalian cells. *Nano Lett.*, **6**(4), 662–668.

90. Anish, C., Khan, N., Upadhyay, A. K., Sehgal, D., Panda, A. K. (2014). Delivery of polysaccharides using polymer particles: Implications on size-dependent immunogenicity, opsonophagocytosis, and protective immunity. *Mol. Pharm.*, **11**(3), 922–937.

91. Mueller, S. N., Tian, S., DeSimone, J. M. (2015). Rapid and persistent delivery of antigen by lymph node targeting PRINT nanoparticle vaccine carrier to promote humoral immunity. *Mol. Pharm.*, **12**(5), 1356–1365.

92. Toy, R., Peiris, P. M., Ghaghada, K. B., Karathanasis, E. (2014). Shaping cancer nanomedicine: The effect of particle shape on the in vivo journey of nanoparticles. *Nanomedicine (Lond)*, **9**(1), 121–134.

93. Agarwal, R., Jurney, P., Raythatha, M., et al. (2015). Effect of shape, size, and aspect ratio on nanoparticle penetration and distribution inside solid tissues using 3D spheroid models. *Adv. Healthc. Mater.*, **4**(15), 2269–2280.

94. Champion, J. A., Mitragotri, S. (2006). Role of target geometry in phagocytosis. *Proc. Natl. Acad. Sci. U. S. A.*, **103**(13), 4930–4934.

95. Agarwal, R., Singh, V., Jurney, P., Shi, L., Sreenivasan, S. V., Roy, K. (2013). Mammalian cells preferentially internalize hydrogel nanodiscs over nanorods and use shape-specific uptake mechanisms. *Proc. Natl. Acad. Sci. U. S. A.*, **110**(43), 17247–17252.

96. Sun, B., Ji, Z., Liao, Y. P., et al. (2013). Engineering an effective immune adjuvant by designed control of shape and crystallinity of aluminum oxyhydroxide nanoparticles. *ACS Nano*, **7**(12), 10834–10849.

97. Niikura, K., Matsunaga, T., Suzuki, T., et al. (2013). Gold nanoparticles as a vaccine platform: Influence of size and shape on immunological responses in vitro and in vivo. *ACS Nano*, **7**(5), 3926–3938.

98. Khisamutdinov, E. F., Li, H., Jasinski, D. L., Chen, J., Fu, J., Guo, P. (2014). Enhancing immunomodulation on innate immunity by shape transition among RNA triangle, square and pentagon nanovehicles. *Nucleic Acids Res.*, **42**(15), 9996–10004.

99. Kumar, S., Anselmo, A. C., Banerjee, A., Zakrewsky, M., Mitragotri, S. (2015). Shape and size-dependent immune response to antigen-carrying nanoparticles. *J. Control. Release*, **220**(pt A), 141–148.
100. Bonner, D. K., Zhao, X., Buss, H., Langer, R., Hammond, P. T. (2013). Crosslinked linear polyethylenimine enhances delivery of DNA to the cytoplasm. *J. Control. Release*, **167**(1), 101–107.
101. Islam, M. A., Park, T. E., Singh, B., et al. (2014). Major degradable polycations as carriers for DNA and siRNA. *J. Control. Release*, **193**, 74–89.
102. Rehman, Z., Zuhorn, I. S., Hoekstra, D. (2013). How cationic lipids transfer nucleic acids into cells and across cellular membranes: Recent advances. *J. Control. Release*, **166**(1), 46–56.
103. Fromen, C. A., Robbins, G. R., Shen, T. W., Kai, M. P., Ting, J. P., DeSimone, J. M. (2015). Controlled analysis of nanoparticle charge on mucosal and systemic antibody responses following pulmonary immunization. *Proc. Natl. Acad. Sci. U. S. A.*, **112**(2), 488–493.
104. Shahbazi, M. A., Fernandez, T. D., Makila, E. M., et al. (2014). Surface chemistry dependent immunostimulative potential of porous silicon nanoplatforms. *Biomaterials*, **35**(33), 9224–9235.
105. Fernandes, R., Smyth, N. R., Muskens, O. L., et al. (2015). Interactions of skin with gold nanoparticles of different surface charge, shape, and functionality. *Small*, **11**(6), 713–721.
106. Chen, H., Li, P., Yin, Y., et al. (2010). The promotion of type 1 T helper cell responses to cationic polymers in vivo via toll-like receptor-4 mediated IL-12 secretion. *Biomaterials*, **31**(32), 8172–8180.
107. Bueter, C. L., Lee, C. K., Rathinam, V. A., et al. (2011). Chitosan but not chitin activates the inflammasome by a mechanism dependent upon phagocytosis. *J. Biol. Chem.*, **286**(41), 35447–35455.
108. Bueter, C. L., Lee, C. K., Wang, J. P., Ostroff, G. R., Specht, C. A., Levitz, S. M. (2014). Spectrum and mechanisms of inflammasome activation by chitosan. *J. Immunol.*, **192**(12), 5943–5951.
109. Pearson, R. M., Juettner, V. V., Hong, S. (2014). Biomolecular corona on nanoparticles: A survey of recent literature and its implications in targeted drug delivery. *Front. Chem.*, **2**, 108.
110. Yu, T., Greish, K., McGill, L. D., Ray, A., Ghandehari, H. (2012). Influence of geometry, porosity, and surface characteristics of silica nanoparticles on acute toxicity: Their vasculature effect and tolerance threshold. *ACS Nano*, **6**(3), 2289–2301.

111. Liu, Y., Hu, Y., Huang, L. (2014). Influence of polyethylene glycol density and surface lipid on pharmacokinetics and biodistribution of lipid-calcium-phosphate nanoparticles. *Biomaterials*, **35**(9), 3027–3034.

112. Akhter, A., Hayashi, Y., Sakurai, Y., Ohga, N., Hida, K., Harashima, H. (2014). Ligand density at the surface of a nanoparticle and different uptake mechanism: Two important factors for successful siRNA delivery to liver endothelial cells. *Int. J. Pharm.*, **475**(1–2), 227–237.

113. Ghaghada, K. B., Saul, J., Natarajan, J. V., Bellamkonda, R. V., Annapragada, A. V. (2005). Folate targeting of drug carriers: A mathematical model. *J. Control. Release*, **104**(1), 113–128.

114. Reuter, K. G., Perry, J. L., Kim, D., Luft, J. C., Liu, R., DeSimone, J. M. (2015). Targeted PRINT hydrogels: The role of nanoparticle size and ligand density on cell association, biodistribution, and tumor accumulation. *Nano Lett.*, **15**(10), 6371–6378.

115. Bandyopadhyay, A., Fine, R. L., Demento, S., Bockenstedt, L. K., Fahmy, T. M. (2011). The impact of nanoparticle ligand density on dendritic-cell targeted vaccines. *Biomaterials*, **32**(11), 3094–3105.

116. Schmidt, N. W., Jin, F., Lande, R., et al. (2015). Liquid-crystalline ordering of antimicrobial peptide-DNA complexes controls TLR9 activation. *Nat. Mater.*, **14**(7), 696–700.

117. Anselmo, A. C., Zhang, M., Kumar, S., et al. (2015). Elasticity of nanoparticles influences their blood circulation, phagocytosis, endocytosis, and targeting. *ACS Nano*, **9**(3), 3169–3177.

118. Huang, W. C., Burnouf, P. A., Su, Y. C., et al. (2016). Engineering chimeric receptors to investigate the size- and rigidity-dependent interaction of PEGylated nanoparticles with cells. *ACS Nano*, **10**(1), 648–662.

119. Guo, Y., Wang, D., Song, Q., et al. (2015). Erythrocyte membrane-enveloped polymeric nanoparticles as nanovaccine for induction of antitumor immunity against melanoma. *ACS Nano*, **9**(7), 6918–6933.

120. Nagy, J. A., Chang, S. H., Shih, S. C., Dvorak, A. M., Dvorak, H. F. (2010). Heterogeneity of the tumor vasculature. *Semin. Thromb. Hemost.*, **36**(3), 321–331.

121. Sakuma, S., Suita, M., Inoue, S., et al. (2012). Cell-penetrating peptide-linked polymers as carriers for mucosal vaccine delivery. *Mol. Pharm.*, **9**(10), 2933–2941.

122. Cruz, L. J., Rosalia, R. A., Kleinovink, J. W., Rueda, F., Lowik, C. W., Ossendorp, F. (2014). Targeting nanoparticles to CD40, DEC-205

or CD11c molecules on dendritic cells for efficient CD8(1) T cell response: A comparative study. *J. Control. Release*, **192**, 209–218.

123. Zheng, Y., Stephan, M. T., Gai, S. A., Abraham, W., Shearer, A., Irvine, D. J. (2013). In vivo targeting of adoptively transferred T-cells with antibody- and cytokine-conjugated liposomes. *J. Control. Release*, **172**(2), 426–435.

124. Akita, H., Kogure, K., Moriguchi, R., et al. (2010). Nanoparticles for ex vivo siRNA delivery to dendritic cells for cancer vaccines: Programmed endosomal escape and dissociation. *J. Control. Release*, **143**(3), 311–317.

125. Kajimoto, K., Sato, Y., Nakamura, T., Yamada, Y., Harashima, H. (2014). Multifunctional envelope-type nano device for controlled intracellular trafficking and selective targeting in vivo. *J. Control. Release*, **190**, 593–606.

126. Reissmann, S. (2014). Cell penetration: Scope and limitations by the application of cell-penetrating peptides. *J. Pept. Sci.*, **20**(10), 760–784.

Chapter 22

Metal-Based Nanoparticles and the Immune System: Activation, Inflammation, and Potential Applications

Yueh-Hsia Luo, PhD,[a] Louis W. Chang, PhD,[b] and Pinpin Lin, PhD[a,b]

[a]*Division of Environmental Health and Occupational Medicine, National Health Research Institutes, Zhunan, Taiwan*
[b]*National Environmental Health Research Center, National Health Research Institutes, Zhunan, Taiwan*

Keywords: metal-based nanoparticles, nanosilver, macrophages, dendritic cells, neutrophils, mast cells, natural killer (NK) cells, T cells, B cells, pattern recognition receptors PRRs, pathogen-associated molecular patterns (PAMPs), immunization, antigen delivery, adjuvanticity, rejection, toll-like receptors (TLRs), major histocompatibility complex (MHC), costimulatory molecules, cytokines, titanium dioxide (TiO_2), zinc oxide (ZnO), silicon dioxide (SiO_2), danger-associated molecular patterns (DAMPs), interleukin (IL), tumor necrosis factor (TNF), interferon (INF), inflammasome, high-density lipoprotein (HDL), lymphoma

22.1 Introduction

Nanotechnology is one of the most exciting industrial innovations of the 21st century. Nanomaterials are used in various industrial applications and products, including sporting goods, tires,

---

Immune Aspects of Biopharmaceuticals and Nanomedicines
Edited by Raj Bawa, János Szebeni, Thomas J. Webster, and Gerald F. Audette
Copyright © 2018 Pan Stanford Publishing Pte. Ltd.
ISBN 978-981-4774-52-9 (Hardcover), 978-0-203-73153-6 (eBook)
www.panstanford.com

sunscreens, cosmetics, electronics, and fuel additives as well for a variety of medical purposes such as diagnostic imaging and drug delivery. Many nanomaterials are metal-based nanoparticles, such as nanosilver, nanometallic oxides (zinc oxide, titanium dioxide, iron oxide, and quantum dots), and are applied for many uses [1]. For example, zinc oxide (ZnO) and titanium dioxide (TiO_2) are used in sunscreens and cosmetic products [2, 3], and nanosilver is used in detergents, antibacterial agents, paints, printer inks, and textiles [4–9].

Nanoparticles frequently have remarkably different physicochemical properties than their conventional bulk materials. These properties can be a "double-edged sword," providing positive advantages for usefulness and negative impacts on health upon exposure. Toxicity due to some metal-based nanoparticles such as silver, gold, and copper increases with decreasing nanoparticle size [10]. Other physicochemical properties such as elemental composition, charge, shape, crystallinity, surface area, solubility, and surface derivatives also influence the toxic potential of the nanoparticles [11–15]. Therefore, metal-based nanoparticles should not be considered a homogeneous population with simple toxic attributes because they act independently to mediate diverse biological reactions.

Many investigators have explored the properties and toxicities of various metal-based nanoparticles. The toxicities of various metal-based nanoparticles, both *in vitro* and *in vivo*, were recently reviewed and summarized by Schrand et al. (Table 22.1) [10].

The engineering of nanoparticles for applications in the immune system is now an exciting, emerging field. Although certain nanomaterials are immunotoxic or immunomodulatory, a concise overview of the interactions between nanoparticles and the immune system would be valuable and indispensable to students and researchers alike. The focus of this review is to outline the interactions of innate and adaptive immune systems with metal-based nanoparticles (Fig. 22.1). We discuss the role of toll-like receptor interactions with nanoparticles and their potential implications. Different effects of nanoparticles on innate immune cells (macrophages, dendritic cells, neutrophils, mast cells, and natural killer cells) and adaptive immune cells (T cells and B cells) are reviewed. This information will enhance the understanding for immunological effects of nanomaterials and help to develop safe metal-based nanoproducts.

Table 22.1 Selected comparative *in vitro* and *in vivo* toxicity studies

Nanoparticle rank for toxicity	Cell line(s)	Dose and time	Comments	References
Cu > Zn > Co > Sb > Ag > Ni > Fe > Zr > Al$_2$O$_3$ > TiO$_2$ > CeO, low toxicity for W	Two human pulmonary cell lines (A549 and THP-1)	0.1–3300 µg/mL, 3 and 24 h	MTT assay on THP-1 cell line exposed to NP for 24 h most sensitive experimental design	Lanone et al. [64]
ZnO > CeO$_2$/TiO$_2$	BEAS-2B	6.125–50 µg/mL, 1–6 h	CeO$_2$ due to particle dissolution to Zn^{2+}	George et al. [65]
ZnO > CeO$_2$/TiO$_2$	BEAS-2B and RAW264.7 macrophages	10–50 µg/mL, 1–24 h	ZnO dissolution in endosomes CeO$_2$ suppressed ROS production and TiO$_2$ did not elicit protective or adverse effects	Xia et al. [66]
ZnO > Fe$_2$O$_3$ > TiO$_2$/CeO$_2$	Human mesothelioma and rodent fibroblast cell line	30 µg/mL, 3–6 days	Human MSTO cells highly sensitive to Fe$_2$O$_3$	Brunner et al. [67]
ZnO > Fe > SiO$_2$	L2 rat epithelial cells and rat primary alveolar macrophages and cocultures	0.0052–520 mg/cm^2, 1–48 h	*In vivo* and *in vitro* measurements demonstrated little correlation	Sayes et al. [68]

(Continued)

Table 22.1 (Continued)

Nanoparticle rank for toxicity	Cell line(s)	Dose and time	Comments	References
ZnO > TiO$_2$, Fe$_3$O$_4$, Al$_2$O$_3$, and CrO$_3$	Neuro-2A cell line	10–200 µg/mL, 2–72 h	ZnO was more toxic compared to other NPs	Jeng and Swanson [69]
CdCl$_2$ > CdSO$_4$ > ZnSO$_4$ > ZnO > CuSO$_4$ > ZnCl$_2$ > V$_2$O$_5$ > CuCl$_2$ > NiSO$_4$ > NiCl$_2$ > Fe$_2$(SO$_4$)$_3$ > CrCl$_2$ > VCl$_2$ > CrK(SO$_4$)2 > FeCl$_2$	A549	0.005–5 mM, 2–48 h	RLE-6TN rat epithelia cells more sensitive than A549 cells	Riley et al. [70]
Ag > Fe$_2$O$_3$ > Al$_2$O$_3$ > ZrO$_2$ > Si$_3$N$_4$ > TiO$_2$ in RAW264.7 and ZrO$_2$ > Al$_2$O$_3$/Fe$_2$O$_3$/Si$_3$N$_4$/Ag > TiO$_2$ in THB-1 and A549	Murine alveolar macrophage (RAW264.7), human macrophage (THB-1), and human epithelial A549	5 µg/mL, 48 h	THB-1 and A549 cells more sensitive than RAW264.7 and no correlation between specific surface area or NP morphology and toxicity	Soto et al. [71, 72]
Ag > MoO$_3$ > Al/Fe$_3$O$_4$/TiO$_2$	Rat cell line (BRL 3A)	5–25 µg/mL, 24 h	Ag produces toxicity through oxidative stress	Hussain et al. [73]
Ag > Mn	PC-12 cells	1–100 µg/mL, 24 h	Ag produced cell shrinkage and irregular membrane borders and Mn dose-dependently depleted dopamine	Hussain et al. [74]

Nanoparticle rank for toxicity	Cell line(s)	Dose and time	Comments	References
Ag > NiO > TiO$_2$	Murine macrophage cell line	5 μg/mL, 48 h	Nanoparticles characterized as aggregates, caution on Ag	Soto et al. [75]
Ag > MoO$_3$ > Al	Mouse spermatogonial stem cells	5–100 μg/mL, 48 h	Concentration-dependent toxicity for all NPs tested	Braydich-Stolle et al. [76]
Cu and Mn > Al	PC-12 cells	10 μg/mL, 24 h	Txnrd1, Gpx1, Th, Maoa, Park2, and Snca genes expression altered	Wang et al. [77]
VOSO$_4$ > TiO$_2$, SiO$_2$, NiO, Fe$_2$O$_3$, CeO$_2$, and Al$_2$O$_3$	BEAS-2B	1–100 μg/mL, 24 h	Manufactured pure oxides less toxic than natural particulate matter derived from soil dust and IL-6 secretion did not correlate with the generation of ROS in cell-free media	Veranth et al. [78]
Mn$_3$O$_4$ > Co$_3$O$_4$ > Fe$_2$O$_3$ > TiO$_2$	Lung epithelial cells A549	30 μg/mL, 4 h	Acellular ROS assay demonstrates catalytic conditions of NPs based on elemental composition	Limbach et al. [79]
Al > Al$_2$O$_3$	Rat alveolar macrophages	25–250 μg/mL, 24 h	Phagocytosis hindered after exposure to Al NPs	Wagner et al. [80]

(Continued)

Table 22.1 (Continued)

Nanoparticle(s)	Animal	Dose/route	Result	References
Ag	Rat	30–1000 mg/kg (subacute oral for 28 days)	Dose-dependent effect on alkaline phosphatase and cholesterol. Twofold more accumulation of NP in kidneys of female than male	Kim et al. [81]
Ag	Rat	1.73×10^4/cm^3 to 1.32×10^6/cm^3 (subacute inhalation, 6 h/day, 5 days/week for 4 weeks)	Liver histopathological effect but no effect in hematology and biochemical parameters	Ji et al. [82]
Ag	Zebrafish	5–100 μg/mL (exposure, 72 h)	Dose-dependent toxicity in embryos Ag NP distributed in brain, heart, yolk, and blood of embryos	Asharani et al. [83]
Ag	Rat	NP was implanted intramuscularly for 7, 14, 30, 90, and 180 days	Inflammation	Chen et al. [84]
Ag	Mice	100–1000 mg/kg (acute oral)	Oxidative stress gene expression alterations	Rahman et al. [85]

Nanoparticle(s)	Animal	Dose/route	Result	References
Ag, Cu, and Al	Mice and rat	30–50 mg/kg (intravenous/intraperitoneal)	BBB penetration	Sharma [86]
Au	Mice	2×10^5 PPB (oral for 7 days)	NP uptake occurred in the small intestine by persorption through single, degrading enterocytes extruded from a villus. Smaller particles cross the GI tract more readily	Hillyer and Albrecht [87]
Cu	Zebrafish	0.25–1.5 mg/L (exposure, 48 h)	Biochemical, histopathological changes and alterations in gene expression	Griffitt et al. [88]
Cu	Mice	108–1080 mg/kg (acute oral)	NP-induced gravely toxicological effects and heavy injuries on kidney, liver, and spleen of treated mice	Chen et al. [89]
Fe_2O_3	Rat	0.8–20 mg/kg (inhalation)	Oxidative stress, inflammation, and pathology	Zhu et al. [90]
TiO_2	Mice	5 g/kg (acute oral)	Biochemical and histopathological effects	Wang et al. [91]
SiO_2 magnetic-NPs	Mice	25–100 mg/kg (intraperitoneal for 4 weeks)	NPs were detected in brain indicating BBB penetration	Kim et al. [92]

Source: Reproduced with permission from ref. [10]. Copyright © 2010 John Wiley & Sons, Inc.

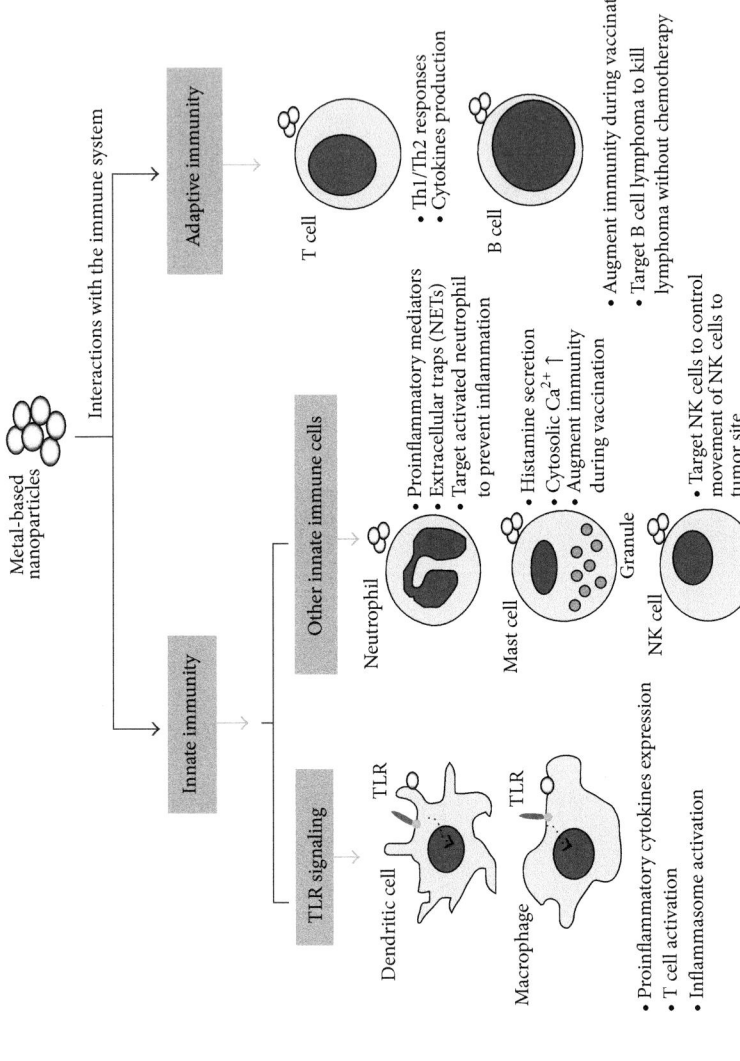

Figure 22.1 Metal-based nanoparticle interaction with the immune system.

22.2 Nanoparticles and Immune System

The immune system can defend against foreign antigens, which has been divided into two general types of immunity: innate immunity and adaptive immunity. Innate immunity is the nonspecific and first line of the body's defense system, which relies on pattern recognition receptors PRRs to recognize broad and conserved molecular patterns found on pathogens (pathogen-associated molecular patterns, PAMPs) [16]. Therefore, the innate immune system plays an essential role in the early recognition and subsequent proinflammatory response. The adaptive immune system is antigen specific and reacts only with the organism that induced the response. Innate and adaptive immunity can be thought of as two equally important aspects of the immune system.

Most nanoparticles are recognized as foreign materials and eliminated by the immune system. However, in the immune system, if the foreign materials are not recognized as a threat, they are ignored or tolerated. Undesirable overwhelming activation of immune responses may lead to harmful consequences. Therefore, the response of the immune system to the nanoparticles must be considered when developing a nanomaterial for *in vivo* application. For example, avoiding immune system detection is crucial if a nanomaterial is to be used for gene or drug delivery [17]. In addition to avoiding immune system detection, nanoparticles also can play an important role in vaccine immunization via antigen delivery and adjuvanticity. Another viewpoint is that nanoparticles targeting immune cells (e.g., macrophages or dendritic cells) can manipulate or control immunological diseases such as infectious diseases or tumor therapy. For example, nanomaterials might be designed to modify effective immune responses of tumor microenvironments as accompanied with anti-inflammatory drugs or specific cytokines.

Three immune related consequences must be considered when a nanomaterial is engineered for applications *in vivo*. The first is immune-mediated destruction or rejection, which could initiate a defensive immune reaction resulting in the elimination of the nanomaterials. Second is immunotoxicity, which could damage the immune system and cause pathological changes. The third is immunocompatibility, which does not interfere with the immune response [18]. Nanoparticle properties such as size, charge, hydrophobicity, hydrophilicity, and the steric effects of nanoparticle coatings direct nanoparticle compatibility with the immune system [17, 19, 20]. For example, nanoparticles that are designed by encapsulation with PEG or other types of polymers provide a hydrophilic environment and shield them from immune recognition [21]. However, some reports showed that the immune system can produce PEG-specific antibodies after administration of PEG-coated liposomes [22, 23]. The research on how and whether nanoparticles trigger antibody production is limited and we need further studies to answer these inconclusive questions.

22.2.1 Nanoparticles and Innate Immunity

The innate immune system consists of different cells and proteins that are nonspecific and are a first line of defense. The main components of the innate immune system include physical epithelial barriers, phagocytic cells (monocyte/macrophages, dendritic cells, and polymorphonuclear leukocytes), phagocytic leukocytes, basophils, mast cells, eosinophils, natural killer (NK) cells, and circulating plasma proteins.

In recent decades, many studies have highlighted rapid progress in understanding toll-like receptors of the innate immune system, which induce expression genes of involved inflammation. Moreover, toll-like receptors activate both the innate and adaptive immune system and play an important role in antiviral and anti-immunity [24]. In this review, we will first discuss the toll-like receptor signaling mechanisms triggered by metal-based nanoparticles and then describe the effects of nanoparticles on other innate immune cells.

22.2.2 The Role of Toll-Like Receptor Signaling in the Innate Immune System

The innate immune system, also known as the non-specific immune system and the first line of defense, relies on the recognition of PAMPs through a limited number of germ line-encoded pattern recognition receptors, belonging to the family of toll-like receptors

(TLRs) [25]. The *Toll* gene was originally discovered in *Drosophila*, responsible for dorsoventricular polarization during embryonic development and antifungal and antibacterial properties of the adult fly [26]. TLR1, TLR2, TLR4, TLR5, TLR6, and TLR10 are present on the cell surface whereas TLR3, TLR7, TLR8, TLR9, TLR12, and TLR13 are localized into intracellular vesicles such as endosomes, lysosomes, and the endoplasmic reticulum (ER). TLR1/TLR2 sense bacterial tri-acylated lipopeptides. TLR2/TLR6 recognize di-acetylated lipopeptides and bacterial lipoteichoic acid or peptidoglycans and mycobacterial cell wall components. TLR3 binds to viral double stranded RNA while TLR4 responds to LPS, and TLR5 senses flagellin. TLR7 and TLR8 respond to the single stranded RNA from viruses, while TLR9 binds to DNA-containing unmethylated CpG motifs which are commonly found in bacterial DNA. TLR12 recognizes profilin, while TLR13 senses bacterial 23S ribosomal RNA (rRNA) [16]. The activation of TLR signaling can not only induce cytokine production but also increase macrophage phagocytosis and natural killer (NK) cells cytolytic activity. Most importantly, TLR signaling activation also can enhance antigen presentation via upregulating the expression of the major histocompatibility complex (MHC) and costimulatory molecules (CD80 and CD86) on dendritic cells leading to adaptive immunity activations. Thus, TLR agonists were believed to be powerful vaccine adjuvant, allergy, infection, and antitumor therapeutics in preclinical studies [24]. The TLR antagonists also have therapeutic value in clinical trials to treat septic shock and autoimmune disorders [27]. For example, TLR agonists or nanoparticles that enhance TLR signaling pathways would be powerful adjuvants [28, 29]. In contrast, TLR antagonists or inhibitors that reduced the inflammatory response would have beneficial therapeutic effects in autoimmune diseases and sepsis [30]. These potential applications may open up innovative directions for the design of nanoparticle conjugates to meet different requirements.

22.2.3 Effects of Nanoparticles on TLR Signaling of Innate Immunity

TLRs are classified as type I transmembrane receptors containing an N-terminal leucine-rich repeat domain (transmembrane region) and a C-terminal cytoplasmic domain. Upon recognition of a PAMP, TLRs recruit a specific set of adaptor molecules that contain the TIR domain, such as MyD88 and TRIF, and initiate downstream signaling events that lead to the secretion of inflammatory cytokines, type I IFN, and chemokines [31]. The TLR signaling cascade results in the activation of transcription factors, nuclear factor κ light chain enhancer of activated B cells (NF-κB), interferon-regulatory factors (IRFs), and mitogen-activated protein kinase; these factors affect the transcription of genes involved in inflammatory and immune responses [32, 33].

Schmidt et al. first reported that Ni^{2+} as an inorganic activator was acting directly through TLR binding to trigger inflammation responses [34]. This interesting finding also makes us think of whether the other chemicals components such as metal-based nanoparticles were also involved in TLR signaling inflammation. Recently, several studies have demonstrated the effects of nanoparticles on innate immunity via TLR signaling pathways [35]. Several nanoparticles (e.g., TiO_2, ZnO, zirconium dioxide (ZrO_2), and silver) modulated immune responses via TLRs. TiO_2 and ZrO_2 nanoparticles increased *TLR7* and *TLR10* mRNA levels in human macrophage U-937 cells and *TLR2* and *TLR4* mRNA levels in the mouse liver cells [36, 37]. N-(2-Mercaptopropionyl) glycine (tiopronin) capped-silver nanoparticles enhanced the TLR3 ligand and TLR9 ligand-induced IL-6 secretion in mouse macrophage Raw264.7 cells [38]. ZnO nanoparticles induced MyD88-dependent proinflammatory cytokines via a TLR signal pathway [39]. Quantum dot 705 activated MyD88-dependent TLRs at the surface or inside of cells, which is a fundamental mechanism for nanoparticle-induced inflammatory responses [40]. TLRs may have important roles not only for different NPs uptake but also for their cellular response [41]. Moreover, the mechanisms of interaction between NPs and TLR are still unclear. There are two possibilities to explain how NPs interact with TLRs. One is that the smaller NPs may just like LPS have cooperated with some small molecules such as the LPS binding protein and then the complex activates further TLRs signaling pathways. The other is that the larger size of NPs may directly associate with TLRs [41]. However, these hypotheses need more studies to confirm.

Proinflammatory cytokines can be induced by TLR signaling pathways. Many cytokines, such as interleukin-(IL-) 1, IL-6, and tumor necrosis factor- (TNF-) α, can activate inflammatory cells, increase vascular permeability, and cause swelling and redness during acute inflammatory responses [42]. IL-1 and IL-6 are important mediators of fever [43]. TNF-α activates endothelial cells leading to hypotension. IL-8 is a chemokine that activates neutrophils or other granulocytes and recruits them to the site of inflammation [44]. Interferon- (IFN-) γ plays an important role in the inflammatory process, recruiting macrophages to the site where an antigen is present [42]. Many studies have reported that NPs can trigger cytokine production associated with inflammatory responses. The levels of proinflammatory cytokines are measured as biomarkers of nanoparticle immunomodulatory effects and immune-mediated toxicity [42]. TiO_2 nanoparticles, nanodiamond, and nanoplatinum also have been reported to trigger proinflammatory cytokine production, dendritic cell maturation, and naïve T cell activation and proliferation [45, 46]. Hanley et al. also reported that ZnO nanoparticles increased the expression of IFN-γ, TNF-α, and IL-12 in primary human immune cells [47]. Gold nanoparticles (10 nm and 50 nm in size) induced IL-1β, IL-6, and TNF-α in rat liver cells after 1 day of acute treatment and then subsided by day 5 of subchronic treatment. The 50 nm gold nanoparticles produced more severe inflammation than the 10 nm gold nanoparticles [48]. However, only limited studies have demonstrated whether or which TLR is involved in the NPs induced proinflammatory cytokines production.

Another interesting field is inflammasomes, which are multiprotein complexes leading to caspase-1 activation, further causing pro-IL-1β and pro-IL-18 maturations and secretions. The IL-1β synthesis and secretion are tightly regulated by TLR signaling and inflammasome activation. A first signal, such as by toll-like receptor activation, triggers the synthesis of pro-IL-1β by transcriptional induction, whereas a second stimulus leads to inflammasome oligomerization, caspase-1 autoactivation, and caspase-1-dependent cleavage and the release of the biologically active, mature IL-1β [49]. The second signal can be triggered by an ever-expanding group of chemically and biologically unrelated danger-associated molecular patterns (DAMPs) or pathogen-associated molecular patterns (PAMPs) [50]. Studies of nanoparticles that induce IL-1β via inflammasome signaling pathway mechanisms is an emerging theme [51, 52].

Some engineered nanoparticles can also activate inflammasome signaling pathways [49, 53, 54]. Among various inflammasomes, nucleotide-binding oligomerization domain- (NOD-) like receptor protein 3 (NLRP3) activation is linked to exposure to various nanoparticles [54, 55]. TiO$_2$ and SiO$_2$ nanoparticles activate the NLRP3 inflammasome and IL-1β release in LPS-primed murine bone marrow-derived macrophages and human macrophage cell lines THP-1 [49, 56]. Peeters et al. [55] recently reported that crystalline silica (SiO$_2$) activated NLRP3 inflammasomes in human lung epithelial cells BEAS-2B and primary human bronchial epithelial cells, which prolonged the inflammatory signal and affected fibroblast proliferation. Silver nanoparticles induced inflammasome formation and triggered IL-1β release and subsequent caspase-1 activation [53]. Inflammasome-activation-associated IL-1β production by dendritic cells in response to particle treatment was size-dependent and maximal at particle diameters between 400 and 1000 nm [57]. Yazdi et al. reported that nano-TiO$_2$ and nano-SiO$_2$, but not nano-ZnO, activate the NLRP3 inflammasome, leading to IL-1β release, and in addition induce the regulated release of IL-1α. Unlike other particulate NLRP3 agonists, nano-TiO$_2$-dependent NLRP3 activity does not require cytoskeleton-dependent phagocytosis and induces IL-1α/β secretions in nonphagocytic keratinocytes. However, the exact mechanism of nano-TiO$_2$ uptake remains elusive, as blocking lipid raft-mediated, caveolin-dependent, or clathrin-dependent endocytosis did not efficiently block IL-1β secretion [49]. The more knowledge we have of cytokine profiles induced by nanoparticles, the better we can utilize the cytokines as biomarkers of immunomodulatory properties of nanoparticles. Moreover, it is also necessary to clarify whether these proinflammatory cytokines were induced by nanoparticle physiochemical properties or by bacterial endotoxin contaminants.

22.2.4 Effects of Nanoparticles on Innate Immune Cells

The innate leukocytes include mast cells, neutrophils, eosinophils, basophils, natural killer (NK) cells, gamma/delta T cells, and the phagocytic cells including macrophages and dendritic cells. We summarized several studies which reported the effects of metal-based nanoparticles on phagocytic cells, neutrophils, mast cells, and NK cells. There are still many challenges to investigate the effects and potential applications of nanoparticles to other innate immune cells such as eosinophils, basophils, and gamma/delta T cells.

22.2.4.1 Phagocytic cells (macrophages, dendritic cells)

Macrophages and dendritic cells play many key roles in the host defense system. They can remove dead cells and pathogens by phagocytosis. They also can shape the inflammatory response by secreting cytokines through the TLR signaling pathway and modulate adaptive immunity by presenting antigens to lymphocytes [58]. In general, macrophages and dendritic cells readily uptake nanoparticles. Therefore, many metal-based nanoparticles (e.g., magnetic nanoparticles and nanoparticles-based PET agents) were commonly used for visualizing of macrophages in human diseases including cancer, atherosclerosis, myocardial infarction, aortic aneurysm, and diabetes [58]. In addition to image applications, targeting tumor-associated macrophages or dendritic cells via nanoparticles for drug, antigen delivery, or vaccine is also a promising tumor therapeutic application. For example, Lin et al. reported that gold nanoparticle delivery of modified CpG can stimulate macrophages and inhibit tumor growth for immunotherapy [59]. Ahn et al. recently demonstrated that gold nanoparticles enable efficient tumor-associated self-antigen delivery to dendritic cells and then activate the cells to facilitate cross-presentation, inducing antigen-specific cytotoxic T cell responses for effective cancer therapy [60].

22.2.4.2 Neutrophils

During acute inflammation, polymorphonuclear neutrophil cells (PMNs) are the first type of leukocytes to migrate to an inflammatory site and then produce several proinflammatory mediators including chemokines, which further attract other PMNs and other cell types like monocytes-macrophages and lymphocytes, corresponding to chronic inflammation. Gold nanoparticles were found trapped by neutrophils in their extracellular traps (NETs), being composed mainly of DNA and a variety of antibacterial proteins [61]. The cell-gold networks were visible after as early as 15 min of treatment of neutrophils with the gold nanoparticles. NETs may contribute to alerting the immune system of a danger signal by activating DNA receptors such as TLR9. This activation might turn out to help in the recruitment of immune cells to mount an acquired immune response or to resolve the inflammation. NETs can either fight inflammatory disease or cause disease depending on the place, time, and dose [62]. However, NETs triggered by nanoparticles need further investigation to figure out their physiological roles. Wang et al. found that the delivery of drugs into inflammatory neutrophils by nanoparticles can prevent vascular inflammation [63]. This study provides a novel nanoparticle-based therapeutic approach for targeting activated neutrophils to treat a range of inflammatory disorders.

22.2.4.3 Mast cells

Mast cells contain many granules in histamine and heparin and have important roles of allergy and anaphylaxis. When activated, mast cells rapidly release histamine and heparin from their granules to dilate blood vessels and recruit neutrophils and macrophages. Chen et al. demonstrated that TiO_2 nanoparticles not only dose-dependently increased histamine secretion, but also increased cytosolic Ca^{2+} concentration in rat mast cells [93]. Their results suggest that systemic circulation of nanoparticles may prompt histamine release without prior allergen sensitization, causing abnormal inflammatory diseases or potential exacerbating manifestations of multiple allergic responses. It was recently reported that the granules of mast cells are powerful enhancers of adaptive immunity when they are released at sites of infection or vaccine administration. John et al. engineered nanoparticles consisting of mast cell granules to augment immunity during vaccination [94]. It is believed that other metal-based nanoparticles also have the possibility of developing this efficient vaccination system.

22.2.4.4 NK cells

NK cells control several types of tumors and microbial infections by limiting their spreading and subsequent tissue damage. NK cells are also regulatory cells which can interact with dendritic cells, macrophages, T cells, and endothelial cells. Therefore, NK cells are believed to limit or exacerbate immune responses [95]. A clinical study has demonstrated that patients with a high level of NK infiltration were found to have a better prognosis than those with a low level of NK infiltration and suggests that enhancement of NK cell infiltration could be a useful antitumor strategy [96]. Lim et al. provided evidence of cell tracking with quantum dots (QD) by labeling NK cells with an anti-CD56 antibody-coated QD705 and tracking the labeled cells for up to 12 days after intratumoral injections [97]. The authors further found a decreased size of tumors treated with NK cells compared with controls [97]. QD labeling was thought as the well-suited imaging technique for tracking different cell populations; however, currently available compounds are not clinically applicable because of their toxic cadmium cores or other nondegradable components; cadmium-free or biodegradable QDs are currently being developed [98]. Jang et al. used magnetic nanoparticles (Fe_3O_4/SiO_2) to control the movement of human natural killer cells (NK-92MI) by an external magnetic field, loading NK-92MI cells infiltrated into the target tumor site and their killing activity is still the same as the NK-92MI cells without the nanoparticles [99]. This study provides an alternative clinical treatment with reduced toxicity of the nanoparticles and enhanced infiltration of immunology to the three-dimensional target site without surgical treatment.

22.2.5 Nanoparticles and Adaptive Immunity

Nanoparticles can be designed to deliver vaccine antigens through specific intracellular pathways such as phagocytosis, macropinocytosis, and endocytosis, allowing better antigen presentation for activating the adaptive immune system [100]. Nanoparticles interact most frequently with APCs in the blood circulation, including B cells, macrophages, and dendritic cells. APCs engulf and digest foreign antigens present on the surface of major histocompatibility complexes of B and T cells [101]. Dendritic cells are the most specialized APCs, which capture and process antigens and migrate to lymphoid tissues leading to T cell or B cell activation. The costimulatory molecules of dendritic cells and the cytokine environment affect the T cell response. T cells including T helper (Th) cells, regulatory T cells (formerly known as suppressor T cells), and cytotoxic T cells express various surface proteins including CD3 and CD4 on Th cells, CD3, and CD8 on cytotoxic T cells. The cytokine environment is produced by dendritic cells via activated CD4+ T cells, neutrophils, and macrophages, which are recruited to the inflammatory site and stromal cells [100]. For example, immature dendritic cells encountered antigens, which are presented to T cells for self-tolerance (T cell anergy) without costimulatory molecule expression. This also occurs for regulatory T (Treg) cells in the presence of transforming growth factor-β1 (TGF-β1) and interleukin- (IL-) 10. Exogenous antigen activates and matures dendritic cells leading to costimulatory molecule expression and Th1, Th2, or Th17 cell activation [102]. Antigen presentation in the IL-6 and IL-23 cytokine microenvironment can also stimulate naïve CD4+ T cells to differentiate into Th17 cells [100]. Th17 cells are potent inducers of inflammation and play key roles in the development of autoimmunity diseases [103]. Th1 cells mediate cellular immunity and further regulate inflammation responses. On the other hand, Th2 cells induce proliferation of master cells and eosinophils and mediate the differentiation of B cells to produce immunoglobulin (Ig) G and IgE, thereby promoting humoral immunity [42].

22.2.6 Effects of Nanoparticles on T Cells

Only a few metal-based nanoparticles have been reported to activate T cell responses or homeostasis. For example, TiO_2 nanoparticles provoke inflammatory cytokines and increase dendritic cell maturation, expression of costimulatory molecules, and prime naïve T cell activation and proliferation [45]. Cd trapped inside fullerene cage nanoparticles (Gd@C82(OH)22) has specific immunomodulatory effects on T cells and macrophages, including polarization of the cytokine balance towards Th1 cytokines, decreasing the production of Th2 cytokines (IL-4, IL-5, and IL-6) and increasing the production of Th1 cytokines (IL-2, IFN-γ, and TNF-α) [104]. One important theory of adaptive immunity is T cell homeostasis (Th1/Th2 balance). Th1 cells drive the cellular immunity to fight viruses and other intracellular pathogens, eliminate cancerous cells, and stimulate delayed-type hypersensitivity skin reactions. Th2 cells drive humoral immunity and upregulate antibody production to fight extracellular organisms. Overactivation of either pattern can cause disease, and either pathway can downregulate the other [105]. Th1 cells secrete large amounts of interferon- (IFN-) γ, IL-2, IL-3, granulocyte macrophage colony-stimulating factor, and a small amount of TNF. Th2 cells produce large amounts of IL-3, IL-4, IL-5, IL-6, and IL-10 and a small amount of TNF. Brandenberger et al. demonstrated that silica nanoparticles promote an adjuvant Th2/Th17 response in murine allergic airway disease [106]. Recently, Tomic´ et al. demonstrated that smaller gold nanoparticles (10 nm) have stronger inhibitory effects on maturation and antitumor functions of DCs, which were induced either by LPS or heat-killed tumor necrotic cells, compared to larger gold nanoparticles (50 nm). Gold nanoparticles (10 nm) can inhibit LPS-induced production of IL-12p70 by dendritic cells and potentiated Th2 polarization, while 50 nm gold nanoparticles promoted Th17 polarization [107]. The authors supposed that the size-dependent immunomodulatory effects of gold nanoparticles could be attributed to different mechanisms of their internalization, levels of accumulation, and intracellular distribution within DCs, leading to different modulation of maturational signaling. Furthermore, these results point to potential adverse effects of smaller gold nanoparticles if used in photothermal therapy and cancer diagnostics. The Th1 or Th2 responses elicited by APCs may be influenced by many factors, such as the maturation states of the APCs and routes of antigen uptake. Nanoparticle size plays a decisive role in determining whether antigen conjugated nanoparticles induce Th1 or Th2 immune responses [108]. Therefore, nanoparticle size may be a critical and fundamental parameter for the induction of specific immunity in vaccine development. The precise selection of nanoparticle size for vaccination can influence the type 1/type 2 cytokine balance after one immunization, and this will be useful in the development of effective vaccines against common human pathogens. However, it is still unclear whether other different physical and chemical properties of nanoparticles, such as charge or chemical stability, can drive the T cell polarization.

22.2.7 Effects of Nanoparticles on B Cells

B cells are another type of lymphocyte in the adaptive immune system. B cells present a unique surface receptor (B cell receptor) to bind with specific antigens. When B cell receptors bind with its specific antigen, an antigen is delivered, degraded, and returned to a surface bound with MHC class II. This antigen, MHC II complex, can be recognized by antigen-specific T helper cells. B cells receive an additional signal from a T helper cell, further differentiating into antibody-secreting B cells. It has been reported that the nanostructure of antigens is used to improve B cell antibody response [109]. Different kinds of synthetic nanoparticles have been designed to carry antigens as an effective vaccination system [101]. Temchura et al. recently reported that calcium phosphate (CaP) nanoparticles coated with protein antigens are promising vaccine candidates for the induction of humoral immunity [110]. In general, it is believed that nanoparticles do not result in the activation of B-cells, unless they are coated with the antigen. In contrast, it was also reported that iron oxide nanoparticles can compromise subsequent antigen-specific immune reactions, including antibody production and T cell responses [111]. The effects of various metal-based nanoparticles on B cell functions are worthy for further and more comprehensive investigations and further development for potential applications.

22.2.8 Therapeutic Approach of Nanoparticles on Lymphoma

Lymphoma is a type of cancer immune cell occurring in B or T lymphocytes which divide faster than normal cells or live longer than they are supposed to. It was reported that engineering nanoparticles have the potential to develop a nontoxic new treatment for lymphoma and other cancers which does not involve chemotherapy [112]. Yang et al. used gold nanoparticle combined with synthetic HDL (high-density lipoprotein) to trick B cell lymphoma, which prefers to eat HDL cholesterol. Once the B cell lymphoma cells start engulfing the gold nanoparticles (or artificial HDL particles), they get plugged up and can no longer feed on any more cholesterol. Deprived of B cell lymphoma's favorite food, the lymphoma cells essentially starve to death. The common treatments of lymphoma are chemotherapy, radiotherapy, or bone marrow transplantation. However, chemotherapy has strong side effects, even leading to possible long-term consequences such as infertility, second cancer risks, and lung damage. Promising and effective nanoparticle drugs may prevent the occurrence of these side effects. While designing novel nanodrugs for cancer therapy, we should consider their molecular mechanisms; for example, Ag nanoparticles have been reported to have antiangiogenic ability [113]. Therefore, Ag nanoparticles are one of the attractive and potential approaches to develop an antitumor effect. Sriram et al. also demonstrated the antitumor activity of silver nanoparticles in Dalton's lymphoma ascites in a tumor model both *in vitro* and *in vivo* by the activation of the caspase-3 enzyme [114]. Moreover, nanodrugs are mainly developed according to their ability to distinguish between malignant and nonmalignant cells, making them a promising alternative to existing drugs. The targeting efficiency of nanoparticles can be accomplished by combining them with the RGD peptide [115] or an antibody against specific tumor markers [116]. In a nutshell, nanoparticles may provide a new way to kill lymphoma without chemotherapy.

22.3   Conclusion and Future Perspectives

Nanoparticles can be used as vaccine carriers, adjuvants, and drug delivery vehicles to target specific inflammation-associated diseases or cancer. Nanoparticles, particularly noble metal nanoparticles, have considerable potential for biomedical applications, such as diagnostic assays, thermal ablation, and radiotherapy enhancement as well as drug and gene delivery. Currently, we are still challenged by the limited knowledge of nanoparticle pharmacokinetics, biodistribution, and immunotoxicity.

The interactions of nanomaterials with the immune system have attracted increasing attention. The physicochemical properties of nanoparticles influence the immunological effects of nanoparticles. Comprehensive studies to explore the effects of physicochemical properties (such as size, shape, and charge) on the immunotoxicity of metal-based nanoparticles are still needed. Assessment of potential adverse effects on the immune system is also a critical component of the overall evaluation of nanodrug toxicity. Further mechanistic studies investigating nanoparticle immunomodulatory effects or inflammatory reactions are required to improve knowledge of the physicochemical properties of nanoparticles, which influences the immune system. A cooperation between materials science and immunology, as well as immunobioengineering, is an emerging field which has great potential to develop prophylactic and therapeutic vaccine applicants.

Disclosures and Conflict of Interest

This chapter was originally published as: Luo, Y.-H., Chang, L. W., and Lin, P. (2015). Metal-based nanoparticles and the immune system: Activation, inflammation, and potential applications, *BioMed Res. Intl.*, **2015**, 143720, DOI: 10.1155/2015/143720, under the Creative Commons Attribution License (https://creativecommons.org/licenses/by/3.0/) and appears here, with edits, by kind permission of the publisher, Hindawi Inc. This work was supported by research Grants NM-103-PP-08 and BN-104-PP-28 from the Institute of Biomedical Engineering and Nanomedicine at the National Health Research Institutes, Taiwan. No writing assistance was utilized in the production of this chapter and the authors have received no payment for its preparation.

Corresponding Author

Dr. Pinpin Lin
National Environmental Health Research Center
National Health Research Institutes
35 Keyan Road, Zhunan 35053, Miaoli County, Taiwan
Email: pplin@nhri.org.tw

References

1. Tourinho, P. S., van Gestel, C. A. M., Lofts, S., Svendsen, C., Soares, A. M. V. M., Loureiro, S. (2012). Metal-based nanoparticles in soil:

Fate, behavior, and effects on soil invertebrates. *Environ. Toxicol. Chem.*, **31**(8), 1679–1692.

2. Mu, L., Sprando, R. L. (2010). Application of nanotechnology in cosmetics. *Pharm. Res.*, **27**(8), 1746–1749.

3. Newman, M. D., Stotland, M., Ellis, J. I. (2009). The safety of nanosized particles in titanium dioxide- and zinc oxide-based sunscreens. *J. Am. Acad. Dermatol.*, **61**(4), 685–692.

4. Durán, N., Marcato, P. D., de Souza, G. I. H., Alves, O. L., Esposito, E. (2007). Antibacterial effect of silver nanoparticles produced by fungal process on textile fabrics and their effluent treatment. *J. Biomed. Nanotechnol.*, **3**(2), 203–208.

5. Benn, T. M., Westerhoff, P. (2008). Nanoparticle silver released into water from commercially available sock fabrics. *Environ. Sci. Technol.*, **42**(11), 4133–4139.

6. Panáček, A., Kvítek, L., Prucek, R., et al. (2006). Silver colloid nanoparticles: Synthesis, characterization, and their antibacterial activity. *J. Phys. Chem. B*, **110**(33), 16248–16253.

7. Baker, C., Pradhan, A., Pakstis, L., Pochan, D. J., Shah, S. I. (2005). Synthesis and antibacterial properties of silver nanoparticles. *J. Nanosci. Nanotechnol.*, **5**(2), 244–249.

8. Kaegi, R., Sinnet, B., Zuleeg, S., et al. (2010). Release of silver nanoparticles from outdoor facades. *Environ. Pollut.*, **158**(9), 2900–2905.

9. Zhang, F., Wu, X. L., Chen, Y. Y., Lin, H. (2009). Application of silver nanoparticles to cotton fabric as an antibacterial textile finish. *Fibers Polym.*, **10**(4), 496–501.

10. Schrand, A. M., Rahman, M. F., Hussain, S. M., Schlager, J. J., Smith, D. A., Syed, A. F. (2010). Metal-based nanoparticles and their toxicity assessment. *Wiley Interdiscip. Rev. Nanomed. Nanobiotechnol.*, **2**(5), 544–568.

11. Dreher, K. L. (2004). Health and environmental impact of nanotechnology: Toxicological assessment of manufactured nanoparticles. *Toxicol. Sci.*, **77**(1), 3–5.

12. Oberdörster, G., Maynard, A., Donaldson, K., et al. (2005). Principles for characterizing the potential human health effects from exposure to nanomaterials: Elements of a screening strategy. *Part. Fibre Toxicol.*, **2**, article 8.

13. Nel, A., Xia, T., Mädler, L., Li, N. (2006). Toxic potential of materials at the nanolevel. *Science*, **311**(5761), 622–627.

14. Nel, A. E., Mädler, L., Velegol, D., et al. (2009). Understanding biophysicochemical interactions at the nano-bio interface. *Nat. Mater.*, **8**(7), 543–557.
15. Tiede, K., Boxall, A. B. A., Tear, S. P., Lewis, J., David, H., Hassellov, M. (2008). Detection and characterization of engineered nanoparticles in food and the environment. *Food Addit. Contam., Part A*, **25**(7), 795–821.
16. Mogensen, T. H. (2009). Pathogen recognition and inflammatory signaling in innate immune defenses. *Clin. Microbiol. Rev.*, **22**(2), 240–273.
17. Moyano, D. F., Goldsmith, M., Solfiell, D. J., et al. (2012). Nanoparticle hydrophobicity dictates immune response. *J. Am. Chem. Soc.*, **134**(9), 3965–3967.
18. Boraschi, D., Costantino, L., Italiani, P. (2012). Interaction of nanoparticles with immunocompetent cells: Nanosafety considerations. *Nanomedicine (London)*, **7**(1), 121–131.
19. Zolnik, B. S., González-Fernández, Á., Sadrieh, N., Dobrovolskaia, M. A. (2010). Nanoparticles and the immune system. *Endocrinology*, **151**(2), 458–465.
20. Dobrovolskaia, M. A., McNeil, S. E. (2007). Immunological properties of engineered nanomaterials. *Nat. Nanotechnol.*, **2**(8), 469–478.
21. Moghimi, S. M. (2002). Chemical camouflage of nanospheres with a poorly reactive surface: Towards development of stealth and target-specific nanocarriers. *Biochim. Biophys. Acta*, **1590**(1–3), 131–139.
22. Wang, X., Ishida, T., Kiwada, H. (2007). Anti-PEG IgM elicited by injection of liposomes is involved in the enhanced blood clearance of a subsequent dose of PEGylated liposomes. *J. Control. Release*, **119**(2), 236–244.
23. Ishida, T., Wang, X., Shimizu, T., Nawata, K., Kiwada, H. (2007). PEGylated liposomes elicit an anti-PEG IgM response in a T cell-independent manner. *J. Control. Release*, **122**(3), 349–355.
24. Engel, A. L., Holt, G. E., Lu, H. (2011). The pharmacokinetics of Toll-like receptor agonists and the impact on the immune system. *Expert Rev. Clin. Pharmacol.*, **4**(2), 275–289.
25. Takeda, K., Akira, S. (2004). TLR signaling pathways. *Semin. Immunol.*, **16**(1), 3–9.
26. Medzhitov, R., Preston-Hurlburt, P., Janeway Jr., C. A. (1997). A human homologue of the *Drosophila* Toll protein signals activation of adaptive immunity. *Nature*, **388**(6640), 394–397.

27. Makkouk, A., Abdelnoor, A. M. (2009). The potential use of Toll-like receptor (TLR) agonists and antagonists as prophylactic and/or therapeutic agents. *Immunopharmacol. Immunotoxicol.*, **31**(3), 331–338.

28. Tamayo, I., Irache, J. M., Mansilla, C., Ochoa-Repáraz, J., Lasarte, J. J., Gamazo, C. (2010). Poly(anhydride) nanoparticles act as active Th1 adjuvants through Toll-like receptor exploitation. *Clin. Vaccine Immunol.*, **17**(9), 1356–1362.

29. Gnjatic, S., Sawhney, N. B., Bhardwaj, N. (2010). Toll-like receptor agonists are they good adjuvants? *Cancer J.*, **16**(4), 382–391.

30. Tse, K., Horner, A. A. (2007). Update on Toll-like receptor-directed therapies for human disease. *Ann. Rheum. Dis.*, **66**(suppl 3), iii77–iii80.

31. Kawai, T., Akira, S. (2010). The role of pattern-recognition receptors in innate immunity: Update on Toll-like receptors. *Nat. Immunol.*, **11**(5), 373–384.

32. Lafferty, E. I., Qureshi, S. T., Schnare, M. (2010). The role of toll-like receptors in acute and chronic lung inflammation. *J. Inflammation (London)*, **7**, article 57.

33. Kawai, T., Akira, S. (2008). Toll-like receptor and RIG-1-like receptor signaling. *Ann. N. Y. Acad. Sci.*, **1143**, 1–20.

34. Schmidt, M., Raghavan, B., Müller, V., et al. (2010). Crucial role for human Toll-like receptor 4 in the development of contact allergy to nickel. *Nat. Immunol.*, **11**(9), 814–819.

35. Smith, D. M., Simon, J. K., Baker Jr., J. R. (2013). Applications of nanotechnology for immunology. *Nat. Rev. Immunol.*, **13**(8), 592–605.

36. Lucarelli, M., Gatti, A. M., Savarino, G., et al. (2004). Innate defence functions of macrophages can be biased by nano-sized ceramic and metallic particles. *Eur. Cytokine Network*, **15**(4), 339–346.

37. Cui, Y. L., Liu, H. T., Zhou, M., et al. (2011). Signaling pathway of inflammatory responses in the mouse liver caused by TiO_2 nanoparticles. *J. Biomed. Mater. Res. Part A*, **96**(1), 221–229.

38. Castillo, P. M., Herrera, J. L., Fernandez-Montesinos, R., et al. (2008). Tiopronin monolayer-protected silver nanoparticles modu-late IL-6 secretion mediated by Toll-like receptor ligands. *Nanomedicine (London)*, **3**(5), 627–635.

39. Chang, H., Ho, C.-C., Yang, C. S., et al. (2013). Involvement of MyD88 in zinc oxide nanoparticle-induced lung inflammation. *Exp. Toxicol. Pathol.*, **65**(6), 887–896.

40. Ho, C.-C., Luo, Y.-H., Chuang, T.-H., Yang, C.-S., Ling, Y.-C., Lin, P. (2013). Quantum dots induced monocyte chemotactic protein-1 expression via MyD88-dependent Toll-like receptor signaling pathways in macrophages. *Toxicology*, **308**, 1–9.

41. Chen, P., Kanehira, K., Taniguchi, A. (2013). Role of Toll-like receptors 3, 4 and 7 in cellular uptake and response to titanium dioxide nanoparticles. *Sci. Technol. Adv. Mater.*, **14**(1), article ID 015008.

42. Elsabahy, M., Wooley, K. L. (2013). Cytokines as biomarkers of nanoparticle immunotoxicity. *Chem. Soc. Rev.*, **42**(12), 5552–5576.

43. Kozak, W., Kluger, M. J., Soszynski, D., et al. (1998). IL-6 and IL-1beta in fever—studies using cytokine-deficient (knockout) mice. *Ann. N. Y. Acad. Sci.*, **856**, 33–47.

44. Struyf, S., Gouwy, M., Dillen, C., Proost, P., Opdenakker, G., van Damme, J. (2005). Chemokines synergize in the recruitment of circulating neutrophils into inflamed tissue. *Eur. J. Immunol.*, **35**(5), 1583–1591.

45. Schanen, B. C., Karakoti, A. S., Seal, S., Drake III, D. R., Warren, W. L., Self, W. T. (2009). Exposure to titanium dioxide nanomaterials provokes inflammation of an in vitro human immune construct. *ACS Nano*, **3**(9), 2523–2532.

46. Ghoneum, M., Ghoneum, A., Gimzewski, J. (2010). Nanodiamond and nanoplatinum liquid, DPV576, activates human monocyte-derived dendritic cells in vitro. *Anticancer Res.*, **30**(10), 4075–4079.

47. Hanley, C., Thurber, A., Hanna, C., Punnoose, A., Zhang, J., Wingett, D. G. (209). The influences of cell Type and ZnO nanoparticle size on immune cell cytotoxicity and cytokine induction. *Nanoscale Res. Lett.*, **4**(12), 1409–1420.

48. Khan, H. A., Abdelhalim, M. A. K., Alhomida, A. S., Al Ayed, M. S. (2013). Transient increase in IL-1beta, IL-6 and TNF-alpha gene expression in rat liver exposed to gold nanoparticles. *Genet. Mol. Res.*, **12**(4), 5851–5857.

49. Yazdi, A. S., Guarda, G., Riteau, N., et al. (2010). Nanoparticles activate the NLR pyrin domain containing 3 (Nlrp3) inflammasome and cause pulmonary inflammation through release of IL-1a and IL-1b. *Proc. Natl. Acad. Sci. U. S. A.*, **107**(45), 19449–19454.

50. Pétrilli, V., Papin, S., Dostert, C., Mayor, A., Martinon, F., Tschopp, J. (2007). Activation of the NALP3 inflammasome is triggered by low intracellular potassium concentration. *Cell Death Differ.*, **14**(9), 1583–1589.

51. Demento, S. L., Eisenbarth, S. C., Foellmer, H. G., et al. (2009). Inflammasome-activating nanoparticles as modular systems for optimizing vaccine efficacy. *Vaccine*, **27**(23), 3013–3021.
52. Reisetter, A. C., Stebounova, L. V., Baltrusaitis, J., et al. (2011). Induction of inflammasome-dependent pyroptosis by carbon black nanoparticles. *J. Biol. Chem.*, **286**(24), 21844–21852.
53. Yang, E.-J., Kim, S., Kim, J. S., Choi, I.-H. (2012). Inflammasome formation and IL-1beta release by human blood monocytes in response to silver nanoparticles. *Biomaterials*, **33**(28), 6858–6867.
54. Yang, M., Flavin, K., Kopf, I., et al. (2013). Functionalization of carbon nanoparticles modulates inflammatory cell recruitment and NLRP3 inflammasome activation. *Small*, **9**(24), 4194–4206.
55. Peeters, P. M., Perkins, T. N., Wouters, E. F. M., Mossman, B. T., Reynaert, N. L. (2013). Silica induces NLRP3 inflammasome activation in human lung epithelial cells. *Part. Fibre Toxicol.*, **10**(1), article 3.
56. Hornung, V., Bauernfeind, F., Halle, A., et al. (2008). Silica crystals and aluminum salts activate the NALP3 inflammasome through phagosomal destabilization. *Nat. Immunol.*, **9**(8), 847–856.
57. Sharp, F. A., Ruane, D., Claass, B., et al. (2009). Uptake of particulate vaccine adjuvants by dendritic cells activates the NALP3 inflammasome. *Proc. Natl. Acad. Sci. U. S. A.*, **106**(3), 870–875.
58. Weissleder, R., Nahrendorf, M., Pittet, M. J. (2014). Imaging macrophages with nanoparticles. *Nat. Mater.*, **13**(2), 125–138.
59. Lin, A. Y., Almeida, J. P. M., Bear, A., et al. (2013). Gold nanoparticle delivery of modified CpG stimulates macrophages and inhibits tumor growth for enhanced immunotherapy. *PLoS One*, **8**(5), article ID e63550.
60. Ahn, S., Lee, I.-H., Kang, S., et al. (2014). Gold nanoparticles displaying tumor-associated self-antigens as a potential vaccine for cancer immunotherapy. *Adv. Healthcare Mater.*, **3**(8), 1194–1199.
61. Bartneck, M., Keul, H. A., Zwadlo-Klarwasser, G., Groll, J. (2010). Phagocytosis independent extracellular nanoparticle clearance by human immune cells. *Nano Lett.*, **10**(1), 59–63.
62. Brinkmann, V., Zychlinsky, A. (2012). Neutrophil extracellular traps: Is immunity the second function of chromatin? *J. Cell Biol.*, **198**(5), 773–783.
63. Wang, Z. J., Li, J., Cho, J., Malik, A. B. (2014). Prevention of vascular inflammation by nanoparticle targeting of adherent neutrophils. *Nat. Nanotechnol.*, **9**(3), 204–210.

64. Lanone, S., Rogerieux, F., Geys, J., et al. (2009). Comparative toxicity of 24 manufactured nanoparticles in human alveolar epithelial and macrophage cell lines. *Part. Fibre Toxicol.*, **6**, article 14.
65. George, S., Pokhrel, S., Xia, T., et al. (2010). Use of a rapid cytotoxicity screening approach to engineer a safer zinc oxide nanoparticle through iron doping. *ACS Nano*, **4**(1), 15–29.
66. Xia, T., Kovochich, M., Liong, M., et al. (2008). Comparison of the mechanism of toxicity of zinc oxide and cerium oxide nanoparticles based on dissolution and oxidative stress properties. *ACS Nano*, **2**(10), 2121–2134.
67. Brunner, T. J., Wick, P., Manser, P., et al. (2006). In vitro cytotoxicity of oxide nanoparticles: Comparison to asbestos, silica, and the effect of particle solubility. *Environ. Sci. Technol.*, **40**(14), 4374–4381.
68. Sayes, C. M., Reed, K. L., Warheit, D. B. (2007). Assessing toxicity of fine and nanoparticles: Comparing in vitro measurements to in vivo pulmonary toxicity profiles. *Toxicol. Sci.*, **97**(1), 163–180.
69. Jeng, H. A., Swanson, J. (2006). Toxicity of metal oxide nanoparticles in mammalian cells. *J. Environ. Sci. Health, Part A*, **41**(12), 2699–2711.
70. Riley, M. R., Boesewetter, D. E., Turner, R. A., et al. (2005). Comparison of the sensitivity of three lung derived cell lines to metals from combustion derived particulate matter. *Toxicol. in Vitro*, **19**(3), 411–419.
71. Soto, K. F., Carrasco, A., Powell, T. G., et al. (2005). Comparative in vitro cytotoxicity assessment of some manufactured nanoparticulate materials characterized by transmission electron microscopy. *J. Nanopart. Res.*, **7**(2–3), 145–169.
72. Soto, K., Garza, K. M., Murr, L. E. (2007). Cytotoxic effects of aggregated nanomaterials. *Acta Biomater.*, **3**(3), 351–358.
73. Hussain, S. M., Hess, K. L., Gearhart, J. M., et al. (2005). In vitro toxicity of nanoparticles in BRL 3A rat liver cells. *Toxicol. in Vitro*, **19**(7), 975–983.
74. Hussain, S. M., Javorina, A. K., Schrand, A. M., et al. (2006). The interaction of manganese nanoparticles with PC-12 cells induces dopamine depletion. *Toxicol. Sci.*, **92**(2), 456–463.
75. Soto, K. F., Carrasco, A., Powell, T. G., et al. (2006). Biological effects of nanoparticulate materials. *Mater. Sci. Eng. C*, **26**(8), 1421–1427.
76. Braydich-Stolle, L., Hussain, S., Schlager, J. J., Hofmann, M. C. (2005). In vitro cytotoxicity of nanoparticles in mammalian germline stem cells. *Toxicol. Sci.*, **88**(2), 412–419.

77. Wang, J. Y., Rahman, M. F., Duhart, H. M., et al. (2009). Expression changes of dopaminergic system-related genes in PC12 cells induced by manganese, silver, or copper nanoparticles. *Neurotoxicology*, **30**(6), 926–933.

78. Veranth, J. M., Kaser, E. G., Veranth, M. M., et al. (2007). Cytokine responses of human lung cells (BEAS-2B) treated with micron-sized and nanoparticles of metal oxides compared to soil dusts. *Part. Fibre Toxicol.*, **4**, article 2.

79. Limbach, L. K., Wick, P., Manser, P., et al. (2007). Exposure of engineered nanoparticles to human lung epithelial cells: Influence of chemical composition and catalytic activity on oxidative stress. *Environ. Sci. Technol.*, **41**(11), 4158–4163.

80. Wagner, A. J., Bleckmann, C. A., Murdock, R. C., et al. (2007). Cellular interaction of different forms of aluminum nanoparticles in rat alveolar macrophages. *J. Phys. Chem. B*, **111**(25), 7353–7359.

81. Kim, Y. S., Kim, J. S., Cho, H. S., et al. (2008). Twenty-eight-day oral toxicity, genotoxicity, and gender-related tissue distribution of silver nanoparticles in Sprague-Dawley rats. *Inhalation Toxicol.*, **20**(6), 575–583.

82. Ji, J. H., Jung, J. H., Kim, S. S., et al. (2007). Twenty-eight-day inhalation toxicity study of silver nanoparticles in Sprague-Dawley rats. *Inhalation Toxicol.*, **19**(10), 857–871.

83. Asharani, P. V., Wu, Y. L., Gong, Z. Y., Valiyaveettil, S. (2008). Toxicity of silver nanoparticles in zebrafish models. *Nanotechnology*, **19**(25), article ID 255102.

84. Chen, J. P., Patil, S., Seal, S., McGinnis, J. F. (2006). Rare earth nanoparticles prevent retinal degeneration induced by intracellular peroxides. *Nat. Nanotechnol.*, **1**(2), 142–150.

85. Rahman, M. F., Wang, J., Patterson, T. A., et al. (2009). Expression of genes related to oxidative stress in the mouse brain after exposure to silver-25 nanoparticles. *Toxicol. Lett.*, **187**(1), 15–21.

86. Sharma, H. S. (2006). Hyperthermia induced brain oedema: Current status & future perspectives. *Indian J. Med. Res.*, **123**(5), 629–652.

87. Hillyer, J. F., Albrecht, R. M. (2001). Gastrointestinal persorption and tissue distribution of differently sized colloidal gold nanoparticles. *J. Pharm. Sci.*, **90**(12), 1927–1936.

88. Griffitt, R. J., Weil, R., Hyndman, K. A., et al. (2007). Exposure to copper nanoparticles causes gill injury and acute lethality in zebrafish (*Danio rerio*). *Environ. Sci. Technol.*, **41**(23), 8178–8186.

89. Chen, Z., Meng, H. A., Xing, G. M., et al. (2006). Acute toxicological effects of copper nanoparticles in vivo. *Toxicol. Lett.*, **163**(2), 109–120.

90. Zhu, M. T., Feng, W. Y., Wang, B., et al. (2008). Comparative study of pulmonary responses to nano- and submicron-sized ferric oxide in rats. *Toxicology*, **247**(2–3), 102–111.

91. Wang, J. X., Zhou, G. Q., Chen, C. Y., et al. (2007). Acute toxicity and biodistribution of different sized titanium dioxide particles in mice after oral administration. *Toxicol. Lett.*, **168**(2), 176–185.

92. Kim, J. S., Yoon, T. J., Kim, B. G., et al. (2006). Toxicity and tissue distribution of magnetic nanoparticles in mice. *Toxicol. Sci.*, **89**(1), 338–347.

93. Chen, E. Y., Garnica, M., Wang, Y.-C., Mintz, A. J., Chen, C.-S., Chin, W.-C. (2012). A mixture of anatase and rutile TiO_2 nanoparticles induces histamine secretion in mast cells. *Part. Fibre Toxicol.*, **9**, article 2.

94. John, A. L. S., Chan, C. Y., Staats, H. F., Leong, K. W., Abraham, S. N. (2012). Synthetic mast-cell granules as adjuvants to promote and polarize immunity in lymph nodes. *Nat. Mater.*, **11**(3), 250–257.

95. Vivier, E., Tomasello, E., Baratin, M., Walzer, T., Ugolini, S. (208). Functions of natural killer cells. *Nat. Immunol.*, **9**(5), 503–510.

96. Ishigami, S., Natsugoe, S., Tokuda, K., et al. (2000). Prognostic value of intratumoral natural killer cells in gastric carcinoma. *Cancer*, **88**(3), 577–583.

97. Lim, Y. T., Cho, M. Y., Noh, Y.-W., Chung, J. W., Chung, B. H. (2009). Near-infrared emitting fluorescent nanocrystals-labeled natural killer cells as a platform technology for the optical imaging of immunotherapeutic cells-based cancer therapy. *Nanotechnology*, **20**(47), article ID 475102.

98. Jha, P., Golovko, D., Bains, S., et al. (2010). Monitoring of natural killer cell immunotherapy using noninvasive imaging modalities. *Cancer Res.*, **70**(15), 6109–6113.

99. Jang, E.-S., Shin, J.-H., Ren, G., et al. (2012). The manipulation of natural killer cells to target tumor sites using magnetic nanoparticles. *Biomaterials*, **33**(22), 5584–5592.

100. Hubbell, J. A., Thomas, S. N., Swartz, M. A. (2009). Materials engineering for immunomodulation. *Nature*, **462**(7272), 449–460.

101. Gregory, A. E., Titball, R., Williamson, D. (2013). Vaccine delivery using nanoparticles. *Front. Cell. Infect. Microbiol.*, **3**, article 13.

102. Capurso, N. A., Look, M., Jeanbart, L., et al. (2010). Development of a nanoparticulate formulation of retinoic acid that suppresses Th17 cells and upregulates regulatory T cells. *Self/Nonself*, **1**(4), 335–340.

103. Bettelli, E., Oukka, M., Kuchroo, V. K. (2007). T(H)-17 cells in the circle of immunity and autoimmunity. *Nat. Immunol.*, **8**(4), 345–350.

104. Liu, Y., Jiao, F., Qiu, Y., et al. (2009). The effect of Gd@C_{82} $(OH)_{22}$ nanoparticles on the release of Th1/Th2 cytokines and induction of TNF-α mediated cellular immunity. *Biomaterials*, **30**(23–24), 3934–3945.

105. Kidd, P. (2003). Th1/Th2 balance: The hypothesis, its limitations, and implications for health and disease. *Altern. Med. Rev.*, **8**(3), 223–246.

106. Brandenberger, C., Rowley, N. L., Jackson-Humbles, D. N., et al. (2013). Engineered silica nanoparticles act as adjuvants to enhance allergic airway disease in mice. *Part. Fibre Toxicol.*, **10**(1), article 26.

107. Tomić, S., Đokić, J., Vasilijić, S., et al. (2014). Size-dependent effects of gold nanoparticles uptake on maturation and antitumor functions of human dendritic cells in vitro. *PLoS One*, **9**(5), article ID e96584.

108. Mottram, P. L., Leong, D., Crimeen-Irwin, B., et al. (2007). Type 1 and 2 immunity following vaccination is influenced by nanoparticle size: Formulation of a model vaccine for respiratory syncytial virus. *Mol. Pharm.*, **4**(1), 73–84.

109. Storni, T., Kündig, T. M., Senti, G., Johansen, P. (2005). Immunity in response to particulate antigen-delivery systems. *Adv. Drug Deliv. Rev.*, **57**(3), 333–355.

110. Temchura, V. V., Kozlova, D., Sokolova, V., Überla, K., Epple, M. (2014). Targeting and activation of antigen-specific B-cells by calcium phosphate nanoparticles loaded with protein antigen. *Biomaterials*, **35**(23), 6098–6105.

111. Shen, C.-C., Wang, C.-C., Liao, M.-H., Jan, T.-R. (2011). A single exposure to iron oxide nanoparticles attenuates antigen-specific antibody production and T-cell reactivity in ovalbumin-sensitized BALB/c mice. *Int. J. Nanomed.*, **6**, 1229–1235.

112. Yang, S., Damiano, M. G., Zhang, H., et al. (2013). Biomimetic, synthetic HDL nanostructures for lymphoma. *Proc. Natl. Acad. Sci. U. S. A.*, **110**(7), 2511–2516.

113. Gurunathan, S., Lee, K. J., Kalishwaralal, K., Sheikpranbabu, S., Vaidyanathan, R., Eom, S. H. (2009). Antiangiogenic properties of silver nanoparticles. *Biomaterials*, **30**(31), 6341–6350.

114. Sriram, M. I., Kanth, S. B. M., Kalishwaralal, K., Gurunathan, S. (2010). Antitumor activity of silver nanoparticles in Dalton's lymphoma ascites tumor model. *Int. J. Nanomed.*, **5**(1), 753–762.

115. Garanger, E., Boturyn, D., Dumy, P. (2007). Tumor targeting with RGD peptide ligands-design of new molecular conjugates for imaging and therapy of cancers. *Anti-Cancer Agents Med. Chem.*, **7**(5), 552–558.

116. Scott, A. M., Allison, J. P., Wolchok, J. D. (2012). Monoclonal antibodies in cancer therapy. *Cancer Immun.*, **12**, article 14.

Chapter 23

Silica Nanoparticles Effects on Hemostasis

Volodymyr Gryshchuk, PhD,[a,b] Volodymyr Chernyshenko, PhD,[b] Tamara Chernyshenko,[b] Olha Hornytska, PhD,[b] Natalya Galagan, PhD,[c] and Tetyana Platonova, DrSc[b]

[a]ESC "Institute of Biology and Medicine" of Kyiv National Taras Shevchenko University, Kyiv, Ukraine
[b]Palladin Institute of Biochemistry of National Academy of Sciences of Ukraine, Kyiv, Ukraine
[c]Chuiko Institute of Surface Chemistry of National Academy of Sciences of Ukraine, Kyiv, Ukraine

Keywords: activated partial thromboplastin time (APTT), anticoagulant, bioengineering, biocompatibility, blood coagulation, blood plasma, coagulation, Factor Xa, immunotoxicity, platelet aggregation, platelet activation, Russel viper venom (RVV), platelet rich plasma (PRP), platelets, procoagulant, prothrombin, prothrombin time (PT), silica nanoparticles (SiNPs), hemostasis, coagulation profile, nanoparticles, blood loss, bleeding prevention, kallikrein–kinin system, contact system of blood coagulation activation

23.1 Introduction

Interactions of nanoparticles with the blood coagulation system can be beneficial or adverse depending on the intended use of a

---

Immune Aspects of Biopharmaceuticals and Nanomedicines
Edited by Raj Bawa, János Szebeni, Thomas J. Webster, and Gerald F. Audette
Copyright © 2018 Pan Stanford Publishing Pte. Ltd.
ISBN 978-981-4774-52-9 (Hardcover), 978-0-203-73153-6 (eBook)
www.panstanford.com

nanomaterial. Nanoparticles can be engineered to be procoagulant or anticoagulant or to carry drugs to intervene in other pathological conditions in which coagulation is a concern [1]. The use of nanoparticles as drug carriers is extremely promising and is of interest for scientists all over the World [2, 3]. Nanoparticles also can be designed as bacteriostatic agents as well [4]. Silica nanoparticles (SiNPs) have been shown to be a promising alternative for biomedical applications [5] as a novel delivery in tumor research vector or tumor-targeting agent [6, 7], and as the carriers of anticancer drugs [8, 9]. For the use of SiNPs in bioengineering, the study of their biocompatibility, including cell toxicity [10], immunotoxicity [11], genotoxic effects [12], is essential. The aim of present study was the estimation of the primary effects of SiNPs on hemostasis *in vitro*.

23.2 Materials and Methods

23.2.1 Materials

23.2.1.1 Chemicals

Chromogenic substrates S2238 (H-D-Phe-Pip-Arg-pNA), S2765 (Z-D-Arg-Gly-Arg-pNA) and S2251 (H-D-Val-Leu-Lys-pNA) were purchased from BIOPHEN, and APTT-reagent was from Renam (Russia). Ecamulin purified from *Echis multisquamatis* venom was kindly donated by Dr. Dar'ya Korolova, PhD, Palladin Institute of Biochemistry of NAS of Ukraine. Factor X activator from *Russel vipera* venom (RVV) and ADP were purchased from Sigma (US).

23.2.1.2 Silica nanoparticles

Amorphous silica nanoparticles (SiNPs) A300 (Kalush, Ukraine) were pre-treated at 400°C for 2 h. The resulting material had the average space of surface as approximately 300 m^2/g and the size of particles ranged from 10 to 40 nm [13].

23.2.1.3 Preparation of fraction of vitamin K-dependent proteins

Ten milliliters of human blood plasma was mixed with 0.6 g of $BaSO_4$ and spinned down at 2000 rpm for 10 min at 4°C. The fraction of vitamin K-dependent proteins was eluted from the pellet by 0.05M tris-HCl buffer pH 7.4 with 0.25M NaCl, 0.001M of benzamidine and 0.02M of EDTA and then desalted using PD-10 column and used *ex temporo*.

23.2.2 Methods

23.2.2.1 Platelet-rich plasma (PRP) and blood plasma preparation

Venous blood of healthy volunteers who had not taken any medication for 7 days was collected into 38 g/liter sodium citrate (9 parts blood to 1 part sodium citrate). PRP was isolated by the centrifugation of blood at 160 g for 20 min at 23°C. For the preparation of platelet-poor blood plasma, PRP was spinned-down at 1500 rpm for 30 min [14].

23.2.2.2 Activated partial thromboplastin time

Activated partial thromboplastin time (APTT) was performed according to the following procedure: 0.1 ml of blood plasma was mixed with an equal volume of APTT-reagent and incubated for 3 min at 37°C. Then the coagulation was initiated by adding 0.1 ml of 0.025M solution of $CaCl_2$. The clotting time was monitored by the Coagulometer Solar CGL-2410 (Belorussia) [15].

23.2.2.3 Amidase activity assay

Hydrolysis of chromogenic substrates (S2238, H-D-Phe-Pip-Arg-pNA; S2765, Z-D-Arg-Gly-Arg-pNA; or S2251, H-D-Val-Leu-Lys-pNA) under the influence of silica nanoparticles (SiNPs) was studied using the reader (Thermo-scientific), E405–E492. The analysis was done in 0.05M Tris-HCl buffer of pH 7.4 solution at 37°C. The concentration of chromogenic substrates was 0.3 mM and that of SiNPs was 0.4 mg/ml [16].

23.2.2.4 Fibrinogen concentration study

Fibrinogen concentration in the blood plasma was determined by the modified spectrophotometric method. Blood plasma (0.2 ml) and PBS (1.7 ml) were mixed in a glass tube. Coagulation was initiated by the addition of 0.1 ml of thrombin-like enzyme from the venom of *Agkistrodon halys halys* (1 NIH/ml), which allowed avoiding fibrin cross-linking [17]. The mixture was incubated for 30 min at 37°C. The fibrin clot was removed and re-solved in 5 ml of 1.5% acetic acid. The concentration of protein was measured using spectrophotometer SF-2000 (Russia) at 280 nm ($\varepsilon = 1.5$).

23.2.2.5 Calculation of the speed of fibrin clot lysis

The clots were formed in the volume of 1 ml containing 0.05M tris-HCl buffer (pH 7.4) with 0.13M NaCl and 0.001M of $CaCl_2$, 70 μl blood plasma, streptokinase at the final concentration of 3 IU/ml, and SiNPs (0.4 mg/ml). The plasma clotting was initiated by adding 1.3 NIH/ml of thrombin-like enzyme from the venom of *Agkistrodon halys halys*. The change of turbidity during the formation and degradation of fibrin clot was monitored constantly at 350 nm. The speed of clot lysis was calculated as a drop-in of the turbidity of the clot per second at the linear stage of hydrolysis [18].

23.2.2.6 Platelet aggregation study

The platelet aggregation measurement was based on the changes in the turbidity of platelet-rich human plasma [19]. In a typical experiment, 0.4 ml platelet-rich plasma was incubated with 0.02 ml of 0.025M $CaCl_2$ and ADP in final concentration 12.5 μM at 37°C. The studied concentrations of SiNPs were 0.001 and 0.01 mg/ml. The aggregation was detected for 10 min with Aggregometer Solar AP2110.

23.2.2.7 Flow cytometry

The shape and granulation of platelets after incubation with silica nanoparticles (SiNPs) vs. platelets activated by thrombin were monitored on COULTER EPICS XL Flow Cytometer [20]. SiNPs (0.01 mg/ml) were added to 1 ml of washed platelets suspension and the samples were incubated for 90 min at 25°C. Scattered and transmitted light were monitored for examining any changes in the platelet granulation and shape.

23.2.2.8 Statistic data analysis

Statistical data analysis was performed using Microsoft Excel. All assays were performed in series of three replicates and the data were fitted with standard errors using "Statistica 7."

23.3 Results

23.3.1 Proteins of Coagulation System with SiNPs

Amorphous silica nanoparticles (SiNPs) were precipitated in the PBS with the final concentration of stock solution 2 mg/ml. All samples of SiNPs were mixed using Vortex *ex temporo*.

The basic coagulation tests APTT and prothrombin time (PT) were performed in the presence of SiNP suspension in the final concentrations 0.2 and 0.4 mg/ml. It was shown that SiNPs distinctly shortened coagulation time in both tests in a concentration dependent manner (Fig. 23.1). The effect was more evident for PT than for APTT.

Specific chromogenic substrates to prothrombin (S2238) and factor Xa (S2765) have been used to avoid the influence of fibrinogen sorption by SiNPs on the results of the tests.

We studied the effect of SiNPs on blood coagulation directly estimating the activation of prothrombin and factor X. Prothrombin was activated by the prothrombin activator from *Echis multisquamatis* venom—ecamulin. The resulting thrombin activity was detected by thrombin-specific chromogenic substrate S2238. It was shown that SiNPs taken in the concentration 0.4 mg/ml did not affect the direct activation of prothrombin (Fig. 23.2A).

Factor X was activated directly by RVV [21], activity of activated factor Xa estimated with specific chromogenic substrate S2765. It was shown that 0.4 mg/ml of SiNPs inhibited the factor X activation (Fig. 23.2B).

Thus, we observed the inhibitory effect of SiNPs on blood coagulation that consisted of prolongation of the time of coagulation initiated by thromboplastin and APTT-reagent as well as the decrease in the activation of factor Xa but not prothrombin.

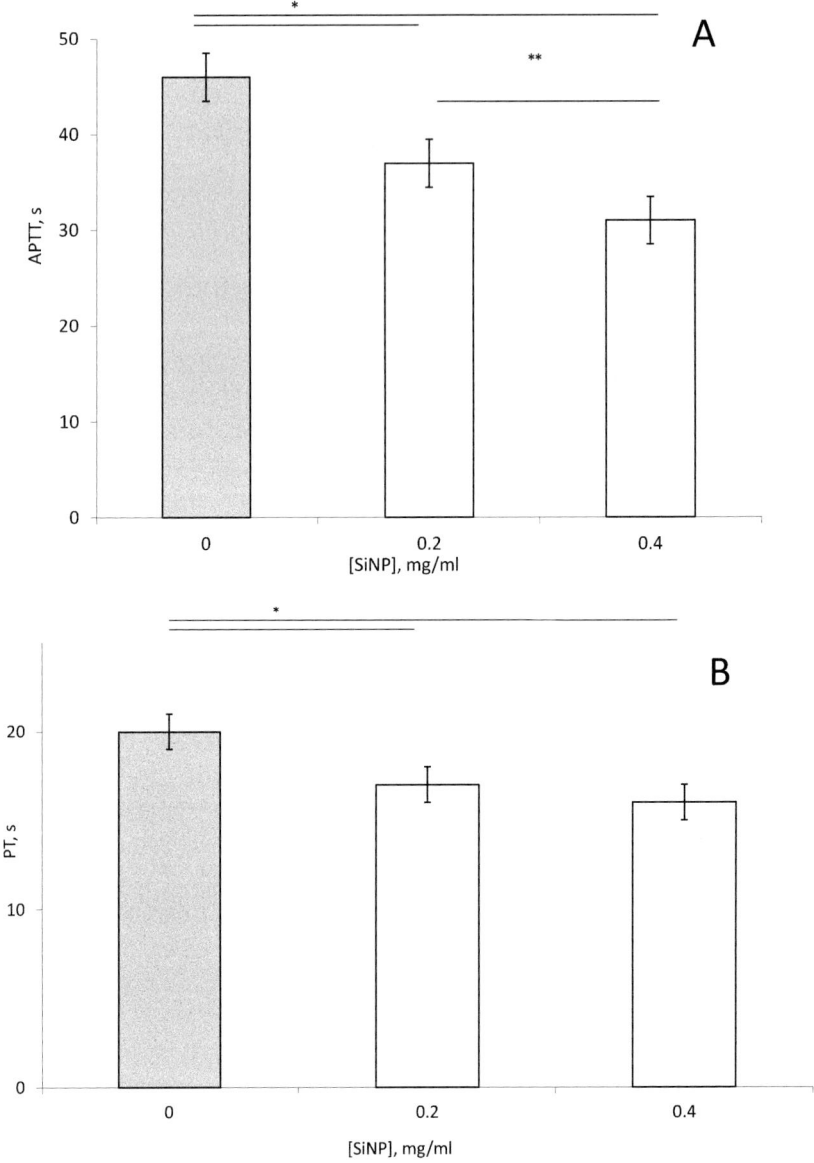

Figure 23.1 Activated partial thromboplastin time (APTT) of human blood plasma (A), and prothrombin time (PT) of human blood plasma (B) in the presence of 0.2 mg/ml and 0.4 mg/ml of SiNPs.

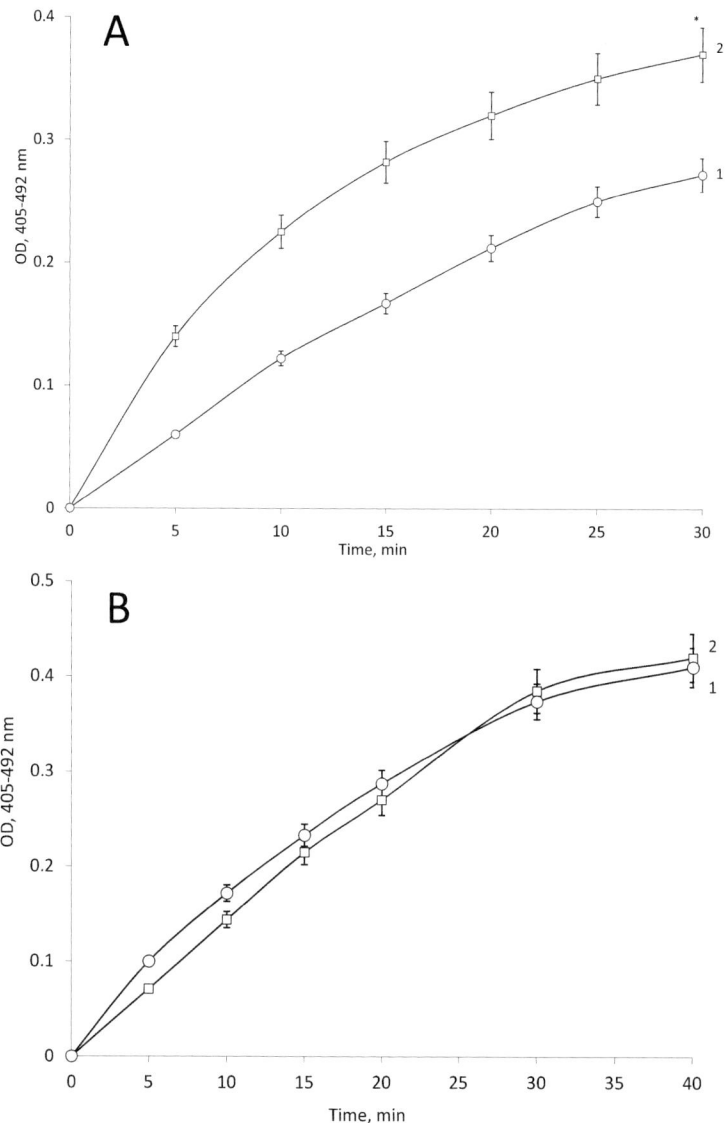

Figure 23.2 Effects of SiNPs on activation of prothrombin and factor X by non-physiologic activators ecamulin and RVV, respectively. (A) Effect of SiNPs (0.4 mg/ml) on prothrombin activation by ecamulin. (B) Effect of SiNPs (0.4 mg/ml) on factor X activation by RVV. 1—control; 2—sample with SiNPs.

23.3.2 Targeting of SiNPs Effects on Coagulation Cascade

To identify the targets of SiNPs in the coagulation cascade that are activated by SiNPs, we compared the activation of factor X by RVV and thromboplastin in the presence of 0.5 mg/ml of SiNPs in human blood plasma (Fig. 23.3A) and in the fraction of vitamin K-dependent proteins (Fig. 23.3B). The fraction of vitamin

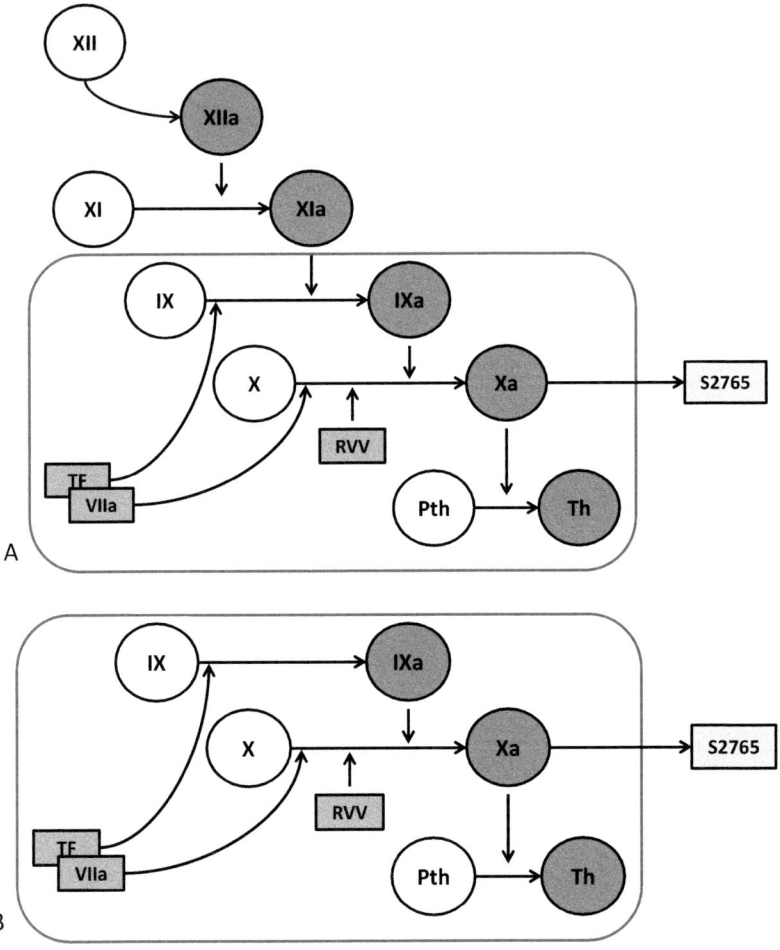

Figure 23.3 Pathway of activation of factor X stimulated by thromboplastin or RVV in human blood plasma (A) or fraction of vitamin K-dependent proteins (B). Pth, prothrombin; Th, thrombin; TF, tissue factor (B).

K-dependent proteins contains prothrombin, factor X, factor IX, factor VII, and protein C and does not contain fibrinogen or components of the kallikrein–kinin contact activation system [22]. Blood plasma or the fraction of vitamin K-dependent proteins were activated by thromboplastin or RVV and the effect of SiNPs on the generation of active factor Xa was measured using specific chromogenic substrate on the microtiter plate spectrophotometer as described above.

We clearly demonstrated the decreasing effect of SiNPs on the factor Xa generation in blood plasma independently on the selected activator. SiNPs stimulated the activation of factor X by thromboplastin and were also effective in the case of RVV-induced activation (Fig. 23.4). However, no activating effect was observed in the fraction of vitamin K-dependent proteins. Thus, we can conclude that SiNPs were ineffective in the absence of the kallikrein–kinin system that was removed in the fraction of vitamin K-dependent proteins.

23.3.3 Fibrinolysis

Fibrinolysis is the main system aimed to remove blood clot from the vessel that is being activated immediately after fibrin formation. It is known that plasminogen is activated on fibrin [23]. A number of preparations that affect the anticoagulation system are known to co-influence the fibrinolytic system [24]. As SiNPs inhibit blood clot formation, we had to study its effect on plasminogen activation and degradation of fibrin clot.

Blood clot was formed in the presence of SiNPs (0.1–0.5 mg/ml) and plasminogen activator (streptokinase). It was demonstrated that the speed of fibrin clot lysis was prolonged by SiNPs distinctly in a concentration-dependent manner (Fig. 23.5A).

However, we did not observe any SiNP-induced changes of plasmin generation in blood plasma (Fig. 23.5B). Plasminogen was activated by streptokinase, and the activity of the resulting plasmin was measured using chromogenic substrate S2251.

Thus, we can assume that SiNPs inhibit blood clot degradation independently of the plasminogen–plasmin system. The observed inhibitory effect can be explained by structural abnormalities of the fibrin clot formed in the presence of SiNPs but not by the impairment of plasminogen activation.

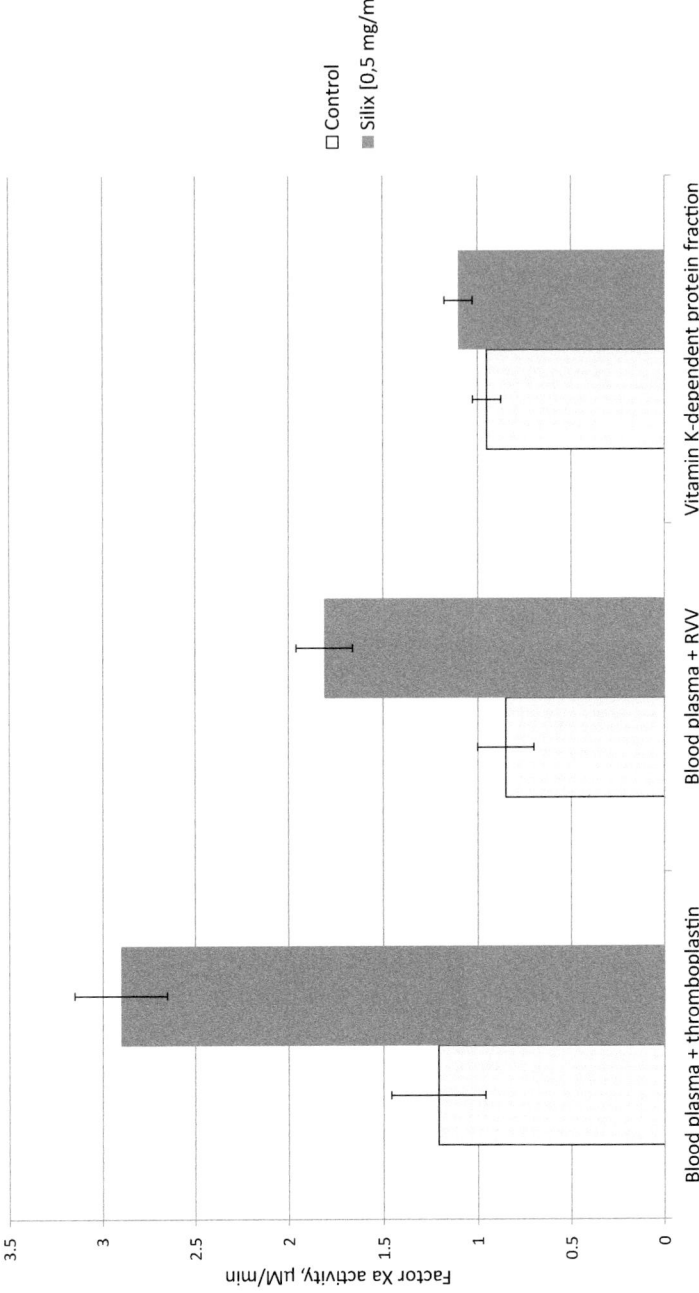

Figure 23.4 Activation of factor Xa by RVV or thromboplastin measured by the hydrolysis of specific chromogenic substrate S2765 in blood plasma and fraction of vitamin K-dependent proteins in the presence or absence of SiNPs (0.5 mg/ml).

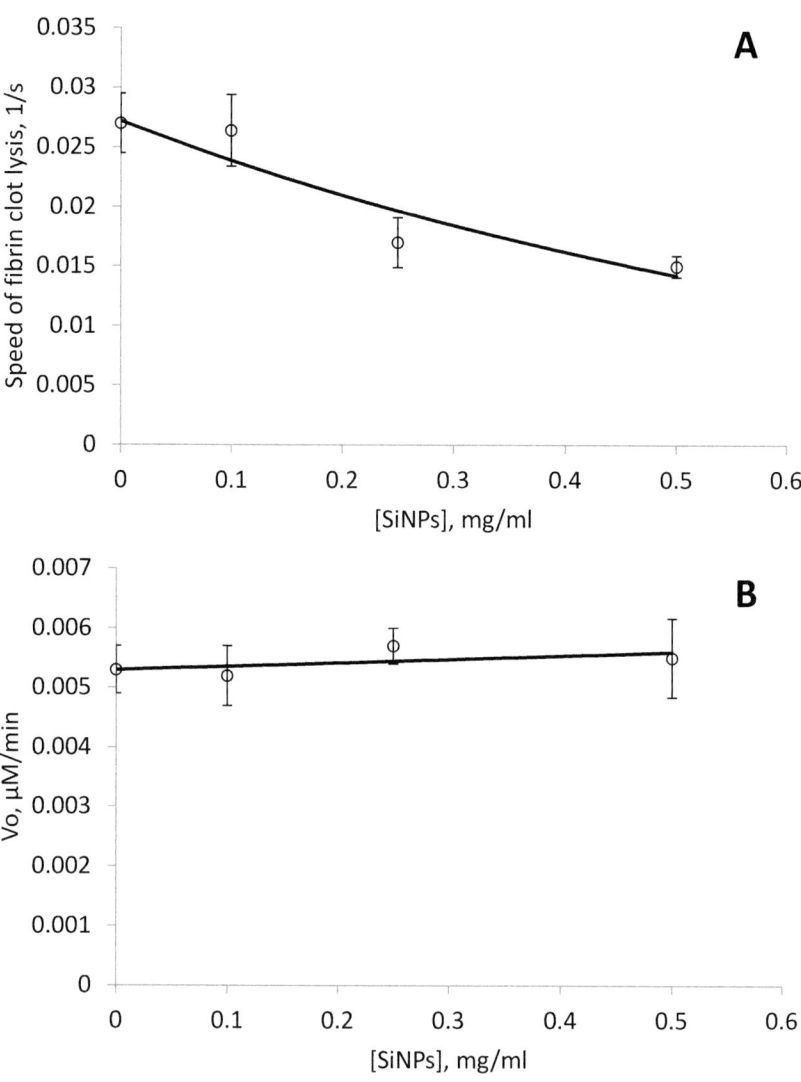

Figure 23.5 The effects of SiNPs on the degradation of fibrin clot by plasmin (A) and on the activation of plasminogen in human blood plasma (B).

23.3.4 Platelet Aggregation and Activation

Platelet aggregation in PRP induced by ADP was studied in the presence of SiNPs or equivalent volume of buffer at the final concentration 0.001 and 0.01 mg/ml. It was shown that these concentrations of SiNPs inhibited the rate and the speed of platelet aggregation (Fig. 23.6). 0.001 mg/ml of SiNPs decreased the aggregation rate twice and provoked huge disaggregation of platelets. These effects were even more evident when 0.01 mg/ml of SiNPs was taken.

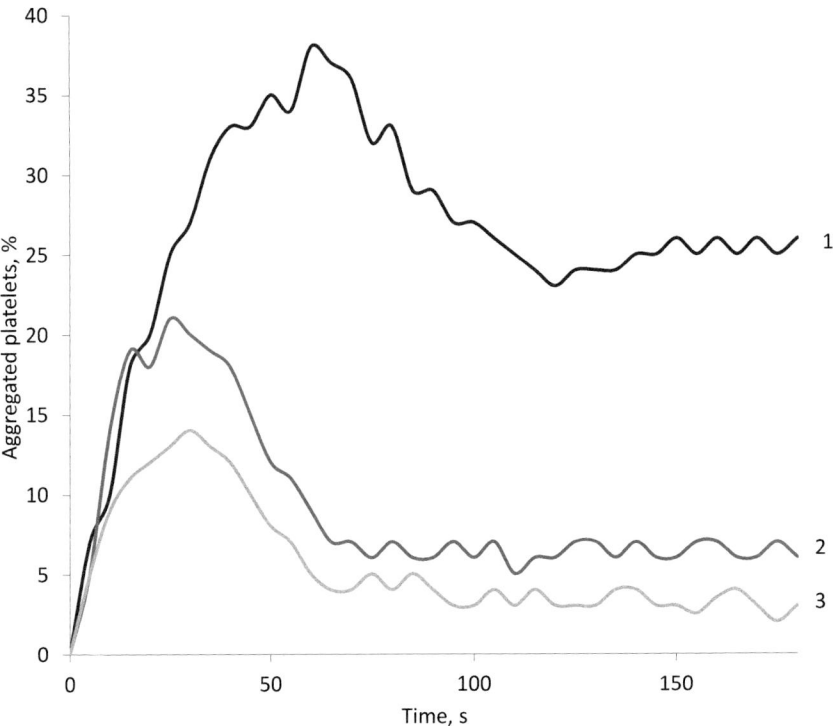

Figure 23.6 Platelet aggregation in platelet rich plasma (PRP) induced by 12.5 μM ADP in the presence of 0.001 mg/ml and 0.01 mg/ml of SiNPs. (1): aggregation of control PRP; (2) aggregation of PRP in the presence of 0.001 mg/ml of SiNPs; (3) aggregation of PRP in the presence of 0.01 mg/ml of SiNPs.

It is known that fibrinogen takes part in platelet aggregation by attracting the platelets to each other and to the newly formed clot. That is why we examined the ability of SiNPs to adsorb fibrinogen. It was shown that 1 mg of SiNPs adsorbed 0.325 mg of fibrinogen of blood plasma (Fig. 23.7). Thus, we can assume that the sorption of fibrinogen by SiNPs taken in the concentrations 0.001 and 0.01 mg/ml could not inhibit platelet aggregation by removing the plasma fibrinogen.

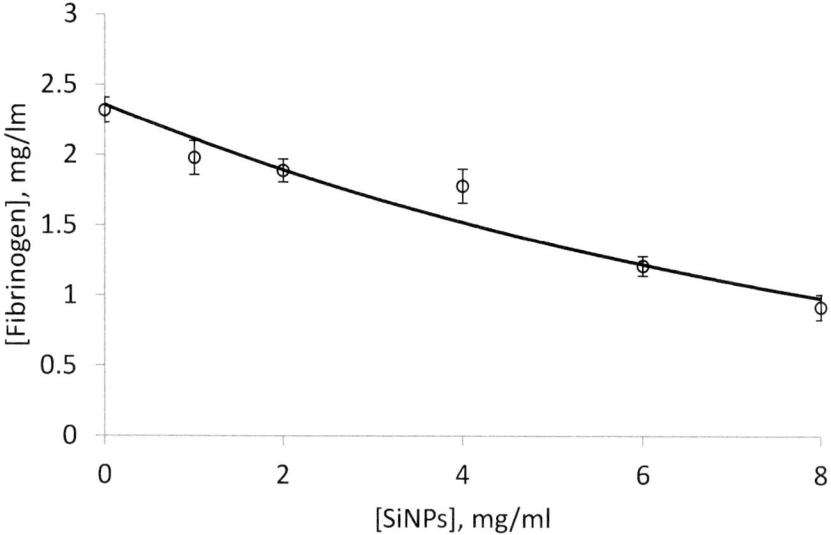

Figure 23.7 Content of fibrinogen in human blood plasma incubated with SiNPs for 30 min at ambient temperature. The drop-in of the fibrinogen concentration is a consequence of its sorption on SiNPs.

Another way to exclude the influence of fibrinogen content on the SiNPs' effect on platelet aggregation was the evaluation of the platelet aggregation rate in the presence of fixed SiNP concentration (0.01 mg/ml) at different concentrations of fibrinogen. It was concluded that being regulated by fibrinogen sorbtion, the rate of platelet aggregation in the presence of SiNPs would be dependent on the level of fibrinogen. Platelets were washed and re-suspended in the HEPES buffer containing different concentrations of fibrinogen (0.5–2.5 mg/ml). It was observed that the average inhibitory effect of SiNPs increased with an increase in the fibrinogen content and did not differ much in the range of fibrinogen concentration from 1.5 to 2.5 mg/ml (Fig. 23.8). This allowed us to assume that the effect of SiNPs on platelet aggregation was mostly fibrinogen-independent.

According to our findings, the inhibitory effect on platelet aggregation could be explained by the action of the studied agent on intracellular signaling or/and on the binding of fibrinogen with its platelet receptor IIbIIIa [25]. So we have to examine whether the SiNPs could inhibit platelet activation. For this purpose, the activation of platelets induced by thrombin was monitored using flow cytometry. The platelet shape and granularity were studied by direct and orthogonal light scattering [18].

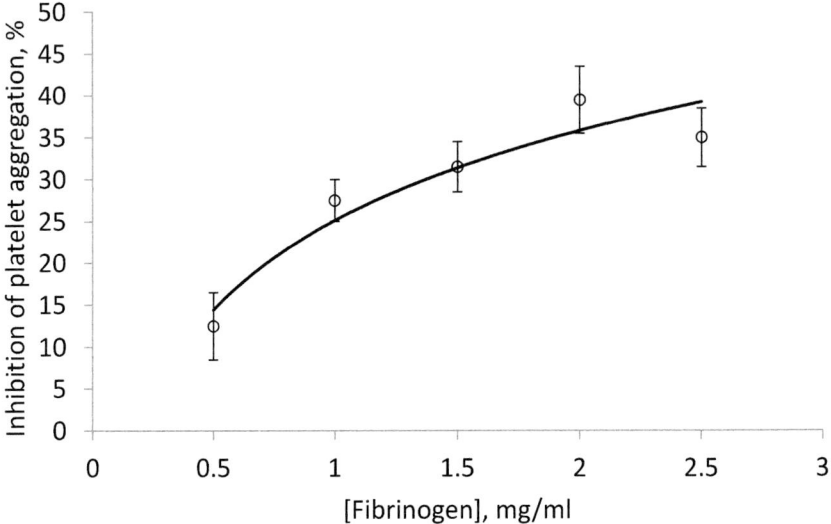

Figure 23.8 The rate of washed platelet aggregation induced by ADP in the presence of 0.01 mg/ml of SiNPs at different concentrations of fibrinogen (0.5–2.5 mg/ml).

To verify any possible effect of SiNPs on human platelets, PRP was analyzed by flow cytometry (Fig. 23.9). Resting human platelets (A) were incubated with 0.001 mg/ml of SiNPs (B) or 0.01 mg/ml of SiNPs (D) for 5 min. It was shown that the population of resting platelets decreased from 80±3% in the control sample (A) to 67±4% and 52±2% in the samples with 0.001 mg/ml of SiNPs (B) and 0.01 mg/ml of SiNPs (D), respectively (Fig. 23.4). In contrast, the activation of platelets by thrombin (C) caused dramatic change to the platelet shape and granularity. Therefore, we concluded that SiNPs were able to induce slight changes in the shape and granularity of platelets that did not cause platelet aggregation but could affect platelet aggregation induced by ADP.

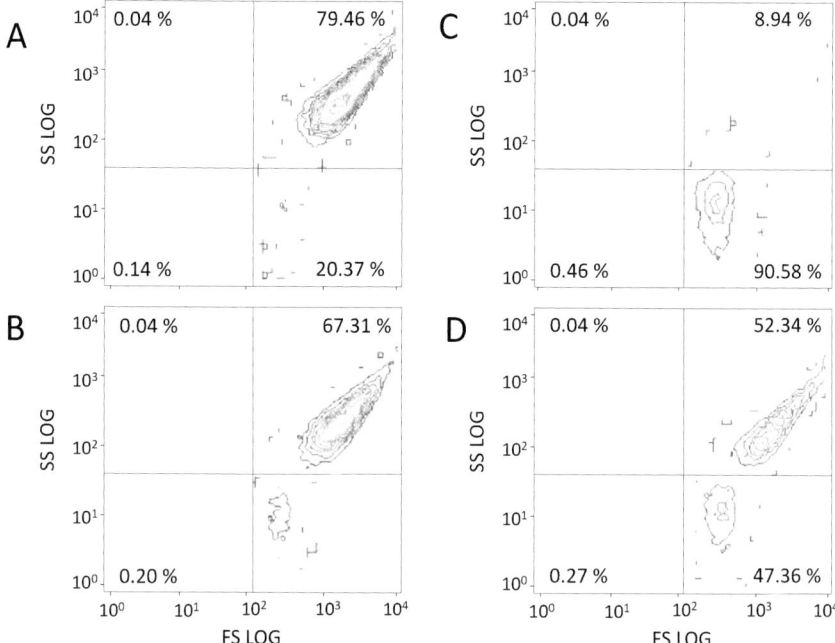

Figure 23.9 Flow cytometry of human platelet rich plasma incubated with 0.01 mg/ml and 0.001 mg/ml of silica nanoparticles (SiNPs). Distribution of platelets according to their shape and granulation. SS LOG—parameter of platelets granulation; FS LOG—parameter of platelets shape. (A) Resting platelets incubated with equal volume of TBS as a control. (B) Resting platelets incubated with 0.001 mg/ml of SiNPs (silica nanoparticles) for 5 min. (C) Platelets activated by thrombin (0.125 NIH/ml) for 2 min. (D) Platelets activated by thrombin, pre-incubated with 0.01 mg/ml of SiNPs for 5 min.

23.4 Discussion

The interaction of silica with blood coagulation proteins is a well-studied area. Previously it was shown that the silica activity depends on the particle diameter and consists of the geometrical relations between the silica and the protein molecules [26]. The conformation of adsorbed proteins on the colloidal silica surfaces plays a role in modulating the amount of their function and cell binding [27]. It was shown that SiNPs effectively adsorbed fibronectin, fibrinogen, etc. Therefore, SiNPs also modulate the processes of cell adhesion by absorbing adhesion molecules [28] and also by direct incorporation in blood cells [29]. It was reported that SiNPs penetrated the platelet plasma membrane and stimulated a rapid and prolonged NO release, IIbIIIa activation, and finally the platelet aggregation [25].

In our study, we showed that the pro-coagulant effects of silica reported in several studies and used in combat gauze development [30, 31] caused mainly by the action of SiNPs on factor Xa activation. Our previous findings reported in [32] showed that this activation did not occur when intrinsic coagulation factors (namely XII and XI) were removed from the incubation volume. These data correspond to the results that showed the sorption of factor XII on the surface of SiNPs, which strongly depends on the size of the particles [33, 34]. However, we did not observe direct coagulant action of SiNPs on unactivated blood plasma or the activating action of SiNPs on platelet aggregation. Despite this, the studied SiNPs were able to change the shape and granularity of the resting platelets studied by flow cytometry, which corresponded to other scientists' data that demonstrated the incorporation of SiNPs in the living cells [25]. We can assume that small SiNPs from the studied samples (10 nm) were incorporated into cells and activated them (as it was shown by Carbalan et al.), but bigger nanoparticles (up to 40 nm) that also were present in the samples inhibited platelet aggregation by absorbing fibrinogen [23–24], which is sufficient for platelets aggregation [21].

Our findings also corresponded to the results obtained recently on *in vivo* models of SiNP administration where systemic activation of the coagulation cascade and platelets was shown [30, 35].

23.5 Conclusions

Amorphous SiNPs are able to increase the activation of the coagulation cascade by adsorbing and stimulating intrinsic pathway coagulation factors. This effect resulted in the shortening of the coagulation time in APTT and PT tests, as well as in the increase of factor X activation by RVV in blood plasma but not in the sample with removed factors XI and XII. SiNPs did not induce platelet aggregation in PRP but changed the shape and granularity of the resting platelets and inhibited their aggregation. The possibility of the use of SiNPs in nanomedicine is strongly dependent on their final concentration in the bloodstream and the size of the particles that are used. However, SiNPs are highly promising as hemostatic agents for preventing blood loss after injury.

Disclosures and Conflict of Interest

This chapter is an updated and revised version that was originally published as: Gryshchuk, V., Galagan, N. (2016). Silica nanoparticles effects on blood coagulation proteins and platelets. *Biochem. Res. Intl.*, 2959414, DOI: 10.1155/2016/2959414, under the Creative Commons Attribution License (https://creativecommons.org/licenses/by/3.0/). It appears here by kind permission of the publisher, Hindawi, Inc. The authors declare that they have no conflict of interest and have no affiliations or financial involvement with any organization or entity discussed in this chapter. No writing assistance was utilized in the production of this chapter and the authors have received no payment for it.

Corresponding Author

Dr. Volodymyr Gryshchuk
ESC "Institute of Biology"
Kyiv National Taras Shevchenko University
64/13 Volodymyrska Street, Kyiv 01601, Ukraine
Email: gryshchukv@gmail.com

About the Authors

Volodymyr Gryshchuk is a junior research fellow of the Protein Structure and Functions department of the Palladin Institute of Biochemistry of the National Academy of Sciences (NAS) of Ukraine, Ukraine. He defended his PhD thesis entitled "Vitamin K-dependent proteins of coagulation system during haemostasis disorders."

Volodymyr Chernyshenko graduated from Kyiv National Taras Shevchenko University and defended his PhD thesis entitled "Fibrinogenase from the venom of *Echis multisquamatis* as an instrument of the study of role of fibrinogen BβN-domain in protein-protein interactions" in the Protein Structure and Functions department of Palladin Institute of Biochemistry of NAS of Ukraine. He is currently a research fellow in the Protein Structure and Functions department of Palladin Institute of Biochemistry of NAS of Ukraine, where his main scientific interests are limited proteolysis of macro-molecules and probing the structure of fibrinogen using specific proteases.

Tamara Chernyshenko is a leading specialist in the laboratory working on hemostasis in the Protein Structure and Functions Department of the Palladin Institute of Biochemistry of the National Academy of Sciences of Ukraine. She studied colloidal chemistry and graduated from the Department of Chemistry, Kyiv National Taras Shevchenko University, Ukraine. She is the author of several scientific patents, including the modification of fibrin glue preparation technique.

Olha Hornytska graduated from the Biological Department of Kyiv National Taras Shevchenko University. She defended her PhD thesis entitled "Isolation, characterization and use of the enzymes from *Agkistrodon halys halys venom*" in the Palladin Institute of Biochemistry of NAS of Ukraine. She is a senior scientist in the Protein Structure and Functions department of Palladin Institute of Biochemistry of NAS of Ukraine. Her main scientific interest is the purification and characterization of snake venom enzymes targeted to hemostasis proteins.

Natalya Galagan is a leading scientist at the Chuiko Institute of Surface Chemistry of the National Academy of Sciences of Ukraine. Her main scientific interest is the development of new nanomaterials based on highly dispersed silica. She holds numerous patents on the application of silica nanoparticles as biological effectors.

Tetyana Platonova is a leading scientist in the Protein Structure and Functions department, Palladin Institute of Biochemistry of the National Academy of Sciences of Ukraine. She defended her Doctor of Sciences thesis entitled "Protein-protein interactions during the formation of supramolecular structures of fibrin clot in blood coagulation." She is studying hemostasis at pathologies, laboratory testing of hemostasis system parameters and molecular mechanisms of fibrin formation and degradation.

References

1. Ilinskaya, A. N., Dobrovolskaia, M. A. (2013). Nanoparticles and the blood coagulation system. Part II: Safety concerns. *Nanomedicine (Lond)*, **8**(6), 969–981.
2. Lu, J., Liong, M., Zink, J. I., Tamanoi, F. (2007). Mesoporous silica nanoparticles as a delivery system for hydrophobic anticancer drugs. *Small*, **3**(8), 1341–1346.
3. Park, K. (2014). Controlled drug delivery systems: Past forward and future back. *J. Control. Release,* **28**(190), 3–8.
4. Jamil, B., Bokhari, H., Imran, M. (2017). Mechanism of action: How nano-antimicrobials act? *Curr. Drug Targets*, **18**(3), 363–373.
5. Hao, N., Li, L., Tang, F. (2014). Shape-mediated biological effects of mesoporous silica nanoparticles. *J. Biomed. Nanotechnol.*, **10**(10), 2508–2538.
6. Wu, X., Wu, M., Zhao, J. X. (2014). Recent development of silica nanoparticles as delivery vectors for cancer imaging and therapy. *Nanomedicine,* **10**(2), 297–312.

7. Milgroom, A., Intrator, M., Madhavan, K., et al. (2014). Mesoporous silica nanoparticles as a breast-cancer targeting ultrasound contrast agent. *Colloids Surf. B: Biointerfaces,* **116**, 652–657.
8. Zhang, P., Kong, J. (2015). Doxorubicin-tethered fluorescent silica nanoparticles for pH-responsive anticancer drug delivery. *Talanta,* **134**, 501–507.
9. Andreani, T., Kiill, C. P., de Souza, A. L., et al. (2014). Surface engineering of silica nanoparticles for oral insulin delivery: Characterization and cell toxicity studies. *Colloids Surf. B: Biointerfaces,* **123**, 916–923.
10. Koch, F., Möller, A. M., Frenz, M., et al. (2014). An in vitro toxicity evaluation of gold-, PLLA- and PCL-coated silica nanoparticles in neuronal cells for nanoparticle-assisted laser-tissue soldering. *Toxicol In Vitro,* **28**(5), 990–998.
11. Lee, S., Kim, M. S., Lee, D., et al. (2013). The comparative immunotoxicity of mesoporous silica nanoparticles and colloidal silica nanoparticles in mice. *Int. J. Nanomed.,* **8**, 147–158.
12. Lin, X., Zhao, N., Yan, P., et al. (2015). The shape and size effects of polycation functionalized silica nanoparticles on gene transfection. *Acta Biomater.,* **11**, 381–392.
13. Pokrovskiy, V., Galagan, N., Chuiko, A. (2006). Interaction of cells with nanoparticles. *Surf. Chem. Biomed. Environ. Sci.,* **2**, 277–286.
14. Korolova, D., Chernyshenko, T., Gornytska, O., et al. (2014). Meizothrombin preparation and its role in fibrin formation and platelet aggregation. *Adv. Biosci. Biotechnol.,* **5**, 588–595.
15. Chernyshenko, V. O., Korolova, D. S., Dosenko, V. E., et al. (2015). Calix[4]arene C-145 effects on plasma haemostasis. *Pharm. Anal. Acta,* **6**, 406.
16. Lottenberg, R., Christensen, U., Jackson, C. M., Coleman, P. L. (1981). Assay of coagulation proteases using peptide chromogenic and fluorogenic substrates. In: Kelman, Z., ed. *Methods in Enzymology,* **80**, 341–361.
17. Castro, H. C., Zingali, R. B., Albuquerque, M. G., et al. (2004). Snake venom thrombin-like enzymes: From reptilase to now. *Cell. Mol. Life Sci.,* **61**(7–8), 843–856.
18. Chernyshenko, V. O., Pirogova, L. V., Cherenok, S. O., et al. (2016). Effects of calix[4]arene C-145 on overall haemostatic potential of blood plasma in vitro and in vivo. *J. Int. Res. Med. Pharm. Sci.,* **10**(61), 146–151.

19. Cattaneo, M., Cerletti, C., Harrison, P., et al. (2013). Recommendations for the standardization of light transmission aggregometry: A consensus of the working party from the platelet physiology subcommittee of SSC/ISTH. *J. Thromb. Haemost.,* **11**(6), 1183–1189.
20. Konokhova, A. I., Yurkin, M. A., Moskalensky, A. E., et al. (2012). Light-scattering flow cytometry for identification and characterization of blood microparticles. *J. Biomed. Opt.,* **17**(5), 057006.
21. Marsh, N., Williams, V. (2005). Practical applications of snake venom toxins in haemostasis. *Toxicon,* **45**(8), 1171–1181.
22. El Asmar, M. S., Naoum, J. J., Arbid, E. J. (2014). Vitamin K dependent proteins and the role of vitamin K2 in the modulation of vascular calcification: A review. *Oman Med. J.,* **29**(3), 172–177.
23. Yakovlev, S., Makogonenko, E., Kurochkina, T., et al. (2000). Conversion of fibrinogen to fibrin: Mechanism of exposure of tPA- and plasminogen-binding sites. *Biochemistry,* **39**(51), 15730–15741.
24. Fareed, J., Bacher, P., Messmore, H. L., Walenga, J. M., Hoppensteadt, D. A., Strano, A., Pifarre, R. (1992). Pharmacological modulation of fibrinolysis by antithrombotic and cardiovascular drugs. *Prog. Cardiovasc. Dis.,* **34**(6), 379–398.
25. Jackson, S. P. (2007). The growing complexity of platelet aggregation. *Blood,* **109**(12), 5087–5095.
26. Kondo, A., Oku, S., Higashitani, K. (1991). Structural changes in protein molecules adsorbed on ultrafine silica particles. *J. Colloid Interface Sci.,* **143**(1), 214–221.
27. Lord, M. S., Cousins, B. G., Doherty, P. J., et al. (2006). The effect of silica nanoparticulate coatings on serum protein adsorption and cellular response. *Biomaterials,* **27**(28), 4856–4862.
28. Wilson, C. J., Clegg, R. E., Leavesley, D. I., Pearcy, M. J. (2005). Mediation of biomaterial-cell interactions by adsorbed proteins: A review. *Tissue Eng.,* **11**(1–2), 1–18.
29. Corbalan, J. J., Medina, C., Jacoby, A., et al. (2012). Amorphous silica nanoparticles aggregate human platelets: potential implications for vascular homeostasis. *Int. J. Nanomed.,* **7**, 631–639.
30. Englehart, M. S., Cho, S. D., Tieu, B. H., et al. (2008). A novel highly porous silica and chitosan-based hemostatic dressing is superior to HemCon and gauze sponges. *J. Trauma,* **65**(4), 884–890.
31. Sambasivan, C. N., Cho, S. D., Zink, K. A., et al. (2009). A highly porous silica and chitosan-based hemostatic dressing is superior

in controlling hemorrhage in a severe groin injury model in swine. *Am. J. Surg.*, **197**(5), 576–580.

32. Chernyshenko, T. M., Gryshchuk, V. I., Galagan, N. P., et al. (2013). Action of high-dispersed silica on blood coagulation factors activation. *Biotechnol. Acta*, **6**(1), 81–85.

33. Yoshida, T., Yoshioka, Y., Tochigi, S. et al. (2013). Intranasal exposure to amorphous nanosilica particles could activate intrinsic coagulation cascade and platelets in mice. *Part. Fibre Toxicol.*, **10**(41).

34. Kushida, T., Saha, K., Subramani, C., et al. (2014). Effect of nano-scale curvature on the intrinsic blood coagulation system. *Nanoscale*, **6**(23), 14484–14487.

35. Liu, X., Sun, J. (2013). Time-course effects of intravenously administered silica nanoparticles on blood coagulation and endothelial function in rats. *J. Nanosci. Nanotechnol.*, **13**(1), 222–228.

Chapter 24

Valproate-Induced Rodent Model of Autism Spectrum Disorder: Immunogenic Effects and Role of Microglia

Prabha S. Awale, PhD,[a] James C. K. Lai, PhD,[a] Srinath Pashikanthi, PhD,[a] and Alok Bhushan, PhD[b]

[a]*Department of Biomedical and Pharmaceutical Sciences, College of Pharmacy, Idaho State University, Pocatello, Idaho, USA*
[b]*Department of Pharmaceutical Sciences, Jefferson College of Pharmacy, Thomas Jefferson University, Philadelphia, Pennsylvania, USA*

Keywords: autism spectrum disorders (ASD), valproic acid (VPA), microglia, immune cells, primary motor cortex, rodent model, prenatal exposure, fetal valproate syndrome, environmental risk, proenkephalin (PENK), brain derived neutrophic factor (BDNF), central nervous system (CNS), γ-aminobutyric acid (GABA), histone deacetylase (HDAC), multi-target drug

24.1 Introduction

Because of their unexpectedly high prevalence (14.7 per 1000 or 1 in 68) [1] in children in the United States alone, autism spectrum disorders (ASD) have raised much alarm, concern, and

---

Immune Aspects of Biopharmaceuticals and Nanomedicines
Edited by Raj Bawa, János Szebeni, Thomas J. Webster, and Gerald F. Audette
Copyright © 2018 Pan Stanford Publishing Pte. Ltd.
ISBN 978-981-4774-52-9 (Hardcover), 978-0-203-73153-6 (eBook)
www.panstanford.com

debate in the public. Not only do they pose as significant public health problems with massive health, social and education costs, the psychological and economic impact on families harboring children with ASD has yet to be fully or comprehensively realized [2].

According to *The Diagnostic and Statistical Manual of Mental Disorders*, 5th edition [3], the formally separate diagnostic entities (i.e., autistic disorder, Asperger's disorder, pervasive developmental disorder not otherwise specified) have been merged into a single dimension, ASD. ASD share two core features: (i) deficits in social behaviors and communications and (ii) restricted interests and repetitive patterns of behavior [3]. Furthermore, a new diagnostic category, namely social communication disorder, has been added to depict patients with deficits in social communications but not exhibiting repetitive behaviors or restricted interests [3]. Because this review chapter primarily focuses on the valproate-induced rodent model of ASD and the immunogenic effects and role of microglia therein, the etiology, pathogenesis, and pathophysiology of ASD as well as the immunogenic mechanisms underlying ASD will only be very summarily mentioned.

24.2 Autism Spectrum Disorders: Etiology and Pathogenesis

As the prevalence of ASD rises, there is also a simultaneous increase in understanding their etiology. Although ASD have a strong genetic component (~25% of the cases have identifiable DNA variants) [4], studies combining multiple genetic approaches with advanced statistical modeling have predicted that some 1000–1500 genes may be associated with ASD, thus demonstrating the genetic heterogeneity of ASD (see [4] and references therein). Currently, although hundreds of genetic mutations have been identified in ASD patients, they are either rare variants, affecting only a few ASD patients or common variants bestowing a small risk for ASD in the general population [4]. This genetic heterogeneity of ASD poses real challenges to the unraveling of ASD pathogenesis and the design of targeted therapy. Nonetheless, recent studies have converged on two possible pathogenically and/

or pathophysiologically relevant entities, namely synaptic and immune function (see later) [5].

Compared with some 25% of ASD cases having a strong genetic origin, the remaining cases (75%) are labeled as "idiopathic." The idiopathic ASD cases have been assumed to have arisen from environmental factors: Key examples of such factors include exposure to infection, pesticides, toxins, and drugs such as thalidomide, ethanol, and valproic acid (VPA) (see [6] and references therein). Relevant to the connection between exposure to VPA and development of ASD is the long recognized co-occurrence of ASD and epilepsies and the increasing recognition of clinical overlap in patients presenting with epilepsies and ASD. (See [3] for further discussion of this connection.)

In addition to the development of ASD subsequent to exposure to drugs such as VPA, maternal exposure to infection (see [7] and references therein) is also associated with a high risk of ASD in the offspring. This hint of the involvement of the immune system in the pathogenesis of ASD has prompted increasing studies supporting the hypothesis that several arms of the immune system are involved in high rates of ASD. While there is a consensus that in the ASD brains, pro-inflammatory mediators (e.g., MCP-1, TGF-β, IL-6, TNF-α, IL-8, etc.) are increased, anti-inflammatory cytokines (e.g., IL-4, IL-5) are not (see [7] and references therein). On the other hand, the findings of the levels of the pro- versus anti-inflammatory cytokines in the plasma of ASD children are somewhat conflicting and await clarification. Clearly, these interesting areas require further investigation. Moreover, studies employing the valproate-induced rodent model of ASD have yielded interesting findings that have provided new avenues for clinical and translational studies in ASD patients as well as exciting opportunities for other translational and drug discovery studies to develop new and/or improved therapies for treating ASD.

24.3 Valproate-Induced Rodent Model of Autism Spectrum Disorders

Although it is primarily employed for treating epilepsy [8], valproic acid (VPA) is a teratogen that induces neural tube defects (mainly spina bifida aperta), causes major congenital anomalies (e.g., congenital heart defects, genital abnormalities, limb defects), and induces the "fetal valproate syndrome," characterized by facial dysmorphisms (see [9] and references therein). Because VPA induces developmental deficits reminiscent of those found in ASD, it is considered an environmental risk factor for autism, consistent with the strong association of VPA with ASD noted in several retrospective and prospective clinical studies (see [6] and references therein). Thus, VPA-induced rodent models can be instrumental in the elucidation of mechanisms underlying the pathogenesis and pathophysiology of ASD. Although both prenatal and post-natal VPA-induced models of ASD exist, many—if not most—of the behavioral and pathological changes are shared between these two types of models [10] [22]. As many of the behavioral abnormalities of human ASD and prenatal rodent VPA model (one dose of 600 mg/kg at embryonic day (ED) 12.5, i.e., the most investigated model) are similar, we will focus our discussion mainly based on this model.

24.3.1 Behavioral Alterations Induced by Prenatal Exposure to VPA

The behavioral abnormalities noted in VPA-induced rodent models of ASD on several post-natal days (PND) after they had been exposed to VPA *in utero* (embryonic day (E)) are summarized in Table 24.1. Because of space limitations, their behavioral abnormalities will not be discussed here. Nonetheless, the noteworthy abnormalities include, but are not limited to, decreased number of social explorations, active lack of social interactions, increased locomotor and repetitive self-grooming and digging behaviors, low sensitivity to pain and higher sensitivity to non-painful stimuli, deficits in attention and processing information, decreased number of pup vocalizations, increased anxiety, delayed maturation and motor development and increased amygdala-dependent fear memories and fear generalizations. For all the above, see [6] and references therein.

Table 24.1 Behavioral deficits induced by maternal valproic acid challenge

Strain/species	VPA dose/administration	Time of exposure	Outcome	Age of outcome assessment
Wistar rats	600 mg/kg; single i.p. injection	E12.5	↓ number of social explorations, ↓ prepulse inhibition, ↓ sensitivity to painful stimuli, ↓ exploratory activity, ↓ place aversion to naloxone, ↑ latency to social behavior, ↑ latency to reach home bedding, ↑ locomotor, repetitive/stereotyped activity, ↑ sensitivity to nonpainful stimuli, ↑ anxiety, delayed maturation and motor development	PND 7–90
Wistar rats	500 mg/kg; single i.p. injection	E12.5	↓ play behavior, ↓ social exploration, active avoidance of social interactions, ↓ sensitivity to thermal pain, ↓ prepulse inhibition, ↓ fear extinction, ↑ repetitive behavior, ↑ anxiety, ↑ amygdala-dependent fear memories, ↑ fear generalization	PND 90
C57BL/6Hsd mice	600 mg/kg; single s.c. injection	E13	↓ pup distress calls, ↓ adult 70-kHz premating vocalizations, ↓ social preference, ↓ prepulse inhibition, ↑ repetitive self-grooming behavior, ↑ anxiety	PND 2–75
Hybrid of C57BL/6, CF-1, Swiss Webster and DBA-2 mouse strains	800 mg/kg; 1.5 g of peanut butter mixed with VPA	E11	lack of sociability, ↑ latency to find home bedding, ↓ general exploration	PND 9–25

(Continued)

Table 24.1 (Continued)

Strain/species	VPA dose/administration	Time of exposure	Outcome	Age of outcome assessment
Sprague-Dawley rats	400 mg/kg; single s.c. injection	E12	lack of sociability and preference for social novelty	PND 30
C57BL/6Hsd mice	600 mg/kg; single s.c. injection	E13	↑ repetitive self-grooming and digging behaviors, ↑ anxiety	PND 42–71
ICR (CD-1) mice	500 mg/kg; single i.p. injection	E12.5	↓ social interactions, ↓ locomotor activity, ↓ general exploratory activity, ↑ anxiety	PND 30 and 60
C57BL/6J mice	600 mg/kg; single i.p. injection	E12.5	lack of sociability, ↓ social interactions, ↓ number of pup vocalizations, ↓ number of complex call types (Complex, Two-syllable, Downward, Composite, Harmonic, Frequency step), ↑ number of Flat call type, ↑ repetitive self-grooming and digging behaviors	PND 8–38
Sprague-Dawley rats	400 mg/kg; single s.c. injection	E12	lack of sociability	PND 30
ICR (CD-1) mice	300 mg/kg; single s.c. injection	E10	lack of sociability, ↓ social interactions, ↓ nest building, ↑ repetitive digging behavior	PND 30–120

i.p., intraperitoneal; s.c., subcutaneous; E, embryonic day.

Source: Adapted and modified from *Exp. Neurol.*, 299, Part A, 217–227, Nicolini, C., Fahnestock, M., The valproic acid-induced rodent model of autism, Copyright (2017). with permission from Elsevier.

Table 24.2 Cellular, molecular and neurochemical alterations induced by VPA model of ASD

Strain/species	VPA dose/administration	Time of exposure	Outcome	Age of outcome assessment
Wistar rats	500 mg/kg; single i.p. injection	E11.5	↑ protein levels of NMDA receptor subunits (NR2A, NR2B) and CaMKII, ↑ NMDA receptor-mediated synaptic currents, ↑ postsynaptic LTP in somatosensory cortex R	PND 12–16
Wistar rats	600 mg/kg; single i.p. injection	E12.5	↓ proenkephalin mRNA expression in the dorsal striatum and nucleus accumbens	PND 60–90
Long-Evans rats	350 mg/kg; single i.p. injection	E12.5	↑ dendritic arborization complexity of apical dendrites in motor cortex layer II pyramidal cells	PND 39–42
Wistar rats	500 mg/kg; single i.p. injection	E11.5	↑ number of direct connections between close neighboring layer 5 pyramidal cells, weaker excitatory synaptic responses, ↓ cell excitability, ↑ network reactivity in somatosensory and medial prefrontal cortices	PND 12–16
Wistar rats	500 mg/kg; single i.p. injection	E12.5	↑ reactivity to stimulation, ↑ LTP, ↓ inhibition in the basolateral amygdala	PND 12–16
Hybrid of C57BL/6, CF-1, Swiss Webster and DBA-2 mouse strain	800 mg/kg; 1.5 g of peanut butter mixed with VPA	E11	↓ neuroligin 3 mRNA in hippocampal subregions (CA1, CA3, DG) and in the somatosensory cortex	PND 60 or 61

(*Continued*)

Table 24.2 (Continued)

Strain/species	VPA dose/administration	Time of exposure	Outcome	Age of outcome assessment
C57BL/6 mice	500 mg/kg; single i.p. injection	E10.5	↓ parvalbumin-positive inhibitory neurons in parietal and occipital cortices	PND 60–90
Hybrid of C57BL/6, CF-1, Swiss Webster and DBA-2 mouse strains	800 mg/kg; 1.5 g of peanut butter mixed with VPA	E11	↓ BDNF mRNA in the somatosensory cortex	PND 60 or 61
C57BL/6Hsd mice	600 mg/kg; single s.c. injection	E13	↓ gamma-phase locking and evoked gamma-power	PND 97
ICR (CD-1) mice	500 mg/kg; single i.p. injection	E12.5	↓ Nissl-positive neurons in layers II-III and V of the prefrontal cortex of female mice	PND >21
ICR (CD-1) mice	500 mg/kg; single i.p. injection	E12.5	↓ Nissl-positive neurons in layers II-III and V of the prefrontal cortex and layers IV-V of the somatosensory cortex of male mice	PND 30 and 60
Sprague-Dawley rats	500 mg/kg; single i.p. injection	E12.5	↓ dendritic spine number in the prefrontal cortex, ↑ dendritic spine number in the ventral hippocampus	PND 21, 35, 70
Wistar rats	500 mg/kg; single i.p. injection	E12.5	↓ total and phosphorylated Akt, mTOR and 4E-BP1, ↓ phosphorylated rpS6	PND 35–38

i.p. = intraperitoneal; s.c. = subcutaneous; E = embryonic day; LTP = long-term potentiation; rpS6 = ribosomal protein S6.
Source: Adapted and modified from *Exp. Neurol.*, 299, Part A, 217–227, Nicolini, C., Fahnestock, M., The valproic acid-induced rodent model of autism, Copyright (2017), with permission from Elsevier.

24.3.2 Neurochemical, Cellular, and Molecular Changes Induced by Prenatal Exposure to VPA

The similarities in behavioral abnormalities of human ASD and prenatal rodent VPA models prompted researchers to examine the underlying cellular and molecular mechanisms. Due to space limitations, the latter findings, summarized in Table 24.2, will not be discussed. Nonetheless, some of neurochemical systems affected and/or implicated are worthy of a brief mention because they may eventually turn out to exhibit pathophysiological interrelationships. The systems altered/implicated in the VPA-induced models include decreased mRNA expression of endogenous opioid proenkephalin (PENK); altered glutamatergic receptors and signaling pathway; decreased expression of brain derived neurotrophic factor (BDNF) and its downstream effectors signaling pathway PI3k-Akt-mTOR in the somatosensory cortex; and decreased expression of postsynaptic adhesion molecule neuroligin 3 whose impairment may lead to shifting of excitatory-inhibitory balance (see [6] and references therein).

At the cellular level, several neuronal losses have been noted in these VPA-induced ASD rodent models: loss of Nissl positive neurons in the middle (II–III) and lower (V) layers of the prefrontal cortex and in the lower (IV–V) layers of the somatosensory cortex. Additionally, abnormal microcircuit connectivity (i.e., local hyper-connectivity and distal hypo-connectivity) among layer V pyramidal cells of somatosensory and medial prefrontal cortex, hyper-plasticity, hyper-reactivity to synaptic stimulation of the basolateral amygdala, and complex dendritic arborization in motor cortex layer II pyramidal cells were also found in such models (see [6] and references therein).

24.3.3 Immune System Changes Induced by Prenatal Exposure to VPA

The consensus that the neuropathology of ASD may be due to the disturbances in early postnatal development has been supported by retrospective studies showing increasing head circumference and postmortem studies revealing a 10–15% increase in brain weight and volume based on MRI of ASD individuals in early childhood compared with those of age-matched controls [11–16]. Similarly, an accelerated overgrowth of neocortical pyramidal neurons, particularly in the temporal and frontal neocortices was found [17, 18]: This observation was hypothesized to be due to either abnormal/dysregulated neurogenesis or abnormal pruning during pre- or postnatal development. Indeed, one major function of microglia is the pruning of the central nervous system (CNS).

Microglia are the resident macrophages (immune cells) of the CNS that colonize the brain during early prenatal development and comprise 5–15% of brain cells [19, 20]. Recently microglia were found to actively remove neural precursor cells, and perform phagocytosis of redundant synapses, neurons, cell debris, as well as facilitate neuronal development by secreting cytokines, neurotrophins, growth factors and regulation of stem cell proliferation during late stages of cortical neurogenesis [21, 22]. Consequently, any factors altering the number or normal physiological states of microglia during the pre- or postnatal period may result in aberrant pruning, ultimately leading to aberrant overgrowth of neurons and developmental disorders such as ASD.

Until recently, few, if any, studies have addressed the effects of prenatal exposure to VPA on early postnatal microglial number. Although ASD is diagnosed early on in development, most behavioral, neurochemical, cellular and molecular studies have been carried out in adult VPA-exposed mice. As we were the first to examine the effects of prenatal exposure to VPA on postnatal microglial number in the brains of these pups and found that the microglial cell numbers in the primary motor cortex of the male mice at postnatal day 6 (P6) and P10 [23]—here neurons exhibit complex arborization in their apical dendrites—were significantly reduced compared to those in age- and sex-matched controls [24]. Although the mechanisms underlying these microglial changes have yet to be elucidated, we hypothesize VPA may inhibit proliferation of microglial cells and thereby reduce their number because VPA is known to inhibit proliferation of neural progenitor cells via the β-catenin-Ras-ERK-p21Cip/WAF1 pathway in both *in vitro* and *in vivo* in the cerebral cortex of rat embryos [25].

24.4 Mechanism of Action of VPA

The mechanisms by which valproic acid (VPA) induces ASD-like behaviors are currently not well understood. The anti-seizure effect of VPA (an established anti-epilepsy drug) is reportedly to elevate brain levels of the inhibitory neurotransmitter, γ-aminobutyric acid (GABA) although the underlying mechanisms have yet to be resolved. In addition, valproate also directly inhibits voltage gated sodium channels, thereby suppressing the high frequency firing of neurons [26]. Furthermore, VPA has histone deacetylase (HDAC) inhibitory activity: Prenatal exposure to VPA causes transient hyperacetylation of H3 and H4 that consequently leads to postnatal behavioral impairments in the offspring [27]. HDAC belongs to the metalloenzyme family: This observed HDAC inhibition by VPA might be due to chelation of carboxylate in VPA with zinc in the active site as observed with classic HDAC inhibitors [28]. Thus, strategies aimed at preventing these electrostatic interactions will lead to the preservation of the HDAC activity and the minimization of the teratogenic effect. By contrast, exposure of mice to valpromide, (a VPA analog lacking histone deacetylase inhibition activity) does not induce transient hyperacetylation or affect their behavior [27]. These studies suggest inhibition of HDAC by VPA during a crucial period of neurogenesis is associated with ASD-like pathologies.

VPA is a multi-target drug. In addition to those discussed above, other mechanism may also contribute to the ASD-like behavioral deficits induced by VPA. VPA can indirectly inhibit GSK3β and this may alter axonal remodeling in developing neurons [29, 30]. Additionally, since VPA inhibits GABA transaminase, the GABA-metabolizing enzyme but also enhances glutamic acid decarboxylase, the GABA-synthesizing enzyme, VPA could increase the levels of GABA during early development [31]. As GABA plays a critical role in maturation of neuronal networks in the developing brain, alteration in the levels of GABA may lead to disruption of neuronal network and ASD-like behaviors in the offspring.

24.5 Conclusions and Future Prospects

In this concise review and position paper, our focus has been centered on valproate-induced rodent models of ASD and the immunogenic effects and role of microglia therein. Despite the scant literature on this role of microglia, a clearer picture is emerging regarding the effect of VPA on microglial cells. We were the first to demonstrate that valproic acid decreases the number of microglia in the primary motor cortex of P6 and P10 male mice. We hypothesize that VPA inhibits the proliferative capacity of the microglial cells, thereby reducing their number. As evident from this concise chapter, there is still much to be discovered concerning the effects of VPA on microglial cells and how that would impact cellular, molecular, neurochemical, and behavioral paradigms in the VPA models of ASD. The availability of the VPA rodent model of ASD, along with other models, greatly facilitates systematic investigation of the cellular and molecular mechanisms underlying ASD. Clearly, these interesting and exciting areas merit further investigation.

Disclosures and Conflict of Interest

The authors declare that they have no conflict of interest and have no affiliations or financial involvement with any organization or entity discussed in this chapter. This includes employment, consultancies, honoraria, grants, stock ownership or options, expert testimony, patents (received or pending) or royalties. No writing assistance was utilized in the production of this chapter and the authors have received no payment for its preparation. The findings and conclusions here reflect the current views of the authors. They should not be attributed, in whole or in part, to the organizations with which they are affiliated, nor should they be considered as expressing an opinion with regard to the merits of any particular company or product discussed herein. Nothing contained herein is to be considered as the rendering of legal advice.

Corresponding Author

Dr. Alok Bhushan
Professor and Chair Department of Pharmaceutical Sciences
Jefferson College of Pharmacy
Thomas Jefferson University
901 Walnut St., Suite 915, Philadelphia, PA 19107, USA
Email: alok.bhushan@jefferson.edu

About the authors

Prabha S. Awale is an assistant professor of pharmacology and toxicology in the Department of Biomedical and Pharmaceutical Sciences, Idaho State University. She received her BPharm degree from Rajiv Gandhi University of Health Sciences, Bangalore, India, and a PhD in cellular and molecular biology from Kent State University/ Northeast Ohio Medical University. She was a postdoctoral research fellow in the laboratory of Mark C. Austin in the Department of Biological Sciences at Idaho State University before joining the College of Pharmacy. During her graduate and postdoctoral tenure, she has worked on several research projects, including identifying high-affinity ligands to the rosiglitazone-binding site of mitoNEET, Structure activity relationship and docking studies of thiazolidinedione-type compounds with monoamine oxidase B, understanding the role of resveratrol in altering proinflammatory cytokine in chemically induced hepatocarcinogenesis, and microglial alteration in valproic acid model of autism. Her research interest includes understanding the effect of environmental agents and drugs like valproic acid in a rodent model of autism. Dr. Awale has published three research articles and seven abstracts.

James C. K. Lai is a professor of pharmacology and toxicology in the Department of Biomedical and Pharmaceutical Sciences, College of Pharmacy, Division of Health Sciences, Idaho State University, Pocatello, Idaho. He received his BSc (honors) in microbiology from Cardiff University, Wales, in 1970, MSc in neurocommunications from the University of Birmingham, England, in 1971, and PhD in biochemistry from the University of London, England, in 1975. Prior to joining Idaho State University in 1991, Dr. Lai had held faculty appointments at Albert Einstein College of Medicine, Bronx, NY, and Cornell Medical College, New York, NY. He was an editorial board member of *Neurotoxicology* from 1985 to 1990; associate editor of *Pharmacy Case Review* from 1993 to 1996; editorial board member of *Metabolic Brain Disease* from 1990 to 2010; and editorial board member of *Neurochemical Research* from 1995 to 2008. His long-standing research interests include neuroscience, neurotoxicology, and regulation of brain metabolism in aging, neurodegenerative diseases, hypoxia, and epilepsy. In the past decade, he has also worked on nanoscience and nanotoxicology as well as cancer pharmacology and anti-cancer drug discovery. He has published over 230 journal articles, proceedings papers, reviews, and book chapters in these and related areas.

Srinath Pashikanthi is an assistant professor of medicinal chemistry in the Department of Biomedical and Pharmaceutical Sciences, College of Pharmacy, Division of Health Sciences, Idaho State University. He received his MSc in organic chemistry with specialization in medicinal chemistry from Kakatiya University, India, MS in chemistry and biochemistry from South Dakota State University, followed by a PhD in medicinal chemistry from the School of Pharmacy, The University of Kansas. He was a postdoctoral research fellow in the laboratory of Webster L. Santos at Virginia Tech. During his graduate and postdoctoral tenure, he has worked on multidisciplinary research projects, including identifying biomarkers for glycation of histone H1, developing natural product–based glycation inhibitors, total synthesis and SAR studies of Jaspine B, developing probes to understand the biochemical aspects of sphingosine kinase I and II, and method development toward the synthesis of vinyl silanes in a regio- and stereoselective fashion using copper (II) in water. His research interests include application of organic medicinal chemistry and protein biochemistry in developing small molecules targeting ceramide metabolizing enzyme and identifying biomarkers of covalently modified proteins under oxidative stress. Dr. Pashikanti has published 7 research articles and co-authored 1 book chapter and 12 abstracts.

Alok Bhushan is the chair and professor in the Department of Pharmaceutical Sciences, Jefferson College of Pharmacy at Thomas Jefferson University, Philadelphia. He received his BSc (honors) and MSc in chemistry from the University of Delhi, India, and his PhD in biochemistry from Punjab Agricultural University, India. He was a postdoctoral fellow at the Johns Hopkins School of Medicine, Medical University of South Carolina, and the University of Vermont. He has held faculty positions at the Department of Pharmacology, School of Medicine, University of Vermont, and Department of Biomedical and Pharmaceutical Sciences, College of Pharmacy, Idaho State University. He has over 15 years of experience in pharmacy as well as graduate education. He successfully completed the Academic Fellows Leadership Program at AACP. He has been trained to serve as a site evaluator for accreditation by ACPE and has had the opportunity to be a site evaluator. Dr. Bhushan's areas of interest include cancer pharmacology and signal transduction focused on treatment and prevention. His research areas are (1) understanding mechanisms to block glioblastoma invasion in the brain and synthesis of novel agents that target the process of invasion, (2) understanding the mechanism of the role of isoflavones in preventing breast and oral cancer, (3) pharmacology of nanomaterials, (4) natural drug discovery for the prevention and treatment of cancer, and (5) characterization and cloning of a novel reduced folate transporter, which plays role in methotrexate and cisplatin resistance. Dr. Bhushan has over 75 journal articles and reviews to his credit. He has served as a major advisor to several PhD and master's students.

References

1. Baio., J. (2014). Prevalence of autism spectrum disorder among children aged 8 years—Autism and Developmental Disabilities Monitoring Network, 11 Sites, United States, 2010. Available at:

https://www.cdc.gov/mmwr/preview/mmwrhtml/ss6302a1.htm (accessed on December 19, 2017).

2. Jarbrink, K., Knapp M. (2001). The economic impact of autism in Britain. *Autism,* **5**, 7–22.

3. Lee, B. H., Smith, T., Paciorkowski, A. R. (2015). Autism spectrum disorder and epilepsy: Disorders with a shared biology. *Epilepsy Behav.,* **47**, 191–201.

4. Ziats, M. N., Rennert, O. M. (2016). The evolving diagnostic and genetic landscapes of autism spectrum disorder. *Front. Genet.,* **7**, 65, 1–6.

5. Voineagu, I., Eapen, V. (2013). Converging pathways in autism spectrum disorders: Interplay between synaptic dysfunction and immune responses. *Front. Hum. Neurosci.,* **7**, 738, 1–5.

6. Nicolini, C., Fahnestock M. (2018). The valproic acid-induced rodent model of autism. *Exp. Neurol.,* **299**, 217–227.

7. Meltzer, A., Van de Water, J. (2017). The role of the immune system in autism spectrum disorder. *Neuropsychopharmacology,* **42**, 284–298.

8. Loscher, W. (2002). Basic pharmacology of valproate: A review after 35 years of clinical use for the treatment of epilepsy. *CNS Drugs,* **16**, 669–694.

9. Ornoy, A. (2009). Valproic acid in pregnancy: How much are we endangering the embryo and fetus? *Reprod. Toxicol.,* **28**, 1–10.

10. Chomiak, T., Turner, N., Hu, B. (2013). What we have learned about autism spectrum disorder from valproic acid. *Patholog. Res. Int.* **2013**, 712758, 1–8.

11. Courchesne, E., Carper, R., Akshoomoff, N. (2003). Evidence of brain overgrowth in the first year of life in autism. *JAMA,* **290**, 337–344.

12. Dawson, G., Munson, J., Webb, S. J., Nalty, T., Abbott, R., Toth, K. (2007). Rate of head growth decelerates and symptoms worsen in the second year of life in autism. *Biol. Psychiatry,* **61**, 458–464.

13. Dementieva, Y. A., Vance, D. D., Donnelly, S. L., Elston, L. A., Wolpert, C. M., Ravan, S. A., DeLong, G. R., Abramson, R. K., Wright, H. H., Cuccaro, M. L. (2005). Accelerated head growth in early development of individuals with autism. *Pediatr. Neurol.,* **32**, 102–108.

14. Hazlett, H. C., Poe, M., Gerig, G., Smith, R. G., Provenzale, J., Ross, A., Gilmore, J., Piven, J. (2005). Magnetic resonance imaging and head circumference study of brain size in autism: Birth through age 2 years. *Arch. Gen. Psychiatry,* **62**, 1366–1376.

15. Courchesne, E., Karns, C. M., Davis, H. R., Ziccardi, R., Carper, R. A., Tigue, Z. D., Chisum, H. J., Moses, P., Pierce, K., Lord, C., Lincoln, A. J., Pizzo, S., Schreibman, L., Haas, R. H., Akshoomoff, N. A., Courchesne, R. Y. (2001). Unusual brain growth patterns in early life in patients with autistic disorder: An MRI study. *Neurology,*. **57**, 245–254.

16. Sparks, B. F., Friedman, S. D., Shaw, D. W., Aylward, E. H., Echelard, D., Artru, A. A., Maravilla, K. R., Giedd, J. N., Munson, J., Dawson, G., Dager, S. R. (2002). Brain structural abnormalities in young children with autism spectrum disorder. *Neurology*, **59**, 184–192.

17. Courchesne, E., Pierce, K., Schumann, C. M., Redcay, E., Buckwalter, J. A., Kennedy, D. P., Morgan, J. (2007). Mapping early brain development in autism. *Neuron*, **56**, 399–413.

18. Courchesne, E., Mouton, P. R., Calhoun, M. E., Semendeferi, K., Ahrens-Barbeau, C., Hallet, M. J., Barnes, C. C., Pierce, K. (2011). Neuron number and size in prefrontal cortex of children with autism. *JAMA*, **306**, 2001–2010.

19. Monier, A., Evrard, P., Gressens, P., Verney, C. (2006). Distribution and differentiation of microglia in the human encephalon during the first two trimesters of gestation. *J. Comp. Neurol.*, **499**, 565–582.

20. Monier, A., Adle-Biassette, H., Delezoide, A. L., Evrard, P., Gressens, P., Verney, C. (2007). Entry and distribution of microglial cells in human embryonic and fetal cerebral cortex. *J. Neuropathol. Exp. Neurol.*, **66**, 372–382.

21. Cunningham, C. L., Martinez-Cerdeno, V., Noctor, S. C. (2013). Microglia regulate the number of neural precursor cells in the developing cerebral cortex. *J. Neurosci.*, **33**, 4216–4233.

22. Boche, D., Perry, V. H., Nicoll, J. A. (2013). Review: Activation patterns of microglia and their identification in the human brain. *Neuropathol. Appl. Neurobiol.*, **39**, 3–18.

23. Awale, P. S., Geldenhuys, W. J., Carroll, R. T. (2011). Immune contribution to valproic acid model of autism. 2011 Neuroscience Meeting planner. Washington DC. Society for Neuroscience, 2011. Available at: http://www.abstractsonline.com/Plan/ViewAbstract#330.13/A43 (accessed on December 19, 2017).

24. Snow, W. M., Hartle, K., Ivanco, T. L. (2008). Altered morphology of motor cortex neurons in the VPA rat model of autism. *Dev. Psychobiol.*, **50**, 633–639.

25. Jung, G. A., Yoon, J. Y., Moon, B. S., Yang, D. H., Kim, H. Y., Lee, S. H., Bryja, V., Arenas, E., Choi, K. Y. (2008). Valproic acid induces differentiation

and inhibition of proliferation in neural progenitor cells via the beta-catenin-Ras-ERK-p21Cip/WAF1 pathway. *BMC Cell Biol.*, **9**, 66.

26. Rho, J. M., Sankar, R. (1999). The pharmacologic basis of antiepileptic drug action. *Epilepsia*, **40**, 1471–1483.

27. Kataoka, S., Takuma, K, Hara, Y, Maeda, Y, Ago, Y, Matsuda, T. (2013). Autism-like behaviours with transient histone hyperacetylation in mice treated prenatally with valproic acid. *Int. J. Neuropsychopharmacol.*, **16**, 91–103.

28. Lombardi, P. M., Cole, K. E., Dowling, D. P., Christianson, D. W. (2011). Structure, mechanism, and inhibition of histone deacetylases and related metalloenzymes. *Curr. Opin. Struct. Biol.*, **21**, 735–743.

29. Chen, G., Huang, L. D., Jiang, Y. M., Manji, H. K. (1999). The mood-stabilizing agent valproate inhibits the activity of glycogen synthase kinase-3. *J. Neurochem.*, **72**, 1327–1330.

30. Hall, A. C., Brennan, A., Goold, R. G., Cleverley, K., Lucas, F. R., Gordon-Weeks, P. R., Salinas, P. C. (2002). Valproate regulates GSK-3-mediated axonal remodeling and synapsin I clustering in developing neurons. *Mol. Cell Neurosci.*, **20**, 257–270.

31. Sernagor, E., Chabrol, F., Bony, G., Cancedda, L. (2010). GABAergic control of neurite outgrowth and remodeling during development and adult neurogenesis: General rules and differences in diverse systems. *Front. Cell Neurosci.*, **4**, 11.

Chapter 25

Accelerated Blood Clearance Phenomenon and Complement Activation-Related Pseudoallergy: Two Sides of the Same Coin

Amr S. Abu Lila,[a,b,c] Janos Szebeni,[d] and Tatsuhiro Ishida[a]

[a]*Department of Pharmacokinetics and Biopharmaceutics,*
Institute of Biomedical Sciences, Tokushima University, Japan
[b]*Department of Pharmaceutics and Industrial Pharmacy,*
Faculty of Pharmacy, Zagazig University, Zagazig, Egypt
[c]*Department of Pharmaceutics, College of Pharmacy,*
Hail University, Saudi Arabia
[d]*Nanomedicine Research and Education Center, Institute of Pathophysiology,*
Semmelweis University and SeroScience Ltd, Budapest, Hungary

Keywords: accelerated blood clearance (ABC) phenomenon, anaphylactoid reactions, anti-PEG IgM, complement activation-related pseudoallergy (CARPA), complement inhibitor, complement activation, hypersensitivity reactions (HSRs), liposomes, polyethylene glycol (PEG), porcine model, pulmonary intravascular macrophages (PIM), spleen, toll-like receptors (TLR), nanocarriers, nanoparticles, mononuclear phagocyte system (MPS), Doxil®, Ambisome®, DaunoXome®, anaphylatoxin receptors (ATR), danger-signaling receptors (DSRs), pathogen-associated molecular pattern (PAMP), large multilamellar vesicle (LMV), small unilamellar vesicle (SUV), tolerance induction

25.1 Introduction

Nanocarriers, such as liposomes and nanoparticles, have been widely explored as promising delivery vehicles in modern pharmacotechnology [1]. They combine the merits of improving bioavailability [2], reducing the incidence of systemic toxicity [3], along with, controlling the release of the encapsulated payload at the site of action [4]. Furthermore, nanocarriers can improve the therapeutic efficacy of encapsulated drug(s) via manipulating their *in vivo* fate [5–7]. These merits have led to the widespread adoption of nanocarrier-based therapeutics in modern pharmacotherapy. Nonetheless, besides their unique therapeutic/diagnostic advantages, nanocarriers seem to prime the immune system, following their intravenous administration, leading to the development of adverse reactions and/or loss of efficacy [8–11].

One major issue in the clinical application and approval of such nanocarriers is the safety of these formulations manifested by the absence of an immunogenic response. Nanocarriers are known to interact with the innate immune system, including the complement system and the mononuclear phagocyte system (MPS) to varying extents. These interactions might, on the one hand, activate complement leading sometimes to an infusion reaction [12–14] and, on the other hand, might trigger an antibody-related immune response which could affect the *in vivo* fate of the administered nanocarriers [15–17].

It is noteworthy that the increased awareness of nanocarrier-induced immunogenic responses has recently been reflected by the fact that testing for immunogenicity of nanocarrier-based therapeutics has been recommended by the US Food and Drug Administration (FDA) to assure the safety of drugs and/or nanocarriers [18].

25.2 Immunogenicity of Liposomal Drug Delivery Systems

25.2.1 The Accelerated Blood Clearance Phenomenon

The interaction of liposomes with the immune system has contributed to the challenges in translation to clinical use. Structural modifications to enhance their utility as delivery vehicles can trigger immunogenic responses against liposomal components [19, 20]. A brilliant example is liposomal surface modification with the hydrophilic polymer polyethylene glycol,

PEG, (PEGylation). As a matter of fact, PEGylation was regarded as a major breakthrough in the field of liposomal-based therapeutics. Surface decoration of liposomes with PEG is reported to prevent their uptake by the cells of MPS and prolong their biological half-lives after intravenous administration [21, 22]. Therefore, PEGylated liposomes are more likely to accumulate in solid tumors via the enhanced permeation and retention (EPR) effect [23, 24]. Nevertheless, in spite of the usefulness and importance of PEGylation, research on PEG-related immune responses has recently emphasized that the injection of PEGylated liposomes into mice, rats, beagle dogs, cynomolgus monkeys, and mini pigs can elicit a PEG-related immune response, exemplified by the extensive production of anti-PEG IgM antibodies, which is responsible for the accelerated clearance of subsequent doses of PEGylated liposomes—the so-called "accelerated blood clearance (ABC)" phenomenon. Therefore, the ABC phenomenon is of clinical concern because it decreases the therapeutic efficacy of encapsulated drugs upon repeated administration.

Since the first report concerning the pharmacokinetic irregularities (the ABC phenomenon) upon repeated administration of PEGylated liposomes by Dams et al. [25], research efforts have focused on elucidating the underlying mechanism of the ABC phenomenon [15–17]. Laverman et al. [15] identified two phases of the ABC phenomenon: the induction phase, in which the host immune system is "primed" by the first injection of PEGylated liposome, and the effectuation phase, in which a subsequent dose of PEGylated liposomes is rapidly opsonized and cleared from systemic circulation, by the Kupffer cells. Later on, in a series of our studies [16, 17, 26–28], we revealed that anti-PEG IgM, which is produced in the spleen in response to the first dose, selectively binds to the PEG upon the second dose of liposomes injected several days later and subsequently activates the complement system, and, as a consequence, the liposomes are taken up by the Kupffer cells (Fig. 25.1).

Thus far, it has been reported that both the occurrence and the magnitude of the ABC induction are influenced by a number of factors including the dose and physicochemical properties of the PEGylated liposomes, the time interval between repeated injections, the animal species, the species of the encapsulated drugs and liposomal composition. Factors affecting the ABC

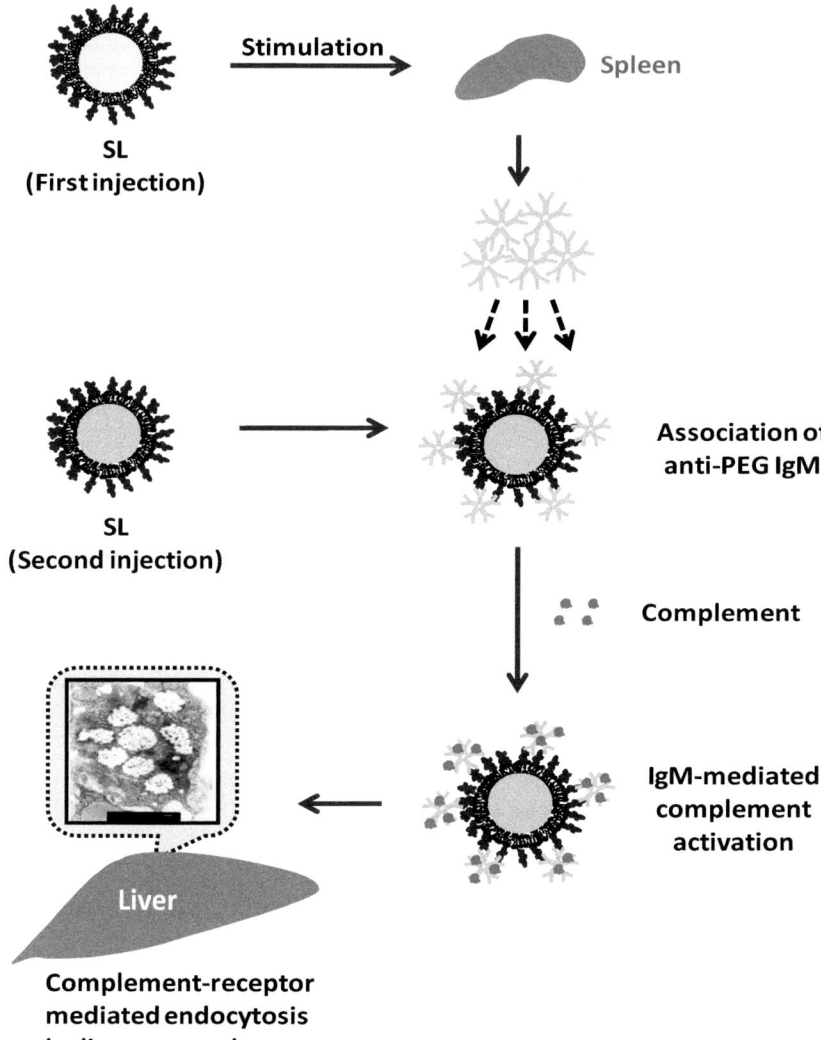

Figure 25.1 Mechanism of the ABC phenomenon.

phenomenon were extensively reviewed in our recent publication [20]. Briefly, the ABC phenomenon is substantially affected by the lipid composition and physicochemical properties (such as size [19, 29]), lipid membrane fluidity [30], PEG density [16, 31] and the terminal group of PEG [31–33]). In addition, the ABC phenomenon is more pronounced at a lower first dose

(0.001–0.1 mmol lipid/kg) than at a higher dose (5 mmol lipid/kg) [34], empty PEGylated liposomes than some drug-loading [35] or special designed gene-loading liposomes [36, 37], slow vascular infusion compared to bolus injection [31] and short time intervals between doses (7 days and 21 days) versus long time intervals (28 days) [38]. Besides, third (or more) injection of PEGylated liposomes has less severe ABC phenomenon due to saturation of MPS cells and/or consumption of anti-PEG IgM in the circulation [39]. Therefore, to attenuate the ABC phenomenon, a bolus injection of higher-dose anticancer drug-loaded PEGylated liposomes with longer dosage intervals is encouraged.

25.2.2 Complement Activation-Related Pseudoallergy

As introduced in other chapters, besides the slow specific immunogenic response manifested in antibody formation against PEGylated liposomes [15, 16, 27], a mounting body of evidence has also revealed that intravenous administration of nanocarriers could provoke acute hypersensitivity reactions (HSR) that are classified as complement (C) activation-related pseudoallergy (CARPA) since they are not initiated/mediated by pre-existing IgE antibodies but rather arise as a consequence of the activation of the complement system [8, 12, 14, 40–42]. Such "hypersensitivity reactions" or "anaphylactoid reactions" typically occur directly at first exposure to the nanocarriers without prior sensitization, and the symptoms usually lessen and/or disappear on later treatment, that is why these reactions have recently been called "pseudoallergic" [40–42]. The symptoms of HSR are mostly minor and transient and include cardiopulmonary distress such as tachypnea, dyspnea, tachycardia, hypertension/hypotension, chest pain and back pain [43]. However, life-threatening or even deadly reactions can occur occasionally in hypersensitive individuals [40–43].

Drugs and agents causing CARPA include radiocontrast media, liposomal drugs (Doxil®, Ambisome®, and DaunoXome®), micellar solvents (e.g., Cremophore EL; the vehicle of Taxol), PEGylated proteins and monoclonal antibodies [40–49]. The first direct evidence for the causal relationship between complement activation and hypersensitivity reactions (CARPA) to PEGylated liposomes was provided by Brouwers et al. [50], who reported that three out of nine patients receiving ^{99m}Tc-labeled PEGylated liposomes for scintigraphic detection of bowel inflammation have developed severe hypersensitivity reactions. Later on, Szebeni and his colleagues, in a series of studies [11, 13, 14, 49], emphasized the potential contribution of complement activation to infusion reactions observed in up to 25% of patients treated with Doxil®. Chanan-Khan et al. [51] also demonstrated that Doxil® activated complement in the majority of patients and induced moderate to severe hypersensitivity reactions in about 50% of cancer patients infused with Doxil® for the first time. Furthermore, CARPA reactions have been elicited with other liposomal formulations regardless their structure or the presence/absence of PEG surface modification [43, 45, 48]. Table 25.1 represents the marketed liposomal drugs that have been reported to cause HSRs along with some of their features and symptoms of hypersensitivity. Currently, CARPA is considered one of the safety issues for nanomedicines, including liposomes, to be evaluated prior the clinical approval of generic intravenous liposomal formulations.

Table 25.1 Hypersensitivity reactions elicited by clinically approved liposomal formulations

Trade name	Encapsulated drug	Pharmacological category	Symptoms
Doxil, Caelyx	Doxorubicin	Cytotoxic agent	Flushing, facial swelling, shortness of breath, headache, chills, back pain, tightness in the chest or throat, hypotension
Ambisome, Abelcet	Amphotericin B	Antifungal agent	Chills, rigors, fever, nausea, vomiting, cardio-respiratory events
DaunoXome	Daunorubicin	Cytotoxic agent	Back pain, flushing, chest tightness
Visudyne	Verteporfin	Photosensitizing agents	Chest pain, syncope, sweating, dizziness, rash, dyspnea, flushing, changes in blood pressure and heart rate, back pain

25.3 Features Distinguishing CARPA from Classical IgE-Mediated Immunity

CARPA symptoms are very similar to those of a common allergic response, but a few features are unique to CARPA, including

   (i) the rise of symptoms upon first exposure;
   (ii) the absence or disappearance of symptoms upon re-exposure;
   (iii) spontaneous resolution of the reaction symptoms;
   (iv) the dependence of reaction magnitude to administration speed; and
   (v) response to steroid and antihistamine premedication.

   Nonetheless, it is notable that CARPA symptoms might be observed in some patients at the second or third treatment. In these cases, immunogenicity of the nanocarrier system itself could be an aggravating factor; immunoglobulin response to the first administration of the nanocarriers.

25.4 Mechanism of CARPA

It has been known for a long time that liposomes can activate the complement system via both innate and acquired immunity [40, 52, 53]. The evidence of a casual role of complement activation in liposome-induced CARPA include, among other observations, that (1) the hemodynamic effects of liposomes were mimicked by the complement activator zymosan; (2) incubation of pig serum with liposomes *in vitro* led to the release of C5a and (3) the specific complement inhibitors, GS1 and sCR1, caused significant inhibition of liposome-induced pulmonary hypertension in pigs [14, 42, 54, 93]. Nevertheless, to date, the exact mechanism that leads to CARPA upon the intravenous administration of liposomal nanocarriers has not been fully explored as the reaction proceeds via a highly complex cascade of molecular and cellular activations and interactions wherein complement activation is an essential, but not necessarily rate-limiting factor [55, 56]. In particular, complement activation can lead to the liberation of anaphylatoxins (such as C3a, C5a), which trigger mast cells, basophils and other phagocytic cells, via their specific receptors, for the secretion of a score of vasoactive mediators, collectively called allergomedins, including PAF, histamine, tryptase, leukotrienes (LTB2, LTC4, LTD4, LTE4, PGD2, TXA2, and TXD4). Some of these allergomedins (e.g., PAF, histamine, tryptase, and TXA2) are preformed and released from the cells immediately upon activation, while others are de novo synthesized and, hence, released gradually. This differential, multistep release of allergomedins from anaphylatoxin-responsive cells may explain the individual variation in the start of clinical symptoms. In the next step of CARPA, allergomedins bind to their respective receptors on endothelial and smooth muscle cells, modifying their function in ways that lead to the symptoms of CARPA (Fig. 25.2) [57].

Nonetheless, in speculating on the nature of triggers of pseudoallergy other than anaphylatoxins, a new hypothesis known as "the double hit hypothesis of CARPA" has been postulated [14]. In this hypothesis, liposome-induced CARPA is thought to arise as a consequence of two hits on allergy-mediating secretory cells (mast cells, basophils, macrophages, etc.); one "hit" being the anaphylatoxin signal, and the second is direct binding of the reaction-trigger drug or particle to these cells via surface receptors that are also linked in the signal transduction network that triggers secretory responses. As shown in Fig. 25.3, following the intravenous administration of liposomes, liposomes trigger the complement system to liberate anaphylatoxins. The binding of anaphylatoxins to their respective receptors (anaphylatoxin receptors, ATR) on mast cells (or pulmonary intravascular macrophages) leads to secretion of vasoactive secondary mediators. Likewise, direct binding of liposomes to these cells via surface receptors or pattern recognition receptors (PRR), possibly Toll-like receptors (TLRs 2 and 4) or other danger-signaling receptors (DSRs) on mast cells and macrophages may trigger a secretory response. As increasingly recognized, TLRs and DSRs recognize molecular arrays that are broadly shared by pathogens (called pathogen-associated molecular patterns, PAMPs), and, hence, they could cross-react with repetitive surface elements on nanomedicines that resemble PAMPs. These "pseudo-PAMPs" could also take part in complement activation, thereby triggering pseudoallergy on two independent pathways. However, the concept of the binding of pseudo-PAMPS on the surface of nanoparticles to TLRs/DSRs and their cooperation with anaphylatoxins in causing pseudoallergy are entirely hypothetical proposals at this time.

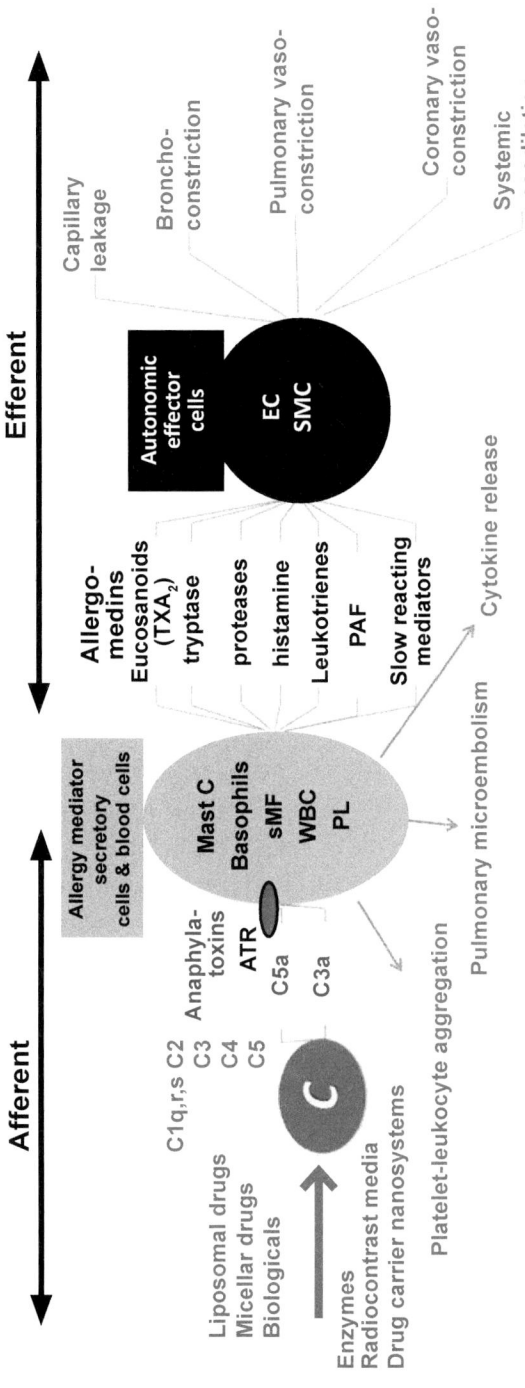

Figure 25.2 The CARPA cascade. The hypothetical scheme illustrates the steps and interactions among a great number of cells and mediators involved in CARPA (modified from [14]).

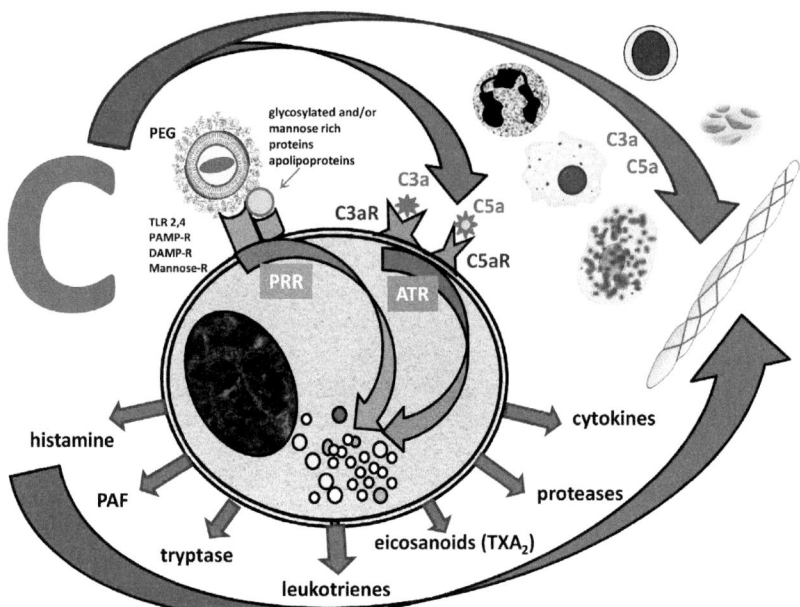

Figure 25.3 The double hit theory of liposome-induced CARPA (modified from [14]).

As mentioned, complement activation per se cannot solely explain liposome-induced pseudoallergy; instead, it may be a trigger or contributing factor, but not the sole cause of HSRs in sensitive individuals. This proposal is based on recent clinical studies showing a nonlinear relationship between complement activation by reactogenic liposomal formulations and HSRs in patients [51] where much higher frequencies of complement activation than of HSR were observed. Such nonlinear relationship between the two phenomena can be rationalized by the complex pathomechanism of CARPA involving numerous redundant activation and intracellular messaging pathways as well as control points (Fig. 25.2). Furthermore, Walsh et al. [58] have reported that Ambisome®, an extremely potent complement activator just like zymosan, induced significant complement activation in all normal human sera *in vitro*, while it elicited CARPA reaction in only 5% of the treated patients. This finding appears consistent with the above concept.

25.5 Factors Affecting Complement Activation by Liposomes

25.5.1 Effect of Liposome Size

Theoretically, antibody binding and complement activation by liposomes should be proportional to the overall surface area of vesicles directly exposed to plasma. Large multilamellar vesicles (LMV) were reported to be highly immunogenic leading to greater activating and pulmonary hypertensive effects, compared to smaller counterparts. Furthermore, small uniform unilamellar vesicles (SUV) are anticipated to reduce the proneness for complement activation and consequently the incidence of CARPA [59].

25.5.2 Effect of Liposome Surface Characteristics

Surface functionalization of liposomes with methoxy-terminated PEG (mPEG) is supposed to substantially suppress/inhibit complement activation [60]. Indeed, such surface modification bestows liposomes with long circulating characteristics via reducing the adsorption of opsonins onto the surfaces of the liposomes, and thus, inhibiting the particle uptake by the cells of the mononuclear phagocyte system (MPS) in the liver and spleen [61–63]. However, in contrast to this general assumption, recent studies have strongly suggested that PEGylated liposomes are capable of activating the complement system via either the classical or the alternative pathways [64, 65]. Complement activation through the classical pathway may be initiated through the binding of naturally occurring antibodies against PEG, which may contribute to the ABC phenomenon; some studies have reported 25% of the general population to be positive for anti-PEG antibodies [66, 67]. In addition to this, PEGylated liposomes may even trigger complement activation independent of anti-PEG antibodies. A recent report by Moghimi et al. [65] demonstrated a crucial role for the anionic phosphate oxygen moiety of the mPEG-phospholipid conjugate incorporated into liposomes in triggering both the alternative and C1q-dependent pathways of the complement system.

25.5.3 Effect of Lipid Composition

The finding that electroneutral large multilamellar vesicles (LMV) composed of only 1,2-dimyristoyl-sn-glycero-3-phosphocholine (DMPC) with and without cholesterol was barely capable of eliciting pulmonary hypertension in pigs [59] is consistent with the general knowledge that charged liposomes are better activators of the complement system than neutral ones [52]. Among the reasons for complement activation by charged and uncharged LMV, the binding of naturally occurring, complement activating anti-phospholipid antibodies was proposed [68], which may also contribute to the ABC of these liposomes. Consequently, such data provided evidence that complement-mediated activation might be an intrinsic property of liposomal phospholipids per se.

Furthermore, cholesterol content has been reported to dictate complement activity potency [59, 69, 70]. The presence of cholesterol in LMV at levels < 45 mol% did not appear to be critical for pulmonary activity [59]. By contrary, Szebeni et al. [59] have reported that 71 mol% cholesterol increased the complement activation and the pulmonary hypertensive effects of liposomes in such a way that small quantities (5 mg) of these vesicles caused immediate death in 50% of tested pigs with symptoms of anaphylactic shock. One possible explanation for this difference is that the excess cholesterol (>45%) in liposomes may become exposed on the membrane surface as patches of microcrystals, and thereby, become increasingly accessible to anti-cholesterol antibodies [59].

25.5.4 Effect of Liposome Morphology

The fact that natural antibodies against liposome-composed lipids including PEGylated lipids and cholesterol show increased binding affinity to liposomes with flat surfaces, due to steric proximity or favorable positioning of epitopes, has been obviously exemplified by the immunogenicity of Doxil® [59]. The finding that Doxil® could elicit a significant increase of SC5b-9 formation, whereas free doxorubicin was not a contributor to this effect, suggests that doxorubicin's complement activating effect might be indirect, via modifying the surface of liposomes [71]. Morphological evidence of the presence of low curvature oval, elongated or irregular liposomes [11], which is considered as an important factor in determining the buildup of multimolecular complexes, such as the C3 convertases (C4b2b or C3bBb) and C5 convertases (C4b2b3b or C3bBb3b), may explain, at least in part, the increased reactogenicity of Doxil®.

25.5.5 Effect of Liposomal Surface Charge

Vesicle surface charge is one of the established factors promoting complement activation by liposomes. As to the molecular details of this effect, Moghimi et al. [65] suggested that the negatively charged phosphor diester moiety of mPEG phospholipid is responsible for enhanced complement activation, at least for the case of mPEGylated dipalmitoyl phosphatidylcholine (DPPC) liposomes. The evidence included the lack of complement activation by non-PEGylated vesicles and the inhibition of complement activation by methylation of this acidic moiety of the mPEG-DPPE [65]. In other studies, it was further shown that the complement activating negative charge needs to be part of anionic phospholipids, as carboxylate-containing negatively charged liposomes, with the same surface charge (zeta potential), failed to induce significant complement activation [72]. These observations suggest that complement activation, which is sensitive to the surface topography of negative charges, and negative surface or zeta potential, per se, may not necessarily imply complement activating potency.

25.5.6 Method and Speed of Intravenous Administration

The finding that administration of small unilamellar vesicles (SUV) by infusion attenuated and/or delayed the pulmonary reaction compared to that observed with bolus injection emphasized the critical role of the method and/or speed of liposomal administration on CARPA. This observation can most easily be rationalized by blood anaphylatoxin levels being rate limiting to pulmonary vasoactivity. As is known, the steady-state levels of functional anaphylatoxins are set by the relative rates of production from plasma C3 and C5, and clearance by cellular receptors and plasma carboxypeptidases [59]. Since clearance is unlikely to be affected by the method of administration, anaphylatoxins level in the blood is likely to depend not only on the potency of vesicles for complement activation but also on the speed by which liposomes enter into blood. This mechanism provides explanation for the reversal of hypersensitivity reactions to liposomes in some patients by slowing down the rate of infusion [73].

25.6 Predictive Tests for CARPA

Because of its potential fatal outcomes, CARPA has been considered as a safety issue in nanopharmacotherapy [44, 57, 64, 65] whose assessment is highly urged. Many *in vitro* testing procedures have been proposed for evaluating the CARPAgenic activity of intravenously administered nanomedicines, similar to those mandated by regulatory agencies for the human application of medical devices [9]. These assays include ELISAs of complement cleavage products (C3a, C5a, C4d, Bb, and SC5b-9), the hemolytic (CH50) complement assay, FACS measurement of basophil activation or cell- or large liposome-bound C3 derivatives, and multiplex bead assays for complement activation by products (Table 25.2). However, none of these *in vitro* tests has been standardized for the assessment of CARPAgenic activity of intravenously administered nanomedicines.

The consequences of CARPA have been studied in many animal models, including rats, dogs, rabbits, and pigs [74–76], as well as non-human primates [76]. Among these models, the porcine CARPA model (Fig. 25.4) appears to most closely mimic human CARPA in terms of reaction kinetics, spectrum of symptoms and the conditions of reaction induction [77–79]. For example, it has been reported that the drug dose of Doxil® that triggers CARPA in pigs corresponds to the infusion reaction-triggering dose in a hypersensitive man [80].

The underlying cause of high sensitivity of pigs has not been clarified. However, the predominance of pulmonary symptoms suggests that the reactions may be related to the high number of pulmonary intravascular macrophages in the microcirculation of porcine lungs [81–83]. Simultaneously, the major hemodynamic changes of CARPA are most likely due to the temporary cardiopulmonary circulation blockade/vasoconstriction caused by the abundant release of thromboxane, other eicosanoids and leukotrienes, histamine, and an additional range of potent vasoactive substances from mast cells and basophil leukocytes [41].

Table 25.2 *In vitro* tests for estimating CARPA

Test	Estimated parameter for CARPA	Advantages	Limitation
ELISA assay of complement activation in normal human serum	Complement activating potency by test material (e.g., nanocarriers) in pooled normal human sera (NHS)	High sensitivity, high specificity as it differentiates between different C activation pathways	There is substantial inter-individual variation in complement response to an activator; the reaction necessitate the testing of multiple sera in order to evaluate the frequency/extent of reaction
ELISA assay of complement activation in whole blood	Complement activating potency by test material (e.g., nanocarriers) in fresh whole blood	Because of the presence of blood cells, it provides a closer model of the human blood than serum/plasma	Care should be taken on the use of anticoagulants as they may affect C activation potential
Hemolytic assay (CH50) of complement activation in animal blood or sera	Complement activating potency by test material (e.g., nanocarriers) in animal blood or serum	Species independent inexpensive assay	Relatively low sensitivity; not applicable for mouse serum
FACS assay of cell- or large liposome-bound C3 derivatives	Measure multiple complement split products	Simultaneous measurement of several complement cleavage products; less time consuming than ELISA	FACS is less accessible technology than ELISA
FACS assay of basophil leukocyte activation in whole blood	Quantitate the basophil leukocyte activating capability by test material (e.g., nanocarriers) in whole blood	High sensitivity	Contribution of IgE-mediated activation to the reaction results; FACS is less accessible technology than ELISA

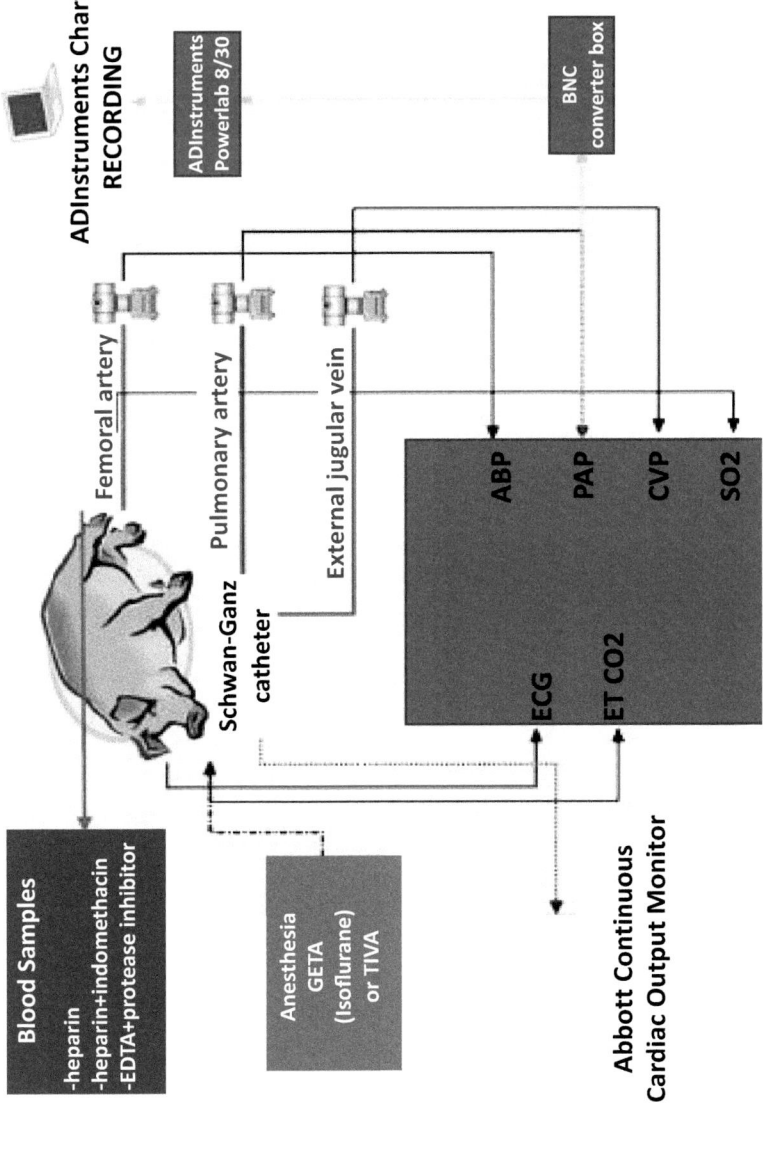

Figure 25.4 Scheme of the porcine CARPA model (modified from [57]).

25.7 Strategies to Attenuate/Abrogate CARPA

25.7.1 Patient Prophylactic Conditioning

Patient pretreatment with anti-inflammatory and anti-allergic agents (such as dexamethasone, ibuprofen, acetaminophen, antihistamines) has been considered one of the most commonly applied approaches to prevent and/or reduce CARPA [84–87]. In addition, it is standard practice to infuse reactogenic medicines slowly, especially at the start of infusion [88, 89]. Nonetheless, such preventive measures do not provide full protection against infusion hypersensitivity reactions, including CARPA.

It is important to raise the possibility that the above strategies of CARPA prevention might also change the extent of the ABC phenomenon. However, to our best knowledge, it has never been shown that CARPA would be associated with ABC, although the two phenomena are likely to be associated, as illustrated in Fig. 25.1. One explanation is that the liposomes that trigger CARPA represent a small fraction of all injected liposomes, and although they clear rapidly from the circulation, this fraction may be too small to detect ABC easily. In contrast, at the time of significant ABC, i.e., after repeated administrations of liposomes when IgM rises against them, ABC should be associated with strong CARPA, at least if the IgM-mediated complement activation background of the ABC phenomenon (Fig. 25.1) holds up in general. Future studies along this line will hopefully address these open questions.

25.7.2 Manipulation of the Physicochemical Properties of Liposomes

The propensity of complement activation by liposomal nanomedicines might be enhanced by certain liposomal characteristics including; increasing size [59], presence of a positive or negative surface charge [65], lack of liposomal homogeneity, presence of aggregates, presence of drugs that can alter liposomal morphology, presence of cholesterol in the bilayer membrane at ≥70% [59], and PEGylation with mPEG-phospholipid conjugate [64, 65]. Based on these findings, neutral small unilamellar vesicles have been shown to be the least reactogenic of the liposomal platforms [90]. Formulation strategies aimed at minimizing the immunogenicity of liposomes include: controlling cholesterol content, methylation of the anionic charge localized on the phosphate oxygen of mPEG-phospholipid conjugate [65] or the use of other non-ionic lipopolymers and lipid conjugates, such as mPEG-substituted synthetic ceramides [91]. However, manipulation of the physicochemical properties of liposomes could lead to altered original pharmacokinetics, loss of efficacy, and the rise of other potential toxicities.

25.7.3 Tolerance Induction

Anaphylaxis prevention by tolerance induction with an empty nanocarrier placebo has been proposed to attenuate/abrogate the incidence of CARPA following the intravenous administration of liposomal platforms [92]. Szebeni et al. [92] have reported that short (15–30 min) infusion of Doxebo (Doxil®-like empty liposomes) to pigs significantly reduced, or entirely prevented the HSRs to subsequently administered Doxil®. In addition, such a protective effect was maintained for at least 24 h post Doxebo administration [92]. They attributed tolerance induction by Doxebo to the possible consumption of an early mediator of CARPA, such as natural anti-PEG antibodies [67], or the down-regulation of a signaling process in cells that mediate the reaction. Accordingly, this proposed approach of tolerance induction might be utilized for other nanomedicine-induced hypersensitivity reactions. In addition, the concept might also be applicable to avoid the ABC phenomenon as antibody decoys (such as Doxebo) before treatment with the real drug (such as Doxil) might clean up anti-PEG IgM from blood circulation and/or saturate the capacity of Kupffer cells for taking up an opsonized second dose of drug-containing PEGylated liposomes. Thus, Doxebo may be available for attenuation of not only the CARPA but also the ABC phenomenon.

25.7.4 Use of Complement Inhibitors

The application of complement inhibitors to ameliorate these HSRs was suggested [93], and strategies to anchor complement inhibitor proteins or their active fragments to liposomal membranes were reported [94, 95]. Most recently, Meszaros et al. [75] investigated the possible use of the natural complement inhibitor factor H (FH) against CARPA. They reported that exogenous FH could inhibit complement activation induced by different CARPAgenic drugs, namely, the antifungal liposomal Amphotericin-B (AmBisome®), the widely used solvent of anticancer drugs, Cremophor EL, and the anticancer monoclonal antibody rituximab *in vitro*. These findings suggested that FH might be considered as a promising strategy for the pharmacological prevention of CARPA. Unfortunately, the function of FH in the ABC phenomenon has not been elucidated yet.

25.8 Conclusions

Besides their unique therapeutic or diagnostic advantages, nanocarrier-based therapeutics seem to share the problem of being recognized by the immune system as foreign materials, which may lead to the rise of adverse reactions and/or loss of efficacy. CARPA and ABC phenomenon are considered two sides of the same coin, both being spin-off phenomena of a vicious cycle of complement activation by nanoparticles, antibody formation and antibody binding to the activator-accelerating complement activation [14]. Therefore, predicting and/or assessing the immunogenic and/or antigenic potential of nanocarrier-based therapeutics could represent an essential safety measure in drug research and development. In addition, exploring certain strategies to attenuate and/or ameliorate the severity of such immunologic reactions; CARPA and ABC phenomenon, will advance the safe use of nanomedicines and biologicals.

Disclosures and Conflict of Interest

This study was supported, in part, by a Grant-in-Aid for Scientific Research (B) 15H04639, the Ministry of Education, Culture, Sports, Science and Technology, Japan and European Union Seventh Framework Program (FP7/2007-2013) grant No. NMP-2012–309820 (NanoAthero) and grant No. NMP-2012–309820 (TheraGlio). The authors declare no conflict of interest. No writing assistance was utilized in the production of this chapter and the authors have received no payment for its preparation.

Corresponding Author

Dr. Tatsuhiro Ishida
Department of Pharmacokinetics and Biopharmaceutics
Institute of Biomedical Sciences, Tokushima University
1-78-1, Sho-machi, Tokushima 770-8505, Japan
Email: ishida@tokushima-u.ac.jp

About the Authors

Amr S. Abu Lila received his master's degree in 2005 in pharmaceutics from the Faculty of Pharmacy, Zagazig University, Egypt, and PhD in pharmaceutical life sciences in 2010 from the Graduate School of Pharmaceutical Sciences, The University of Tokushima, Japan. From 2011 to 2013, he was a JSPS postdoctoral fellow at the Graduate school of Pharmaceutical Sciences, The University of Tokushima. Currently, he is an associate professor in the Faculty of Pharmacy, Zagazig University. He has published more than 30 peer-reviewed scientific journal papers, 5 review articles, and 5 book chapters. His current research focuses on anticancer therapeutics using liposomal drug delivery systems for the selective targeting of either chemotherapeutic agents or genes to tumor tissues.

Janos Szebeni is an immunologist and director of the Nanomedicine Research and Education Center at Semmelweis University, Hungary. He is also founder and CEO of a contract research SME "SeroScience," and full professor of (immune) biology at Miskolc University, Hungary. He has held various academic positions in Hungary and abroad, living in the United States for 22 years before returning to Hungary in 2006. His research on various themes in hematology, membrane biology, nanotechnology, and immunology resulted 150+ peer-reviewed papers, book chapters, patents, etc. (citations:

≈6000, H index: 39), and a book entitled *The Complement System: Novel Roles in Health and Disease* (Kluwer, 2004). Three fields stand out where he has been most active: artificial blood, liposomes, and the complement system. His original works in the 1990s led to the "CARPA" concept delineated in this chapter.

Tatsuhiro Ishida graduated from Tokushima University, Japan, in 1993 and then received his master's degree in 1995 and PhD in 1998 from the Faculty of Pharmaceutical Sciences, Tokushima University, Japan. From 1998 to 2000, he was a postdoctoral fellow at the University of Alberta (Pharmacology), Canada. In 2000, he became a lecturer in Pharmaceutical Sciences at Tokushima University and was promoted to associate professor in 2003. He has been a full professor there since 2014. He has published more than 120 peer-reviewed papers, more than 10 review articles, and 5 book chapters. He has given more than 30 presentations as an invited speaker at international conferences. He is interested in immunological responses against PEGylated therapeutics and the mechanisms behind the response.

References

1. Doi, Y., Abu Lila, A. S., Matsumoto, H., Okada, T., Shimizu, T., Ishida, T. (2016). Improvement of intratumor microdistribution of PEGylated liposome via tumor priming by metronomic S-1 dosing. *Int. J. Nanomed.*, **11**, 5573–5582.

2. Wacker, M. (2013). Nanocarriers for intravenous injection–the long hard road to the market. *Int. J. Pharm.*, **457**, 50–62.

3. Soloman, R., Gabizon, A. A. (2008). Clinical pharmacology of liposomal anthracyclines: Focus on pegylated liposomal Doxorubicin. *Clin. Lymphoma Myeloma*, **8**, 21–32.

4. Gulati, N., Gupta, H. (2011). Parenteral drug delivery: A review. *Recent Pat. Drug Deliv. Formul.*, **5**, 133–145.

5. Mross, K., Niemann, B., Massing, U., Drevs, J., Unger, C., Bhamra, R., Swenson, C. E. (2004). Pharmacokinetics of liposomal doxorubicin (TLC-D99; Myocet) in patients with solid tumors: An open-label, single-dose study. *Cancer Chemother. Pharmacol.*, **54**, 514–524.

6. Slingerland, M., Guchelaar, H. J., Gelderblom, H. (2012). Liposomal drug formulations in cancer therapy: 15 years along the road. *Drug Discov. Today*, **17**, 160–166.
7. Yamamoto, Y., Kawano, I., Iwase, H. (2011). Nab-paclitaxel for the treatment of breast cancer: Efficacy, safety, and approval. *Onco Targets Ther.*, **4**, 123–136.
8. Bedocs, P., Capacchione, J., Potts, L., Chugani, R., Weiszhar, Z., Szebeni, J., Buckenmaier, C. C. (2014). Hypersensitivity reactions to intravenous lipid emulsion in swine: Relevance for lipid resuscitation studies. *Anesth. Analg.*, **119**, 1094–1101.
9. Dezsi, L., Fulop, T., Meszaros, T., Szenasi, G., Urbanics, R., Vazsonyi, C., Orfi, E., Rosivall, L., Nemes, R., Kok, R. J., Metselaar, J. M., Storm, G., Szebeni, J. (2014). Features of complement activation-related pseudoallergy to liposomes with different surface charge and PEGylation: Comparison of the porcine and rat responses. *J. Control. Release*, **195**, 2–10.
10. Orfi, E., Szebeni, J. (2016). The immune system of the gut and potential adverse effects of oral nanocarriers on its function. *Adv. Drug Deliv. Rev.*, **106**, 402–409.
11. Szebeni, J., Bedocs, P., Rozsnyay, Z., Weiszhar, Z., Urbanics, R., Rosivall, L., Cohen, R., Garbuzenko, O., Bathori, G., Toth, M., Bunger, R., Barenholz, Y. (2012). Liposome-induced complement activation and related cardiopulmonary distress in pigs: Factors promoting reactogenicity of Doxil and AmBisome. *Nanomedicine*, **8**, 176–184.
12. Szebeni, J. (2005). Complement activation-related pseudoallergy caused by amphiphilic drug carriers: The role of lipoproteins. *Curr. Drug Deliv.*, **2**, 443–449.
13. Szebeni, J. (2005). Complement activation-related pseudoallergy: A new class of drug-induced acute immune toxicity. *Toxicology*, **216**, 106–121.
14. Szebeni, J. (2014). Complement activation-related pseudoallergy: A stress reaction in blood triggered by nanomedicines and biologicals. *Mol. Immunol.*, **61**, 163–173.
15. Laverman, P., Carstens, M. G., Boerman, O. C., Dams, E. T., Oyen, W. J., van Rooijen, N., Corstens, F. H., Storm, G. (2001). Factors affecting the accelerated blood clearance of polyethylene glycol-liposomes upon repeated injection. *J. Pharmacol. Exp. Ther.*, **298**, 607–612.
16. Ishida, T., Harada, M., Wang, X. Y., Ichihara, M., Irimura, K., Kiwada, H. (2005). Accelerated blood clearance of PEGylated liposomes

following preceding liposome injection: Effects of lipid dose and PEG surface-density and chain length of the first-dose liposomes. *J. Control. Release*, **105**, 305–317.

17. Ishida, T., Ichihara, M., Wang, X., Kiwada, H. (2006). Spleen plays an important role in the induction of accelerated blood clearance of PEGylated liposomes. *J. Control. Release*, **115**, 243–250.

18. Hastings, K. L. (2002). Implications of the new FDA/CDER immunotoxicology guidance for drugs. *Int. Immunopharmacol.*, **2**, 1613–1618.

19. Kaminskas, L. M., McLeod, V. M., Porter, C. J., Boyd, B. J. (2011). Differences in colloidal structure of PEGylated nanomaterials dictate the likelihood of accelerated blood clearance. *J. Pharm. Sci.*, **100**, 5069–5077.

20. Abu Lila, A. S., Kiwada, H., Ishida, T. (2013). The accelerated blood clearance (ABC) phenomenon: Clinical challenge and approaches to manage. *J. Control. Release*, **172**, 38–47.

21. Allen, T. M., Cullis, P. R. (2013). Liposomal drug delivery systems: From concept to clinical applications. *Adv. Drug Deliv. Rev.*, **65**, 36–48.

22. Barenholz, Y. (2012). Doxil(R)–the first FDA-approved nano-drug: Lessons learned. *J. Control. Release*, **160**, 117–134.

23. Yuan, F., Dellian, M., Fukumura, D., Leunig, M., Berk, D. A., Torchilin, V. P., Jain, R. K. (1995). Vascular permeability in a human tumor xenograft: Molecular size dependence and cutoff size. *Cancer Res.*, **55**, 3752–3756.

24. Maeda, H. (2001). The enhanced permeability and retention (EPR) effect in tumor vasculature: The key role of tumor-selective macromolecular drug targeting. *Adv. Enzyme Regul.*, **41**, 189–207.

25. Dams, E. T., Laverman, P., Oyen, W. J., Storm, G., Scherphof, G. L., van Der Meer, J. W., Corstens, F. H., Boerman, O. C. (2000). Accelerated blood clearance and altered biodistribution of repeated injections of sterically stabilized liposomes. *J. Pharmacol. Exp. Ther.*, **292**, 1071–1079.

26. Ishida, T., Maeda, R., Ichihara, M., Mukai, Y., Motoki, Y., Manabe, Y., Irimura, K., Kiwada, H. (2002). The accelerated clearance on repeated injection of pegylated liposomes in rats: Laboratory and histopathological study. *Cell. Mol. Biol. Lett.*, **7**, 286.

27. Ishida, T., Wang, X., Shimizu, T., Nawata, K., Kiwada, H. (2007). PEGylated liposomes elicit an anti-PEG IgM response in a T cell-independent manner. *J. Control. Release*, **122**, 349–355.

28. Shimizu, T., Ishida, T., Kiwada, H. (2013). Transport of PEGylated liposomes from the splenic marginal zone to the follicle in the induction phase of the accelerated blood clearance phenomenon. *Immunobiology*, **218**, 725–732.
29. Koide, H., Asai, T., Hatanaka, K., Urakami, T., Ishii, T., Kenjo, E., Nishihara, M., Yokoyama, M., Ishida, T., Kiwada, H., Oku, N. (2008). Particle size-dependent triggering of accelerated blood clearance phenomenon. *Int. J. Pharm.*, **362**, 197–200.
30. Xu, H., Ye, F., Hu, M., Yin, P., Zhang, W., Li, Y., Yu, X., Deng, Y. (2015). Influence of phospholipid types and animal models on the accelerated blood clearance phenomenon of PEGylated liposomes upon repeated injection. *Drug Deliv.*, **22**, 598–607.
31. Li, C., Cao, J., Wang, Y., Zhao, X., Deng, C., Wei, N., Yang, J., Cui, J. (2012). Accelerated blood clearance of pegylated liposomal topotecan: Influence of polyethylene glycol grafting density and animal species. *J. Pharm. Sci.*, **101**, 3864–3876.
32. Sherman, M. R., Williams, L. D., Sobczyk, M. A., Michaels, S. J., Saifer, M. G. (2012). Role of the methoxy group in immune responses to mPEG-protein conjugates. *Bioconjug. Chem.*, **23**, 485–499.
33. Shimizu, T., Ichihara, M., Yoshioka, Y., Ishida, T., Nakagawa, S., Kiwada, H. (2012). Intravenous administration of polyethylene glycol-coated (PEGylated) proteins and PEGylated adenovirus elicits an anti-PEG immunoglobulin M response. *Biol. Pharm. Bull.*, **35**, 1336–1342.
34. Wang, X., Ishida, T., Kiwada, H. (2007). Anti-PEG IgM elicited by injection of liposomes is involved in the enhanced blood clearance of a subsequent dose of PEGylated liposomes. *J. Control. Release*, **119**, 236–244.
35. La-Beck, N. M., Zamboni, B. A., Gabizon, A., Schmeeda, H., Amantea, M., Gehrig, P. A., Zamboni, W. C. (2012). Factors affecting the pharmacokinetics of pegylated liposomal doxorubicin in patients. *Cancer Chemother. Pharmacol.*, **69**, 43–50.
36. Tagami, T., Nakamura, K., Shimizu, T., Yamazaki, N., Ishida, T., Kiwada, H. (2010). CpG motifs in pDNA-sequences increase anti-PEG IgM production induced by PEG-coated pDNA-lipoplexes. *J. Control. Release*, **142**, 160–166.
37. Tagami, T., Uehara, Y., Moriyoshi, N., Ishida, T., Kiwada, H. (2011). Anti-PEG IgM production by siRNA encapsulated in a PEGylated lipid nanocarrier is dependent on the sequence of the siRNA. *J. Control. Release*, **151**, 149–154.

38. Li, C., Zhao, X., Wang, Y., Yang, H., Li, H., Tian, W., Yang, J., Cui, J. (2013). Prolongation of time interval between doses could eliminate accelerated blood clearance phenomenon induced by pegylated liposomal topotecan. *Int. J. Pharm.*, **443**, 17–25.

39. Suzuki, T., Ichihara, M., Hyodo, K., Yamamoto, E., Ishida, T., Kiwada, H., Ishihara, H., Kikuchi, H. (2012). Accelerated blood clearance of PEGylated liposomes containing doxorubicin upon repeated administration to dogs. *Int. J. Pharm.*, **436**, 636–643.

40. Szebeni, J. (2001). Complement activation-related pseudoallergy caused by liposomes, micellar carriers of intravenous drugs, and radiocontrast agents. *Crit. Rev. Ther. Drug Carrier Syst.*, **18**, 567–606.

41. Szebeni, J., Baranyi, L., Savay, S., Bodo, M., Milosevits, J., Alving, C.R., Bunger, R. (2006). Complement activation-related cardiac anaphylaxis in pigs: Role of C5a anaphylatoxin and adenosine in liposome-induced abnormalities in ECG and heart function. *Am. J. Physiol. Heart Circ. Physiol.*, **290**, H1050–H1058.

42. Szebeni, J., Storm, G. (2015). Complement activation as a bioequivalence issue relevant to the development of generic liposomes and other nanoparticulate drugs. *Biochem. Biophys. Res. Commun.*, **468**, 490–497.

43. Alberts, D. S., Garcia, D. J. (1997). Safety aspects of pegylated liposomal doxorubicin in patients with cancer. *Drugs*, **54**(Suppl 4), 30–35.

44. Andersen, A. J., Hashemi, S. H., Andresen, T. L., Hunter, A. C., Moghimi, S. M. (2009). Complement: Alive and kicking nanomedicines. *J. Biomed. Nanotechnol.*, **5**, 364–372.

45. Sculier, J. P., Coune, A., Brassinne, C., Laduron, C., Atassi, G., Ruysschaert, J. M., Fruhling, J. (1986). Intravenous infusion of high doses of liposomes containing NSC 251635, a water-insoluble cytostatic agent. A pilot study with pharmacokinetic data. *J. Clin. Oncol.*, **4**, 789–797.

46. Levine, S. J., Walsh, T. J., Martinez, A., Eichacker, P. Q., Lopez-Berestein, G., Natanson, C. (1991). Cardiopulmonary toxicity after liposomal amphotericin B infusion. *Ann. Intern. Med.*, **114**, 664–666.

47. Ringden, O., Andstrom, E., Remberger, M., Svahn, B. M., Tollemar, J. (1994). Allergic reactions and other rare side-effects of liposomal amphotericin. *Lancet*, **344**, 1156–1157.

48. Laing, R. B., Milne, L. J., Leen, C. L., Malcolm, G. P., Steers, A. J. (1994). Anaphylactic reactions to liposomal amphotericin. *Lancet*, **344**, 682.

49. Uziely, B., Jeffers, S., Isacson, R., Kutsch, K., Wei-Tsao, D., Yehoshua, Z., Libson, E., Muggia, F. M., Gabizon, A. (1995). Liposomal doxorubicin: Antitumor activity and unique toxicities during two complementary phase I studies. *J. Clin. Oncol.*, **13**, 1777–1785.

50. Brouwers, A. H., De Jong, D. J., Dams, E. T., Oyen, W. J., Boerman, O. C., Laverman, P., Naber, T. H., Storm, G., Corstens, F. H. (2000). Tc-99m-PEG-Liposomes for the evaluation of colitis in Crohn's disease. *J. Drug Target.*, **8**, 225–233.

51. Chanan-Khan, A., Szebeni, J., Savay, S., Liebes, L., Rafique, N. M., Alving, C. R., Muggia, F. M. (2003). Complement activation following first exposure to pegylated liposomal doxorubicin (Doxil): Possible role in hypersensitivity reactions. *Ann. Oncol.*, **14**, 1430–1437.

52. Szebeni, J. (1998). The interaction of liposomes with the complement system. *Crit. Rev. Ther. Drug Carrier Syst.*, **15**, 57–88.

53. Szebeni, J., Baranyi, L., Savay, S., Milosevits, J., Bodo, M., Bunger, R., Alving, C. R. (2003). The interaction of liposomes with the complement system: In vitro and in vivo assays. *Methods Enzymol.*, **373**, 136–154.

54. Jackman, J. A., Meszaros, T., Fulop, T., Urbanics, R., Szebeni, J., Cho, N. J. (2016). Comparison of complement activation-related pseudoallergy in miniature and domestic pigs: Foundation of a validatable immune toxicity model. *Nanomedicine*, **12**, 933–943.

55. Moghimi, S. M., Hamad, I. (2008). Liposome-mediated triggering of complement cascade. *J. Liposome Res.*, **18**, 195–209.

56. Moghimi, S. M., Andersen, A. J., Hashemi, S. H., Lettiero, B., Ahmadvand, D., Hunter, A. C., Andresen, T. L., Hamad, I., Szebeni, J. (2010). Complement activation cascade triggered by PEG-PL engineered nanomedicines and carbon nanotubes: The challenges ahead. *J. Control. Release*, **146**, 175–181.

57. Szebeni, J., Muggia, F., Gabizon, A., Barenholz, Y. (2011). Activation of complement by therapeutic liposomes and other lipid excipient-based therapeutic products: Prediction and prevention. *Adv. Drug Deliv. Rev.*, **63**, 1020–1030.

58. Walsh, T. J., Finberg, R. W., Arndt, C., Hiemenz, J., Schwartz, C., Bodensteiner, D., Pappas, P., Seibel, N., Greenberg, R. N., Dummer, S., Schuster, M., Holcenberg, J. S. (1999). Liposomal amphotericin B for empirical therapy in patients with persistent fever and neutropenia. National Institute of Allergy and Infectious Diseases Mycoses Study Group. *N. Engl. J. Med.*, **340**, 764–771.

59. Szebeni, J., Baranyi, L., Savay, S., Bodo, M., Morse, D. S., Basta, M., Stahl, G. L., Bunger, R., Alving, C. R. (2000). Liposome-induced pulmonary hypertension: Properties and mechanism of a complement-mediated pseudoallergic reaction. *Am. J. Physiol. Heart Circ. Physiol.*, **279**, H1319–H1328.

60. Bradley, A. J., Devine, D. V., Ansell, S. M., Janzen, J., Brooks, D. E. (1998). Inhibition of liposome-induced complement activation by incorporated poly(ethylene glycol)-lipids. *Arch. Biochem. Biophys.*, **357**, 185–194.

61. Allen, C., Dos Santos, N., Gallagher, R., Chiu, G. N., Shu, Y., Li, W. M., Johnstone, S. A., Janoff, A. S., Mayer, L. D., Webb, M. S., Bally, M. B. (2002). Controlling the physical behavior and biological performance of liposome formulations through use of surface grafted poly(ethylene glycol). *Biosci. Rep.*, **22**, 225–250.

62. Torchilin, V. P. (1998). Polymer-coated long-circulating microparticulate pharmaceuticals. *J. Microencapsul.*, **15**, 1–19.

63. Lasic, D. D., Martin, F. J., Gabizon, A., Huang, S. K., Papahadjopoulos, D. (1991). Sterically stabilized liposomes: A hypothesis on the molecular origin of the extended circulation times. *Biochim. Biophys. Acta*, **1070**, 187–192.

64. Moghimi, S. M., Szebeni, J. (2003). Stealth liposomes and long circulating nanoparticles: Critical issues in pharmacokinetics, opsonization and protein-binding properties. *Prog. Lipid Res.*, **42**, 463–478.

65. Moghimi, S. M., Hamad, I., Andresen, T. L., Jorgensen, K., Szebeni, J. (2006). Methylation of the phosphate oxygen moiety of phospholipid-methoxy(polyethylene glycol) conjugate prevents PEGylated liposome-mediated complement activation and anaphylatoxin production. *FASEB J.*, **20**, 2591–2593.

66. Armstrong, J. K., Hempel, G., Koling, S., Chan, L. S., Fisher, T., Meiselman, H. J., Garratty, G. (2007). Antibody against poly(ethylene glycol) adversely affects PEG-asparaginase therapy in acute lymphoblastic leukemia patients. *Cancer*, **110**, 103–111.

67. Richter, A. W., Akerblom, E. (1984). Polyethylene glycol reactive antibodies in man: Titer distribution in allergic patients treated with monomethoxy polyethylene glycol modified allergens or placebo, and in healthy blood donors. *Int. Arch. Allergy Appl. Immunol.*, **74**, 36–39.

68. Olds, L. C., Miller, 3rd, J. J. (1990). C3 activation products correlate with antibodies to lipid A in pauciarticular juvenile arthritis. *Arthritis Rheum.*, **33**, 520–524.

69. Moein Moghimi, S., Hamad, I., Bunger, R., Andresen, T. L., Jorgensen, K., Hunter, A. C., Baranji, L., Rosivall, L., Szebeni, J. (2006). Activation of the human complement system by cholesterol-rich and PEGylated liposomes-modulation of cholesterol-rich liposome-mediated complement activation by elevated serum LDL and HDL levels. *J. Liposome Res.*, **16**, 167–174.

70. Wassef, N. M., Johnson, S. H., Graeber, G. M., Swartz, Jr., G. M., Schultz, C. L., Hailey, J. R., Johnson, A. J., Taylor, D. G., Ridgway, R. L., Alving, C. R. (1989). Anaphylactoid reactions mediated by autoantibodies to cholesterol in miniature pigs. *J. Immunol.*, **143**, 2990–2995.

71. Pedersen, M. B., Zhou, X., Larsen, E. K., Sorensen, U. S., Kjems, J., Nygaard, J. V., Nyengaard, J. R., Meyer, R. L., Boesen, T., Vorup-Jensen, T. (2010). Curvature of synthetic and natural surfaces is an important target feature in classical pathway complement activation. *J. Immunol.*, **184**, 1931–1945.

72. Sou, K., Tsuchida, E. (2008). Electrostatic interactions and complement activation on the surface of phospholipid vesicle containing acidic lipids: Effect of the structure of acidic groups. *Biochim. Biophys. Acta*, **1778**, 1035–1041.

73. Dezube, B. J. (1996). Clinical presentation and natural history of AIDS-related Kaposi's sarcoma. *Hematol. Oncol. Clin. North Am.*, **10**, 1023–1029.

74. Wu, J., Wu, Y. Q., Ricklin, D., Janssen, B. J., Lambris, J. D., Gros, P. (2009). Structure of complement fragment C3b-factor H and implications for host protection by complement regulators. *Nat. Immunol.*, **10**, 728–733.

75. Meszaros, T., Csincsi, A. I., Uzonyi, B., Hebecker, M., Fulop, T. G., Erdei, A., Szebeni, J., Jozsi, M. (2016). Factor H inhibits complement activation induced by liposomal and micellar drugs and the therapeutic antibody rituximab in vitro. *Nanomedicine*, **12**, 1023–1031.

76. Goicoechea de Jorge, E., Caesar, J. J., Malik, T. H., Patel, M., Colledge, M., Johnson, S., Hakobyan, S., Morgan, B. P., Harris, C. L., Pickering, M. C., Lea, S. M. (2013). Dimerization of complement factor H-related proteins modulates complement activation in vivo. *Proc. Natl. Acad. Sci. U. S. A.*, **110**, 4685–4690.

77. Kopp, A., Hebecker, M., Svobodova, E., Jozsi, M. (2012). Factor h: A complement regulator in health and disease, and a mediator of cellular interactions. *Biomolecules*, **2**, 46–75.

78. Szebeni, J., Bedocs, P., Csukas, D., Rosivall, L., Bunger, R., Urbanics, R. (2012). A porcine model of complement-mediated infusion reactions to drug carrier nanosystems and other medicines. *Adv. Drug Deliv. Rev.*, **64**, 1706–1716.

79. Kajander, T., Lehtinen, M. J., Hyvarinen, S., Bhattacharjee, A., Leung, E., Isenman, D. E., Meri, S., Goldman, A., Jokiranta, T. S. (2011). Dual interaction of factor H with C3d and glycosaminoglycans in host-nonhost discrimination by complement. *Proc. Natl. Acad. Sci. U. S. A.*, **108**, 2897–2902.

80. Tortajada, A., Yebenes, H., Abarrategui-Garrido, C., Anter, J., Garcia-Fernandez, J. M., Martinez-Barricarte, R., Alba-Dominguez, M., Malik, T. H., Bedoya, R., Cabrera Perez, R., Lopez Trascasa, M., Pickering, M. C., Harris, C. L., Sanchez-Corral, P., Llorca, O., Rodriguez de Cordoba, S. (2013). C3 glomerulopathy-associated CFHR1 mutation alters FHR oligomerization and complement regulation. *J. Clin. Invest.*, **123**, 2434–2446.

81. Schneberger, D., Aharonson-Raz, K., Singh, B. (2012). Pulmonary intravascular macrophages and lung health: What are we missing?. *Am. J. Physiol. Lung Cell. Mol. Physiol.*, **302**, L498–L503.

82. Gaca, J. G., Palestrant, D., Lukes, D. J., Olausson, M., Parker, W., Davis, Jr., R. D. (2003). Prevention of acute lung injury in swine: Depletion of pulmonary intravascular macrophages using liposomal clodronate. *J. Surg. Res.*, **112**, 19–25.

83. Sierra, M. A., Carrasco, L., Gomez-Villamandos, J. C., Martin de las Mulas, J., Mendez, A., Jover, A. (1990). Pulmonary intravascular macrophages in lungs of pigs inoculated with African swine fever virus of differing virulence. *J. Comp. Pathol.*, **102**, 323–334.

84. Szebeni, J. (2004). Hypersensitivity reactions to radiocontrast media: The role of complement activation. *Curr. Allergy Asthma Rep.*, **4**, 25–30.

85. Small, P., Satin, R., Palayew, M. J., Hyams, B. (1982). Prophylactic antihistamines in the management of radiographic contrast reactions. *Clin. Allergy*, **12**, 289–294.

86. Greenberger, P. A. (1984). Contrast media reactions. *J. Allergy Clin. Immunol.*, **74**, 600–605.

87. Lasser, E. C., Lang, J. H., Lyon, S. G., Hamblin, A. E. (1979). Complement and contrast material reactors. *J. Allergy Clin. Immunol.*, **64**, 105–112.

88. Kang, S. P., Saif, M. W. (2007). Infusion-related and hypersensitivity reactions of monoclonal antibodies used to treat colorectal cancer–

identification, prevention, and management. *J. Support. Oncol.*, **5**, 451–457.

89. Lenz, H. J. (2007). Management and preparedness for infusion and hypersensitivity reactions. *Oncologist*, **12**, 601–609.

90. Sercombe, L., Veerati, T., Moheimani, F., Wu, S. Y., Sood, A. K., Hua, S. (2015). Advances and challenges of liposome assisted drug delivery. *Front. Pharmacol.*, **6**, 286.

91. Webb, M. S., Saxon, D., Wong, F. M., Lim, H. J., Wang, Z., Bally, M. B., Choi, L. S., Cullis, P. R., Mayer, L. D. (1998). Comparison of different hydrophobic anchors conjugated to poly(ethylene glycol): Effects on the pharmacokinetics of liposomal vincristine. *Biochim. Biophys. Acta*, **1372**, 272–282.

92. Szebeni, J., Bedocs, P., Urbanics, R., Bunger, R., Rosivall, L., Toth, M., Barenholz, Y. (2012). Prevention of infusion reactions to PEGylated liposomal doxorubicin via tachyphylaxis induction by placebo vesicles: A porcine model. *J. Control. Release*, **160**, 382–387.

93. Szebeni, J., Fontana, J. L., Wassef, N. M., Mongan, P. D., Morse, D. S., Dobbins, D. E., Stahl, G. L., Bunger, R., Alving, C. R. (1999). Hemodynamic changes induced by liposomes and liposome-encapsulated hemoglobin in pigs: A model for pseudoallergic cardiopulmonary reactions to liposomes. Role of complement and inhibition by soluble CR1 and anti-C5a antibody. *Circulation*, **99**, 2302–2309.

94. Smith, R. A. (2002). Targeting anticomplement agents. *Biochem. Soc. Trans.*, **30**, 1037–1041.

95. Smith, G. P., Smith, R. A. (2001). Membrane-targeted complement inhibitors. *Mol. Immunol.*, **38**, 249–255.

Chapter 26

Current and Rising Concepts in Immunotherapy: Biopharmaceuticals versus Nanomedicines

Matthias Bartneck, PhD

Department of Medicine III, Medical Faculty, RWTH Aachen, Germany

Keywords: macrophages, immune cells, leukocytes, biopharmaceuticals, nanomedicine, drug delivery, small molecules, nanocarriers, polymers, vascular endothelial growth factor a (VEGFA), natural killer (NK) cells, myeloid-derived suppressor cells, T helper 2 (Th2) cells, regulatory T (T_{Reg}) cells, nanoantibodies, aptamers, micro-RNAs (miR), reticuloendothelial cells (RES), interlukin 1 β (IL1β), tumor necrosis factor (TNF), dendritic cells (DC), toll-like receptor (TLR)

26.1 Immunity in Inflammatory Disease and Cancer

Immunity is the ability of an organism to repel disease by recognizing and neutralizing foreign substances and pathogens. It is made of organs, cells, and soluble compounds, which protect the individual from injury, which is in large part based on the

---

Immune Aspects of Biopharmaceuticals and Nanomedicines
Edited by Raj Bawa, János Szebeni, Thomas J. Webster, and Gerald F. Audette
Copyright © 2018 Pan Stanford Publishing Pte. Ltd.
ISBN 978-981-4774-52-9 (Hardcover), 978-0-203-73153-6 (eBook)
www.panstanford.com

recognition of "self" and "non-self." Its cells and molecules are also referred to as cellular and humoral immunity, respectively. For decades, the origin of immune cells was thought to be the bone marrow only. Notably, during the past decade, scientists have discovered that tissue-resident macrophages (MΦ) originate in large parts from local tissue stem cells, but not from circulating monocytes [1]. Yet, it is clear that MΦ play crucial roles in initiating and resolving inflammation in different organs. One of the most prominent mediators generated by MΦ is the tumor necrosis factor (TNF), which covers a broad line of biological functions and, in many cases, is involved in cell and organ injury [2].

In many cases, TNF leads to necroptotic parenchymal cell death; for instance, keratinocytes form skin barriers, whereas hepatocytes form the parenchyma of the liver. Throughout different organs, many parenchymal cells are sensitive to MΦ-derived TNF, which fuels many types of inflammatory disease, including liver inflammation [3], rheumatoid arthritis [4], autoimmune disease [5], and psoriasis [6]. Biopharmaceuticals such as infliximab have changed the course of many chronic inflammatory diseases but are among the most expensive pharmaceuticals [7]. Under normal conditions, acute inflammation is terminated by a healing phase, whereas in chronic inflammation, immune cell mediators such as TNF lead to continuous tissue injury resulting in organ fibrosis and loss of function [8]. Fibrotic diseases with excessive extracellular matrix (ECM) growth are exerted by fibroblast-like cells. In the liver, these are hepatic stellate cells [9], and corresponding cell types also mediate skin [10] and cardiac scarring [11]. Fibroblasts produce different types of ECM proteins, the most abundant and popular ones being collagen. During fibrosis, collagen I is abundantly expressed [3], whereas in cancer, specific collagens, such as collagen IV, are generated, which stimulate cancer cell proliferation [12]. In summary, all organs have specific types of parenchymal, fibroblast-like, and immune cell subsets, which interact in disease.

Owed to the fact that they infiltrate injured organs, leukocytes of different organs share major similarities [13]. Generally, one can basically differentiate lymphoid and myeloid immune cells and both have certain major modes of activation. Three main types of lymphocytes have been distinguished: natural killer (NK) cells,

B lymphocytes, and T lymphocytes. NK cells belong to the innate immune cells and one of their major functions is killing abnormal or cancer cells that are missing MHC I markers and that cannot be recognized as abnormal by other cell types. T and B lymphocytes can be activated toward a specific antigen only and thus exhibit a high level of specificity.

In contrast to B and T lymphocytes, which augment their functions based on clonal expansion and modifications of somatic genes, myeloid cells such as MΦ exhibit a high level of plasticity encoded in their DNA and can be polarized into diverse distinct subtypes. Importantly, human MΦ only proliferate under very specific conditions. Despite this fact, there are many reports from studies based on mouse experiments that suggest human MΦ are also highly proliferative [14]. These "big eaters" are strongly influenced by their microenvironment and it is advisable to classify the different subpopulations using gene expression, surface markers, and specifically, cytokine expression to understand their functions in disease [15]. The most popular distinction in MΦ dichotomy is to classify them as inflammatory macrophages (M1-MΦ) or anti-inflammatory M2-MΦ, based on their expression of specific markers [16]. Outbalances between the M1 and the M2 subtype appear in different diseases: M1-MΦ are overrepresented in inflammatory diseases, whereas certain subtypes of M2-MΦ support tumor growth.

Immune cells produce a huge spectrum of compounds named cytokines acting on other cell types and tissues, which constitutes a major part of humoral immunity. Serum cytokine levels are important in diagnostics as they represent potential indicators of organ injury. There is evidence that cancerogenesis is influenced by different MΦ subsets, but the roles of the subsets are not fully understood. Several reports demonstrated that M2-polarized MΦ are frequent in cancer, which can be designated as type 2 tumor-associated macrophages (TAM2). The TAM2 and other immune cells exhibit immunosuppressive functions, such as myeloid-derived suppressor cells, T helper 2 (Th2) cells, and regulatory T (T_{Reg}) cells, which are overrepresented in many types of cancer allowing for immune evasion of tumors [17]. Similar to M2-MΦ and TAM2, also T_{Reg} produce large amounts of interleukin 10 (IL10) [18]. M2-MΦ additionally produce many other anti-inflammatory cytokines such as the CC chemokine ligand 18

(CCL18), transforming growth factor β (TGF-β), or interleukin 1 receptor antagonist (IL1RA), depending on the subset of M2-MΦ [19]. Currently our unpublished research in experimental liver cancer has demonstrated that in addition to the TAM2, there are also inflammatory TAM, which one might term TAM1, which are highly distinctive from TAM2 and from MDSC in mRNA profiles.

Furthermore, certain MΦ subtypes generate proangiogenic factors such as the vascular endothelial growth factor a (VEGFA). Our earlier studies have demonstrated that infiltrating inflammatory monocytes express VEGFA in chronic liver disease and thereby generate new blood vessels. It is obvious that the simplified scheme in Fig. 26.1 is just an excerpt of a few aspects of inflammatory diseases and cancer since there are indeed several exceptions and additional specifications from the simplified scheme, considering, for example leukemia, which is a type of cancer driven by proliferating immune cells. Yet, there are many basic similarities of inflammatory diseases representing important basic concepts for immunomodulation mostly independent of the targeted organ (Fig. 26.1). This review shows selected current immunotherapies for inflammatory disease and cancer, addressing selected biopharmaceuticals, small molecules, and novel nanomedicines (NMs).

26.2 Nanomedicine

26.2.1 Design of Nanomedicine

The tools of nanotechnology allow for a huge variety of novel drugs. Virtually all biological macromolecules (carbohydrates, lipids, proteins, and nucleic acids) are based on polymeric building units. Carbohydrates play important roles in energy saving and also represent important constituents of biological matrices, in being part of glycoproteins. NMs have adopted several miniaturizations from larger macromolecules. In order to miniaturize glycoproteins, functional sugar carbohydrate groups might be used separately from their protein part. As an example for reducing size but retaining functionality, selectins (glycoproteins) can be reduced to their carbohydrate part and retain biological functionality [20]. In a recent study, we analyzed the efficiency of selectin-mimetic

NM by using a selectin-derived tetrasaccharide sialyl-Lewis^X anchored to a polymer backbone. The construct was able to inhibit the migration of human primary MΦ based on the carbohydrate part only [20] and exhibited therapeutic effects in inflammatory liver disease where it targeted resident hepatic macrophages (Kupffer cells) [21].

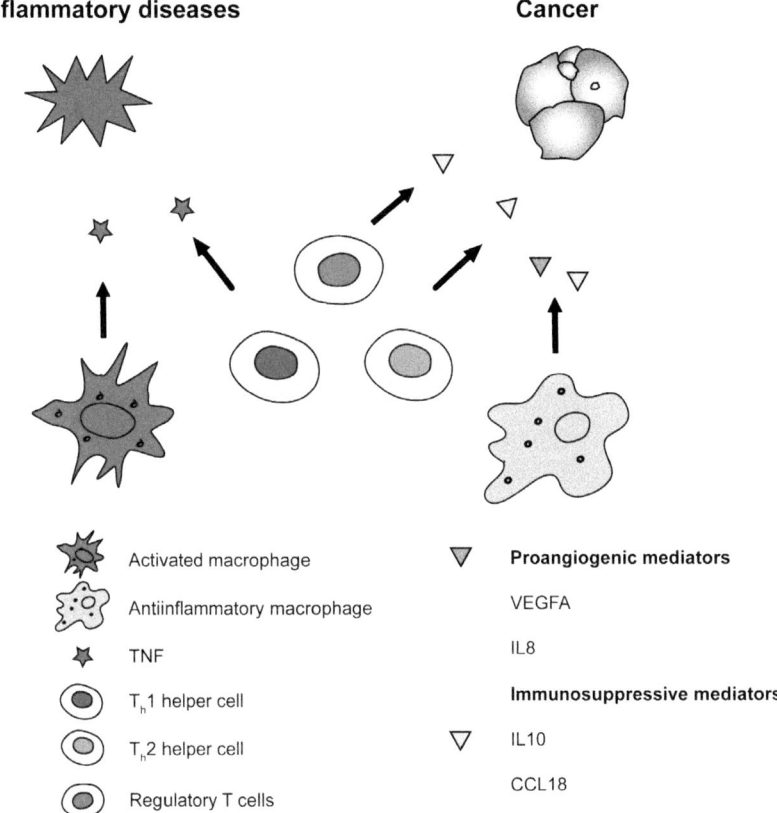

Figure 26.1 Leukocytes are drivers of inflammation and cancer. Based on their secretion of mediators such as cytokines, leukocytes have a major impact on the progression of disease. Lymphocytes exert their influence based on clonal proliferation whereas macrophages exhibit a high level of plasticity.

Lipid-based drugs are an important transport vehicle, for instance, for chemotherapeutics, with phospholipids being among the most common lipid molecule in cell membranes. Owing to this property, phospholipids are useful in designing fully biocompatible drug delivery systems [22, 23], or as parts of the membranes of artificial cells. Using lipids for drug encapsulation dramatically changes the pharmacokinetics of drugs such as the significantly increased circulation of doxorubicin, pioneered by Gabizon and colleagues in the 1990s [24], a method allowing for the enhanced permeability and retention (EPR) effect of NM [25].

Protein-based biopharmaceuticals have reached a quarter trillion market volume in 2017 and by 2020, are assumed to generate US$290 billion in revenue and to comprise 27% of the pharmaceutical market [111]. There are diverse strategies to replace the complex functionality of antibodies by using smaller versions of proteins or molecules with similar properties. Some of these small structures are peptides, nanoantibodies, and aptamers. The earliest attempt to miniaturize antibodies was to extract isolated short peptide sequences. The tripeptide arginine-glycine-asparagine (RGD) is derived from the protein fibronectin, with a size of 350 Da only, and probably is the most prominent short bioactive peptide. Similar to its mother protein, it binds to integrins at the cell surface and thereby mediates the attachment of many cell types, which otherwise cannot attach to cell-repellant materials such as PEGylated, poly acrylic acid (PAA)-coated, or fluorinated and hydrophobic substrates. Notably, MΦ exhibit an outstanding capability of adhesion to various substrates [26, 27], whereas many cell types, including fibroblasts, require specific motifs such as RGD for proper attachment and stretching [28]. While peptides have been studied for many years, current attempts increasingly focus on nanobodies and aptamers to replace monoclonal antibodies for structural targeting.

Miniaturized motifs are of interest for decorating NMs to evoke targeting specificity and already enrich many microscopical and imaging methods [29]. Nanobodies have a size ranging 2–4 nm and molecular weight of about 12–15 kDa, comprising about 100–120 amino acids [29]. They are composed of the functional part of antibodies. Aptamers have some major advantage over conventional proteins and antibody-based targeting because they are more heat-resistant and less sensitive to contaminations with bacteria than proteins. They are chemically produced based on self-folding oligonucleotides with 15–100 nucleotides [30] of size 2–3 nm and weight 10–15 kDa. Size reductions represent a great advantage in microscopical and imaging methods, since some small antigens only have 10 binding sites [29]. In contrast to their condensed versions (peptides, nanoantibodies, many natural antibodies exhibit a molecular weight of about 150 kDa and a size of approximately 10 nm [31].

Nucleic acids are of major importance for conserving and transmitting genetic material and allow NM to transmit information. In the classical dogma of molecular biology, genetic information encoded by DNA is transcribed into mRNA, which is translated into a protein. During the beginning of this century, small non-coding RNAs were discovered. Among these, specifically, micro-RNAs (miR) are tiny but strong molecules usually of size about 20 nucleotides only, but exhibit strong, regulatory effects on mRNA [32]. Bioinformaticians have demonstrated that miR regulates approximately 60% of protein-coding genes [33]. The action of miR is based on binding to the 3′ untranslated region of the target mRNA. However, due to the fact that the complementarities of miR to mRNA sequences occurs only in parts, some miR can bind to multiple different mRNAs [32]. MiR was shown to play an important role in the regulation of inflammatory liver disease, as demonstrated by Roy and colleagues [34], and thus the modulation of miR activity is a useful tool for versatile NM, both as a target and as a component to deliver.

26.2.2 Smart Nanomedicines

In order to increase therapeutic efficiency, smart features should be adopted into novel NM. The first nanomedical drug, liposomal doxorubicin (Doxil®), had several smart features already in 1995 when it was approved by the FDA. Its PEGylation evoked prolonged circulation and avoidance of reticuloendothelial cell (RES) clearance. The loading principle allowed for drug release to nearby tumors and, having a liposome lipid bilayer in a liquid ordered phase (liquid crystalline exhibiting a melting temperature of 53°C) [35]. Doxil is "passively targeted" to tumors, based on the EPR effect, and its cargo is released and becomes available to tumor cells by sequestration in the lysosome.

Importantly, disease conditions severely affect temperature and pH: At inflammatory sites and during fever, the temperature increases [36]. In wounds [37], and in the skin [37], the pH is below the neutral pH of 7. Therefore, NM should be adapted to evoke a controlled release at its site of action, mediated for example, by temperature and pH-dependent induction of its release from the carrier substance.

Depending on the targeted disease, NM can further be modified to be responsive to many other stimuli. Responsiveness can be achieved toward redox potential, for example, to glutathione, which increases in cancer cells [38], to enzymes which catalyze a specific reaction near the site of the lesion [39], or they can further be photosensitive, electric pulse-sensitive, or magnetic field-sensitive [40]. In gel-based systems, smart features of materials might also influence the swelling behavior and a potential drug release [41]. Importantly, every cell membrane–penetrating NM which has to be delivered into the cytoplasm requires a certain mechanism to assure endosomal escape [42]. The cargo has to leave the carrier before it is degraded by enzymes of the lysosome, especially in the case of targeting MΦ which are rich in lysosomes. Endosomal escape can, in the case of lipid-based nanocarriers, be realized via incorporation of pH-sensitive lipids, leading to enhanced drug release close to the lysosome [43].

Currently, smart features of NMs mostly are at a preclinical stage, but it is assumed that properties such as those mentioned above will enhance the efficiency of future formulations.

26.3 Therapeutic Modulation of Immunity

26.3.1 Targeting Immune Cell-Related Disease

The presence of inflammatory cytokines in serum secreted by immune cells is known to affect reactions in the whole body. Interleukin 1 β (IL1β) and TNF are two essential cytokines that affect many inflammatory diseases. Blocking of TNF represents a promising therapeutic option for inflammatory disorders such as hepatitis, rheumatoid arthritis, inflammatory bowel disease, or graft versus host disease [44]. Anti-inflammatory cytokines directly suppress inflammatory mediators; for example, IL10 suppresses

IL6 and TNFα production [45]. Cytokines of M1-MΦ, specifically TNF, are targeted by pharmaceuticals such as etanercept, which has been approved since 1998. Many other pharmaceuticals inhibit TNF to overcome the negative side effects of the cytokine on healthy tissue which occurs in dysregulated inflammatory disease such as arthritis [46]. It is a matter of fact that the TNF blockers indirectly target MΦ (and neutrophil)-derived cytokines [4]. Recently, novel small molecule-based drugs which block the route of action of cytokines have been marketed. In particular, tyrosine kinase inhibitors have entered the sector of anti-inflammatories. The novel class of Janus kinase (JAK) inhibitors intend to block cytokine signaling, including that of TNF. For instance, Tofacitinib, a pan-JAK inhibitor, received approval for several inflammatory diseases such as inflammatory bowel diseases [47], psoriasis [48], and arthritis [49].

Attributable to their outstanding plasticity, other cytokines of MΦ are also pharmaceutical targets. For instance, the vascular endothelial growth factor a (VEGFA), which has a strong impact on angiogenesis by inducing the proliferation of endothelial cells, is also blocked by the biopharmaceuticals bevacizumab or sorafenib in cancer treatment [50]. A treatment which facilitates VEGFA inhibition has a 10-year tradition and it has become clear during this period that tumor resistance toward inhibitors such as bevacizumab is an obstacle for the success of anti-VEGFA therapies [51]. Drugs can also affect angiogenesis indirectly by acting on cells which induce the formation of novel blood vessels based on their secretion of angiogenic factors such as VEGFA: The inhibition of circulating CC-chemokine ligand 2 (CCL2) by a neutralizing aptamer reduces the migration of monocytes into inflamed liver and consequently also decreases hepatic angiogenesis in chronic inflammation [52]. However, chemokines and correspondingly their inhibitors have a low level of specificity, and thus, these drugs might act on many cell types in the body. Thus, there is an urgent need for novel therapies on cancer and inflammatory disease with increased specificity for cell types.

A targeted depletion of MΦ can be done by using particulate carriers such as micron-sized clodronate liposomes which MΦ scavenge actively. The clodronate of these micro-sized lipid carriers is, upon uptake, released intracellularly and induces apoptosis of MΦ. A recent study employed liposomal clodronate to treat sarcoma, a disease which is also strongly promoted by MΦ [53]. Subsequent strategies to target MΦ make use of biopharmaceuticals which target receptors involved in MΦ differentiation. Scientists have reported the inhibition of TAM using antibodies directed against macrophage colony stimulating factors (MCSF, CSFR1) or CC chemokine receptor 2 (CCR2) in cancer treatment [54]. TAM are currently being engaged in phase I clinical trials using the monoclonal antibody IMC-CS4 which blocks CSFR1 (study identifier NCT01346358) [55].

However, blocking of TAM infiltration into tumors has a limited level of success, because tumors recruit an alternative type of myeloid cell when TAM are blocked. Recently, scientists discovered a mechanism that leads to the infiltration of tumors with TAM precursors, which normally suppresses their infiltration with tumor-associated neutrophils (TAN) [56], also referred to as polymorphonuclear myeloid-derived suppressor cells (PMN-MDSC). Upon inhibition of CSF1R in cancer, an increased amount of TAN is recruited into the tumors, which compensates for a lack of TAM or monocytic MDSC. The solution to overcome this myeloid cancer immune escape might be to combine inhibitors for TAM (CSF1R-inhibitors) and TAN (CXCR2-inhibitors) [56]. Targeting of toll-like receptor (TLR) signaling has been approved for melanoma treatment [57]. Importantly, TLR are differentially expressed by subsets of TAM as recently discovered in our own investigations in liver cancer, and the cellular effects on myeloid cells remain to be understood in more detail.

Dendritic cells (DC) exhibit many similarities with MΦ and share some markers [58]. In contrast to MΦ, DC are specialized on presenting an antigen to T cells and hence migrate more rapidly to the lymph node than macrophages and thus play a central role for vaccination. Researchers have therefore aimed to target DC in order to vaccinate against cancer. There are many possibilities to perform vaccinations based on targeting DC: *ex vivo* priming of DC can be done to trigger these cells to present a specific tumor-associated antigen to lymphocytes. Lymphocytes in turn kill cells expressing a certain antigen [59]. Vaccination can be performed by using different molecules for vaccination: electroporation-enhanced immunization is explored using DNA-based vaccine such as GX-188E [60], and cancer treatment in the future might be based on tumor antigen delivery to DC *in vivo* as explored in hepatocellular carcinoma patients [61].

Targeting of non-phagocytic immune cells works can be realized by antibodies against cell-type specific markers, mostly those present on the cellular surface. In therapy of B cell lymphoma, the CD20 depleting antibody rituximab has been approved by the FDA in 1998 and was the first antibody against cancer to be approved. It is still in extensive use, but with the patent of rituximab expiring in 2013 in the EU and in 2016 in the USA, novel strategies, such as improved versions of the inhibitor or biosimilars to replace the drug, are under development. Monospecific antibodies such as rituximab act by labeling the designated cell type to be eliminated by other immune cells, based on binding of the Fc-part of the antibody to the Fc-receptor of immune cells. In order to increase the efficiency of killing CD20 expressing cells, bispecific antibodies with binding sites for CD20 and also for FAS were explored. After binding of both sites of the bi-specific antibody to single B cells, FAS triggers apoptosis. Interestingly, binding of FAS alone does not have this effect [62]. Scientists have followed on optimizing such constructs and CD20×CD95 antibodies have opened novel perspectives for the treatment of B-cell-mediated autoimmune disease and lymphoma [63], and this concept also opens new perspectives for advanced pharmaceuticals.

Cytotoxic T cells can actively kill cancer cells, but during autoimmune disease, they destroy the body's-own cells. In the course of therapies, they can be depleted using anti-CD8 antibodies [64]. In order to use their killing capabilities to fight cancer, innovative strategies already make use of cytotoxic CD8 cells. Many fields of research have been inspired a lot by the inhibitors of programmed cell death 1 (PD1), such as pembrolizumab. In this regard, immunomodulatory therapies utilize the killing capabilities of T cells for killing tumor cells [65]. The so-called chimeric antigen receptor (CAR)-T cells represent another major innovation which has been approved in 2017 by the FDA for children and young adults with relapsed or refractory B-cell acute lymphoblastic leukemia [66]. CAR T cells are generated based on extracting a patient's immune cells and genetically engineering and reinjecting them into the patient where they shall seek and destroy cancer cells [67].

Unfortunately, there were tragic episodes in earlier attempts to engage T cells. For instance, the drug labeled as JCAR015 of the company Juno Therapeutics failed in B cell acute lymphoblastic leukemia, based on excess fluids accumulated in the brains of patients and was suspended in 2015 (trial NCT02535364). Earlier failed immunotherapies (in 2006) that made use of a superagonist of CD28 (TGN1412) led to similar tragic results with six people becoming seriously sick within minutes. Yet, data on blocking CD28 are promising despite the kickback and the strategy now has come back for arthritis therapy [68]. Thus, immunotherapies bear risks, specifically cellular hyperactivation with cytokine storm. Regulatory T cells (T_{Reg}) are another cell type focused on for immunotherapeutics. In inflammatory ear disease, inflammation is prolonged when T_{Reg} are depleted by a CD25 depleting antibody [69], demonstrating their role in suppressing immunity, which is desired in the therapy of inflammatory disease.

Similar to other infiltrating immune cells, neutrophils can also be targeted in cancer therapy. Acharyya and Massague suggested interfering with downstream targets of the MAPK cascade such as leukotriene biosynthesis, CXCR2, Anti-Bv8, and G-CSF antibodies [70]. Importantly, a combined inhibition of TAN and TAM precursors (PMN-MDSC/mMDSC) might open a novel avenue for cancer therapy [56]. Scientists have also started to explore the therapeutic potential of platelet targeting, which is very promising due to the very high number of platelets in the body. The platelet-expressed collagen receptor glycoprotein VI was shown by Boilard and colleagues to be a key trigger for platelet microparticle generation in arthritis pathophysiology [71]. Their glycoprotein (GP) Ib-V-IX receptor affects their adhesion, activation, and procoagulant activity and has been explored as a novel preventive treatment of thrombus formation using therapeutic antibodies [72]. Furthermore, platelets are targeted in chronic HIV infection where different anti-inflammatory drugs are currently compared at a stage II clinical trial (NCT02578706) (Table 26.1). There were also reports that platelets are critically involved in liver disease [73].

Table 26.1 Selected biopharmaceutic, microcarrier, and small molecule-based immunotherapies for inflammatory diseases and cancer

Cell type	Disease	Drug, outcome	Stage	Reference
Macrophages	Rheumatoid arthritis, AID	Infliximab (anti-TNF mAB), ameliorates inflammation	Approved	[4]
	Sarcoma	Clodlips, depletes macrophages and ameliorates disease	Preclinical	[53]
	Breast and prostate cancer	Anti-CSF1R mAB (IMC-CS4), reduced cancer therapy	Phase I	[55]
	Pancreatic cancer	mAB vs. CSF1R or CCR2, improved chemotherapy	Preclinical	[54]
	Cancer	Imiquimod (TLR7 agonist), reduces cancer	Approved	[57]
	Tofacitinib	Inflammatory diseases, inhibiting cytokine signaling	Approved	[47–49]
	Cancer (EMT)	Encapsulation of macrophages	Preclinical	[81]
Dendritic cells	Cancer	Pulsed tumor antigen	Clinical	[61]
B cells	Leukemia	CAR T cells, kill CD20 expressing B cells	Approved	[66]
	Leukemia, AID	CD20×CD95-mAB, depletes B cells	Preclinical	[63]
	Leukemia, AID, rheumatoid arthritis	Rituximab depletes B cells, anti CD20 mAB	Approved	[109]

(*Continued*)

Table 26.1 (*Continued*)

Cell type	Disease	Drug, outcome	Stage	Reference
Cytotoxic T cell	Lymphoma, rheumatoid arthritis	Anti-CD20 mAB, increases T_H1 and reduces T_{Regs}	Preclinical	[110]
	Cancer	PD1 inhibitor	Approved	[65]
Regulatory T cell	Ear inflammation	CD25 (T_{Reg}-depleting mAB), prolonged inflammation	Preclinical	[69]
Neutrophils	Cancer	Inhibition of CXCR2, Bv8, G-CSF	Suggestion	[56, 70]

Abbreviations: AID, Autoimmune disease; AuNR, Gold nanorods; Bv8, Prokineticin 2 (PROK2); CaP, Calcium phosphate; CAR T cells, Chimeric antigen receptor T cells; CCR2, Chemokine receptor 2; CD, Cluster of differentiation; CSF1R, Colony-stimulating factor receptor 1; CXCR2, CXC-chemokine receptor 2; Dex, Dexamethasone; EMT, Epithelial to mesenchymal transition; G-CSF, Granulocyte colony-stimulating factor; GLF, The peptide sequence GLF; LNC, Lipid-based nanocarriers; mAB, Monoclonal antibody; MTC, Mannose-modified trimethyl chitosan–cysteine conjugate nanoparticles; NASH, Non-alcoholic steatohepatitis; PD1, Programmed cell death; pPB, "C∗SRNLIDC∗" (peptide vs. PDGFβR); RGD, The peptide sequence RGD; RNAi, RNA interference; siRNA, Silencing RNA; SSL, Sterically stabilized liposomes; TAA, Thioacetamide; T_H1 cells, Type I T helper cell; TLR, toll-like receptor; TNF, Tumor necrosis factor; T_{Reg}, Regulatory T cells.

26.3.2 Immunomodulatory Nanocarriers

Virtually any drug and NM can have immunomodulatory effects. There is a large number of NM available of either organic or inorganic nature and all of their properties such as material type, size, and functionalization have an impact on biodistribution in organs and cells. Particles of size 10–250 nm mostly distribute to the liver and spleen, whereas those smaller than 10 nm additionally also translocate to other organs such as the kidney, testis, and brain [74]. The most prominent inorganic NM probably are gold nanoparticles (AuNP), which can be easily modified in size, shape, and functional end groups, allowing, for example, for peptide-decorated gold nanorods [75]. There have been a few clinical trials facilitating gold nanoparticles: One was on CYT-6091, a gold nanocarrier for TNF with phase I clinical studies completed in cancer therapy [76]. However, inorganic nanoparticles accumulate in the body since they are not biodegradable, reflected by our own studies on gold nanorods which are retained at similar levels in the liver one week after injection [13].

Considering the fact that many biopharmaceuticals target immune cell-derived cytokines in circulation, it would be much more useful to target TNF and other compounds even before they are produced by the cells. This is possible using NMs which enter the cells and modify their status of activation. Based on interfering with inflammatory pathways, the production of TNF and similar mediators can be reduced. For instance, this can be done by delivering anti-inflammatory substances such as glucocorticoids [22, 23]. Immune cells can be targeted selectively based on their expression of receptors, by using pharmaceuticals decorated with targeting ligands functioning as the key to biological structures. Importantly, there are phagocytic immune cells which are quite easy to target, as a consequence of their unspecific uptake activities: Monocytes and MΦ are probably the most efficient mobile cellular scavengers of nearly any kind of material that they can internalize physically. They are central in the removal of materials from blood, such as gold nanoparticles [13, 16, 77]. The physicochemical properties of any given material influences the reaction of the phagocytes which secrete a large spectrum of cytokines that act on other cells [78] and tissues [79]. The liver is, due to its large number of MΦ, probably the most important internal organ for the accumulation of NM [13]. Therefore, any kind of serious nanomaterial studied should, after cytotoxicity tests, be monitored for the response of phagocytes to the material and assess biodistribution. Interactions with materials alter the state of MΦ activation and thereby affect the release profile of their cytokines *in vitro* and *in vivo* [13, 22, 23, 77]. Thus, targeting of phagocytes is comparatively easy, and the challenge is rather to avoid targeting them if, for instance, the drugs are directed against lymphocytes, which are not phagocytically active.

Our own research with human primary immune cells demonstrated an M2 polarization of MΦ by two tripeptides (RGD or GLF)-functionalized gold nanorods *in vitro* [77]. However, *in vivo* studies facilitating models for acute and chronic liver inflammation did not obtain therapeutic effects at comparable dosage and the coupled peptides even increased hepatic injury (Table 26.2) [13]. Subsequent studies of our research group put the focus on delivering anti-inflammatory drugs such as corticosteroids to achieve M2-polarization of MΦ [22]. Using liposomal dexamethasone (Lipodex), we were able to cure Concanavalin A-induced autoimmune hepatitis and found a tendency toward decreased fibrosis in the carbon tetrachloride (CCl_4)-based model of chronic liver injury [23]. Treatment of inflammatory activation has been demonstrated using small interfering RNA (siRNA) administered by polymeric nanoparticles which inhibit M1 polarization of MΦ [80].

Table 26.2 Selected nanomedicine-based immunotherapies for inflammatory diseases and cancer

Cell type	Disease	Drug, outcome	Stage	Reference
Macrophages	Liver inflammation	AuNR with RGD or GLF, increased ConA-based injury and no effect on fibrosis	Preclinical	[13]
	Liver inflammation	Liposomal Dex, significantly reduced hepatitis and slightly diminished fibrosis	Preclinical	[22, 23]
	Acute hepatitis	MTC anti-TNF siRNA, inhibition of M1-macrophages	Preclinical	[80]
	Cancer	Feraheme, induction of M1-macrophages, growth arrest	Preclinical	[84]
	Cancer	MiR155 nanovectors, repolarize TAM	Preclinical	[82]
Dendritic cells	Cancer	RNA-based lipoplexes	Preclinical	[90]

Cell type	Disease	Drug, outcome	Stage	Reference
B cells	Cancer (vaccination)	CaP-TLR nanoparticle, stronger immunization	Preclinical	[92]
	B cell lymphoma	CD38-targeted LNC anti Cyclin D1 RNAi, amelioration	Preclinical	[93]
Regulatory T cell	AID	Nanoparticles with AID-peptides, T_{Reg} reprogramming	Preclinical	[94]
Myeloid blasts	Acute myeloid leukemia	Liposomal combination drug CPX-351	Stage III	[86]
Neutrophils	Inflammation	Albumin-encapsulated piceatannol	Preclinical	[95]

Abbreviations: AID, Autoimmune disease; AuNR, Gold nanorods; Bv8, Prokineticin 2 (PROK2); CaP, Calcium phosphate; CAR T cells, Chimeric antigen receptor T cells; CCR2, Chemokine receptor 2; CD, Cluster of differentiation; CSF1R, Colony-stimulating factor receptor 1; CXCR2, CXC-chemokine receptor 2; Dex, Dexamethasone; EMT, Epithelial to mesenchymal transition; G-CSF, Granulocyte colony-stimulating factor; GLF, The peptide sequence GLF; LNC, Lipid-based nanocarriers; mAB, Monoclonal antibody; MTC, Mannose-modified trimethyl chitosan–cysteine conjugate nanoparticles; NASH, Non-alcoholic steatohepatitis; PD1, Programmed cell death; pPB, "C∗SRNLIDC∗" (peptide vs. PDGFβR); RGD, The peptide sequence RGD; RNAi, RNA interference; siRNA, Silencing RNA; SSL, Sterically stabilized liposomes; TAA, Thioacetamide; T_H1 cells, Type I T helper cell; TLR, Toll-like receptor; TNF, Tumor necrosis factor; T_{Reg}, Regulatory T cells.

In a very recent innovative study, macrophages were, together with a polarizing stimulus, encapsulated in an alginate matrix to be used as a production site for cytokines. These complexes were shown to inhibit the epithelial-to-mesenchymal (EMT) transition [81]. Repolarizing macrophages have been shown to work efficiently to induce tumor regression based on the delivery of micro-RNA155 by polypeptide nanovectors which increased the expression of miR155 up to 400-fold in TAM and triggers them to express M1 and to repress M2 markers [82].

Cellular delivery of small non-coding RNA such as siRNA or micro-RNA requires nanocarriers or alternative vehicles, because their stability *in vivo* is limited. Liposomal dexamethasone (Lipodex) serves as another representative example for improving the pharmacokinetics of a drug based on nanotechnology what has been explored in our studies: While free dexamethasone acts on a systemic level on virtually all accessible parts of the body, encapsulation protects many cells from its cytotoxicity. Specifically, human fibroblasts and MΦ were demonstrated to be protected from the cytotoxic effects of free dexamethasone shown in our own *in vitro* studies [22]. *In vivo*, Lipodex efficiently targets immune cells resulting in therapeutic effects which cannot be obtained by using the free drug: At a concentration of 1 mg/kg body weight, autoimmune-mediated liver disease was cured, whereas this was not achieved by free dexamethasone at identical concentration [23]. The benefits of drug encapsulation are even more pronounced for doxorubicin, whose half-life is extremely short (few minutes) and which can be prolonged for several days using a lipid-based carrier [83].

Recently, it was discovered that ferumoxytol, a drug normally used to treat anemia, also repolarizes M2 into M1-MΦ, based on the iron overload of MΦ [84]. However, the iron supplementation of feraheme might lead to side effects such as inflammation, which shows that it is a complex issue to modulate immune cells. Similar to the M1-shift induced by ferumoxtol, also Lipodex [22] and nanocarrier-coupled stavudine were discovered to lead to inflammatory activation of MΦ [85], possibly due to the accumulation of the drug inside the cells, making exact dosing important. While many strategies exist to cure lymphoid leukemia based on engaging the cells by depleting antibodies (Rituximab, CAR-T cells), there are few strategies to cure myeloid leukemia. Yet, one promising attempt for an improved therapy of acute myeloid leukemia is represented by CPX-351, which contains the two drugs cytarabine and daunorubicin in a single liposomal formulation. CPX-351 was shown to efficiently target myeloid cells, with greater efficiency than the single administration of both drugs [86], and the drug is now approved by the FDA for the treatment of acute myeloid leukemia (AML) [87]. The mechanism of action is hypothesized to be based on phagocytosis of the myeloid blasts.

There is huge interest in using nanocarriers for improved methods of DC vaccination by exploiting the specific properties of nanocarriers [88]. Nano-sized delivery systems enable improved vaccination since they can protect biologically sensitive molecules such as antigen-bearing peptides, DNA or mRNA encoding for antigens, or immunostimulatory oligonucleotides, from degradation [89]. Gold-based nanoparticles accumulate in the MHCII processing compartment of DC [77], which might explain the efficient manipulation of DC by NM. The physicochemical properties of nanocarriers play important roles for the activation of the DC, and conjugated peptides alter the maturation of DC [77]. RNA-based lipoplexes appear to be promising nanocarriers for vaccination against cancer [90].

Targeting non-phagocytic cells using NM requires cell-specific markers, such as CD molecules, in order to target different leukocyte populations, similar to biopharmaceuticals such as rituximab which target CD20 expressed by B cells. Compared to biopharmaceuticals, NM offers many more options for improved pharmacokinetics, for example, based on drug encapsulation. In case of a limited targeting capability for non-phagocytes such as lymphoid cells by encapsulated drugs, improved options might arise from using the TAT-peptide, a short peptide which increases uptake by lymphocytes [91]. In order to directly target B cells for an optimized method of vaccination, calcium phosphate nanoparticles coated with TLR ligands were studied in vaccination. The delivery system was shown to be more efficient than conventional methods for immunization [92]. Delivery of interfering RNA (RNAi) neutralizing Cyclin D1 via CD38-directed targeting lipid nanocarriers has been explored in order to target B cell lymphoma [93].

T_{Reg} hold great potential as therapeutic targets in treating autoimmunity, but currently there are no strategies to expand antigen-specific T_{Reg} *in vivo*. Clemente-Casares and colleagues have demonstrated that by using nanoparticles decorated with AID-relevant peptides bound to MHCII complex molecules, antigen-specific type I $CD4^+$ T cells (TR1) are induced. This strategy makes use of the existing T_{Reg} and turns them from disease-primed autoreactive T cells into TR1-like cells, which consequently suppress autoantigen-loaded antigen-presenting cells, and further differentiate into cognate disease-suppressing regulatory B cells [94].

Nanocarriers were also exploited to target neutrophils. Administration of albumin nanoparticles loaded with the spleen tyrosine kinase inhibitor, piceatannol, inactivates the inflammatory function of activated neutrophils (Table 26.2) [95]. NM or perhaps picomedicine are also predestined to be used to manipulate platelets in targeted therapies, probably using miniaturized platelet-targeting structures for these small cells of size 2–3 μm only.

26.3.3 Strategies to Modulate or Make Use of Immune Cell Migration

There are two key elements which distinguish immune cells from other (healthy) cell types: their migration and secretory activity. The combination of both facilitates a mobile production site of cytokines and makes them important for virtually all types of disease. Immunomodulatory drugs can act via targeting the whole process of leukocyte extravasation which comprises their migration from the endothelium into different organs. This process is based on a coordinated interplay between endothelial cells and immune cells: Endothelial cells activate leukocytes and direct them to the point where they extravasate, and leukocytes in turn instruct endothelial cells to open a path for transmigration [96]. Selectins control the first step in the adhesion of leukocytes to the endothelium and to inflamed sites of the body. We have demonstrated recently that selectin-binding glycopolymers anchored to polymer backbones strongly impact the migration of human primary MΦ [20], and that selectin-binding glycopolymers are potential novel therapeutics for liver disease [21].

Chemokines function among others in the activation of leukocytes during their process of rolling along the endothelium, and we have noted only modest effects on inflammatory disease progression by inhibiting CCL2, but a reduction in angiogenesis due to the reduced number of monocytes [52]. Until now, there have been few attempts in targeting chemokine receptors in liver cancer with nanocarriers, despite the few drugs available for the disease. The CXC chemokine receptor 4 (CXCR4) exerts many important functions at different stages of HCC and can be targeted by using ligands for CXCR4 such as CXCL12, which is expressed by monocytes and MΦ [97]. Liu et al. have used CXCR4-directed nanocarriers loaded with siRNA against VEGF as an alternative antiangiogenic therapy and received a potent antitumor response by their carriers [98]. However, chemokines and their receptors generally exhibit a low level of cell type specificity, and more specific modulation of immune cell subpopulations might increase the efficiency of therapeutics.

Integrins appear at the final stage of leukocyte extravasation, just upfront to transmigration. Yet, only 3 of 24 known human integrins are currently engaged by antibody therapies [99]. The process of immune cell extravasation therefore has three major points of intervention given by molecules targeting selectins, chemokines, or integrins (Fig. 26.2). However, it has to be taken into account that not all immune cells originate from the blood stream, but that portions of them originate from local stem cells [1]. In fact, very little is known about the role of MΦ which originate from organ-resident stem cells, in inflammatory disease and cancer, and how they might interact with infiltrating monocytes.

The outstanding mobility of immune cells can also be exploited for another compelling concept which uses the high number and mobility of immune cells to transport nanoparticles to the desired site of action—using immune cells as Trojan horses. A promising study shown by Choi and colleagues illustrated that monocytes might be used to destroy hypoxic tumor regions by their delivery of cytotoxic compounds [100]. The immunological synapse, the process in which antigen-presenting cells instruct CD4 expressing helper cells, has also been engaged by nanoparticles which attach to free thiol groups on proteins at the surface of T cells: Using maleimide-functionalized nanoparticles, Stephan and coworkers used T cell-linked synthetic nanoparticles as drug delivery vehicles into the immunological synapse [101], which might be useful to instruct T cells against cancer antigens.

26.3.4 Unintended Interactions of Nanomedicine with Myeloid Immunity

NM which is intended for therapeutic use will, in virtually any case, be faced with cells of the immune system and endothelial cells [102]. It is known from biomaterial sciences that unintended interactions of materials with innate immune cells might result in chronic inflammation, tissue damage, and implant rejection [103]. Similarly, other materials that are considered as foreign to the body and as a threat to the body might be attacked by immune cells. Endothelial cells, monocytes, and MΦ can be defined as the reticuloendothelial system (RES). There are different attempts to reduce unspecific interactions with the RES, the most frequently performed strategy is to reduce unspecific binding to proteins and cells by PEGylation [104]. However, many researchers are ignoring the fact that PEG accumulates in parts in the body and that immunity generates antibodies against PEG, a fact that is responsible for the accelerated blood clearance (ABC) effect where liposomes are cleared upon repeated administration [105].

Alternative strategies to inhibit undesired reactions with the RES are optimizing the immune response using coatings that reduce the acute immune reaction, such as poly(N-isopropylacrylamide) hydrogel particles cross-linked with poly(ethylene glycol) diacrylate [106]. Others have proposed usage of ultrasmall nanogels sizing only 2 nm, which is below the size that allows the clearance of these particles by the RES [107]. Studies by our group have shown that endothelial cells internalize large amounts of nanoparticles *in vitro* where they are cultured as a monolayer [77], yet, *in vivo*, endothelial cells did not contribute to the high uptake of nanoparticles into the liver, and nanoparticles were found in hepatic MΦ only [13].

Importantly, specialized leukocytes such as monocytes, MΦ, and DC recognize pathogens and process antigens of novel pathogens and present them via their major histocompatibility complex II to T helper cells [77]. In turn, T cells generate complex molecular sites targeting the pathogen-derived antigen (T cell-receptor) which they present on their surface and use to target the pathogen. B cells generate antibodies which share many similarities with the T cell receptor. Our earlier studies have demonstrated that material composition, surface chemistry, and peptide modifications affect the maturation of DC [77] and, thus, might also affect T cell instruction by DC. The MHCII complex serves as a route of material transmission to the cells of the adaptive immune system, specifically to T helper cells. In addition to their direct effects on phagocytic myeloid immune cells, NM can potentially also interact with antigen presenting cells and influence immunizations. Clemente-Casares and colleagues have shown that a directed manipulation of the immunological synapse might be used to treat autoimmune disease [94]. However, NM-based vaccination might also interfere with normal functions of the immune system.

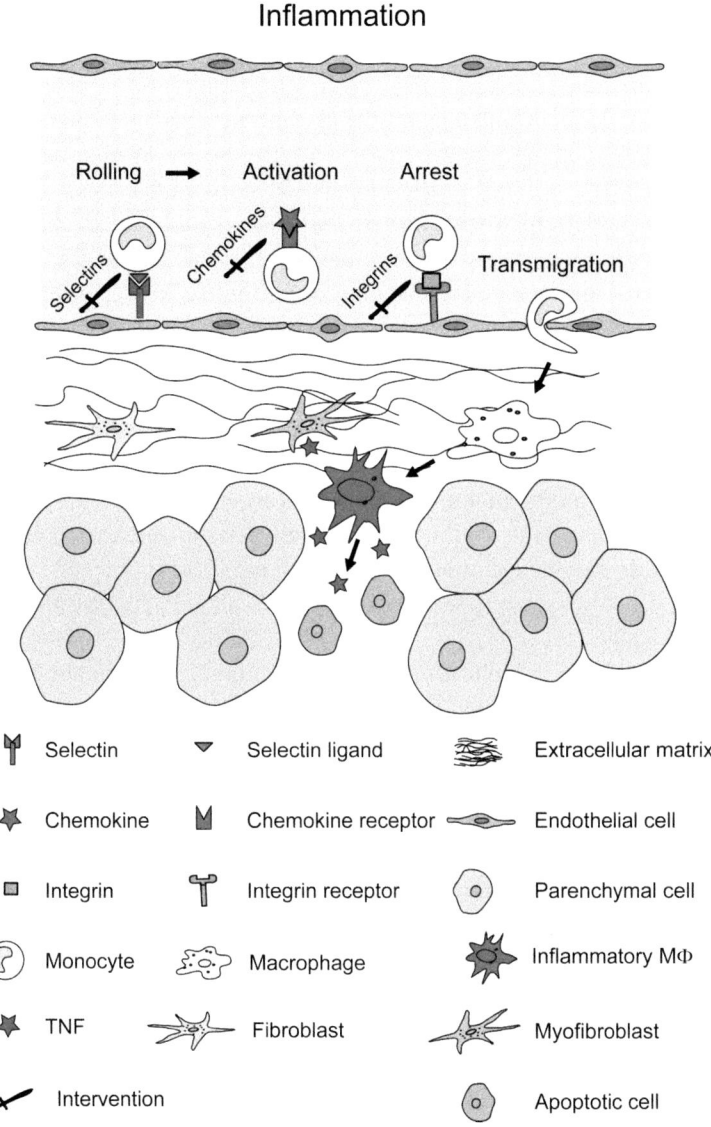

Figure 26.2 Therapeutic modulation of leukocyte migration as treatment of inflammatory disease. Upon inflammation, monocytes and other immune cells migrate to inflammatory sites through the bloodstream. Monocytes differentiate into specific macrophages which have a major impact on the progression of inflammation. Following inflammatory activation, certain mediators such as the tumor necrosis factor (TNF), may harm parenchymal cells.

26.4 Conclusions

The inevitable interactions with immunity will probably remain a major challenge of novel therapeutics including NM. The unintended targeting of the RES, specifically of MΦ, cannot be solved and assured easily but should be controlled for every novel formulation. Inappropriate carriers might induce chronic inflammation or impact the immunological synapse. Compared to biopharmaceuticals, NM can inhibit the production of pathogenic factors one step ahead than biopharmaceuticals do and inhibit the production of the mediators intracellularly. NM can further combine the properties of different drugs into a single formulation such as CPX-351, which contains two chemotherapeutic drugs in a single formulation. NM surface groups can target receptors of cells and might in the future even target multiple receptors, similar to bispecific antibodies. Controlled drug release enabled by nanotechnological delivery systems can transport sufficient drugs to a desired site of action, instead of distributing it systemically.

NM, or the application of nanotechnology to healthcare, should make increasing use of the novel insights into human molecular biology retrieved from the different "omics" analytics such as genomics, transcriptomics, proteomics, and metabolomics which enabled the discovery of various biomarkers useful for early stage detection of diseases, specifically in oncology [108]. NM holds great promise for significantly improving therapies based on enhanced drug delivery, reduced side effects, molecular imaging, and theranostics, which is the ability to use carriers for combined therapeutic and diagnostic purpose. Nevertheless, even with targeting groups and smart features incorporated, production methods for NM have to be simple and affordable to have the chance of being successful in the clinics [108]. Thus, the magic bullet of targeted NM requires continuous interdisciplinary efforts for successful clinical translation.

Disclosures and Conflict of Interest

This chapter is an updated and revised version of a manuscript originally published as: Bartneck, M. Immunomodulatory nanomedicine. *Macromol. Biosci.* **17**, 1700021 (2017) (DOI: 10.1002/mabi.201700021), and appears here with kind permission of the author and the publisher, Wiley WILEY-VCH Verlag GmbH & Co. The author gratefully acknowledges financial support from the medical faculty of the RWTH Aachen University, of the Wilhelm Sander Foundation, and of the COST Action BM1404 Mye-EUNITER (http://www.mye-euniter.eu). COST is part of the European Union Framework Program Horizon 2020. The author declares that he has no conflict of interest and has no affiliations or financial involvement with any organization or entity discussed in this chapter. This includes employment, consultancies, honoraria, grants, stock ownership or options, expert testimony, patents (received or pending) or royalties. No writing assistance was utilized in the production of this chapter and the author has received no payment for its preparation.

Corresponding Author

Dr. Matthias Bartneck
Department of Medicine III, Medical Faculty
RWTH Aachen
Pauwelsstr. 30, 52074 Aachen, Germany
Email: mbartneck@ukaachen.de

About the Author

Matthias Bartneck graduated from the University of Bielefeld with a diploma in biology in 2007, where he performed metabolite and gene expression profiling of human primary breast cancer at the Institute of Genetics. Immune cell signatures, specifically of macrophages, in cancer samples raised his interest in immunology. He pursued his research at RWTH Aachen University, where he studied the effects of nanomaterials on human primary immune cells and obtained a doctorate in natural sciences in 2010. His current research at Medical Clinic III at the University Hospital Aachen covers targeted immunomodulatory nanomedicine facilitating lipid, micellar, polymer, and metal-based nanocarriers and combined molecular imaging and cell sorting. He is a member of the European network Mye-EUNITER, which is devoted to triggering exploratory research on myeloid regulatory cells.

References

1. Epelman, S., Lavine, K. J., Randolph, G. J. (2014). Origin and functions of tissue macrophages. *Immunity*, **41**(1), 21–35.
2. Duprez, L., et al. (2011). RIP kinase-dependent necrosis drives lethal systemic inflammatory response syndrome. *Immunity*, **35**(6), 908–918.
3. Bartneck, M., et al. (2016). Histidine-rich glycoprotein promotes macrophage activation and inflammation in chronic liver disease. *Hepatology*, **63**(4), 1310–1324.
4. Seymour, H. E., et al. (2001). Anti-TNF agents for rheumatoid arthritis. *Br. J. Clin. Pharmacol.*, **51**(3), 201–208.
5. Chatzantoni, K., Mouzaki, A. (2006). Anti-TNF-alpha antibody therapies in autoimmune diseases. *Curr. Top. Med. Chem.*, **6**(16), 1707–1714.
6. Kircik, L. H., Del Rosso, J. Q. (2009). Anti-TNF agents for the treatment of psoriasis. *J. Drugs Dermatol.*, **8**(6), 546–559.
7. Martelli, L., et al. (2017). Cost-effectiveness of drug monitoring of anti-TNF therapy in inflammatory bowel disease and rheumatoid arthritis: A systematic review. *J. Gastroenterol.*, **52**(1), 19–25.
8. Arango Duque, G., Descoteaux, A. (2014). Macrophage cytokines: Involvement in immunity and infectious diseases. *Front. Immunol.*, **5**, 491.
9. Bartneck, M., et al. (2015). Isolation and time lapse microscopy of highly pure hepatic stellate cells. *Anal. Cell. Pathol. (Amst)*, **2015**, 417023.
10. Kendall, R. T., Feghali-Bostwick, C. A. (2014). Fibroblasts in fibrosis: Novel roles and mediators. *Front. Pharmacol.*, **5**, 123.
11. Travers, J. G., et al. (2016). Cardiac fibrosis: The fibroblast awakens. *Circ. Res.*, **118**(6), 1021–1040.
12. Ohlund, D., et al. (2013). Type IV collagen stimulates pancreatic cancer cell proliferation, migration, and inhibits apoptosis through an autocrine loop. *BMC Cancer*, **13**, 154.
13. Bartneck, M., et al. (2012). Peptide-functionalized gold nanorods increase liver injury in hepatitis. *ACS Nano*, **6**(10), 8767–8777.

14. Jenkins, S. J., et al. (2011). Local macrophage proliferation, rather than recruitment from the blood, is a signature of TH2 inflammation. *Science*, **332**(6035), 1284–1288.
15. Murray, P. J., et al. (2014). Macrophage activation and polarization: Nomenclature and experimental guidelines. *Immunity*, **41**(1), 14–20.
16. Bartneck, M., et al. (2010). Rapid uptake of gold nanorods by primary human blood phagocytes and immunomodulatory effects of surface chemistry. *ACS Nano*, **4**(6), 3073–3086.
17. Mantovani, A., Sica, A. (2010). Macrophages, innate immunity and cancer: Balance, tolerance, and diversity. *Curr. Opin. Immunol.*, **22**(2), 231–237.
18. Butt, A. Q., Mills, K. H. (2014). Immunosuppressive networks and checkpoints controlling antitumor immunity and their blockade in the development of cancer immunotherapeutics and vaccines. *Oncogene*, **33**(38), 4623–4631.
19. Yona, S., et al. (2013). Fate mapping reveals origins and dynamics of monocytes and tissue macrophages under homeostasis. *Immunity*, **38**(1), 79–91.
20. Moog, K. E., et al. (2017). Polymeric selectin ligands mimicking complex carbohydrates: From selectin binders to modifiers of macrophage migration. *Angew. Chem. Int. Ed. Engl.*, **56**(5), 1416–1421.
21. Bartneck, M., et al. (2017). Immunomodulatory therapy of inflammatory liver disease using selectin-binding glycopolymers. *ACS Nano*, **11**(10), 9689–9700.
22. Bartneck, M., et al. (2014). Liposomal encapsulation of dexamethasone modulates cytotoxicity, inflammatory cytokine response, and migratory properties of primary human macrophages. *Nanomedicine*, **10**(6), 1209–1220.
23. Bartneck, M., et al. (2015). Fluorescent cell-traceable dexamethasone-loaded liposomes for the treatment of inflammatory liver diseases. *Biomaterials*, **37**, 367–382.
24. Gabizon, A., et al. (1994). Prolonged circulation time and enhanced accumulation in malignant exudates of doxorubicin encapsulated in polyethylene-glycol coated liposomes. *Cancer Res.*, **54**(4), 987–992.
25. Gabizon, A. A., Shmeeda, H., Zalipsky, S. (2006). Pros and cons of the liposome platform in cancer drug targeting. *J. Liposome Res.*, **16**(3), 175–183.
26. Bartneck, M., et al. (2012). Inducing healing-like human primary macrophage phenotypes by 3D hydrogel coated nanofibres. *Biomaterials*, **33**(16), 4136–4146.

27. Bartneck, M., et al. (2010). Induction of specific macrophage subtypes by defined micro-patterned structures. *Acta Biomater.*, **6**(10), 3864–3872.
28. Groll, J., et al. (2005). A novel star PEG-derived surface coating for specific cell adhesion. *J. Biomed. Mater. Res. A*, **74**(4), 607–617.
29. Opazo, F., et al. (2012). Aptamers as potential tools for super-resolution microscopy. *Nat. Methods*, **9**(10), 938–939.
30. Schulz, C., et al. (2016). Generating aptamers interacting with polymeric surfaces for biofunctionalization. *Macromol. Biosci.*, **16**(12), 1776–1791.
31. Szeitner, Z., et al. (2014). A rational approach for generating cardiac troponin I selective Spiegelmers. *Chem. Commun. (Camb)*, **50**(51), 6801–6804.
32. Lam, J. K., et al. (2015). siRNA versus miRNA as therapeutics for gene silencing. *Mol. Ther. Nucleic Acids*, **4**, e252.
33. Friedman, R. C., et al. (2009). Most mammalian mRNAs are conserved targets of microRNAs. *Genome Res.*, **19**(1), 92–105.
34. Roy, S., et al. (2015). The role of miRNAs in the regulation of inflammatory processes during hepatofibrogenesis. *Hepatobiliary Surg. Nutr.*, **4**(1), 24–33.
35. Barenholz, Y. (2012). Doxil(R)–the first FDA-approved nano-drug: Lessons learned. *J. Control. Release*, **160**(2), 117–134.
36. Andral, M. (1841). On the physical alterations of the blood and animal fluids in disease: Alterations of the blood in fever and inflammation. *Prov. Med. Surg. J. (1840)*, **2**(47), 419–421.
37. Schneider, L. A., et al. (2007). Influence of pH on wound-healing: A new perspective for wound-therapy? *Arch. Dermatol. Res.*, **298**(9), 413–420.
38. Torchilin, V. P. (2014). Multifunctional, stimuli-sensitive nanoparticulate systems for drug delivery. *Nat. Rev. Drug Discov.*, **13**(11), 813–827.
39. de la Rica, R., Aili, D., Stevens, M. M. (2012). Enzyme-responsive nanoparticles for drug release and diagnostics. *Adv. Drug Deliv. Rev.*, **64**(11), 967–978.
40. Mura, S., Nicolas, J., Couvreur, P. (2013). Stimuli-responsive nanocarriers for drug delivery. *Nat. Mater.*, **12**(11), 991–1003.
41. Topuz, F., et al. (2017). One-step fabrication of biocompatible multifaceted nanocomposite gels and nanolayers. *Biomacromolecules*, **18**(2), 386–397.

42. Varkouhi, A. K., et al. (2011). Endosomal escape pathways for delivery of biologicals. *J. Control. Release*, **151**(3), 220–228.
43. Hafez, I. M., Cullis, P. R. (2004). Tunable pH-sensitive liposomes. *Methods Enzymol.*, **387**, 113–134.
44. Dinarello, C. A. (2000). Proinflammatory cytokines. *Chest*, **118**(2), 503–508.
45. Trindade, M. C., et al. (2001). Interleukin-10 inhibits polymethylmethacrylate particle induced interleukin-6 and tumor necrosis factor-alpha release by human monocyte/macrophages in vitro. *Biomaterials*, **22**(15), 2067–2073.
46. Pope, R. M. (2002). Apoptosis as a therapeutic tool in rheumatoid arthritis. *Nat. Rev. Immunol.*, **2**(7), 527–535.
47. Danese, S., Vuitton, L., Peyrin-Biroulet, L. (2015). Biologic agents for IBD: Practical insights. *Nat. Rev. Gastroenterol. Hepatol.*, **12**(9), 537–545.
48. Kwatra, S. G., et al. (2012). JAK inhibitors in psoriasis: A promising new treatment modality. *J. Drugs Dermatol.*, **11**(8), 913–918.
49. Yamaoka, K. (2016). Janus kinase inhibitors for rheumatoid arthritis. *Curr. Opin. Chem. Biol.*, **32**, 29–33.
50. Sullivan, L. A., Brekken, R. A. (2010). The VEGF family in cancer and antibody-based strategies for their inhibition. *MAbs*, **2**(2), 165–175.
51. Ferrara, N., Adamis, A. P. (2016). Ten years of anti-vascular endothelial growth factor therapy. *Nat. Rev. Drug Discov.*, **15**(6), 385–403.
52. Ehling, J., et al. (2014). CCL2-dependent infiltrating macrophages promote angiogenesis in progressive liver fibrosis. *Gut*, **63**(12), 1960–1971.
53. Guth, A. M., et al. (2013). Liposomal clodronate treatment for tumour macrophage depletion in dogs with soft-tissue sarcoma. *Vet. Comp. Oncol.*, **11**(4), 296–305.
54. Mitchem, J. B., et al. (2013). Targeting tumor-infiltrating macrophages decreases tumor-initiating cells, relieves immunosuppression, and improves chemotherapeutic responses. *Cancer Res.*, **73**(3), 1128–1141.
55. Crusz, S. M., Balkwill, F. R. (2015). Inflammation and cancer: Advances and new agents. *Nat. Rev. Clin. Oncol.*, **12**(10), 584–596.
56. Kumar, V., et al. (2017). Cancer-associated fibroblasts neutralize the anti-tumor effect of CSF1 receptor blockade by inducing PMN-MDSC infiltration of tumors. *Cancer Cell*, **32**(5), 654–668.e5.

57. Vacchelli, E., et al. (2012). Trial watch: FDA-approved Toll-like receptor agonists for cancer therapy. *Oncoimmunology*, **1**(6), 894–907.
58. Guilliams, M., et al. (2016). Unsupervised high-dimensional analysis aligns dendritic cells across tissues and species. *Immunity*, **45**(3), 669–684.
59. Pyzer, A. R., Avigan, D. E., Rosenblatt, J. (2014). Clinical trials of dendritic cell-based cancer vaccines in hematologic malignancies. *Hum. Vaccin. Immunother.*, **10**(11), 3125–3131.
60. Kim, T. J., et al. (2014). Clearance of persistent HPV infection and cervical lesion by therapeutic DNA vaccine in CIN3 patients. *Nat. Commun.*, **5**, 5317.
61. Lee, J. H., et al. (2015). A phase I/IIa study of adjuvant immunotherapy with tumour antigen-pulsed dendritic cells in patients with hepatocellular carcinoma. *Br. J. Cancer*, **113**(12), 1666–1676.
62. Jung, G., et al. (2001). Target cell-restricted triggering of the CD95 (APO-1/Fas) death receptor with bispecific antibody fragments. *Cancer Res.*, **61**(5), 1846–1848.
63. Nalivaiko, K., et al. (2016). A recombinant bispecific CD20×CD95 antibody with superior activity against normal and malignant B-cells. *Mol. Ther.*, **24**(2), 298–305.
64. Clement, M., et al. (2016). Targeted suppression of autoreactive CD8+ T-cell activation using blocking anti-CD8 antibodies. *Sci. Rep.*, **6**, 35332.
65. Hoos, A. (2016). Development of immuno-oncology drugs–from CTLA4 to PD1 to the next generations. *Nat. Rev. Drug Discov.*, **15**(4), 235–247.
66. [No authors listed] (2017). Panel OKs CAR T therapy for leukemia. *Cancer Discov.*, **7**(9), 924.
67. Jackson, H. J., Rafiq, S., Brentjens, R. J. (2016). *Driving CAR T-cells forward. Nat. Rev. Clin. Oncol.*, **13**(6), 370–383.
68. Tyrsin, D., et al. (2016). From TGN1412 to TAB08: The return of CD28 superagonist therapy to clinical development for the treatment of rheumatoid arthritis. *Clin. Exp. Rheumatol.*, **34**(4 Suppl 98), 45–48.
69. Christensen, A. D., et al. (2015). Depletion of regulatory T cells in a hapten-induced inflammation model results in prolonged and increased inflammation driven by T cells. *Clin. Exp. Immunol.*, **179**(3), 485–499.
70. Acharyya, S., Massague, J. (2016). Arresting supporters: Targeting neutrophils in metastasis. *Cell Res.*, **26**(3), 273–274.

71. Boilard, E., et al. (2010). Platelets amplify inflammation in arthritis via collagen-dependent microparticle production. *Science*, **327**(5965), 580–583.
72. Maurer, E., et al. (2013). Targeting platelet GPIbbeta reduces platelet adhesion, GPIb signaling and thrombin generation and prevents arterial thrombosis. *Arterioscler. Thromb. Vasc. Biol.*, **33**(6), 1221–1229.
73. Kondo, R., et al. (2013). Accumulation of platelets in the liver may be an important contributory factor to thrombocytopenia and liver fibrosis in chronic hepatitis C. *J. Gastroenterol.*, **48**(4), 526–534.
74. Muller, R. H., et al. (1996). Phagocytic uptake and cytotoxicity of solid lipid nanoparticles (SLN) sterically stabilized with poloxamine 908 and poloxamer 407. *J. Drug Target.*, **4**(3), 161–170.
75. Nikoobakht, B., et al. (2002). The quenching of CdSe quantum dots photoluminescence by gold nanoparticles in solution. *Photochem. Photobiol.*, **75**(6), 591–597.
76. Libutti, S. K., et al. (2010). Phase I and pharmacokinetic studies of CYT-6091, a novel PEGylated colloidal gold-rhTNF nanomedicine. *Clin. Cancer Res.*, **16**(24), 6139–6149.
77. Bartneck, M., et al. (2012). Effects of nanoparticle surface-coupled peptides, functional endgroups, and charge on intracellular distribution and functionality of human primary reticuloendothelial cells. *Nanomedicine*, **8**(8), 1282–1292.
78. Zimmermann, H. W., et al. (2010). Functional contribution of elevated circulating and hepatic non-classical CD14CD16 monocytes to inflammation and human liver fibrosis. *PLoS One*, **5**(6), e11049.
79. Mullarky, I. K., et al. (2006). Tumor necrosis factor alpha and gamma interferon, but not hemorrhage or pathogen burden, dictate levels of protective fibrin deposition during infection. *Infect. Immun.*, **74**(2), 1181–1188.
80. He, C., et al. (2013). Multifunctional polymeric nanoparticles for oral delivery of TNF-alpha siRNA to macrophages. *Biomaterials*, **34**(11), 2843–2854.
81. Sola, A., et al. (2018). Microencapsulated macrophages releases conditioned medium able to prevent epithelial to mesenchymal transition. *Drug Deliv.*, **25**(1), 91–101.
82. Liu, L., et al. (2017). Tumor associated macrophage-targeted microRNA delivery with dual-responsive polypeptide nanovectors for anti-cancer therapy. *Biomaterials*, **134**, 166–179.

83. Gabizon, A., et al. (1998). Development of liposomal anthracyclines: From basics to clinical applications. *J. Control. Release*, **53**(1–3), 275–279.

84. Zanganeh, S., et al. (2016). Iron oxide nanoparticles inhibit tumour growth by inducing pro-inflammatory macrophage polarization in tumour tissues. *Nat. Nanotechnol.*, **11**(11), 986–994.

85. Zazo, H., et al. (2017). Gold nanocarriers for macrophage-targeted therapy of human immunodeficiency virus. *Macromol. Biosci.*, **17**(3), doi:10.1002/mabi.201600359.

86. Cortes, J. E., et al. (2015). Phase II, multicenter, randomized trial of CPX-351 (cytarabine:daunorubicin) liposome injection versus intensive salvage therapy in adults with first relapse AML. *Cancer*, **121**(2), 234–242.

87. Nikanjam, M., et al. (2018). Persistent cytarabine and daunorubicin exposure after administration of novel liposomal formulation CPX-351: Population pharmacokinetic assessment. *Cancer Chemother. Pharmacol.*, **81**(1), 171–178.

88. Zentel, R. (2016). Nanoparticles and the immune system: Challenges and opportunities. *Nanomedicine (Lond)*, **11**(20), 2619–2620.

89. Grabbe, S., et al. (2016). Nanoparticles and the immune system: Challenges and opportunities. *Nanomedicine (Lond)*, **11**(20), 2621–2624.

90. Grabbe, S., et al. (2016). Translating nanoparticulate-personalized cancer vaccines into clinical applications: Case study with RNA-lipoplexes for the treatment of melanoma. *Nanomedicine (Lond)*, **11**(20), 2723–2734.

91. Dodd, C. H., et al. (2001). Normal T-cell response and in vivo magnetic resonance imaging of T cells loaded with HIV transactivator-peptide-derived superparamagnetic nanoparticles. *J. Immunol. Methods*, **256**(1–2), 89–105.

92. Zilker, C., et al. (2016). Nanoparticle-based B-cell targeting vaccines: Tailoring of humoral immune responses by functionalization with different TLR-ligands. *Nanomedicine*, **13**(1), 173–182.

93. Weinstein, S., et al. (2016). Harnessing RNAi-based nanomedicines for therapeutic gene silencing in B-cell malignancies. *Proc. Natl. Acad. Sci. U. S. A.*, **113**(1), E16–E22.

94. Clemente-Casares, X., et al. (2016). Expanding antigen-specific regulatory networks to treat autoimmunity. *Nature*, **530**(7591), 434–440.

95. Wang, Z., et al. (2014). Prevention of vascular inflammation by nanoparticle targeting of adherent neutrophils. *Nat. Nano*, **9**(3), 204–210.
96. Vestweber, D. (2015). How leukocytes cross the vascular endothelium. *Nat. Rev. Immunol.*, **15**(11), 692–704.
97. Sanchez-Martin, L., et al. (2011). The chemokine CXCL12 regulates monocyte-macrophage differentiation and RUNX3 expression. *Blood*, **117**(1), 88–97.
98. Liu, J. Y., et al. (2015). Delivery of siRNA using CXCR4-targeted nanoparticles modulates tumor microenvironment and achieves a potent antitumor response in liver cancer. *Mol. Ther.*, **23**(11), 1772–1782.
99. Ley, K., et al. (2016). Integrin-based therapeutics: Biological basis, clinical use and new drugs. *Nat. Rev. Drug Discov.*, **15**(3), 173–183.
100. Choi, M. R., et al. (2007). A cellular Trojan Horse for delivery of therapeutic nanoparticles into tumors. *Nano Lett.*, **7**(12), 3759–3765.
101. Stephan, M. T., et al. (2012). Synapse-directed delivery of immunomodulators using T-cell-conjugated nanoparticles. *Biomaterials*, **33**(23), 5776–5787.
102. Almeida, J. P., et al. (2011). In vivo biodistribution of nanoparticles. *Nanomedicine (Lond)*, **6**(5), 815–835.
103. Griffiths, M. M., Langone, J. J., Lightfoote, M. M. (1996). Biomaterials and granulomas. *Methods*, **9**(2), 295–304.
104. Bazile, D., et al. (1995). Stealth Me.PEG-PLA nanoparticles avoid uptake by the mononuclear phagocytes system. *J. Pharm. Sci.*, **84**(4), 493–498.
105. Abu Lila, A. S., Kiwada, H., Ishida, T. (2013). The accelerated blood clearance (ABC) phenomenon: Clinical challenge and approaches to manage. *J. Control. Release*, **172**(1), 38–47.
106. Bridges, A. W., et al., Reduced acute inflammatory responses to microgel conformal coatings. *Biomaterials*, **29**(35), 4605–4615.
107. Goh, S. L., et al. (2004). Cross-linked microparticles as carriers for the delivery of plasmid DNA for vaccine development. *Bioconjug. Chem.*, **15**(3), 467–474.
108. Rosenblum, D., Peer, D. (2014). Omics-based nanomedicine: The future of personalized oncology. *Cancer Lett.*, **352**(1), 126–136.
109. Lim, S. H., et al. (2010). Anti-CD20 monoclonal antibodies: Historical and future perspectives. *Haematologica*, **95**(1), 135–143.

110. Deligne, C., et al. (2015). Anti-CD20 therapy induces a memory Th1 response through the IFN-gamma/IL-12 axis and prevents protumor regulatory T-cell expansion in mice. *Leukemia*, **29**(4), 947–957.

111. Nambisan, P. (2017). *An Introduction to Ethical, Safety and Intellectual Property Rights Issues in Biotechnology*, 1st ed., Academic Press, ISBN 9780128092316 (Paperback), 9780128092514 (eBook).

Supplementary References

1. Bronte, V., et al. (2016). Recommendations for myeloid-derived suppressor cell nomenclature and characterization standards. *Nat. Commun.*, **7**, 12150.

2. Singh, G. (2018). Biosimilars. In: Vohora, D, Singh, G., eds. *Pharmaceutical Medicine and Translational Clinical Research*, chapter 22, Academic Press, Boston, pp. 355–367, ISBN: 9780128021033.

3. Saldanha, M., et al. (2018). A regulatory perspective on testing of biological activity of complex biologics. *Trends Biotechnol.*, **36**(3), 231–234.

4. Egeberg, A., et al. (2018). Safety, efficacy and drug survival of biologics and biosimilars for moderate-to-severe plaque psoriasis. *Br. J. Dermatol.*, **178**(2), 509–519.

5. Pallardy, M. J., et al. (2017). Why the immune system should be concerned by nanomaterials? *Front. Immunol.*, **8**, 544.

6. Scott, E. A., et al. (2017). Overcoming immune dysregulation with immunoengineered nanobiomaterials. *Annu. Rev. Biomed. Eng.* **19**, 57–84.

7. Aguado, B. A., et al. (2018). Engineering precision biomaterials for personalized medicine. *Sci. Transl. Med.*, **10**(424), eaam8645.

8. Bawa, R., Audette, G. F., Rubinstein, I. (eds.) (2016). *Handbook of Clinical Nanomedicine: Nanoparticles, Imaging, Therapy, and Clinical Applications*, Pan Stanford Publishing, Singapore, 1662 p., 55 chapters, ISBN 9789814669207 (Hardcover), 9789814669214 (eBook).

9. Bawa, R. (ed.); Audette, G. F., Reese, B. E. (assist. eds.) (2016). *Handbook of Clinical Nanomedicine: Law, Business, Regulation, Safety, and Risk*, Pan Stanford Publishing, Singapore, 1448 p., 60 chapters, ISBN 9789814669221 (Hardcover), 9789814669238 (eBook).

10. Tinkle, S., McNeil, S. E., Mühlebach, S., Bawa, R., Borchard, G., Barenholz, Y., Tamarkin, L., Desai, N. (2014). Nanomedicines: Addressing the scientific and regulatory gap. *Ann. N. Y. Acad. Sci.*, **1313**, 35–56.

Chapter 27

Characterization of the Interaction between Nanomedicines and Biological Components: *In vitro* Evaluation

Cristina Fornaguera, MSc, PhD

Group of Materials Engineering (GEMAT), Bioengineering Department, IQS School of Engineering, Ramon Llull University, Barcelona, Spain

Keywords: nanotechnology, nanoscience, nanomedicine, nanobiotechnology, nanomaterials, nanoparticles, nanosystems, personalized therapies, characterization techniques, interaction with biological components, safety, efficacy, technology transfer, biocompatibility, cytotoxicity, blood proteins interaction, blood cells, blood components, albumin, fibrinogen, C3 complement protein, hemolysis, coagulation, cell uptake, phagocytosis, cell transfection, target cells, erythrocytes

27.1 Introduction

Nanosystems and nanomaterials are general terms to designate any entity with at least one dimension having sizes ranging in the nanometric scale, which includes a huge variety of nanoentities with different properties, such as nano-emulsions, nanoparticles, polyplexes, dendritic structures, micelles and liposomes, among others [1–3]. The interest of colloidal nanosystems has witnessed an exponential increase since the first reports appeared in the 1990s (Fig. 27.1A). Specifically, the interest in their use for biomedical applications, the field in which they are called nanomedicine, has increased notably during the past 20 years (Fig. 27.1B) [4–6], because current treatments have not yet solved some drawbacks, such as the controlled release of therapeutic compounds and their appropriate biodistribution. In parallel, there has been an exponential increase in the impact of personalized medicine on nanotherapies (Fig. 27.1C). Since the beginning of gene therapy, this type of therapy has arisen as an opportunity to treat each individuals with the specific requirements defined in their genome so it must be taken into account when designing nanosystems. It is thought that by combining the advantages of nanotechnology and personalized medicine, very efficacious treatments will be developed in the next few years.

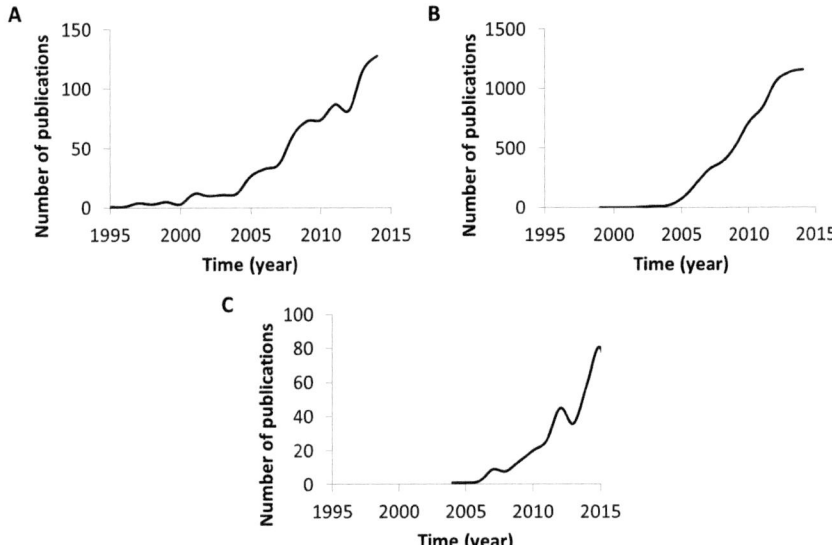

Figure 27.1 Number of publications per year indexed in Scopus that contain the following terms: (A) "nanosystem," (B) "nanomedicine," and (C) "personalized nanomedicine."

In conventional treatments, high doses of biological actives are required, which usually produce severe side effects. In some cases, the treatment proposed is useful for one person but not for others. In this context, novel delivery nanosystems are advantageous in numerous aspects for personalized therapies: (1) They protect the actives they encapsulate; (2) they enable the controlled and sustained release of actives, allowing a decrease of therapeutic concentrations; (3) they can be tuned to reach their targeting organs, thus enabling a localized active release and reduction of side effects, and (4) their surface can be tailored with a variety of chemical moieties to achieve multifunctional therapeutics specific for each individual [4, 6–9]. The numerous advantages that nanomedicines possess have been perceived by the pharmaceutical industry, thus resulting in the industrial production and commercialization of a variety of nanomedicines, such as the widely known Abraxane®, a nanomedicine approved in 2005 for its demonstrated efficacy in various cancers, which consists of the nanostructured conjugate of paclitaxel [6, 10–12]. Remarkably, not only the therapeutic applications of nanomedicines but their applications as diagnostic nanosystems and even for theranostic purposes (combination of therapy and diagnosis in the same nanosystem) are of interest [6, 13, 14].

Prior to colloidal nanosystems, transference to pharmaceutical industries and their commercialization, and after a wide characterization of their physicochemical properties, a deep study of their interaction with biological components is required (Fig. 27.2) to ensure safe nanomedicines are obtained and to clearly define their behavior in physiological conditions. Specifically, the interaction of nanosystems with blood components (e.g., proteins and cells) is of the utmost importance, since most administered nanomaterials will end up being distributed by the bloodstream. Although huge efforts have been made in the field of nanomedicine characterization [6, 15], for example, by the Nanotechnology Characterization Laboratory in the United States, currently, standardized protocols, assays, and/or methodologies for the adequate characterization of colloidal nanomaterials before preclinical assays do not yet exist, and the Food and Drug Administration has not yet formulated guidelines [3, 12]. However, several reviews have appeared with the objective to summarize characterization techniques for nanomaterials, mainly for nanoparticles [2, 3, 6, 15, 16]. Each of them presents a particular point of view of the most relevant techniques to study nanomaterials. Most of them have pointed out the importance of characterizing nanomedicines in the physiological desired conditions to avoid misleading results, since environmental conditions affect the properties of nanomaterial [2, 15, 16], but the interactions of nanomaterials with biological components were only described in a few and with a brief description, without specifying useful techniques [15, 16].

In addition, concerning the type of nanomedicine studied, most of these reviews focused mainly on nanoparticles [2, 3, 6, 15, 16]. In fact, the available information is dispersed in the current literature and a single review representing a complete guide to follow in the development of novel nanomaterials is missing. In this context, this chapter aims to summarize the main techniques concerning the interaction of nanomedicines with biological systems. Its objective is to review the currently available techniques for a complete characterization of the interaction of nanomaterials with biological components, which must be addressed when designing nanomedicines, before starting clinical and preclinical studies. For the study of each parameter, different techniques are described, highlighting their advantages and disadvantages and indicating which nanomaterials can be properly studied by each technique. From now on, the terms nanosystems, nanomedicines, and colloidal nanomaterials will be used as general synonyms to name nanostructures with nanometric dimensions, independent of the specific entity they represent (e.g., nanorods or nanoparticles), their composition (e.g., polymer or metal), and their use (e.g., drug delivery systems or non viral gene delivery systems) (Fig. 27.3).

As stated above, the translation of knowledge from novel designed nanosystems at a research laboratory scale to real human therapies is usually a limiting or even a final point due to the lack of systematic studies. Therefore, this book chapter will be a useful support for those nanotechnology scientists aiming to develop nanosystems for real nanomedicine applications.

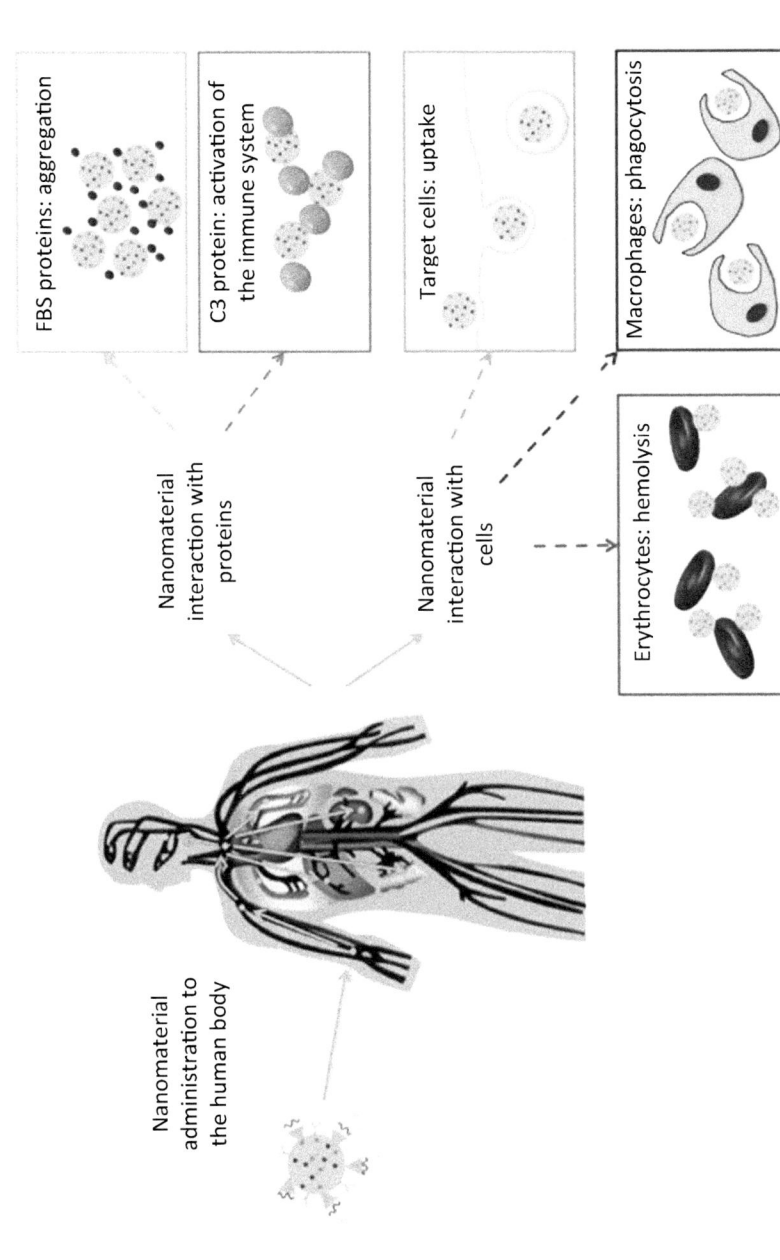

Figure 27.2 Schematic representation of possible interactions of some nanosystems with biological components, namely cells and proteins.

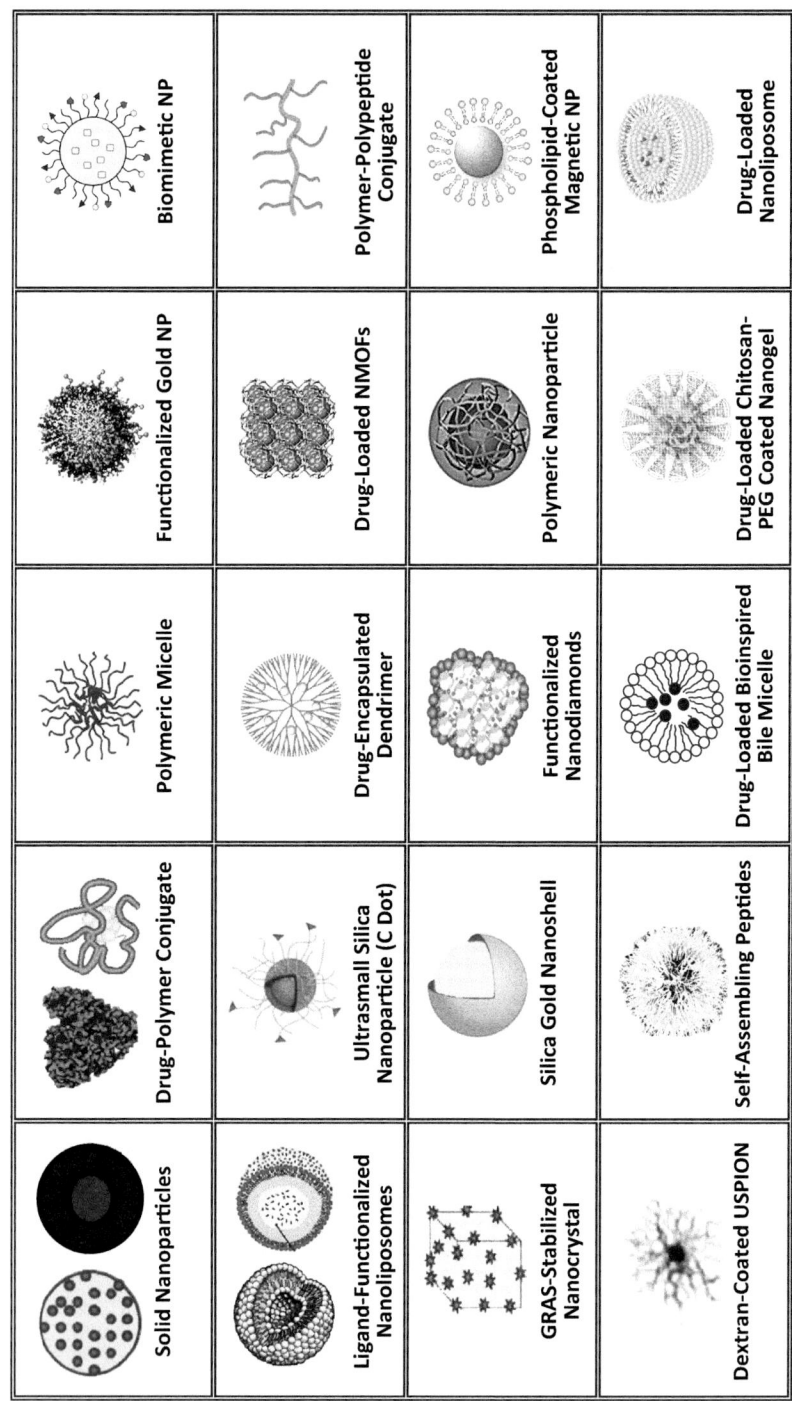

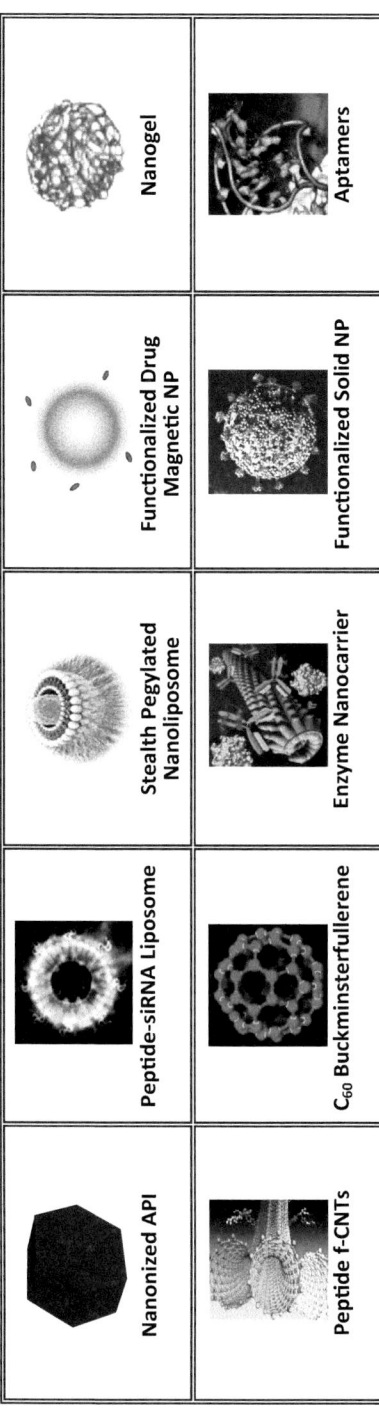

Figure 27.3 Nanoscale Drug delivery system platforms (nanodrug products). Schematic representation of selected engineered nanoparticles (NPs) used in drug delivery that are either approved by regulatory bodies, are in preclinical development or are in clinical trials. In most cases shown above, they are considered as first or second generation multifunctional engineered NPs, ranging in average diameter from one nanometer (1 nm) to a micron (1000 nm). Active bio-targeting of the NP is often achieved via conjugating ligands or functional groups (antibodies, peptides, aptamers, folate, hyaluronic acid). These molecules are tagged to the NP surface with or without spacers/linkers such as PEG. Many nanodrugs depicted above (e.g., metal-based NPs, f-CNTs, NMOFs, etc.), although extensively advertised for drug delivery, will pose enormous drug approval and commercialization challenges and will not appear in the clinic this century. Non-engineered antibodies, biological motors (e.g., sperms), engineered nanomotors, and naturally occurring NPs (natural protein nanotubes) are specifically excluded here. Antibody–drug conjugates (ADCs) are encompassed by the cartoons labeled "Polymer-Polypeptide Conjugate" or "Drug-Polymer Conjugate." Therapeutic monoclonal antibodies (TMAbs), polymer-polypeptide conjugates, and aptamers shown above are classic biologics but they are also nanodrugs as they fall within the widely accepted standard definition of nanodrugs. The list of NPs depicted here is not meant to be exhaustive, the illustrations do not reflect precise three-dimensional shape or configuration, and the NPs are not drawn to scale. *Abbreviations*: NPs, nanoparticles; PEG, polyethylene glycol; GRAS, Generally Recognized As Safe; C dot, Cornell dot; API, active pharmaceutical ingredient; ADCs, antibody–drug conjugates; NMOFs, nanoscale metal organic frameworks; f-CNTs, functionalized carbon nanotubes; siRNA, small interfering ribonucleic acid; USPION, ultrasmall superparamagnetic iron oxide nanoparticle. Copyright 2018 Raj Bawa. All rights reserved.

27.2 Experimental Techniques for the Analysis of Nanoparticle Interaction with Biological Components

A deep physicochemical characterization of designed nanomaterials, intended for biomedical applications, is a must before *in vitro* efficacy studies [3]. The rigorous characterization required for nanosystems is noteworthy, since they are usually composed of various parts that must be fully characterized individually and together in nanomedicine (e.g., multifunctional nanoparticles with encapsulated compounds and various functionalization moieties) [6]. In general, the key parameters to study can be classified in three types of techniques: analytical, colloidal, and biological techniques [2, 3, 12, 15, 17]. In this review, a deep description of the third type of technique is detailed: those aimed to define the interaction of nanomaterials with biological components in a physiological environment.

27.2.1 Techniques to Study the Interaction of Nanomaterials with Proteins

Most therapeutic nanosystems are designed for their intravenous systemic administration (parenteral route), targeting the desired organ. Therefore, prior to reaching the target tissue, they get in contact with many proteins in the blood. For this reason, the characterization of the interaction of nanomaterials with proteins is of the utmost importance, since they must be specifically designed to avoid undesirable interactions with any kind of protein. Techniques to study these interactions are specified in the following subsections. In brief, the stability of nanomaterials in a physiological environment is described first. Then, electrophoretic techniques to study the interaction of nanomaterials with proteins (a cocktail of different kinds of proteins or a specific individual protein) are detailed, and finally, the specific case of interactions with blood components is also discussed.

27.2.1.1 Nanomaterial stability in a physiological environment

Prior to studying specific protein interactions, it is important to confirm nanomaterial stability in a biological environment, simulating physiological conditions, such as the widely known fetal bovine serum (FBS) protein solution. Therefore, a preliminary study of nanomaterial interactions with proteins can be performed without the need of any biological system, only using the purified proteins. There are many techniques that have been used to assess the stability of nanosystems in physiological conditions—most of them summarized in a previous study by Fornaguera et al. [18], who studied the stability of PLGA nanoparticles, with different modifications, in a protein solution. In general, they are recommended as a first step in the characterization of nanomaterial–biological component interactions because since they do not use a biological material (except for the purified proteins solutions), they are less costly, safe, rapid, and easy to perform than *in vitro* and *in vivo* studies.

First, it is recommended to measure a key parameter of the nanomaterial, such as droplet size or surface charge, as a function of time, under incubation at 37°C with FBS, since this is the temperature of physiological systems and FBS is representative of the whole blood protein solution. If changes in the measured parameter take place, it is also recommended to deeply study individual proteins to find out which one produces the change and which would be the one interacting with the nanomaterial. In addition, the study of interactions with a single protein is recommended because it must be taken into account that the use of serum from different origins (e.g., different species or different individuals of a same specie) can result in divergent results. Serum is a biological component, obtained from animals or individuals. Therefore, the reproducibility of the results could be difficult when changing the serum origin because of the inter-individual variability that could result in the identification of different interactions. For this reason, the testing of sera from different species (e.g., human, bovine) and from different commercial brands is strongly recommended, since the whole results can give a complete view of the expected *in vivo* behavior of designed nanomaterials in a protein environment.

Albumin is the most abundant protein in the blood, so it is the first candidate to test. In parallel, fibrinogen can be also tested, since it represents a different group of proteins, due to its elongated shape, compared with the spherical 3D conformation of albumin [9, 18]. When incubating nanoparticles with albumin or fibrinogen, the droplet size of nanoparticles could notably increase due to the formation of microscopic aggregates, thus not enabling measuring by light scattering techniques (Fig. 27.4). However, in some cases, the size remains in the nanometric range and can be measured by DLS (Fig. 27.4A), although in most cases, the presence of aggregates is so evident that measurements overpass the nanometric range and can be confirmed under optical microscopy (see Fig. 27.4B as an example). If the aggregation is even huge, macroscopic sediment will also appear (Fig. 27.4C).

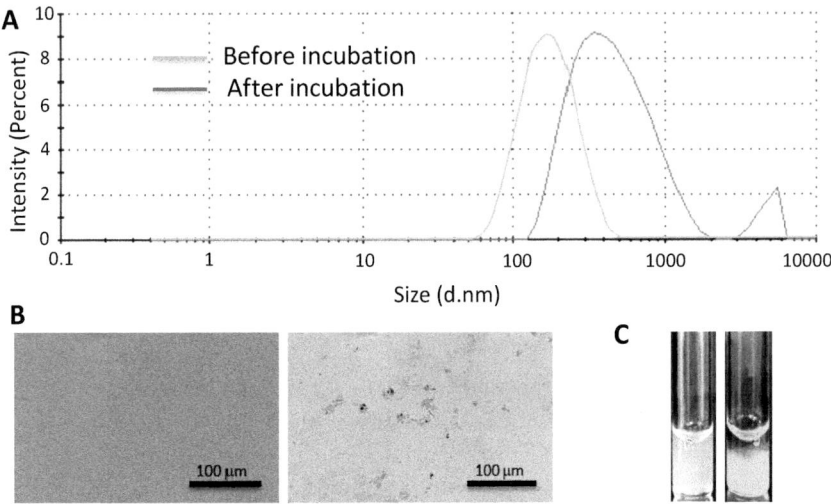

Figure 27.4 (A) Hydrodynamic size distribution of a PLGA nanoparticle dispersion before (left) and after (right) incubation with fibrinogen; (B) optical micrographs; (C) macroscopic appearance. The change in the visual aspect, microscopic structure, and nanometric size indicates the presence of aggregates.

It is worth noting that physiological conditions may vary in patients (diseased individuals), compared with healthy individuals. For this reason, to design a personalized therapy for each patient, physiological conditions to test the stability of nanomaterials must be tuned in order to simulate the environment of each patient. However, achieving commercial protein mixtures that simulate disease conditions is usually much more difficult than using healthy materials, which could make the study difficult and even result in difficult extrapolation of results obtained in research laboratory to what would happen in disease physiological conditions. In this case, after the laboratory study of the interactions with conditions similar to those of diseased individuals, it is recommended to study protein interactions in an animal model of the disease, to find out if the expected behavior of the designed nanomaterial takes place.

27.2.1.2 Electrophoretic techniques

Electrophoretic techniques (summarized in Table 27.1) are also used to study the interactions between nanomaterials and proteins. They consist of the migration of charged molecules (DNA, RNA or proteins) through a polymeric matrix under an

Table 27.1 Summary of main electrophoretic techniques

Technique	Characteristics of analyses	Advantages	Disadvantages	References
SDS-PAGE electrophoresis	Proteins present in a sample separated as a function of their molecular weight	Simple and affordable	Requires preparative steps	Manabe (2000) Hames (2002)
Rocket immuno-electrophoresis and radial immuno-diffusion	Quantification of individual proteins Study of BSA interaction with nanoparticles	Simple, quick, and reproducible High accuracy Quantitative results Use of little equipment No preparative separation steps	Require a calibration curve	Mancini (1965) Laurell (1966) Weeke (1973) Vauthier (2009) Vauthier (2011)
2D immuno electrophoresis	Quantification of individual proteins Study of the activation of the complement system	Simple, quick, and reproducible Use of little equipment Quantitative results No preparative separation steps	Require a calibration curve Difficult to extrapolate to *in vivo* results Low sensitivity	Bertholon (2006) Vauthier (2011) d'Addio (2012)

electric field provided by submerged electrodes, the migration being dependent on either the size or the isoelectric point of the studied material [19]. The polymeric matrix is usually composed of agarose, to separate macromolecules, such as nucleic acids or polyacrylamide, to separate smaller molecules such as proteins [19]. They have been widely used in the nanomedicine field to determine hydrodynamic sizes, study the formation of the protein corona onto the nanomaterial surface and the adsorption of specific proteins onto the nanomaterial surface [3, 18, 20–22]. Their main uses are described in the following subsections.

27.2.1.2.1 SDS-PAGE Electrophoresis

Sodium dodecyl sulfate–polyacrylamide gel electrophoresis (SDS-PAGE) is the most commonly used electrophoresis to study proteins. It is a denaturing electrophoresis that, performed in a single dimension (the most common application), separates proteins as a function of their apparent molecular mass, since after the denaturing process and incubation with SDS reagent, all proteins are equally negatively charged, without their 3D structure [19, 23]. The size of the pores formed in the polymeric matrix depends on the percentage of the polyacrylamide used [19].

SDS-PAGE is a simple and affordable technique that has been used, prior to *in vitro* and *in vivo* studies, to find out and quantify the proteins attached to the nanoparticle surface, for example see Fig. 27.5 [19, 24]. It is recommended as a first approximation to find out the level of protein corona formation on a nanomaterial surface, which is called opsonization, and represents the first step of a nanosystem phagocytosis, which is not desired, because it neutralizes the nanomaterial and then it cannot reach the target organ, thus unable to perform the therapeutic action. When analyzing a library of formulations, this technique can be useful to select the one that is less opsonized. However, since it is a basic technique, the information that it gives is limited, and after this preliminary analysis of opsonization, more specific techniques (detailed below) are recommended. It has some disadvantages, such as the number of preparative steps required prior to electrophoresis and the only semi-quantitative result that it gives [23].

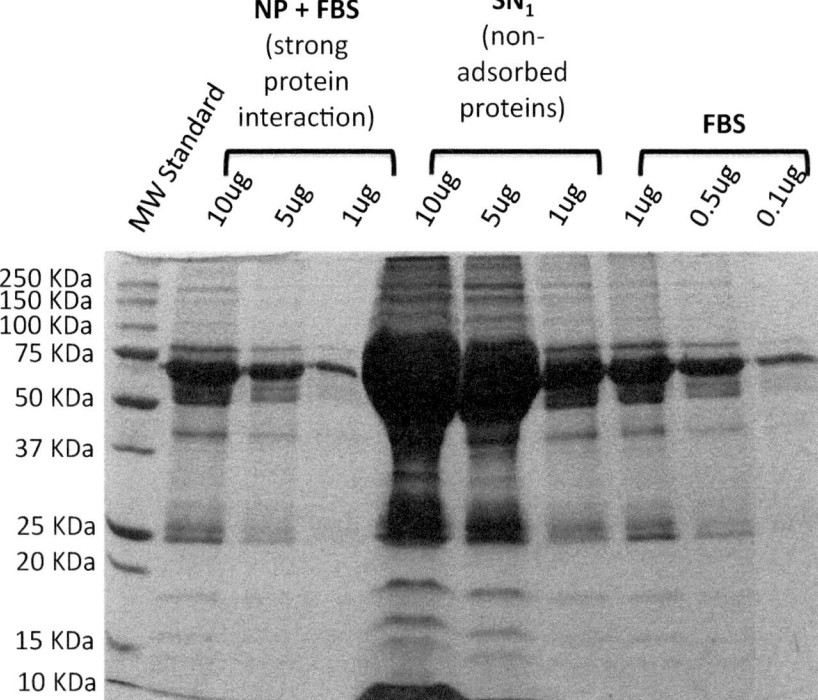

Figure 27.5 Example of SDS-PAGE of PLGA nanoparticles incubated with fetal bovine serum (FBS). NP + FBS fraction corresponds to those proteins strongly adsorbed onto the nanoparticle surface, while the SN1 fraction corresponds to the non-adsorbed proteins. As it can be easily seen, the amount and kind of adsorbed protein is lower than the non-adsorbed ones [25].

27.2.1.2.2 Immunomethods

The term immunomethods or immunochemical methods comprises a variety of immunoelectrophoretic methods, which could be defined as the use of antibodies to detect the results obtained by electrophoresis [9]. Therefore, the high sensitivity of immunomethods is an advantage, in contrast to the high cost that antibodies involve. Different kinds of agarose electrophoresis are performed, where proteins migrate under various conditions. After the time left to run, proteins are detected through gel incubation with specific antibodies, which are further stained, for example, with Comassie Blue. In this review, three kinds of immunomethods are described—radial immunodiffusion, rocket immune-electrophoresis, and 2D-immunoelectrophoresis—which were considered of interest to study the interactions of proteins with nanomedicines in colloidal dispersion (examples in Fig. 27.6).

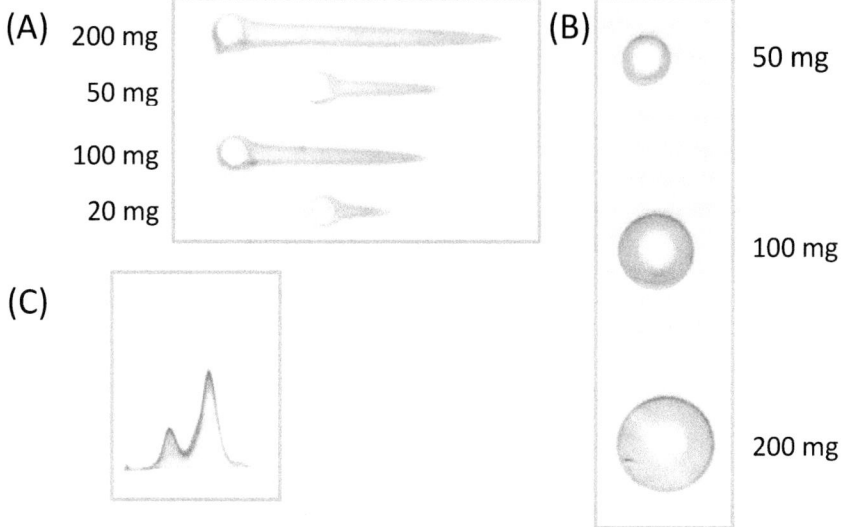

Figure 27.6 Examples of the gels obtained using described techniques. (A) Rocket immunoelectrophoresis of the free bovine serum albumin (BSA) after being incubated with polymeric nanoparticles at increasing concentrations. (B) Radial immunodiffusion of the free bovine serum albumin (BSA) after being incubated with polymeric nanoparticles at increasing concentrations. (C) 2D-immunoelectrophoresis of the C3 proteins after being incubated with polymeric nanoparticles. Adapted from [25].

27.2.1.2.2.1 Rocket immunoelectrophoresis

Rocket immunoelectrophoresis is a rapid, simple, and reproducible immunochemical technique to determine and quantify a single protein contained in a sample in which other components, even other proteins, can be present. Samples are loaded in circular wells at one edge of the agarose gels and migrate in the gel under an electric field. The protein migration results in a rocket-shape precipitation (Fig. 27.6A), whose height is proportional to the protein concentration [9, 26].

This method is widely used for the determination of the bovine serum albumin (BSA) protein that remains in solution, after its incubation with different kinds of polymeric nanoparticles, although it could be used for other types of nanosystems, since BSA is the major protein of the blood and its interaction with nanomaterials will influence their fate and biodistribution [9, 27]. It can be also used for the study of other kinds of proteins. Nevertheless, since BSA is the main protein in blood, when studying other proteins, it must be taken into account that the behavior of individual minor isolated proteins could vary when major proteins are present. To circumvent this problem, a strategy is to incubate nanomaterials with FBS or a protein cocktail but detect specifically the desired protein.

27.2.1.2.2.2 Radial immunodiffusion

Radial immunodiffusion is an immunochemical method for the quantification of a single protein contained in a sample with other components. Like the others, it consists of an agarose gel. In this method, however, the samples are loaded in circular wells at a central region of the gel and diffuse through the gel, forming a sedimentation ring (Fig. 27.6B) whose diameter is proportional to the amount of free protein that has diffused [9].

As rocket immunoelectrophoresis, radial immunodiffusion has been widely used for the quantification of free BSA after its incubation with polymeric nanoparticles [9]. The difference between them is the concentration of proteins that they can detect: While rocket immunoelectrophoresis can detect protein concentrations as low as 5 µg/mL, radial immunodiffusion requires at least 50 µg/mL of proteins or higher to be detected. Therefore, it is recommended to use both techniques for the same study to have a wider concentration range to study.

Since both radial immunodiffusion and rocket immunoelectrophoresis are similar techniques, their advantages and drawbacks are summarized together. An important advantage of both techniques is the possibility to obtain quantitative results with little equipment and a rapid, easy, sensitive, and highly accurate method [9, 22, 28]. Another relevant advantage is that they can be performed directly with bioconjugates, without a separation step of proteins from nanomaterials, which is advantageous because separation steps could change the balance of attached/nonattached proteins and perturb the results. A drawback of these techniques is the requirement of running a calibration curve, necessary for the quantification of the studied proteins. The calibration curve is usually composed of known concentrations of the proteins to test. For radial immunodiffusion, the calibration curve consists of a linear relationship between the square of the diameter of the sedimentation rings and the concentration of the protein in the well, while for rocket immunoelectrophoresis, it is a linear relationship between the height of the immunoprecipitation peak and the protein concentration in the well [9, 25]. Another important drawback is the difficulty to perform these techniques with surfactant containing samples, since some surfactants, such as the widely used polysorbate 80, were described to perturb the quantification of the results [29].

27.2.1.2.3 2D-immunoelectrophoresis to detect nanomaterial interaction with C3 complement protein

2D immunoelectrophoresis is a very versatile technique that can be used for a multiplicity of purposes, all of them involving protein migration, such as the careful study of protein opsonization (using a nondenaturalizing gel): running of a first dimension as a function of protein molecular weight and running of the second dimension as a function of the isoelectric point (pI). Although many proteins are present in FBS, each spot of the 2D gel will represent a protein of a characteristic MW and a characteristic pI, which can be attributed to a single protein. Since this technique has been used for many purposes for a long time, its use in the nanomedicine field is not extended, and its description is beyond the scope of this chapter.

However, in this chapter, 2D immunoelectrophoresis is presented as a useful tool to study the activation of the complement system, as an indication of the influence of nanomedicine on the immune system, which is a key parameter to study nanomedicines before *in vivo* administration, since it will determine the lifetime of the nanomedicine, its fate, and biodistribution [30, 31]. This technique specifically detects the interaction of the nanosystems with the C3 protein. This protein, when activated, is broken into diverse fragments of different shapes and sizes (Fig. 27.7A). 2D immunoelectrophoresis of C3 protein consists of the use of a horizontal agarose protein electrophoresis in two dimensions. In the first dimension, the C3 and its subunits are separated as a function of their molecular weight (the smaller the fragment, the further it migrates), while in the second dimension, they are separated as a function of their concentration (the higher the concentration, the further they run) (Fig. 27.6C shows an example of a C3 activated material).

To study the complement system activation, the C3 protein is the best choice, since it is a key protein and the major protein of the complement cascade [30–32]. Therefore, it has to be incubated with the samples to test. After incubation, samples are loaded into the well of the gel and then migrate through the two dimensions. If the C3 protein is not activated, which means that nanomaterials do not influence the complement cascade, a single peak at slow migration in the first dimension is obtained, corresponding to the entire C3 protein (Fig. 27.7B, left). In contrast, if a sample activates the C3 protein, it breaks into fragments that—since they are smaller—migrate longer in the first dimension and are detected as two peaks (Fig. 27.7B, right) [31].

This methodology is a rapid, versatile, and facile technique that can be applied with simple equipment. However, its sensitivity is low and if the nanosystems is composed of protein components, it will give a higher interaction of the C3 protein, although in some cases that does not mean that a higher activation of the immune system is produced (false positive) [18, 30]. When designing nanomaterials for therapeutic applications, in many cases, it is not desired to activate the immune system, since if it detects the administered nanosystems, it will activate the degradation of the component detected as exogenous. As a consequence, the nanosystems will be cleared, not enabling their arrival to the target organ. An exemption of this assumption is the case of immunotherapy, where the activation of the immune system is required. Nevertheless, in any case, complement activation is an important parameter to study.

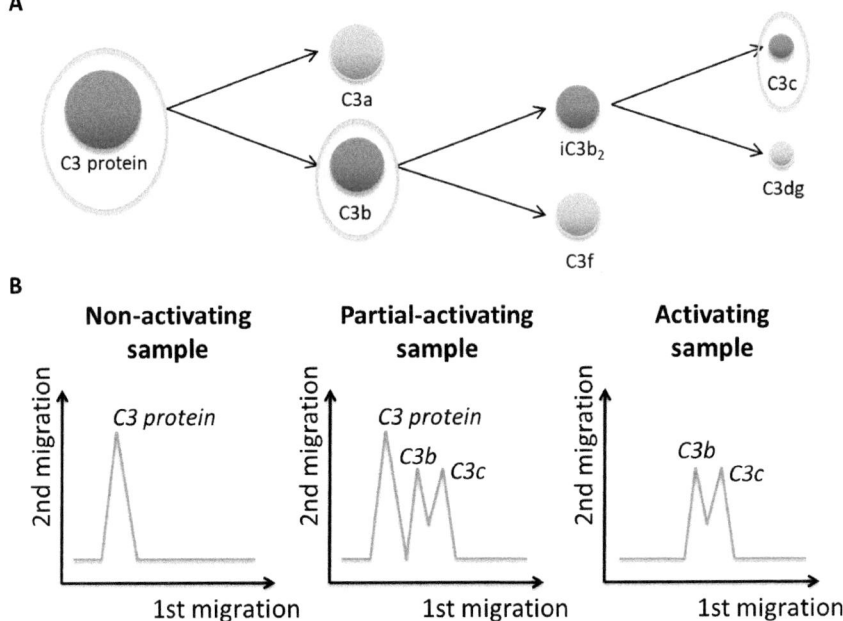

Figure 27.7 (A) Schematic representation of the classical pathway of the complement activation. Circles represent the proteins analyzed by this technique. (B) The expected 2D-immunoelectrophoresis gels when studying different types of samples.

Therefore, the application of this technique is recommended together with the study of protein corona formation (described in Section 27.2.1.2) for a preliminary study of the interaction of nanomaterials with proteins and their further detection and elimination by the immune system.

27.2.1.3 Blood coagulation study

The coagulation or clotting time is another important parameter to study before preclinical studies, which is related to the interaction of the nanosystems with proteins of the coagulation cascade. This interaction could elongate or shorten the coagulation time, but in most cases, it is not desired to modify coagulation time. Although standard values for the coagulation time exist, they could slightly vary for each individual. As stated earlier, these slight variations are especially important when designing personalized therapies, specifically in the case of diseased individuals, since some diseases can modify the blood coagulation times (e.g., hemophilia), and controls of diseased blood are required to study the effect of nanomaterials.

Blood coagulation times can be assessed by performing two tests that correspond to the two pathways of coagulation cascade activation (Fig. 27.8). It is important to study both the extrinsic and the intrinsic coagulation times, since nanomaterials could produce a specific effect only in one of the pathways that could be underestimated if only the other one is studied.

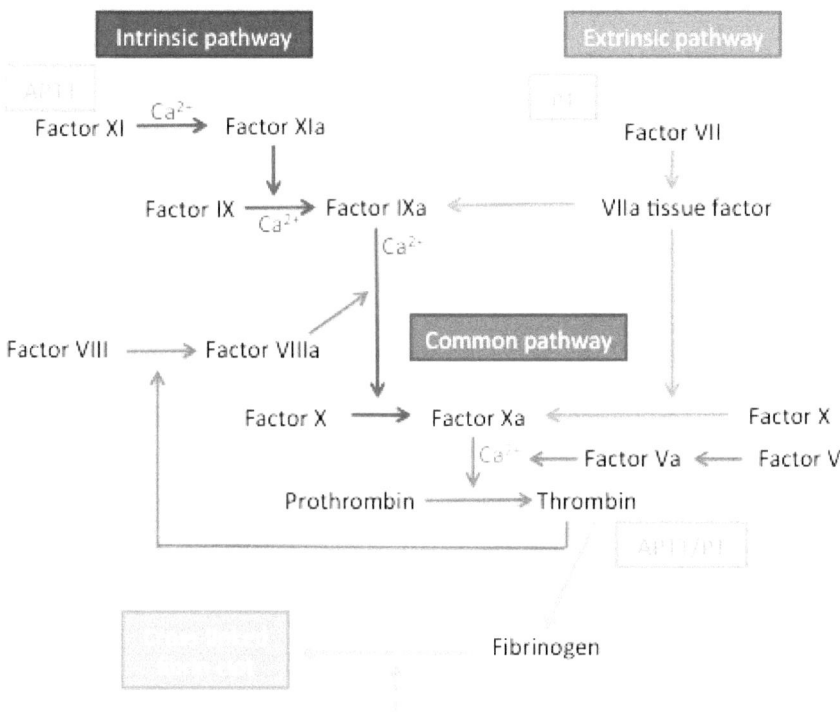

Figure 27.8 Schematic representation of the blood coagulation cascade.

The prothrombin time (PT) test measures the activation of coagulation by the extrinsic or tissue factor. For the assessment of the coagulation time, once a sample is placed into the coagulometer, phospholipid calcium thromboplastin is added and the coagulation time is automatically assessed. Normal values for healthy humans are in the range 12–15 s.

The activated partial thromboplastin time (APTT) test consists of the measure of the activation of the contact or intrinsic coagulation pathway. To calculate this parameter, cephalin, a negatively charged phospholipid acting as a contact activator, is added to the sample previously placed in the coagulometer. The mixture is incubated for 2 min, followed by the addition of calcium chloride to activate the clot formation and measurement of the coagulation time. For this experiment, the normal values for healthy humans are 25–35s [33].

To perform this technique, it is important to obtain blood (human blood if possible) of high quality and maintained in a noncoagulation environment. In addition, the use of fresh blood is strongly recommended, since after some days, the blood is damaged. As indicated above for other techniques, this technique is based on the use of a biological material, the blood; so variations between individuals, and more with diseased individuals, are expected. Therefore, it is a must to study the effect of nanomaterials on the coagulation of blood from different individuals.

It must be taken into account that one of the key proteins of the coagulation cascade is fibrinogen (Fig. 27.8). Therefore, the interaction of nanomaterials with this protein could influence the coagulation cascade [33]. For this reason, it is recommended to study the fibrinogen aggregation if alterations on the coagulation cascade are found, using techniques stated above, such as a nanosystems' stability in the presence of fibrinogen and fibrinogen electrophoresis (instead of BSA, as described above) after incubation with the nanosystems.

27.2.2 Techniques to study interactions of nanomaterials with cells

Apart from interactions with proteins, when nanomaterials enter the body, they also get in contact with the cells. The interaction of nanomaterials with one type of cells can be desired, when aiming to transfect them using genes, for example, but it may also be undesired, when this interaction is with immune cells (with the exception of immune therapies). In addition, it must be taken into account that the presence of the nanomaterial or the presence of its degradation products can produce cell toxicity.

Techniques to study both cell interactions and cell toxicity are summarized in the following subsections.

27.2.2.1 Desired interaction with cell surface: nanomaterial uptake

Nanomaterial uptake is a key point to further achieve pharmacological action of the drug or the transfection of genetic material. The most common mechanism of nanoparticle uptake is by endocytosis, using different pathways (e.g., caveolae or clathrin) that depend on the properties of the nanomaterial. Therefore, nanosystems must be carefully designed in order to target the desired receptor. In addition, cell type–specific receptors must be also used, which need nanomaterials to be previously functionalized with an active target moiety.

It is important to remark on the use of fluorescent dyes to follow a nanosystems' uptake, labeling both the nanosystem and cell components. The colocalization of the nanosystem dye with the subcellular structure labeled means that the nanosystem penetrates the cell through this route.

In this case, in contrast to the techniques used to study the nanomaterials' interaction with proteins, to study the interactions of nanomaterials with cells, cells are required. Similar to all techniques that use cell cultures, the results will depend not only on the properties of the nanomaterial studied but also on the cell type used. In most studies, immortalized cell lines are used, due to their ease of use, commercial availability, and immortality. However, the results using primary cell lines are more reliable, since they come directly from biological individuals or animals and they have suffered fewer modifications than immortalized cell lines. Therefore, when translating into *in vivo* studies, the results are expected to be consistent. However, obtaining primary cell lines can be difficult for nonspecialized groups as working with them requires well-trained researchers, together with access to these cells.

27.2.2.2 Phagocytosis assessment

Phagocytosis is a specific kind of nanomaterial uptake, but it can be only performed by the so-called phagocytic cells, macrophages being the most representative. In contrast to endocytosis, phagocytosis is not desirable in most cases, since it represents the first step in the elimination of the nanosystem by the reticuloendothelial system (RES). Many parameters of the nanomaterial influence the phagocytic rate. For example, the PEGylation of the nanomaterial, together with an elongated shape, is a factor that decreases phagocytosis [16, 33]. Similar to cell penetration, phagocytosis can be studied *in vitro* taking advantage of fluorescence microscopy.

27.2.2.3 Toxicity assessment techniques

Colloidal nanomedicines have to be always designed to ensure the use of biocompatible and biodegradable materials, thus achieving a safe therapy. However, the properties of nanomaterials sometimes differ from those of the materials they are made of and physiological interactions of nanomaterials with biological components may differ from those of isolated materials [2, 3]. For this reason, before starting any *in vitro* and further *in vivo* assays, nanomaterial toxicity has to be studied in depth using various cell lines and diverse toxicity tests, since each test measures different parameters of the cells and they have different sensitivity [34]. These tests, in general, have the advantage of a facile, rapid, and economic performance, and more important, the use of animals is reduced, which should be done only when *in vitro* alternatives do not exist [35]. However, it is also remarkable that some results obtained *in vitro* cannot be extrapolated to the *in vivo* behavior [35]. In the following sections, the most commonly used tests to assess nanomaterials toxicity *in vitro*, namely hemolysis and cytotoxicity, are discussed.

27.2.2.3.1 Hemolysis determination

Erythrocytes or red blood cells (RBCs) represent the most abundant cells in the blood (around 45%). They are responsible for oxygen transport to the tissues and the removal of carbon dioxide, also contributing to the acid–base blood balance. Therefore, a nanomaterials' effects on erythrocyte integrity represents a widely used measure of their toxicity in the blood [18, 33]. The term hemolysis refers to the potential of a substance to damage erythrocytes, which may lead to their loss. It occurs when the hemoglobin is released from a compromised or ruptured erythrocyte plasma membrane. Not only is erythrocyte hemolysis a dangerous process for an individual's survival, but also the released hemoglobin represents a toxic component in the blood.

Hemolysis is one of the most used techniques to study the effects that injected materials produce on the blood, since in this case, these *in vitro* results usually correlate quite well with *in vivo* results [33]. To measure the produced hemolysis, spectroscopic measures are performed in nanomaterials incubated with erythrocytes during different times, as schematically described in Fig. 27.9.

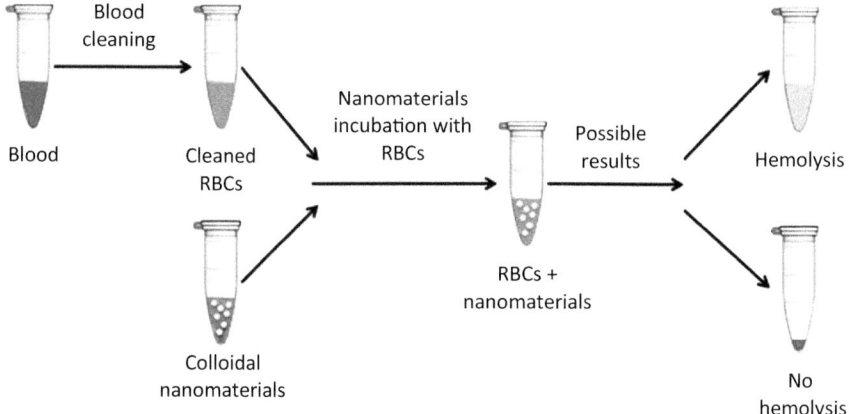

Figure 27.9 Schematic representation of the hemolysis assay.

It is worth noting that the interaction of nanomedicines with erythrocytes are largely determined by their physicochemical characteristics: mainly size, surface charge, and shape [33].

27.2.2.3.2 Cytotoxicity assessment

Colloidal nanosystem cytotoxicity is a key parameter that must be studied when initiating *in vitro* assays, before any preclinical studies. The cytotoxic character of a nanomaterial can be given not only by its components but also by its physicochemical properties. Remarkably, although a bulk material is biocompatible and biodegradable, in the nanomaterial form, it could produce some toxicity [3]. In addition, the compatibility of nanosystems with biological tissues depends also on the tissue type as well as on the assay performed, since different assays measure different parameters related with cell viability. For this reason, it is strongly recommended to study colloidal nanomaterial toxicity in different cell lines, performing various toxicity tests, which are described in the following subsections [34, 35].

Currently many fluorescent dyes are available that can be useful in determining cell viability under confocal microscopy or by flow cytometry, such as DAPI. However, they are beyond the scope of this chapter.

27.2.2.3.2.1 MTT test

The 3-(4,5-dimethylthiazol-2-yl)-2,5-diphenyltetrazolium bromide (MTT) colorimetric assay is by far the most widely used cytotoxicity test *in vitro*, together with the MTS assays, which is a modification of the MTT but both give the same results [36–38]. This test assesses the cell viability by means of measurement of the mitochondrial cell activity [38, 39]. Since mitochondrial dehydrogenase enzymes cleave the tetrazolium ring, tetrazolium salts (the MTT reagent) are quantified to perform this test. After their formation, they are dissolved in dimethyl sulfoxide and their amount is quantified by spectroscopic colorimetric absorbance. The highest the absorbance, the higher the number of MTT crystals formed and, consequently, the highest the cell viability. Together with the neutral red assay (detailed in the following text), they are the most sensitive cytotoxicity tests [17, 34].

Although MTT has been used extensively, in some cases, it does not give reliable results. Since it measures the mitochondrial activity of cells, when working with very slow dividing cells or even in confluence conditions (e.g., to reproduce the BBB, a body structure that does not divide in normal conditions), metabolic cell activity is very low, thus resulting in a limited formation of crystals and resulting in a low absorbance that can be barely quantified. Therefore, in this case, MTT is not recommended.

27.2.2.3.2.2 LDH assays

The lactate dehydrogenase assay (LDH assay) measures LDH activity in the extracellular media, which gives an indication of the rupture of the cell membrane [33, 39, 40]. It was first developed to test the cytotoxicity in neuronal cells [41]. Specifically, this test measures the production of lactate and NAD+ from pyruvate and NADH, reaction that only takes place in the presence of the LDH enzyme, by measuring the absorbance of NADH [34, 41]. Therefore, only in the presence of LDH, the absorbance should decrease due to the oxidation of NADH [41]. Since it is a measure of the extracellular LDH, the highest the LDH activity, the more damaged the cell membrane and, consequently, the lowest the viability of cells [17].

27.2.2.3.2.3 Neutral red

The neutral red assay is another spectrophotometric test to study nanomaterial cytotoxicity in cell cultures. It is based on the incubation of cell cultures with neutral red (toluene red), which is internalized by live cells and accumulated in lysosomes. Therefore, the higher the neutral red uptake, the higher the cell viability [17, 34].

27.2.2.3.2.4 Trypan Blue

The trypan blue assay is based on a principle similar to that of the neutral red assay. In this case, however, trypan blue, as a diazo dye, is only permeable to compromised cell membranes; therefore, only dead cells are stained in blue. The amount of cell death is quantified via optical light microscopy measurements and calculation of viability percentage [17]. This assay can also be performed with other similar dyes, such as the Evans blue assay.

Summarizing cytotoxicity assessment techniques *in vitro*, a combination of at least two techniques is recommended when performing a safety profile of nanomaterials *in vitro* to have complete and reliable results. Although in the vast majority of studies, only the MTT (or the alternative MTS) test is used, as described above, there are some specific cases in which it does not give reliable results. For this reason, it is recommended to perform the MTT but in combination with another technique. Among the other three assays described in this chapter, trypan blue is used in most studies only for the qualitative assessment of the cells when passaging cultures, due to its easy and rapid performance, to ensure that the number of cells counted are alive. Neutral red is less used. Therefore, for comparison with previous studies, it is recommended to test the MTT together with the LDH.

27.3 Conclusions and Future Prospects

During the design of nanosystems for medical and pharmaceutical purposes, it is a key factor to study their interaction with biological components, since it will determine the nanomaterial half-life, fate, and biodistribution in the body.

As described in this chapter, there are many techniques to study nanomaterial interaction with proteins. Since most nanomaterials are designed to be administered through the parenteral route, it is important to apply a technique for the study of the general interaction with all blood proteins, as well as to specific techniques for the study of key proteins. Among blood proteins, interactions with albumin should be studied, because it is the most abundant protein in the blood and, consequently, nanomaterials will rapidly find themselves surrounded by albumin when administered. In addition, the interactions with the C3 complement system protein are also encouraged, since this study enables the first indication of the reaction of the immune system through a designed nanomaterial.

On the other hand, it is also important to study the interaction of nanosystems with cells. To achieve a personalized therapy, nanomaterials must be designed with a specific surface, targeting the desired receptor of a target cell. A first approximation of the study of nanomaterial uptake can be performed *in vitro*. However, there are also many undesired interactions with other cells, which could result in cell and/or blood toxicity, which must be known before *in vivo* studies.

Nevertheless, most current studies focus on the therapeutic activity of nanomaterials, while they forget to study the safety profile. Therefore, the translation of the results to clinical studies is very difficult and nonexistent in most cases. To overcome this problem in future studies, a deep study of the interaction of nanomaterials with any kind of biological protein and/or cell is strongly recommended before translation to clinical trials to establish novel and personalized therapies.

Disclosures and Conflict of Interest

This work was supported by Sagetis Biotech SL. The author declares that she has no conflict of interest and has no affiliations or financial involvement with any organization or entity discussed in this chapter. No writing assistance was utilized in the production of this chapter and no payment was received for its preparation. The findings and conclusions here reflect the current views of the author and should not be attributed, in whole or in part, to the organization to which she is affiliated.

Corresponding Author

Associate Professor Cristina Fornaguera
Group of Materials Engineering (GEMAT)
Bioengineering Department
IQS School of Engineering, Ramon Llull University
Via Augusta 390, 08017 Barcelona, Spain
Email: cristina.fornaguera@iqs.url.edu

About the Author

Cristina Fornaguera is an associate professor of Institut Químic de Sarrià (IQS), from Ramon Llull University, since September 2018. She obtained her BSc in biotechnology, specializing in biomedicine, from the Autonomous University of Barcelona and PhD in development and control of medicines with honors, cum laude and awarded with the Extraordinary Thesis Award of the University of Barcelona during the 2014–2015 academic year. The main aim of her thesis was the preparation and characterization of polymeric nanoparticles to cross the blood–brain barrier. During this period and thanks to the multidisciplinary character of her PhD, she started acquiring expertise in colloidal chemistry, nanomedicine, biotechnology, pharmaceutical technology, advanced delivery systems formulation and characterization, *in vitro* studies of safety and efficacy of nanosystems, and *in vivo* experimentation to test the designed compounds. She is now consolidating her experience in all these research fields. Her interest in nanomedicine led her to industrial research, specifically to Sagetis Biotech, where she worked as a postdoctoral fellow from 2015 to 2018, on the design of innovative polymeric nanosystems capable of overcoming a fundamental engineering challenge: the development of safe and effective delivery vectors for currently unmet medical needs. Simultaneously, she developed an academic career as a tutor and docent in the IQS; work that continuous currently with her recent position as associate lecturer and researcher at IQS. Her research interests are focused on nanomedicine, biomaterials, bioengineering and drug delivery systems for immunotherapeutic applications. She has presented her research results at national and international conferences, authored many original research publications and review papers, and co-edited a special issue on personalized nanomedicine in the *Journal of Personalized Medicine* of the MDPI group (ISSN: 2075-4426) and collaborated as a reviewer for many peer-reviewed journals.

References

1. Duncan, R., Gaspar, R. (2011). Nanomedicine(s) under the microscope. *Mol. Pharmaceutics*, **8**, 2101–2141.
2. Sapsford, K. E., et al. (2011). Analyzing nanomaterial bioconjugates: A review of current and emerging purification and characterization techniques. *Anal. Chem.*, **83**, 4453–4488.
3. Lin, P. C., et al. (2014). Techniques for physicochemical characterization of nanomaterials. *Biotechnol. Adv.*, **32**(4), 711–726.
4. Pinto Reis, C., et al. (2006). Nanoencapsulation I. Methods for preparation of drug-loaded polymeric nanoparticles. *Nanomed. Nanotechnol. Biol. Med.*, **2**, 8–21.
5. Kreuter, J., et al. (2007). Covalent attachment of apolipoprotein A-I and apolipoprotein B-100 to albumin nanoparticles enables drug transport into the brain. *J. Control. Release*, **118**, 54–58.
6. Hall, J. B., et al. (2007). Characterization of nanoparticles for therapeutics. *Nanomedicine (Lond)*, **2**, 789–803.
7. Kreuter, J. (2014). Drug delivery to the central nervous system by polymeric nanoparticles: What do we know? *Adv. Drug Deliv. Rev.*, **71**(2013), 2–14.
8. Sahni, J. K., et al. (2011). Neurotherapeutic applications of nanoparticles in Alzheimer's disease. *J. Control. Release*, **152**(2), 208–231.
9. Vauthier, C., Bouchemal, K. (2009). Methods for the preparation and manufacture of polymeric nanoparticles. *Pharm. Res.*, **26**(5), 1025–1058.
10. Wang, A. Z., Langer, R., Farokhzad, O. C. (2012). Nanoparticle delivery of cancer drugs. *Annu. Rev. Med.*, **63**, 185–198.

11. Miele, E., et al. (2009). Albumin-bound formulation of paclitaxel (Abraxane® ABI-007) in the treatment of breast cancer. *Int. J. Nanomed.*, **4**, 99–105.
12. Dobrovolskaia, M. A. McNeil, S. E. (2007). Immunological properties of engineered nanomaterials. *Nat. Nanotechnol.*, **2**, 469–478.
13. Brigger, I., Dubernet, C., Couvreur, P. (2002). Nanoparticles in cancer therapy and diagnosis. *Adv. Drug Deliv. Rev.*, **54**, 631–651.
14. Mura, S., Couvreur, P. (2012). Nanotheranostics for personalized medicine. *Adv. Drug Deliv. Rev.*, **64**(13), 1394–1416.
15. Crist, R. M., et al. (2012). Common pitfalls in nanotechnology: Lessons learned from NCI's Nanotechnology Characterization Laboratory. *Integr. Biol.*, **5**(1), 66–73.
16. Cho, E. J., et al. (2013). Nanoparticle characterization: State of the art, challenges, and emerging technologies. *Mol. Pharmaceutics*, **10**, 2093–2110.
17. Lewinski, N., Colvin, V., Drezek, R. (2008). Cytotoxicity of nanoparticles. *Small*, **4**(1), 26–49.
18. Fornaguera, C., Calderó, G., et al. (2015). Interactions of PLGA nanoparticles with blood components: Protein adsorption, coagulation, activation of the complement system and hemolysis studies. *Nanoscale*, **7**(14), 6045–6058.
19. Hames, B. (2002). *Gel Electrophoresis of Proteins: A Practical Approach*. The Practical Approach Series, 3rd ed., Oxford University Press, USA.
20. Gaucher, G., et al. (2009). Effect of poly(V-vinyl-pyrrolidone)-block-poly(D,L-lactide) as coating agent on the opsonization, phagocytosis, and pharmacokinetics of biodegradable nanoparticles. *Biomacromolecules*, **10**, 408–416.
21. Cerdevall, T., et al. (2007). Detailed identification of plasma proteins adsorbed on copolymer nanoparticles. *Angew. Chem. Int. Ed.*, **46**, 5754–5756.
22. Vauthier, C., et al. (2011). Protein adsorption and complement activation for di-block copolymer nanoparticles. *Biomaterials*, **32**(6), 1646–1656.
23. Manabe, T. (2000). Combination of electrophoretic techniques for comprehensive analysis of complex protein systems. *Electrophoresis*, **21**, 1116–1122.
24. Mamedova, N. N., et al. (2001). Albumin–CdTe nanoparticle bioconjugates: Preparation, structure, and interunit energy transfer with antenna effect. *Nano Lett.*, **1**, 281–286.

25. Fornaguera, C. (2015). Development of multifunctional polymeric nanoparticles by nano-emulsion templating as advanced nanocarriers targeting the blood-brain barrier. Doctoral Thesis. University of Barcelona, Barcelona, Spain.
26. Weeke, B. (1973). Rocket immunoelectrophoresis. *Scand. J. Immunol.*, **2**(S1), 37–46.
27. Vroman, L., Adams, A. (1986). Why plasma proteins interact at interfaces. In: Horbett, T., Brash, J., eds. *Proteins at Interfaces: Physicochemical and Biochemical Studies.* ACS Symposium, Washington, pp. 154–164.
28. Mancini, G., Carbonara, A., Heremans, J. (1965). Immunochemical quantification of antigens by single radial immunodiffusion. *Immunochemistry*, **2**(3), 235–254.
29. Fornaguera, C., et al. (2015). Protein-nanoparticle interactions evaluation by immunomethods: Surfactants can disturb quantitative determinations. *Eur. J. Pharm. Biopharm.*, **94**, 284–290.
30. D'Addio, S. M., et al. (2012). Effects of block copolymer properties on nanocarrier protection from in vivo clearance. *J. Control. Release*, **162**(1), 208–217.
31. Bertholon, I., Vauthier, C., Labarre, D. (2006). Complement activation by core-shell poly(isobutylcyanoacrylate)-polysaccharide nanoparticles: Influences of surface morphology, length, and type of polysaccharide. *Pharm. Res.*, **23**(6), 1313–1323.
32. Zandanel, C., Vauthier, C. (2012). Characterization of fluorescent poly-(isobutylcyanoacrylate) nanoparticles obtained by copolymerization of a fluorescent probe during Redox Radical Emulsion Polymerization (RREP). *Eur. J. Pharm. Biopharm.*, **82**(1), 66–75.
33. Dobrovolskaia, M. A., McNeil, S. E. (2013). *Handbook of Immunological Properties of Engineered Nanomaterials*, 1st ed., Frontiers in Nanobiomedical Research series, World Scientific Publishing.
34. Fotakis, G., Timbrell, J. A. (2006). In vitro cytotoxicity assays: Comparison of LDH, neutral red, MTT and protein assay in hepatoma cell lines following exposure to cadmium chloride. *Toxicol. Lett.*, **160**, 171–177.
35. Hanks, C. T., Wataha, J. C., Sun, Z. (1996). In vitro models of biocompatibility: A review. *Dent. Mater.*, **12**, 186–193.
36. Grijalvo, S., Eritja, R. (2012). Synthesis and in vitro inhibition properties of oligonucleotide conjugates carrying amphipathic

proline-rich peptide derivatives of the sweet arrow peptide (SAP). *Mol. Diversity*, **16**(2), 307–317.

37. Fornaguera, C., Llinàs, M., et al. (2015). Design and in vitro evaluation of biocompatible dexamethasone-loaded nanoparticle dispersions, obtained from nano-emulsions, for inhalatory therapy. *Colloids Surf., B*, **125**, 58–64.

38. Putnam, D., et al. (2001). Polymer-based gene delivery with low cytotoxicity by a unique balance of side-chain termini. *Proc. Natl. Acad. Sci. U. S. A.*, **98**, 1200–1205.

39. Posadas, I., et al. (2009). Highly efficient transfection of rat cortical neurons using carbosilane dendrimers unveils a neuroprotective role for HIF-1alpha in early chemical hypoxia-mediated neurotoxicity. *Pharm. Res.*, **26**(5), 1181–1191.

40. Weber, N., et al. (2008). Characterization of carbosilane dendrimers as effective carriers of siRNA to HIV-infected lymphocytes. *J. Control. Release*, **123**(1), 55–64.

41. Koh, J., Choi, D. (1987). Quantitative determination of glutamate mediated cortical neuronal injury in cell culture by lactate dehydrogenase efflux assay. *J. Neurosci. Methods*, **1**(20), 83–90.

Supplementary References

1. Lundqvist, M., Augustsson, C., Lilja, M., Lundkvist, K., Dahlbäck, B., Linse, S., et al. (2017). The nanoparticle protein corona formed in human blood or human blood fractions. *PLoS ONE*, **12**(4), e0175871.

2. Hadjidemetriou, M., Kostarelos, K. (2017). Nanomedicine: Evolution of the nanoparticle corona. *Nat. Nanotechnol.*, **12**, 288–290.

3. Tran, S., DeGiovanni, P. J., Piel, B., Rai, P. (2017). Cancer nanomedicine: A review of recent success in drug delivery. *Clin. Trans. Med.*, **6**, 44.

4. Boraschi, D., Swartzwelter, B. J., Italiani, P. (2018). Interaction of engineered nanomaterials with the immune system: Health-related safety and possible benefits. *Curr. Opin. Toxicol.*, **10**, 74–83.

5. Greish, K., Mathur, A., Bakhiet, M., Taurin, S. (2018). Nanomedicine: Is it lost in translation? *Ther. Deliv.*, **8**(4).

6. Bawa, R., Audette, G. F., Rubinstein, I. (eds.) (2016). *Handbook of Clinical Nanomedicine: Nanoparticles, Imaging, Therapy, and Clinical Applications*, Pan Stanford Publishing, Singapore, 1662 p., 55 chapters, ISBN 978-981-4669-20-7 (Hardcover), 978-981-4669-21-4 (eBook).

7. Bawa, R. (ed.); Audette, G. F., Reese, B. E. (assist. eds.) (2016). *Handbook of Clinical Nanomedicine: Law, Business, Regulation, Safety, and Risk*, Pan Stanford Publishing, Singapore, 1448 p., 60 chapters, ISBN 978-981-4669-22-1 (Hardcover), 978-981-4669-23-8 (eBook).

8. Bawa, R. (2017). A practical guide to translating nanomedical products. In: Cornier, J., et al., eds. *Pharmaceutical Nanotechnology: Innovation and Production*, 1st ed., Wiley-VCH Verlag, chapter 28, pp. 663–695.

9. Tinkle, S., McNeil, S. E., Mühlebach, S., Bawa, R., Borchard, G., Barenholz, Y., Tamarkin, L., Desai, N. (2014). Nanomedicines: Addressing the scientific and regulatory gap. *Ann. N. Y. Acad. Sci.*, **1313**, 35–56.

Chapter 28

Unwanted Immunogenicity: From Risk Assessment to Risk Management

Cheryl Scott

BioProcess International, Boston, Massachusetts, USA

Keywords: vaccines, immunotherapy, immunogenicity, antidrug antibody (ADA), biotherapeutics, personalized medicine, therapeutic proteins, immunotoxicity, biologics, biosimilars, tolerization, tolerogenicity, biomanufacturing, protein aggregates, neoantigens, neutralizing antibodies (NABs), polysorbate, pharmacodynamics activity, postmarket surveillance, epitope, clinical markers, formulation erythropoietin, European Medicines Agency (EMA), US Food and Drug Administration (FDA)

28.1 Introduction

Although vaccines and immunotherapies are designed to engage the human immune system in fighting disease, unwanted immunogenicity can be a major problem for protein-based therapeutics. Some patients produce antidrug antibodies (ADAs), which might lead to drug inactivation or adverse effects. Even human and humanized proteins have proven to be surprisingly immunogenic

---

Immune Aspects of Biopharmaceuticals and Nanomedicines
Edited by Raj Bawa, János Szebeni, Thomas J. Webster, and Gerald F. Audette
Copyright © 2018 Pan Stanford Publishing Pte. Ltd.
ISBN 978-981-4774-52-9 (Hardcover), 978-0-203-73153-6 (eBook)
www.panstanford.com

in some cases, suggesting that immune tolerance requires careful consideration in biologic product design. In rushing to deliver new drugs to market, some biotherapeutics developers have overlooked factors that contribute to protein immunogenicity. Fortunately, the parameters influencing vaccine efficacy have been thoroughly studied for years, which allows biopharmaceutical companies to draw parallels when addressing immunogenicity of their protein therapeutics. Individual variations in patient tendencies to develop ADAs are probably genetic; so this is another area where personalized medicine and pharmacogenomics may help the industry progress.

The immune response to biologic products often involves B or T cells. The former produce antibodies that bind to proteins and thus reduce or eliminate their therapeutic effects. Potential complications can be life threatening. Thus, measuring the tendency to trigger antibody formation is an important part of determining the clinical safety and efficacy of protein-based drugs. T cells help activate B cells, especially for disease cases in which a patient's natural protein is defective in some way. That patient's T cells could treat protein therapeutics as if they were foreign invaders because they are different from the native protein. Such a response has been noted, for example, in some hemophilia patients, whose blood factor VIII is genetically defective. They may develop ADAs when infused with a correct factor VIII therapeutic protein (e.g., Bayer's Kogenate or Baxter's Advate products), presenting a significant impediment to such treatment. It is, in fact, considered to be "the most important problem in hemophilia A care today" [1]. Rheumatoid arthritis is another condition for which treatments are complicated by immunogenicity [2].

28.1.1 Regulators Respond

Regulations often come about in response to specific events that highlight the need for government oversight of a specific business activity. Late in the 20th century, notable occurrences of pure red-cell aplasia (PRCA) associated with erythropoietin treatments [3] led to closer scrutiny by regulators in several countries, and Box 28.1 lists some of the results. Meanwhile, biotherapeutics developers began looking closer at the subject, too [4]. Immunogenicity is now considered to be a basic aspect of biologic product safety. It is generally accepted that repeated administration of protein therapeutics can lead to ADA production in patients. So there is no dearth of guidance available to biopharmaceutical companies embarking on immunogenicity assessment of their products in development [5].

Box 28.1 International guidelines

European Medicines Agency

EMEA/CHMP/BMWP/14327/2006: *Guideline on Immunogenicity Assessment of Biotechnology-Derived Therapeutic Proteins.* European Medicines Agency: London, UK, 13 December 2007; www.ema.europa.eu/docs/en_GB/document_library/Scientific_guideline/2009/09/WC500003946.pdf.

EMA/CHMP/BMWP/86289/2010: *Guideline on Immunogenicity Assessment of Monoclonal Antibodies Intended for In Vivo Clinical Use.* European Medicines Agency: London, UK, 24 May 2012; www.ema.europa.eu/docs/en_GB/document_library/Scientific_guideline/2012/06/WC500128688.pdf.

EMA/275542/2013: *Concept Paper on the Revision of the Guideline on Immunogenicity Assessment of Biotechnology-Derived Therapeutic Proteins (CHMP/BMWP/42832/2005)* (Draft). European Medicines Agency: London, UK, 20 February 2014; www.ema.europa.eu/docs/en_GB/document_library/Scientific_guideline/2014/03/WC500163623.pdf.

US Food and Drug Administration

CBER/CDER. *Draft Guidance for Industry: Assay Development for Immunogenicity Testing of Therapeutic Proteins.* US Food and Drug Administration: Rockville, MD, December 2009; www.fda.gov/downloads/Drugs/.../Guidances/UCM192750.pdf.

CDER. *Draft Guidance for Industry: Immunogenicity-Related Considerations for the Approval of Low Molecular Weight Heparin for NDAs and ANDAs.* US Food and Drug Administration: Rockville, MD, April 2014; www.fda.gov/downloads/Drugs/GuidanceComplianceRegulatoryInformation/Guidances/UCM392194.pdf.

CBER/CDER. *Guidance for Industry: Immunogenicity Assessment for Therapeutic Protein Products.* US Food and Drug Administration: Rockville, MD, August 2014; www.fda.gov/downloads/Drugs/GuidanceComplianceRegulatoryInformation/Guidances/UCM338856.pdf.

For example, guideline S6 from the International Conference on Harmonisation of Technical Requirements for the Registration of Pharmaceuticals for Human Use (ICH) describes approaches to preclinical testing, including that for immunogenicity [6]. The guideline was adopted by the European Medicines Agency's (EMA's) Committee for Medicinal Products for Human Use (CHMP) in 2011 and by both the US Food and Drug Administration (FDA) and the Japanese Ministry of Health, Labour, and Welfare (MHLW) in 2012. Note that immunogenicity is not to be confused with *immunotoxicity* (the unwanted side effect of immune-system suppression), for which ICH has another specific guidance (ICH S8).

28.2 Cause and Effect

Many factors contribute to the immunogenicity of recombinant biologics, some related to the route and frequency of administration or a patient's own circumstances, but most directly involving the product itself [7]. Nonhuman sequences of amino acids in a protein's make-up are well known to elicit an immune response; that is the main reason the industry pushed for "humanizing" and developing "fully human" antibodies some two decades ago. Contaminants and adventitious agents are likely to do the same. Immunogenic glycosylation patterns have limited the use of plant- and yeast-based expression systems. Some stabilizing agents and storage media can make products more immunogenic (even acting like vaccine adjuvants), especially in combination with out-of-specification temperatures or rough handling. Protein oxidation and aggregation also are likely to cause trouble; the former can lead to the latter, which also can be caused by other means. Often no single underlying factor can be pointed to, but rather several factors will interact to cause a problem.

In addition to the erythropoietin events referred to above, ADA issues have arisen over time with blood factors VIII and IX, interferons α and β, interleukin, hormones, and monoclonal antibodies (MAbs)—with clinical consequences ranging from none at all (for growth hormone) to loss of product efficacy (the most common result of immunogenicity), allergic anaphylaxis, and PRCA as mentioned above. In the latter case, patients' immune systems went on to attack their own native protein as well as the introduced recombinant product. Regulatory guidance for immunogenicity continues to be revised, and class-specific guidance is emerging as research brings more cause-and-effect understanding to light.

Protein aggregates are often immunogenic. However, "in the absence of adjuvants, proteins usually are not immunogenic, in fact, often they are tolerogenic" [1]. That is the basis of allergy immunotherapy, for example, through which controlled exposure to small amounts of a given allergen builds up a patient's tolerance to it. "Tolerogenicity" is a never-ending source of consternation for vaccine developers—but something that biotherapeutics companies love to see. So immunogenicity of therapeutic protein products—especially fully human and humanized products—often can be attributed to copurified impurities (e.g., from animal-sourced products used in biomanufacturing), aggregation and conformation changes that attract T-cell attention, and even simple product storage and handling. One common cause of unwanted immunogenicity is inclusion of polysorbate (Tween) surfactants in biotherapeutic formulations [8]. All these are the types of problems that manufacturing changes can solve.

"Little is known about the mechanism of aggregate induction of immunogenicity," says Ed Maggio (CEO of Aegis Therapeutics). The most immunogenic type seems to be large, multimeric protein aggregates made up of native-like molecules with repetitive epitopes on their surfaces—leading to an assumption that more available binding sites increase risk [9]. Aggregates of denatured protein also can induce antibody production, but the results should not be product-neutralizing antibodies because protein conformation is lost through denaturation. That is, ADAs that recognize unfolded linear sequences are not likely to latch onto properly folded versions of those same sequences; it happens, but it is rare.

So the occurrence of aggregation itself does not guarantee trouble. ADAs against aggregates may not bind to single monomers and, even if they do, they do not always cause problems. Different immunological mechanisms have been hypothesized based on the different types of immunogenic threat, some involving T cells and some not. The biopharmaceutical industry is actively seeking reliable predictive models to address immunogenicity of protein aggregates, but doing so requires improved understanding of how it works.

When it comes to formulation-based immunogenicity, Maggio says that polysorbates 80 and 20 are the main offenders. To complicate matters, these nonionic surfactants are included in most biotherapeutic drug products to prevent aggregation and extend shelf life. They are mixtures, themselves (of fatty-acid esters) that inevitably include spontaneously generated oxidative contaminants (e.g., protein-damaging peroxides, epoxy acids, and reactive aldehydes that react with methionines, histidines, lysines, cysteines, tryptophans, tyrosines, and primary amino groups). Those impurities can form immunogenic "neoantigens." Lot variability in the extent of polysorbate degradation can cause subsequent lot variability in biologics immunogenicity. At the very least, that can complicate bioequivalence comparisons required for regulatory approval and postmarket surveillance—or lead to lot failures, recalls, and even necessitate reformulation and requalification that can involve expensive human trials. Maggio points to studies that have found a broad range of hydroperoxide content in commercially available polysorbate products [10]. The level of reactive contaminants in those preparations varies over time because auto-oxidation of polysorbates is spontaneous and progressive.

In response to such findings, suppliers have made available specifically "high-purity" polysorbate preparations. These are treated to remove peroxides and packaged with oxygen excluded from their containers (in the headspace, it is replaced with nitrogen or argon). But oxidation begins as soon as the contents come into contact with the air, Maggio points out, and such reactions will accelerate in aqueous solution [11]. Again, this may or may not cause a problem—polysorbates may help or harm—so risk assessment must be based on real data from laboratory analyses.

28.3 Evaluating Immunogenicity

Because therapeutic proteins have different properties and aggregation profiles, each poses a unique immunological risk. So experts recommend a case-by-case approach to assessing such risk. That posed by the presence of aggregates, for example, comes from not only their tendency to trigger ADAs, but also the clinical consequences that those antibodies might have [9].

> The amount, size, and type of aggregates necessary to trigger immune responses are a major concern for pharmaceutical companies and regulatory agencies. Current US Pharmacopeia (USP) particulate requirements state that particles >10 μm in size should be controlled below 6,000 particles per container, but no regulations have been established for the smaller sizes. Therefore, it is possible that immunogenically relevant protein aggregates have been routinely ignored by these regulations. For every aggregate >10 μm present in a formulation, there can be considerable amounts of slightly smaller aggregates, and each may contain hundreds or thousands of protein units. Thus, given the lack of knowledge regarding the most immunogenic aggregate sizes, it is important to develop instruments, protocols, and regulations that properly address a broader range of particle sizes, especially in the subvisible range [9].

Circulating ADAs are the primary measurement used in defining an immune response to recombinant protein products. As mentioned above, both patient-related and product-related factors come into play, and those provide a starting point for immunogenicity risk assessment. Ideally, all potential factors should be taken into consideration early in drug development.

Information about patient immune responses to protein therapeutics is required for marketing applications. And limited data may be available at the time of product launch, so postmarket surveillance is often necessary. Methods for evaluating immunogenicity include *in silico* screening and protein structure analysis, analytical characterization, preclinical testing and bioanalytical assays, and ultimately clinical trials. During early development, preclinical data can help developers design later clinical studies and interpret their results.

ICH S6 suggests measuring ADAs in nonclinical studies when there is evidence of altered pharmacodynamic activity or immune-mediated reactions in animal models or human subjects [6]. "Samples should be preemptively collected for potential analysis." And when ADAs are detected, companies should assess their effects on study interpretation. The guideline says that determining the potential for neutralizing antibodies (NAbs) is warranted when ADAs are detected but there is no identified PD marker to demonstrate their sustained activity.

The EMA's 2007 guideline cautions that "the predictive value of nonclinical studies for evaluation of immunogenicity of a biological medicinal product in humans is low due to inevitable immunogenicity of human proteins in animals" [13]. Although such experiments normally are not required, they can "be of value in evaluating the consequences of an immune response." Companies must develop "adequate screening and confirmatory assays" to measure immune responses against their products. Such assays should distinguish neutralizing from non-neutralizing antibodies, both in pivotal clinical trials and postmarket studies. Data should be systematically collected from "a sufficient number of patients." A sampling schedule is determined for each product by considering risks associated with an unwanted immune response in patients. Thus, immunogenicity issues should be addressed in a product's overall risk management plan.

The classic immunological assay is a sandwich type of enzyme-linked immunosorbent assay (ELISA). Other approaches involved in immunogenicity testing include aggregation and particle analysis [14]. Experience in applying immunochemical analysis to clinical diagnostics and disease research has led supplier companies such as Cygnus Technologies into the immunogenicity field. That company offers reagents and test kits developed to overcome many nonspecific and false-positive reactions that can confuse analysis of patient immune responses. According to the Cygnus website, well-characterized, immunospecific reagents can be used to establish baseline (pretherapy) immune response; detect titer increases after therapy; screen patient populations to identify those who may have neutralizing antibodies; correlate antibody titers to therapies; map immunogenic epitopes on protein structures; distinguish between recombinant (drug) and the natural protein forms and between immune response to a drug or associated process contaminants; and establish isotypes and subtypes of antibodies involved.

Other providers of immunogenicity assay reagents, kits, and other technologies include Antitope (an Abzena company), Meso Scale Discovery, Pall ForteBio, Perkin Elmer, and ProZyme. But like viral safety, immunogenicity testing is an area of such specialized knowledge that many drug-sponsor companies prefer to outsource the work to contract testing service providers such as those listed in Box 28.2. And some contract manufacturing and development organizations—e.g., Covance, Fujifilm Diosynth Biotechnologies, and Lonza—make a point of touting their own expertise in this area. Immunogenicity analysts use instruments such as the Biacore system from GE Healthcare, high-performance liquid chromatography (HPLC) systems from Agilent Technologies, Cirascan microarrayers from Aushon Biosystems, the Optim 2 fluorescence and static light-scattering system from Avacta Analytical, Bruker's AVANCE-II nuclear magnetic resonance (NMR) spectrometer, Gyrolab automation from Gyros, the Sysmex FPIA3000 particle-sizing system and MicroCal microcalorimetry from Malvern Instruments, and dual-layer multiplexing technology from SQI Diagnostics.

Box 28.2 Some immunogenicity testing services

Accuro Biologics (www.accurobio.com) in Scotland

Advanced Biodesign (www.a-biodesign.com) in France

Anabiotec (www.anabiotec.com) in Belgium

BioAgilytix (www.bioagilytix.com) in North Carolina

Charles River Laboratories (www.criver.com) in many locations across Europe and the United States

Eurofins Bioanalytical Services (www.eurofins.com) in several locations around the world

Haemtech Biopharma Services (www.haemtechbiopharma.com) in Vermont

ImmunXperts (www.immunxperts.com) in Belgium

Intertek (www.intertek.com) in many locations worldwide

Pacific BioLabs (www.pacificbiolabs.com) in California

ProImmune (www.proimmune.com) in Florida and in Oxford, UK

Quintiles (www.quintiles.com) in multiple locales worldwide

SGS (www.sgs.com) in many locations around the world

TNO Triskelion (www.tnotriskelion.com) in The Netherlands

28.3.1 The Tiered Strategy

Many experts endorse a tiered approach to immunogenicity determination [15]. For example, Jo Goodman (senior R&D manager at MedImmune's Cambridge, UK, site) said at a 2013 conference that "it is essential to adopt an appropriate strategy for development of adequate screening and confirmatory assays to measure an immune response against a therapeutic protein." Assays, she said, must be sensitive and detect responses at clinically relevant concentrations. Developers need to consider possible interference from "matrix factors" in patient sera, for example. Immunogenicity assays should be able to detect different types of responses (e.g., IgG or IgM) and be validated and standardized to distinguish neutralizing from non-neutralizing antibodies. An assay should be designed for the clinical population to which it will be applied and ideally capable of detecting low-affinity antibodies. All these criteria help prevent false-negative results. Use of relevant positive and negative controls will help.

The tiered approach begins with patient samples taken and screened at specified times during clinical trials [15]. Positive results go through confirmatory testing, and negative results are rejected (which explains the importance of preventing false negatives). Samples that are confirmed positive go through further analysis for characterization and titer, as well as neutralization assay development. Using assays for clinical markers and assessment of clinical response in patients, analysts then assess the correlation of characterized ADAs with clinical responses.

Goodman describes immunogenicity assay technology as "a changing landscape." With the issue first brought to light just over a decade ago, assay technology and strategies for determining immune responses have evolved over time. "Early assay types mainly involved ELISA, surface plasmon resonance (SPR), fluorescence, and radioimmunoprecipitation (RIP) techniques. And those technologies are still viable approaches. But new technology has become available, and each platform has its own advantages and disadvantages." Most new approaches are variations on older themes rather than wholly novel analytical technologies.

Starting before clinical testing begins, service providers such as Lonza stress immunogenicity assessment by addressing T-cell epitope mapping throughout drug development. By managing potential drug immunogenicity at the earliest possible stage,

Lonza says that companies can save time and money by creating the safest possible protein products.

A number of issues can arise in immunogenicity assay development and execution:

- cell-related problems in cell-based assays (e.g., cell-line stability, passaging, media/sera changes, mycoplasma or other contamination, cell banking)
- reagent trouble (e.g., stability and activity, changes in positive controls or detection antibodies, microplate and detection kit variability, and biological activity of specialized reagents)
- instrumental issues (e.g., maintenance/calibration, pipette changes, varying parameters, introduction of new brands or models of instrument)
- problems related to assay format and execution (e.g., different analysts, method drift over time, potential differences among patient samples)

The tiered approach may not work for biosimilar product development [16]. Small differences in production processes can lead to protein conformational or folding changes—potentially causing aggregation and subsequent immunogenicity. Screening and confirmation assays in the tiered approach tend to measure ligand binding. Neutralizing assays are cell based and may involve proliferation or gene reporters or measure potency (e.g., antibody-dependent, cell-mediated cytotoxicity). Biosimilar developers have different questions to answer: Is my biosimilar as immunogenic as the innovator? Might it even be "biosuperior?" What about interchangeability?

Immunogenicity assays for biosimilar products may involve two different positive control antibodies: one against the innovator product and one against its biosimilar counterpart. Some developers wonder whether they can simply use an innovator's assay to detect ADAs against a biosimilar. However, that single assay can reveal only relative immunogenicity rates between biosimilar and innovator products. "A second assay may be necessary to reveal true differences" [16]. In preclinical testing, immunogenicity rates may differ because of the small number of animals used. Even if a biosimilar product appears to have lower immunogenicity than its original counterpart, the developer will need to determine the biological relevance of such results in combination with other data (e.g., pharmacokinetics, pharmacodynamics). All assays used should have comparable sensitivity, selectivity, and precision to those used by the innovator. Here, as everywhere in biopharmaceutical development, a knowledge-based risk-management approach is advised:

> The more we learn regarding the factors triggering immunogenicity, and how they affect and activate the immune system, the better we can implement assays to predict and assess immunogenicity at a preclinical stage, with improved clinical translational value [16].

28.4 Predict and Prevent

Before a product ever reaches clinical-phase studies and associated bioanalysis, immunogenicity can and should be considered. Even before preclinical testing, some companies are getting proactive. Some even believe they may be able to ward off trouble at the protein-design stage. Doing so could turn potential "biosimilars," into "biobetters"—or give innovators the kind of patent edge they are looking for. Although as yet no reliable, straightforward models for widespread prediction have been put forth, a number of approaches are in different stages of development. They may be *in vitro* (cell-based analysis), *in vivo* (animal testing), or *in silico* (computer modeling) in nature.

28.4.1 Planning and Predicting

In silico profiling is gaining interest for a number of reasons. Modeling proteins based on their amino-acid sequences aids in product characterization and helps companies make process and formulation decisions early on. And some are using it at the product discovery and lead-optimization stage to find problematic sequences that are likely to induce immunogenicity in circulation. T cells bind to specific epitopes (small linear fragments of protein antigens) displayed on the surface of antigen-presenting cells [17]. Computer algorithms have been created to map the locations of such epitopes in protein three-dimensional structures using frequency analysis, support-vector machines, hidden Markov models, and neural networks.

With programs such as the EpiMatrix system from EpiVax and Antitope's iTope software and T Cell Epitope Database information, companies can quickly screen large genomic sequences for putative epitopes—and successes among analysts involved in vaccine design and studying autoimmune disorders are naturally leading others to apply them in assessment of unwanted immunogenicity. Some developers say their software can measure the potential immunogenicity of whole proteins and their subregions. An "epitope-density approach" is gaining acceptance for comparing protein therapeutics, such as in drug discovery and biosimilar development. It may reduce the risk of failure due to immunogenicity in the clinical setting [17].

In silico approaches are still a new technology. "They are potentially useful in R&D," says Maggio of Aegis Therapeutics, "but they may not impress regulatory agencies" when incorporated into investigational new-drug (IND) applications for going into clinical trials. Robin Thorpe (scientific advisor at ImmunXperts and an expert in the regulatory field) says that regulators consider them a way to guide further investigation. "But for product development, *in silico* analysis can aid in selection of most appropriate versions of products for development. To prevent overestimation, it is almost always advisable to confirm results using *in vitro* T-cell assays."

Many scientists may feel more comfortable with *in vitro* results—especially those already using cell-based assays to measure product potency, for example. Peptides some 9-25 amino acids long can be used to measure T-cell responses *in vitro* [17]. Animal testing, too, can provide results that may be predictive of human immunogenicity—especially that associated with protein aggregation. But an important caveat here is that predictive information is used more to guide choices made in product design, formulation, and clinical testing—not to serve as safety data by itself.

"Given the number of other factors that may equally influence immunogenicity, it seems very unlikely that [*in vitro* or *in silico*] methods may one day fully predict an immune response to either monomeric or aggregated therapeutic proteins" [17]. The EpiScreen assay from Antitope is one success story: A preclinical *in vitro* assay that has been shown to correlate with clinical data [18]. Companies such as Merck and Amgen have used it in development as well as in reformulation of problem drugs [19–21]. Box 28.3 goes into more detail.

Box 28.3 Expert commentary: predictive analysis

I spoke with Chloé Ackaert, a project manager at ImmunXperts SA in Belgium, about *in silico* and *in vitro* immunogenicity analysis. Here is part of our conversation:

Q: How does *in silico* immune profiling fit in with the biobetter concept?

A: *In silico* analysis and sequence optimization is the first step, but still every optimized protein needs to undergo other analyses before it can be concluded that the changes made resulted in a better protein. Sequence optimization might need other optimizations, as well. First, one must always ensure that the protein is not impaired or altered functionally. Second, changes in amino-acid sequence may imply different structure, stability, and/or pharmacokinetics and pharmacodynamics. Finally, the formulation might need to be adapted to suit a new sequence better. This approach has a different aim and consequences from screening a large number of candidates in discovery. But both provide information based on calculations and statistics, which benefits from subsequent *in vitro* analysis to select the best candidate.

Q: How do you distinguish between predictive analysis and immunogenicity assessment?

A: *Predictive* is a strong word that we avoid with any *ex vivo* analyses. We prefer to say "early immunogenicity risk assessment." We don't have a crystal ball with which to predict what a certain product will do in patients. But we can generate a risk profile that puts a product into perspective with others for which the immunogenic potential is known. By combining all information from available analyses, you can make an informative profile of a candidate molecule. But unknown factors remain that can lead to a different profile once a drug is in the clinic. So we don't *predict* how a molecule will behave; we assess its immunogenic risk in preclinical models. *Immunogenicity assessment* is generally used to describe antidrug antibody (ADA) measurement, which is currently possible only once a product is administered to a living organism.

Q: What are the most promising methods?

A: We advise customers to combine *in silico* and *in vitro* analysis, the first to screen a large panel of molecules for the best candidates. With those selected candidates, the best way to proceed is to perform an *in vitro* analysis, in most cases a T-cell proliferation assay with human immune cells. Use of those cells is very important to provide a strong surrogate marker for the potential of drug candidates to induce an ADA response later on. That puts their immunogenic potential of in perspective with that of known products (benchmark molecules). The key is to screen early to avoid problems later on.

> **Q:** What's the most problematic aspect of this work?
> **A:** Looking at data without the right context can lead to wrong conclusions. The assays and technologies available today are of a very high quality. But the importance of using the right controls and benchmark molecules is becoming clear and urgent. A key concept that could improve the science is standardizing the currently used assays and how they present results. Currently, every service provider has its own approach, which makes assay comparison difficult. At ImmunXperts, we aim to centralize all assays available and gather information in a structured, standardized, and coherent way. This helps us merge results provided by different experts with our final aim of providing the most complete immunogenicity profile for a given drug candidate.

In vivo testing, on the other hand, involves the entire immune system of a complex organism and thus can "better simulate the extremely complex scenario that results in complete immune responses, especially when protein aggregates are involved." But animal models must be well chosen because most modern recombinant proteins are humanized or fully human in nature, making them foreign to animals by nature. Some transgenic and knock-in/knock-out strains have been developed, as a result, and they do offer promise as predictive immunogenicity models. This science, however, is still in its early stages.

28.4.2 Preparation and Prevention

Risk management follows from risk assessment. Biopharmaceutical companies have many options for addressing the problem of immunogenicity when it arises. Factors associated with drug administration and patient-specific conditions can be difficult for biomanufacturers to control. Often such issues must be addressed through changes at the clinical level. For example, the erythropoietin problem led to a change in the product's mode of administration from subcutaneous to intravenous as well as revised storage and handling protocols, along with increased patient monitoring and maintenance efforts [12]. This may be an area where pharmacogenomics and personalized medicine can make a difference: by identifying patients who are more or less likely to have a problem as well as those inclined to benefit from a given therapy.

Box 28.4 Expert commentary: predictive analysis, deimmunization, and humanization

I spoke with Neil Butt, vice president of business development at Abzena plc in England, about immunogenicity analysis and antibody humanization. Here is part of our conversation:

Q: Tell me about your company's EpiScreen assay.
A: It is an *ex vivo* assay that quantifies T-cell responses to protein therapeutics based on stimulating peripheral blood mononuclear cells (PBMCs) isolated from human donors with test samples. The assay is performed as a service and is not available in a kit format. Three whole proteins run in parallel can be assessed for immunogenicity in eight weeks, with additional proteins (run in parallel) taking a little longer. The assay has not been automated to date.

Q: How will EpiScreen data fit in with *in silico* information?
A: The EpiScreen assay simulates the four key events that lead to a clinical immunogenic response: antigen uptake by antigen-presenting cells (APCs), antigen processing within each APC, presentation of processed antigen in the context of the major histocompatibility complex (MHC), and recognition of the peptide–MHC complex by T-cell receptors. *In silico* tools take into account only the third stage of that process and are thus very over-predictive of overall immunogenicity. They can be used as a low-resolution screen, but accurate assessment requires the use of a T-cell assay. The two can be combined for deimmunizing proteins. Once a T cell epitope is identified, *in silico* tools can be used to design protein variants that do not contain those MHC-binding peptides.

Q: And how would you say those results complement whole-organism studies?
A: Currently there are no ideal animal models for predicting the immunogenicity of protein therapeutics in man. HLA class II transgenic mice can be used to identify T-cell epitopes derived from therapeutic proteins. But they are limited by the HLA diversity that can be tested. An alternative to transgenic models is engraftment of neonatal NOD *scid* IL2Rγ^{null} mice with human CD34+ stem cells derived from PBMCs. But their MHC class II repertoire is limited and they are not tolerized against all human proteins. Without germline transfer of genes encoding human immune cells, each mouse has to be engrafted individually (which is expensive and time consuming). The EpiScreen assay provides an alternative, more robust technology that has been demonstrated to correlate with clinical immunogenicity.

Q: How important is it for companies to approach immunogenicity analysis from all those different angles?
A: Immunogenicity assessment can be incorporated at all stages of the development process, and each tool has its place. *In silico* tools can be used

to narrow down a large number (10–100) of lead sequences to a manageable number for T cell assays. *Ex vivo* time-course assays can be used to assess the overall immunogenicity of a smaller subset of preferred proteins (1–10) during early lead selection. T-cell epitope mapping can be used to map the precise location, number, and potency of T-cell epitopes within a protein sequence—allowing for potential deimmunization. *In silico* tools then can be used for redesigning that sequence either to remove T-cell epitopes or design new therapeutic proteins without them. Later, an EpiScreen time-course assay can again be used to assess the success of deimmunization as well as the effects of formulation and aggregation on therapeutic proteins.

Q: Can you elaborate on the idea of deimmunization? Does it apply only to new products/biosimilars? Or can it help turn marketed proteins into biobetters?

A: Deimmunization technologies can apply to antibodies and other proteins. Our Composite Human Antibody technology combines humanization and deimmunization technologies into one platform to make fully humanized antibodies without T-cell epitopes. For other proteins, our Composite Protein technology incorporates EpiScreen technology for accurate identification of T-cell epitopes and validation of their absence in deimmunized proteins, with iTope technology and the TCED database for designing deimmunized sequences. Both technologies apply to development of both new products and biobetters (the latter based on molecules that exhibit high levels of clinical immunogenicity).

Q: Does "humanization" not apply to nonantibody glycoproteins?

A: The concept is based on the adaptive nature of the immune response, allowing an antibody to be derived that will bind to virtually any antigen with high affinity. Because humans cannot be immunized experimentally with antigens to isolate a specific antibody, the procedure is performed in model animals to create and isolate hybridomas that secrete antibody with the desired specificity. Because such antibodies are nonhuman, they require engineering to make them "human-like" before they can be used clinically.

For most other classes of protein, a direct homologue exists in humans that matches the activity of a protein identified in animal models. So that can simply be expressed and used as a therapeutic without humanization. An exception to this rule are replacement therapies for certain diseases in which the protein is absent due to genetic mutation. When administered to patients, it would appear foreign to their immune systems, giving such proteins the potential to be immunogenic. Similarly, some other proteins that are clinically useful and derived from microorganisms (e.g., bacterial toxins, for which no human homologue exists) tend to are highly immunogenic and must be deimmunized by removal of T-cell epitopes.

For product-related factors, biomanufacturers can make formulation changes to prevent aggregation or even change the drug substance itself. When predictive analysis has identified T-cell epitope issues, deimmunization by epitope modification is one strategy for reducing protein immunogenicity. Some proteins may benefit from fusion or conjugation with other molecules (e.g., polyethylene glycol (PEG), which is also used as a formulation excipient) to help make them more "invisible" to patients' immune systems—primarily by improving their solubility and making them less prone to aggregation.

28.4.3 Epitope Modification

When predictive methods identify potential immunogenicity issues early in product development, companies have more options available for addressing such problems than if they were not discovered until clinical testing. Humanization is such a common answer for antibodies that it is almost a given part of product development now. Most companies add amino-acid sequence segments derived from variable regions of human antibodies, taking care not to use known T-cell epitopes. If nonhuman glycosylation patterns are the problem, then an expression system can be chosen that performs the correct posttranslational modifications.

Modern predictive methods can identify problematic T-cell epitopes on recombinant proteins in development, allowing companies to use selective mutation methods that remove them. Once a library of sequence variants has been designed and expressed, those can be screened for retention of their therapeutic activity as well as reduced immunogenicity using the same *in silico*, *in vitro*, and/or *in vivo* methods.

28.4.4 Formulation Changes

Unwanted immunogenicity is not always attributable to a protein's amino acid sequence. If problems arise later in development, the drug product's formulation is often to blame. And that usually means that aggregation is involved. Protein aggregation has consequences beyond immunogenicity, as well. Aggregated proteins lose their bioactivity and stability, often irreversibly, and they waste valuable product. Aggregation-prone proteins need to be protected carefully throughout bioprocessing, especially during shaking and shipping; freezing and thawing; drying and reconstitution; and formulation, fill, and finish. Structural modifications can help, but the range of formulation options should be considered first. When aggregation in processing is inevitable, the last resort option is to remove aggregates before drug-product formulation—but that wastes valuable product and thus cannot be very cost effective [22].

Maggio of Aegis focuses on polysorbate problems [8]. "Replacing polysorbates with surfactants that do not cause progressive protein degradation (and increased immunogenicity) represents a critical need and a significant opportunity for creation of better innovator and biosimilar biotherapeutics." He points to alkylsaccharides as a promising alternative. Historically used in both food and cosmetics, these nonionic surfactants are each made of a sugar coupled to an alkyl chain. "They are being adopted by innovator biotherapeutic companies such as Hoffmann-La Roche, which has recently licensed Aegis's ProTek dodecylmaltoside for stabilization of products in development. Results have been published for an interferon product" [23]. A second "big pharma" company has entered into a ProTek license similar to Roche's, Maggio reports, and two more are in early discussion with his company regarding both MAbs and non-MAb biotherapeutics.

28.4.5 Tolerization

Even fully human antibodies can be immunogenic. A new approach to reducing the problem introduces regulatory T-cell (Treg) epitopes to the protein sequence that induce immune tolerance to it. This may well complement humanization (replacing foreign sequences with human ones) and deimmunization (reducing T-effector epitopes) approaches to product design. The concept brings us back to one expert's claim that most proteins are actually tolerogenic rather than immunogenic [1]. It leads us to the question: How do we convey that property on others? Immune responses are controlled by several different biological mechanisms, some of which could be exploited for induction of tolerance to protein therapeutics [24]. *Natural tolerance* is the basic recognition of "self"; *adaptive tolerance* (the basis of allergy immunotherapy) happens with certain types of repeated exposure—when the body recognizes that a given "invader" is not harmful.

Tolerization may offer an option for some products that are already on the market. For example, humanized alemtuzumab (Genzyme's Campath) antibodies treat leukemia and potentially multiple sclerosis. The product was humanized because the original rat-derived version induced neutralizing antibodies in many clinical trial participants [24]. With up to 75% of patients still developing antibody responses to the humanized form (especially after several doses), Genzyme is considering administration of a nonbinding version to tolerize patients to the drug before they begin a course of therapy. Some experts are saying this approach has "significant potential to accelerate development of biobetter products."

28.5 From Start to Finish—and Beyond

Immunogenicity is a key metric of product safety that makes it a key metric of product quality. A biotherapeutic product's life cycle can be broadly described as moving from discovery/design through lead optimization, product characterization and preformulation, process development, preclinical testing, and clinical trials to market authorization—and from there on to postmarket surveillance. Companies often think of the biologics license (BLA) or new drug application (NDA) as the final goal. And they often consider quality, safety, and efficacy to be a "three-legged stool" of separate but interdependent parameters. But risk management continues beyond market authorization, and quality could be said to include both safety and efficacy. After all, no product can be called "high quality" if it does not do the job it needs to do without doing any harm—no matter its purity.

Once a bit of an afterthought, immunogenicity is now an essential part of biotherapeutic development from start to finish. The industry's (and its regulators') growing knowledge of this subject is making proving, predicting, and preventing it an ever more integral part of quality by design.

Disclosures and Conflict of Interest

This chapter is an edited and updated version that was originally published in *BioProcess International*, November 2014, Vol. 12, No. 10, and appears here with permission of the copyright holder (http://www.bioprocessintl.com/analytical/downstream-validation/unwanted-immunogenicity-risk-assessment-risk-management). The author declares that she has no conflict of interest and has no affiliations or financial involvement with any organization or entity discussed in this chapter. No writing assistance was utilized in the production of this chapter and the author has received no payment for its preparation.

Corresponding Author

Cheryl Scott
BioProcess International
PO Box 70
Dexter, OR 97431 USA
Email: cscott@knect365.com

About the Author

Cheryl Scott is cofounder and senior technical editor of *BioProcess International*, a monthly, controlled-circulation magazine devoted to the development, scale-up, and manufacture of biotherapeutics. With experience as a medical/psychological transcriptionist and in aerospace engineering support, she began covering the biopharmaceutical industry in 1996 as an editor for *BioPharm* after earning her BA in journalism from the University of Arkansas at Fayetteville. In 2002, she worked with three colleagues to develop a new publication, *BioProcess International*, and contributed to production of the BPI Conference series for Informa. Also a science fiction author (as CA Scott, known for the *Racing History* series), she continues to learn on the job every day from the best advisors, colleagues and authors in the business every day.

References

1. Lollar, P. (2013). The immune response to blood coagulation factor VIII. *14th Annual Immunogenicity for Biotherapeutics*. Baltimore, MD, 18 March 2013. Institute for International Research, New York, NY.
2. Ryan, S. (2014). Explaining immunogenicity. *Irish Med. Times*, 10 September 2014, Available at: www.imt.ie/clinical/2014/09/explaining-immunogenicity.html (accessed on February 7, 2018).
3. Thorpe, R., Wadhwa, M. (2003). Unwanted immunogenicity of therapeutic biological products: problems and their consequences. *BioProcess Int.*, **1**(9), 60–62.
4. Schellekens, H. (2002). Immunogenicity of therapeutic proteins: Clinical implications and future prospects. *Clin. Ther.*, **24**(11), 1720–1740.
5. Cai, X.-Y., Cullen, C. (2013). Current regulatory guidelines for immunogenicity assays. *Immunogenicity Assay Development, Validation, and Implementation*. Future Medicine, London, UK, pp. 6–17.
6. ICH S6(R1) (2011). Preclinical safety evaluation of biotechnology-driven pharmaceuticals. *International Council on Harmonisation of Technical Requirements for the Registration of Pharmaceuticals for Human Use*, Geneva, Switzerland, Available at: www.ich.org/fileadmin/Public_Web_Site/ICH_Products/Guidelines/Safety/S6_R1/Step4/S6_R1_Guideline.pdf (accessed on February 7, 2018).
7. Purcell, R. T., Lockey, R. F. (2008). Immunologic responses to therapeutic biologic agents. *J. Invest. Allergol. Clin. Immunol.*, **18**(5), 335–342.
8. Maggio, E. T. (2012). Polysorbates, immunogenicity, and the totality of the evidence. *BioProcess Int.*, **10**(10), 44–49. Available at: www.bioprocessintl.com/sponsored-content/preclinical-immunogenicity-assessment (accessed on February 7, 2018).
9. Filipe, V., et al. (2010). Aggregation and immunogenicity of therapeutic proteins. In: Wang, W., Roberts, C. J. eds. *Aggregation of Therapeutic Proteins*, John Wiley & Sons, Hoboken, NJ, chapter 10, pp. 416–433.
10. Wasylaschuk, W. R., et al. (2007). Evaluation of hydroperoxides in common pharmaceutical excipients. *J. Pharm. Sci.*, **96**, 106–116.
11. Ha, E., Wang, W., Wang, Y. J. (2002). Peroxide formation in polysorbate 80 and protein stability. *J. Pharm. Sci.*, **91**, 2252–2264.

12. Schellekens, H. (2005). Immunologic mechanisms of EPO-associated pure red cell aplasia. *Best Pract. Res. Clin. Haematol.*, **18**(3), 473–480.
13. EMEA/CHMP/BMWP/14327/2006 (2007). Guideline on immunogenicity assessment of biotechnology-derived therapeutic proteins. European Medicines Agency, London, UK, 13 December 2007. Available at: www.ema.europa.eu/docs/en_GB/document_library/Scientific_guideline/2009/09/WC500003946.pdf (accessed on February 7, 2018).
14. Mire-Sluis, A., et al. (2011). Analysis and immunogenic potential of aggregates and particles: A practical approach, Part 1. *BioProcess Int.*, **9**(10), 38–47; Part 2. *BioProcess Int.*, **9**(11), 38–43.
15. Goodman, J. (2013). Using the Aushon Cirascan™ to develop sensitive and drug tolerant immunogenicity assays. *14th Annual Immunogenicity for Biotherapeutics*, Baltimore, MD, 19 March 2013. IIR USA, New York, NY.
16. Sauerborn, M. (2013). Is the tiered immunogenicity testing of biologics the adequate approach in preclinical development? *Bioanalysis*, **5**(7), 743–746.
17. De Groot, A. S., Martin, W. (2009). Reducing risk, improving outcomes: bioengineering less immunogenic protein therapeutics. *Clin. Immunol.*, **131**(1), 189–201.
18. Baker, M. P., Jones, T. D. (2007). Identification and removal of immunogenicity in therapeutic proteins. *Curr. Opin. Drug Discov. Dev.*, **10**(2), 219–227.
19. Jaber, A., Baker, M. (2007). Assessment of the immunogenicity of different interferon beta-1α formulations using ex vivo T-cell assays. *J. Pharm. Biomed. Anal.*, **43**(4), 1256–1261.
20. Jaber, A., et al. (2007). The Rebif® new formulation story: It's not trials and error. *Drugs R&D*, **8**(6), 335–348.
21. Joubert, M. K., et al. (2012). Highly aggregated antibody therapeutics can enhance the in vitro innate and late-stage T-cell immune responses. *J. Biol. Chem.*, **287**(30), 25266–25279.
22. Wang, W., Warne, N. W. (2010). Approaches to managing protein aggregation in product development. In: Wang, W., Roberts, C. J. eds. *Aggregation of Therapeutic Proteins*, John Wiley & Sons, Hoboken, NJ, chapter 8, pp. 333–365.
23. Rifkin, R. A., et al. (2011). *n*-Dodecyl-β-D-maltoside inhibits aggregation of human interferon-β-1b and reduces its immunogenicity. *J. Neuroimmune Pharmacol.*, **6**(1), 158–162.

24. De Groot, A. S., et al. (2013). Beyond humanization and de-immunization: Tolerization as a method for reducing the immunogenicity of biologics. *Exp. Rev. Clin. Pharmacol.*, **6**(6), 651–662.

For Further Investigation

- Brinks, V. (2013). Immunogenicity of biosimilar monoclonal antibodies. *GaBI J.*, **2**(4), 188–193.
- Carr, F. (2013). Fully human antibodies: Myths and magic bullets. *MedNous*, 18.
- Dobrovolskaia, M. A. (2013). Nanoparticles and immunogenicity. *14th Annual Immunogenicity for Biotherapeutics*. Baltimore, MD, 18 March 2013. Institute for International Research, New York, NY.
- Gunsior, M. (2010). Implications of immunogenicity in drug development (Global bioanalytical services white paper). Covance Laboratories, Chantilly, VA. Available at: www.covance.com/content/dam/covance/assetLibrary/whitepapers/adawhitepaperpdf.pdf (accessed on February 7, 2018).
- Jawa, V., et al. (2013). T-cell dependent immunogenicity of protein therapeutics: Preclinical assessment and mitigation. *Clin. Immunol.*, **149**(3), 534–555.
- Liang, B. A., Mackey, T. (2011). Emerging patient safety issues under health care reform: Follow-on biologics and immunogenicity. *Ther. Clin. Risk Manage.*, **7**, 489–493.
- Nielsen, M., et al. (2010). NetMHCIIpan-2.0: Improved pan-specific HLA-DR predictions using a novel concurrent alignment and weight optimization training procedure. *Immunome Res.*, **6**, 9.
- Parenky, A., et al. (2014). New FDA draft guidance on immunogenicity. *AAPS J.*, **16**(3), 499–503.
- Perry, L. C. A., Jones, T. D., Baker, M. P. (2008). New approaches to prediction of immune responses to therapeutic proteins during preclinical development. *Drugs R&D*, **9**(6), 385–396.
- Rup, B. (2011). Biotherapeutic immunogenicity risk factors: The science, reliability, and concepts for implementing predictive tools to improve their reliability. *Mastering Immunogenicity Summit*, Boston, MA, 13 September 2011. ProImmune Ltd., Oxford, UK.
- Sauna, Z. E. (2011). The assessment of immune responses against biological drugs. *Mastering Immunogenicity Summit*, Boston, MA, 13 September 2011. ProImmune Ltd., Oxford, UK.

- Sauna, Z. E. (2012). The immunogenicity of protein therapeutics: Time to get personal? *Mastering Immunogenicity Summit*, Boston, MA, September 2012. ProImmune Ltd., Oxford, UK. Available at: www.proimmune.com/ecommerce/pdf_files/sauna1.pdf (accessed on February 7, 2018).
- Shankar, G., et al. (2014). Assessment and reporting of the clinical immunogenicity of therapeutic proteins and peptides: Harmonized terminology and tactical recommendations. *AAPS J.*, **16**(4), 658–673.
- Stas, P., Lasters, I. (2009). Strategies for preclinical immunogenicity assessment of protein therapeutics. *IDrugs*, **12**(3), 169–173.
- Van Beers, M. M. C., Bardor, M. (2012). Minimizing immunogenicity of biopharmaceuticals by controlling critical quality attributes of proteins. *Biotechnol. J.*, **7**(12), 1473–1484.
- Van de Weert, M., Møller, E. H., eds. (2008). *Immunogenicity of Biopharmaceuticals*, Springer, New York, NY.
- Weber, C. A., et al. (2009). T cell epitope: Friend or foe? Immunogenicity of biologics in context. *Adv. Drug Deliv. Rev.*, **61**(11), 965–976.

Chapter 29

Emerging Therapeutic Potential of Nanoparticles in Pancreatic Cancer: A Systematic Review of Clinical Trials

Minnie Au, MBBS,[a,b,*] Theophilus I. Emeto, PhD,[a,*] Jacinta Power, MBBS,[b] Venkat N. Vangaveti, PhD,[c] and Hock C. Lai, MBBS[b]

[a]*Public Health and Tropical Medicine, College of Public Health, Medical and Veterinary Sciences, James Cook University, Townsville, Australia*
[b]*Townsville Cancer Centre, The Townsville Hospital, Townsville, Australia*
[c]*College of Medicine and Dentistry, James Cook University, Townsville, Australia*

Keywords: pancreatic cancer, nanoparticles, clinical trials, cancer therapy, fluorodeoxyglucose, intravenous, nanoparticle albumin bound-paclitaxel (nab-paclitaxel), randomized control trials, positron emission tomography, enhanced permeability and retention (EPR), toxicity, recognition motifs, randomized control trial (RCT), Jadad score, nanomedicine, pathotrophic nanoparticle gene delivery, Response Evaluation Criteria in Solid Tumours (RECIST), personalized dosing regimen, gold nanoparticles, micelle nanoparticles, liposomal nanoparticles, safety profile, efficacy, RNA interference, photodynamic therapy, photothermal therapy

Pancreatic cancer is an aggressive disease with a five-year survival rate of less than 5%, which is associated with late presentation. In recent years, research into nanomedicine and the use of

---

*These authors contributed equally to this work.

Immune Aspects of Biopharmaceuticals and Nanomedicines
Edited by Raj Bawa, János Szebeni, Thomas J. Webster, and Gerald F. Audette
Copyright © 2018 Pan Stanford Publishing Pte. Ltd.
ISBN 978-981-4774-52-9 (Hardcover), 978-0-203-73153-6 (eBook)
www.panstanford.com

nanoparticles as therapeutic agents for cancers has increased. This chapter describes the latest developments on the use of nanoparticles and evaluates the risks and benefits of nanoparticles as an emerging therapy for pancreatic cancer. The Preferred Reporting Items of Systematic Reviews and Meta-Analyses checklist was used. Studies were extracted by searching the Embase, MEDLINE, SCOPUS, Web of Science, and Cochrane Library databases from inception to March 18, 2016, with no language restrictions. Clinical trials involving the use of nanoparticles as a therapeutic or prognostic option in patients with pancreatic cancer were considered. Selected studies were evaluated using the Jadad score for randomized control trials and the Therapy CA Worksheet for intervention studies. Of the 210 articles found, 10 clinical trials, including one randomized control trial and nine phase I/II clinical trials, met the inclusion criteria and were analyzed. These studies demonstrated that nanoparticles can be used in conjunction with chemotherapeutic agents increasing their efficacy while reducing their toxicity. Increased efficacy of treatment with nanoparticles may improve the clinical outcomes and quality of life in patients with pancreatic cancer, although the long-term side effects are yet to be defined. The study registration number is CRD42015020009.

29.1   Introduction

Pancreatic cancer is a rare but aggressive disease that is plagued by a myriad of problems, including late diagnosis often when the cancer has metastasized, no early warning symptoms, and inadequate therapeutic options on diagnosis [1]. The incidence rate of pancreatic cancer for gender is close to 1, with approximate rates of 8 per 100,000 in men and 6 per 100,000 in women globally [2]. Worldwide, it is responsible for 331,000 deaths annually [2]. It is the sixth most common cause of cancer-related death in Australia and the fourth globally [3]. Despite years of research, the five-year survival rate remains at approximately 5% [1]. The median age of diagnosis has been reported to range between 66 and 68 years [4]; however, early onset pancreatic cancer occurring in patients under 50 years of age is associated with more advanced disease at presentation and a poorer prognosis [4, 5]. Currently 97% of the burden of disease from pancreatic cancer is due to years of life lost to premature death [6] with a median survival time of six to ten months for locally advanced disease, and three to six months for metastatic disease [7, 8]. Established risk factors for pancreatic cancer include a family history of the disease and smoking, which account for 5% to 10% of cases. Other weaker associations include obesity, diabetes mellitus, chronic pancreatitis, periodontal disease, *Helicobacter pylori*, and gallstones [4]. A challenge to the management of pancreatic cancer is the drug-resistant nature of pancreatic tumor cells to gemcitabine, a pyrimidine antagonist used as the first-line chemotherapeutic agent [9]. Unlike many other cancers, pancreatic cancer is characterized by several pathophysiological complications that make it hard to treat, specifically with drugs. Traditionally, complete surgical resection provides the most recognized form of treatment [10]. A complete analysis of the difficulties in treating pancreatic cancer is aptly reviewed by Oberstein and Olive [11].

Nanoparticles are 100 to 10,000 times smaller than human cells and can interact with biological molecules intra- and extracellularly [12]. Nanomedicine is the use of nanoparticles in medicine, and they can be attached to lipids or form polymers to encapsulate drugs to increase drug solubility, permeability, and delivery to target cells leading to higher therapeutic efficiency [13]. Their unique properties include the ability to remain stable in the physiological environment and passively target pancreatic cancer cells via the enhanced permeability and retention effect (EPR). EPR is due to the size of the nanoparticles, which allows them to extravasate from leaky blood vessels, supplying the carcinoma and targeting it. Due to the poor lymphatic drainage in tumors, nanoparticles are able to accumulate within tumor capillaries and are large enough to escape filtration by the kidney and small enough to evade phagocytic removal by Kupffer cells and splenocytes. However, the non-physiological surface chemistry of nanoparticles may cause non-specific cellular targeting and precipitation leading to cell damage [14]. Alternatively, nanoparticles can be used to actively target tumor cells by combination of specific recognition motifs such as antibodies and sugar molecules within nanomedicine formulations [15]. Evidence suggests that active targeting by nanoparticles is efficient for poorly leaky tumors, whereas passive targeting is better for highly leaky tumors [15].

Toxicity from nanoparticles may occur as a result of composition, size, or charge of the nanoparticles [16]. For example, cationic liposomal nanoparticles can interact with the extracellular matrix, serum proteins, and lipoproteins, with consequent aggregation and or oxidative stress resulting in non-target tissue damage [17, 18]. Gold nanoparticles are able to cross the placenta and damage the developing fetus [19]. Gold particles are also implicated in the induction of reactive oxygen formation and the initiation of autoimmunity [20]. Given the large diversity of materials used in the construction of nanoparticles, there is an infinite number of combinations of interactions with a high potential of negative interactions that should be taken into consideration to ensure patient safety [21].

A range of *in vitro* and *in vivo* animal studies have shown promising results using a variety of nanoparticles as nanocarriers or in combination with standard chemotherapeutic agents [22–32]. There has been a surge of interest in the use of nanoparticles as therapeutic agents for various cancers in recent years. For example, a number of clinical trials have been conducted using nanoparticles as nanocarriers for a range of solid organ tumors such as colorectal cancer [33], non-small cell lung cancer [34], gastric cancer [35], breast cancer [36] and adenocarcinomas of the esophagus and gastroesophageal junction [37]. Therefore, the aim of this systematic review is to synthesize available literature on clinical trials performed up to March 2016 on the latest developments in the use of nanoparticles as an emerging therapy for pancreatic cancer.

29.2 Methods Section

29.2.1 Literature Search

This systematic review was performed in accordance to the Preferred Reporting Items of Systematic Reviews and Meta-

Analyses (PRISMA) statement [38]. The study protocol can be found on the PROSPERO international prospective register of systematic review (PROSPERO 2015: CRD42015020009). Briefly, a literature search to identify studies investigating the use of nanoparticles in the management of pancreatic cancer was conducted. The Embase (1980), MEDLINE (1966), SCOPUS (1996), Web of Science (1965), and Cochrane Library databases (1992) were searched from inception to March 2016 with no language restrictions. Search terms applied included: "nanoparticles" OR "nanomedicine," [Title/Abstract] AND "pancreatic cancer management" OR "pancreatic cancer therapy," AND/OR "clinical trials" OR "clinical studies" OR "human participants." Titles and abstracts were independently screened by two authors (M.A and J.P) to identify possibly relevant studies. The full texts for articles that appear ambiguous were assessed to determine their suitability for inclusion. Database searches were supplemented by scanning the reference lists of included studies and employing the related articles function in PubMed. Subsequently, the full texts of all potentially eligible studies were evaluated in detail for inclusion by the two authors. Discrepancies were resolved at a consensus meeting between the two authors. If the two authors failed to reach a consensus, a third author (T.I.E.) was involved in making a final decision.

29.2.2 Inclusion/Exclusion Criteria

The studies included in this chapter are clinical trials involving human participants diagnosed with pancreatic cancer. Interventions used in the studies must include at least one group of participants being treated with nanoparticles for pancreatic cancer, and the impact of nanoparticle treatment on the outcome of disease progression or overall survival must be measured. Studies excluded were studies not involving human participants, studies evaluating the use of nanoparticles in the imaging/diagnosis of pancreatic cancer and not the treatment, and studies evaluating the use of nanoparticles in patients without pancreatic cancer.

29.2.3 Data Collection

Two investigators (M.A and J.P) extracted data using the aforementioned strategy. Data extracted included specific details about the population, interventions, comparison, outcome (PICO) and study methods of significance to the review question and specific objectives. Authors of eligible studies were contacted where additional information was required. Data were cross-checked in a consensus meeting and again, discrepancies were resolved through discussion and mutual agreement between the two authors. The third author (T.I.E.) was available to make a final decision if required.

29.2.4 Quality of Methods Assessment

Two independent reviewers (M.A and J.P) assessed the validity of the studies using the Jadad score [39] for randomized control trials (RCT) and the Therapy CA Worksheet [40], for intervention studies. If there was any disagreement, the third reviewer (T.I.E.) interceded to make a final decision. The Jadad score assesses randomization, blinding, and attrition to derive a score ranging from 0 (low quality) to 5 (high quality). For this review, a Jadad score greater than 2 was deemed to be of sound methodology. The Therapy CA Worksheet assesses whether the study was randomized, whether there was sufficient and complete follow-up, and whether groups were analyzed according to their random allocations, blinding, group characteristics and outcome (mean survival). Articles were categorized as "low," "moderate," or "high" according to analysis.

29.3 Results

29.3.1 Study Selection

We identified 210 potentially eligible studies from initial database searches after removing duplicates (Fig. 29.1). A total of 157 articles were excluded following review of their titles and abstract. The most frequent reasons for exclusion were:

not being clinical trials, not involving nanoparticles, and not involving patients with pancreatic cancer. After appraising 53 full text articles, a further 50 were excluded because they were not clinical trials or involved the diagnosis or investigation of pancreatic cancer but not the management. Six additional studies met the inclusion criteria on hand searching the reference lists of included studies; therefore 10 studies were included in this study (Table 29.1). Ten clinical trials were found from the search strategy, including one randomized controlled trial and nine phase I/II clinical trials. The types of nanoparticles evaluated include nanoparticles containing a retroviral gene, gold nanoparticles, micelle nanoparticles, liposomal nanoparticles and albumin nanoparticles conjugated with chemotherapeutics [7, 8, 41–48]. In addition to evaluating the effects of the drug on the progression of pancreatic cancer, the maximum tolerated dose and adverse effects were also investigated.

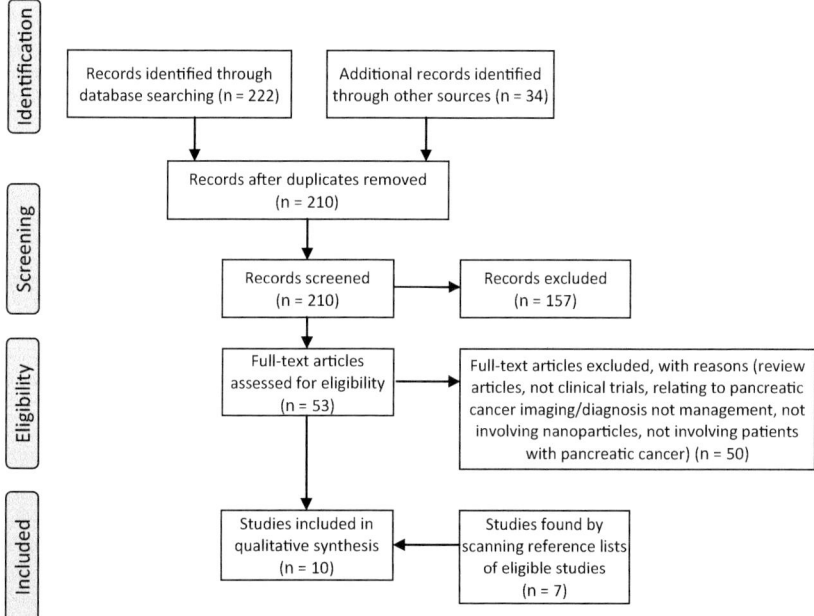

Figure 29.1 Flow diagram illustrating data collection protocol employed in this study.

Table 29.1 Characteristics of included studies

Country/region	Sample size	Age range (years)	% Males	Previous treatment	Follow-up (months)	Mortality (%)	Assessment	Reference
North America, Eastern Europe, Australia, Western Europe	n = 861	27–88	58	None	24	692 total deaths (80) 333 in the treatment group (77) 359 in the gemcitabine group (83)	Nab-paclitaxel plus gemcitabine vs. gemcitabine monotherapy	Von Hoff et al. 2013 [41]
United States	n = 19	24–80	47.4	Chemotherapy (gemcitabine containing regimen)*	36	16 at 16 months (84)	Determine the effectiveness of nab-paclitaxel monotherapy as a second line agent	Hosein et al. 2013 [8]
United States	n = 67	30–72	48	None	18	32 at 12 months (48)	Identify the safety and maximum tolerated dose of nab-paclitaxel plus gemcitabine	Von Hoff et al. 2011 [42]
United States Philippines	Trial 1 n = 6 Trial 2 n = 3 Trial 3 n = 1	Trial 1 45–64 Trial 2 53–68 Trial 3 Not stated	Not stated	Chemotherapy (gemcitabine containing regimen)	Trial 1: 13 Trial 2: 6 Trial 3: 6	Trial 1: 6 (100) Trial 2: 1 (33) Trial 3: 1 (100)	Trial 1: Determine the safety of Rexin-G at varying doses Trial 2: Determine the safety of Rexin-G at varying doses Trial 3: Determine the effectiveness of a personal dosing regimen for Rexin-G	Gordon et al. 2006 [7]

Country/region	Sample size	Age range (years)	% Males	Previous treatment	Follow-up (months)	Mortality (%)	Assessment	Reference
Philippines	n = 3	47–56	33	Surgical resection, chemotherapy (gemcitabine containing regimen) and external beam radiotherapy	14	1 (33)	Evaluate the safety and efficacy of Rexin-G	Gordon et al. 2004 [43]
United States	n = 13	50–83	46	Chemotherapy (gemcitabine containing regimen)	12	13 (87)	Determine the effectiveness and most appropriate dose of Rexin-G	Chawla et al. 2010 [48]
Unites States	n = 12	42–71	75	Chemotherapy (gemcitabine containing regimen)	6	11 (92)	Determine the effectiveness and most appropriate dose of Rexin-G	Galanis et al. 2008 [47]
United States	n = 3	Not stated	Not stated	Chemotherapy	Not analyzed	Not analyzed	Evaluate the efficacy and safety of CYT6091	Libutti et al. 2010 [46]
Japan	n = 11	43–72	Not stated	Chemotherapy	Not analyzed	Not analyzed	Determine the maximum tolerated dose, safety and efficay of NK105	Hamaguchi et al. 2007 [44]
Greece	n = 24	47–80	46	Chemotherapy	8	17 (71)	Evaluate the safety and efficacy of lipoplatin	Stathopolous et al. 2006 [45]

*Two patients received non-gemcitabine-based frontline therapy.

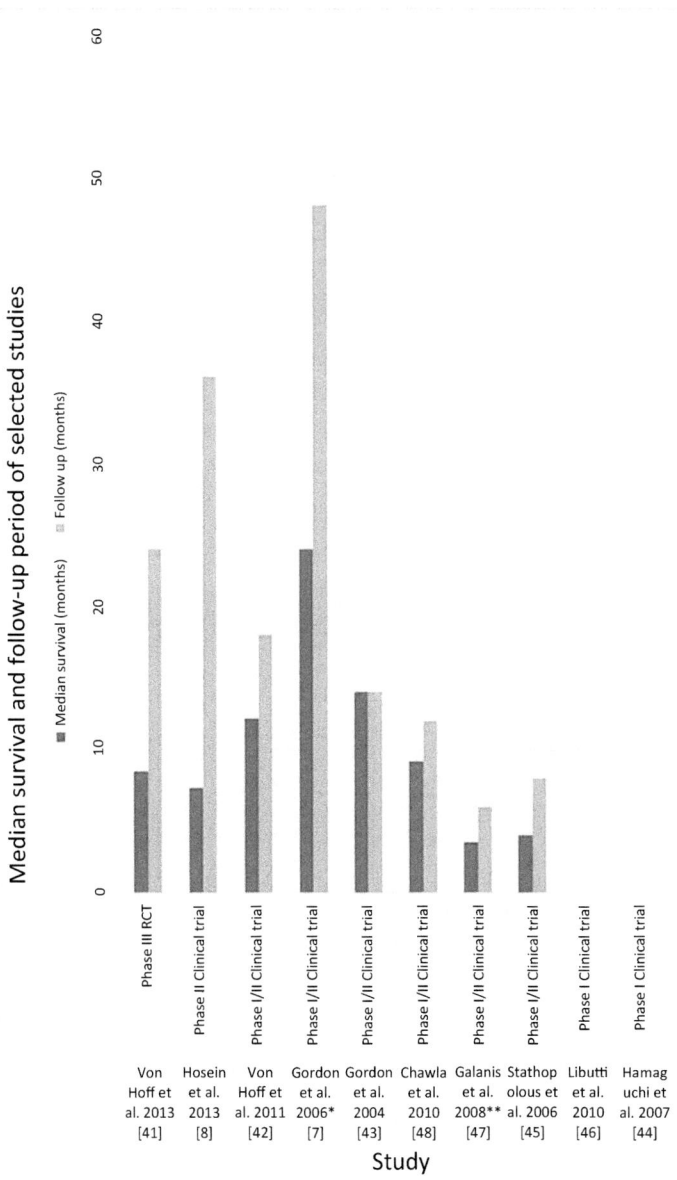

Figure 29.2 Median survivals and follow-up period of selected studies. Abbreviation: RCT, randomized controlled clinical trial.

Table 29.2 Summary of findings associating nanoparticles with pancreatic cancer

Authors	Study design	Nanoparticle formulation	States (from the criteria	Main objective	Participants[*]	Overall median survival/ outcome	Adverse reactions[†]	Conclusion
Von Hoff et al. [41]	Phase III Randomized control trial	Nanoparticle albumin bound paclitaxel (nab-paclitaxel)	Metastatic pancreatic cancer Karnofsky performance status score of 70+	Evaluate the safety and efficacy of nab-paclitaxel plus gemcitabine vs. gemcitabine monotherapy in patients with metastatic pancreatic cancer	n = 861; age 27–88 years; metastatic pancreatic cancer	8.5 months (95% CI, 7.89 to 9.53)	Major: neutropenia Minor: fatigue, nausea, vomiting, anorexia and neuropathy	Increased overall survival; adverse effects of peripheral neuropathy and myelosuppression increased
Hosein et al. [8]	Phase II clinical trial	Nanoparticle albumin bound paclitaxel (nab-paclitaxel)	Pre-treated advanced pancreatic cancer	Evaluate the safety and efficacy of nab-paclitaxel monotherapy in patients with advanced pancreatic cancer	n = 19; age 22–80 years; stage III and IV pancreatic cancer	7.3 months (95% CI, 2.8–15.8)	Major: sepsis and neutropenia Minor: fatigue and neuropathy	—
Von Hoff et al. [42]	Phase I/II clinical trial	Nanoparticle albumin bound paclitaxel (nab-paclitaxel)	Untreated advanced pancreatic cancer	Identify the safety and maximum tolerated dose of nab-paclitaxel plus gemcitabine in patients with untreated advanced pancreatic cancer	n = 67; age 30–72 years	12.2 months (95% CI, 9.8 to 17.9)	Major: sepsis and neutropenia Minor: fatigue and neuropathy	Increased overall survival; slightly higher occurrence of febrile neutropenia (3% vs. 1%)

(Continued)

Table 29.2 (Continued)

Authors	Study design	Nanoparticle formulation	States (from the criteria)	Main objective	Participants*	Overall median survival/outcome	Adverse reactions†	Conclusion
Gordon et al. [7]	(A) Phase I/II clinical trial	Rexin-G	Trial A: Locally advanced pancreatic cancer	Trial A: Determine the safety of Rexin-G at varying doses in patients with locally advanced pancreatic cancer	Trial A, n = 6 Age 45–64 years	Trial A: 24 months (95% CI, 11.1 to 39.5)	Trial A: nil minor or major side effects	Trial A: Increased overall survival; symptom relief
	(B) Phase I/II clinical trial	Rexin-G	Trial B: Metastatic cancer	Trial B: Determine the safety of Rexin-G at varying doses in patients with various types of metastatic cancer	Trial B, n = 3 Age 53–68 years	Trial B: 9 months (95% CI, 2.4 to 14.9)	Trial B: nil minor or major side effects	Trial B: Increased overall survival; symptom relief
	(C) Expanded access clinical trial	Rexin-G	Trial C: Solid organ cancer	Trial C: Determine the effectiveness of a personal dosing regimen for Rexin-G in solid tumors; nanoparticle: Rexin-G (non-replicating retroviral vector expressing a cytocidal gene)	Trial C, n = 1, Age (not stated)	Trial C: Unknown	Trial C: Major anemia requiring red cell transfusions and sporadic thrombocytopenia	Trial C: Reduction in size of metastatic lesions

Authors	Study design	Nanoparticle formulation	States (from the criteria	Main objective	Participants*	Overall median survival/outcome	Adverse reactions†	Conclusion
Gordon et al. [43]	Phase I/II clinical trial	Rexin-G	Stage 4 pancreatic cancer	Evaluate the safety and efficacy of Rexin-G in patients with stage 4 pancreatic cancer; nanoparticle: Rexin-G	n = 3 Stage 4 pancreatic cancer	14 months** (95% CI, −5.8 to 31.8)	Major: nil Minor: nil	Increased overall survival; no adverse events
Chawla et al. [48]	Phase I/II clinical trial	Rexin-G	Gemcitabine resistant metastatic cancer	Determine the effectiveness and most appropriate dose of Rexin-G in patients with gemcitabine resistant metastatic cancer; nanoparticle: Rexin G	n = 13 Age 50–83 years Gemcitabine refractory Metastatic disease	2.6 months at dose 0–1, n = 6; 9.3 months at dose 2, n = 7	Major: nil Minor: fatigue, chills and headache	Increased overall survival; low severity of adverse events
Galanis et al. [47]	Phase I/II clinical trial	Rexin-G	Gemcitabine resistant metastatic disease	Determine the effectiveness and most appropriate dose of Rexin-G in patients with gemcitabine resistant metastatic cancer; nanoparticle: Rexin G	n = 12 Age 42–71 years Gemcitabine refractory Metastatic disease	3.5 months from treatment initiation	Major: nil Minor: nausea, fever, diarrhea, hypermagnesemia and raised liver enzymes (alanine aminotransferase (ALT), aspartate aminotransferase (AST), alkaline phosphate (ALP))	Significant increase in tumor size; low severity of adverse events

(Continued)

Table 29.2 (*Continued*)

Authors	Study design	Nanoparticle formulation	States (from the criteria	Main objective	Participants[*]	Overall median survival/outcome	Adverse reactions[†]	Conclusion
Libutti et al. [46]	Phase I clinical trial	Colloid gold nanoparticle PEGlycated with recombinant TNF	Solid organ cancer	Evaluate the efficacy and safety of CYT6091 in patients with advanced stage cancer	n = 3 with Pancreatic cancer	Not specified	Major: nil Minor: lymphopenia, hypoalbuminemia, hypokalemia, hypophosphatemia and deranged liver function tests (bilirubin and AST)	Nanoparticle CYT6091 preferentially targets tumor tissue
Hamaguchi et al. [44]	Phase I clinical trial	NK105 (micelle nanoparticle)	Refractory solid organ cancers	Determine the maximum tolerated dose, safety and efficacy of NK105 in 19 patients with refractory solid organ cancers	n = 11 Age 43–72 years (range for all participants)	Not specified; antitumor response of 1 year for 1 patient, one had stable disease for 4 weeks	Major: neutropenia Minor: fever; nausea, fatigue, stomatitis, rash, alopecia (for all participants with a solid organ cancer)	Decrease in size of metastatic lesions; low severity of adverse events
Stathopolous et al. [45]	Phase I/II clinical trial	Lipoplatin	Refractory pancreatic cancer	Evaluate the safety and efficacy of lipoplatin and gemcitabine in patients with refractory pancreatic cancer	n = 24 Age 47–80 years; refractory pancreatic cancer	4 months from beginning of treatment (range 2–8 months)	Major: no neurological/renal toxicity Minor: self-resolving abdominal pain; myelotoxicity (grade 3)	Treatment resulted in symptom relief and a partial response/stable disease; low severity of adverse events

[*]Only includes participants with pancreatic cancer.
[**]Including 1 patient still alive after 20 months.
[†]Major reactions include clinically significant neurotoxicity, hemotoxicity, and renal/liver toxicity. Minor reactions include non-life threatening symptoms that resolve with minimal or no intervention.

29.3.2 Study Characteristics

All studies were prospectively performed and conducted in hospitals, mostly in tertiary centers in the United States, Australia, Greece, and Europe. Three studies (30%) were conducted in Japan and the Philippines [7, 43, 44]. All studies stated that informed consent was obtained from the participants and were granted ethics approval. All participants had a formal diagnosis of pancreatic cancer confirmed by histology, imaging and tumor markers; the majority had metastatic disease refractory to conventional chemotherapy. The sample size of studies included ranged from 1 to 861 with a median sample size of 12 participants (interquartile range (IQR), 3–23). Participant age was not stated in the two studies, in the rest of the studies, participant age ranged from 27 to 88 years (Table 29.1). The median survival for participants ranged from 3.5 to 24 months, with an overall median of 8.9 months (IQR, 3.5–13.6), adverse effects ranged from minor ones such as headaches to major effect such as neutropenia and sepsis (Table 29.2). Two studies (20%) did not state the median survival time, or the follow-up period [44, 46] (see Tables 29.1 and 29.2). For the remaining studies (80%), the follow-up period ranged from six to 48 months, with an overall median of 16 months (IQR, 9.0–33.0), (Table 29.2 and Fig. 29.2).

29.4 Synthesis of Study Results

29.4.1 Nanoparticle Albumin Bound Paclitaxel

Paclitaxel is a plant chemotherapeutic alkaloid that is mixed with human serum albumin in an aqueous solvent under high pressure to form a 100–200 nm drug nanoparticle albumin bound paclitaxel (nab-paclitaxel) [49]. One phase I/II study and one phase II study were found on investigating the effect of nab-paclitaxel [8, 42]. Promising beneficial effects of a combination of nab-paclitaxel and gemcitabine were reported in the first study [42]. The second study involved patients with advanced pancreatic cancers and failed to show convincing therapeutic effect of this medication [8]. In a phase I/II study involving 67 patients randomized into three groups, 20 receiving

100 mg/m², 44 receiving 125 mg/m² and three receiving 150 mg/m² of nab-paclitaxel, followed by 1000 mg of gemcitabine on three days in every 28-day cycle. Von Hoff and colleagues reported that the maximum tolerated dose of nab-paclitaxel was 125 mg/m² once a week for three weeks plus 1000 mg/m² gemcitabine every 28 days [42]. They found that the dose-limiting adverse reactions were neutropenia and sepsis; the progression-free survival was 7.9 months (95% CI 5.8–11 months) with a median overall survival of 12.2 months (95% CI 9.8–17.9 months) and a one-year survival rate of 48%. Positron emission tomography (PET) analysis of patients showed a median decrease in metabolic activity of 79% in all three treatment groups with a higher reduction in metabolic activity in the group receiving 125 mg/m² nab-paclitaxel-gemcitabine compared to those receiving 100 mg/m², 68% vs. 53%, respectively (p = 0.044). However, in 19 patients with stage III/IV pancreatic cancer who progressed on gemcitabine-based treatment, and recruited into a single-arm, open-label phase II clinical trial of nab-paclitaxel, Hosein and colleagues reported similar side effects to the Van Hoff study above, a progression-free survival of 1.7 months (95% CI, 1.5–3.5 months), good overall tolerance, and median overall survival of 7.3 months (95% CI, 2.8–15.8 months) [8].

In a more recent phase III RCT involving 861 participants with metastatic pancreatic cancer randomly assigned to a treatment regimen involving nab-paclitaxel and gemcitabine or gemcitabine alone, the same authors reported a significant increase in median overall survival in the group receiving nab-paclitaxel compared to the group receiving gemcitabine alone of 8.5 months and 6.7 months, respectively (p < 0.001) [41]. At the one-year mark, the survival was 5.5 months in the nab-paclitaxel-gemcitabine group compared to 3.7 months in the gemcitabine group (p < 0.001). The adverse events associated with treatment were more prominent in patients receiving nab-paclitaxel plus gemcitabine; these include neutropenia (38% in the nab-paclitaxel plus gemcitabine group, 27% in the gemcitabine group), fatigue (17% in the nab-paclitaxel plus gemcitabine group, 7% in the gemcitabine group) or neuropathy (17% in the nab-paclitaxel plus gemcitabine group, 1% in the gemcitabine group). However, the rates of myelosuppression and neuropathy were also increased [41]. Taken together, these studies suggest that nab-paclitaxel may serve as a promising treatment modality in the future.

29.4.2 Pathotrophic Nanoparticle Gene Delivery

Rexin-G is a pathotropic retroviral based nanoparticle/gene delivery vector produced by transient co-transfection of human embryonic kidney 293T cells with the Moloney murine leukemia virus, and encodes a dominant negative mutant construct of the human *cyclin G1* gene [43]. The first clinical trial using Rexin-G in the treatment of pancreatic cancer in the Philippines was performed by Gordon et al. [43], who reported tumor stabilization in doses ranging from 2.7 × 10^{10} to 3 × 10^{11} colony-forming units; tumor growth was arrested in three of three patients with no experience of dose-limiting toxicity. Two patients were stable 5 and 14 months from diagnosis, respectively. There were no adverse events such as bone marrow suppression, significant alterations in liver and kidney function, nausea or vomiting, mucositis, or hair loss. In a further multicenter/country study, the same performed a series of clinical trials investigating the use of Rexin-G in patients with locally advanced or metastatic pancreatic cancer [7]. Clinical trial A assessed the use of Rexin-G in six patients with pancreatic cancer. Five patients showed a partial response and one had stable disease. Half of the participants had a >30% reduction in tumor size by Response Evaluation Criteria in Solid Tumours (RECIST) or by tumor volume measurement. Progression-free survival ranged from two to nine months with a mean of 3.8 months. The median overall survival of patients treated with Rexin-G from diagnosis was 24 months, whereas that for patients on conventional therapy was 4.4 months. Clinically, all six participants had no associated nausea, vomiting, diarrhea, mucositis, hair loss, or neuropathy, although three participants had symptomatic relief of pain. The only adverse reactions association with treatment were a generalized rash and urticaria in two participants [7].

Clinical trial B investigated the effectiveness of Rexin-G in patients with metastatic cancer, and it involved three patients with metastatic pancreatic cancer. For the patients with metastatic pancreatic cancer, two had a partial response with a >30% reduction in tumor size, necrosis of the primary tumor and decrease in number and size of metastatic nodules. One patient had progressive disease. All three had symptomatic relief of pain. These patients did not suffer from any treatment-related adverse reactions [7].

Clinical trial C investigated the effectiveness of using a personalized dosing regimen (Calculus of Parity) to calculate the dose of Rexin-G in patients with metastatic cancer. This trial involved two patients with metastatic pancreatic cancer. Both patients responded to therapy with one demonstrating necrosis and cystic conversion of an unresectable pancreatic tumor, while the other patient showed significant reduction in the primary pancreatic tumor and a reduction from 28 to 12 pulmonary nodules. None of the patients experienced nausea, vomiting, diarrhea, mucositis, hair loss, or neuropathy. However, two patients developed anemia requiring packed red cell transfusions, which was potentially due to bleeding into the necrotic tumors [7].

In a similar study involving 13 patients with metastatic pancreatic cancer resistant to standard chemotherapy containing gemcitabine, Chawla et al. reported that four patients left the trial due to complications related to their disease or personal reasons after less than one cycle of therapy [48]. They found that the median overall survival was 2.6 months for six patients at dose level 0–1 (1×10^{11} colony-forming units, 2–3 times a week) and 9.3 months for seven patients at dose level 2 (2×10^{11} colony-forming units, thrice a week for four weeks). Treatment-related grade 1 adverse events were experienced by three participants; two experienced fatigue and one experienced chills with a headache [48].

Galanis et al. [47] carried out a study to determine the dose of Rexin-G that provided the best response in 12 patients with gemcitabine refractory metastatic pancreatic cancer. The investigators found that at a dose level between 1×10^{11} to 6×10^{11} colony-forming units per cycle, the treatment was mostly well tolerated with only one participant experiencing a dose-limiting toxicity of raised serum transaminases at a dose of 1.5×10^{11} colony-forming units. The median survival was

3.5 months with 11 participants showing progressive disease and one showing radiographically stable disease with clinical deterioration. Although the treatment was well tolerated, there was no evidence of clinical anti-tumor activity; CT and PET scans pre-treatment at day 28 showed significantly increased tumor volume with a mean increase of 204.5% (p = 0.001), increase in CA 19.9 by a mean of 204.5% (p = 0.001), median increase in PET standardized uptake of fluorodeoxyglucose (FDG) was 36.3% (p = 0.0244).

Overall, Rexin-G is reported to selectively target metastatic cancer sites with associated angiogenesis and increase mean survival in patients with pancreatic cancer.

29.4.3 Gold Nanoparticles

Libutti et al. conducted a clinical trial using CYT-6091 in 30 patients with advanced solid organ cancer, including three participants with pancreatic cancer [46]. CYT-6091 consists of colloid gold nanoparticles with surface-bound recombinant tumor necrosis factor and thiolyated polyethylene glycol. They found that CYT9061 selectively targeted tumor tissue in the three patients with pancreatic adenocarcinoma. Electron microscopy examination of biopsies of the tumor and the adjacent healthy tissue showed that particles in normal tissues were between 0 and 2 in the three participants with pancreatic adenocarcinoma and 5–6 particles in tumor tissue. There were minor adverse effects reported, including lymphopenia, hypoalbuminemia, electrolyte disturbances, and derangement in hepatic enzymes but did not specify any overall survival [46]. This study suggests that colloid gold nanoparticles combined with recombinant tumor necrosis factor selectively target pancreatic cancer sites, aiding the delivery of chemotherapeutic agents to pancreatic cancer tissue.

29.4.4 Micelle Nanoparticles

Micelle nanoparticles are constructed by using polyethylene glycol as the hydrophilic component and modified polyaspartate as the hydrophobic component which entraps the drug paclitaxel [44]. Paclitaxel is an antimicrotubule chemotherapeutic agent for a range of solid organ cancers; however, its efficacy is limited by poor water solubility [44]. The use of a micelle nanoparticle formulation overcomes this by encapsulating paclitaxel in a "core-shell" that is water soluble and has been shown to have enhanced anti- tumor activity due to the EPR effect [44, 50].

Hamaguchi et al. performed a phase I clinical trial to determine the maximum tolerated dose, dose-related toxicities, and pharmacokinetics of NK105, a micelle carrier system for paclitaxel [44]. Nineteen cancer patients were recruited, including 11 patients with pancreatic cancer who received IV infusion of NK105. NK105 was generally well tolerated; six patients developed peripheral neuropathy, and none of the patients developed clinically significant hematological toxicities. A partial response was seen in a patient with metastatic pancreatic cancer who received 150 mg/m^2; their liver metastases reduced in size by 90%, although the effect on pancreatic cancer was not specifically reported. Hence, it is unclear whether micelle nanoparticles would be useful in pancreatic cancer management.

29.4.5 Liposomal Nanoparticles

A liposomal-cisplatin nanoparticle (lipoplatin) is constructed from cisplatin and liposomes composed of dipalmitoyl phosphatidyl glycerol, methoxy-polyethylene glycol-distearoyl phosphatidylethanolamine, and soy phosphatidyl choline [51]. Stathopoulos et al. investigated the efficacy and safe dose of lipoplatin with gemcitabine in 24 patients with refractory pancreatic cancer [45]. The response to treatment was determined by CT (computed tomography) measurement of the tumors. A partial response (>50% reduction in the sum of products of the perpendicular diameters of lesions lasting for at least four weeks) was seen in two patients. Stable disease (<50% reduction and <25% increase in the size of the products of two perpendicular diameters of lesions for at least eight weeks) was seen in 14 patients. Median survival from the beginning of treatment was four months. The treatment dose of fortnightly administration of up to 100 mg/m² of lipoplatin and 1000 mg/m² of gemcitabine was well tolerated by the participants with no evidence of neurotoxicity or renal toxicity.

29.4.6 Quality of Methods of Included Studies

The quality of methods assessment of 10 studies included is outlined in Table 29.3. With a Jadad score of 3, the one RCT included is of reasonably sound methodology, (Table 29.3a). In the other nine non-randomized clinical trials included, the Therapy CA Worksheet indicates that included studies ranged from low to moderate quality of methodology (Table 29.3b). Common weaknesses identified were failure to blind, small sample sizes and/or failure to justify sample size, and failure to identify and account for all confounders.

29.5 Discussion

Pancreatic cancer remains a devastating cause of death globally [3] and is plagued by limited therapeutic options on diagnosis [1]. A significant challenge in the management of pancreatic cancer is the drug-resistant nature of first-line chemotherapy [9]. The ability of nanoparticles to bypass some of these difficulties due to their unique characteristics has enabled their trials as putative therapeutic agents for pancreatic cancers in recent years. This chapter appraised available literature on clinical trials performed up to March 2015 on the use of nanoparticles as therapeutic agents for pancreatic cancer.

Overall, clinical trials have demonstrated that nanoparticles can improve the efficacy of anticancer agents [7, 8, 41–48]. For example, nanoparticles were shown to increase the delivery, cellular targeting of gemcitabine the current first-line chemotherapy for pancreatic cancer, while reducing associated adverse effects [14]. Gemcitabine is known to be plagued by issues such as low solubility and poor expression of intracellular gemcitabine-uptake regulating nucleoside transporters on pancreatic cells [27]. Additionally, multidrug resistance proteins, the anti-tumor microenvironment such as epithelial-mesenchymal transition cells with migratory and invasive properties, and the hypoxic stroma in pancreatic cancers also play a role as a physical barrier preventing chemotherapeutic agents from targeting pancreatic cancer cells [4]. The evidence reviewed in this chapter suggests that these barriers are broken by nab-paclitaxel which increases drug bioavailability and delivery to the malignant tissue [31, 48, 49]. Nab-paclitaxel, for example, has been reported to not only enhance the effect of paclitaxel by increasing its activity and reducing toxicity, but to also acts synergistically with gemcitabine [42]. Since the development of gemcitabine in 1996, eight phase III clinical trials involving chemotherapeutic [52–57] or biologic agents [58, 59, 60] have failed to show an improvement in survival. Improvement was seen in 2006 when a phase III randomized controlled trial demonstrated that erlotinib and gemcitabine lead to an overall survival of 6.42 months, which was significantly prolonged compared to gemcitabine and a placebo [61].

Rexin-G was the first targeted genetic medicine reported to show an increase in overall survival with no organ related toxicity [48]. None of the studies reviewed reported any systematic toxicity [7, 47, 48]. However, one study failed to show any evidence of vector specific- or neutralizing antibodies in the sera of the participants, and no evidence of vector DNA integration or recombination events in non-target organs, including lymphocytes [48]. Collectively, these studies suggest that Rexin-G is superior to standard chemotherapy in terms of safety profile, efficacy in the management of gemcitabine-resistant pancreatic cancer, and improving quality of life.

Libutti et al. performed the first clinical trial involving CTY-6091 and reported potential tumor reducing effects with a moderate safety profile [46]. This outcome is supported by previously reported data on the safety of colloid gold in medicine such as in the treatment for rheumatoid arthritis [62]. In support, pre-clinical studies employing CYT-6091 suggest increased accumulation in solid tumors and a reduction in systemic toxicity [63]. Similarly, the studies using liposomal nanoparticles were reported to exhibit a high safety profile, low toxicity, adequate tumor targeting ability, low immunogenicity, and no renal or neurological toxicity [45].

Table 29.3 Quality assessment of included studies

Table 29.3a Quality assessment of included randomized controlled trial using the JADAD score

Author and year	Randomization	Blinding	An account of all patients	Total score
Von Hoff et al. 2013 [41]	2	0	1	3

Table 29.3b Quality assessment of included studies using the Therapy CA Worksheet

Author and year	Randomization	Sufficient and complete follow-up	Groups analyzed as per randomization	Blinding	Groups treated equally apart from intervention	Groups have similar characteristics at the start	Median survival (months)	95% CI
Hosein et al. 2013 [8]	N	Y	N/A	N	N/A	N/A	7.3	2.8–15.8
Von Hoff et al. 2011 [42]	N	Y	N/A	N	N/A	N/A	12.2	9.8–17.9
Gordon et al. 2006 [7]	Trial A: N Trial B: N Trial C: N	Trial A: Y Trial B: Y Trial C: N	Trial A: N/A Trial B: N/A Trial C: N/A	Trial A: N Trial B: N Trial C: N	Trial A: N/A Trial B: N/A Trial C: N/A	Trial A: N/A Trial B: N/A Trial C: N/A	Trial A: 25 Trial B: 9 Trial C: N/A	12.36–38.30* 3.58–13.76* N/A
Gordon et al. 2004 [43]	N	Y	N/A	N	N/A	N/A	13	−2.30–28.30*

(Continued)

Table 29.3 (Continued)

Author and year	Randomization	Sufficient and complete follow-up	Groups analyzed as per randomization	Blinding	Groups treated equally apart from intervention	Groups have similar characteristics at the start	Median survival (months)	95% CI
Chawla et al. 2010 [48]	N	Y	N/A	N	N/A	N/A	Dose 0-1:4.3 Dose 2:9.2	N/A†
Galanis et al. 2008 [47]	N	Y	N/A	N	N/A	N/A	3.5	2.66–4.34*
Libutti et al. 2010 [46]	N	N	N/A	N	N/A	N/A	N/A	N/A†
Hamaguchi et al. 2007 [44]	N	N	N/A	N	N/A	N/A	N/A	N/A†
Stathopolous et al. 2006 [45]	N	Y	N/A	N	N/A	N/A	4	3.37–4.63*

Abbreviations: CI, Confidence interval; Y, Yes; N, No; N/A, Not applicable.
*Calculated based on values in paper.
†Unable to calculate based on information in paper.

29.5.1 Current Progress

As demonstrated in this systematic review, there is indeed ongoing research into the development of nanotechnology based on the unique tumor microenvironment, which is able to deliver clinically pertinent doses of active formulations to the tumor site while evading various physiological barriers in the fight against pancreatic cancers. Evidence suggests that there is progress in developing nanoparticles able to increase the efficacy per dose of a therapeutic agent by increasing its bioavailability, and that can also be modified for targeted specificity toward cancer cells with negligible damage to non-target tissues, which is generally associated with current chemotherapy [7, 41, 43–48, 64–67]. With the establishment of the Alliance for Nanotechnology in Cancer responsible for fostering innovation and collaboration among researchers to expedite the use of nanotechnology for cancer diagnosis and therapy by the United States National Cancer Institute in September 2004, there has been some success in the design and synthesis of nanoparticles that can encapsulate and deliver a diverse suite of cancer-targeting therapeutic formulations such as nanoparticles delivering chemotherapy drugs or RNA interference inhibitors [68, 69, 70, 71], and nanoparticles co-delivering two chemotherapeutic drugs at a fraction of the dose with minimal side effects and with the potential to reduce cost [64, 65]. There are many other emerging strategies such as the use of nanoparticles (e.g., magnetic nanoparticles) synergistically to improve photodynamic therapy (use of specific wavelength irradiation to selectively kill cancer cells via oxidative stress and caspase-dependent apoptotic mediated mechanisms) [72, 73], and photothermal therapy (use of near-infrared light of longer wavelengths to ablate cancer cells) [74], or both [75], or by employing nanoparticles composed of high atomic numbers such as gold nanoparticles [76], titanium oxide nanotubes [77], or gadolinium-based nanoparticles [78] to enhance radiation therapy.

29.5.2 Limitations

When nanoparticles enter the biological environment, the surface proteins associated with the nanoparticle interact with biological molecules; this interaction depends highly on the composition of proteins on the nanoparticle. Inappropriate surface chemistry of nanoparticles has the potential to cause unwanted reactions, reduction in efficacy, and adverse effects [21].

Clinical trials in this study involve participants who have refractory pancreatic cancer; further studies need to be done to ascertain the effects of nanoparticles on patients with less localized pancreatic cancer.

Studies included in this review were heterogeneous precluding a meta-analysis. Variability was identified in the way the dosage of nanoparticles for administration was determined; since dosage is related to toxicity, this may be a confounder in the frequency and severity of side effects found. In order for the studies to be comparable, a standardized form of dosing should be used in future studies.

Nanoparticles can be generated in many forms, and only a few of them have been investigated in clinical trials as demonstrated by this study. Many other nanoparticle types have been investigated in *in vivo* studies with promising results [64, 65]. In the future, it is expected that many more clinical trials will be published on these emerging therapies such as quantum dots, carbon nanotubes, paramagnetic nanoparticles, metallic nanoparticles, and silver nanoparticles. Since great diversity exists in the form that nanoparticles can take, this study is only representative of gold nanoparticles, micelle nanoparticles, Rexin-G, and liposomal nanoparticles. The nanoparticles used in the clinical trials identified in this study vary greatly among themselves, and the results cannot be generalized to all the forms of nanoparticles available.

Multiple cell lines of origin for pancreatic cancer exist; the trials included in this chapter did not identify the cell line of pancreatic cancer for the participants. This is a limitation, as the types of mutation present in the cell line provide information on the growth characteristics, tumorigenicity, and chemosensitivity of the tumor [79]. For example, panc-1 cells have a 5× greater ability to invade compared to BxPC-3 cells and Capan-1 cells have a higher angiogenic potential compared to Panc-1 cells [80–83]. Although there is limited evidence on the best method to obtain cell line information, further research in this area can enhance the interpretation of results from the use of nanomedicine.

29.5.3 Future Research

These studies highlight the potential of nanoparticles to be used in human participants; the results demonstrate a safe toxicity profile and ability to increase overall survival. Despite promising research showing the efficacy and safety of nanoparticles in *in vitro* and *in vivo* studies in animal models, more research is required to determine the clearance mechanisms of nanoparticles and their molecular interactions in human participants [14]. The long-term side effects of using nanoparticles are yet to be defined. More randomized controlled trials are required to determine implications of nanomedicine on the quality of life of patients with pancreatic cancer.

29.6 Conclusions

Clinical trials have been performed involving a retroviral vector, albumin, colloid gold, micelles, and liposomes. The clinical trials have demonstrated that nanoparticles can be used in conjunction with chemotherapeutic and other agents increasing their efficacy while reducing their toxicity. Increased efficacy of treatment with nanoparticles may improve the clinical outcomes and quality of life in patients with pancreatic cancer, although the long-term side effects of these agents remain unknown.

Abbreviations

%	Percentage
CI	Confidence interval
FDG	Fluorodeoxyglucose
IV	Intravenous
mg/m^2	Milligram per meter square
nab-paclitaxel	Nanoparticle albumin bound-paclitaxel
RCT	Randomized control trials
PET	Positron emission tomography

Disclosures and Conflict of Interest

This chapter was originally published as an open-access article distributed under the terms and conditions of the Creative Commons Attribution (CC BY) license (http://creativecommons.org/licenses/by/4.0/) and originally appeared in *Biomedicines* 2016, 4, 20; doi:10.3390/biomedicines4030020. The copyright permission of the author and publisher has been obtained. This chapter is an unaltered version of the original with minor reformatting and editing. The authors thank the Research Team, Public Health and Tropical Medicine, James Cook University for technical guidance. Minnie Au and Theophilus I. Emeto designed and wrote the chapter. Minnie Au, Jacinta Power, and Theophilus I. Emeto critically appraised the literature. Venkat N. Vangaveti and Hock C. Lai reviewed and edited the chapter. All authors read and approved the final manuscript for submission. The authors declare no conflict of interest. The authors have not received any payment for the preparation of this chapter.

Corresponding Author

Dr. Theophilus I. Emeto
Public Health and Tropical Medicine
College of Public Health, Medical and Veterinary Sciences
James Cook University
James Cook Drive, Douglas, Townsville QLD 4811, Australia
Email: Theophilus.emeto@jcu.edu.au

References

1. Robotin, M. C., Jones, S. C., Biankin, A. V., Waters, L., Iverson, D., Gooden, H., Barraclough, B., Penman, A. G. (2010). Defining research priorities for pancreatic cancer in Australia: Results of a consensus development process. *Cancer Causes Control,* **21**, 729–736.
2. Ferlay, J., Soerjomataram, I., Dikshit, R., Eser, S., Mathers, C., Rebelo, M., Parkin, D. M., Forman, D., Bray, F. (2015). Cancer incidence and mortality worldwide: Sources, methods and major patterns in globocan 2012. *Int. J. Cancer,* **136**, E359–E386.
3. Hariharan, D., Saied, A., Kocher, H. (2008). Analysis of mortality rates for pancreatic cancer across the world. *HPB,* **10**, 58–62.

4. Ansari, D., Chen, B.-C., Dong, L., Zhou, M.-T., Andersson, R. (2011). Pancreatic cancer: Translational research aspects and clinical implications. *World J. Gastroenterol.*, **18**, 1417–1424.
5. Tingstedt, B., Weitkamper, C., Andersson, R. (2011). Early onset pancreatic cancer: A controlled trial. *Ann. Gastroenterol.*, **24**, 206–212.
6. Australian Institute of Health and Welfere, Australasian Association of Cancer Registries (2012). *Cancer in Australia: An Overview 2012*, Australian Institute of Health and Welfare, Canberra, Australia.
7. Gordon, E. M., Lopez, F. F., Cornelio, G. H., Lorenzo, C. C., Levy, J. P., Reed, R. A., Liu, L., Bruckner, H. W., Hall, F. L. (2006). Pathotropic nanoparticles for cancer gene therapy Rexin-G™ IV: Three-year clinical experience. *Int. J. Oncol.*, **29**, 1053–1064.
8. Hosein, P. J., de Lima Lopes, G., Jr., Pastorini, V. H., Gomez, C., Macintyre, J., Zayas, G., Reis, I., Montero, A. J., Merchan, J. R., Rocha Lima, C. M. (2013). A phase II trial of nab-paclitaxel as second-line therapy in patients with advanced pancreatic cancer. *Am. J. Clin. Oncol.*, **36**, 151–156.
9. Yu, X., Zhang, Y., Chen, C., Yao, Q., Li, M. (2010). Targeted drug delivery in pancreatic cancer. *Biochim. Biophys. Acta*, **1805**, 97–104.
10. Shaib, Y. H., Davila, J. A., El-Serag, H. B. (2006). The epidemiology of pancreatic cancer in the united states: Changes below the surface. *Aliment. Pharmacol. Ther.*, **24**, 87–94.
11. Oberstein, P. E., Olive, K. P. (2013). Pancreatic cancer: Why is it so hard to treat? *Ther. Adv. Gastroenterol.*, **6**, 321–337.
12. Buzea, C., Pacheco, I. I., Robbie, K. (2007). Nanomaterials and nanoparticles: Sources and toxicity. *Biointerphases*, **2**, 17–71.
13. Jain, K. (2010). Advances in the field of nanooncology. *BMC Med.*, **8**, 83.
14. Malekigorji, M., Curtis, A., Hoskins, C. (2014). The use of iron oxide nanoparticles for pancreatic cancer therapy. *J. Nanomed. Res.*, **1**, 00004.
15. Kunjachan, S., Pola, R., Gremse, F., Theek, B., Ehling, J., Moeckel, D., Hermanns-Sachweh, B., Pechar, M., Ulbrich, K., Hennink, W. E., et al. (2014). Passive versus active tumor targeting using RGD- and NGR-modified polymeric nanomedicines. *Nano Lett.*, **14**, 972–981.
16. Sharma, A., Madhunapantula, S. V., Robertson, G. P. (2012). Toxicological considerations when creating nanoparticle based drugs and drug delivery systems? *Expert Opin. Drug Metab. Toxicol.*, **8**, 47–69.

17. Dokka, S., Toledo, D., Shi, X., Castranova, V., Rojanasakul, Y. (2000). Oxygen radical-mediated pulmonary toxicity induced by some cationic liposomes. *Pharm. Res.*, **17**, 521–525.
18. Lv, H., Zhang, S., Wang, B., Cui, S., Yan, J. (2006). Toxicity of cationic lipids and cationic polymers in gene delivery. *J. Control. Release*, **114**, 100–109.
19. Keelan, J. A. (2011). Nanotoxicology: Nanoparticles versus the placenta. *Nat. Nanotechnol.*, **6**, 263–264.
20. Chang, C. (2010). The immune effects of naturally occurring and synthetic nanoparticles. *J. Autoimmun.*, **34**, J234–J246.
21. Laurent, S., Mahmoudi, M. (2011). Superparamagnetic iron oxide nanoparticles: Promises for diagnosis and treatment of cancer. *Int. J. Mol. Epidemiol. Genet.*, **2**, 367–390.
22. Wang, C., Zhang, H., Chen, B., Yin, H., Wang, W. (2011). Study of the enhanced anticancer efficacy of gambogic acid on Capan-1 pancreatic cancer cells when mediated via magnetic Fe_3O_4 nanoparticles. *Int. J. Nanomed.*, **6**, 1929–1935.
23. Papa, A.-L., Basu, S., Sengupta, P., Banerjee, D., Sengupta, S., Harfouche, R. (2012). Mechanistic studies of gemcitabine-loaded nanoplatforms in resistant pancreatic cancer cells. *BMC Cancer*, **12**, 419.
24. Gannon, C. J., Patra, C. R., Bhattacharya, R., Mukherjee, P., Curley, S. A. (2008). Intracellular gold nanoparticles enhance non-invasive radiofrequency thermal destruction of human gastrointestinal cancer cells. *J. Nanobiotechnol.*, **6**, 2.
25. Ristorcelli, E., Beraud, E., Verrando, P., Villard, C., Lafitte, D., Sbarra, V., Lombardo, D., Verine, A. (2008). Human tumor nanoparticles induce apoptosis of pancreatic cancer cells. *FASEB J.*, **22**, 3358–3369.
26. Kudgus, R. A., Szabolcs, A., Khan, J. A., Walden, C. A., Reid, J. M., Robertson, J. D., Bhattacharya, R., Mukherjee, P. (2013). Inhibiting the growth of pancreatic adenocarcinoma in vitro and in vivo through targeted treatment with designer gold nanotherapeutics. *PLoS One*, **8**, e57522.
27. Lu, J., Liong, M., Zink, J. I., Tamanoi, F. (2007). Mesoporous silica nanoparticles as a delivery system for hydrophobic anticancer drugs. *Small*, **3**, 1341–1346.
28. Yoshida, M., Takimoto, R., Murase, K., Sato, Y., Hirakawa, M., Tamura, F., Sato, T., Iyama, S., Osuga, T., Miyanishi, K. (2012). Targeting anticancer drug delivery to pancreatic cancer cells using a fucose-bound nanoparticle approach. *PLoS One*, **7**, e39545.

29. Patra, C. R., Bhattacharya, R., Wang, E., Katarya, A., Lau, J. S., Dutta, S., Muders, M., Wang, S., Buhrow, S. A., Safgren, S. L. (2008). Targeted delivery of gemcitabine to pancreatic adenocarcinoma using cetuximab as a targeting agent. *Cancer Res.*, **68**, 1970–1978.

30. Glazer, E. S., Zhu, C., Massey, K. L., Thompson, C. S., Kaluarachchi, W. D., Hamir, A. N., Curley, S. A. (2010). Noninvasive radiofrequency field destruction of pancreatic adenocarcinoma xenografts treated with targeted gold nanoparticles. *Clin. Cancer Res.*, **16**, 5712–5721.

31. Frese, K. K., Neesse, A., Cook, N., Bapiro, T. E., Lolkema, M. P., Jodrell, D. I., Tuveson, D. A. (2012). Nab-paclitaxel potentiates gemcitabine activity by reducing cytidine deaminase levels in a mouse model of pancreatic cancer. *Cancer Discov.*, **2**, 260–269.

32. Khan, J. A., Kudgus, R. A., Szabolcs, A., Dutta, S., Wang, E., Cao, S., Curran, G. L., Shah, V., Curley, S., Mukhopadhyay, D. (2011). Designing nanoconjugates to effectively target pancreatic cancer cells in vitro and in vivo. *PLoS One*, **6**, e20347.

33. Batist, G., Gelmon, K. A., Chi, K. N., Miller, W. H., Chia, S. K., Mayer, L. D., Swenson, C. E., Janoff, A. S., Louie, A. C. (2009). Safety, pharmacokinetics, and efficacy of CPX-1 liposome injection in patients with advanced solid tumors. *Clin. Cancer Res.*, **15**, 692–700.

34. Seymour, L. W., Ferry, D. R., Kerr, D. J., Rea, D., Whitlock, M., Poyner, R., Boivin, C., Hesslewood, S., Twelves, C., Blackie, R. (2009). Phase II studies of polymer-doxorubicin (PK1, FCE28068) in the treatment of breast, lung and colorectal cancer. *Int. J. Oncol.*, **34**, 1629–1636.

35. Matsumura, Y., Gotoh, M., Muro, K., Yamada, Y., Shirao, K., Shimada, Y., Okuwa, M., Matsumoto, S., Miyata, Y., Ohkura, H. (2004). Phase I and pharmacokinetic study of MCC-465, a doxorubicin (DXR) encapsulated in peg immunoliposome, in patients with metastatic stomach cancer. *Ann. Oncol.*, **15**, 517–525.

36. Gradishar, W. (2005). Superior efficacy of albumin-bound paclitaxel, ABI-007, compared with polythylated castor oil-based pclitaxel in women with metastatic breast cancer: Results of a phase III trial. *J. Clin. Oncol.*, **23**, 5983–5992.

37. Valle, J. W., Armstrong, A., Newman, C., Alakhov, V., Pietrzynski, G., Brewer, J., Campbell, S., Corrie, P., Rowinsky, E. K., Ranson, M. (2011). A phase 2 study of SP1049C, doxorubicin in P-glycoprotein-targeting pluronics, in patients with advanced adenocarcinoma of the esophagus and gastroesophageal junction. *Investig. New Drugs*, **29**, 1029–1037.

38. Liberatti, A., Altman, D. G., Tetzlaff, J., Mulrow, C., Gøtzsche, P., Ioannidis, J. (2009). The PRISMA statement for reporting systematic review and meta-analysis of studies that evaluate healthcare interventions: Explanation and elaboration. *Ann. Intern. Med.*, **151**(4).

39. Halpern, S. H., Joanne Douglas, M. (2005). Jadad scale for reporting randomized controlled trials. In: Halpern, S. H., Joanne Douglas, M., eds. *Evidence-Based Obstetric Anesthesia*, Blackwell Publishing, Malden, MA, USA, pp. 237–238.

40. Centre for Evidence-Based Medicine (2015). Therapy critical appraisal worksheet. Canadian Institute of Health Research. Available online: http://ktclearinghouse.ca/cebm/teaching/worksheets/therapy (accessed on 30 March 2015).

41. Von Hoff, D. D., Ervin, T., Arena, F. P., Chiorean, E. G., Infante, J., Moore, M., Seay, T., Tjulandin, S. A., Ma, W. W., Saleh, M. N. (2013). Increased survival in pancreatic cancer with nab-paclitaxel plus gemcitabine. *N. Engl. J. Med.*, **369**, 1691–1703.

42. Von Hoff, D. D., Ramanathan, R. K., Borad, M. J., Laheru, D. A., Smith, L. S., Wood, T. E., Korn, R. L., Desai, N., Trieu, V., Iglesias, J. L. (2011). Gemcitabine plus nab-paclitaxel is an active regimen in patients with advanced pancreatic cancer: A phase I/II trial. *J. Clin. Oncol.*, **29**, 4548–4554.

43. Gordon, E. M., Cornelio, G. H., Lorenzo, C. C., Levy, J. P., Reed, R. A., Liu, L., Hall, F. L. (2004). First clinical experience using a 'pathotropic' injectable retroviral vector (Rexin-G) as intervention for stage IV pancreatic cancer. *Int. J. Oncol.*, **24**, 177–185.

44. Hamaguchi, T., Kato, K., Yasui, H., Morizane, C., Ikeda, M., Ueno, H., Muro, K., Yamada, Y., Okusaka, T., Shirao, K. (2007). A phase I and pharmacokinetic study of NK105, a paclitaxel-incorporating micellar nanoparticle formulation. *Br. J. Cancer*, **97**, 170–176.

45. Stathopoulos, G. P., Boulikas, T., Vougiouka, M., Rigatos, S. K., Stathopoulos, J. G. (2006). Liposomal cisplatin combined with gemcitabine in pretreated advanced pancreatic cancer patients: A phase I-II study. *Oncol. Rep.*, **15**, 1201–1204.

46. Libutti, S. K., Paciotti, G. F., Byrnes, A. A., Alexander, H. R., Gannon, W. E., Walker, M., Seidel, G. D., Yuldasheva, N., Tamarkin, L. (2010). Phase I and pharmacokinetic studies of CYT-6091, a novel PEGylated colloidal gold-rhTNF nanomedicine. *Clin. Cancer Res.*, **16**, 6139–6149.

47. Galanis, E., Carlson, S. K., Foster, N. R., Lowe, V., Quevedo, F., McWilliams, R. R., Grothey, A., Jatoi, A., Alberts, S. R., Rubin, J. (2008). Phase I trial of a pathotropic retroviral vector expressing a

cytocidal cyclin G1 construct (Rexin-G) in patients with advanced pancreatic cancer. *Mol. Ther.*, **16**, 979–984.

48. Chawla, S. P., Chua, V. S., Fernandez, L., Quon, D., Blackwelder, W. C., Gordon, E. M., Hall, F. L. (2010). Advanced phase I/II studies of targeted gene delivery in vivo: Intravenous Rexin-G for gemcitabine-resistant metastatic pancreatic cancer. *Mol. Ther.*, **18**, 435–441.

49. Kratz, F. (2008). Albumin as a drug carrier: Design of prodrugs, drug conjugates and nanoparticles. *J. Control. Release*, **132**, 171–183.

50. Torchilin, V. P. (2006). Multifunctional nanocarriers. *Adv. Drug Deliv. Rev.*, **58**, 1532–1555.

51. Stathopoulos, G. P., Boulikas, T., Vougiouka, M., Deliconstantinos, G., Rigatos, S., Darli, E., Viliotou, V., Stathopoulos, J. G. (2005). Pharmacokinetics and adverse reactions of a new liposomal cisplatin (lipoplatin): Phase I study. *Oncol. Rep.*, **13**, 589–595.

52. Berlin, J. D., Catalano, P., Thomas, J. P., Kugler, J. W., Haller, D. G. (2002). Phase III study of gemcitabine in combination with fluorouracil versus gemcitabine alone in patients with advanced pancreatic carcinoma: Eastern cooperative oncology group trial E2297. *J. Clin. Oncol.*, **20**, 3270–3275.

53. Lima, C. M. R., Green, M. R., Rotche, R., Miller, W. H., Jeffrey, G. M., Cisar, L. A., Morganti, A., Orlando, N., Gruia, G., Miller, L. L. (2004). Irinotecan plus gemcitabine results in no survival advantage compared with gemcitabine monotherapy in patients with locally advanced or metastatic pancreatic cancer despite increased tumor response rate. *J. Clin. Oncol.*, **22**, 3776–3783.

54. Louvet, C., Labianca, R., Hammel, P., Lledo, G., De Braud, F., Andre, T., Cantore, M., Ducreux, M., Zaniboni, A., De Gramont, A. (2004). GemOx (Gemcitabine + Oxaliplatin) versus Gem (Gemcitabine) in non resectable pancreatic adenocarcinoma: Final results of the GERCOR/GISCAD intergroup phase III. In *Proceedings of the ASCO Annual Meeting*, New Orleans, LA, USA, 5–8 June 2004, p. 4008.

55. Oettle, H., Richards, D., Ramanathan, R., van Laethem, J., Peeters, M. (2005). A randomized phase III study comparing gemcitabine pemetrexed versus gemcitabine in patients with locally advanced and metastatic pancreas cancer. *Ann. Oncol.*, **16**, 1639–1645.

56. Abou-Alfa, G. K., Letourneau, R., Harker, G., Modiano, M., Hurwitz, H., Tchekmedyian, N. S., Feit, K., Ackerman, J., De Jager, R. L., Eckhardt, S. G., et al. (2006). Randomized phase III study of exatecan and

gemcitabine compared with gemcitabine alone in untreated advanced pancreatic cancer. *J. Clin. Oncol.*, **24**, 4441–4447.

57. Poplin, E., Levy, D., Berlin, J., Rothenberg, M., O'Dwyer, P., Cella, D. (2006). Phase III trial of gemcitabine (30-min infusion) versus gemcitabine (fixed-dose rate infusion) versus gemcitabine plus oxaliplatin (GEMOX) in patients with advanced pancreatic cancer. *J. Clin. Oncol.*, **24**, 933s.

58. Van Cutsem, E., van de Velde, H., Karasek, P., Oettle, H., Vervenne, W., Szawlowski, A., Schoffski, P., Post, S., Verslype, C., Neumann, H. (2004). Phase III trial of gemcitabine plus tipifarnib compared with gemcitabine plus placebo in advanced pancreatic cancer. *J. Clin. Oncol.*, **22**, 1430–1438.

59. Bramhall, S., Schulz, J., Nemunaitis, J., Brown, P., Baillet, M., Buckels, J. (2002). A double-blind placebo-controlled, randomised study comparing gemcitabine and marimastat with gemcitabine and placebo as first line therapy in patients with advanced pancreatic cancer. *Br. J. Cancer*, **87**, 161–167.

60. Watson, S. A., Gilliam, A. D. (2001). G17DT: A new weapon in the therapeutic armoury for gastrointestinal malignancy. *Expert Opin. Biol. Ther.*, **1**, 309–317.

61. Moore, M. J., Goldstein, D., Hamm, J., Figer, A., Hecht, J. R., Gallinger, S., Au, H. J., Murawa, P., Walde, D., Wolff, R. A. (2007). Erlotinib plus gemcitabine compared with gemcitabine alone in patients with advanced pancreatic cancer: A phase III trial of the national cancer institute of Canada clinical trials group. *J. Clin. Oncol.*, **25**, 1960–1966.

62. Graham-Bonnalie, F. E. (1971). Gold for rheumatoid arthritis. *Br. Med. J.*, **2**, 277.

63. Myer, L., Jones, D., Tamarkin, L., Paciotti, G. (2008). Nanomedicine-based enhancement of chemotherapy. *Cancer Res.*, **68**, 5718.

64. Meng, H., Mai, W. X., Zhang, H., Xue, M., Xia, T., Lin, S., Wang, X., Zhao, Y., Ji, Z., Zink, J. I., et al. (2013). Codelivery of an optimal drug/sirna combination using mesoporous silica nanoparticles to overcome drug resistance in breast cancer in vitro and in vivo. *ACS Nano*, **7**, 994–1005.

65. Meng, H., Wang, M., Liu, H., Liu, X., Situ, A., Wu, B., Ji, Z., Chang, C. H., Nel, A. E. (2015). Use of a lipid-coated mesoporous silica nanoparticle platform for synergistic gemcitabine and paclitaxel delivery to human pancreatic cancer in mice. *ACS Nano*, **9**, 3540–3557.

66. Ferrari, M. (2005). Cancer nanotechnology: Opportunities and challenges. *Nat. Rev. Cancer*, **5**, 161–171.
67. McCarroll, J., Teo, J., Boyer, C., Goldstein, D., Kavallaris, M., Phillips, P. A. (2014). Potential applications of nanotechnology for the diagnosis and treatment of pancreatic cancer. *Front. Physiol.*, **5**, 2.
68. Singh, S., Sharma, A., Robertson, G. P. (2012). Realizing the clinical potential of cancer nanotechnology by minimizing toxicologic and targeted delivery concerns. *Cancer Res.*, **72**, 5663–5668.
69. Schroeder, A., Heller, D. A., Winslow, M. M., Dahlman, J. E., Pratt, G. W., Langer, R., Jacks, T., Anderson, D. G. (2012). Treating metastatic cancer with nanotechnology. *Nat. Rev. Cancer*, **12**, 39–50.
70. Namiki, Y., Fuchigami, T., Tada, N., Kawamura, R., Matsunuma, S., Kitamoto, Y., Nakagawa, M. (2011). Nanomedicine for cancer: Lipid-based nanostructures for drug delivery and monitoring. *Acc. Chem. Res.*, **44**, 1080–1093.
71. Blanco, E., Hsiao, A., Mann, A. P., Landry, M. G., Meric-Bernstam, F., Ferrari, M. (2011). Nanomedicine in cancer therapy: Innovative trends and prospects. *Cancer Sci.*, **102**, 1247–1252.
72. Robertson, C. A., Evans, D. H., Abrahamse, H. (2009). Photodynamic therapy (PDT): A short review on cellular mechanisms and cancer research applications for PDT. *J. Photochem. Photobiol. B*, **96**, 1–8.
73. Piktel, E., Niemirowicz, K., Waţek, M., Wollny, T., Deptuła, P., Bucki, R. (2016). Recent insights in nanotechnology-based drugs and formulations designed for effective anti-cancer therapy. *J. Nanobiotechnol.*, **14**, 1–23.
74. Han, J., Li, J., Jia, W., Yao, L., Li, X., Jiang, L., Tian, Y. (2014). Photothermal therapy of cancer cells using novel hollow gold nanoflowers. *Int. J. Nanomed.*, **9**, 517–526.
75. Fan, Z., Dai, X., Lu, Y., Yu, E., Brahmbatt, N., Carter, N., Tchouwou, C., Singh, A. K., Jones, Y., Yu, H., et al. (2014). Enhancing targeted tumor treatment by near IR light-activatable photodynamic–photothermal synergistic therapy. *Mol. Pharm.*, **11**, 1109–1116.
76. Hainfeld, J. F., Dilmanian, F. A., Slatkin, D. N., Smilowitz, H. M. (2008). Radiotherapy enhancement with gold nanoparticles. *J. Pharm. Pharmacol.*, **60**, 977–985.
77. Townley, H. E., Kim, J., Dobson, P. J. (2012). In vivo demonstration of enhanced radiotherapy using rare earth doped titania nanoparticles. *Nanoscale*, **4**, 5043–5050.

78. Le Duc, G., Miladi, I., Alric, C., Mowat, P., Bräuer-Krisch, E., Bouchet, A., Khalil, E., Billotey, C., Janier, M., Lux, F., et al. (2011). Toward an image-guided microbeam radiation therapy using gadolinium-based nanoparticles. *ACS Nano*, **5**, 9566–9574.
79. Deer, E. L., González-Hernández, J., Coursen, J. D., Shea, J. E., Ngatia, J., Scaife, C. L., Firpo, M. A., Mulvihill, S. J. (2010). Phenotype and genotype of pancreatic cancer cell lines. *Pancreas*, **39**, 425–435.
80. Eibl, G., Bruemmer, D., Okada, Y., Duffy, J. P., Law, R. E., Reber, H. A., Hines, O. J. (2003). PGE 2 is generated by specific COX-2 activity and increases VEGF production in COX-2-expressing human pancreatic cancer cells. *Biochem. Biophys. Res. Commun.*, **306**, 887–897.
81. Eibl, G., Reber, H. A., Wente, M. N., Hines, O. J. (2003). The selective cyclooxygenase-2 inhibitor nimesulide induces apoptosis in pancreatic cancer cells independent of COX-2. *Pancreas*, **26**, 33–41.
82. Molina, M. A., Sitja-Arnau, M., Lemoine, M. G., Frazier, M. L., Sinicrope, F. A. (1999). Increased cyclooxygenase-2 expression in human pancreatic carcinomas and cell lines growth inhibition by nonsteroidal anti-inflammatory drugs. *Cancer Res.*, **59**, 4356–4362.
83. Yip-Schneider, M. T., Sweeney, C. J., Jung, S.-H., Crowell, P. L., Marshall, M. S. (2001). Cell cycle effects of nonsteroidal anti-inflammatory drugs and enhanced growth inhibition in combination with gemcitabine in pancreatic carcinoma cells. *J. Pharmacol. Exp. Ther.*, **298**, 976–985.

Chapter 30

SGT-53: A Novel Nanomedicine Capable of Augmenting Cancer Immunotherapy

Joe B. Harford, PhD,[a] Sang-Soo Kim, PhD,[a,b] Kathleen F. Pirollo, PhD,[b] Antonina Rait, PhD,[b] and Esther H. Chang, PhD[b]

[a]SynerGene Therapeutics, Inc., Potomac, MD, USA
[b]Department of Oncology, Georgetown University Medical Center, Washington, DC, USA

Keywords: immunotherapy, gene therapy, targeted therapy, nanomedicine, nanocomplex, targeted nanocomplex, liposome, drug delivery, DNA delivery, P53, tumor suppressor, PD1, Anti-PD1, immune checkpoint, checkpoint inhibitor, checkpoint blockade, transferrin receptor, endocytosis receptor-mediated endocytosis, transcytosis, receptor-mediated transcytosis, blood–brain barrier, combination therapy, mouse tumor models, syngeneic tumor models, antibody targeting, breast cancer, glioblastoma, lung cancer, head and neck cancer

30.1 Introduction

For decades, cancers have been treated with surgery, chemotherapy, radiotherapy, or some combination of these modalities. As our understanding of the molecular pathways operative in cancer cells has grown, the field of targeted therapeutics (i.e., "precision medicine") has emerged [1]. The term "targeted therapy" in oncology generally refers to the use of agents that block the growth and spread of cancer cells by interfering with specific molecules ("molecular targets") that are involved in the cancer growth, progression, and spread [2]. Targeted therapeutics can be designed with an intended target in view or be based on lead compounds that were detected *via* screening assays based on the particular molecular target. Underlying the field of targeted therapeutics is the assumption that if a molecular target is preferentially found in the cancer cells or if it is in some fashion distinct from its counterpart in normal cells, the targeted therapeutic may have the ability to impede the growth or survival of cancer cells with less harm to normal cells than generally seen with traditional chemotherapy or radiotherapy. Targeted therapeutics can be small molecules e.g., imatinib mesylate (Gleevec®/Glivec®, formerly called STI571) inhibits the BCR-ABL tyrosine kinase that is found specifically in chronic myeloid leukemia but not in normal cells [3, 4]. Targeted therapeutics can also take the form of a monoclonal antibody (mAb) that binds to its cognate antigen on cancer cells to exert a therapeutic effect. For example, rituximab (Rituxan®), the first therapeutic antibody approved for cancer by the US Food and Drug Administration (FDA), binds to the B-cell antigen CD20 [5, 6]. Other examples of therapeutic mAbs include trastuzumab (Herceptin®), which recognizes the HER2 molecules found on some breast cancer and certain other malignancies [7, 8], and bevacizumab (Avastin®), which recognizes VEGF involved in the formation of new blood vessels (angiogenesis) that supply tumors [9, 10].

Cancer gene therapy and cancer immunotherapy are also considered to be targeted therapies. In gene therapy, the aim is often to restore a particular gene function that is lost or defective in cancer cells or to induce cancer cell death by introducing foreign genetic materials into patient's cells [11, 12]. We will herein describe SGT-53, our novel nanomedicine for p53, gene therapy that is "double targeted" in that its payload targets the cellular network regulated by p53 and the delivery system targets tumors *via* a receptor the expression of which is elevated on cancer cells. Cancer immunotherapy can take various forms, but we will focus on the manipulation of immune checkpoints using antibodies that has now become an important component of cancer treatment. Agents that are now approved by the FDA for this form of immunotherapy include mAbs targeting the programmed cell death-1 (PD1) pathway (i.e., antibodies against PD1 or PD-L1, the ligand for PD1) that block negative regulation of T-cell immunity [13–16]. This class of therapeutic agents are among those known as "checkpoint inhibitors" (CPIs) and can be thought of as releasing the "brakes" on the immune system that are naturally present to prevent its attack of normal cells (autoimmunity).

Cancer cells evade immune attack by both rendering cytotoxic T cells (CTLs) dysfunctional and impeding recognition of tumor-specific antigens. The overarching goal of cancer immunotherapy is to develop interventions that enable the body's own immune system to eradicate the cancer with minimal side effects. Despite exciting results in cancer patients receiving therapy based on CPIs, a considerable fraction of patients either do not respond or become refractory/resistant to these agents [17–19]. Moreover, in some instances, treatment with CPIs is accompanied by unacceptable toxicities which are the result of increased immune-related adverse events [20–24].

Much effort is now being expended to explore combining multiple CPIs or combining CPIs with other classes of therapeutic agents to improve clinical response and/or minimize toxicities [25, 26]. In this chapter, we will discuss results combining cancer gene therapy and cancer immunotherapy. Specifically, we will describe SGT-53, a novel nanomedicine that is now in human clinical trials for multiple cancer indications [27, 28]. SGT-53 is a nanocomplex that targets tumor cells and results in relatively high expression of wild-type p53 in cultured cancer cells and in tumors *in vivo* [29–31]. We have employed four syngeneic models in immunocompetent mice, all of which are relatively refractory to checkpoint blockade. These models represent head and neck (H&N) cancer [32], lung cancer, metastatic breast cancer, and glioblastoma [33]. In all of these syngeneic models, our results indicate that SGT-53 treatment results in p53 expression in the tumor cells and markedly augments immunotherapy based on an anti-PD1 mAb. In addition, in one of the syngeneic tumor models

(BALB/c mice bearing 4T1 metastatic breast tumors), we have demonstrated that addition of SGT-53 to the anti-PD1 treatment regimen ameliorates the toxicity associated with anti-PD1 monotherapy [33].

30.2 The Role of p53 in Cancer

The tumor suppressor p53, encoded by the *TP53* gene, is among the most studied molecules with more than 80,000 publications in nearly 40 years of research [34]. The potential of p53 as a therapeutic target has been highlighted both in scientific publications and in popular media. In 1993, p53 was dubbed "molecule of the year" by *Science* in an article that stated: "...p53 and its fellow tumor suppressors are generating an excitement that suggests prevention now and hope for a cure of a terrible killer in the not-too-distant future" [35]. In 1996, p53 made the cover of *Newsweek* under the headline "The Cancer Killer" suggesting that it may be "the key to the cure" [36]. The definition of "not-too-distant future" can be debated, but cancer is still with us, and investigators are still seeking to understand how p53 functions as a tumor suppressor and how best to translate this understanding into therapeutic interventions to the benefit of cancer patients.

The beginning of the p53 story can be traced to 1979 (see Fig. 30.1) when it was identified in a complex with the SV40 tumor-virus oncoprotein [37–40]. Initially, p53 was described as an oncogene [40, 41] but as a result of research in the 1980s was reclassified as a tumor suppressor gene. About 10 years before the actual identification of p53, four families were described with a strong propensity to cancer that would come to be known as classical Li-Fraumeni syndrome (LFS) [42]. LFS families are characterized by the occurrence of a constellation of tumor types with relatively early onset in affected family members. Estimates of the increased risk of cancer have revealed that a staggering 50% of affected LFS family members develop cancer before age 40 years, and 90% before age 60 years [43]. In 1990, two laboratories simultaneously published data that identified germline alterations in the *TP53* gene as the inherited defects in the LFS families [44, 45].

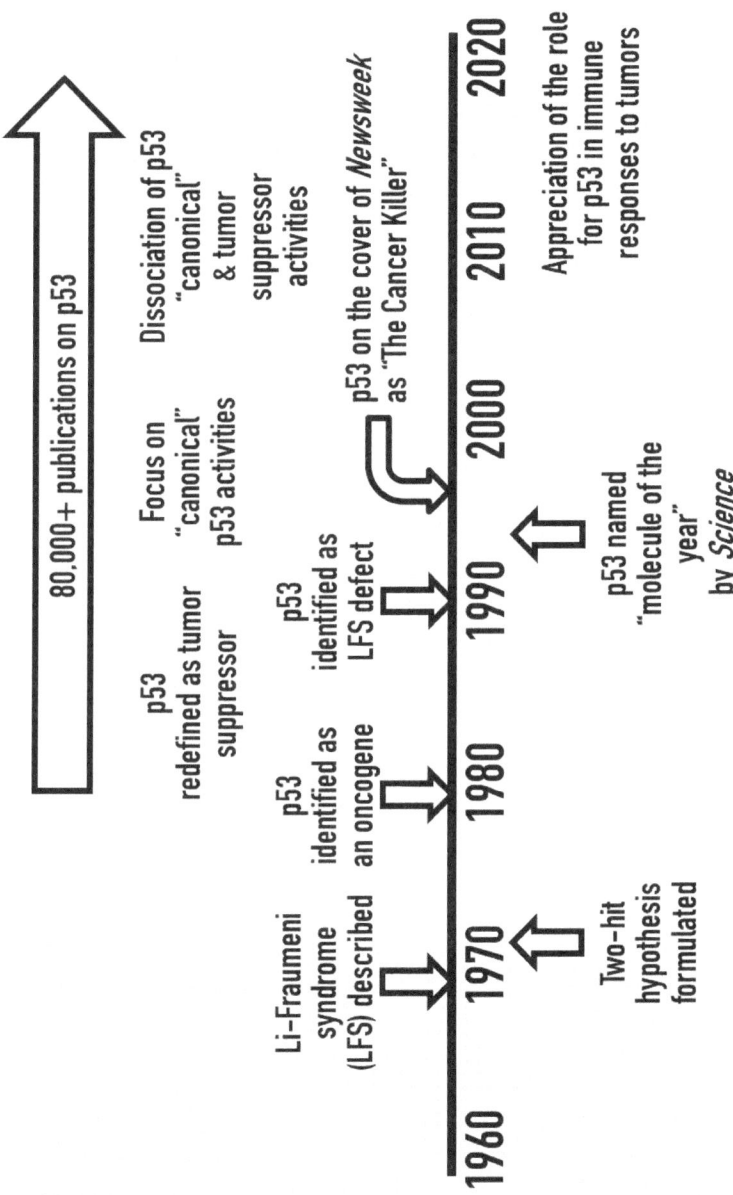

Figure 30.1 The timeline of p53. See text for additional details.

The familial cancers in LFS family members were due to inheritance of only one functional allele of the *TP53* gene and subsequent loss of this heterozygosity in somatic cells (i.e., loss of or mutation in their one functional *TP53* allele). This explanation for LFS was in accord with the "two-hit hypothesis" that had been formulated by Al Knudson based on statistical analysis comparing sporadic and inherited forms of retinoblastoma [46]. The two-hit hypothesis is now accepted with cases of inherited retinoblastoma involving a germline defect in one allele of the *RB1* gene with subsequent somatic loss of the remaining *RB1* allele. In this regard, *RB1* in inherited retinoblastoma is analogous to *TP53* in LFS.

Understanding the genetic basis for tumorigenesis in cancer-prone families has proven to be informative in identifying genes that are also involved in non-inherited (i.e., sporadic) cancers. Over 50% of all human cancers have either deletion, mutational inactivation of p53 or alterations in the pathway downstream of p53 [47–49]. Indeed, *TP53* is the single most frequently mutated gene in cancer [50–53]. It has even been asserted that "direct mutation or aberration in one of its many regulatory pathways is a hallmark of VIRTUALLY EVERY TUMOR" [34] (*emphasis added*). Mutation of p53 can lead to loss of its normal functions, but the mutant p53 proteins that are often overexpressed in tumors exhibit new functions that can contribute to oncogenesis [34, 54–55]. These gain-of-function mutations provide potential targets of therapeutics, but utility of these therapeutics may be limited to cancers with particular mutations in p53.

Substantial effort has been devoted to the understanding of how p53 functions as a tumor suppressor [34, 54–56]. The *TP53* gene encodes a transcription factor that regulates the transcription of hundreds of targets in the human genome—some in a tissue-specific manner [57–61]. Given the large number of genes under the influence of p53 and the intricacy of cellular pathways involving the products of these genes, it is perhaps not surprising that the p53 story is quite complicated. Early studies revealed that p53 is involved in response to acute DNA damage (e.g., by radiation or chemicals) by triggering cell cycle arrest, apoptosis and/or senescence [54, 62–64]. The roles of p53 in cell cycle arrest, apoptosis, and senescence in response to DNA damage have been referred to as the "canonical functions" of p53, and it was envisioned that these functions were responsible for the tumor suppressor activity of p53. By arresting the cell cycle in response to DNA damage, p53 in effect was giving the cell opportunity to engage in DNA repair to preclude accumulation and propagation of mutations that could give rise to tumors. If the DNA damage was too extensive to be repaired, the cell could be eliminated *via* p53-driven programmed cell death (apoptosis), and this too would suppress tumor formation. Senescence, characterized by a permanent cessation of cell division, was seen as yet another p53-mediated alternative to tumor formation. The canonical functions of p53 provided a plausible explanation for how p53 functions to suppress tumors, and p53 was described as "guardian of the genome" based on its canonical roles in cell cycle arrest, apoptosis and senescence [65].

A very interesting observation has been made in large mammals e.g., elephants. Given the fact that elephants have 100-times more cells than humans, it would be anticipated that they would be approximately 100-times MORE susceptible to the somatic events leading to tumors, and yet elephants are estimated to be 5-times LESS susceptible to cancers than humans. This observation is related to what is referred to as "Peto's paradox" named for Richard Peto, who noted that cancer incidence in various species does not correlate with the number of cells in the organism as might be expected [66]. In exploring the molecular basis for the lower cancer susceptibility in elephants, it was found that their genome contains at least 20 copies (40 alleles) of genes for p53 rather than the two alleles found in humans [67–69]. The "extra" copies of *TP53*-related genes found in elephants appear to contribute to enhanced response of elephant cells to DNA damage induced by radiation or chemicals [67–68]. In contrast to elephants having excess copies of p53-related alleles, strains of mice engineered to lack the *TP53* gene (termed p53-null or p53$^{-/-}$ mice) are highly predisposed to development of various tumors [70]. In this regard, p53 null mice resemble affected LFS family members in that both are afflicted with multiple types of cancers.

Although p53 certainly does trigger cell cycle arrest, apoptosis, and senescence in response to DNA damage, its tumor suppressor function is clearly more complicated than just responding to DNA damage as originally envisioned. Something of a paradigm shift in thinking about p53 occurred when research indicated that the protein's ability to trigger a response to acute DNA damage could be lost in certain p53 mutants that nonetheless retained tumor suppressor activity [71, 72]. Similarly, the cell cycle arrest and apoptosis triggered by p53 involve downstream effectors p21, *Puma* and *Noxa* and in mice genetically lacking these effectors, p53 can no longer trigger response to acute DNA damage. Yet even in the absence of ability to respond to DNA damage, p53 is a tumor suppressor [73]. This divorcing of DNA damage response from tumor suppression has led to research that seeks to understand the range of activities associated with the tumor suppressor activity of p53 that might be exploited therapeutically [54–56].

One of the most influential descriptions of cancer described six "hallmarks of cancer" [74]. This perspective on cancer was subsequently expanded to include 10 hallmarks [75] that included cancer-specific alterations in metabolism, angiogenesis, genetic instability, immune evasion, cell death, replicative immortality, sustained proliferation, invasion/metastasis, inflammation, and the tumor microenvironment. Evidence exists for an involvement of p53 in each of the 10 hallmarks [76]. Moreover, a Cancer Hallmarks Analytics Tool (CHAT) has been developed for text mining of the scientific literature on cancer to organize and evaluate it in the context of the 10 hallmarks of cancer [77]. When CHAT was used on a large scale i.e., using "p53" to query 150 million sentences extracted from 24 million PubMed Abstracts, hits on p53 were found to be related to each of the 10 hallmarks of cancer. It has been suggested that tumor suppression by p53 is not related to a single role but to some summation of its functions. This would perhaps account for the fact that mutations in p53 in cancers are much more common than mutations in any of the downstream components of the p53 network [56].

30.3 Cancer Therapeutics Based on p53

Given the abundance of data linking the loss of p53 to oncogenesis, restoring the missing p53 function(s) has long been recognized as an attractive cancer therapeutic strategy [78–81]. Attempts at restoration of p53 activity have included gene therapy whereby a functional copy of p53 is transferred by one means or another into tumor cells [82]. One advantage to *TP53* gene therapy is that it allows one to be agnostic on just how p53 works as a tumor suppressor. All functions of p53 that are affected by the loss of p53 function should be restored by wild-type *TP53* gene therapy. Viral delivery strategies have been employed to bring functional *TP53* gene into tumors to trigger apoptosis [83, 84]. Our own nonviral delivery of functional *TP53* using a tumor-targeted nanocomplex will be described in more detail below, but we have extensive data in mice bearing a variety of human tumor xenografts that nonviral delivery of a functional *TP53* gene to tumors sensitizes these cells to chemotherapy and radiotherapy by rendering cells expressing exogenous p53 more prone to apoptosis [29, 31, 85–88].

Indirect restoration of p53 functions has also been attempted by aiming at other components of the p53 network [54–56, 89]. It has long been appreciated that alterations in proteins other than p53 can influence p53 functions. Perhaps the most extensively studied are MDM2 and MDMX, two related negative regulators of p53 activity that have ubiquitin ligase activity [90–92]. In response to DNA damage, phosphorylation events inhibit the interaction of p53 with MDM2/MDMX resulting in p53 stabilization with consequent activation of downstream transcriptional targets [93, 94]. Because MDM2/MDMX are negative regulators of p53, overexpression of MDM2/MDMX results in loss of p53 function that can result in oncogenesis [95–97]. In those instances wherein p53 function is lost in a tumor through overexpression of its negative regulators, interventions that down-modulate these regulators or interfere with their interaction with p53 would be expected to restore p53 function for tumor suppression. Such molecules are under study as cancer therapeutics [96, 98–100], but their usefulness may be limited to that subset of cancers wherein wild-type p53 is inhibited by MDM2/MDMX overexpression.

30.4 The Role of p53 as Guardian of Immune Integrity

A linkage between p53 and the immune system was evident in early studies with p53 null mice. These mice were engineered to lack *TP53* develop tumors, but approximately one-quarter of these animals die of unresolved infections prior to tumor development suggestive of a severely compromised immune system [101]. The p53 null mice are also more susceptible to autoimmune diseases as demonstrated in a streptozotocin-induced diabetes model with elevated proinflammatory cytokines [102] indicating dysfunction in immune regulation. Other studies have shown that p53 serves as a host antiviral factor that enhances both innate and adaptive immune responses to infections by influenza virus [103]. In experiments designed to assess the impact of losing and restoring p53 function in tumors, RNA interference (RNAi) was used to conditionally regulate endogenous p53 in a doxycycline-dependent fashion [104]. In this mouse model, p53 expression is suppressed in the absence of doxycycline and increased when mice are exposed to doxycycline. Using this murine model, the absence of p53 was found to be required for maintenance of an aggressive hepatocellular carcinoma. However, when p53 expression was restored, the tumor regressed, and this regression was not due to apoptosis. Rather, tumor regression was hampered when macrophages, neutrophils, or natural killer cell functions were blocked individually indicating that the tumor regression triggered by restoring p53 expression was linked to the innate immune system.

It has become clear that, in addition to its well-documented role in response to genotoxic and oncogenic stresses by inducing cell cycle arrest, apoptosis, and senescence, p53 also participates in immune regulation at various levels. Compelling evidence links p53 dysfunction to immunological consequences that contribute to tumorigenesis and tumor progression [105, 106]. Given that DNA binding sites for p53 as a transcription factor are found in a large number of genes and that immune responses also involve complex interplay among many gene products, it is perhaps not surprising that p53 has been implicated in immunity. The recent studies show that p53 is involved in controlling genes involved in immune signaling, autoimmunity, post-apoptotic dead cell clearance, immune tolerance, and immune checkpoint regulation. The mechanisms by which p53 modulates the immune system include direct regulation of the expression of immune-relevant molecules *via* p53 functioning as a master transcription factor. Immune-relevant parameters may also be altered *via* indirect downstream effects emanating from one or more components of the many p53-regulated cellular pathways. In the aforementioned analysis using CHAT to analyze the scientific literature, immune destruction of tumors was the hallmark of cancer with the least hits on the query "p53" [77] perhaps reflective of the fact that this is a relatively new area of p53 research. Nonetheless, it has been suggested that the title "guardian of immune integrity" be added to its title "guardian of the genome" in describing p53 [107].

30.5 SGT-53, A Novel Nanomedicine for *TP53* Gene Therapy

We are developing a platform technology for delivery of therapeutic agents that is based on a nanocomplex termed scL (for single chain Liposome; see Fig. 30.2). The scL nanocomplex consists of: (a) a cationic liposome which can encapsulate various types of therapeutic molecules; (b) the therapeutic payload itself; and (c) a targeting moiety consisting of a single chain mAb Fv fragment (scFv) that recognizes the transferrin receptor (TfR). The scL nanocomplexes are capable of encapsulating diverse therapeutic payloads ranging from plasmid DNAs for gene therapy [30, 87, 108, 109] to antisense oligonucleotides, siRNAs, miRNAs to modulate cellular gene expression [111–114] to small molecule therapeutic agents like temozolomide (TMZ) [115] or other small chemotherapeutics [116].

Virtually all cancer cells examined display increased expression of TfRs compared to normal cells and the TfR pathway has long been recognized as promising for targeting therapeutics to tumors [117, 118]. The TfR engages in endocytosis for delivery

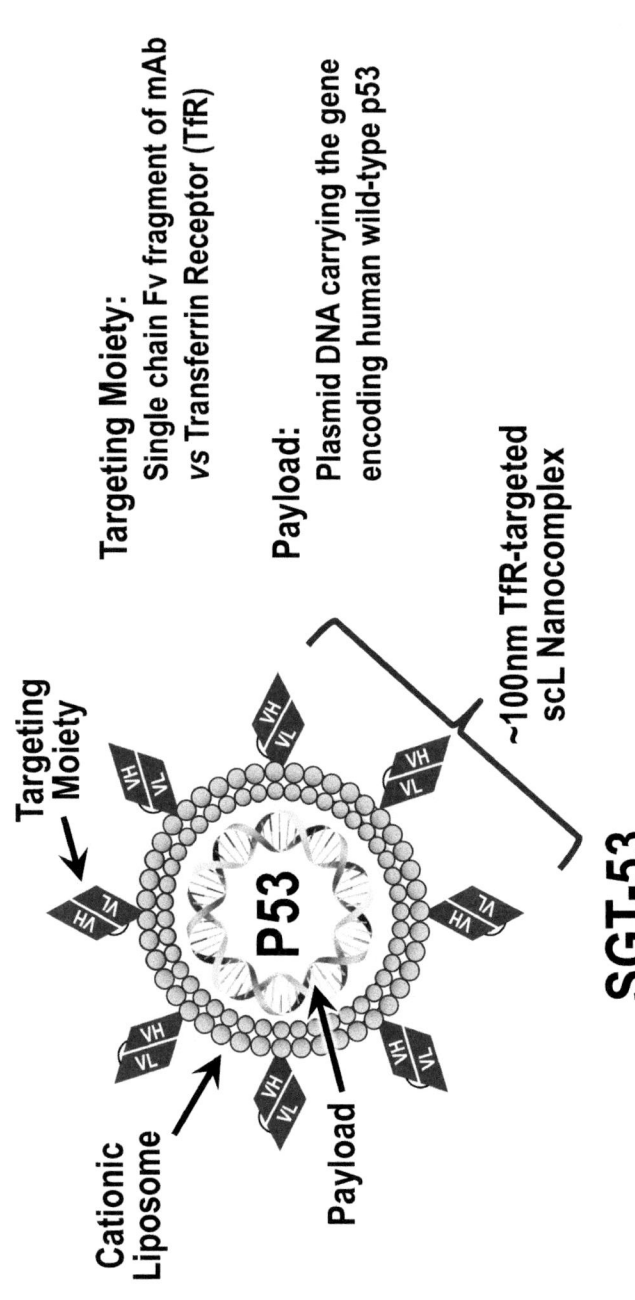

Figure 30.2 Schematic representation of SGT-53. The scL (single chain Liposome) nanocomplex has three components: a payload (a plasmid DNA carrying the wild-type human *TP53* gene in the case of SGT-53); a cationic liposome; and a targeting moiety in the form of a single chain Fv (scFV) fragment of a monoclonal antibody raised against the human transferrin receptor (TfR).

of iron that is found in serum bound to transferrin (Tf). Cells acquire iron by TfR-mediated endocytosis that involves delivery of iron into cells with recycling of the TfR. The TfR on the surface of tumor cells undergo high cycling rates reflective of increased iron consumption [119]. Our scL nanocomplexes are taken into cells *via* this normal physiological pathway (see Fig. 30.3). When the payload of the scL nanocomplex is a plasmid DNA carrying a gene of interest, the gene is released from endosomes, transcribed in the nucleus, and its mRNA translated in the cytoplasm where the protein encoded by the gene can be detected. When the scL nanocomplex is injected intravenously (i.v.), the complex "homes" to tumors with exquisite specificity. Two features of the scL nanocomplex appear to contribute to its specificity in targeting of tumors. The size of the nanocomplex (~100 nm) allows them to extravasate into the interstitial space of the tumor *via* the newer, leakier angiogenic vessels that supply blood to tumors (see Fig. 30.4A). This feature is related to the enhanced permeability and retention (EPR) effect that accounts for a degree of accumulation of even untargeted nanomaterials in tumors. The EPR effect has been reviewed [120, 121]. The scL nanocomplexes engage in the enhanced permeability component of the EPR effect. With untargeted nanomaterials, the retention component of the EPR effect reflects only a small difference between material entering and exiting the tumor i.e., the retention is not an active process mediated by a specific interaction with the tumor cells. In contrast, retention of scL nanocomplexes (and their payloads) within in the tumor is NOT passive. Rather, elevated TfR expression on tumors (see Fig. 30.4B) and the nature of its intracellular trafficking pathway that includes receptor recycling [122, 123] allow the payload to accumulate in the tumor cells.

An example of the specificity of this nanodelivery system is shown in Fig. 30.4C. This experiment involves a mouse bearing a primary pancreatic tumor that has metastasized to multiple organs. The scL nanocomplex injected into the tail vein of this animal had as its payload a plasmid DNA carrying the *LacZ* gene. This reporter gene allowed assessment of where the β-galactosidase encoded by *LacZ* is being expressed. As can be seen in Fig. 30.4C, only tumor tissue shows the characteristic blue color of *LacZ* expression whereas adjacent normal tissue

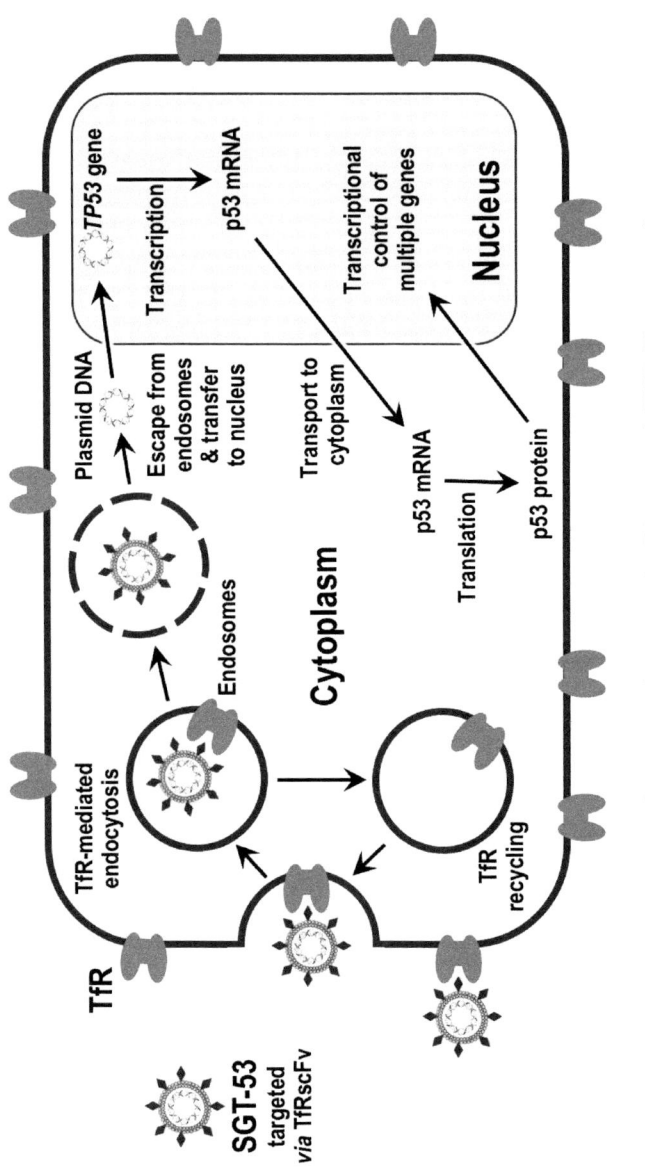

Figure 30.3 Uptake of SGT-53 and expression of p53. SGT-53 enters cancer cells with elevated TfR expression *via* TfR-mediated endocytosis after which the plasmid DNA carrying the *TP53* gene escapes the endosomal compartment and is transcribed into p53 mRNA in the nucleus. The p53 protein is translated from p53 mRNA in the cytoplasm and acts in the nucleus as a transcription factor that regulates expression of many genes.

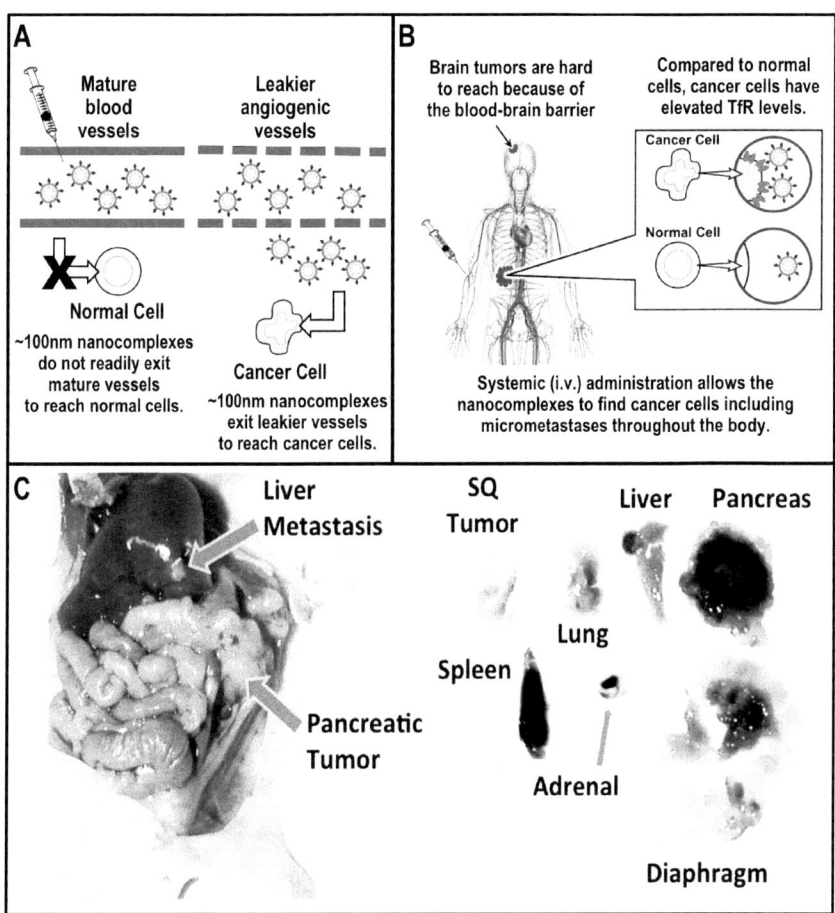

Figure 30.4 The basis of the tumor-targeting by scL nanocomplexes. The selective delivery of scL payloads to tumors derives from preferential exit of the nanocomplexes from the newer, leakier blood vessels that supply tumors (Panel A) and overexpression of the TfR on cancer cells (Panel B). In Panel C is shown a mouse bearing a primary pancreatic tumor with metastases to multiple organs. The mouse received three intravenous injections of scL-LacZ over 24 h and 60 h later, organs were stained for β-galactosidase activity (modified, with permission, from ref. 163).

remains uncolored. This can perhaps best be seen where tumor has invaded a portion of the diaphragm or with the tumor nodule seen on the surface of the liver. It should be noted that the payload in scL-LacZ is not itself colored. The blue staining of the tumor cells requires uptake of the nanocomplex, escape from the endosomal compartment, transfer of DNA into the nucleus, transcription into mRNA, transport of mRNA back into the cytoplasm and translation into the enzyme β-galactosidase responsible for the blue color development through cleavage of the colorimetric substrate X-gal.

Another feature of the scL nanocomplex is its ability to actively cross the blood–brain barrier (BBB) to deliver payloads into the brain. The BBB comprises cerebral endothelial cells that line the brain's blood vessels and blocks entry into the brain parenchyma of many therapeutic agents, including most macromolecules [124]. In the treatment of brain tumors and other neurological diseases, a delivery vehicle that ferries payloads from the bloodstream into the brain would be highly desirable. Such a physiological process naturally exists by which diferric Tf (MW ~80 kDa) is transported across the brain endothelial cells by a process called transcytosis to maintain iron homeostasis in the brain. Iron is an important molecule for brain function but must be imported across the BBB for use by brain cells, and Tf is the primary iron transporting protein [125, 126]. Brain endothelial cells by are unique among endothelial cells in that they exclusively overexpress TfR compared to peripheral endothelium to facilitate iron acquisition by the brain [127]. The Tf/TfR system has long been recognized as having potential to be exploited for delivery of therapeutic agents into the brain [124, 127]. We are developing the scL nanocomplex as a versatile delivery system capable of being loaded with a range of therapeutic and diagnostic molecules to traverse the BBB. These scL nanocomplexes are somewhat akin to the "Trojan horse" of ancient Greece in that they are able to enter the brain by a normal physiological process carrying a payload that would not otherwise be granted entry (see Fig. 30.5A). Endothelial cell TfR molecules pick up diferric Tf on the blood side of the endothelial cells that line brain capillaries. Once on the brain side of the BBB, the nanocomplex enters individual brain cells *via* TfR-mediated endocytosis (see Fig. 30.3), the physiological

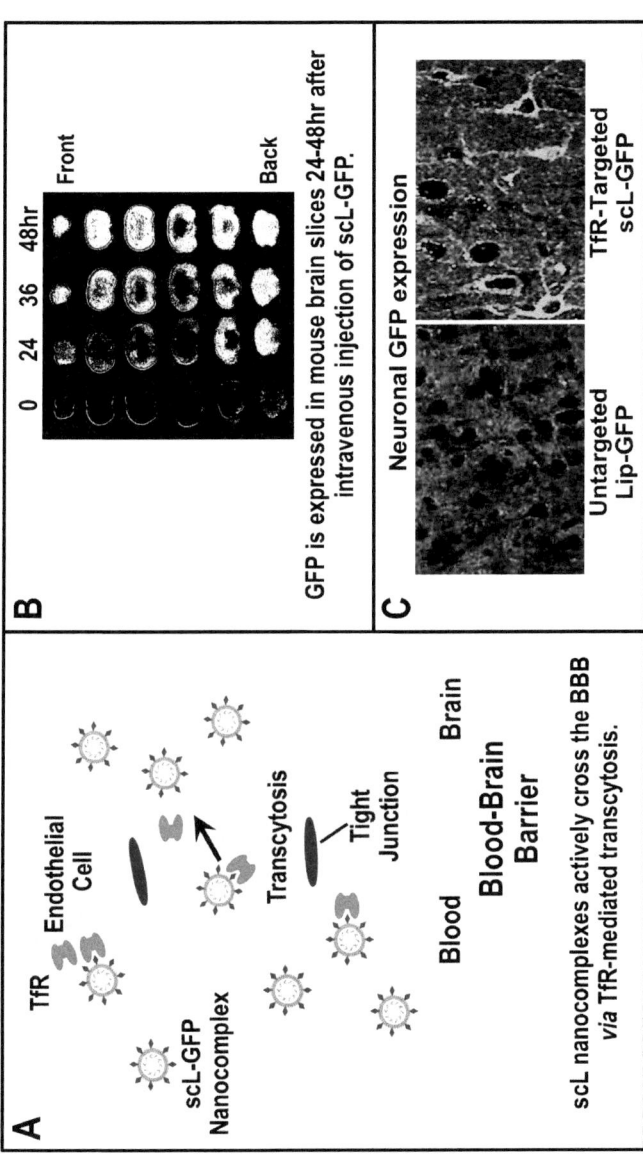

Figure 30.5 The crossing of the blood–brain barrier (BBB) by scL nanocomplexes. The BBB consists of brain microvascular endothelial cells with tight junctions. Diferric transferrin is normally shuttled across the BBB by TfR-mediated transcytosis to supply the brain with iron for metabolism. The scL nanocomplexes hijack this process to enter the brain after intravenous administration (Panel A). Following injection of mice with a scL nanocomplex carrying gene for green fluorescent protein (GFP), the fluorescent gene product can be seen in brain slices (Panel B) and in the cytoplasm of neurons in the cortex (Panel C).

pathway by which these cells take up diferric Tf to acquire iron for metabolism. To demonstrate the validity of this approach, we have encapsulated a plasmid DNA carrying the gene for green fluorescent protein (scL-GFP) and administered this nanocomplex by tail vein injection. When brains are sliced and visualized for GFP signal, we observed the signal in the deep brain (see Fig. 30.5B). High-resolution microscopy revealed the fluorescent GFP in the cytoplasm of in the cells in the brain cortex (see Fig. 30.5C). The fluorescent signal in the neuronal cell cytoplasm arises from the GFP produced *in situ* by transcription of the GFP plasmid DNA and subsequent translation of GFP mRNA in the cytoplasm. This finding confirms that the nanocomplex's payload is not trapped in the endothelial cells of the BBB or in the endosomes of brain cells.

Various alternative strategies have been employed to address this lack of access to brain by therapeutics. These include direct intraventricular injection [128], intrathecal or intranasal administration to avoid the BBB [129–131], or invasive strategies involving mechanical or chemical disruptions of the BBB [132–135]. However, disruption of the BBB to allow entry of therapeutic agents would concurrently compromise the normal protective role of the BBB with potential for brain toxicity or secondary infections [136–138]. Moreover, these approaches are not suitable for the multiple doses of medicines to treat chronic neurological or neuromuscular disorders that may require longer-term interventions [138–142].

We have also used scL nanocomplexes to deliver siRNA payloads into the brain [113, 114]. The potential of RNAi-based therapeutics has been recognized since its discovery in the late 1990s [143]. RNAi encompasses several variations on the theme of RNA-directed gene regulation including siRNA. Successful siRNA-based therapeutics require delivery of the siRNA with complementarity to a specific cellular mRNA target, the silencing of which would be of importance in treating a disease or condition [144, 145]. However, difficulties have been encountered in attempts to develop RNAi-based therapeutics due to RNA instability and/or the inability of the putative RNAi therapeutic to reach the appropriate target cells. Protection of RNA from degradation and the ability to deliver to a desired site of action are considered to be the two grand challenges in RNAi-based therapeutics, and our platform technology addresses both of these issues. Delivery of siRNA having therapeutic potential is a multifaceted challenge, but one key aspect relates to traversing the various tissue barriers that exist in the body [146]. This issue is perhaps nowhere more pronounced than in treatments aimed at brain tumors or other neurological diseases, because the BBB is impervious to most systemically administered therapeutics including nucleic acids and their analogs [148, 148]. We have shown that the scL nanocomplex can ferry siRNA payloads of therapeutic relevance into the brains of mice after systemic administrations. The therapeutic benefit of scL-delivered siRNA was demonstrated in experiments wherein the death of mice and neuronal apoptosis triggered by bacterial lipopolysaccharide were prevented [113, 114].

We have also utilized scL nanocomplexes to deliver small molecule therapeutics into the brain [115, 149]. The chemotherapeutic agent TMZ is the current standard of care for GBM, but as currently used, TMZ improves median survival by only ~2.5 months, and even this modest effect is seen in only about 30% of patients. When brain tumors recur after initial treatment, these tumors are more often TMZ-resistant. Although a relatively small molecule (MW = 194), TMZ does not readily cross the intact BBB. Access of TMZ to the GPM in the brain occurs primarily as a result of disruption of the BBB by the growing tumor meaning that smaller foci of cancer cells that have not as yet disrupted the BBB are not effectively reached by TMZ. We have produced a novel targeted nanocomplex formulation of TMZ that actively crosses the BBB and then actively targets cancer cells in the brain [115]. This nanocomplex formulation (termed scL-TMZ) has been shown to be markedly more active than free TMZ both *in vitro* and in animal models for human GBM including those that are highly TMZ-resistant. Encapsulation of TMZ in the nanocomplex also enhances the agent's activity against other types of tumor cell lines (lung, prostate, pancreas, breast, melanoma and colorectal). Based on these findings, there is every reason to suspect that scL-TMZ will prove effective against a range of cancer types beyond GBM.

Two products that utilize the scL platform delivery system for gene therapy are now in human clinical trials. SGT-53 (see Fig. 30.2), our lead oncology product for p53 gene therapy, has completed phase Ia and Ib clinical trials [27, 28] and is currently in multiple human Phase II clinical trials. SGT-53 has been shown to be well tolerated with some indications of anti-cancer activity in the safety trials. SGT-53 is now being tested against recurrent GBM in combination with TMZ (ClinicalTrials.gov Identifier: NCT02340156), against metastatic pancreatic cancer in combination with gemcitabine/nab-paclitaxel (ClinicalTrials.gov Identifier: NCT02340117), and in combination with topotecan/ cyclophosphamide for refractory or recurrent non-CNS solid tumors in children (ClinicalTrials.gov Identifier: NCT02354547). In addition, two children with brain tumors have been treated with SGT-53 under an extended access (aka compassionate-use) investigational new drug application. No serious safety issues arose in these patients, and evidence of tumor responses was apparent in CT scans (unpublished data).

A second investigational anti-cancer product (SGT-94) based on *RB94*, a version of the *RB1* gene that encodes RB94 that has an N-terminal truncation compared to full-length RB110 [109, 110]. SGT-94 has also completed a Phase I clinical trial (ClinicalTrials.gov Identifier: NCT01517464) with an excellent safety trial [150]. SGT-53 and SGT-94 are among a small number of receptor-targeted nanomedicines now undergoing clinical evaluation [151]. The exceptional safety profiles in trials for SGT-53 and SGT-94, both of which utilize the scL delivery system, suggest that the "regulatory risk" of all products utilizing this system has been effectively reduced. We are currently exploring other applications for scL-mediated nanodelivery.

30.6 SGT-53 Augments Cancer Immunotherapy Based on an Anti-PD1 Monoclonal Antibody

Despite clinical success of antibodies against PD1, many patients are nonresponsive and only a subset experiences durable responses. A large number of trials are now under way that combine inhibitors of the PD1/PD-L1 pathway with a wide range of other agents [26]. These trials aim to expand the percentage of patients that respond to checkpoint blockade and/or improve the anti-tumor activity in patients who do respond. In preclinical studies, we have utilized several syngeneic murine tumor models including a breast tumor (4T1), a non-small cell lung carcinoma (LL2), and glioblastoma (GL261) to explore combining our investigational nanomedicine SGT-53 with a checkpoint inhibitor [33]. The combination of SGT-53 and anti-PD1 antibody resulted in enhanced inhibition of tumor growth compared to either agent individually (see Fig. 30.6). The inhibition of tumor growth results in a corresponding extension of lifespan in the mice receiving SGT-53 plus an anti-PD1 antibody. Similar observations, i.e., decreased tumor growth and increased survival, were also observed in mice bearing MOC1, a H&N tumor treated with the combination of SGT-53 plus an anti-PD1 antibody [32]. It should be noted that all four of these murine tumors are relatively refractory to immune checkpoint inhibitors as monotherapy and, in this regard, resemble patients with tumors that do not respond to checkpoint blockade.

To assess how increased p53 expression emanating from SGT-53 treatment was modulating the immune system to inhibit tumor growth and improve survival, a number of immune-relevant markers were examined in mice bearing 4T1 tumors that were treated with SGT-53 [33]. FACS analysis of harvested tumors revealed significantly increased surface expression of immune cell recognition molecules (e.g., calreticulin (CRT), FAS, PD-L1, CD80, CD86, ICAM1, and MHC class I) after SGT-53 treatment. Although the elements involved in tumor response to anti-PD1 antibodies are complex [26], studies seeking biomarkers that might be used to predict patient response have found expression of PD-L1 on tumors to be the single feature most highly correlated with response [152]. It has been suggested that cancer patients who do not respond to treatment with anti-PD1 antibodies are those having tumors with relatively low expression of PD-L1 [74] although a lack of response of patients to CPIs is likely more complicated. In addition to the observation of elevated PD-L1 on 4T1, we have shown that treatment of mice with SGT-53 elevated the expression of PD-L1 on MOC1 H&N tumor cells [32] as well as on GL261 glioblastoma and LL2 lung cancer [33 and unpublished data]. It is our hypothesis that SGT-53 treatment would also elevate expression of PD-L1 on human tumors, and this effect could result not only in enhanced response to anti-PD1 antibodies in the subset of patients who currently respond but might also expand the percentage of patients who respond. Conversion of non-responders to responders could have profound effects on the use of immunotherapy based on PD1/PD-L1. We have used the NanoString technique to assess gene expression in tumors treated with SGT-53 and found increased mRNA levels corresponding to transporter associated with antigen processing 1 (TAP1) and TAP2 after SGT-53 treatment. These molecules have been implicated as important p53-dependent components of antigen presentation and immunogenic cell death (ICD). These data strongly suggest that p53 expression from SGT-53 alters the expression of immunogenic markers on the surface of tumor cells and induces ICD of tumor cells. We also examined infiltration of tumors by cells of the innate and adaptive immune system. The number of dendritic cells, macrophages, and NK cells associated with 4T1 tumors were each increased by SGT-53 treatment. When mice bearing 4T1 breast tumors were treated with anti-PD1 mAb plus SGT-53, we observed a significant increase in activated $CD4^+$ and $CD8^+$ T cells associated with the tumors. Neither the anti-PD1 antibody nor SGT-53 alone led to this increase in tumor-associated T cells [33]. Comparable observations were made in mice with MOC1 tumors [32].

Triggering apoptosis is considered a canonical p53 function, and we have observed that SGT-53 treatment enhances apoptosis. More recently, the involvement of p53 in post-apoptotic clearance of dead cell bodies through regulation of DD1α has come into focus [153]. Homophilic interaction between DD1α molecules on apoptotic cells and DD1α on phagocytes mediates phagocytic engulfment of the apoptotic cells. DD1α-deficient mice experience autoimmune disorders indicating that DD1α is involved in self-tolerance. It appears that in addition to its involvement in the clearance of post-apoptotic cells, DD1α upregulation contributes to suppression of autoimmunity by p53. When mice bearing 4T1 breast tumors were treated with SGT-53, a significant increase in DD1α on the cancer cells was observed (unpublished data). Based on cell surface markers and gene expression analysis, the tumors in mice treated with SGT-53 appear to be more immunologically "hot." This change would be expected to result in an increased immune response to tumor cells expressing the p53 encoded by the DNA plasmid carried by SGT-53.

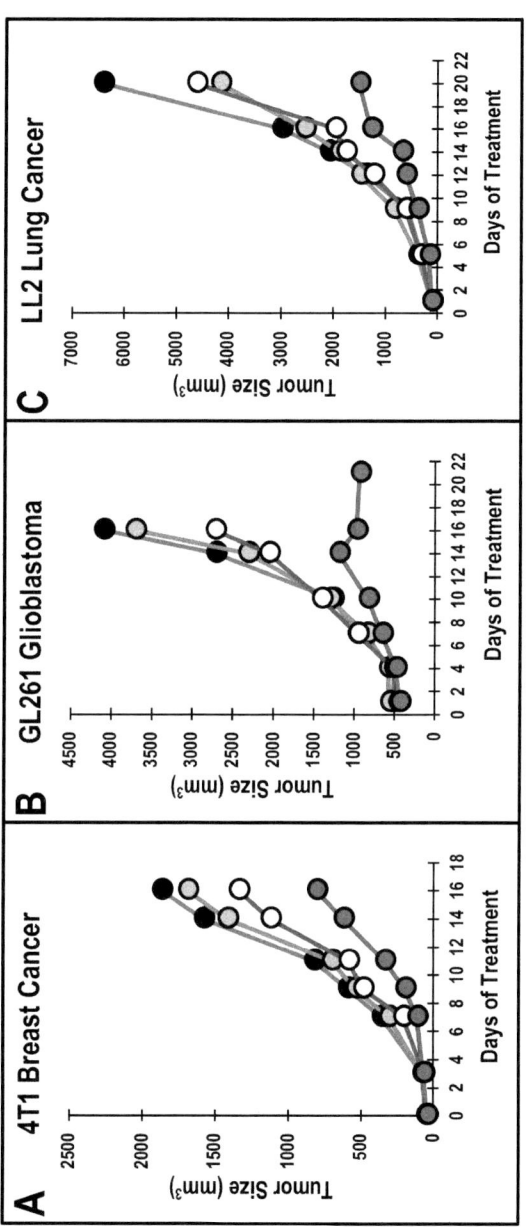

Figure 30.6 Augmentation by SGT-53 of cancer immunotherapy based on an anti-PD1 monoclonal antibody. Tumor growth is compared for 4T1 breast cancer (Panel A), GL261 glioblastoma (Panel B) and LL2 lung cancer (Panel C). All of these syngeneic murine tumor models are relatively refractory to immunotherapy based on an anti-PD1 antibody monotherapy and to SGT-53 monotherapy. The combination of SGT-53 plus anti-PD1 antibody is markedly more inhibitory of tumor growth than either monotherapy.

The vast majority of cancer deaths are due to metastatic disease. Mice having 4T1 breast tumors experience metastases to the lung akin to what is seen in some breast cancer patients. The 4T1 metastases appear as nodules of cells that can be quantified microscopically in sections of the lungs of mice. We have observed that SGT-53 alone is able to substantially reduce 4T1 lung metastases [33]. Treatment of the mice with an anti-PD1 antibody alone was essentially ineffective in blocking lung metastases. However, in mice treated with the combination of SGT-53 plus an anti-PD1 antibody, 4T1 lung nodules were essentially undetectable.

Although PD1/PD-L1 checkpoint blockade is normally well tolerated in patients, three patients receiving nivolumab died from pneumonitis while participating in trials [20]. We observed that BALB/c mice bearing 4T1 tumors were all killed by repeated administration (~4–5 doses) of anti-PD1 antibody. These deaths occurred significantly before tumors were large enough to kill the mice. Fatal hypersensitivity after repeated injections of anti-PD1 or anti-PD-L1 has also been observed in 4T1-bearing mice by others [154]. Autopsies of our mice killed by anti-PD1 revealed lung abnormalities that reflected an increase in neutrophil and macrophage infiltration of the lungs. This infiltration was seen with anti-PD1 antibody treatment alone but not with either SGT-53 alone or in the combination-treatment group. The finding that the deaths of mice seen after treatment with an anti-PD1 antibody was abrogated if the checkpoint inhibitor was given in conjunction with SGT-53 [33] suggests that SGT-53 when added to anti-PD1 immunotherapy may render the CPI not only more effective but also safer for patients.

To understand the molecular basis of the ability of SGT-53 to protect mice from anti-PD1-induced death, we performed gene expression profiling of tumor tissues from the mice employing the NanoString technique. We specifically looked for candidate genes that were modulated by anti-PD1 treatment alone but not when SGT-53 was added to anti-PD1. The logic here was that some change(s) in gene expression might be linked to the death of 4T1-bearing mice, and that if SGT-53 prevents this death, then SGT-53 might prevent the underlying alteration in gene expression. Multiple genes involved in immune hypersensitivity and neutrophil recruitment, priming, and activation (Ccl17, Cd74,

Prg2, Nod1, and GM-CSF) were identified as being upregulated by anti-PD1 monotherapy, while genes involved in inhibiting excessive neutrophil infiltration (e.g., Gzmm) were downregulated. The changes in expression of this set of genes were not observed with the combination treatment involving anti-PD1 plus SGT-53. Of note is that GM-CSF has been directly linked to lung-damaging neutrophil accumulation [155]. GM-CSF mRNA was upregulated with anti-PD1 treatment, but when SGT-53 was added to the checkpoint blockade, GM-CSF mRNA expression in tumors was similar to that seen in the tumors of untreated animals. A similar result was observed with GM-CSF in sera as assessed by ELISA that was significantly increased (~4-fold) in 4T1 tumor-bearing mice after anti-PD1 treatment alone, but was restored to normal levels when SGT-53 was added to the treatment regimen. Anti-PD1 treatment also increased the serum level of the pro-inflammatory cytokine TNFα, in this case by ~13-fold. TNFα is known to be related to the cytokine-release syndrome seen after infusion of certain mAbs [156]. SGT-53 treatment added to anti-PD1 treatment resulted in TNFα levels equivalent to those seen in untreated animals. We have thus identified a set of genes that are candidates in the fatal hypersensitivity reaction that we see in 4T1-bearing mice. SGT-53 treatment prevents the modulation of the expression of these genes by anti-PD1 monotherapy and "rescues" the mice from the otherwise fatal consequences of treatment with this checkpoint inhibitor. Taken together with the above efficacy data, these results support the notion that SGT-53 may render checkpoint blockade not only more effective but also safer.

Tumors engage in immunosuppression as part of evading the body's immune system. The cytotoxicity of $CD8^+$ tumor-infiltrating lymphocytes (TILs) against the tumor can be influenced by multiple immunosuppressive factors such as suppressive cytokines, suppressor cells e.g., regulatory T cells (Tregs) and myeloid-derived suppressor cells (MDSCs), and signaling through inhibitory immune ligands [157–159]. Tregs and MDSCs are crucial populations in enforcing immunosuppression in the tumor microenvironment [160]. We found that elevated p53 expression resulting from SGT-53 treatment reduced both Tregs and MDSCs associated with tumors, and so SGT-53 can be exploited as a new means of reducing or eliminating these immunosuppressive cells. Gene expression profiling revealed that following SGT-53 treatment, a significant down-modulation of IDO1, an enzyme known for its key immunosuppressive role in many human cancers, was observed. It has been previously shown that tryptophan depletion by IDO1 could lead to the T cell anergy and activation of immunosuppressive Tregs and MDSCs [161, 162]. Currently, drugs targeting the IDO1 pathway are in clinical trials to reverse the tumor-induced immunosuppression [162]. In accordance with our data, a recent report demonstrated that restoration of p53 activity *via* nutlin-3a was able to induce ICD and promote CD8$^+$ T cell-dependent anti-tumor immunity in mice bearing EL4 tumors [160]. In that study, nutlin-activated p53 was able to eliminate immunosuppressive MDSCs. However, reactivation of endogenous p53 *via* nutlin-3a requires tumor cells harboring wild-type p53, and many tumor cells with TP53 deletions or otherwise not expressing mutant p53 would be expected to be unresponsive to nutlin-3a. In contrast, our tumor-targeted gene therapy approach to restore functional p53 and subsequently induce tumor cell immunogenicity and anti-tumor immunity would not be expected to be dependent on the p53 status of the tumor. Indeed, in human tumor cell lines, we have observed that SGT-53 can push cells harboring either wild-type or mutated p53 into apoptotic death [31, 88].

30.7 Summary and Perspectives

We have produced a nanodelivery system capable of carrying a wide range of payload types. The salient features of our scL nanocomplex are: (a) an ability to carry a variety of payloads into tumor cells with a high degree of specificity based on augmented access to tumors and overexpression of TfR on tumor cells; (b) an ability to carry payloads across the BBB *via* transcytosis mediated by the TfR and into CNS cells *via* TfR-mediated endocytosis; and (c) acceptable safety profiles (based on the trials for SGT-53 and SGT-94, both of which utilize this delivery system). Our most advanced product, SGT-53, is now in multiple Phase II trials.

We have demonstrated that p53 gene therapy delivered in the form of a tumor-targeted nanomedicine (SGT-53) enhances the efficacy of immunotherapy in several mouse models of syngeneic tumors including H&N, breast, lung, and brain [32, 33]. The involvement of p53 in immune regulation is clearly multifaceted. We have examined the impact of SGT-53 treatment on a number of immune-relevant markers in animals bearing syngeneic tumors. We observed that surface expression of the ER protein CRT and release of the nuclear protein HMGB1 are elevated on tumor cells after SGT-53 treatment. Both of these surface markers are hallmarks of ICD. We also observed a significant increase in PD-L1 expression in mice treated with SGT-53 and an apparent enhancement in innate immunity as evidenced by an increase in tumor-associated dendritic cells and macrophages [32, 33]. Tumor expression of CD80 and CD86 was also elevated by SGT-53. The change in expression of CD80 and CD86 was particularly pronounced in mice bearing MOC2 H&N tumors [32]. Total CD8+ T cells and activated cytotoxic (granzyme B+) T cells were also increased in tumors after SGT-53 treatment. In all tumor types examined, a significantly enhanced inhibition of tumor growth was achieved with a systemic administration of SGT-53 plus anti-PD1 antibody in mouse models of tumors with inherent therapeutic resistance to anti-PD1 therapy (see Fig. 30.6). Moreover, treatment of the mice with SGT-53 alone was able to substantially reduce metastases of 4T1 breast tumor cells in the lungs, while treatment with anti-PD1 alone was essentially ineffective in reducing metastases. In mice treated with the combination of SGT-53 plus anti-PD1 antibody, virtually no 4T1 metastatic lung nodules were detected. Collectively, SGT-53 treatment modified immunogenicity of tumors (i.e., made the tumor "hot" immunologically), increased both innate and adaptive immune activity, and reduced immunosuppression in the tumor microenvironment (see Fig. 30.7).

In BALB/c mice bearing 4T1 breast tumors, we have observed mortality (anaphylaxis) after the mice were given multiple injections of an anti-PD1 antibody that was due to the fatal hypersensitivity to xenogeneic anti-PD1 antibody mediated by neutrophils. Surprisingly, we observed that this fatal hypersensitivity was not seen when PD1 blockade was done in the context of combination therapy with SGT-53. Using gene expression analysis, a number of candidate genes that may be responsible for the neutrophil invasion of the lungs have been identified. Collectively, our data suggest that SGT-53, when added to anti-PD1 immunotherapy, may augment checkpoint blockade not only in terms of rendering checkpoint inhibitors more effective but also making them safer for patients.

A large number of trials are now under way that combine inhibitors of the PD1/PD-L1 pathway with a wide range of other agents [26]. In many cases, these combinations appear to be empirical in nature and lacking a strong mechanistic rationale for the particular combination being tested. The ability of SGT-53 to modulate a number of relevant immune marker, including tumor PD-L1 in multiple syngeneic mouse models, provides such rationale for combining SGT-53 with an anti-PD1 antibody in a cancer trial. This trial should be relatively easy to mount given that a number of anti-PD1 antibodies are already on the market, and SGT-53 is already in Phase II clinical trials in combination with conventional chemotherapeutic agents.

In summary, we herein describe a novel nanomedicine that not only increased the immunogenicity of tumor cells and the number of tumor-infiltrating immune cells but also alleviated immunosuppression and improved anti-tumor activity when used in combination with an anti-PD1 antibody. This improved efficacy of the combination therapy was observed in mice bearing 4T1 breast cancer, LL2 non-small cell lung cancer, or GL261 glioblastoma [33]. A similar enhancement of anti-tumor activity was also observed in a mouse syngeneic model of MOC1 H&N cancer [32]. Given that SGT-53 could alleviate fatal hypersensitivity associated with an anti-PD1 antibody in 4T1 breast cancer, this nanomedicine may be able to reduce immune-related adverse events that are sometimes seen with cancer immunotherapies. Collectively, our data suggest that SGT-53, representing tumor-targeted p53 gene therapy, has potential to augment significantly immune checkpoint blockade agents for improved outcomes in a variety of malignancies. It is possible that the SGT-53 would not only improve outcomes in patients that already respond to checkpoint blockade, but also increase the percentage of patients who respond. SGT-53 has completed a first-in-man Phase I and Ib trials with favorable safety profiles [27, 28] and is now being evaluated in Phase Ib and Phase II trials as combination therapy with currently approved chemotherapeutic agents. Our data here provide a strong mechanistic rationale for combining SGT-53 and PD1/PD-L1-based immune checkpoint blockade in a clinical trial setting.

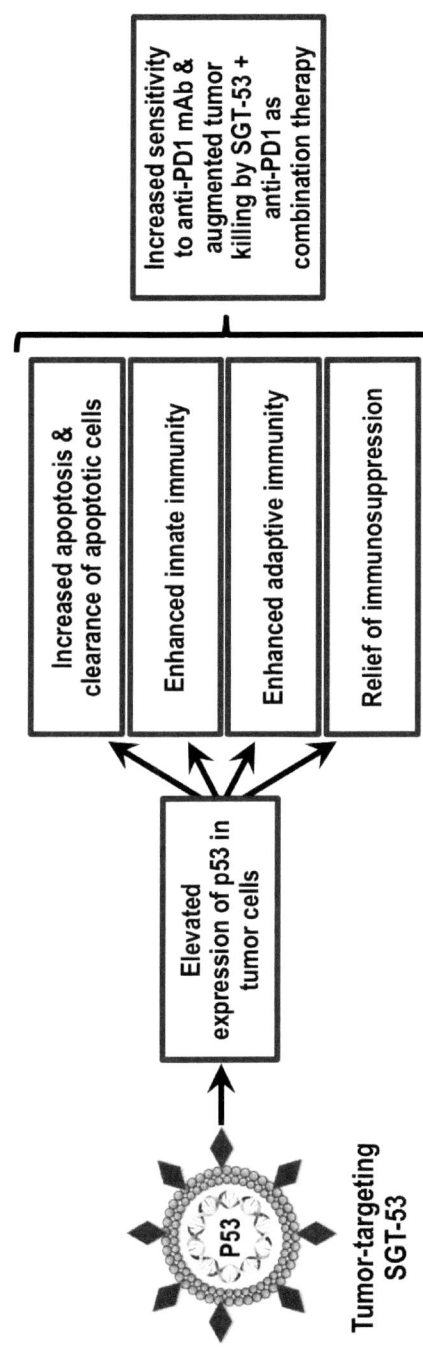

Figure 30.7 Mechanism of the augmentation of cancer immunotherapy by SGT-53. The elevation of p53 expression emanating from SGT-53 treatment has been shown to increase apoptosis, clearance of apoptotic cells by macrophages, to enhance both innate and adaptive immune responses and to relieve tumor immunosuppression.

Disclosures and conflict of interest

Drs. Chang and Pirollo are inventors of the described technology, for which several patents owned by Georgetown University have been issued. The patents have been licensed to SynerGene Therapeutics for commercial development. Dr. Chang is a professor at Georgetown University who owns an equity interest in SynerGene Therapeutics, Inc., and serves as a non-paid scientific consultant to the company. Dr. Harford serves as salaried president & CEO of SynerGene Therapeutics, Inc., and owns stock in the same. Dr. Kim is a salaried employee of SynerGene Therapeutics, Inc., and owns stock options in the company. Dr. Rait serves as a consultant for SynerGene Therapeutics, Inc., and owns stock options in the company. Dr. Pirollo has received research support from SynerGene Therapeutics, Inc.

Corresponding Author

Dr. Joe B. Harford
SynerGene Therapeutics, Inc.
9812 Falls Rd., Suite 114, Potomac, MD 20854, USA
Email: harfordj@synergeneus.com

References

1. Wu, H-C., Chang, D-K., Huang, C-T. (2006). Targeted therapy for cancer. *J. Cancer Mol.*, **2**, 57–66.
2. Targeted Cancer Therapies. Available at: https://www.cancer.gov/about-cancer/treatment/types/targeted-therapies/targeted-therapies-fact-sheet (accessed on June 25, 2018).
3. Druker, B. J. (2002). STI571 (Gleevec) as a paradigm for cancer therapy. *Trends Mol. Med.*, **8**, S14–S18.
4. Lambert, G. K., Duhme-Klair, A. K., Morgan, T., Ramjee, M. K. (2013). The background, discovery and clinical development of BCR-ABL inhibitors. *Drug Discov. Today*, **18**, 992–1000.
5. Reff, M. E., Carner, K., Chambers, K. S., Chinn, P. C., Leonard, J. E., Raab, R., et al. (1994). Depletion of B cells *in vivo* by a chimeric mouse human monoclonal antibody to CD20. *Blood*, **83**, 435–445.
6. Maloney, D. G., Liles, T. M., Czerwinski, D. K., Waldichuk, C., Rosenberg, J., Grillo-Lopez, A., et al. (1994). Phase I clinical trial using escalating single-dose infusion of chimeric anti-CD20

monoclonal antibody (IDEC-C2B8) in patients with recurrent B-cell lymphoma. *Blood*, **84**, 2457–2466.

7. Hudziak, R. M., Lewis, G. D., Winget, M., Fendly, B. M., Shepard, H. M., Ullrich, A. (1989). p185HER2 monoclonal antibody has antiproliferative effects *in vitro* and sensitizes human breast tumor cells to tumor necrosis factor. *Mol. Cell. Biol.*, **9**, 1165–1172.

8. Slamon, D., Pegram, M. (2001). Rationale for trastuzumab (Herceptin) in adjuvant breast cancer trials. *Semin. Oncol.*, **28**, 13–19.

9. Willett, C. G., Boucher, Y., di Tomaso, E., Duda, D. G., Munn, L. L., Tong, R. T., et al. (2004). Direct evidence that the VEGF-specific antibody bevacizumab has antivascular effects in human rectal cancer. *Nat. Med.*, **10**, 145–147.

10. Ferrara, N., Hillan, K. J., Gerber, H. P., Novotny, W. (2004). Discovery and development of bevacizumab, an anti-VEGF antibody for treating cancer. *Nat. Rev. Drug Discov.*, **3**, 391–400.

11. McCormick, F. (2001). Cancer gene therapy: Fringe or cutting edge? *Nat. Rev. Cancer*, **1**, 130–141.

12. Sato-Dahlman, M., Wirth, K., Yamamoto, M. (2018). Role of gene therapy in pancreatic cancer: A review. *Cancers*, **10**(4), 103.

13. Mellman, I., Coukos, G., Dranoff, G. (2011). Cancer immunotherapy comes of age. *Nature*, **480**, 480–489.

14. Pardoll, D. M. (2012). The blockade of immune checkpoints in cancer immunotherapy. *Nat. Rev., Cancer,* **12**, 252–264.

15. Zou, W., Wolchok, J. D., Chen, L. (2016). PD-L1 (B7-H1) and PD-1 pathway blockade for cancer therapy: Mechanisms, response biomarkers and combinations. *Sci. Translational Med.*, **8**(328), 328rv4.

16. Li, Y., Li, F., Jiang, F., Lv, X., Zhang, R., Lu, A., et al. (2016). A mini-review for cancer immunotherapy: Molecular understanding of PD-1/PD-L1 pathway & translational blockade of immune checkpoints. *Int. J. Mol. Sci.*, **17**(7), E1151.

17. Vinay, D. S., Ryan, E. P., Pawelec, G., Talib, W. H., Stagg, J., Elkord, E., et al. (2015). Immune evasion in cancer: Mechanistic basis and therapeutic strategies. *Semin. Cancer Biol.*, **35 Suppl**, S185–S198.

18. Kim, J. M., Chen, D. S. (2016). Immune escape to PD-L1/PD-1 blockade: Seven steps to success (or failure). *Ann. Oncol.*, **27**, 1492–1504.

19. Sharma, P., Hu-Lieskovan, S., Wargo, J. A., Ribas, A. (2017). Primary, adaptive, and acquired resistance to cancer immunotherapy. *Cell*, **168**, 707–723.

20. Gangadhar, T. C., Vonderheide, R. H. (2014). Mitigating the toxic effects of anticancer immunotherapy. *Nat. Rev. Clinical Oncol.*, **11**, 91–99.
21. Naidoo, J., Page, D. B., Li, B. T., Connell, L. C., Schindler, K., Lacouture, M. E., et al. (2015). Toxicities of the anti-PD-1 and anti-PD-L1 immune checkpoint antibodies. *Ann. Oncol.*, **26**, 2375–2391.
22. Topalian, S. L., Hodi, F. S., Brahmer, J. R., Gettinger, S. N., Smith, D. C., McDermott, D. F., et al. *J. of Medicine*, **366**, 2443–2454.
23. Nishijima, T. F., Shachar, S. S., Nyrop, K. A., Muss, H. B. (2017). Safety and tolerability of PD-1/PD-L1 inhibitors compared with chemotherapy in patients with advanced cancer: A meta-analysis. *Oncologist*, **22**, 470–479.
24. Wu, J., Hong, D., Zhang, X., Lu, X., Miao, J. (2017). PD-1 inhibitors increase the incidence and risk of pneumonitis in cancer patients in a dose-independent manner: A meta-analysis. *Sci. Rep.*, **7**, 44173.
25. Kyi, C., Postow, M. A. (2016). Immune checkpoint inhibitor combinations in solid tumors: Opportunities and challenges. *Immunotherapy*, **8**, 821–837.
26. Chen, D. S., Mellman, I. (2017). Elements of cancer immunity and the cancer-immune set point. *Nature*, **541**, 321–330.
27. Senzer, D., Nemunaitis, J., Nemunaitis, D., Bedell, C., Edelman, G., Barve, M., et al. (2013). Phase I study of a systemically delivered p53 nanoparticle in advanced solid tumors. *Mol. Ther.*, **21**, 1096–1103.
28. Pirollo, K. F., Nemunaitis, J., Leung, P. K., Nunan, R., Adams, J., Chang, E. H. (2016). Safety and efficacy in advanced solid tumors of a targeted nanocomplex carrying the p53 gene used in combination with docetaxel: A phase 1b study. *Mol. Ther.*, **24**, 1697–1706.
29. Xu. L., Pirollo, K. F., Chang, E. H. (2001). Tumor-targeted p53 gene therapy enhances the efficacy of conventional chemo/radiotherapy. *J. Control. Release*, **74**, 115–128.
30. Xu, L., Tang, W., Huang, C., Alexander, W., Xiang, L., Pirollo, K. F., et al. (2001). Systemic p53 gene therapy of cancer with immunolipoplexes targeted by anti-transferrin receptor ScFv. *Mol. Med.*, **7**, 723–734.
31. Kim, S. S., Rait, A., Kim, E., Pirollo, K. F., Nishida, M., Farkas, N., et al. (2014). A nanoparticle carrying the p53 gene targets tumors including cancer stem cells, sensitizes glioblastoma to chemotherapy and improves survival. *ACS Nano*, **8**, 5494–5514.
32. Moore, E. C., Sun, L., Clavijo, P. E., Friedman, J., Harford, J. B., Saleh, A. D., et al. (2018). Nanocomplex-based TP53 gene therapy

promotes anti-tumor immunity through TP53- and STING dependent mechanisms. *Oncoimmunology,* **7**, e1404216.

33. Kim, S. S., Harford, J. B., Moghe, M., Rait, A., Chang, E. H. (2018). Combination with SGT-53 overcomes tumor resistance to a checkpoint inhibitor. *Oncoimmunology.,* in press.
34. Joerger, A. C., Fersht, A. R. (2016). The p53 pathway: Origins, inactivation in cancer, and emerging therapeutic approaches. *Annu. Rev. Biochem.,* **85**, 375–404.
35. Koshland, D. E., Jr. (1993). Molecule of the year. *Science,* **262**, 1953.
36. *Newsweek,* December 23, 1996.
37. Lane, D. P., Crawford, L. V. (1979). T antigen is bound to a host protein in SV40-transformed cells. *Nature,* **278**, 261–263.
38. Linzer, D. I., Levine, A. J. (1979). Characterization of a 54K dalton cellular SV40 tumor antigen present in SV40-transformed cells and uninfected embryonal carcinoma cells. *Cell,* **17**, 43–52.
39. Kress, M., May, E., Cassingena, R., May, P. (1979). Simian virus 40-transformed cells express new species of proteins precipitable by anti-simian virus 40 tumor serum. *J. Virol.,* **31**, 472–483.
40. DeLeo, A. B., Jay, G., Appella, E., Dubois, G. C., Law, L. W., Old, L. J. (1979). Detection of a transformation-related antigen in chemically induced sarcomas and other transformed cells of the mouse. *Proc. Natl. Acad. Sci. U. S. A.,* **76**, 2420–2424.
41. Rotter, V. (1983). p53, a transformation-related cellular-encoded protein, can be used as a biochemical marker for the detection of primary mouse tumor cells. *Proc. Natl. Acad. Sci. U. S. A.,* **80**, 2613–2617.
42. Li, F. P., Fraumeni, J. F., Jr. (1969). Soft-tissue sarcomas, breast cancer, and other neoplasms. A familial syndrome? *Ann. Intern. Med.,* **71**, 747–752.
43. Lustbader, E. D., Williams, W. R., Bondy, M. L., Strom, S., Strong, L. C. (1992). Segregation analysis of cancer in families of childhood soft-tissue-sarcoma patients. *Am. J. Hum. Genet.,* **51**, 344–356.
44. Srivastava, S., Zou, Z. Q., Pirollo, K., Blattner, W., Chang, E. H. (1990). Germ-line transmission of a mutated p53 gene in a cancer-prone family with Li-Fraumeni syndrome. *Nature,* **348**, 747–749.
45. Malkin, D., Li, F. P., Strong, L. C., Fraumeni, J. F., Jr., Nelson, C. E., et al. (1990). Germ line p53 mutations in a familial syndrome of breast cancer, sarcomas, and other neoplasms. *Science,* **250**, 1233–1238.
46. Knudson, A. G., Jr. (1971). Mutation and cancer: Statistical study of retinoblastoma. *Proc. Natl. Acad. Sci. U. S. A.,* **68**, 820–823.

47. Vogelstein, B., Lane, D., Levine, A. J. (2000). Surfing the p53 network. *Nature*, **408**, 307–310.
48. Leroy, B., Anderson, M., Soussi, T. (2014). TP53 mutations in human cancer: Database reassessment and prospects for the next decade. *Hum. Mutat.*, **35**, 672–688.
49. Hainaut, P., Pfeifer, G. P. (2016). Somatic TP53 mutations in the era of genome sequencing. *Cold Spring Harb. Perspect. Med.*, **6**(11), a026179.
50. Lawrence, M. S., Stojanov, P., Mermel, C. H., Robinson, J. T., Garraway, L. A., Golub, T. R., et al. (2014). Discovery and saturation analysis of cancer genes across 21 tumour types. *Nature*, **505**, 495–501.
51. Kandoth, C., McLellan, M. D., Vandin, F., Ye, K., Niu, B., Lu, C., et al. (2013). Mutational landscape and significance across 12 major cancer types. *Nature*, **502**, 333–339.
52. Olivier, M., Hollstein, M., Hainaut, P. (2010). TP53 mutations in human cancers: Origins, consequences, and clinical use. *Cold Spring Harb. Perspect. Biol.*, **2**(1), a001008.
53. Robles, A. I., Harris, C. C. (2010). Clinical outcomes and correlates of TP53 mutations and cancer. *Cold Spring Harb. Perspect. Biol.*, **2**(3), a001016.
54. Bieging, K. T., Mello, S. S., Attardi, L. D. (2014). Unravelling mechanisms of p53-mediated tumour suppression. *Nat. Rev. Cancer*, **14**, 359–370.
55. Tanaka, T., Watanabe, M., Yamashita, K. (2018). Potential therapeutic targets of TP53 gene in the context of its classically canonical functions and its latest non-canonical functions in human cancer. *Oncotarget*, **9**, 16234–16247.
56. Mello, S. S., Attardi, L. D. (2018). Deciphering p53 signaling in tumor suppression. *Curr. Opin. Cell Biol.*, **51**, 65–72.
57. el-Deiry, W. S., Kern, S. E., Pietenpol, J. A., Kinzler, K. W., Vogelstein, B. (1992). Definition of a consensus binding site for p53. *Nat. Genet.*, **1**, 45–49.
58. Hoh, J., Jin, S., Parrado, T., Edington, J., Levine, A. J., Ott, J. (2002). The p53MH algorithm and its application in detecting p53-responsive genes. *Proc. Natl. Acad. Sci. U. S. A.*, **99**, 8467–8472.
59. Bieging, K. T., Attardi, L. D. (2012). Deconstructing p53 transcriptional networks in tumor suppression. *Trends Cell Biol.*, **22**, 97–106.
60. Wei, C. L., Wu, Q., Vega, V. B., Chiu, K. P., Ng, P., Zhang, T., et al. (2006). A global map of p53 transcription-factor binding sites in the human genome. *Cell*, **124**, 207–219.
61. Joerger, A. C., Fersht, A. R. (2008). Structural biology of the tumor suppressor p53. *Annu. Rev. Biochem.*, **77**, 557–582.

62. Levine, A. J. (1997). p53, the cellular gatekeeper for growth and division. *Cell*, **88**, 323–331.
63. Vousden, K. H., Lane, D. P. (2007). p53 in health and disease. *Nat. Rev. Mol. Cell Biol.*, **8**, 275–283.
64. Lane, D., Levine, A. (2010). p53 Research: The past thirty years and the next thirty years. *Cold Spring Harb. Perspect. Biol.*, **2**(12), a000893.
65. Lane, D. P. (1992). Cancer. p53, guardian of the genome. *Nature*, **358**, 15–16.
66. Peto, R. (1977). Epidemiology, multistage models and short-term mutagenicity tests. In: Hiatt, H. H., Watson, J. D., Winsten, J. A., eds. *Origins of Human Cancer.* Cold Spring Harbor Publications, New York, NY, pp. 1403–1428.
67. Abegglen, L. M., Caulin, A. F., Chan, A., Lee, K., Robinson, R., Campbell, et al. (2015). Potential mechanisms for cancer resistance in elephants and comparative cellular response to DNA damage in humans. *JAMA*, **314**, 1850–1860.
68. Sulak, M., Fong, L., Mika, K., Chigurupati, S., Yon, L., Mongan, N. P., et al. (2016). TP53 copy number expansion is associated with the evolution of increased body size and an enhanced DNA damage response in elephants. *eLife*, **5**, e11994.
69. Haupt, S., Haupt, Y. (2017). P53 at the start of the 21st century: Lessons from elephants. *F1000Research*, **6**, 2041.
70. Garcia, P. B., Attardi, L. D. (2014). Illuminating p53 function in cancer with genetically engineered mouse models. *Semin. Cell Dev. Biol.*, **27**, 74–85.
71. Brady, C. A., Jiang, D., Mello, S. S., Johnson, T. M., Jarvis, L. A., Kozak, M. M., et al. (2011). Distinct p53 transcriptional programs dictate acute DNA-damage responses and tumor suppression. *Cell*, **145**, 571–583.
72. Li, T., Kon, N., Jiang, L., Tan, M., Ludwig, T., Zhao, Y., et al. (2012). Tumor suppression in the absence of p53-mediated cell-cycle arrest, apoptosis, and senescence. *Cell*, **149**, 1269–1283.
73. Valente, L. J., Gray, D. H., Michalak, E. M., Pinon-Hofbauer, J., Egle, A., Scott, C. L., et al. (2013). p53 efficiently suppresses tumor development in the complete absence of its cell-cycle inhibitory and proapoptotic effectors p21, Puma, and Noxa. *Cell Rep.*, **3**, 1339–1345.
74. Hanahan, D., Weinberg, R. A. (2000). The hallmarks of cancer. *Cell*, **100**, 57–70.
75. Hanahan, D., Weinberg, R. A. (2011). Hallmarks of cancer: The next generation. *Cell*, **144**, 646–674.

76. Nahta, R., Al-Mulla, F., Al-Temaimi, R., Amedei, A., Andrade-Vieira, R., Bay, S. N., et al. (2015). Mechanisms of environmental chemicals that enable the cancer hallmark of evasion of growth suppression. *Carcinogenesis*, **36 Suppl 1**, S2–18.
77. Baker, S., Ali, I., Silins, I., Pyysalo, S., Guo, Y., Hogberg, J., et al. (2017). Cancer Hallmarks Analytics Tool (CHAT): A text mining approach to organize and evaluate scientific literature on cancer. *Bioinformatics*, **33**, 3973–3981.
78. Ventura, A., Kirsch, D. G., McLaughlin, M. E., Tuveson, D. A., Grimm, J., Lintault, L., et al. (2007). Restoration of p53 function leads to tumour regression *in vivo*. *Nature*, **445**, 661–665.
79. Yu, X., Vazquez, A., Levine, A. J., Carpizo, D. R. (2012). Allele-specific p53 mutant reactivation. *Cancer Cell*, **21**, 614–625.
80. Cheok, C. F., Lane, D. P. (2017). Exploiting the p53 Pathway for Therapy. *Cold Spring Harb. Perspect. Med.*, **7**(3), a026310.
81. Valente, J. F., Queiroz, J. A., Sousa, F. (2018). p53 as the focus of gene therapy: Past, present and future. *Current Drug Targets* (Jan 15, Epub ahead of print.
82. Ajith, T. A. (2017). Strategies used in the clinical trials of gene therapy for cancer. *J. Exp. Ther. Oncol.*, **11**, 33–39.
83. Tazawa, H., Kagawa, S., Fujiwara, T. (2013). Advances in adenovirus-mediated p53 cancer gene therapy. *Expert Opin. Biol. Ther.*, **13**, 1569–1583.
84. Fan, X., Lu, H., Cui, Y., Hou, X., Huang, C., Liu, G. (2018). Overexpression of p53 delivered using recombinant NDV induces apoptosis in glioma cells by regulating the apoptotic signaling pathway. *Exp. Ther. Med.*, **15**, 4522–4530.
85. Chang, E. H., Pirollo, K. F., Bouker, K. B. (2000). Tp53 gene therapy: A key to modulating resistance to anticancer therapies? *Mol. Med. Today*, **6**, 358–364.
86. Xu, L., Pirollo, K. F., Tang, W. H., Rait, A., Chang, E. H. (1999). Transferrin-liposome-mediated systemic p53 gene therapy in combination with radiation results in regression of human head and neck cancer xenografts. *Hum. Gene Ther.*, **10**, 2941–2952.
87. Camp, E. R., Wang, C., Little, E. C., Watson, P. M., Pirollo, K. F., Rait, A., et al. (2013). Transferrin receptor targeting nanomedicine delivering wild-type p53 gene sensitizes pancreatic cancer to gemcitabine therapy. *Cancer Gene Ther.*, **20**, 222–228.
88. Kim, S. S., Rait, A., Kim, E., Pirollo, K. F., Chang, E. H. (2015). A tumor-targeting p53 nanodelivery system limits chemoresistance to

temozolomide prolonging survival in a mouse model of glioblastoma multiforme. *Nanomedicine*, **11**, 301–311.

89. Wasylishen, A. R., Lozano, G. (2016). Attenuating the p53 pathway in human cancers: Many means to the same end. *Cold Spring Harb. Perspect. Med.*, **6**(8), a026211.

90. Kussie, P. H., Gorina, S., Marechal, V., Elenbaas, B., Moreau, J., Levine, A. J., et al. (1996). Structure of the MDM2 oncoprotein bound to the p53 tumor suppressor transactivation domain. *Science*, **274**, 948–953.

91. Schon, O., Friedler, A., Bycroft, M., Freund, S. M., Fersht, A. R. (2002). Molecular mechanism of the interaction between MDM2 and p53. *J. Mol. Biol.*, **323**, 491–501.

92. Popowicz, G. M., Czarna, A., Holak, T. A. (2008). Structure of the human Mdmx protein bound to the p53 tumor suppressor transactivation domain. *Cell Cycle*, **7**, 2441–2443.

93. Canman, C. E., Lim, D. S., Cimprich, K. A., Taya, Y., Tamai, K., Sakaguchi, K., et al. (1998). Activation of the ATM kinase by ionizing radiation and phosphorylation of p53. *Science*, **281**, 1677–1679.

94. Khosravi, R., Maya, R., Gottlieb, T., Oren, M., Shiloh, Y., Shkedy, D. (1999). Rapid ATM-dependent phosphorylation of MDM2 precedes p53 accumulation in response to DNA damage. *Proc. Natl. Acad. Sci. U. S. A.*, **96**, 14973–14977.

95. Kubbutat, M. H., Jones, S. N., Vousden, K. H. (1997). Regulation of p53 stability by Mdm2. *Nature*, **387**, 299–303.

96. Wade, M., Li, Y. C., Wahl, G. M. (2013). MDM2, MDMX and p53 in oncogenesis and cancer therapy. *Nat. Rev. Cancer*, **13**, 83–96.

97. Chen, X., Gohain, N., Zhan, C., Lu, W. Y., Pazgier, M., Lu, W. (2016). Structural basis of how stress-induced MDMX phosphorylation activates p53. *Oncogene*, **35**, 1919–1925.

98. Vassilev, L. T. (2007). MDM2 inhibitors for cancer therapy. *Trends Mol. Med.*, **13**, 23–31.

99. Li, Q., Lozano, G. (2013). Molecular pathways: Targeting Mdm2 and Mdm4 in cancer therapy. *Clin. Cancer Res.*, **19**, 34–41.

100. Wang, S., Zhao, Y., Aguilar, A., Bernard, D., Yang, C. Y. (2017). Targeting the MDM2-p53 protein-protein interaction for new cancer therapy: Progress and challenges. *Cold Spring Harb. Perspect. Med.*, **7**(5), a026245.

101. Donehower, L. A., Harvey, M., Slagle, B. L., McArthur, M. J., Montgomery, C. A., Jr., Butel, J. S., et al. (1992). Mice deficient for p53 are developmentally normal but susceptible to spontaneous tumours. *Nature*, **356**, 215–221.

102. Zheng, S. J., Lamhamedi-Cherradi, S. E.,Wang, P., Xu, L., Chen, Y. H. (2005). Tumor suppressor p53 inhibits autoimmune inflammation and macrophage function. *Diabetes*, **54**, 1423–1428.

103. Munoz-Fontela, C., Pazos, M., Delgado, I., Murk, W., Mungamuri, S. K., Lee, S. W., et al. (2011). p53 serves as a host antiviral factor that enhances innate and adaptive immune responses to influenza A virus. *J. Immunol.*, **187**, 6428–6436.

104. Xue, W., Zender, L., Miething, C., Dickins, R. A., Hernando, E., Krizhanovsky, V., et al. (2007). Senescence and tumour clearance is triggered by p53 restoration in murine liver carcinomas. *Nature*, **445**, 656–660.

105. Cui, Y., Guo, G. (2016). Immunomodulatory function of the tumor suppressor p53 in host immune response and the tumor microenvironment. *Int. J. Mol. Sci.*, **17**(11), E1942.

106. Lowe, J., Shatz, M., Resnick, M. A., Menendez, D. (2013). Modulation of immune responses by the tumor suppressor p53. *BioDiscovery*, **8**, e8947.

107. Munoz-Fontela, C., Mandinova, A., Aaronson, S. A., Lee, S. W. (2016). Emerging roles of p53 and other tumour-suppressor genes in immune regulation. *Nat. Rev. Immunology*, **16**, 741–750.

108. Xu, L., Huang, C., Huang, W., Tang, W., Rait, A., Yin, Y. Z., et al. (2002). Systemic tumor-targeted gene delivery by anti-transferrin receptor scFv-immunoliposomes. *Mol. Cancer Ther.*, **1**, 337–346.

109. Pirollo, K. F., Rait, A., Zhou, Q., Zhang, X. Q., Zhou, J., Kim, C. S., et al. (2008). Tumor-targeting nanocomplex delivery of novel tumor suppressor RB94 chemosensitizes bladder carcinoma cells *in vitro* and *in vivo*. *Clin. Cancer Res.*, **14**, 2190–2198.

110. Xu, H. J., Xu, K., Zhou, Y., Li, J., Benedict, W. F. Hu, S. X. (1994). Enhanced tumor cell growth suppression by an N-terminal truncated retinoblastoma protein. *Proc. Natl. Acad. Sci. U. S. A.*, **91**, 9837–9841.

111. Pirollo, K. F., Rait, A., Sleer, L. S., Chang, E. H. (2003). Antisense therapeutics: From theory to clinical practice. *Pharmacol. Ther.*, **99**, 55–77.

112. Pirollo, K. F., Chang, E. H. (2008). Targeted delivery of small interfering RNA: Approaching effective cancer therapies. *Cancer Res.*, **68**, 1247–1250.

113. Kim, S. S., Rait, A., Garrido-Sanabria, E., Pirollo K. F., Harford, J., Chang, E. H. (2018). Nanotherapeutics for gene modulation that prevents apoptosis in the brain and fatal neuroinflammation. *Mol. Ther.*, **26**, 1–11.

114. Kim, S. S., Harford, J. B., Moghe, M., Rait, A., Pirollo, K. F., Chang, E. H. (2018). Targeted nanocomplex carrying siRNA against MALAT1 sensitizes glioblastoma to temozolomide. *Nucleic Acids Res.*, **46**, 1424–1440.

115. Kim, S. S., Rait, A., Kim, E., DeMarco, J., Pirollo, K. F., Chang, E. H. (2015). Encapsulation of temozolomide in a tumor-targeting nanocomplex enhances anti-cancer efficacy and reduces toxicity in a mouse model of glioblastoma. *Cancer Lett.*, **369**, 250–258.

116. Hwang, S. H., Rait, A., Pirollo, K. F., Zhou, Q., Yenugonda, V. M., Chinigo, G. M., et al. (2008). Tumor-targeting nanodelivery enhances the anticancer activity of a novel quinazolinone analogue. *Mol. Cancer Ther.*, **7**, 559–568.

117. Daniels, T. R., Bernabeu, E., Rodriguez, J. A., Patel, S., Kozman, M., Chiappetta, D. A., et al. (2012). The transferrin receptor and the targeted delivery of therapeutic agents against cancer. *Biochim. Biophys. Acta*, **1820**, 291–317.

118. Tortorella, S., Karagiannis, T. C. (2014). Transferrin receptor-mediated endocytosis: A useful target for cancer therapy. *J. Membr. Biol.*, **247**, 291–307.

119. Hopkins, C. R., Trowbridge, I. S. (1983). Internalization and processing of transferrin and the transferrin receptor in human carcinoma A431 cells. *J. Cell Biol.*, **97**, 508–521.

120. Fang, J., Nakamura, H., Maeda, H. (2011). The EPR effect: Unique features of tumor blood vessels for drug delivery, factors involved, and limitations and augmentation of the effect. *Adv. Drug Deliv. Rev.*, **63**, 136–151.

121. Maeda, H. (2015). Toward a full understanding of the EPR effect in primary and metastatic tumors as well as issues related to its heterogeneity. *Adv. Drug Deliv. Rev.*, **91**, 3–6.

122. Dautry-Varsat, A. (1986). Receptor-mediated endocytosis: The intracellular journey of transferrin and its receptor. *Biochimie*, **68**, 375–381.

123. Mayle, K. M., Le, A. M., Kamei, D. T. (2012). The intracellular trafficking pathway of transferrin. *Biochim. Biophys. Acta*, **1820**, 264–281.

124. Pardridge, W. M. (2012). Drug transport across the blood-brain barrier. *J Cereb Blood Flow Metab*, **32**, 1959–1972.

125. Mills, E., Dong, X. P., Wang, F., Xu, H. (2010). Mechanisms of brain iron transport: Insight into neurodegeneration and CNS disorders. *Future Med. Chem.*, **2**, 51–64.

126. Paterson, J., Webster, C. I. (2016). Exploiting transferrin receptor for delivering drugs across the blood-brain barrier. *Drug Discov. Today Technol.*, **20**, 49–52.
127. Johnsen, K. B., Burkhart, A., Melander, F., Kempen, P., Vejlebo, J., Siupka, P., et al. (2017). Targeting transferrin receptors at the blood-brain barrier improves the uptake of immunoliposomes and subsequent cargo transport into the brain parenchyma. *Sci. Rep.*, **7**(1), 10396.
128. Jones, A. R., Shusta, E. V. (2007). Blood-brain barrier transport of therapeutics via receptor-mediation. *Pharm. Res.*, **24**, 1759–1771.
129. Miller, T. M., Pestronk, A., David, W., Rothstein, J., Simpson, E., Appel, S., et al. (2013). An antisense oligonucleotide against SOD1 delivered intrathecally for patients with SOD1 familial amyotrophic lateral sclerosis: A phase 1, randomised, first-in-man study. *Lancet Neurol.*, **12**, 435–442.
130. Xiao, C., Davis, F. J., Chauhan, B. C., Viola, K. L., Lacor, P. N., Velasco, P. T., et al. (2013). Brain transit and ameliorative effects of intranasally delivered anti-amyloid-beta oligomer antibody in 5XFAD mice. *J. Alzheimers Dis.*, **35**, 777–788.
131. Huang, M., Hu, M., Song, Q., Song, H., Huang, J., Gu, X., et al. (2015). GM1-modified lipoprotein-like nanoparticle: Multifunctional nanoplatform for the combination therapy of Alzheimer's disease. *ACS Nano*, **9**, 10801–10816.
132. Miyagami, M., Tsubokawa, T., Tazoe, M., Kagawa, Y. (1990). Intra-arterial ACNU chemotherapy employing 20% mannitol osmotic blood-brain barrier disruption for malignant brain tumors. *Neurol. Med. Chir.*, **30**, 582–590.
133. Allard, E., Passirani, C., Benoit, J. P. (2009). Convection-enhanced delivery of nanocarriers for the treatment of brain tumors. *Biomaterials*, **30**, 2302–2318.
134. Kreuter, J. (2001). Nanoparticulate systems for brain delivery of drugs. *Adv. Drug Deliv. Rev.*, **47**, 65–81.
135. McDannold, N., Arvanitis, C. D., Vykhodtseva, N., Livingstone, M. S. (2012). Temporary disruption of the blood-brain barrier by use of ultrasound and microbubbles: Safety and efficacy evaluation in rhesus macaques. *Cancer Res.*, **72**, 3652–3663.
136. Abbott, N., Ronnback, L., Hansson, E. (2006). Astrocyte-endothelial interactions at the blood-brain barrier. *Nat. Rev. Neurosci.*, **7**, 41–53.
137. Chen, X., Gawryluk, J., Wagener, J., Ghribi, O., Geiger, J. (2008). Caffeine blocks disruption of blood brain barrier in a rabbit model of Alzheimer's disease. *J. Neuroinflamm.*, **5**, 12.

138. Zlokovic, B. V. (2008). The blood-brain barrier in health and chronic neurodegenerative disorders. *Neuron*, **57**, 178–201.
139. Goldsmith, M., Abramovitz, L., Peer, D. (2014). Precision nanomedicine in neurodegenerative diseases. *ACS Nano*, **8**, 1958–1965.
140. Evers, M. M., Toonen, L. J., van Roon-Mom, W. M. (2015). Antisense oligonucleotides in therapy for neurodegenerative disorders. *Adv. Drug Deliv. Rev.*, **87**, 90–103.
141. Khorkova, O., Wahlestedt, C. (2017). Oligonucleotide therapies for disorders of the nervous system. *Nat. Biotechnol.*, **35**, 249–263.
142. Sardone, V., Zhou, H., Muntoni, F., Ferlini, A., Falzarano, M. S. (2017). Antisense oligonucleotide-based therapy for neuromuscular disease. *Molecules*, **22**(4), E563.
143. Fire, A., Xu, S., Montgomery, M. K., Kostas, S. A., Driver, S. E., Mello, C. C. (1998). Potent and specific genetic interference by double-stranded RNA in *Caenorhabditis elegans*. *Nature*, **391**, 806–811.
144. Bobbin, M. L., Rossi, J. J. (2016). RNA interference (RNAi)-based therapeutics: Delivering on the promise? *Annu. Rev. Pharmacol. Toxicol.*, **56**, 103–122.
145. Tam, C., Wong, J. H., Cheung, R. C. F., Zuo, T., Ng, T. B. (2017). Therapeutic potentials of short interfering RNAs. *Appl. Microbiol. Biotechnol.*, **101**, 7091–7111.
146. Juliano, R. L. (2016). The delivery of therapeutic oligonucleotides. *Nucleic Acids Res.*, **44**, 6518–6548.
147. Hawkins, B. T., Davis, T. P. (2005). The blood-brain barrier/neurovascular unit in health and disease. *Pharmacological Rev.*, **57**, 173–185.
148. Pardridge, W. M. (2005). The blood-brain barrier: Bottleneck in brain drug development. *NeuroRx*, **2**, 3–14.
149. Kim, S. S., Harford, J. B., Pirollo, K. F., Chang, E. H. (2015). Effective treatment of glioblastoma requires crossing the blood–brain barrier and targeting tumors including cancer stem cells: The promise of nanomedicine. *Biochem. Biophys. Res. Commun.*, **468**, 485–489.
150. Siefker-Radtke, A., Zhang, X. Q., Guo, C. C., Shen, Y., Pirollo, K. F., Sabir, S., et al. (2016). A phase 1 study of a tumor-targeted systemic nanodelivery system, SGT-94, in genitourinary cancers. *Mol. Ther.*, **24**, 1484–1491.
151. van der Meel, R., Vehmeijer, L. J., Kok, R. J., Storm, G., van Gaal, E. V. (2013). Ligand-targeted particulate nanomedicines undergoing clinical evaluation: Current status. *Adv. Drug Deliv. Rev.*, **65**, 1284–1298.

152. Taube, J. M., Klein, A., Brahmer, J. R., Xu, H., Pan, X., Kim, J. H., et al. (2014). Association of PD-1, PD-1 ligands, and other features of the tumor immune microenvironment with response to anti-PD-1 therapy. *Clin. Cancer Res.*, **20**, 5064–5074.
153. Yoon, K. W., Byun, S., Kwon, E., Hwang, S. Y., Chu, K., Hiraki, M., et al. (2015). Control of signaling-mediated clearance of apoptotic cells by the tumor suppressor p53. *Science*, **349**(6247), 1261669.
154. Mall, C., Sckisel, G. D., Proia, D. A., Mirsoian, A., Grossenbacher, S. K., Pai, C. S., et al. (2016). Repeated PD-1/PD-L1 monoclonal antibody administration induces fatal xenogeneic hypersensitivity reactions in a murine model of breast cancer. *Oncoimmunology*, **5**, e1075114.
155. Kudlak, K., Demuro, J. P., Hanna, A. F., Brem, H. (2013). Acute lung injury following the use of granulocyte-macrophage colony-stimulating factor. *Int. J. Crit. Illn. Inj. Sci.*, **3**, 279–281.
156. Descotes, J. (2009). Immunotoxicity of monoclonal antibodies. *mAbs*, **1**, 104–111.
157. Hirayama, Y., Gi, M., Yamano, S., Tachibana, H., Okuno, T., Tamada, S., et al. (2016). Anti-PD-L1 treatment enhances antitumor effect of everolimus in a mouse model of renal cell carcinoma. *Cancer Sci.*, **107**, 1736–1744.
158. Harter, P. N., Bernatz, S., Scholz, A., Zeiner, P. S., Zinke, J., Kiyose, M., et al. (2015). Distribution and prognostic relevance of tumor-infiltrating lymphocytes (TILs) and PD-1/PD-L1 immune checkpoints in human brain metastases. *Oncotarget*, **6**, 40836–40849.
159. Fridman, W. H., Pages, F., Sautes-Fridman, C., Galon, J. (2012). The immune contexture in human tumours: Impact on clinical outcome. *Nat. Rev. Cancer*, **12**, 298–306.
160. Guo, G., Yu, M., Xiao, W., Celis, E., Cui, Y. (2017). Local Activation of p53 in the Tumor Microenvironment Overcomes Immune Suppression and Enhances Antitumor Immunity. *Cancer Res.*, **77**, 2292–2305.
161. Holmgaard, R. B., Zamarin, D., Li, Y., Gasmi, B., Munn, D. H., Allison, J. P., et al. (2015). Tumor-expressed IDO recruits and activates MDSCs in a Treg-dependent manner. *Cell Rep.*, **13**, 412–424.
162. Prendergast, G. C., Malachowski, W. P., DuHadaway, J. B., Muller, A. J. (2017). Discovery of IDO1 inhibitors: From bench to bedside. *Cancer Res.*, **77**, 6795–6811.
163. Pirollo, K. F., Dagata, J., Wang, P., Freedman, M., Vladar, A., Fricke, S., et al. (2006). A tumor-targeted nanodelivery system to improve early MRI detection of cancer. *Mol. Imaging*, **5**, 41–52.

Index

ABC, *see* accelerated blood clearance
Abelcet 2, 46, 776
accelerated blood clearance (ABC) 2, 44, 51, 83, 89, 97, 111, 249–250, 252–256, 258, 260, 262, 264–266, 268, 270, 272, 274, 276, 278, 289–290, 292, 771–773, 822
activated partial thromboplastin time (APTT) 731, 733, 736, 854
active pharmaceutical ingredient (API) 2, 11, 29, 87, 187, 189, 362, 841
acute lymphoblastic leukemia 155–156, 173, 296, 811–812
acute myelogenous leukemia 155–156, 173
adaptive immune response 92, 123, 126–127
adaptive immune system 125–126, 260, 453, 627, 707–708, 715, 718, 822, 950
ADAs, *see* antidrug antibodies
ADC, *see* antibody-drug conjugates
adjuvant 54, 123–124, 135–137, 140–142, 144, 266, 289, 291, 300, 313, 507, 519, 522, 555, 562, 564–565, 646, 678, 685, 709, 717
adjuvanticity 137–138, 699, 707
adoptive cell therapies 643, 646, 648–649
adoptive immunotherapy 643–644, 646
ADR, *see* adverse drug reaction
adverse drug reaction (ADR) 2
adverse events 33–34, 40, 85, 88, 96, 492, 537, 539–540, 542–543, 546, 548–549, 552, 567, 905–906, 908–910, 931, 956
aggregates 2, 36, 41–42, 92, 270, 398, 400–401, 404–407, 449, 537, 555, 558–560, 566, 636, 703, 787, 844, 867, 871–873, 881, 885
aggregation of biologic nanodrugs 83, 92
aHUS, *see* atypical hemolytic uremic syndrome
AI, *see* artificial intelligence
albumin 26, 161, 258, 300, 347, 354, 448, 489, 537, 563, 817, 820, 835, 843, 848–849, 860, 893, 899, 903, 907, 919
allergen 194, 537, 568, 714, 871
allergy 44, 97, 103, 109–110, 289, 294, 361–362, 369, 374, 376–377, 389, 396, 537, 550, 709, 714, 778–779, 871, 886
alternative pathway 94, 298, 311–313, 329, 348, 351, 404, 419, 421, 522
Ambisome 2, 26, 46, 106–107, 112, 325, 370, 374, 390, 392–393, 422–426, 771, 775–776, 780, 789

aminoacyl-tRNA synthetase 155, 161, 173
γ-aminobutyric acid (GABA) 753, 763
anaphylactic shock 84, 376, 389, 399, 417, 782
anaphylactoid reactions 2, 83, 297, 389, 403, 405, 407, 771, 775
anaphylatoxin receptors (ATR) 109–110, 377, 394, 396, 399, 771, 778
anaphylatoxins 93, 110, 113, 311, 313, 316, 328, 341, 348, 354, 356, 361, 363, 369, 377, 389, 396, 399, 405, 417, 777–779, 783
anaphylaxis 41, 46, 83–84, 92, 98, 100, 113, 197, 200–201, 207, 361, 389–390, 405–406, 417, 537, 539, 542–543, 547, 551, 559, 567–568, 572, 575, 591–592, 714, 788, 870, 955
animal models 55, 68, 115, 210, 290–291, 295, 322–323, 326, 330, 346, 356–357, 361, 363–364, 366, 380, 389, 514, 520, 569, 574, 784, 873, 881–883, 919, 947
anti-inflammatory 46, 66, 183–184, 192–195, 324, 342, 347, 355, 357, 375, 707, 755, 787, 803, 808, 812, 815–816
anti-PD1 658, 929, 931–932, 948–953, 955–956
anti-PEG antibodies 83, 250, 276, 290–301, 319, 561, 598, 781, 788
anti-PEG IgG 289, 292
anti-PEG IgM 51–52, 89–90, 249, 251–255, 257–269, 274–276, 289, 291–297, 299, 771, 773–775, 788

antibodies 2, 16–17, 29, 34–38, 46, 66–67, 87–88, 125–126, 155–164, 166–167, 197–198, 229–236, 238–244, 250, 252, 275–276, 290–301, 317–319, 347–349, 515–516, 540–542, 544–545, 547–548, 550–553, 555–556, 559, 561, 568–569, 571–574, 585, 588, 590–603, 605–610, 614–615, 617–621, 646–647, 670–671, 674, 775, 781–782, 806–807, 810–812, 822, 841, 847, 867–871, 873–874, 876–877, 883–886, 948–949
antibody-antibiotic conjugates 155, 171, 174
antibody-drug conjugates (ADCs) 2, 26, 29, 87, 155–166, 168–169, 173, 585, 608, 841
antibody targeting 26, 571, 929
antibody–polymer conjugation 155, 171, 174
anticancer agents 913
anticoagulant 348, 350, 595, 598, 603, 622, 731–732
antidrug antibodies (ADAs) 2, 34–35, 38, 44, 83, 88, 97, 111, 867, 880
antigen 33–34, 55, 85, 88, 23–127, 130–131, 138–145, 155–157, 160, 162, 167–168, 170, 191–192, 194, 229–233, 237–243, 312–313, 316, 355, 453, 510, 516, 518, 520–523, 537, 539–540, 550, 552, 557, 559, 562, 592, 601, 603, 619, 623, 630–631, 657–660, 674–675, 678–682, 707, 713, 715–718, 810–811, 813–814, 819, 821–822, 882–883, 930, 950

antigen delivery 138, 192, 675, 699, 707, 713, 811
antigen-presenting cells (APCs) 33, 85, 124–125, 191, 453, 562, 630, 659, 819, 821, 878, 882
antiphospholipid syndrome (APS) 341–342, 347, 349
antiretroviral 183, 192–193
APCs, *see* antigen-presenting cells
API, *see* active pharmaceutical ingredient
APS, *see* antiphospholipid syndrome
aptamers 29, 87, 449, 491, 682, 684, 801, 806, 841
APTT, *see* activated partial thromboplastin time
arrhythmia 361, 365, 417
arrhythmia blood cells 361
artificial intelligence (AI) 2, 72, 628, 633
ASD, *see* autism spectrum disorders
assay specificity 585, 596–597, 614, 620–621
assessment 20, 35, 51, 53–55, 59, 68–69, 103, 209, 229–230, 233, 322, 456, 462, 537–538, 540–542, 544–546, 548–552, 554, 556, 558–562, 564, 566, 568, 570, 572–574, 586–589, 594–595, 599–600, 602–603, 605–606, 610, 619, 621, 720, 757–760, 784, 853, 855–857, 859, 867, 869, 872–873, 876, 879–882, 887, 898, 900–901, 913, 915, 941
atherosclerosis 171–172, 316, 331, 341–343, 345–347, 353, 355–356, 393, 401, 406, 713
ATR, *see* anaphylatoxin receptors
atypical hemolytic uremic syndrome (aHUS) 50, 102, 311, 315
autism spectrum disorders (ASD) 753–755, 757, 759, 761
auto-antibodies 229–236, 238–240, 242–244
autoimmune diseases 171–172, 229–230, 234, 240, 243, 330, 341, 553, 631, 658, 671, 686, 709, 811, 938
autonomous nanomedical devices 627, 631, 638

B-cells 23, 125, 127, 171, 191, 231, 234, 249, 251–252, 257–259, 261–264, 267, 273–274, 298, 327, 453, 485, 519, 537, 540, 552, 562, 606, 627, 630, 674, 699–700, 710, 715–716, 718, 811, 813, 817, 819, 822, 868
bacterial 126, 139, 186, 205, 207, 314, 508, 510–514, 519, 524, 537, 555, 606, 628, 636, 709, 712, 883, 947
Bayh–Dole Act 2, 13
BDNF, *see* brain derived neutrophic factor
binding antibodies 585
biocompatibility 42, 89, 200, 452, 462, 519, 731–732, 835
bioconjugation 441, 489
biodegradation 183, 196
biodistribution 168, 323, 326, 445, 488, 657, 659–662, 664, 669, 674, 681–682, 719, 814–815, 836, 849, 851, 860
bioengineering 686–687, 731–732
bioequivalent 2, 56
Biologic License Application (BLA) 2, 10–12
biological 1–5, 7, 10, 16–21, 23–24, 29, 35, 40–43, 45, 47, 56–60,

63, 72–75, 83, 85, 87, 95–96, 103, 113–114, 138, 147, 188, 205, 209, 234, 239, 289–290, 312, 316, 318, 328, 330, 443–444, 476–477, 538–540, 544, 573–575, 586–588, 594–595, 661, 765, 773, 802, 804, 835–860, 862, 874, 877, 885, 917
biological half-life 289–290
biological products 2–5, 16–21, 23–24, 41, 59, 586–587
biologics 1–8, 10–14, 16–24, 26, 28–30, 32–56, 58, 60, 62–68, 70–72, 74–76, 78, 83–85, 87–88, 93–94, 96, 100, 155, 175, 301, 463, 537–538, 585, 841, 867, 870, 872, 875, 886
Biologics Price Competition and Innovation Act (BPCI Act) 2, 13, 23, 538
biomanufacturing 62, 867, 871
biomarkers 39, 47, 64, 210, 229–232, 234–238, 240, 242–244, 367, 537, 541, 562, 644, 685, 711–712, 766, 824, 949
biomaterial 136, 311, 316–317, 821
biomedical engineering 330, 432, 643, 651, 657, 687, 720
biomolecular drugs 2–3
biopharmaceuticals 2–3, 40, 95, 801–802, 804, 806, 809–810, 815, 819, 824
biosimilars 3, 11, 13, 23, 56–63, 65, 71, 75, 84–85, 538, 549, 590, 811, 867, 877–879, 883, 885
biotherapeutics 2–3, 5, 40, 867–869, 871, 885, 887
bispecific antibody 155, 169–170
BLA, see Biologic License Application

bleeding prevention 731
blood cells 93, 110, 315, 321, 361, 377, 389, 393, 396, 417, 427, 671, 746, 779, 785, 835, 856
blood coagulation 351, 403, 731, 735, 745, 747, 852–853
blood components 43, 316, 835, 837, 842
blood loss 731, 747
blood plasma 188, 731–734, 736, 738–743, 746
blood proteins interaction 835
blood–brain barrier 485, 861, 929, 944–945
BPCI Act, see Biologics Price Competition and Innovation Act
brain derived neutrophic factor (BDNF) 753
brain tumors 643–648, 650, 944, 947–948
breast cancer 156, 162, 164, 242, 682, 825, 896, 929–931, 951–952, 956

C-reactive protein 537, 569
C-type lectin-like molecule-1 155, 169, 174
C3 complement protein 835, 850
C3a 84, 93, 109–110, 115, 312–313, 319–322, 327, 341, 343, 345–346, 351–356, 361, 363, 377, 389, 394, 396, 399, 405, 419–420, 426–430, 682, 777, 779, 784
C3b 49, 84, 101, 109, 312–314, 319, 355, 395, 402–403, 522
C5a 84, 93, 108–110, 115, 312–313, 321, 341–343, 345–346, 348, 353–356, 361, 363, 377, 389, 394, 396, 398–399, 405,

419–420, 426–430, 777, 779, 784
C5b-9 93, 109, 320, 341, 343, 345–346, 352–356, 395, 404, 406
cancer 30, 38, 46, 155–156, 159–160, 162, 164, 167–171, 183, 185, 190, 195, 208–210, 229–230, 235, 237, 241–242, 317, 327, 331, 476, 479, 487, 491–494, 550, 631, 634, 643–644, 646–649, 657–659, 666, 670, 681–682, 685–686, 713, 717–719, 766–767, 776, 801–805, 808–814, 816–817, 819–821, 825, 893–900, 902–914, 916–920, 929–940, 942–944, 946–954, 956–958
cancer immunotherapy 195, 643–644, 657, 681, 929–932, 934, 936, 938, 940, 942, 944, 946, 948–954, 956–958
cancer therapy 713, 719, 812–814, 893, 897
cancer vaccines 643, 646–647, 649, 659
CAR T-cells, *see* chimeric antigen receptor T-cells
cardiac anaphylaxis 84
cardiovascular therapeutics 473
CARPA, *see* complement activation-related pseudoallergy
cationic lipids 441, 450–452, 455–456, 460, 464, 478, 483–484, 486, 494
CCV, *see* Coxiella-containing vacuole
CD4+ cells 123, 127
CD4+ T cell 54, 130, 183, 347, 507, 522
CD8+ cells 123, 127, 143
CD8+ T cell 143, 183, 191–192, 195, 507, 954
CD20 657–658, 811, 813–814, 819, 930
CD28 537, 543, 812
CDC, *see* Centers for Disease Control and Prevention
cell-mediated immunity 126, 130, 141, 231, 507, 514–515
cell transfection 835
cell uptake 444, 449, 669, 681, 835
Centers for Disease Control and Prevention (CDC) 183, 507–508
central nervous system (CNS) 240, 473, 544, 571, 643–644, 753, 762
characterization techniques 19, 835, 838
checkpoint blockade 675, 681, 929, 931, 948–949, 952–953, 956
checkpoint inhibitors 643, 646–647, 649–650, 931, 949, 956
chemical modifications 36, 62, 445–446, 477, 537, 557–558, 671
chemotherapy 46, 156, 249, 261, 488, 643–644, 648–649, 658, 671, 685, 718–719, 813, 900–901, 907, 910, 913–914, 917, 929–930, 937
chimeric antigen receptor T-cells (CAR T-cells) 657–658
chronic kidney disease (CKD) 34, 88, 289, 296, 434
circulating immune complexes 537, 544, 571
CKD, *see* chronic kidney disease
class-2 thymus-independent (TI-2) antigen 289, 298
classical pathway 94, 311–313, 327, 348, 403–404, 421, 781, 852

clinical markers 867, 876
clinical status 491-493
clinical trials 2, 14, 17, 21, 29, 38, 40, 56, 65, 71-72, 87, 41-144, 197, 244, 276, 294-295, 301, 445, 450-451, 491-492, 494, 520, 546-547, 549, 586-587, 591, 607, 643-644, 647-650, 709, 810, 814, 841, 860, 873-874, 876, 879, 886, 893-894, 896-897, 899, 909, 913-914, 918-919, 931, 947-948, 954, 956
clinically relevant 51, 61, 98, 243, 395, 519, 537, 548, 562, 876
CNS, *see* central nervous system
coagulation 17, 94, 188-190, 197, 207, 342-344, 346, 351, 354, 393, 403, 555, 731-733, 735, 738, 745-747, 835, 852-854
combination therapy 486, 929, 955-956
commercialization 2, 21, 29, 84, 87, 837, 841
companion diagnostics 229, 233, 235, 244
complement 2, 44, 46-47, 49, 52, 56, 90, 92-93, 96-98, 100-101, 103, 117, 187, 189-190, 197-198, 200-201, 207, 249-250, 252-256, 258, 292-293, 297-298, 311-320, 322, 324, 326, 328-330, 341-343, 345-349, 351-353, 355-357, 361-362, 379-380, 389-390, 392, 394, 396, 398, 402-408, 417-420, 426-430, 432-434, 456, 510, 515, 771-778, 780-785, 787-789, 791, 850-852
complement activation 2, 44, 46, 90, 93, 96-98, 100, 187, 189-190, 253-256, 258, 289, 292, 311-312, 314, 316, 318, 320, 322, 324, 326, 328-330, 343, 345-346, 348-349, 351-352, 356-357, 379-380, 389-390, 392, 394, 396, 398, 400, 402-404, 406-408, 417-418, 420, 422, 424, 426, 428, 430, 432-434, 771, 774-778, 780-785, 787, 789, 851-852
complement activation-related pseudoallergy (CARPA) 2, 44, 46-47, 93, 96-98, 100, 103-104, 106, 108-109, 111, 113-115, 187, 189, 255, 289, 297-298, 311, 317, 361-374, 376, 378, 380, 389-392, 394-396, 398-408, 417-418, 420-426, 428, 430-432, 434, 771-772, 774-782, 784-790
complement cascade 311, 313, 348, 851
complement inhibitor 349, 771, 777, 788-789
complement system 92-93, 117, 197-198, 250, 252, 254-255, 293, 298, 312-313, 329, 341, 345, 352-353, 379, 418, 434, 456, 515, 682, 772-773, 775, 777-778, 781-782, 791, 845, 850-851, 860
conceptual 627-628, 631, 633, 635, 637
confirmatory assay 457, 585-587, 591, 599, 602, 606, 609-610, 614, 619, 874, 876
conjugated proteins 2, 35, 608
contact system of blood coagulation activation 731
Copaxone 2, 5, 61-67, 83, 96

corona effect 657, 668
costimulatory molecules 136, 699, 709, 715–716
counterpart protein 585
Coxiella burnetii 507–508, 517, 524
Coxiella-containing vacuole (CCV) 507, 511
Cremophor EL 2, 46, 186, 789
CRIM, *see* cross-reactive immunologic material
CRISPR-Cas9 2, 74
cross-linking 36, 159–160, 252, 261, 298, 488, 537, 543, 559, 569–570, 675, 734
cross-reactive immunologic material (CRIM) 537, 554
cross-reactivity 67, 299, 537, 544, 547, 585, 589–591, 605–606
cryo-electron microscopy (cryo-EM) 2, 74
cryo-EM, *see* cryo-electron microscopy
cryptic epitopes 537, 555
CTL, *see* cytotoxic T lymphocyte
cut point 585, 593–595, 597, 602, 604–605, 609–610, 614, 616, 618–620, 622–623
cytokines 17, 21, 110, 113, 25–127, 130, 134–136, 190, 195, 198, 200–201, 203–204, 206–207, 231, 351, 354, 454–455, 457, 479–480, 485, 512, 515, 518–519, 521, 523, 539, 542–544, 552–554, 569–570, 612–613, 657–660, 666, 669, 676, 679, 686, 707, 709–713, 715–717, 755, 762, 765, 803, 805, 808–809, 812–813, 815, 817, 820, 953
cytotoxic agents 249, 262, 647
cytotoxic T lymphocyte (CTL) 123, 681

cytotoxicity 26, 139–140, 156, 162, 169–171, 205, 450–451, 484, 669, 678, 815, 818, 835, 856–859, 877, 953

DAMPs, *see* danger-associated molecular patterns
danger-associated molecular patterns (DAMPs) 679, 699, 711
danger-signaling receptors (DSRs) 771, 778
DaunoXome 2, 46, 771, 775–776
DCs, *see* dendritic cells
delivery platform 507, 522
dendrimers 112, 186, 188, 194, 200, 202, 205, 441, 450, 456, 674
dendritic cells (DCs) 123, 231, 660, 663, 666, 669, 678, 681, 711, 716
dermatomyositis 341, 355
desensitization 115, 380, 537, 548
diagnostic assay 229–230, 242, 719
dialysis membrane 311, 316
diapedesis 627, 632
disease dosing 537
DNA and RNA immunotherapeutics 657, 659
DNA delivery 929
docking 441, 460–463, 475, 765
Doxil 2, 26–27, 30, 46, 48, 99, 107, 112, 114, 198, 200, 250, 255–256, 261, 263–264, 275, 289, 297–298, 317, 327, 364, 370, 372, 374–375, 390, 398, 771, 775–776, 782–784, 788, 807

doxorubicin 26, 30, 107, 164–165, 189–190, 250, 293, 317, 374, 672, 685, 776, 782, 806–807, 818
drug delivery 2, 24, 26–27, 29, 43, 77, 87, 95, 117, 138, 183, 192–194, 206, 249–250, 270, 275, 277, 301–302, 331, 379, 422, 441, 463, 478, 496, 639, 657, 659–660, 685, 700, 707, 719, 772–773, 775, 790, 801, 806, 821, 824, 838, 841, 929
drug delivery systems 2, 24, 26, 250, 270, 275, 277, 302, 379, 478, 772–773, 775, 790, 806, 838
drug-like molecule 2, 73
drug resistance 261, 627, 634
drug tolerance 585, 594–596
drug toxicity 361, 389
druggable genome 2, 73–74
DSRs, see danger-signaling receptors

environmental risk 753, 756
Epibase 2, 55
epidermal growth factor receptor (EGFR) 155–156, 162, 173–174, 643, 647
EpiMatrix 2, 55, 878
EpiScreen 2, 55, 879, 882–883
epitope 2, 34, 55, 74, 88, 252, 257, 298–299, 541, 556, 572, 585, 589, 591–592, 601, 867, 876, 878–879, 882–884
epitope mapping 2, 74, 585, 589, 876, 883
epitope mapping analysis 2, 74
EPR, see enhanced permeability and retention
erythrocytes 45, 322, 351–352, 835, 839, 856–857
European Medicines Agency (EMA) 2, 57, 96, 233, 362, 867, 869–870

effectuation phase 249, 251, 773
EGFR, see epidermal growth factor receptor
EMA, see European Medicines Agency
endogenous protein 537, 540, 544–545, 552–554, 556–557, 572–575, 591
endotoxin 88, 183, 186, 190, 194, 199, 201, 203–205, 207, 209–210, 323, 419, 712
engineered nanomaterials 184, 187, 191, 197, 207, 209–210, 627, 629
enhanced permeability and retention (EPR) 806, 893, 895, 941

Fab fragment 169, 341, 355
factor replacement therapy 537, 551
Factor Xa 94, 731, 735, 739–740, 746
Fc fragment 341, 349, 354
FDA, see US Food and Drug Administration
FD&C Act, see Federal Food, Drug, and Cosmetic Act
Federal Food, Drug, and Cosmetic Act (FD&C Act) 2, 4, 11, 587
fetal valproate syndrome 753, 756
FGF-2, see fibroblast growth factor 2
fibrinogen 207, 345, 395, 401–402, 406–407, 733, 735, 739, 742–746, 748, 835, 843–844, 854

fibroblast growth factor 2 (FGF-2) 341, 344
fluorodeoxyglucose 893, 911, 919
formulation erythropoietin 867
formylglycine-generating enzyme 155, 166, 174
functional excipients 441, 455
functionalized antibodies 2, 35

GABA, *see* γ-aminobutyric acid
gene silencing 171, 441–443, 451, 475, 483–484, 486–487, 491, 494
gene therapy 4, 265, 464, 473, 478, 660, 836, 929–931, 937, 939, 941, 943, 945, 947, 954, 956
generic drugs 2, 4, 56, 62
GHCE, *see* Global Health Care Equivalency
glatiramer acetate 2, 64–66
glioblastoma 767, 929, 931, 949, 951, 956
gliomas 629, 643–646, 648–650, 652
Global Health Care Equivalency (GHCE) 627–628, 637
glycans 155, 158, 164–165, 167–168, 173
glycosylation 36, 41, 537, 561, 870, 884
GM-CSF, *see* granulocyte-macrophage colony-stimulating factor
gold nanoparticles 27, 197, 202, 520, 631, 679, 711, 713–714, 717–718, 814–815, 893, 896, 899, 906, 911, 917–918
granulocyte-macrophage colony-stimulating factor (GM-CSF) 537, 550

hapten 33, 85, 171, 289, 291, 299, 608
HSA, *see* human serum albumin
Hatch–Waxman Act 2, 13
HDAC, *see* histone deacetylase
HDL, *see* high-density lipoprotein
head and neck cancer 929
hemodynamic changes 106–107, 113, 325, 361, 784
hemodynamics 47, 104, 367, 389
hemolysis 45, 188, 200, 311, 321–322, 328, 352–353, 598, 835, 839, 856–857
hemolysis assay 311, 321–322, 328, 857
hemostasis 731–732, 734, 736, 738, 740, 742, 744, 746, 748
heterogeneous 64, 157, 349, 537, 560, 918
HGG, *see* high grade gliomas
high-density lipoprotein (HDL) 487, 699, 718
high grade gliomas (HGG) 643
high mobility group protein B1 (HMGB1) 537
histamine 47, 84, 108, 110, 202, 367, 377–378, 389, 396, 420, 426–427, 429–430, 484, 542, 714, 778–779, 784
histone deacetylase (HDAC) 753, 763
HLA, *see* human leukocyte antigen
HMGB1, *see* high mobility group protein B1
HO-PEG, *see* hydroxyl PEG
homogeneous 157, 159, 161, 165, 172–173, 537, 560, 700
HSRs, *see* hypersensitivity reactions
human epidermal growth factor receptor 2 155–156, 174
human immune system 35, 71, 85, 123, 553, 627–629, 631–633, 635, 867

human leukocyte antigen (HLA) 55, 537, 552
human serum albumin (HSA) 347, 537, 563, 907
Humira 2, 21, 38
humoral immunity 123, 126–127, 131, 231, 507, 716–718, 802–803
Humulin 2, 23
hydroxyl PEG (HO-PEG) 289, 300
hypersensitivity 2, 33, 35, 39, 46, 83–84, 92–93, 96, 194, 249–250, 255–256, 276, 289, 296–298, 327, 361, 363–364, 374, 389–390, 417, 430, 456, 537, 542–543, 570, 585, 588, 717, 771, 775–776, 784, 787–788, 952–953, 955–956
hypersensitivity reactions (HSRs) 2, 33, 39, 46, 84, 93, 194, 249–250, 255–256, 276, 289, 296–298, 327, 361, 363, 374, 389, 417, 430, 456, 542, 771, 775–776, 784, 787–788, 953
hypertension 107–108, 324–325, 350, 361, 365, 367, 371, 373, 378, 393, 395, 398–399, 401, 405, 407, 417, 419–421, 427–428, 431, 775, 777, 782
hyperviscosity 341, 351
hypotension 197, 317, 361, 373, 378, 399, 405, 407, 419, 421–422, 426–429, 569, 711, 775–776

ICs, *see* immune complexes
IFN-beta 537, 552
Ig, *see* immunoglobulin
IgE 46, 98, 133–134, 297–298, 317, 407, 537, 540, 542, 568, 585, 592, 622, 716, 775, 777, 785
IgG 126, 130–132, 156, 158–159, 168, 171–172, 174, 232, 252, 289, 292–293, 295, 341, 348–349, 353, 537, 540, 559, 585, 591, 607, 621, 665, 672, 680, 876
IgM 51–52, 89–90, 249, 251–255, 257–269, 273–276, 289, 291–299, 319, 515, 585, 591, 601, 607, 621, 771, 773–775, 787–788, 876
IL, *see* interleukin
IL1β *see* interlukin 1β
immune augmentation 627
immune cells 43, 68, 71, 127–128, 134, 136, 140–141, 172, 183, 188, 195, 201, 203–204, 256–257, 264, 294, 313, 512, 542, 628–629, 646, 648, 663, 668–669, 674, 700, 707–708, 711–714, 753, 762, 801–804, 808, 811–812, 815, 818, 820–823, 825, 854, 880, 882, 956
immune checkpoint 650, 929, 939, 949, 956
immune complexes (ICs) 2, 35, 38–39, 83, 92, 312, 537, 544, 570–571
immune-deficient 537, 550, 574
immune enhancement 627
immune reactions 33, 38, 70–71, 84, 128, 312, 441–442, 444, 446, 448, 450, 452–458, 460, 462, 464, 718
immune response 33–35, 39–42, 85, 88, 92–93, 111, 123–127, 130–132, 134, 136, 138–141, 143–145, 191, 193, 229–230, 232, 254, 262–263, 268, 273, 291–292, 294–295, 299, 477–481, 512, 515–516,

518–523, 537–541, 544–545, 548–563, 567, 570–572, 574–575, 586–588, 592, 601, 603, 607–608, 644, 646–647, 660–661, 664–669, 679–683, 707–708, 710, 714–715, 772–773, 873–874, 876, 938
  adaptive 39, 92, 123, 125–127, 938, 957
  unwanted 34, 88, 540, 874
immune response modulating impurities (IRMI) 537, 562
immune-stimulating complexes 123, 137
immune system 34–35, 40–41, 46, 71–73, 75, 85, 93, 96, 123–129, 137–138, 140–142, 145, 147, 183–186, 188–190, 192, 194, 196, 198, 200, 202, 204, 206, 208–210, 229–231, 257, 260–262, 289–290, 407–408, 452–453, 456, 512, 550, 553, 627–633, 635, 638, 644, 646–649, 699–700, 702, 704, 706–720, 755, 772–773, 821–822, 850–852, 931, 938–939, 949–950
immunity 55, 64, 98, 123, 126–127, 130–131, 138, 141, 143, 192, 210, 231, 289–290, 292, 294, 296, 298–302, 330, 361, 389, 463, 507, 509, 514–516, 518–519, 521, 554, 658, 679, 686, 707–710, 713–718, 777, 801–803, 808–809, 811–813, 815, 817, 819, 821–824, 931, 939, 954–955
immunization 92, 130, 144, 231, 252, 298, 522, 668, 699, 707, 717, 810, 817, 819
immuno-suppressive 202
immunodeficiency 341, 476, 491

immunogenic epitopes 2, 72, 276, 874
immunogenicity 20, 34–35, 37, 39–42, 44, 51–56, 59–61, 64, 66–69, 73–75, 83, 85, 88–89, 92, 95, 97, 129–130, 135–136, 138, 143–145, 147, 197, 203–204, 250, 270–271, 275–276, 289–291, 293, 295–297, 456–457, 462–463, 537–542, 544–568, 570, 572–575, 585–590, 592, 594, 596–600, 602, 606–610, 614–618, 620, 622–624, 658–659, 678, 686, 772–773, 867–888, 954–956
immunoglobulin (Ig) 16, 156, 174, 234, 241, 341–342, 350–351, 353–357, 461, 591, 716, 777
immunology 33, 71, 84, 116–117, 183, 209–210, 356, 379, 434, 507, 523–524, 644, 650, 715, 720, 790, 825
immunomodulation 341, 627, 629, 679, 804
immunomodulator 2, 64, 66
immunomodulatory effects 2, 40, 64, 473–474, 476, 478, 480, 482, 484, 486, 488, 490, 492, 494, 496, 711, 716–717, 720, 814
immunopharmacology 2, 40, 210
immunotherapy 123–124, 126, 128, 130, 132, 134, 136, 138, 140, 142, 144, 146, 148, 183, 195, 627, 643–650, 652, 657–666, 668, 670–686, 713, 801–802, 804, 806, 808, 810, 812, 814, 816, 818, 820, 822, 824, 826, 851, 867, 871, 886, 929–932, 934, 936, 938, 940, 942, 944, 946, 948–954, 956–958

immunotoxic effects 2, 33, 71, 84
immunotoxicity 40, 54, 66,
    183–188, 190, 197–203,
    205–210, 364–365, 379–380,
    418, 707, 719, 731–732, 867,
    870
in silico tools 441, 458, 882–883
induction phase 249, 251, 773
INF, *see* interferon
infectious disease 123, 507, 524,
    586, 659
inflammasome 137, 190, 512, 666,
    668, 679, 699, 711–712
inflammation 35, 44, 97, 136–137,
    172, 183, 190, 195, 205, 207,
    294, 297, 311, 313, 316, 328,
    341–343, 345–346, 353, 356,
    394, 430, 456, 473, 480, 485,
    487, 657, 664, 668, 686, 699,
    704–705, 708, 710–711,
    713–714, 716, 719–720, 776,
    802, 805, 809, 812–818, 821,
    823–824, 936
inflammatory mediators 325, 328,
    361, 369, 402, 417, 455, 480,
    755, 808
inflammatory response 123, 136,
    139–140, 194, 202–203, 210,
    342, 345, 456, 512, 518, 554,
    709, 711, 713
infusion reactions 41, 92, 255, 297,
    364, 538, 543, 776
inhibition 50, 102, 107, 115, 160,
    163, 202, 293, 298, 348–349,
    356, 405, 419, 456, 480, 482,
    515, 544–545, 574, 596–597,
    610, 612–615, 619, 757, 759,
    763, 777, 783, 809–810, 812,
    814, 816, 949, 955
innate immune system 96, 123,
    125–126, 137–138, 142, 260,
    266, 447, 453, 512, 627, 647,
    707–708, 772, 938

innate immunity 98, 126, 138, 210,
    330, 361, 389, 463, 679,
    707–710, 955
interaction with biological
    components 43, 835, 837,
    842, 860
interchangeable product 2, 58, 60
interferon (INF) 38, 67, 130, 191,
    275, 293, 295–296, 300–301,
    442, 448, 454, 456–457, 476,
    479, 491, 513, 518, 538, 563,
    606, 674, 676, 699, 710–711,
    717, 885
interferon-alpha 538, 563
interleukin (IL) 108, 131, 189–190,
    341, 344, 456, 518, 538, 563,
    699, 711, 716, 803–804, 808,
    870
interleukin-2 538, 563
interlukin 1β (IL1β) 801
intra-assay variability 585
intracellular delivery 445, 447, 449,
    451, 487, 657, 659, 661, 663,
    675–676, 678–679, 685
intracellular pathogens 141,
    507–512, 514, 516–520, 522,
    524
intracranial lymphatic vessels 643
intravascular macrophages (PIM)
    cells 84, 108, 323, 367, 429,
    771, 778, 784
intravenous 11, 26, 37, 49, 56, 101,
    103, 113, 193, 257–259,
    266–268, 291, 296, 324–326,
    341, 350, 357, 380, 417–419,
    421–422, 427, 429–430, 484,
    486, 551, 559, 673, 705,
    772–773, 775–778, 783, 788,
    842, 881, 893, 919, 943, 945
intravenous immunoglobulin (IVIG)
    341, 357

intrinsic tolerance mechanisms 643
IRMI, *see* immune response modulating impurities
isotyping 585, 589–590
iTope 2, 55, 878, 883
IVIG, *see* intravenous immunoglobulin

Jadad score 893–894, 898, 913, 915

kallikrein–kinin system 731, 739

large cell variants (LCV) 507, 509
large multilamellar vesicle (LMV) 771
lectin pathway 94, 311–313, 327
leukocytes 108, 342–344, 348, 351, 353, 394–395, 400–401, 404–406, 430–431, 708, 712–713, 784, 801–802, 805, 820, 822
ligand density 657, 659, 662, 664, 669
lipid dose 249, 256, 260, 267, 293
lipid nanoparticles 100, 137–138, 194, 198, 254, 257, 265–266, 476, 483–484, 492, 495, 660, 668, 670–671, 676
lipid-polymer hybrid nanoparticles 441, 452
lipid recognition 441, 460
lipoplexes 265–266, 273, 441, 450, 481, 484–485, 487, 816, 819
lipopolysaccharide (LPS) 172, 174, 200, 202, 430, 455, 485, 507, 510, 562, 947

liposomal nanoparticles 893, 896, 899, 912, 914, 918
liposome-encapsulated hemoglobin 389, 392, 407, 420
liposomes 26–27, 30, 51, 53, 89, 91, 95, 98, 103, 106–107, 131, 133, 137, 184, 186, 194, 197–198, 200, 250–265, 267–269, 271–273, 276, 289–293, 297–299, 317–318, 325, 361–362, 364, 370–375, 389, 391–393, 398–399, 417–423, 425–426, 430–431, 434, 452–453, 477, 479, 482–486, 493, 669–670, 672–676, 771–773, 775–778, 780–785, 787–788, 807, 809, 939–940
LMV, *see* large multilamellar vesicle
LN, *see* lymph node
LPS, *see* lipopolysaccharide
lung cancer 896, 929, 931, 949, 951, 956
lymph node (LN) 123, 364, 680–682, 810
lymphoma 156, 682, 684, 699, 718–719, 811, 814, 817, 819

mABs, *see* monoclonal antibodies
MAC, *see* membrane attack complex
macrophages 45, 52, 84, 90, 93, 108–110, 113, 123, 125–130, 135–136, 139, 172, 190–196, 199, 202, 251, 253–254, 342, 344–346, 348, 354, 367, 389, 396, 399, 419, 428–429, 453, 486, 511–514, 666, 669–671, 674, 681, 699–703, 707–717, 762, 771, 774, 778, 784, 801–803, 805, 810, 813,

816–817, 823, 950, 952, 955, 957
major histocompatibility complex (MHC) 37, 55, 127, 518, 699, 709, 822, 882
mannose-binding lectin (MBL) 109, 311, 313, 318
mast cells 108–111, 316, 342, 346, 363, 374, 377, 389, 396, 430, 453, 699–700, 708, 712–714, 777–778, 784
maternal 144, 538, 545, 755, 757
matrix interference 585, 598, 613–614
matrix metalloproteinases (MMPs) 341, 345
MBL, see mannose-binding lectin
MD2 441
medical sciences 209, 433, 441
medicine 4, 24, 27, 42, 47, 57, 63, 67, 74, 77, 83, 89, 95, 116–117, 156, 184, 209, 229, 233–234, 311, 329–331, 379–380, 389, 417, 434, 463, 474, 507, 523–524, 639–640, 686, 699, 731, 766–767, 801, 825, 836, 862, 867–868, 881, 893, 895, 914, 920, 930
membrane attack complex (MAC) 93–94, 311, 313–314, 352, 400–402, 406–407, 510
metal-based nanoparticles 699–700, 702, 704, 706, 708, 710, 712–714, 716, 718–720
metastases 643, 645–647, 649, 912, 943, 952, 955
methoxypoly(ethylene glycol) (mPEG) 289, 299
metronomic chemotherapy 249, 261
MHC, see major histocompatibility complex
MHC class I 123, 127, 949
MHC class II 55, 123, 127, 718, 882
micelle nanoparticles 893, 899, 911–912, 918
micelles 83, 112, 196, 250, 256, 268–270, 289, 293, 299, 681, 836, 919
micro-RNAs (miR) 801, 807
microarrays 229, 235–237, 239–241, 243
microbial transglutaminase 155, 160, 173
microemulsions 250, 268, 289, 293
microglia 753–754, 762, 764
microRNA 473, 475
minimal required dilution (MRD) 585, 594, 599
miR, see micro-RNAs
mitigation strategies 538, 540, 542, 545, 548, 554, 561, 564, 568
MMPs, see matrix metalloproteinases
molecular disassembly array 627, 633
molecular manufacturing 627–628, 638
molecular structure 22, 402, 474, 538, 556
monitoring 47, 54, 68, 104, 230, 319, 367, 538, 547–549, 574, 603, 881
monoclonal antibodies (mABs)
human 555
therapeutic 29, 62, 658, 841
monodisperse particles 507, 520
monomethyl auristatin E 155, 159, 173
mononuclear phagocyte/phagocytic system (MPS) 183, 189–190, 311–312, 444–445
mononuclear phagocyte system (MPS) 250, 289–290, 771–772, 781

mouse model   164, 194, 273, 311, 327, 330, 353, 938
mouse tumor models   929
mPEG, *see* methoxypoly (ethylene glycol)
MPS, *see* mononuclear phagocyte/phagocytic system
MRD, *see* minimal required dilution
multi-target drug   753, 763
multispecific antibodies   538, 555
myeloid-derived suppressor cells   658, 801, 803, 810, 953
myocardial infarction   105, 316, 341, 343, 347, 393, 399, 713

nab-paclitaxel, *see* nanoparticle albumin bound-paclitaxel
NABs, *see* neutralizing antibodies
nanoantibodies   801, 806–807
nanocarriers   42, 196, 201, 208, 249–250, 252, 254, 256–258, 260, 262–264, 266, 268–270, 272, 274–276, 278, 290, 292–293, 299, 362, 451, 480, 487, 678, 680, 771–772, 775, 777, 785, 801, 808, 814, 817–820, 825, 896
nanocomplex   929, 931, 937, 939–941, 944–947, 954
nanodrugs   1–6, 8, 10, 12, 14, 16, 18, 20, 22, 24–56, 58, 60, 62–64, 66–68, 70–72, 74–76, 78, 83–85, 87–89, 91–94, 96–99, 101, 103, 105, 107, 109, 111, 362, 380, 643, 719–720, 841
nanoemulsion   123, 138, 202
nanoformulations   71, 183, 193, 198, 201, 205

nanomaterials   5, 32, 45, 54, 88, 184–188, 191, 196–197, 199–200, 203, 205, 207, 209–210, 328, 496, 627, 629, 699–700, 707, 719, 767, 825, 835, 837–839, 842–846, 849–860, 941
nanomedical device   627, 632, 635, 639
nanomedical platform   627–628, 631, 633, 635, 637
nanomedicines   3, 5, 83–84, 116, 316–319, 321–323, 325–328, 331, 341–342, 344, 346, 348, 350, 352, 354, 356, 361–362, 370–371, 376, 378–380, 389–390, 417–418, 420, 431–432, 776, 778, 784, 788–790, 801, 804, 835–838, 840, 842, 844, 846, 848, 850, 852, 854, 856–858, 860, 862, 929–932, 934, 936, 938–950, 952, 954, 956, 958
nanoparticle albumin bound-paclitaxel (nab-paclitaxel)   893, 919
nanoparticle asymmetry   657, 665
nanoparticle surface chemistry   657, 668
nanoparticles   27, 29, 32, 83–84, 87–88, 95, 100, 103–104, 111–112, 123, 128–131, 134–135, 137–144, 183–200, 202–210, 250–252, 254, 259, 262, 264–266, 311–312, 317–318, 323, 330, 361–362, 407–408, 451–452, 476, 482–484, 491–492, 507–508, 519–520, 629–631, 657–662, 664–686, 699–700, 706–720, 731–736, 744–748, 771, 814–817, 819–822, 835–836, 838, 841–843, 847–849,

893–900, 902–904, 910–914, 916–920
nanoparticulate drug formulations 2–3, 5, 84
nanopharmaceuticals 2–3, 5, 27, 84, 380
nanoscale 2, 5, 24, 26, 29, 31, 36, 42, 87, 94, 137, 141, 210, 390, 633, 841
nanoscience 27, 31, 766, 835
nanosilver 699–700
nanosimilars 2–3, 62–63, 65, 71, 75, 84–85
nanosystems 27, 45, 100, 110, 362, 377, 396, 779, 835–839, 842–843, 849, 851–852, 854–855, 857, 860–861
nanotechnology 2, 24, 27, 31–32, 84, 95, 116–117, 124, 128–129, 131, 133, 135, 137, 139, 141–143, 145–146, 183–185, 190, 192, 195, 197–199, 201, 206, 208–209, 330–331, 361, 376, 379, 433, 508, 627–628, 637, 639, 651, 699, 790, 804, 818, 824, 835–838, 917
nanotherapeutics 5, 330, 627, 629
nanotoxicities 98, 311
nanovaccine 507–510, 512, 514, 516–522, 524
natural killer (NK) cells 123, 126, 231, 630, 679, 699–700, 708–709, 712, 715, 801–802, 938
NBCD, *see* nonbiologic complex drug
NBCD similar 2, 75
NBEs, *see* New Biological Entities
NCEs, *see* New Chemical Entities
NDA, *see* New Drug Application
near-infrared fluorescent 155, 167, 174

neoantigens 555, 557, 867, 872
neonatal 232, 350, 538, 544–545, 882
neuro-oncology 643–644, 648–649, 652
neutralizing antibodies (NABs) 34, 66, 88, 252, 291, 327, 516, 538, 540–541, 545, 547–548, 550–552, 571–574, 585, 588, 615, 867, 871, 873–874, 876, 886, 914
neutralizing assay 585
neutrophils 123, 126, 136, 196, 316, 348, 352, 399, 420, 453, 699–700, 711–714, 716, 810, 812, 814, 817, 820, 938, 955
New Biological Entities (NBEs) 2, 5, 10
New Chemical Entities (NCEs) 2, 5
New Drug Application (NDA) 2, 5, 10–13, 17, 589, 886, 948
non-acute immune response 538, 570
nonbiologic complex drug (NBCD) 2, 63, 83, 95
nonspecific uptake 168, 476, 657, 662–663, 670
NP, *see* nanoparticle
nucleic acid recognition 441, 458
nucleic acids 170, 190, 197, 201, 249, 265, 273, 483, 562, 661, 804, 807, 846, 947

oncolytic virus therapies 643, 646–647
opsonins 84, 109, 135, 249–250, 451, 781
opsonization 84, 93, 109, 123, 135, 252, 254, 313, 315, 327–328, 660, 846, 850

optical imaging 155, 166, 174
ovalbumin 131, 275, 289, 291, 428, 665

P53 230, 929–939, 942, 947, 949–950, 953–957
PAI-1, see plasminogen activator inhibitor
PAMPs, see pathogen-associated molecular patterns
pancreatic cancer 168, 813, 893–900, 902–914, 916, 918–920, 948
particle-mediated epidermal delivery (PMED) 123, 143
particle replication in nonwetting templates (PRINT) 657, 665
particle size 128–134, 190, 192, 249, 256, 270, 293, 318, 664–665
particulate 108, 112, 124, 137, 140, 142, 145, 203–204, 369, 390, 538, 703, 712, 809, 873
PASylation 289, 301
patents 2, 11, 24, 31, 70, 84, 95, 117, 146–147, 331, 379, 434, 495, 523, 748, 764, 790, 825, 958
pathogen-associated molecular patterns (PAMPs) 109, 123, 126, 441, 452, 507, 512, 699, 707, 771, 778
pathophysiology 83, 116, 352, 389, 417–418, 433–434, 538–539, 542, 568, 754, 756, 771, 812
pathotrophic nanoparticle gene delivery 893, 909
patient history 538, 551
pattern recognition receptors (PRRs) 123, 126, 318, 647, 686, 699, 707–708, 778

PCBylation 249
PD, see pharmacodynamic
PD1 658, 811, 814, 817, 929, 931–932, 948–953, 955–956
PEG, see polyethylene glycol
PEG-ADA, see PEG-modified bovine adenosine deaminase
PEG-ASNase, see PEGylated asparaginase
PEG-chain length 249, 256–258, 293
PEG-INF-α 289, 295
PEG-modified bovine adenosine deaminase (PEG-ADA) 289, 295
PEG-PAL, see PEGylated phenylalanine ammonia lyase
PEG-surface density 249, 257–258
PEG terminal group 249, 258
PEGinesatide 53, 91, 275, 289, 296
Pegloticase, see PEGylated uricase
PEGylated asparaginase (PEG-ASNase) 289, 296
PEGylated liposomes 2, 51, 89, 106–107, 200, 250–256, 259–265, 267–269, 271–272, 291–293, 297, 773, 775–776, 781, 788
PEGylated nanoparticle 251, 254, 259, 262, 311, 665
PEGylated phenylalanine ammonia lyase (PEG-PAL) 289, 296
PEGylated proteins 2, 46, 99, 250, 268, 289–291, 295, 317, 775
PEGylated uricase (Pegloticase) 289, 297
PEGylation 51, 89, 249–250, 256, 270, 275–276, 289–293, 300, 389, 450, 538, 561, 592, 773, 787, 807, 822, 856
PENK, see proenkephalin

peripheral blood mononuclear cells 155, 169, 174, 202, 513, 569–570, 882
personalized dosing regimen 893, 910
personalized medicine 77, 117, 229, 233–234, 836, 862, 867–868, 881
personalized therapies 835, 837, 852, 860
phagocyte 123, 130, 134, 250, 289–290, 314, 444, 515, 771–772, 781, 815
phagocytosis 45, 93, 108–109, 128, 135, 137, 191, 254, 256, 445, 670–671, 703, 709, 712–713, 715, 762, 818, 835, 839, 846, 855–856
pharmacodynamic (PD) 2, 9, 11, 59, 71, 541, 873
pharmacodynamics activity 867
pharmacokinetics (PK) 2, 157, 160, 165, 197, 249, 259, 262, 277, 289, 293, 302, 326, 448, 538, 541, 547, 585, 599, 622, 664–665, 682, 719, 771, 788, 790, 806, 818–819, 878, 880, 912
photodynamic therapy 893, 917
photothermal therapy 679, 717, 893, 917
pig model 107, 311, 325
PIM, see pulmonary intravascular macrophages
PK, see pharmacokinetics
plasminogen activator inhibitor (PAI-1) 341, 345
platelet activation 44, 97, 188, 344, 346, 351, 354, 393, 395, 400–406, 731, 744
platelet aggregation 207, 394, 400–401, 405, 430, 731, 734, 741–744, 746–747

platelet rich plasma 731, 742, 745
platelets 108, 110, 315, 341, 344–346, 351, 377, 389–391, 393–398, 400–407, 427, 731, 734, 742–747, 812, 820
PLGA, see polymeric poly(lactic-co-glycolic acid)
PMED, see particle-mediated epidermal delivery
poly(carboxybetaine) 289, 300
polyethylene glycol (PEG) 29, 83, 87, 123, 134, 189, 249, 268, 275, 289–290, 389, 449, 488, 538, 561, 598, 665, 771–772, 841, 884, 911–912
polyglycerol 249, 273, 289, 300
polymeric poly(lactic-co-glycolic acid) (PLGA) 130, 132–133, 136, 194, 452, 657, 671
polymers 112, 136, 185, 272–273, 289, 300–301, 318–319, 325, 374, 450, 452, 456, 478, 486, 491, 519–521, 663, 668–670, 675, 678, 708, 801, 895
polymorphism 513, 538, 554
polyplexes 441, 450, 481, 836
polysorbate 138, 186, 296, 564–565, 598, 850, 867, 871–872, 885
polystyrene nanoparticles 84
PolyXen 289, 301
porcine model 361, 367, 418, 771
positron emission tomography 155, 166, 174, 893, 908, 919
postmarket surveillance 867, 872–873, 886
posttranslational modifications 538, 556, 884
precision 24, 114, 462, 585, 590, 599–600, 611, 614–617, 621, 878, 930

preclinical 11, 14, 29, 47, 50, 54, 62, 72, 87, 100, 102–103, 142, 144, 161, 172, 183, 198, 200, 205, 207, 209, 323, 326, 328, 331, 362, 376, 380, 493, 574, 586, 631, 647, 668, 709, 808, 813–814, 816–817, 837–838, 841, 852, 857, 870, 873, 877–880, 886, 948
preexisitng (natural) antibodies 585
pregnancy 347–350, 420, 538, 545
prenatal exposure 753, 756, 761–763
primary motor cortex 753, 762, 764
PRINT, see particle replication in nonwetting templates
procoagulant 188–189, 349, 395, 731–732, 812
product custody 538, 566
proenkephalin (PENK) 753, 761
prophylactic 517, 538, 573, 720, 787
prostate-specific membrane antigen 155, 162, 174
protein aggregates 2, 42, 558–560, 867, 871–873, 881
protein aggregation 2, 41, 203–204, 559–561, 565, 879, 884
protein corona 42–43, 83, 89, 183, 207, 846, 852
protein engineering 155, 158, 175, 538, 558
protein products 2–3, 16–17, 20, 291, 537–550, 552–554, 556–572, 574, 585–590, 592, 594, 596, 598, 600–602, 604, 606–608, 610–612, 614, 616, 618, 620, 622, 624, 869, 871, 873, 877

prothrombin time (PT) 731, 735–736, 853
PRRs, see pattern recognition receptors
pseudoallergic reactions 311, 317, 327, 380
pseudoallergy 2, 44, 46, 83, 93, 96–98, 105, 111, 187, 189, 255, 289, 297, 311, 317, 325, 361–363, 389–390, 392, 394, 396, 398, 400, 402, 404, 406–408, 417–418, 420, 422, 424, 426, 428, 430, 432–434, 771, 775, 778, 780
PT, see prothrombin time
PTO, see US Patent and Trademark Office
Public Health Service (PHS) Act 2, 12
pulmonary 47, 84, 104, 107–110, 112, 139, 203, 323–326, 347, 366–367, 369–371, 373, 377–378, 390, 395–396, 398–399, 405–407, 419–420, 427–431, 463, 514, 521, 668, 701, 771, 777–779, 781–784, 786, 910
pulmonary intravascular macrophages (PIM) 84, 108, 323, 367, 429, 771, 778, 784
pyrrolobenzodiazepine 155, 159, 173
pyrrolysyl-tRNA synthetase 155, 162, 173

Q fever 507, 509–510, 513–514, 517, 524
QC, see quality control
quality control (QC) 585, 602
Qvax 507, 514–515, 517

randomized control trial (RCT) 893–894, 903
RCA, *see* regulators of complement activation
RCT, *see* randomized control trial
R&D, *see* research and development
reactive oxygen species (ROS) 123, 139, 455
receptor-mediated transcytosis 929
RECIST, *see* Response Evaluation Criteria in Solid Tumours
recognition motifs 893, 896
recombinant 3, 17, 23, 34, 38, 49, 88, 101, 144, 197, 203–204, 229, 236–237, 243, 349, 427, 520, 553, 658, 870–871, 873–874, 881, 884, 906, 911
regulators of complement activation (RCA) 311, 329
regulatory T (TReg) cells 716, 801, 803
rejection 623, 660, 683, 699, 707, 821
release kinetics 657, 663, 678
repeated administration 51, 89, 249–251, 262–263, 265, 269–271, 273, 292, 297, 301, 426, 429, 773, 822, 869, 952
replacement therapy 342, 537–538, 550–551, 553–554, 573
research and development (R&D) 2–3, 639, 685, 789
Response Evaluation Criteria in Solid Tumours (RECIST) 893, 909
reticuloendothelial cells 801, 807
reticuloendothelial system 2, 51, 89, 323, 476, 669–670, 822, 856
RF, *see* rheumatoid factor

rheumatoid factor (RF) 585, 593, 607
risk assessment 54, 229–230, 538, 545, 561, 564, 589, 872, 881
risk-based approach 52, 538, 567, 588
risk-benefit assessment 538, 541, 551
RNA interference 441–442, 444, 446, 448, 450, 452, 454, 456, 458, 460, 462–464, 657, 661, 814, 817, 893, 917, 938
rodent model 326, 433, 753–762, 764–766
rodents 324–326, 417–419, 431
ROS, *see* reactive oxygen species
route of administration 130, 347, 538, 547, 551–552, 560
Russel viper venom (RVV) 731
RVV, *see* Russel viper venom

safety profile 859–860, 893, 914
scavenging 341, 356–357
SCD, *see* sickle cell disease
screening assay 585, 591, 606, 609–610, 618, 620–621
SCV, *see* small cell variants
Sec incorporation sequence 155, 163, 174
selectivity 11, 72, 167, 171, 475, 585, 590, 596–597, 601, 603–604, 609, 611, 616, 648, 878
sensitivity 63, 92, 103, 167–169, 229, 233, 239, 320–323, 325, 363, 393, 431, 488, 568, 585, 589–590, 593–595, 598–599, 601–604, 609, 611, 614–616, 618–620, 623, 756–757, 784–785, 845, 847, 851, 856, 878

sentinel 627–628, 630–638, 640
shelf-life stability 123
short peptide tags 155, 158, 165–166, 173
shRNA 442, 473–475, 482
sickle cell anemia 341–342, 351–352
sickle cell disease (SCD) 341, 351–352
silica nanoparticles (SiNPs) 670, 717, 731–736, 738, 740, 742, 744–748
silicon dioxide (SiO_2) 699
single-cell genomics 2, 73
SiNPs, see silica nanoparticles
SiO_2, see silicon dioxide
siRNA 29, 87, 195, 201, 265–267, 273, 442–453, 455, 458, 461, 473–494, 496, 657, 659–660, 668, 672–673, 675–676, 678, 681, 685, 814, 816–818, 820, 841, 946–947
site-specific conjugation 155, 160–162, 165–167, 169, 172
site-specific delivery 2, 478, 685
small cell variants (SCV) 507, 509
small interfering RNA 265, 441, 816
small-molecule drug (SMD) 2, 8, 12–13, 22, 32, 38, 61, 83–85, 95–96
small molecules 21, 33, 85, 312, 328, 658, 676, 710, 766, 801, 804, 930
small unilamellar vesicle (SUV) 771
SMC, see smooth muscle cells
SMD, see small-molecule drug
smooth muscle cells (SMC) 110, 341, 344, 377, 396, 398–399, 778
specific amino acids 155, 158–159, 166, 168, 173
SPIONs, see superparamagnetic iron oxide nanoparticles

spleen 52, 90, 190, 251–253, 257, 262–264, 293, 323–325, 351, 476, 482, 521, 664, 674, 705, 771, 773–774, 781, 814, 820
splenic B cells 249, 257–258, 264, 267, 273
SRS, see stereotactic radiosurgery
stable nucleic acid lipid particles 441
stealth polymers 289, 300
stereotactic radiosurgery (SRS) 643, 646, 649
strain-promoted azide-alkyne cycloaddition 155, 173
stress reaction 112, 361–362, 364, 376, 407
structure-activity relationship 455
superparamagnetic iron oxide nanoparticles (SPIONs) 123, 141, 195
surface charge 36, 42, 85, 109, 123, 128, 134–135, 190, 249, 256, 293, 318, 491, 521, 668, 783, 787, 843, 857
surface modification 257, 271, 273, 275, 422, 489–490, 519, 772, 776, 781
SUV, see small unilamellar vesicle
syngeneic tumor models 929, 931
system suitability control 585, 595, 603

T cell independent antigens 249, 252
T cells 36, 125, 127, 131, 136, 141–144, 169–171, 298, 346, 453, 456, 485, 514–516, 518, 521–522, 538, 543, 550, 557, 570, 606, 627, 631, 648–649, 657–659, 668, 674–675, 679, 682–683, 699–700, 712–713,

715–716, 810–814, 817–819, 821–822, 868, 871, 878, 931, 950, 953, 955
T helper (Th) cells   123, 127, 538, 557, 716
T helper 2 (Th2) cells   518, 801, 803
T4SS, *see* type IV secretion system
tachyphylaxis   107, 111, 361, 373–375, 417, 426
TAFI, *see* thrombin-activated fibrinolysis inhibitor
target cells   45, 135, 141–142, 190, 443, 445, 477, 481, 572, 835, 839, 895, 946
target mediated drug disposition (TMDD)   2, 11
targeted nanocomplex   929, 937, 947
targeted nanoparticles   657
targeted therapy   233, 754, 929–930
TCED   2, 55, 883
terminal complex   255, 298, 389, 395
therapeutic protein product   3, 538, 540–547, 550–556, 558–559, 561–566, 568, 571–574, 585–589, 591–593, 595–599, 601–602, 605–615, 620–622, 624
therapeutic proteins   3, 17, 20, 197, 203–204, 301, 342, 457, 537–574, 585–598, 600–602, 604–616, 618–622, 624, 867, 869, 871–872, 879, 882–883
THIOMAB   155, 159, 163, 171
third dose   249, 262
thrombin-activated fibrinolysis inhibitor (TAFI)   341, 345
thrombo-embolic events   341
thromboxane   84, 104, 324, 344, 378, 389, 417, 421, 427, 429, 784

time interval   249, 259, 275, 773
TiO$_2$, *see* titanium dioxide
tissue permeation   657, 662–663
tissue-type plasminogen activator (tPA)   341, 345
titanium dioxide (TiO$_2$)   629, 699–700
titer   54, 144, 313, 538, 541, 547–548, 554, 571, 585, 588, 595, 605, 610, 617, 874, 876
titer(ing)   585
TLR reporter cell lines   441
TLR3   266, 441, 447, 453, 458, 709–710
TLR4   441, 454, 456, 460–461, 463, 512–513, 522, 709–710
TLR4 signalling pathways   441
TLR7   266–267, 441, 447, 453, 458–459, 709–710, 813
TLR8   441, 447, 455, 709
TLR9   202, 265–266, 274, 441, 519, 669, 709–710, 714
TLRs, *see* toll-like receptors
TMDD, *see* target mediated drug disposition
TNFα, *see* tumor necrosis factor-α
tolerance   107, 145, 261, 290, 374, 521, 538, 548, 552–555, 557, 567, 571–573, 585, 594–596, 643, 646, 658–659, 716, 771, 788, 868, 871, 885–886, 908, 939, 950
tolerance induction   374, 548, 554, 571–573, 771, 788
tolerization   867, 885–886
tolerogenicity   867, 871
toll-like receptors (TLRs)   140, 142, 249, 265–266, 419, 441, 447, 454, 457, 479, 491, 512–513, 562, 643, 647, 657–658, 661, 699–700, 708–711, 771, 778, 801, 810, 814, 817

total protein binding 183, 206–207
toxicity 8, 10, 14, 33–34, 42, 84–85, 146, 160, 189–190, 196, 205, 209, 265, 289–290, 301, 323, 326, 361–364, 376, 418, 426, 449–451, 455–456, 463, 476, 479, 482–483, 487, 629–630, 636, 644, 650, 658, 664, 668–669, 678, 681, 685, 700–704, 711, 715, 720, 854–858, 860, 893–894, 896, 906, 909–910, 913–914, 918–919
tPA, see tissue-type plasminogen activator
transcytosis 929, 944–945, 954
transferrin receptor 452, 485, 491, 929, 939–940
translation 2–3, 25, 35, 76, 84, 93, 124, 185, 190, 203, 207, 237, 322, 330, 442, 455, 475, 772, 824, 838, 860, 944, 946
tumor infiltrating lymphocytes 643–644
tumor necrosis factor-α (TNFα) 183, 192, 341, 344, 485, 518, 699, 711, 801–802, 814, 817, 823, 911
tumor suppressor 929, 932, 934–937
type IV secretion system (T4SS) 507

UNIarray 229, 236, 238, 241–242
uniform size 507
unnatural amino acids 155, 158, 161, 163, 168–170, 173
uricase 270, 289, 291, 293, 297, 300

US Food and Drug Administration (FDA) 2–3, 27, 85, 136, 772, 867, 869–870, 930
user fees 2, 70

vaccines 3–4, 16–17, 33–34, 74, 85, 88, 123–124, 126–148, 184, 190–192, 195, 390, 464, 507–508, 514–516, 519–524, 538–539, 586, 636, 643, 646–647, 649–650, 657–661, 663, 665, 669, 671–675, 680–681, 707, 709, 714–715, 717–720, 810, 867–868, 870–871, 879
validation 20, 200, 209, 230, 235, 237–238, 241–243, 321, 474, 546, 585–586, 589–590, 593–594, 596–598, 600–601, 603–604, 612, 614–620, 622–624, 883, 887
valproic acid (VPA) 753, 755, 757–758, 760, 763–765
Vascular Cartographic Scanning Nanodevice (VCSN) 627, 635
vascular endothelial growth factor (VEGF) 657–658, 801, 804, 809
vascular endothelial growth factor a (VEGFA) 801, 804, 809
vasoactive mediators 108, 366, 369, 373–374, 389
VCSN, see Vascular Cartographic Scanning Nanodevice
VEGF, see vascular endothelial growth factor
VEGFA, see vascular endothelial growth factor a
virus-like nanoparticles 123, 142
visionary 627, 638

Visudyne 2, 46, 776
VPA, *see* valproic acid
vulnerable plaque 341

white blood cells 110, 377, 389, 396, 427
WHO, *see* World Health Organization
World Health Organization (WHO) 47, 57, 100, 643, 645

xenogenic 538
XTEN 289, 300–301

zinc oxide (ZnO) 139, 202, 699–700
ZnO, *see* zinc oxide